Blessed are the poor in spirit

Blessed are they who mourn

Nothing but a Bully

Blessed are the pure of heart

Blessed are the merciful

A Tale of Power
Genocide
and
Green Beards

Blessed are the peacemakers

Blessed are the meek

Blessed are they who hunger and thirst for justice

Blessed are they who suffer persecution for justice's sake

Ptera Hunter

Nothing but a Bully

A Tale of Power, Genocide, and Green-Beards

Ptera Hunter

Northeastern Illinois University

Benjamin Ross & Lane Publishers

First Published in the United States of America

Edition: September 2025
ISBN: 9781-9557-8003-2
Published by Benjamin Ross & Lane
Melbourne Beach, FL
www.BenjaminRossLane.com

Dedicated to Dr. Roger Hansel, a long-time friend

and graduate advisor who

guided me through

times

marred by a belligerent

and self-serving intellectual bully,

Thank you, Roger, for encouraging me through tribulations.

"[There are those who] little dream of the ages of tyranny, mental, moral, and physical, that has reduced the image of God to a mere beast of burden; the emblem of the Divine Mother, to a slave to bear children; and life itself, a curse. But there are others who see, feel, and shed tears of blood in their hearts, who think that there is a remedy for it, and who are ready to apply this remedy at any cost, even to the giving up of life. And 'of such is the kingdom of Heaven.'"

— Swami Vivekananda, letter to Alasinga, 1883 [1]

"And this again, that that insurgent horror was knit to him closer than a wife, closer than an eye; lay caged in his flesh, where he heard it mutter and felt it struggle to be born; and at every hour of weakness, and in the confidence of slumber, prevailed against him, and deposed him out of life."

— R. L. Stevenson, *The Strange Case of Dr. Jekyll and Mr. Hyde* [2]

"Human kindness has never weakened the stamina or softened the fiber of a free people. A nation does not have to be cruel in order to be tough.

— Franklin D. Roosevelt, Oct 13, 1940, radio address for
The Mobilization for Human Needs [3]

"Ah! How the slave yearns for it, that he may be secure from the lash, that he may enjoy his family, and no more be tortured with the worst fear of slavery, the separation of friends and families. The whip we can bear without a murmur, compared to the idea of separation. Oh, my friends, you cannot feel the slave's misery, when he is separated from his kindred. The agony of the mother parting from her children cannot be told."

— Frederick Douglass, *I Have Come to Tell You Something About Slavery: An Address, Lynn, Massachusetts, October 1841* [4]

"Thus you find enemies in all those whom you have injured by occupying that dominion, and you cannot maintain the friendship of those who have helped you to obtain this possession, as you will not be able to fulfil their expectations, nor can you use strong measures with them, being under an obligation to them; for which reason, however strong your armies may be, you will always need the favour of the inhabitants to take possession of a province."

— Niccolò Machiavelli, *The Prince,* [5]

ACKNOWLEDGMENTS

I recognize and give thanks to Northeastern Illinois University for providing a semester of leave to finish this book, which has taken years to flesh out. Without this time, I would still be working on it.

I am grateful for the input of Dr. Roger Hansel, who introduced me to the study of behavior and animal spite, encouraged my work, and read an early draft. He has been an advisor, a mentor, and a true friend for many years.

I also extend thanks to Angela Pettera for her extensive help with copy editing. I extend heartfelt gratitude to Dr. Terrence Puryear and Dr. M. Jones for reviewing my endocrinology and entomology materials. I extend my appreciation to Mr. K. Kelling for his comments on my literary analysis and his assistance with Latin translations. I thank Mr. J. Trowbridge and Mr. N. A. Habib for their aid with languages and linguistics. Their critical comments improved this manuscript.

I appreciate the work of Ms. Mary Hansell, who read it as a representative target audience. I am also indebted to Ms. Criage Althage and Mr. Jonathan Gronli for reviewing my work with fresh eyes, checking it for readability.

I could not have completed this work without the input of Dr. Linda Nedielkof, who reviewed the social psychology. She also introduced me to the quip "jerkalepsy" and the case of the punishing pigeon (Chapter 3).

My students have served as a bottomless source of inspiration. Their constant questions have kept me young and forced me to stay current with the literature. I extend a special recognition to my former student and now colleague, Sarah France, who suggested I include a unit on the microbiome in my animal behavior class. That addition influenced my writing.

I am grateful to the librarians at Northeastern Illinois University in Chicago, the Chicago Public Library, and the Quatrefoil Library in Minneapolis for their invaluable aid with my never-ending requests for information and obscure papers. I extend my appreciation to Elena Cisneros, Assistant Archivist at Oglala Lakota College in Kyle, South Dakota, for her aid in tracking down information on the Wounded Knee Massacre and the changing signage that has described it.

I thank the University of Kolkata, the Ramakrishna Mission Institute of Culture, Golpark, Kolkata, and the Noble Institute for Environmental Peace, Toronto, where I presented papers on my work while preparing the first drafts of this book.

I could never have written this book if not for the myriad of scientists, reporters, authors, theologians, lawyers, and philosophers upon whose work I have relied for support and information and upon whose shoulders I have wobbled. (And Hitachi, my feline typing assistant who regularly inserted extra keystrokes into my drafts, inspected the monitor, and kept me sane.)

And, of course, you, the readers. I hope this book provides insight into why our species, which is capable of incredible generosity, love, and cooperation, too often takes the path of the bully.

CONTENTS

LIST OF FIGURES

PLEASE DON'T SKIP THIS SHORT (TWO-PAGE) PREFACE

How To Use This Book

To the casual reader: Don't get flustered by the vocabulary. If you run across a word you don't know, you probably don't have to know it to get the gist. You don't need to read all the preceding chapters to understand a later one. To paraphrase what they say in 12-step programs, read what you like and leave the rest. You can skip the footnotes. Instead, just remember: The statements are generalities and trends, not absolutes.

To the more involved reader: I apologize for the copious footnotes. I slipped the more technical stuff and personal asides into these. If something seemed nit-picky or too technical, I moved it there to improve the flow. I strove to cite original, peer-reviewed sources whenever possible, using books and review articles to summarize compiled research.

To everyone: Science is one way of knowing the world around us. I used literary quotes as chapter openers and pulled examples from court cases and news articles. I have simplified the biology for general consumption. I limited the number of hormones, neurotransmitters, and brain regions to a few well-studied and recognized key ones. I condensed other discussions that could have run for pages, such as the impact of the microbiome.

This book runs the gamut from workplace and schoolyard bullying to financial bullying, recognized torture, enslavement, and genocide. These sit on a continuum of dehumanization and abuse. I am not equating the

Shoa with one child held down and urinated on by his peers. There is a quantitative difference. Still, for that one child, the humiliation and dehumanization at the hands of the bully will haunt them for life.

PART I:

Definitions and Scope, Biology and Impact of Bullying and Violence

"Any interaction which is aggressive and asymmetric (not on equal footing) is ragging."

— Rajesh Garg [6],
Ragging: A public health problem in India. *Indian Journal of Medical Science*, 63:263-71

Bullying Is Violence

"Your worldly failure, if you have failed, is to your credit rather than to your blame. Remember that the best and greatest among mankind are those who do themselves no worldly good. Every successful man is more or less a selfish man. The devoted fail... 'Charity seeketh not her own.'"

— Thomas Hardy [7, p. 443], *Jude the Obscure* (Part VI, Chapter 4)

CHAPTER 1:

Always with Us

"Cum in servum omnia liceant, est aliquid, quod in hominem licere commune ius animantium vetet." [1]

(The laws of man permit a slave to be abused in any way. Nevertheless, the laws of nature, common to all life, forbid some acts to a human being.)

— Seneca the Younger, Book 1, Chapter 18

Figure 1-1. The most beloved of Asgard's Gods was not Odin or Thor, but Balder, the God of purity, peace, and light. This public domain Illustration by Jacques Reich appears in Sara Power's Old Norse Stories, p 112 (1900) [3075].

Odin's Quest and Thor's Hammer: A Parable

Odin thirsted for knowledge. He saw the discord surrounding him and understood the mayhem would destroy Asgard, the home of the Gods. From Mimir, the wisest of the Gods, he learned he could gain all

---

1 http://www.perseus.tufts.edu/hopper/text?doc=Perseus%3Atext%3A2007.01.0015%3Abook%3D1%3Achapter%3D18%3Asection%3D2

knowledge if he sacrificed all. Odin hung inverted on the world tree, Yggdrasil, to drink from the well of understanding. He swallowed as though downing the finest mead. He gouged out one eye to gain wisdom and forethought. That eye remains within the well, ever seeking more knowledge. Through his insights and his two messengers, Thought and Memory, Odin ruled the Gods.

He had two sons, Balder and Thor. Balder, the just and beautiful beloved of the Gods, had a peaceful nature. Thor, whose troubled spirit could strike with force, wielded his war hammer to guard both Asgard and the peaceful Balder. Thor could protect the gentle Balder. Though Balder could not protect Thor, his gentle soul nurtured harmony among the Gods, making life in Asgard idyllic.

Odin adopted a third son, Loki, the shape-shifting trickster, from the giants who had sworn to undermine the Gods. Through Loki, the Gods mastered deception, the skill needed to complete the triumvirate Asgard required to thrive: The Wisdom of Odin, the Power of Thor, and the Deceit of Loki. Under their protection, Frigg watched over families and the Earth. Freyja, the embodiment of love and beauty, guarded mortal children and brought the peaceful dead to her hall, reuniting them with their treasured ones after death.

The Gods loved Balder, the peaceful, brave, and just one. They extracted oaths from all creation, each creature, rock, and plant pledging to keep the peaceful one safe and never to act against him. Believing Balder's peace could survive all attacks, they tested his resilience with good-natured assaults. He deflected every blow and resisted every poison until Loki's treachery destroyed him. That viper in Asgard gave the blind God, Höd, an arrow of mistletoe, the symbol of goodwill. He could not see the concealed threat. Through treason, Loki's lies, and the false peace of the mistletoe, Balder's death unleashed the hammer of Thor and brought war to the Gods.

Words Shape the World

> *"A 16-year-old from Newcastle was charged under the same Public Order law, for saying 'woof' to a dog in front of police officers. Both were charged under Section 5 of Britain's Public Order Act. The law, which covers both the spoken and written word, stated that a person could be found guilty of an offense if he or she used threatening, abusive or insulting words or behavior."* — Locker, *Time Magazine*, 2013 [3408]

A young man, J, celebrated the birth of his child with a bit too much to drink. While riding home after a night with his friends, his full bladder protested, as often happens after imbibing. The driver pulled over so J could pee on a telephone pole. For most people, the story would end there. If caught in the act, we might get lectured about being stupid or drunk. But J had a misfortune: someone spotted him and took offense. The police charged him with open lewdness. The courts sentenced him to two years and put him on a sex offender registry [8, 9, 10]. The laws placed peeing in the same category as sexual assault.

The words "sex offender" and the laws surrounding it stigmatized J. Because he drank too much and had to pee, the state now limits where he can live, with whom he can associate, and how he can make his living. The term has become a boogyman, haunting not only dangerous predators but countless people, turning their lives into living hells. Beyond the state's sanctions, the phrase conjures images of sexual abuse and condemns those labeled, often without a conviction. Names of people on public lists of offenders are driven from their homes by self-appointed citizens protecting their neighborhoods. Most of us stand by and let it happen, either agreeing with the vigilantes or fearing we shall become tarred by defending the target.

Our vocabulary shapes our world. Words matter. "Woman," "lady," and "lass" all refer to adult female humans, but their richness varies, evoking different emotions and memories. "Man," "gent," and "dude" also paint pictures. What holds true in our vocabulary about people holds true in how we describe violence. Just as the "lady" and "lass" shape our perceptions, so do words used to describe mistreatment. "Bully," "assault," "intimidate," "denigrate," and "dehumanize" overlap in meaning, and the word we choose reflects our values and attitudes toward situations, victims, and perpetrators.

Consider this fictional account. *Jealousy over a promotion prompted three coworkers to toss their twenty-five-year-old colleague, Hank, into a urinal. They verbally assaulted the man while holding him down and peeing on him.* Virtually anyone reading this account would describe the fictional incident as criminal assault and battery.

When it happens to children, we devalue the experience and impact and dismiss it as bullying, or worse, as deserved. J. N. spent his middle and high school years peed on. His classmates beat him to the point of requiring surgery. He was subjected to a mock public rape in front of classmates.[2] The monsters did psychological and physical damage, resulting in multiple suicide attempts. J. dropped out of school and ran away from home. The school did nothing to stop the assaults; some personnel sided with the assailants. Their complacency led this courageous person to return and sue the school for failing to intervene. He lost the case in 1995 but won on appeal in federal court the following year [11].

When we classify what happened to Mr. N. as "bullying," it becomes easier to dismiss the repeated criminal assaults he endured. Dismissing assailants as "nothing but bullies" minimizes the victim's trauma and the violence suffered. The verbal and physical assaults and robberies, the intimidations and microaggressions, the character assassinations and casual dismissal of humanity, denial of individuality, and personal dignity that happen under the radar, under the minimizing cloak of "bullying," are what they are: verbal abuse, assault, theft, intimidation, aggression, character assassination, and dehumanization that would otherwise outrage us.

[2] Following the sexual assault, Jamie was disciplined for leaving the school. He was forced to return and attend classes with his assailants, who received no punishment. After an attempted suicide and a stay with relatives out of state, he returned after a promise he would not need to revisit the site of the assault. The Department of Social Services forced his return to hell.

Dehumanizing violence takes on many intersecting forms. They form a continuum from pushing one's way through a grocery store line to genocide. Violence strikes in many guises and affects different aspects of our lives. We long to label it, hoping to tame the beast with its name. Often, we conspire to hide it.

> *"McDonald's president who made $7.4 million last year says proposal to pay fast-food workers $22 an hour is 'costly and job-destroying' "* — MSN article, 2023 [3166]

When we discuss bullying, violence, and abuse, we blend overlapping forms of maltreatment and fail to recognize others [12, 13]. As a species, we abuse each other physically, sexually, emotionally, temporally, and financially. We dehumanize, dismiss, and demand conformity. We also extend kindness and compassion that bolsters self-esteem and offsets the damage [14]. Yet, in an age and culture that increasingly acknowledges human diversity, we expect victims to respond uniformly to abuses.

CHAPTER 2:
The Impact and Scope; Nothing but a Bully?

"Si vis pacem, para bellum." (If you want to see peace, prepare for war.)

— Latin adage

To Trust or Not to Trust:

"It is equally faulty to trust everyone and to trust no one. Yet the former fault is, I should say, the more ingenious, the latter the more safe."

— Seneca's Letters to Lucilius, III, IV, The Letters on Morality [15]

Humans, like other animals, categorize. Animal brains use quick-sorting algorithms because we cannot always spend enough time to understand the character of every threat or everyone before judging them. In nature, unquestioning tolerance leads to near-certain death. Honeybees attack invaders who do not smell like their nest. Squirrels chase neighboring squirrels from their trees, and chimpanzees declare open warfare on neighboring bands who impinge on their territory. Even clones of corals and sea anemones, which lack nervous systems, attack members of their own species that threaten their expansion. Without the benefit of a central nervous system, they sacrifice themselves for the good of their close genetic relatives [16, 17].

Our brains divide the world into broad categories of "dangerous" and "safe." We take this action with food, animals, inanimate objects, and people. Classifying things and people based on experience and cultural learning kept our ancestors safe throughout our evolutionary history. Today, children learn about "stranger danger." Our "sixth sense" about a person can save us from being mugged, or it can result in bias and discrimination. Avoiding potential threats costs little; dashing unawares toward the arms of danger gets one killed. We evolved our classification system biased toward safety, assuming the identified "outsider," the "unknown," will harm us.

Often, we must evaluate a person's character with minimal information, and our innate tendency to classify and stereotype plays a role in this safety mechanism. Our social learning alerts us to dangers; often, the brain's filter includes identifiable "others" as potential dangers. Our Pleistocene ancestors gained a survival advantage by stereotyping those who looked, behaved, or spoke differently. Strangers could harm our bands or take our resources. Strangers posed risks; strangers could kill you. We have proven unable to fine-tune our adaptive suspicion to exclude *only* the untrustworthy. Primate brains prefer the safer approach, assuming danger until assured of safety. This supposition enabled our ancestors to react to potential threats. It also resulted in ongoing conflicts with neighbors and laid a foundation for xenophobia, nepotism, and discrimination.

> *On Thursday, June 12, 2025, Senator Alex Padilla of California attended a media event hosted by Homeland Security Secretary Kristi Noem. When he asked a question, men in plainclothes pushed him from the room. In the hallway, the NBC video shows people wearing FBI vests forcing him to the ground and cuffing him. "If this is how the Department of Homeland Security responds to a Senator with a question, you can only imagine what they're doing to farmworkers, to cooks, to day laborers... We will hold this administration accountable."*
> – Senator Alex Padilla [3234]

As a species, we cannot agree on what constitutes "*them*." Some populations consider skin tone to be a distinguishing trait. A small but vocal minority of Euro-Americans expect a "race" war to emerge. To this end, the most extreme elements form white "Christian" militias, dreaming of a world order in which white skin and "Christian values" dominate. This vocal subgroup believes that if people would listen and understand their vision of society, it could provide peace, harmony, and freedom for everyone—at least for everyone who counts within their worldview.

People will preach hatred and violence in the name of God, usurping religious symbolism to rationalize their anger. The star and crescent moon of Islam, the Star of David of Judaism, or the cross of Christ in Christianity can turn from symbols of love to a message of dread. As a dear friend of mine said, worshiping hatred violates the first pillar of Islam: There is only one God. Some Hindus will use the trident of Shiva to incite inter-caste violence [e.g., [18] like some Christians instigate inter-racial or sectarian hatred behind a flaming cross. They might insult or taunt to provoke a reaction; others do not care whom they injure or ponder how their actions cascade.

Dismissing bullies and terrorists as sadistic flukes of nature oversimplifies the story [19, 20, 21]. Persecutions by the empowered subjugate victims. Violence can be direct or indirect; it can happen in a quest for power, status, resources, or revenge. Physical, psychological, or economic bullying can occur alone or interlinked. Although the disempowered can fight back and roles can reverse, the powerful seldom foresee themselves as the victims. From ancient markets to modern enslavement, freeborn people have rarely stood on the auction block without a sense of complete shock and disbelief.

Computers provide a modern platform where tormenters can distribute their poison while remote from the quarry, spreading their venom further [22, 23]. The inability to escape the barrage of cyberbullying contributes markedly to social isolation and youth suicide, the population in which this phenomenon is best studied [24, 25, 26, 27, 28], and contributes to depression, social exclusion, and stress in cyberbullied adults [29, 30, 31]. No matter the etiology or its platform, bullying serves a function for the tormentor; it provides him or her with a shortcut to power.

Bullying

"The condition upon which God hath given liberty to man is eternal vigilance."

— John Philpot Curran, 1790 [32, p. 200]

Bullying and violence are fundamental aspects of human nature. Until the late 20th century, we minimized the damage inflicted by the schoolyard and workplace bullies, compartmentalizing their damage and isolating it from "real" threats of "overt" violence. This approach created stigmas. In silence, abused people endured shame, disgust, and sadism. Victims accepted the voices of parents, friends, and coworkers who gaslighted them with well-intentioned advice: ignore the pain and be the better person. Bullying shouldn't hurt because the assailant is "nothing but a bully." Claims that "everyone gets bullied" minimized their distress. However, the degree and frequency of this abuse vary, and the intensity, type, and consistency of victimization alter brain structure and increases the probability of later depression and adjustment, just as with other forms of torture [33, 34, 35, 36].

Cross-cultural reviews of the research on childhood bullying and suicide indicate self-harm, suicidal ideations, and suicide increase in bullied children, with a more significant impact on girls than boys [37, 38]. Childhood victimization leads to an impaired quality of life that includes depression, suicidality, loss of self-esteem, and reduced cognitive performance extending into adulthood [39, 40, 41, 42, 43]. Ignoring the aggressor doesn't work any better than ignoring cancer. You cannot treat cancer without understanding it; you cannot address oppressors without comprehending why people terrorize.

To grasp the mindset of a bully, one must appreciate his vantage point. Still, a willingness to do so arouses condemnation and rage. We resist understanding the phenomenon as if it were tantamount to excusing the behavior. Comprehending the origins of degrading violence, be it a schoolyard tyrant, a manipulative boss, or a political dictator, no more condones it than understanding forest fires leads to arson. Knowing how and why people torment will not eliminate the problem, but it can open avenues for intervention.

Bullying[3] appears across cultures, not just in hyper-competitive WEIRD[4] cultures such as Anglo-North America. Globally, 25% of children are bullied [44]. A Korean study of middle school children demonstrated up to 40% of students participated in bullying, and at least 23% of students in that survey were victims [45]. Estimates of childhood victimization in India range from 25% [46] to 60% [47]. Only about 40% of parents recognized the problem. The rate of childhood victimization by "bullies"[5] varies among countries, with a worldwide average of 36% [48].

Despite widespread misconceptions, nasty, tyrannical behavior does not always spring from abuse. Bullies are not universally victims first. Bully-victims occur less frequently than "pure" bullies who have not been habitual victims and who have not suffered a high level of abuse.[6] Bully-victims have different profiles than people who are only bullies or only victims. Pure bullies often enjoy a high social status among their classmates, but peers readily reject bully-victims [49]. Both types of tormentors have an increased risk of perpetrating more violence as adults. Bullies, bully-victims, and victims show consistent differences in children. Less is known about bullied adults or those exposed to intimidation and violence [50].

Definitions of abuse and acceptable behavior vary among cultures and subcultures. Many dismiss verbal abuse as ephemeral, less consequential than sticks and stones.[7] Verbal assaults are the most common form

---

[3] At some point, everyone gets cast as a bully, though we often fail to realize we played the role. I will use the term to describe a person who habitually engages in degrading, humiliating, dehumanizing, physically damaging, or otherwise harmful behavior. I will not minimize their damage by isolating them from other tormentors.

[4] WEIRD = Western, Educated, Industrialized, Rich, and Democratic. WEIRD cultures differ from traditional cultures, representing neither a normal basal state of thinking for humankind nor a distinct culture. Much available research is done on WEIRD populations. Although cultural norms differ, there is no strict binary distinction between WEIRD and non-WEIRD peoples. Where possible, I will include non-WEIRD research.

[5] These studies limited their cohorts of perpetrators to children, not to adults who might engage in abuse.

[6] Bully-victims are a diverse group who have experienced different forms of bullying. Lumping them into one group here may be an oversimplification [3368].

[7] Many children in the United States are taught the rhyme, "Sticks and stones may break my bones, but names will never hurt me." This short poem demeans the experience of the bullied child and should not be seen as an adequate or effective defense. It is not. Verbal attacks leave no physical scars but can inflict as much long-term pain as physical ones.

of reported abuse; their pain lasts long after physical scars heal. Verbal humiliation predisposes the target to self-doubt and turns the sufferer into a focus of repeated attacks extending into adulthood. When pernicious verbal violence dons the mantel of "harmless" or "bonding" banter," gaslighting can compound the aggression [51].

Adults: The Ignored Victims

> *"Lasciate ogne speranza, voi ch'intrate." (All hope abandon, ye who enter in!)*
>
> — *Dante's Inferno*, Henry W. Longfellow, translator (1871) [52]

Intimidation extends beyond children. Victimization of adults by adults happens, especially in rank-driven situations such as employment settings and nursing facilities. Journalist Andrea Adams recognized the similarities to school bullying and coined the term "workplace bullying" in 1988 [53]. Workplace intimidation and abuse are now recognized as a worldwide phenomenon [54]. Given that most people spend more waking hours engaged in work than other activities or settings, workplace bullying or "mobbing" presents a clear danger to the victim's health and identity [55, 56].

> *"If you can't make it, you're not tough enough to do medicine."* — on the bullying of interns, Colenbrander, Causer & Haire [62]

Some professions consider intimidation and mistreatment acceptable professional hazing or a rite of passage. Physician and nursing trainees in many countries[8] report rates of maltreatment hovering between 25-50% victimization [57, 58, 59, 60, 61, 62, 63, 64] with some estimates exceeding 58%. [65]. In the United States, interns report bullying by staff at levels topping 93% [66]. Additional bullying by clients has increased as cyberspace has rendered online bullying more common [67]. In an American study, the sleep deprivation

[8] These studies included competitive and cooperative cultures, including India, Pakistan, The United Kingdom, Australia, Korea, and Greece.

forced upon medical interns yielded a 16% increase in the likelihood of a car crash [68]. Still, supervisors often dismiss these torments as part of the internship process.

Adult bullying occurs beyond the medical profession. It runs rampant in jobs and society and is trending upward [69, 70]. One finds status-based mistreatment in any place where a recognized person—or class of people—has less power. It often takes the form of threats to professional or personal status. Workers dismissed as "short term" or expendable report similar problems, such as oil-rig workers [71]. High rates of bullying inflict volunteers and workers at nonprofit organizations, where the power and status differences are huge [72]. Hospitality workers describe micro or macro aggressions by guests and management, leading to high turnover rates, stigmatization, and reduced job efficiency [73, 74, 75]. This link between bullied employees and high attrition rates is not limited to the hospitality industry; one also sees it in higher-status and skilled jobs, including information technology, business schools, and health care [76, 77]. Presumably, the unreported instances exceed these.

Academia inflicts one of the highest rates of workplace bullying, with over 50% of university personnel reporting victimization [78, 79, 80], with much of the violence directed at the usual targets (i.e., under-represented groups based on gender and gender identity, race, disability) plus non-tenured faculty who often lack minimal workplace protection [79]. Despite the high figure, an estimated 60%-70% of cases remain unreported [81]. Nontenured faculty regularly suffer from financial bullying in university and college faculties, receiving significantly less pay for the same or similar work [82]. In higher education, non-tenured faculty report higher levels of intimidation than tenured faculty [83].

Recognized bullying tactics include shunning, dismissal of accomplishments, isolation, shaming, overwork, and the victim's physical, psychological, or financial destabilization [84, 85, 86]. Such abuses account for up to 50% of stress-related disorders in the U.K. [87]. In the United States, the number of workers who meet the definitional criterion for experiencing at least one bullying event per week approaches half [88].

Like children shamed by their status as victims, bullied adults often fail to report the intimidation to their superiors, many of whom are their abusers. Adult victims experience an increased risk of stress-related disorders, including PTSD, high blood pressure, coronary heart disease, and immune impairment. Prolonged intimidation psychologically and physically impairs people and leads to premature aging [89, 90, 91, 92, 93] and depression [94]. Many become too disabled to work. Many develop traumatic stress disorders requiring protracted rehabilitation before returning to the workforce—if ever [95, 96].[9]

Incidents of bullying escalate as people pass from youth and middle age to old age. Elder abuse, neglect, and discrimination, although long recognized as problematic, are often left uninvestigated. Social neglect, leading to isolation and loneliness, impacts between 5% and 43% of elders in Europe and the United States [97]. Seniors in nursing homes and elder care facilities report aggressions that can spark legal action, such as physical, sexual, and financial abuse. In the name of profit, organizational maltreatment in nursing homes can extend to the neglect of basic human needs and dignity, such as leaving elders to lay in their own filth and other lack of personal care, activity deprivation, denial of therapy, and insufficient time allotted for eating [98, 99, 100].

Even outside institutionalized settings, elder abuse flourishes. Resource competition contributes to this disgrace. Offspring compete for attention and resources as children. For a wolf cub or baby bird, competing with one's siblings can be a matter of survival. In many species, young, spry upstarts compete with older, weaker ones for mates, food, and territory. Younger male chimpanzees displace older ones as troop leaders, and younger, healthier birds will drive weaker, older ones from bird feeders. Paradoxically, this competition

[9] These medical costs and lost labor harm the worker and create a large drain on societal resources. The bully benefits by instilling fear and suppressing opposition.

increases with the quality of a territory or resource and decreases when the added advantage gained by the competition just isn't worth the effort.

Humans, too, compete for resources, including inheritance. As adults, we may guard against parents who spend too much of "our" expected inheritance. And we do not confine our worries to our siblings getting "our" resources. (A lawyer once told me he'd seen people "go to war over pots and pans.") As in other animals, where the prospective gain is high, resource competition increases, and wealthy families experience proportionally more disputes over estates than fiscally challenged ones.

We also worry about losing to newcomers, such as new spouses and love interests. Economic abuse of the elderly leads not just to a sense of betrayal, but also to emotional distress and a perception of isolation [101]. Financial exploitation ranks on par with neglect as the most common forms of elder mistreatment, with about 3.5 to 5% of seniors reporting the trauma [102]; for each reported case, an estimated five more cases remain undetected [103].

The growing animosity toward baby boomers in the United States has at least some roots in this generational dominance and resource struggle, positioning elder abuse as one of the increasing crimes of the 21st century. The harm from this abuse spills into the general society as older people, often left destitute by exploitation, become dependent upon state aid.

Common complaints by seniors about bullying include bullying about gender, sexual interest, and decreased cognitive capacity [104]. An LGBTQ identity amplifies the terror of institutionalization [105]. Many people and institutions dismiss the complaints of older people as signs of dementia, similar to how the protestations of psychiatric patients were treated in the 19th century, e.g., [106]. Compounding this, entering one's twilight years doesn't automatically bestow sainthood. The staff at nursing facilities can be abused by clients, and the elderly can endure abuse from their age-mates just as children can endure it from both adults and other children [107].

Most people have received memorable physical or emotional scars from a tormentor. The disconcerting aspect of these studies lies not in the quantity of victims but in the number of bullies. The figures suggest most of us, given the right circumstances, will participate in the dehumanization and violence or stand by while someone suffers at another's hand. We also mistreat people, failing to realize or admit our transgression. We differ by how often we engage in these power plays, the degree of the abuse, and how we deal with the consequences. We fall prey to the "fundamental attribution error": ascribing our violence to unavoidable circumstances but our mistreatment by others to inherent flaws in their characters.

The Impact: Nothing But a Bully?

"Have imperator, morituri te salutant!" (Hail Caesar! They who are about to die salute you!)

— Suetonius, *The Twelve Caesars*, 21:4 Gladiator's salute [108, p. 44]

> *"After a fellow basketball player sexually assaulted DP in the locker room, the coach let the assailant play so they wouldn't need to forfeit the game. He later told the team not to 'joke around with guys who can't take a man joke'."* — Sawyer, Carroll & Connaughton, 2010 [3454]

Playgrounds are brutal gladiatorial fields. Children attack, humiliate, and degrade each other, instilling a sense of powerlessness in their targets. When a bully gains control of a victim, nearby children will watch passively or even join the fun. We think of small children as the ones who pounce upon emotionally vulnerable combatants. Older children are more eager to eat their victims alive [109], especially during the middle-school years when the hormones of adolescence spike. Audiences participate in the culture of degradation [110], laughing until the sufferers crumble, stabbed through the heart with pain. Most avoid victims, fearing potential stigmatization. Aiding someone at the bottom of the social ladder can impose a devastating social cost [111, 112, 113].[10] Our primate drive for dominance makes behaviors that risk a loss of our status about as appealing as a root canal without Novocain.

Fledgling oppressors who humiliate their classmates appear to fight a senseless and remote battle. In truth, their stakes are as significant as life itself. Early social status sticks in primates, and these childhood encounters shape our lives. From a biologist's perspective, they battle to gain social rank, social integration, and the ability to reproduce [114, p. 412–431, 115, 116]. By damaging their competition, they increase their *relative* standing among their peers [117]. By destroying authentic, healthy pride in the victims, by removing their sense of control and replacing it with shame, the bully achieves dominance, if not prestige [118, 119]. Bullying can also alter one's lifelong ability to feel pleasure and derive a sense of reward from accomplishments [120, 121, 122]. It can alter the size and shape of the brain in its victims, leaving them with a reduced ability to think and understand the world (cognition) and an impaired ability to survive social stress; these results further reduce the ability of the victim to compete with the bully for social status [123, 36].

Despite its importance, childhood violence received little public scrutiny until a string of suicides and shootings brought the problem to world attention in the 1980s [124]. Public cognizance of bullying runs in cycles. It peaks with news reports of bullying leading to mass murder [125, 126, 127, 128], but awareness fades as the latest headlines wane. Our attention ignores the most common form of school-associated violent deaths: single-victim murders in the disenfranchised populations of urban schools [129].

---

[10] The social cost incurred by intervention can be reduced by manipulating the background culture away from acceptance of bullying behavior [3154] and toward a perception of fairness and an expectation of a positive response by authorities [3160]. For example, peer-buddy programs for autistic students can change the social image of people with autism and other aneurotypical humans, reducing the levels of intimidation and ostracism [3279] both through modeling by high-status students and by decreased opportunities for abuse during unstructured activities such as recess.

[10] For a list of the people directly denied treatment and followed by this study, see Gray [3143, pp. 6-12].

Fewer people have paid attention to the psychic and emotional damage done to tormented adults. Until recently, even less research has explored the psychological damage of bullying on bullies, although many people have speculated on the possible effects. Discoveries about the fates of bullies proved counterintuitive and surprising.

Initially, psychologists defined bullying as a narrow range of behavior. They portrayed it as an isolated form of violence in which one or more children perpetrated physical or psychological abuse on another child over repeated attacks [130]. In time, society recognized that bullying could wear additional faces. We learned about the intent to bully, how people recognize bullying, that adults can bully, and that bullies can attack adults [131]. The concept of "bullying" extended beyond the playground, and the boundaries that separate it from other forms of violence and maltreatment began to blur.

"Bullying" includes the malicious and repeated abuse of power over and degradation of underlings in the home and workplace [132, 133, 134, 135, 63]. Its meaning now extends to malicious hazing (ragging) of new students or under-classmates [136]. Individuals can bully. Intimidation can also arise as a vicious mob or institutional attack on a person's or group's physical, intellectual, financial, or emotional well-being [137, 138, 139, 95, 140]. Schools can bully. Laws can bully. Corporations can bully.

In corporate bullying, the problem becomes depersonalized and systemic. Schools, agencies, and businesses may tacitly or explicitly encourage bullying behaviors and policies to drive profits or agendas [141]. Institutionalized intimidation can take any of these forms, engage in retaliatory behaviors, or simply fail to protect employees from discrimination, shaming, and other oppressive or coercive behaviors [142].

Passive Bullying: Ignorance in the Foreground

> *"Ignorance, the root and stem of every evil."* — Attributed to Plato

Bullying can happen subtly over time by withholding information. Uninformed consent facilitates the purposeful continuation of harmful activities that can result in life-changing disabilities. The concept of abuse through "uninformed consent" has emerged as an issue. Hidden and unspoken conditions permeate social contracts from athletics to medicine and construction, and their impact has expanded in our digital world. Most of us escape the worst physical and psychological repercussions, but there remain winners and losers among the bullies, victims, and onlookers.

Uninformed Dissent: Medical Bullying and Disenfranchisement

> *"One of the most tenacious beliefs was that Blacks did not feel pain or anxiety, which excused painful surgical explorations without anesthesia on Blacks."*
>
> — H. Washington in Medical Apartheid [143]

Western medicine professes an ideal of giving patients information about their treatments, providing them with dignity, honoring their wishes, and not using them as experimental animals. Institutions and physicians have regularly and systematically performed unnecessary or unjustifiable medical procedures despite these values, including lethal treatments and invasive surgeries. This concern became more prominent with the disclosures at the Nuremberg trials of prominent Nazi physicians who had treated human beings as instruments for research without regard for their humanity [144]. It took longer to recognize the medical abuse and oppression of African Americans as unconsenting medical test subjects [145, 146, 147, 143].

As part of our increased focus on the patient, the concept of "informed consent" has grown in importance, as has disdain for unnecessary procedures [148]. In the United States, a routine lack of medical consent and the medical abuse of African Americans led to a widespread distrust of doctors and hospitals. The most cited example of this medical abuse is the Tuskegee syphilis study that started in 1932 and continued through 1972, the longest-known non-therapeutic "health study" on humans in medical history [149, 145, 150]. The United States Public Health Service allowed syphilis to remain untreated and run rampant in an African-American community so medical researchers could see how it developed. Even after treatment became feasible, doctors denied treatment and continued to let the disease ooze through the patients and community [151].

Sadly, the exploitations of the Tuskegee experiments do not stand in isolation, and the history of exploitation has left a legacy of distrust [152]. Abuses wrought upon African Americans have ranged from lethal experimentation and studies to forced sterilizations and substandard medical care. African Americans are less likely than whites to receive adequate pain-control medications and, when treated for addiction, they receive fewer treatment referrals [153]. Experiments subjected people to Mengel-esque procedures. Pneumoencephalography replaced the fluid surrounding the brain with air to take a picture of the brain [154, 155]. Children received injections of heart-damaging drugs in protocols calling for African Americans and Latinos while excluding Whites [156, 143, pp. 271-298]. Prisoners have served as test rabbits, their skin covered in patches of chemicals to test their reactions [146].

> *"It was cheaper to use Niggers than cats, because they were everywhere, and cheap experimental animals."*
> — Dr. H. Bailey 1977 speech on the misuse of African American subjects [143, pp. 10, 347, 357]

The treatments of ailments differed between Whites, Blacks, and Latinx, contributing to higher morbidity and mortality rates in disenfranchised communities. This differential treatment includes a disregard for infants [157]. The forced display of black bodies and other medical abuses spanned from slavery through to the last decade of the 20th century [143] and beyond. Even after death, the sanctity of the body could fall prey to the degradation of grave robbing and non-consensual dissections.

The fallout of the ongoing medical abuse of African Americans includes an earned distrust of medical authorities and procedures. A suspicion of vaccines, including those for annual influenzas, flows from these past abuses [158]. During the early COVID-19 epidemic, a disproportionately large percentage of African Americans suffered and died from the disease due to this distrust and to the lack of treatment access [159, 160, 161, 162]. The distrust of doctors, which sprouted from a history of abuse and medical exploitation, now contributes to people's reticence about vaccines, including ones for COVID-19.

African Americans may have suffered the worst medical bullying in the United States, but other disenfranchised people have also suffered. Orphans, emotionally and intellectually challenged, the poor, and the captive and institutionalized have become unwilling trial rabbits [163]. American corporations, universities, and hospitals continued to use the incarcerated as subjects far beyond Tuskegee. Prisoners served as test-rabbits for household products, disabling psychogenic and carcinogenic drugs, and Phase-1 drug studies [146, 147]. The practice was officially discontinued in 1975, long after the trials of Nazi doctors at Nuremberg [164]. Washington [143] cites numerous examples beyond this supposed expiration date. It is easy to point to the US and decry the abuses; neither the US nor capitalist countries stand alone in such misdeeds. Cruelty is the dominion of all.

The concepts of "necessary" and "informed consent" remain fuzzy. The idiosyncratic perception of risks and benefits by doctors, patients, hospital administrations, lawyers, and accountants contribute to the confusion. Some procedures, such as the purposeful exposure of humans to high-dose full-body radiation [165, pp. 227-263] or the non-consensual irradiation of children's heads for "ringworm" [166] are Mengel-esque in nature. Theoretically, disallowed procedures include ones for which the risks and the costs exceed the perceived therapeutic benefits—or at least a significant level of relief for a patient [167].

The position of authority held by physicians and researchers amplifies the human illusion of personal, intellectual, and moral superiority [168, 169]. This inner confidence can lead doctors to conclude that their patients either share their views of a procedure or have a flawed perception [144]. The position of power held by administrators over physicians has a similar effect. Our other natural human tendency, to obey or be influenced by perceived authorities [170], can lead doctors to adopt the biases of their administrators. The repercussions of such power differentials differ between regions and countries [171].

Doctors can fall victim to a financial and legal system that forces them to conduct unnecessary procedures [172]. Many feel obligated to perform redundant tests. Demands of patients can instill a fear of litigation should they fail to take the most aggressive approach to treatment. These fears can drive physicians to overuse tests and therapies despite potential negative consequences [173].

Reasons for overtreatment in the United States extend beyond the physicians' fear of liability. As additional factors, doctors cite pressure from patients and difficulty accessing the medical records of prior tests and procedures [172]. Malpractice lawsuits contribute to the increased use of defensive medicine and the cost of malpractice insurance for American physicians [174, 175]. Much litigation contends negligence in diagnostics, and many unnecessary surgeries in the United States result from physicians' uncertainty about the effectiveness of specific procedures [176]. Administrators may pressure physicians to perform redundant tests to increase profit margins [177][11] despite the cost-savings to the patients and insurance companies of eliminating duplicate procedures [178]. The American College of Physicians has called to end this financial bullying by reforming the payment and delivery system for healthcare, thus weakening the link between remuneration and excess testing [179].

---

[11] A physician I know refers to doctors who engage in these behaviors as "dollar docs." Doctors who succumb and do unwarranted procedures remain a minority. Far from engaging in an accepted practice, dollar docs suffer a bad reputation among their peers because they spread the stigma. This counterpressure from their peers acts as a partial check against this exploitation.

Student-Athletes

"Je préfère crever de passion que de crever d'ennui!"
(I would rather die from passion than from boredom!)

— Zola [180, p. 291]

Other fields can also dehumanize and objectify people by withholding or minimizing information. Student sports have a long history of imposing physically damaging activities on participants while presenting them as "high-status" activities. For example, student football players average from approximately 240 head collisions per season for those aged 9-12 years to 650 head impacts for high school players [181, 182, 183]. Estimates of how many youthful players will develop concussions from playing the game range from 5% to 20%. Among collegiate players, almost half of the concussions (48.5%) happen during preseason training, without any violence being inflicted by the other team. [184].

Short-term and long-term cognitive problems often follow concussions [185]. Young brains respond more intensely to trauma than the adult brains of professional and college players who have caught the public's attention. However, damage occurs, and the aftermath can stay with the players throughout their lives. Despite this information, we promote and idolize contact sports, such as football, soccer, and hockey [186], without providing this vital information to the families and players [187]. Students who endure head trauma often fail to report the injury, fearing losing playing time or disappointing their team [188].

In the United States, colleges consider players amateurs engaged in club activities [187]. This classification exempts them from employee medical coverage. Student-athletes earn millions of dollars for their schools but receive minimal compensation for their efforts and images. The College Athletics Associations have rules that micro-manage what they can and cannot accept, down to types of food. Colleges make thousands of dollars by marketing the image of an athlete but can suspend a hungry player for receiving a bag of food [187, pp. 68, 71, 352].

The demands made upon them prevent many student-athletes from graduating on time or with high grades—if they graduate. Students who participate in sports that generate revenue for their schools lag behind their peers in academic progress. African-American football players have a graduation rate of only 50.2% [189].

The shoddy treatment extends beyond financial and educational abuse to physical violence. Spectators must include themselves in the bullying scenario [190]. We thrill at the moves that cause head injuries. Our willingness to accept the costs to the student-athletes encourages schools to market this exploitation.

Academic Bullying

> *"[An open access policy] that leads to great opportunities, but it is too difficult for many to traverse, is not really granting access at all, but a cruel hoax."*
>
> — Scherer & Anson [191, p. 2]

School athletics, such as football and hockey, are not the only fields that objectify students under the guise of increasing their status. Higher education, in general, has encouraged students to enroll in programs and colleges that can leave them woefully ill-equipped to obtain a job. Nearly 50% of American higher-education students will not earn a degree [192]. Colleges knowingly take underqualified students under the guise of opening the door to success. Often, the school admits them into expensive remedial programs, knowing that few will succeed. Too many leave the institution with no degree, no further qualifications, no education on alternatives to college, and an increased debt burden [191].

Reduced enrollment standards at schools allow unprepared students to begin college studies. Institutions have made such changes in futile attempts to fill seats as their pool of academically capable entrants erodes. These policies lead to qualified students avoiding those schools and the schools admitting a higher percentage of less-able students to fill those same seats. Fearing drops in retention and graduation rates,[12] administrators may concurrently pressure faculty and departments not to fail too many students, even though many entrants lack the fundamental skills to succeed.

When the act of failing students invites retribution against teachers, grades carry less meaning, and institutions reduce the value of their degrees for all their graduates. Students exit with few prospects for finding college-level work based on diluted academic credentials. Graduates who gain college-level employment carry burdens of high student debt, diminishing the lifetime financial value of their educations. As universities have played their educational Ponzi game, many North American high schools have de-emphasized programs in the trades. They treat these professions as inferior choices for the student and a failure for their school. Personally, I'll survive longer without understanding string theory than with a clogged toilet or leaky roof. Coming from a family of tradesmen, I cannot understand the condescending attitudes of some guidance counselors who denigrate the trades.

The correlation between college degrees and higher salaries is reported in confusing ways. In 2013, the median income for full-time workers, ages 25 to 32, with a bachelor's degree or higher outstripped the median wage of someone without some college or a two-year degree by about 52% [$45.5K vs. $30K; [193]. But financial stagnation and the theft of opportunities for advancement contribute to disconnection and alienation [194]. Cited figures often give the "median" income (half above and half below), do not specify the salary range, and include people with advanced degrees [193]. That study did not address the lack of training alternatives to college. Nor did it criticize the preference of many employers for college-educated workers for jobs that should not require a college-level education.

---

[12] The retention rate is how many students complete their freshman year. The graduation rate is the percentage of students who complete the coursework for their degree, typically within 1-1.5 times the predicted time for completion. Open admissions policies tend to drop both values because underqualified students, who are less likely to finish, enter with the qualified ones.

Many graduates leave school without the skills to earn a position that compensates them for the time and money invested in college. In the 2015-2016 school year, 84% of American students completed high school [195]. The percentage of these graduates who enrolled in college varied from 66% to 70% between 2009 and 2019[13] [196, 195, p. 462, 197]. At its ebb, the probability of attending college remained over 50%.[14] A smaller portion of the population finishes college with a bachelor's degree or higher, moving from 27.5% to 32.1% between 2005 and 2019 [198, 199].

This flood of graduates exceeds the economy's ability to place students into financially rewarding college-level positions. High-paying employers prefer students who come from more prestigious schools. Competition decreases the odds that students from second-tier schools will find engaging or lucrative post-college posts. Yet, as a society, we continue to denigrate the trades, a traditional means of financial advancement that drove the rise of the middle class.

Many students have transformed college into a stop-gap—a place to kill time and accumulate debt until deciding what to do with one's life. To keep their doors open and to maintain bloated bureaucracies, universities encourage this practice, using their influence to engage in systematic monetary abuse of these lost souls. We produce far more college graduates than there are college-level jobs, including entry-level jobs. Many graduates and non-graduates will spend much of their working lives repaying college debt, which often blossoms with interest. They will work at jobs paying near minimum wage, often without benefits, which should not require a college education.

---

[13] Some estimates put the low end of these attendance statistics as low as 56% following the real estate crash. Even though the difference is about 10%, the figures remain above 50% attendance (not graduation). Many students wrack up debt through attendance but never graduate.

[14] Graduation rates are more difficult to estimate because they usually do not account for transfer students who start at one institution but graduate from another school.

A learned expectation of failure leads to depression and doubts about personal competence [200, 201]. Society intimates that under-employed graduates have squandered their college education. Graduates lose self-efficacy and experience learned helplessness that can paralyze future endeavors.[15] Moreover, it reduces personal "grit," the motivation and effort to succeed [202].

Accrued educational debt negates much of the remaining salary gains. Despite this knowledge, we market college education to people with a promise of a road to advancement. Student debt in the United States topped 1.31 trillion dollars as of December 31, 2016 [203, pp. 2, 5]. As of 2021-2022, that figure had risen to $1.58 - $1.77 trillion in student loans. This debt is spread unevenly across 45 million borrowers [204, 205].

Personal debt due to student loans surpasses credit card debt and is second to mortgage debt for most Americans.[16] Under US law, bankruptcy does not discharge student debt, and late payments on student loans range between 10-11% [206]. Overdue payments do more than damage credit scores. They trigger collection fees, placing further financial burdens upon former students. High schools and colleges use the lure of a high-status future to engage ill-prepared people in a high-risk monetary activity that can have long-term negative consequences. Often, the university will not provide adequate counseling or discuss alternative paths to success, streaming ill-equipped students into courses with insufficient background, positioning them to fail. Although the school benefits from having a paying butt sitting in a seat, the student may never recover from the incurred debt. If such continuing financial abuse by an institution is not an act of bullying, then nothing is.

---

[15] The findings on self-efficacy are not limited to Western industrialized (WEIRD) cultures but appear across cultures [3452, 3405, 3171]. Populations showing this effect include Turkish [3377], Saudi [292], Japanese [3414], Chinese [3480], and Indian populations [3175].

[16] This figure is also misleading. Educational loans exceed mortgage loans for people ages 18-29. By age 30, mortgage loans exceed educational loans and remain the primary debt source across age categories [3294]. However, older people attended college at a time when less debt accrued.

The Fallout

> *"Why died I not from the womb? Why did I not give up the ghost when I came out of the belly?"*
>
> — *The Bible*, The Book of Job, 3:11 [207]

Victims extend beyond the direct subjects of violence. Casualties include families and friends impacted by the injury to loved ones [208]. Damages embrace people affected by the lowered social position that can last through life [209, 210, 211]. Long-term consequences of victimization include increased risks for suicide, alcohol dependency, inability to form relationships, adult obesity, and mood disorders [212, 213, 214, 215]. Childhood maltreatment survivors report reduced joy in life [216], an impaired ability to break free of psychoactive drugs such as alcohol and tobacco [217, 218, 219], and less success in combating some physical and mental health disorders [220, 221, 222, 223].

Victims of childhood violence and intimidation experience reduced school performance and damage to their adult financial outcomes [224, 225]. Oxytocin, a hormone that feeds into our feelings of love and acceptance, can turn against the victim to promote stress avoidance and damage relationships[17] [226, 227]. Long-term harm extends to employers and corporations that experience lost income and expertise from bullied workers.

An unvarnished truth became lost in the initial skirmish over the definition: Bullying Is Violence. Bullying exerts power, control, and dehumanization over another through emotional, physical, sexual, verbal, financial, or social violence. It seeks to dominate and degrade. Attempts to segregate bullying, defining it as a side nook, do not respect the experience and pain of the bullied. No natural boundary exists between

---

[17] Oxytocin is far from being the "love hormone" as proclaimed by popular literature. Its impacts depend on context and can range from trust and love to distrust and joy in the harm that befalls others (Schadenfreude).

bullying and other maltreatments that intimidate, degrade, disgust, inflict pain, or brandish physical, financial, temporal, or psychological control.

As children fight their battles, they learn their places in society. Winners develop confidence in their abilities. They walk taller and straighter and experience a higher standing than they would otherwise receive. Winning conflicts alters body chemistry. Stress hormones go down, and bodies increase the production of rewarding chemicals, such as dopamine, oxytocin, and endorphins. The *interactions* of hormones allow the victors to appreciate their successes. Higher chemical payoffs for behaviors increase the likelihood of a person repeating it, be that activity a STEM project, painting, or holding another down and spitting in their mouth. Rewarding physiological changes create feedback cycles that reinforce the actions that produce them.

The Winners

> *"Nemo reside"*
> *(No one left behind)* — US Marine motto

Social mammals compete for standing within their society. In humans, most seek encounters to raise their status and reputation [228, pp. 4-42]. Aggressive battles designed to handicap a rival range from microaggressions to genocides.

When someone wins a power contest, they[18] receive a physiological reward. Four major hormone classes and their receptors bolster our reward pathway: dopamine, serotonin, endorphin, and oxytocin [229].[19] In primates, high-stakes payoffs include improved social rank, which can trigger these rewarding chemicals.

[18] Reminder: I use "he" or "she" interchangeably unless otherwise stated. Most statements will hold across genders. At the time of writing, society has not accepted a universal, well-established common gender pronoun.

[19] There are multiple subtypes of many subtypes and receptors. For simplicity, I will usually collapse them into one.

At school, winners often are or become the cool kids who thumb their noses at those they deem beneath them. On the job, they can achieve rank by usurping credit or demeaning their coworkers' contributions. They exude an attitude of superiority and develop a circle of minions who further reinforce their ego.

Contrary to popular belief, victimizers do not automatically suffer from low self-esteem.[20] On the contrary, they often display narcissism and the favorable self-image accompanying power and station [230, 231, 232]. They often attack weaker individuals, not through a lack of confidence but because these targets provide low-risk, effective displays of power and strength to rivals [233]. As we age, these spectacles can attract sexual partners [234, 235].

Across most studied cultures, victimizers and defenders occupy and maintain high social standing among their peers. On average, the bully holds a higher social rank and becomes more popular with time [236, 237, 238]. The prey occupy the lowest rungs of the hierarchy and continue along this trajectory. They often become bullied adults, have ongoing mental health issues, reduced self-esteem, and impaired satisfaction with life [239, 240, 241, 242, 243].

We want to believe the world is fair. We trust that predators will receive their comeuppance and fail in life or become prey themselves. We enjoy tales of fiction in which the bully or the evil perpetrator of violence gets his just desserts. Too often, reality dashes these dreams. Like our fellow primates, we engage in social ranking. Much of our aggression focuses on reputation and standing within a community. Like other social animals, humans learn to choose their battles and use differences in power to maintain or increase their position relative to their group [244, 245, 246].

---

[20] These data refer to traditional bullying, not cyberbullying. Some evidence suggests that the link between elevated self-esteem and active cyberbullying is weaker than other forms of bullying [3202].

People who achieve status early have an advantage over those who learn they are losers. Adults share this pattern, but childhood traumas shape our lives [241, 247]. It differentially damages women because girls experience more socially condoned and intimate personal onslaughts than men [248, 249].[21] Future behaviors and a sense of enfranchisement for high-ranking males also take shape during this period [250].

Young bullies move from grade school to high school and college to the workplace. As accomplished tyrants, they enter these new institutions with the honed senses and skills of a predator. They often associate with fellow victimizers and peers who reinforce their rank and predatory behaviors [251].[22] In this foreign land, these seasoned hunters spot new prey. They gravitate toward others who show signs of high social rank and learn the Machiavellian politics that determine who will rule among the alphas and the size of their new kingdom.

These individuals know how to survive and succeed when they emerge into adulthood. The bully child becomes a tyrannical adult [242]. They have developed aggressive strategies to bolster and defend their self-esteem [252, 253]. Through status and self-confidence, both males and females achieve goals and attract a choice selection of potential mates [119].

In humans and other primates, male power and success attract females. These attributes provide resources and standing for their children. By selecting successful males, women improve their offspring's chances of survival, social rank, and reproductive success. (Sorry, but as stereotypical as the image may be, all else being equal, CEOs will get more dates than low-level grunts.) She treads carefully, choosing a mate with the highest possible rank without choosing someone who could harm her children. Dominant men can exploit

---

[21] Other disenfranchised groups also suffer from early-life victimization [3433].

[22] There is little consensus about the underlying personality of bullies other than they are disagreeable and often engage in Machiavellian behaviors [3185]. Positive self-images are cited as part of the complex, but other studies dismiss this feature as sub-clinical narcissism. It's hard to distinguish between these descriptions; some of the distinctions may spring from the mind of the beholders.

success to attract a bevy of women across their lifetimes, a mechanism designed to leave more of their genes in the next generation [254, 255, 256].[23]

As a strategy for success, tormenting can incur high costs, especially from other alphas who can form coalitions to maintain their status. These costs limit the benefits of bullying as an adaptive strategy [233]. Both men and women form coalitions to reduce the success of degrading and bullying tactics as paths to reproductive victory.

> *"These people just made these tokens up and sold them for money. You will say to yourself, 'How do I get in on that?' "* — *How Not to Play the Game* by Matt Levine [258]

Although bullying behavior *appears* maladaptive for society, it can prove lucrative for the individual [e.g., relational bullies [257]. With an aggressive skill set, successful bullies do more than cope; they flourish. They traverse the social maze, and their self-confidence can produce a resiliency their victims will lack. That confidence also inspires trust in subordinates and explains the loyalty bestowed on dictators and the willingness of people to invest in "confidence schemes." Cryptocurrency, for example, depends heavily on investors' confidence in the coin and the people running the business [258].

When faced with traumatizing situations, a successful individual will cope better than someone with a history of failure. Self-confident people have more psychological tools and use their ordeals to strengthen their self-image and reframe tribulations as positive growth experiences [259]. Bullies benefit from this process. Victims lose.

---

[23] Bateman's principles have drawn criticism for being sexist. As much as I want them to fail for humans, statistical analyses support them. Mulder and Ross [254] found a possible exception among the Pimbwe. Pimbwe women gained fitness when they had more spouses, and men lost fitness, possibly due to mate-switching by women to improve access to resources (trading up for a higher quality male). However, the study did not account for extra-pair matings (i.e., infidelity) or how the prolonged male fertility relative to females interacted with these factors.

The Losers

> *"I'm starting to feel good about myself, and every time that happens, you have to tear me down. I'm not gonna be your fat sidekick anymore."... "It doesn't even matter if you keep the weight off. You'll always be a fat girl. It's just who you are."*
>
> — *Britany Runs a Marathon*, Colzizzo [260]

In contrast to the perpetrators, victims develop a perceptual framework of pessimism and hopelessness that reduces their ability to perceive success [261, 262, 263]. Victims often come from a disenfranchised or lower socio-economic background than their oppressors [264], making them social targets even before the first attack. These losers of the playground arenas experience social devaluation. They have learned shame[24] and their place before the first confrontations begin [265, 85, 266]. Hunched shoulders and a withdrawn posture may signal their low status. They learn to watch the ground rather than to meet the gaze of others. Eye contact invites discord.

The damage extends far beyond changes to posture. It is a significant risk factor for suicide, the tenth leading cause of death in the United States [267, 268, 269]. In the habitually bullied, several brain regions change their activity or show changes in volume [270]. Some areas grow; others shrink. The white matter, those cables of neurons tasked with connecting our little grey cells, mature earlier and reduce their density [271, 272]. The regions deep in the brain, the subcortical structures associated with emotions and reward recognition, are particularly sensitive to bullying [273], as is the "dorsal striatum," an area related to automatic and habitual responses [36, 274].

Altered brain structure reduces the ability to acquire, store, and retrieve knowledge [123]. The stress hormones elevated by chronic bullying can reduce the size of the hippocampus, a memory-processing unit

[24] In traditional small societies, the shame a person feels for committing a taboo correlates with an understood devaluation of that person. Our species uses this "shame system" as part of normal social functioning, reinforcing ranks [266]. The study did not include how the shame system interacts with the "act" of being victimized.

in the brain critical to learning. Children who experience repeated bouts show smaller volumes in this brain region, which can lead to anxiety, stress, and memory impairment [275]. Areas and networks involved in attention, navigation, and memory show abnormal structural changes, as do other networks involved in attention [276].[25]

A weakness can paint a target on a victim or identify an alpha [277]. People who differ from the norm—such as people with psychological, intellectual, or social disabilities—are targeted for bullying. People with Asperger's syndrome [278, 279, 280] and other neurodivergent conditions [281] become targets because of how they interact with neurotypicals [282, 283, 278], as do those who stray from heteronormative behaviors [284, 285, 286].

Casual observers often miss the damage from bullying on these people. Their usually low-key reactions to abuse fail to trigger empathic behaviors in others.[26] The downward-shifted spectrum of responses can mask their pain [287]. Muted responses to bullying make people with divergent behaviors safer targets for spite than people whose social demeanors fall closer to societal norms.[27] 94% of adults with autism report a history of victimization [288].

By choice or lack of choice, targets associate with the other wounded or people who might defend them [251]. The ability to concentrate and work declines in the quarry, while the bully's performance remains on par with her peers [289, 88, 290].

---

[25] Posterior cingulate cortex, dorsal attention network, fronto-parietal network, ventral attention network, cingulo-opercular network, cingulo-parietal network, and sensory-motor network (Wen, et al., 2023). Not all changes are found in every victim, and age, gender, and type of abuse influence the trauma.

[26] Their low-key reactions fail to trigger mirror neurons, a key component of expressed empathy. I will cover mirror neurons and empathy in a later chapter.

[27] The low-intensity response of many people with autistic traits has led to the mistaken belief that ridicule and social ostracism do not bother people with autism. About 45% of people with Asperger's Syndrome suffer from *gelotophobia* (fear of ridicule) as compared to only 6% of people without Asperger's Syndrome [3456, 3313]. Despite this personal fear, there is no apparent difference between these two groups in the enjoyment derived from *katagelasticism*, laughing at others [3470].

Victims of predictable violence dread their place of abuse. For children, school is often the central location of torment, and they learn to despise it [291, 292]. Victims do not raise their hands in class because mistakes cause humiliation. They feel they have neither the right nor the ability to be heard. Initiative creates censure. Speaking invites blame. They grow into adults with diminished self-confidence and take few chances for fear of doing the wrong thing [293, 294, 295]. Despite its crippling effects, the study of gelotophobia, the fear of humiliation or ridicule, did not begin in earnest until this century [296]. Although our desire for group approval decreases as we reach adulthood, peer victimization can entrench this need at childhood levels. This self-doubt fosters troubled social interactions and future victimization [297]. Here, boys and girls may differ. Cis-gendered boys typically blame others for their victimization and externalize their pain. Girls blame themselves and internalize it [298].[28]

Reduced social rank creates a feedback effect that can spiral out of control. Stress hormones increase as humiliation increases [299], leading to changes in brain chemistry and neurodevelopment that can last a lifetime [300, 301, 302]. These wounded develop maladaptive thought processes, distorted views of social situations, and emotional trauma [303, 304, 305]. They become haunted by ongoing self-blame and a sense of abandonment [306]. Their faith in their decisions diminishes, as does their confidence in the ability of authorities and loved ones to limit or control the abuse [307, 308].[29] Their reduced faith in their skills even impacts the choice of a safe mate, associates, and financial decisions.

Years after the victimization, the socially or financially successful adult may not recognize or emotionally register personal successes. They carry an epigenetic memoir—a molecular bookmark—of what happened

---

[28] These gender differences are not necessarily hormonal or innate. They could arise from learned gender expectations. Less is known about transgendered and nonconforming youth compared to their cis-gendered female peers. One study [3461] found that TGNC individuals may resemble cis-females in their reactions, but that study's sample was biased toward biological females.

[29] Cyberbullying enhances the inability to control or escape the impact of a verbal bully or rumors. With cyberbullying, the victim has a perceived inability to escape the tormentor's domain. Online harassment can follow a person to new schools, jobs, or communities while the bully remains safe behind a veil of anonymity.

during the bullying phase of his or her life [309, 310].[30] This history forms by turning specific genes on and silencing others [311, 312]. These epigenetic changes alter behavior, thinking, and immune function [313]. He may produce insufficient rewarding neurotransmitters and neuroregulators, such as dopamine or endorphins, to provide the rewards experienced by an alpha upon minor accomplishments [300]. Thus, early victimization correlates with an increased danger of substance abuse in adolescents and adults [314, 315, 316, 317].

In contrast, their childhood tormentors learn to frame experiences to bolster their egos. Minor successes can reinforce their affirming self-image. They can get a thrill of accomplishment by eight in the morning, just because they completed morning chores before going to work. High-power individuals draw inspiration from themselves more readily, choose social interactions that provide rewards, and prioritize their social needs over others [318]. As a result, they express themselves more freely and have more rewarding emotional lives [319]. Social dominance leads high-ranking individuals toward further success and power. Subjugation leads to failure, expectations of failure, and reduced quality of life [320].

The harm exceeds the psychological and sociological effects. Various adverse childhood experiences, including bullying, witnessing violence, or the death of a parent, increase the probability of multiple medical symptoms [321, 322, 243, 323]. These adults have at least a doubled threat for emphysema, autoimmune disorders, suicide attempts, depression and anxiety disorders, cancer, and obesity [324, 325]. Even after accounting for other risk factors, such as obesity and diabetes,[31] people with seven or more adverse childhood experiences had a chance of heart disease 360% over those with none of these markers.

---

[30] A person's "epigenetics" is the record of events created by turning on and off sequences of DNA. Identical twins may be virtually identical in DNA at separation, but their epigenetics diverge through life as they experience different events.

[31] Obesity, depression, and diabetes correlate, and obesity is often cited as a risk factor for diabetes. They also relate to chronic inflammation [3088]. Correlation does not imply causation. Long-term inflammation or depression could

Survivors of childhood bullying have a higher rate of body-wide (systemic) low-grade inflammation, an indicator of reduced health and shortened lifespan [326]. The incidences of autoimmune disorders, cancer, and emphysema increase by at least 100% [324]. These data fit with animal studies showing the quality of our immune system improves with social status and rank-associated behaviors [327, 328, 313].

Physiological data dovetail with molecular studies. Stress, including stress from bullying, damages our organs, tissues, and the expression of our genes. Long-term molecular damage from victimization includes changes at the tips of the chromosomes, the telomeres. Telomeres shorten as we age (telomere uncapping). Children with stressful lives have increased levels of a "stress" hormone (cortisol) and decreased telomere length [329, 330, 331]. Adult survivors can have telomere (chromosome tip) reduction corresponding to an abridged lifespan of five to ten years and elevated susceptibility to age-related diseases [332, 333, 334].[32] Such changes in physiology create a biological foundation for the social and psychological changes manifested by the victims.

Bullied children do not respond to social and educational stimuli like other children [335, 336]. Praise ceases to work as an effective behavioral reinforcement. If it draws attention to the victim, it endangers him. Tormentors will seize the opportunity to belittle the accomplishment. They may attack on the playground, in the study hall, or in the neighborhood. Some pounce in front of a teacher or other authority. The bully could be an adult [337, 338] or receive a tacit or overt adult sanction. In one infamous case, a principal refused to take corrective measures against ongoing abuse, even after one boy pushed his victim down and urinated on him [339].

---

trigger weight gain, or weight gain could create conditions to foster the former. Social inequality and victimization, including childhood victimization, may play a causative role in the current obesity epidemic [3519, 3135].

[32] The relationship may not hold for witnesses of traumatic events without intrusive symptoms such as nightmares, distress, or flashbacks (Criterion A trauma) [90].

Figure 2-1. Poverty does not always harm to virtue (or Struggle of the Rich Against the Poor), Otto Vaenius, ca 1646. The artist and humanist satirized the behavior and fears of the upper classes long before "Occupy Wall Street" became a cry. In the foreground, a woman threatens to beat a man with her purse to take possession (beehive?). Behind her, a rich man dehumanizes a peasant, riding him like a horse. [3463]

With noble intentions, parents can feed this cycle. They can worsen the situation by criticizing the child for "introversion" or for somehow being responsible for his state. They push their child to be more gregarious, more outgoing with peers without appreciating or acknowledging the perceived and real dangers. Parents belittle the threat, assuring their children that their persecutor is "nothing but a bully." This dismissal implies terrorization holds minimal importance.

Those who have not suffered under the yoke of a bully underestimate the vulnerable state of the victim [340]. For example, women who fail to fend off the humiliation and torture of rape by an offender find themselves victimized again, portrayed as sluts, complicit or deserving of the attack, and society tends to minimize their shame and post-traumatic stress [341, 342, 343]. The cowering and apparent acquiescence to dehumanization by their terrorized slaves convinced many Nazi concentration camp guards of the natural inferiority of their victims, separating them from humanity, making subsequent degradation easier to justify [344, 345, 346].[33] By denying the humanity of another, we can gain carte blanche to use "it" as an object devoid of moral concerns. If, as the old saying goes, nothing succeeds like success, then nothing creates a downward spiral like a history of failure.[34] Through repeated bouts of persecution, the victim learns that resisting yields more grief. Many expect these

[33] Despite their debasement, prisoners maintained covert resistance cells and engaged in active sabotage despite the mass reprisals. These acts included opportunistic attacks on the guards and the destruction of confiscated goods.

[34] Many experience this phenomenon when trying to break a habit or an addiction. A history of failure makes the cycle harder to crack.

walking wounded, be they children or adults, to respond to torments with more aggression than they can muster [347, 348, 349, 350].

Parents want the problem to disappear. Victimized children cannot meet the perceived expectations of their parents. Demeaning their oppressor as "nothing but a bully"—words meant to comfort—further lowers the dominant persecutor's acknowledged social rank and thus the status of the tormented. The persecuted hear that their pain lacks validity. Attracting attention can reinforce his role as a target for further abuse. Her brain understands that she has again failed to meet society's expectations. The oppression has begun to rewire his brain to recognize success as failure and praise as condemnation.

Victimization by bullies was recognized in children long before we acknowledged it in adults. Therefore, much of this data refers to children. Childhood experiences form the template that crafts the adult nervous and hormonal systems. Employers, neighbors, and family members may substitute for supports and bullies in the same paradigm, but the injuries follow a similar pattern.

Did I Earn It?

The human brain processes undeserved recognition and praise differently than rewards perceived as earned [351, 352]. A history of humiliation and defeat makes it challenging to recognize successes as earned. The belittled individual responds to praise as if it holds little worth. When people no longer see themselves as having control or merit, they cease believing they can exert a constructive force on their world[35] and place little value on their accomplishments.

[35] Self-perception as a positive force has a strong cultural bias. In cultures that focus more on a person's role as part of society rather than on their efficacy as individuals, the bullied person will see their role in the social units as diminished.

Disempowerment continues long after the degradation ends. Even on their most productive day, survivors think they accomplished nothing. If the parents of a non-bullied child put their child's drawing on the refrigerator door, the child will gaze at it with pride. The brain rewards the gesture with a good dose of reinforcing chemicals such as oxytocin and dopamine. Habitually victimized children can feel gaslighted and experience the parent's recognition with shame and a hormonal stress response (cortisol increase), amplifying self-doubt. Even when not intended, this inadvertent gaslighting leads adults to self-doubt, a rejection of praise, and conformation to expected subservient or conformist behaviors [353].

The power differential magnifies this condition for children. A child, faced with inconsistencies in experiences, learns to distrust reassurances by authorities, believing instead that they are humoring her. The parents may feel pride, but the victim sees more humiliation. He might test the sincerity of the caregivers by giving them flawed work. If the parents fail the test and praise the shoddy creation, the child gains evidence of personal worthlessness and will no longer trust their praise and begin to internalize a belief: *My parents must feel pity to lie this boldly. I am a greater failure than I thought.*

In groups as diverse as soccer players [354], children [355, 356] and nurses [357, 358], bullying leads to a lost sense of accomplishment, physical exhaustion, and personal devaluation. This burnout triad appears across society in workers who feel undervalued and intimidated, including children. A sense of betrayal and loss of support accompanies psychological and physical burnout [359]. Businesses engaging in the practice or who fail to control it should measure their costs in terms of loss of talent, initiative, the silencing of employees, training hours, reputation, and dollars. Industry suffers from the stress contagion caused by those who seek power by intimidation through the use of abusive and uncivil behaviors. Fiscal costs and psychological damage mount as the effects of burnout spread from supervisors or workers to their subordinates and families [360, 361, 362, 71].

As the victimized move to high school and college, bullied grade-schoolers enter with reinforced behaviors that push them toward the social ladder's bottom rungs. They have learned their place in society and know the dangers attention can bring. If fortune smiles upon them, they find refuge in a niche without competition from the alphas. Some victims develop a vocation or avocation in a subject requiring more work and time than the socially engaged bullies are willing to invest. Males labeled" geeks" and "nerds" often retreat into science, art, and economics clubs—activities not perceived as archetypically masculine. Because their social lives travel alternative paths, some find refuge in books.[36] Still, on average, their academic scores fall [363].

The Onlookers

> *"A species must possess some adaptive trait that allows it to negotiate the environment. Birds have their wings, the turtle has its shell; mankind has the tribe, the city and the civilization."*
>
> — Crescioni and Baumeister [364]

Bullies employ audiences as tools for humiliation and dehumanization. Although they can potentially turn against the alpha and support the target, this risk is minimal. The predator, not the prey, chooses when to pounce. He influences who becomes the audience; she determines the course of the attack.

Onlookers may side with the downtrodden, but their courage gains little social currency, and intervention may decrease their social standing. Our preference for supporting high-status individuals (including dominant social classes) is an engrained threat-deterrent device also occurring in our closest relatives, the

[36] "Nerd" no longer carries the pejorative label it once had. With the introduction of personal computers in the 1980s, nerd pride developed, and academically oriented people came out of the closet, meeting each other [3162]. Nerds became more mainstream, and people who participate in "nerd culture" became less labeled and bullied. However, the moniker of "nerd," a term traditionally carried a low status, can still sting. For boys, the word bore the connotation of being effeminate [3065]. Not all nerds were bullied, but nerd culture accepted and provided refuge to many who were.

chimps and bonobos [365, 366, 367]. Defending the oppressed entails risk unless one has a secure position within the community. Without social support, most people will back the oppressor. Integrated groups can create alliances to stand against an aggressor more easily than a random audience. Still, the popularity of the harasser typically outstrips that of the target, preventing most power shifts.

When a high-ranking individual slaps down a bully, he can elevate his status. Here, I give a personal shout-out and thank you to Neil Gaiman (author), who defended an 18-month-old child when a cyberbully demeaned him for wearing a costume. The bully did not think the child's costume was good enough [368]. Thank you, Mr. Gaiman, for putting the so-called "true fan" in his place. You turned the tables on this person and likely curtailed more than this abusive behavior. (NB: Academy Awards: Take note, think twice, and reflect on punishing Will Smith but defending Chris Rock, who bullied Mr. Smith's wife under the guise of comedy.)

Primates crave power, and we take our first positions in status hierarchies in childhood. Our experiences with dominance and submission tweak our hormonal levels and protein expression to survive within these social structures [369, 370]. Aggression feeds directly into these hierarchies and our desires for power but does not always yield stable dominance. We turn on bullies.[37]

Complex pecking orders of leaders and followers establish coalitions to obtain and maintain standing and influence [371, 372]. Our primate-based responses to dominance vary among situations. A perception of the fairness and legitimacy of leaders improves reactions and loyalty [373]. So do our bonds and coalitions with group members [374, 375], and the perceived authority and social currency of troop members.

---

[37] At the liberations of the Nazi concentration camps, the cries for food could have less force than the demands to be given the SS men. At some locations the liberating allies either could not stop or turned a blind eye to the thirst for vengeance, e.g., Dachau [3302].

Honesty, humility, and fairness form better predictors of long-term popularity and sustained social dominance than intimidation and humiliation [376].

Exploitive, impulsive, goal-directed leadership styles do not reduce a ruler's popularity [234]. They do decrease his likability, legitimacy, and likelihood of long-term obedience. Tyrants may not benefit from the fairness factor but can exert dominance in many social situations, even when disliked. The tormentors and their allies enjoy more popularity than those who defend the ill-treated [234]. Bullies and onlookers are more popular than the targets or the heroes who protect them. Despite our social condemnation of "hitting girls," males may suffer a loss of status for intervention on behalf of a female victim [377].[38] These forces combine, making it challenging to side with the low-ranking individual as we blame them for their status.

The bystander effect contributes to the passive response in the audiences garnered by bullies. As crowds grow, intervention becomes less likely [378]. The obligation to stop the attack diffuses among more people, diluting the responsibility to stop it. "Neutral" bystanders are far from neutral. Bullies integrate audiences into their craft [379]. Practiced perpetrators use them to enhance their intimidation, spread shame, and silence opposition [380, 381].

Humiliation before an audience magnifies torment by destroying social standing [382]. Part of the damage from cyberbullying [383] emanates from this spectator effect. Online bullying expands the size of the audience and its location. A discrete set of venues no longer serves as a focal point for danger and humiliation. Instead, the region of torment and subjugation extends to the world and across time. Audiences transform from a few onlookers to the potential inclusion of thousands. The public nature

[38] Even outsiders are held in less disdain than victims. For example, consider people treated as lower castes within societies, such as African Americans in the United States. New immigrants often receive higher caste status and get treated better than long-term members of a group perceived as low-caste [3266].

(perceived and potential onlooker size) and the anonymity (safety for the perp) appear to be the primary contributors to the impact of this cyberbullying and trolling, not the medium itself [384].

Public degradations during cyberbullying, as in traditional attacks, include belittling, relational (e.g., social exclusion, sexual control, or the destruction of social bonds), or rumors and sextortion, e.g. [385, 386, 387]. Sheltered by the net, the persecutor can feign friendship, making the betrayal less expected and more painful. Online bullying carries an increased danger of relational bullying, often heaped onto pre-existing forms of emotional degradation and violence [388].

Taking Out the Competition

"The enemy of my enemy is my ally." — ascribed to Kautilya, an Indian tactician

High-status and intelligent peers pose a social hazard for the less capable [389, 390].[39] By targeting and humiliating them before an audience, a status seeker can take down future competition by more than a peg. In an adult workplace, removing the efficient, productive, or well-educated competition contributes to immediate financial survival. This spiteful strategy [117] exposes the bully to minimal risk while neutralizing future competitors for mating and fiscal success. Such malicious behaviors have a strong presence in the animal kingdom as an evolutionary adaptation to reduce the rank of potential rivals [391].[40] Victims who cannot overcome the damage done by childhood demons continue to show telltale signs of depression and trauma (PTSD) through adulthood and into old age [42, 392, 393]. On average, they earn less than the untraumatized, suffer poorer health, and, in socially mobile societies, marry into a lower social

[39] Targeting need not be conscious; it can follow the innate pattern seen in primates: take out the competition when you can.

[40] Remember, these behaviors did not evolve for a playground or modern society. As adaptations, they would have benefited us when we were still barely-human creatures in the Pleistocene era. When focused on non-kinship groups, these behaviors may still have adaptive value in many non-WEIRD societies.

status rather than an equal or higher one. As lower-rank individuals, they can experience impaired working memory [394], tend to occupy lower-status jobs, and are more likely to be unable to attain or maintain employment [243]. Doubting their accomplishments and contributions, many become chronic workaholics, desperately seeking validation. Despite hard work, they retain their low social rank, and their body language broadcasts their damaged self-esteem. They take fewer chances and receive fewer rewards with less recognition.

Victims assume learned submissive roles, while bullies take dominant stances that further limit or regulate the victims' initiative [395]. When they do initiate a project, the chronically intimidated will lack experience in how to proceed or the incentive to succeed [396]. Social blunders plague them, but they acknowledge their insecurities and ask for aid more often than their bullying alphas.[41]

Survivors incorporate the experiences into their bodies through changes in gene expression [309, 397, 398, 399]. The high level of stress hormones and reduced glutamate function experienced during the early development of the endocrine and nervous systems alter their psychological and physiological development [400, 401]. Their reactions are primed and fired by hair triggers later in life.

Through the lifecycle, victimization changes the neural network and its levels of receptors, hormones, and neurotransmitters [300, 402].[42] These systems grow, develop, and mature differently from those in higher-ranking persons. Radical, permanent changes occur in the chemistry of behavior. Proteins and pathways

[41] Personal Note: As a teacher, I love students who ask questions, especially when lost. The question directs me in how to help. I know my love of questions is not universal among faculty. That difference, too, may reflect our individual histories as victims, bystanders, and bullies.

[42] These regulatory chemicals have a cascading effect, influencing diverse body systems. Changes in the levels of testosterone alter our ability to reproduce. Stress even changes our levels of histamines, compromising our ability to fight infection. The impact on histamine responses dates far back in our evolutionary history and can be found in fish and stressed-out insects [585].

that enhance confidence turn silent. The production of stress hormones triggers more easily. The bullies instill biochemical ghosts that haunt their victims for years [311, 403, 404].

The force of these internalized bullies slowly destroys the fragile ego of the wounded. Victims can develop a hair-trigger response to future bullying and stress, leading them to repeat their learned pattern of oppression. To succeed, take the initiative, or accept credit for an original thought upsets the social order into which they have inserted themselves. They have learned through the conditioning of their brains, epigenetics, and endocrine systems that they are unworthy of anything beyond a subservient role.

CHAPTER 3:

The Chemistry of Bullying: Hormones 101

"The biology of context is vastly more complicated than the biology of the behavior itself."

— Robert Sapolsky in an interview with *Psychology Today* [405]

The Car and Its Drivers

People drive their cars, take the bus, or walk to work or school, maintaining their routes along well-trodden paths. Thanks to our interactive control systems, our autopilot functions better than it does in most machines. While our brains focus on entertainment, the news, or the latest problem we need to solve, we reach our destination, often without remembering how we got there. When they work right in a positive environment, we enjoy the benefits of sound decision-making and a flexible social life. When our autopilot malfunctions, we can become depressed, rigid, startle easily, or live lives filled with anxiety or depression.

Many things can send our feedback systems spiraling out of control. Genes can hypersensitize a person to abuse or leave them resilient. Cultural norms can oppress some while exalting others. Social stressors, such as bullying, can modify the physical structures and delivery systems that modulate our behavior and emotions, making us more attuned to threats by increasing our social sensitivity and altering the paths we use to avoid danger [36].

Several of these regulators are in the brain. Far from being a universal dumping ground for information, the brain parcels out tasks to specific zones. Each of these districts has specialized functions shared and interlaced with other areas. For example, the large lump just above the brain stem, the cerebellum, specializes in motor control, but it dips its lobes into different activities, such as emotional regulation, a job associated with a deep almond-shaped structure, the amygdala. An island on the side of the head, the insula, recognizes bodily changes, including changes in breathing, pain levels, and vomiting; it also functions in social intelligence and contains neurons that let us understand what other people are thinking. The prefrontal cortex (PFC) manages much of our "voluntary" behaviors and thoughts, including planning, paying attention, and changing our minds (flexibility); it also regulates our emotional lives and social behavior. No region functions alone. They form an interdependent grid that regulates both automatic and "voluntary" thoughts, emotions, and behaviors. Each plays a significant role in some aspect of our lives, but one region's main job may be likened to another one's hobby.

The networks of the brain, in turn, interact with the body's other feedback system, the chemical transmitters. Each region of the brain has different sensitivities to specific hormones and transmitters. Some drive us toward action or withdrawal by stimulating regions that kick in our survival instincts. A compound that activates one brain region may down-regulate another.

The torture of chronic bullying correlates with changes in hormonal activity and the volume of specific parts [123, 36]. Some areas grow, others shrink, and these changes linger, detectable years after the trauma. The type, chronicity, and severity of bullying make a difference [36, 274].

Although some stress is typical, a constant bombardment creates a continuous pressure wash. In the developing brain of a child, it is the concentrated neurotoxic cocktail of stress hormones that trigger changes in gene activity, physically altering specific brain regions, thickening and hypersensitizing some regions while paring back the veneer of others, leaving them less functional and vulnerable to future attacks. Areas that control emotions and social reactions experience some of the most radical developmental changes, providing a structural explanation for behavioral and cognitive changes induced by bullying [274], changes that can plague victims for life [213, 214, 406, 243]. The limbic system, our emotional core, grows, leaving us vulnerable to social assaults, while our cerebellar cortex and insula can shrink, reducing our ability to cope with the trauma [123, 36]. The PTSD of female genocide survivors also correlates with changed gene expression that can be identified in their blood and the blood of their children [407].

You Have My Permission to Skip This Stuff

> *"He has only half learned the art of reading who has not added to it the more refined art of skipping and skimming."*
>
> — attributed to Arthur Balfour, British Prime Minister from 1902-1905

> *"Rising testosterone levels foster aggression only during challenges to status. Finally, crucially, the rise in testosterone during a status challenge does not necessarily increase aggression; it increases whatever is needed to maintain status."* — Robert Sapolsky, Professor of Neurology Neuroscience & Neurosurgery, Stanford University [3238]

Everything I tell you about hormones is a lie. Technically, everything is an oversimplification. Our defined hormones and glands have changed from the simplistic descriptions of years ago. As biologists learned how signaling mechanisms work, they widened their definitions, often blurring the neat, compartmentalized distinctions between signaling systems. Newly discovered sources have joined the organs long believed to be the sole suppliers of hormones.

How they impact behavior also remains open to some interpretation. You needn't understand everything in this section to get the gist. It's the most technical portion of the book and the hardest to wrap your mind around. If you find yourself mired in the details, shrug your shoulders and skip it. Although endocrine responses explain much of our behavior, I've striven to make each chapter reasonably self-contained. So, if your head feels like a shaken snow globe, close the cover, apologize to the nice book, and skip to the next chapter.

An Introduction to Hormones

Two interlinked control mechanisms dominate our behavior: the nervous and hormonal (endocrine) systems.[43] The nervous system responds to the environment in a flash. The hormones trigger our reactions. When fire service workers awaken to an alarm, the nervous system engages first. The hormonal system maintains arousal as the firefighters remain on high alert for hours on adrenaline. The timing, duration, and history of exposure (experience) contribute to the efficiency of these interactions.

When we study behaviors, we observe the end products of a myriad of interacting biochemical reactions. The natural curiosity of scientists drives their search for the roots of these chemical correlates of actions like physical violence and cowering fear. The hormones and neurotransmitters regulating our bodies provide insights into our programming. Like our physiology, our behaviors develop an optimal range of operation (homeostasis).

To help our puny primate brains understand how the body works as an integrated whole, we study it as interlinking subsystems. Like a complex jigsaw puzzle, we assemble our ideas about the whole by interconnecting the bits. Unlike the game, removing or placing a piece in the wrong position can affect how other pieces fit together, even if they are on the far side of the board. Make extra progesterone or testosterone, and you can mess with cortisol levels.

When approaching the hormones and glands, we separate them by their strongest interconnections, like separating puzzle pieces by color. We then look at the edges of each piece, the complex inputs from each hormone, receptor, and enzyme that control their function and build the foundation for behavior. And just

---

[43] This hand-wave is my first oversimplification in the chapter, but not my last. The nervous and endocrine systems give a broad picture that suffices (for now). The truth is complex. Diet, including proteins, and fatty acids, isozymes (enzymes that share function but have different efficiencies), and isoreceptors (molecules that receive messages with the same or altered function) also play roles. So do genetic, translational variations in the structures of the hormones, and the populations of microbes in our gut.

as a puzzle component may get stuck in the box or fall on the floor, endocrinologists and molecular biologists check for missing pieces stuck in corners we haven't explored. For example, testosterone, once believed to be the product of just the ovaries and testes, is also produced by the adrenal gland [408, 409] and the brain [410]. The production of testosterone by the brain is like the puzzle piece that slipped under the carpet; its presence does not show in standard blood tests, and it could go unnoticed, leaving a gap in our understanding.

Timing, duration, and previous exposure to hormones form part of the puzzle's edge. These interactions lay down the initial contours, shaping the neural pathways that follow. The hypothalamus and pituitary glands flesh out the scaffold. Their hormones link their function to those of other glands and hormones. Together with the adrenal gland, they regulate our stress response. With the gonads, they modulate reproductive changes.

The third "axis," the RAAS,[44] provides additional pieces of the puzzle by regulating blood pressure. This last pathway involves the adrenal cortex but is more diffuse than the first two. It includes secretions by the liver and the small arteries (arterioles) of the kidneys.

When our nervous or endocrine systems fall out of balance, the revisions can tilt us toward maladaptive behavior. Changes in hormones, their receptors, or their regulation can alter our sex drive and activity level.[45] Genetic and epigenetic inclinations that sweep us from the norm can predispose us toward aggression or meekness. They can impose risk-taking or risk-aversion patterns. When penchants interact with the environment, the combination can nudge people to become torturers or victims. Trip the right

---

[44] RAAS stands for Renin-Angiotensin-Aldosterone System. This system is particularly sensitive to cholesterol levels [3067, 3506].

[45] Technical note: regulation can include changes in miRNAs and the switching on or off of genes, including genes for isozymes and isoreceptors.

chemical switch, and the meekest person can become a brutal killer [e.g. [411, 412]. Push a different button, and a well-balanced person will quake in terror.

Throughout our lives, our genetics, personal history, and environments distort the plane of experience, creating folds and cliffs that let us jump from one action to the next. These natural boundaries pose ridges and barriers that protect us. Without them, we would fall from these behavioral precipices into maladaptive zones of conduct. Some teeter on the cliff edges of their chemical Armageddons.

Arsenic is a Natural Food (But You Still Shouldn't Eat It)

> *"Man can be cruel, but nature is indifferent. It is the unrivaled psychopath."*
>
> — James Renner [413]

WEIRD cultures encourage people to assume something natural is good and unnatural things are bad. A moment of reflection demonstrates the caca nature of the argument. For a primate, excreting as you go through the day and pooping over the side of your bed is natural. They empty their bowels and keep on moving. However, I doubt anyone advocates for a return to natural defecation in the streets. Arsenic is naturally found in many water supplies, yet I recommend against taking it as a dietary supplement. Asbestos is a natural fiber. Polio is natural. Formaldehyde and cyanide are found in nature, as are death-cap mushrooms and hemlock. Castor beans contain ricin, another lethal poison. Botulinum poison, salmonella, and anthrax are all "natural."

Violence, from bullying to genocide, is also natural, as are cooperation and compassion. Being natural or artificial does not make something good or bad. The *impact* of a food, fiber, or behavior renders it "good" or "bad" in our mind's eye. Nature designed us to favor people close to us, those we perceive as like us. Few would consider giving a beloved child's college fund to a stranger within the realm of normal behavior. Yet rampant nepotism and resource hoarding harm others and exaggerate inequalities.

Appeals to nature, arguments claiming something natural is inherently good, don't work. What is natural is not necessarily even normal. In industrialized countries, few people follow the "natural pattern" of heading to bed at sundown followed by most diurnal animals.[46]

We also fall prey to believing that if we want something to hold true, it must be natural. We crave explanations that relegate violent crimes to domains we can manipulate and control. We want to see rape, murder, and genocide as aberrant behaviors caused by a social context. We can exert control over our cultural context and reprogram programmed behavior; genes pose a daunting problem. Our biological inheritance creates a framework upon which culture, family, and learning impose their final form. Genetics can provide a built-in explanation for intolerable behavior, but the behavior remains unacceptable. Hormones link to tenderness and violence, to love and rape. However, remember, "natural" does not mean "desirable," and because we see a behavior as beneficial does not make it natural.

Testosterone and The Mencken Principle

> *"There is always a well-known solution to every human problem—neat, plausible, and wrong."*
>
> — Mencken [414]

Remember, like all humans, I am a liar. Everything I tell you in this chapter is false. Everything I tell you in this chapter is true. I'm not trying to mess with your head by presenting a paradox. The data on testosterone, a.k.a. "T," is inconsistent and contradictory. For every study that says testosterone induces violence (sort of), another disputes it (kind of). These contradictions are not the results one wishes to see in a science like biology. We like clean, neat, simple data. To get consistent results, one must at least have

---

[46] Our primary wake happens during the day, classifying us as diurnal. That does not mean we are only active during the day. The sleep pattern for most of human history was biphasic. We'd wake up for a few minutes or hours during the night (the watch), then resume rest until sunrise [3511]. Our expectation of a "good," uninterrupted night's sleep is a post-industrial concept.

clear input, understand the parts of the system and how they interact, and create unbiased experimental designs. One must put aside prejudices about how things *should* behave and see what *is* happening without cherry-picking the bits and pieces that support our preconceptions.

The testosterone story is complex and muddled by biases. T itself is a relatively weak hormone. The body must convert some of it to a related hormone, DHT, for decent activity. This transformation occurs in multiple organs, including the testes, ovaries, liver, skin, and fat cells. Each system has its own collection of controlling molecules, turning specific genes on or off at its own speed, so simply measuring T-levels in the blood provides limited insight.

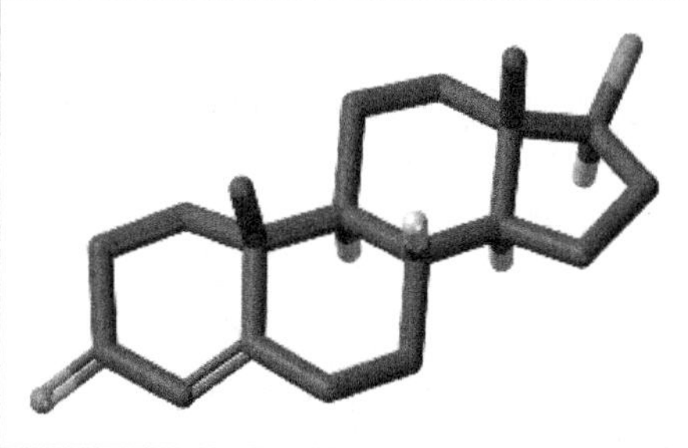

Figure 3-1. A "skeletal stick" model of testosterone. Hormones act like keys, fitting into molecular locks to open the gateways to specific processes in the body. If you make one small change to a "tooth" on the key, it acts differently. — (Public Domain image from Wikimedia courtesy of Jynto, 2018)

An additional preconception hindered our understanding of how testosterone is tied to dominance, status, and aggression. When in doubt, we fall back on a neat, plausible explanation: males outdo females in violence because they have more testosterone.

That understanding has flaws. But science's most beautiful and frustrating power lies in its penchant for exposing its own errors. Science self-corrects as it learns. The strength of science grows from a willingness to change and explore new dimensions, even if it frustrates the outsider. The early studies often ignored data that contradicted what everyone knew: males were aggressive. Some oversimplify the problem, ignoring the interactions of testosterone with other hormones, neurotransmitters, and receptors. Blaming T for violence is akin to watching a movie and crediting its success to a film editor while ignoring the directors, the cast, and the writers.

As we learn about the interplay of hormones, we converge on a complex understanding of testosterone and its supporting cast that together regulate our behavior. The starring players include estrogen, progesterone, oxytocin, and cortisol. Both biological males and females have these hormones and their crews of receptors and regulatory enzymes that tell them when and where to act and when to blend into the background when not needed. The crew sizes and compositions differ, but like actors, the signaling hormones couldn't function without them. And just as in filmmaking, the timing of different players—especially acknowledged stars like testosterone—and their interactions with other crew members can create massive changes in the final product [415].

For years, we have confused different but related phenomena such as aggression, dominance, rank, power, status, leadership, and sway. These social aspects overlap, but they are not identical, and they are fluid. We recognized allied divides in the concept of "dominance" years ago. Still, we continued to muddle them in our minds and in experimental designs, often assuming they are linked. An individual, be it a human, chimp, or raven, may occupy a high rank in one situation but not another. A dominant animal may follow the leadership of a different troop member with superior abilities or knowledge. As a teacher, I am the ranking alpha in my classroom, but I occupy a subordinate role with respect to my chair and my plumber. One parent may dominate a household (making most decisions), but others may exert a more powerful impact on the children.

Rank, power, and dominance shift with situations. In numerous mammals, males lead during intergroup conflicts, but females make decisions about the group's movements [416]. A silverback male gorilla may appear to rule his troop. He cannot retain power without the cooperation and support of the female hierarchy, which can drain him of control and expose the leader to a challenge from a preferred male. Being an alpha has little to do with bombastic gestures or aggression; it's about getting your way, more often than not, within your sphere of influence.

Humans can mistake cowering for respect and aggression for social status. Think about the domineering, bullying people you've encountered. They may exercise power and demand respect, but I doubt you held them in much esteem. You might even fantasize about unhooking their life support to charge your phone battery.[47] Aggressive, browbeating primates have trouble keeping their power, and their long-suffering underlings will enact retribution if given a chance. An alpha cannot rule without the support of empowered coalitions. Once a bullying leader falls, they find few friends among their victims.[48]

Our muddling of dominance and aggression also takes root in our cultural, either-or biases about gender and our simplistic equating of sex and gender. Behavior patterns vary among and within genders [417]. Our binary, either-or language reduces our ability to see gradations. Species can exhibit female social dominance, such as ringtail lemurs, bonobos, and spotted hyenas; others have aggressive alpha males, like chimpanzees and hamadryads. Co-dominant species may share power or exercise leadership in separate domains.

Some species have more than two common body types, often with different body forms behaving differently. Some organisms have multiple sexes [418, 419], with some fungi tipping the scales at 23,000 mating types [420].[49] Hermaphroditic and gender-changing species (serial hermaphrodites), like the clown fish and chalk bass, teach us that brain organization does not always match the genitals on the body [421,

---

[47] In 2017, a report circulated that a teen unplugged her grandfather's life support to charge her phone. The story began as a satire in a Spanish-language article on *El Mundo Today*. It was revived in an English-language satirical magazine (8shit.net), from where it went viral as fake news. The story spread without question because of adult prejudices about teens. An older version goes back to South Africa, where the culprit was a janitor who needed to plug in a floor polisher [3367].

[48] To maintain their leadership, a dictator does not need the support of all or most people. They need strategic backing from powerful groups. Keeping their "essential" constituency satisfied provides their strength. Often, those required to maintain supremacy are not the same as those who helped the alpha to gain it. Those two populations of betas pose threats [1079].

[49] The slime mold *Physarum polycephalum* has a mating system controlled by genes with multiple variants for each site. Each combination yields a different reproductive sex, expanding their ability to recombine genes. One bark fungus, *Schizophyllum commune,* may be the current winner, with an estimated 23K mating types. Not everyone gets to mate with everyone. It's an excellent system for shuffling your genes around.

422].[50, 51] As they switch genders, animals often experience brains and body chemistry out of sync with their physical changes.

In other species, like the nematode *C. elegans*, an organism used to model disease and the immune system, the environment can induce "gender-atypical" behaviors [423]. Like machines with fewer moving parts, less complex organisms often have a smaller set of proteins (proteome) and fewer metabolic interactions (metabolome); they serve as models with simplified signaling and response systems that let us tease out an understanding of our complexity, physiology, and behavior.

Despite the flood of information about gender, sex, aggression, and hormones, we crave neat, plausible answers and strive to put the consequences of each in separate boxes. We isolate factors to create a simple, understandable, and palatable story. The following section is plagued with orthodoxy and studies that explored the influence of hormones, including the almighty "T," without taking our current understanding of gender and sex into account. Think of the hormones as puzzle pieces laid on a table but not assembled into a coherent picture. It's hard to see the image in the scattered bits, and it is even more challenging when we have biased preconceptions of what the picture should look like. It's further complicated when the pieces morph into different shapes and colors over time.

Everyone comes with baggage, even scientists. Most studies assume the world consists of two biological sexes in which the brain and gonads function in harmony. Within that framework, studies may presume a dichotomy or ignore differences. Many equate power, aggression, social rank, and leadership with

---

[50] Did I mention that some ciliates switch gender with the time of day [3282, 3066]?

[51] Hermaphrodism goes beyond simultaneous and serial. Some species have hermaphrodites and males, e.g., Kryptolebias killifishes [3455]. Some flip once (typically female to male), but a few switch back and forth. At least 500 fish species show some form of hermaphroditism. In humans, there are rare cases of people born with female genitalia converting to male organs as they enter puberty. These young men have a standard sensitivity to androgens. They cannot develop into males in the uterus because one enzyme (5-alpha-reductase) doesn't work right or is deficient [3466, 3427]. They cannot convert testosterone to its pumped-up, longer-lasting version, DHT. At puberty, testosterone levels rise, and the testes begin cranking out enough DHT to finish the job they started.

masculinity. Craving simplicity and a desire to finish their work before their grants run dry, they often treat one hormone without considering its molecular ecosystem in the body. With that proviso, you can either tear out your hair in frustration, give up, and skip the chapter, or plow through the data and try to make sense of it. I recommend a good cry before moving forward. That's what I did. It helps.

T. Rex

> *"Godi, slowly raising himself, screaming with fear and anguish, watched his tormentors go... He bled from dozens of gashes, cuts, and punctures. He was never seen again."*
>
> — *Demonic Males: Apes and the Origins of Human Violence* [424, p. 6]

I want to clarify one thing: the popular idea that testosterone is the primary cause of violence is contested [425, 426]. Think of T and DHT as contributing factors rather than the root cause of aggression. Although involved, it does *not* constitute the sole root of violence.[52] Nor does T promote a cold-hearted willingness to sacrifice others for the greater good [427]. An elegant, regulated system integrates this hormone.[53] Additional hormones and neurotransmitters, e.g., MAOA, GABA, serotonin [428] and different receptors [429] contribute to our susceptibility to aggression, forced dominance, and unwilling oppression. These complex and integrated team players regulate various behaviors, not just sex and power drives. Males and females both produce testosterone.

---

[52] Estrogen has a much better public relations team than testosterone, but it, too, plays a role in potentiating violence [528]. Because the birth spike in testosterone quickly converts to estrogens, the resulting estrogen spike may masculinize the male brain [3115]. I will discuss estrogen and progesterone later in this book.

[53] Increasing a person's adult testosterone level does not always increase aggression. Female-to-male transgendered men report a short-term increase in anger (3-7 months), and some studies report a decrease [3179]. Studies show no correlation between blood levels of T and anger [3374]. Some short-acting intramuscular injections cause T fluctuations and may trigger aggression. Other injections counter this problem [3429]. Testosterone can improve social functioning or increase aggression in cis-gendered people. The current data for transgendered people is sparse, and the potential for bias is high, as is the pain from discrimination and ostracism.

Males have about ten times as much T as females, but testosterone is a weak hormone compared to these precursors, especially androstenedione (A_4).[54] Some of these differences result from the different body-fat ratios of men and women. Fat tissue is the primary location of p450 aromatase, a key enzyme that converts androgens into estrogen. For this same reason, obese men can produce quaffing amounts of estrogen and suffer infertility [430, 431]. Females retain less free testosterone than males, converting more of it to estrogen, but their bodies are exquisitely sensitive to T and its precursors, androstenedione and DHEA.[55]

With a half-life of about ten to fifteen minutes,[56] a free, active molecule of T does not survive long in the body. After that, most of the circulating T either gets bound to a blood protein such as albumin[57] or converted to another form. We associate free T with behavioral impacts, not its constrained forms. However, its turnover is high. The balance between free testosterone production and its removal (conversion to less active forms and sequestering) helps determine the wallop it creates.

Testosterone permits the expression of aggression but does not raise the social standing of subordinate individuals nor dictate their level of belligerence. The *interaction* of T or A_4 with different brain chemicals can make an animal susceptible to committing violence. The receptors' number, quality, and location limit testosterone's and A's effect. These interactions influence the downstream processing of the information in the brain.

---

[54] A precursor is a compound in the body that is used to make a different compound. For example, tryptophane is an amino acid used to make melatonin (a sleepy-time chemical), and serotonin (a mood and seasonal regulator).

[55] If you must know, DHEA = dehydroepiandrosterone and is a precursor for testosterone and estrogens. Progesterone, another "female" hormone, split from the others earlier in the enzyme chain. Progesterone can transform into corticosterone. The path to cortisol diverges between these two highways. They all derive from the same parent molecule, cholesterol.

[56] The half-life of a molecule is how long it takes for the body to metabolize and get rid of half of it. Two precursors of testosterone have longer half-lives. DHEA can last for half an hour, but androstenedione can last from 1.7 to 77 hours, depending on conditions. These compounds may be responsible for many of the effects assigned to testosterone in early studies.

[57] Other binding proteins include SHBG, HAS, CBG, and orosomucoid (ORM). In blood chemistry, a high SHBG level means you have sequestered (taken out of active circulation) a lot of testosterone.

Like other hormones, T acts in a complex social structure and physiological environment. I chose to treat testosterone first because it has suffered from vicious slander. T drives a plethora of behaviors. The typical male body fine-tunes its testosterone levels to the social sphere, altering hormonal signals and facilitating pro-social behaviors as well as well-publicized antisocial ones [432]. It contributes to our mood, self-perception, psychological well-being, and willingness to take risks [433, 434, 435]. Testosterone promotes inquisitiveness.[58] It reduces stress and some forms of anxiety [436, 437, 438]. Female white-footed mice (*Peromyscus leucopus*) who receive testosterone after an ovariectomy (eliminating their gonadal hormones) investigate their surroundings more than ordinary females[59] [439].

Testosterone does more than facilitate "aggression." High concentrations link to selfish and goal-oriented behaviors and reduced concern for the welfare of others [440]. It also helps maintain health and well-being. T influences our cognitive abilities, including attention, memory, spatial ability, and social behavior. Low testosterone increases our chance of dementia as we age [441, 442]. If it rises too high in younger men, the imbalance can negatively impact social and physical awareness and cognition [443, 444], and the mega-doses of T-like compounds associated with anabolic steroid abuse increase the risk of dementia [445, 446, 447].

Testosterone levels correlate with dominance and mating in most vertebrates.[60] However, an increase doesn't raise hostility toward those above you on the social food chain. Instead, as a colleague of mine

---

[58] The female hormone, progesterone, also has this effect, promoting both learning and memory [3117].

[59] In this species, the males will travel farther than females in search of mates. Thus, the dosing of these females with testosterone was described as a trigger for this stereotypic masculine behavior (travel). An experiment in which testosterone-treated and untreated females were released into a highway cloverleaf showed higher survivorship for the testosterone-laden females [3478]. This pattern suggests they do not direct this inquisitiveness toward Wanderlust (the stereotypically masculine behavior) but toward territoriality. It could link to an activation of the NA, which responds with a pleasant reward to stimulation (learning).

[60] In guinea pigs, males in the higher echelons of their society have higher baseline testosterone than lower ranking males [3352]. Testosterone can influence specific traits that send honest signals of mate and social quality. The levels of testosterone, which rise after successful dominance contests, influence the comb size in roosters [480].

expressed it, the rise leaves you with a case of "jerkalepsy,"[61] increasing aggression toward subordinates, the low-cost targets upon which dominance can be flaunted.[62] One would thus expect active bullies to have high testosterone. Victims should have reduced levels, skewing their abilities to compete for social status and giving the jerk a competitive edge. Unfortunately, that alluring scenario is too simple.

This hormone decreases in men right after they become fathers [448]. This drop acts as a safety mechanism to reduce aggression toward new mothers and infants during this time. The negative correlation of testosterone with generosity would threaten the newborn's and mom's resources [440]. New fathers also experience a rise in estradiol linked to nurturing.

In most species, the identity of one's father remains murky. Social rank springs from the mother, even among male-dominant species. In spotted hyenas, a matriarchal animal, the mom's status also passes to her daughters, influencing their social rank and fitness[63] [449]. High-ranking females form coalitions to maintain power through both intimidation and social support. They cannot rule alone and must form alliances (friendships) to maintain their standing in the troop. All spotted hyenas have comparable testosterone levels at birth, but males soon outpace females in their production. Female ovaries take a different hormonal path, building the reserves of a precursor, A_4,[64] which mirrors the level found in males when they reach adulthood. These developmental differences lay the foundation for mature hyena-female aggression. Many matriarchal species exhibit high levels of testosterone or A_4 in females during their life

---

[61] Jerkalepsy: the sudden, acute transformation of a reasonable person into a jerk (a quip borrowed from Dr. Linda Nidelkoff, Northeastern Illinois University, College of Education). Sadly, in the DSM-5, the diagnosis does not exist.

[62] When a social species does not follow this roadmap, it suggests that factors besides dominance determine who has access to females. Species with dominant females, like the spotted hyenas, show a different hormonal pattern in males [3497, 526].

[63] In biology, "fitness" means the ability to reproduce.

[64] A_4 = androstenedione. It is considered a "weak" androgen that the body uses to synthesize testosterone and estrone, a "weak" female sex hormone. Human children also experience a rise in A_4 before puberty. It plays a role in brain maturation, social and cultural learning, and aggression [3395]. This precursor also lasts longer than testosterone, and reserves can thus be built.

cycle, including ringtail lemurs, marmosets, and hamsters [450].[65] They lay a chemical foundation for social cooperation, coalition building, and aggressive competition.

Our closest genetic relatives are bonobos and chimpanzees. Chimps and humans engage in higher proactive (goal-directed) aggression levels than bonobos. Bonobos and humans like to "get even," following a slight and engage in more reactive (retaliatory) aggression than chimps [451]. Unlike bonobos, aggressive, mature chimp males rule their troops, harassing and browbeating the females.[66] From our human perspective, female bonobos fare better. They have an earlier surge in their testosterone levels than their chimpanzee counterparts. They experience puberty earlier and build stronger female coalitions than female chimps, letting them dominate the social group [452, 453]. Bonobo females will interfere with breeding attempts by lower-ranking females and may abuse the offspring of their underlings. These assaults ensure the alpha's progeny will form a disproportional percentage of the next generation and its leaders.

In species where aggression improves a male's chances of mating, high-ranking males have high testosterone levels. The link between this hormone and aggression remains tight in natural populations with social males. In many species, a male's ability to dominate translates into successful reproduction. He can also keep his rivals from producing babies that will compete with his offspring. The neighbor's reduced fitness (ability to reproduce) may not benefit the species, but it boosts the *relative fitness* of the spiteful animal compared to his neighbor.

It is tempting to draw a neat picture of sexually successful bellicose males. But, ferociousness and proactive aggression have reproductive downsides, including social and sexual rejection by females and potential

---

[65] Female bonobos have an earlier increase in testosterone levels than do female chimpanzees, experience puberty earlier, and are more likely to form dominant female coalitions than female chimps [452].

[66] Juvenile male chimps will form coalitions with their mother or join adult males in low-aggression situations. The powerful adults seldom seek to join the younger, less powerful males [3097].

danger for mates and offspring. The correlations between testosterone levels, reproductive success, and rank break down in species like hyenas and bonobos, where females reject aggressive males.

In male doves, a successful courtship releases an enzyme (aromatase) that converts testosterone into an estrogen analog. It transforms their aggressive, head-bobbing pursuit of a female into a gentle invitation to join them in the nest [454], an adaptation to the joint chick-rearing activities to follow. In male bonobos, mobs of females will attack antagonistic males. Aggression, androgens, and rank remain correlated in these cousins. Still, only low-ranking males show spikes in their testosterone levels around fertile females, and their high testosterone inversely relates to their status. Friendly social behavior rather than aggression increases their chances of mating [455].[67]

Titers of testosterone decline following social defeats, such as victimization [456, 457]. At least in males,[68] the drop instigates social withdrawal [458].[69] This hormonal shift reduces the loser's chance of another aggressive encounter that could further reduce his social standing.[70]

When a person indulges in testosterone-mimics (anabolic steroids or "roids") to build muscles, it often triggers a behavioral metamorphosis. Problems include increased reactive aggression ("roid rage"), paranoia, hypomania, and confrontational behaviors. Steroids can leave lifelong transformations in their wake [459]. Men with histories of steroid abuse have more violent altercations than comparable men who

---

[67] This disproportional reproduction ensures that the genes that support the behavior are also disproportionally represented (= reproductive fitness).

[68] Females remain understudied. Too often, researchers treated them as a scaled-down version of the male, leading to little data about them.

[69] This specific study looked at males. It did not study hormonal changes in females.

[70] A retreat also stabilizes the social structure. Most small-group societies, like hunters and gatherers, consist of related males with females that have joined the band. Stabilizing society and keeping strife low benefits the genes of the individuals by improving the chances of group survival. Because the males are related, they pass on some of each other's genes during procreation.

have not imbibed in them. Barring a severe metabolic disease, steroid users experience more metabolic and behavioral troubles than people who have never used artificial supplementation.[71]

Not every steroid user becomes an aggressive, chronic jerkaleptic or commits a crime. Multiple factors must be in play for violence to emerge. Testosterone does not create aggression, but under the right circumstances, high levels potentiate violent tendencies [460, 461]. For those who express verbal and physical violence or are less self-restrained than most, steroids aggravate such predispositions [462], and long-term anabolic steroid abuse can recircuit the brain and enlarge the right amygdala [automatic emotional processing [463, 464, 465].

Like several hormones contributing to social dominance and aggression, testosterone creates positive sensations by stimulating a primary reward center in the brain, the *nucleus accumbens (NA)*.[72] In short, we feel pumped. Activities that increase testosterone elevate dopamine and decrease serotonin activity in this area. In males, levels climb in anticipation of aggression and reach their peak and nadir after a confrontation [466, 467]. Defeated rats also show altered dopamine activity in a deeper, older part of the brain, the VTA (ventral tegmental area). This site also responds to rewarding and noxious stimulation. It regulates anxiety and situational fears, and it coordinates the responses of the other reward centers [468, 469, 470].[73] Hormones beyond T and A_4 feed into behavioral and psychological differences, but their interactions with receptors and hormones in different brain regions relate directly to both the bullies and the bullied.

---

[71] One must be careful about drawing conclusions here. The population who imbibes steroids may not be the same as those who do not. Users may start with a predisposition toward rage not present in the non-users.

[72] The brain divides its function into physical compartments. Some regions direct movement. Others control sensation or thought. Still others function as data-processing units or as the seat of our feelings. The nucleus accumbens helps us feel good about life.

[73] These confrontations were territorial in nature, with a defender attacking an intruder. This aggression would be goal oriented.

In humans, young victimized boys show higher testosterone levels than non-bullied boys; victimized girls[74] show a drop in this hormone [471].[75] The reason for this dichotomy remains unknown; it could link to the different styles of intimidation experienced by different genders. Boys endure more physical bullying, requiring an aggressive bodily response. Girls undergo greater psychological terror, for which no adequate physical defense exists, leading to the discrepancy.[76] Or, bullied girls may direct precursors into the progesterone to cortisol pathway or deplete their T and precursors faster.

Animals enjoy dopamine releases in the nucleus accumbens, even when it's artificial. Electrical stimulation of the NA creates a sense of contentment and well-being. A drug (or behavior) can stimulate the accumbens by releasing dopamine into its synapses (neural connections). Within a wide range, the higher the amount of dopamine released, the higher the levels of pleasure and the more appealing their memory. Aggression also activates specific receptors in this brain area, e.g., [472, 473].[77] Steroids and cocaine both stimulate this region and produce similar rewarding feelings. Rats will choose such arousal over sex or food.

Anabolic steroids induce the same good feelings in the NA because they mimic the testosterone created by the body. They attach to the same receptors on the surface of the cells. The accumbens recognizes the drugs and gives us a good feeling of power and status. The agreeable stimulation from steroids goes beyond the NA. It links to parts of the brain that control our emotional life (the limbic system). Our happiness, desires, and anger travel through these interconnections.

---

[74] A quick reminder: Most literature simplifies the data to two biological sexes and ignores gender. For simplicity, assume a dichotomy.

[75] The bullied boys typically do not become bullies themselves. Their pre-existing high levels of stress hormones (cortisol) may help inoculate them against becoming bullies. Aggressive individuals like bullies tend to have high testosterone, low cortisol levels, and a higher social rank [3467]. The increased cortisol levels in victims correlate with anxiety and depression, not aggression.

[76] It will be interesting to see if future research shows a similar decrease in testosterone in boys for cyberbullying, a type of attack with no apparent physical defense.

[77] Animals with a history of successful aggression engage in more encounters than those who experience chronic defeat. A rewarding dose of dopamine here creates feedback encouraging the behavior.

Not all brains respond to testosterone or aggression in the same way. Social history plays a role. Long-tailed macaques (*Macaca fascicularis*) that receive T supplementation increase their aggressive behavior. This aggression increases disproportionately in monkeys that already show dominant behaviors [474]. Monkeys without a history of dominance did not respond as powerfully to increased levels of steroids.

In most species, males show predictable aggression patterns when testosterone rises. Still, minor blips in this hormone do not cause them to go berserk and start a rampage. Natural increases in testosterone-related aggression have an adaptive purpose, such as procuring mates and resources. Spiteful hostility intentionally harms others. Provocations may impose a cost on the aggressor, but spite could be a winning strategy if it benefits his genes.

The spiteful suppression of reproduction in another organism happens in many mammals, including rodents, e.g., naked mole-rats [475]; colonial rodents [476] and some female primates [477, 478]. In some species, subordinate males access fertile females after the dominant male has had his turn at sex.[78] Dominant females may suppress the reproduction of subordinates through social stress [477, 479]. In other species, testosterone lets males attract females with sexy displays—honest markers of status or quality they cannot fake [480].

In many social species, dominant males prevent low-ranking ones from mating altogether. However, in organisms as diverse as squid, fish, birds, and mammals, a few pushy males cannot monopolize females. The ladies and subdominant gentlemen may sneak sex when their rival isn't watching [481, 482, 483].

---

[78] Far from being passive recipients of sexual advances, females play an active role in male courtship behaviors, choosing males and, in some species, directing the sperm of specific males toward their eggs. They use both precopulatory mechanisms, often involving communication, outsmarting the males (guppies), or postcopulatory strategies, such as sending sperm deposits down blind corridors (waterfowl and dolphins), removing or killing the sperm (dragonflies), or having another male do it for them (also dragonflies).

The preceding studies demonstrate that testosterone *links* with aggression. However, they *do not show* that it *causes* antagonism. Correlation is not the same as causation. Active confrontations increase this hormone. When male primates are lumped together to form new groups, testosterone levels are poor indicators of who will dominate. Instead, group members act as behavioral pacemakers. Those males who dominate in hierarchical species soon have a higher testosterone level *after* their ascendency.

The Social Ladder

"Two bears cannot live in one den." — Russian Proverb

Studies of talapoin monkeys clarify how *not* to climb the social ladder [484]. Researchers injected castrated[79] talapoin males with a high testosterone dose, changing its ratio with other hormones and receptors. It had little impact on high and low-status troop members. Monkeys with an intermediate status increased their aggression but not their social rank. They remained subservient to the dominant males and females. They aimed their violent behavior toward subservient troop mates. In short, they became unrepentant bullies. Species with the same pattern, including red deer, show the underlying mechanism extends beyond the primates [485]. Even pissed-off pigeons can unleash a plague of terror on their fellows [486, 487].

This *potentiation* of aggression by testosterone goes beyond the pleasure center of the *NA* to receptors in a different brain region, the limbic system. The limbic system controls our emotional impulses. Testosterone decreases the time between neuronal firings in this region, increasing their speed. As the limbic system processes more emotional signals per unit of time, our emotional brain outpaces the rational one. These

[79] gonadectomized = removed testicles. The use of castrated males controlled for hormonal fluctuations.

testosterone-mediated signals exaggerate their firing in a person whose other body chemistry supports violent outbursts [488, 489, 490].

Dominance and testosterone hold hands, linked together in many vertebrates. This cozy relationship extends beyond primates and beyond males. In many species, dominant individuals have higher T or A_4 levels than their subordinates. In contrast, social defeats reduce these values and increase the relative amount of cortisol (C), a hormone associated with stress [491].[80] Women often hold lower ranks and experience more social defeats than men. In an American study, compared to men, they suffered twice the level of clinical depression [492]. They had greater vulnerability to cocaine abuse [493].[81] They are susceptible to dopamine depletion in the *nucleus accumbens* and changes in serotonin levels, another mood regulator.

Matriarchies Don't Differ Much

Like high-ranking males, dominant females have higher testosterone and/or androstenedione levels than submissive ones. In some species, like spotted hyenas, females have elevated androgen levels throughout their reproductive years. Others express increased T or A_4 at a particular phase of the breeding cycle. In many species of birds, females show aggression related to breeding sites before laying eggs, e.g., juncos [494]. Birds depend on this critical aggression to compete for rare breeding sites [495, 496].

Female tree swallows (*Tachycineta bicolor*) compete for cavities in trees that they need for their nests. Exposure to high testosterone levels while still in the egg increases aggression and T levels in adult females

---

[80] In Syrian hamsters, males who experienced territorial losses showed drops in testosterone and more submissive behavior. This decrease did not immediately impact their ability to mate when presented with a female. This study did not address the potential long-term impact on sperm count and fertility, nor the effect on reproduction due to loss of territory.

[81] Like testosterone, cocaine improves mood, provides a feeling of power, and stimulates the *nucleus accumbens*.

[497]. Before nesting, these mothers-to-be show an increase in both traits [498]. Their sensitivity to this androgen provides the edge needed to gain one of the rare breeding sites [497, 499, 496]. High-T females attack any other lady who wants their piece of real estate. After successfully defending their territory, their bodies prepare to lay their eggs. An enzyme (aromatase) transforms their testosterone into estrogen, promoting nurturing rather than territorial aggression.

Like many vertebrate mothers-to-be, female tree swallows have another hormonal trick under their wings. After ovulation, progesterone levels rise [500]. Like testosterone, progesterone correlates positively with the mama bird's protective behavior. In females, progesterone plays a pivotal role in resource-defense aggression [501].

Like men, women enjoy winning. Before a competition, women show increased testosterone and estradiol (an estrogen). In women, estradiol links to dominance. However, testosterone (a precursor for estradiol) increases during a competitive sports match, and estradiol declines [502]. The body makes estradiol from testosterone, suggesting the female body keeps testosterone high by inhibiting testosterone's conversion during competition.

Beyond any other mammal, the spotted hyena[82] shows the link between social standing and androgens in females. Both genders have high T levels. Unlike most social mammals, the females control this complex carnivore society. A dominant female runs a clan that can include multiple motherlines. Young males leave their natal pack as they approach adulthood. Both biological genders are aggressive, but in this family of cats, the females show more aggression than males. The matriarchy is so strong among hyenas that an erect penis signals male submission rather than belligerence.

[82] There are four species of hyena. The others are solitary and do not have a known dominance hierarchy.

The increased sensitivity of female biology to T and A_4 gives her this aggressive edge. During development, female spotted hyenas mime and exaggerate traits of the male. They have a clitoris that forms a pseudopenis and pseudoscrotum [503].[83] Females use the "fake penis" for urination, signaling status, and delivering their young [504].

The hyenas pay a cost for birthing cubs through their long pseudopenis. Mothers and offspring run an increased mortality risk during these deliveries compared to similar mammals that lack this feature. The high status gained by the androgen-laden mother improves the health, survival, and reproductive potential of her kittens. The rewards are high enough to offset the costs and gain a smidge of reproductive potential. Their social standing at birth correlates positively with their social rank in the clan [505]. High-status offspring will have better functioning immune systems and an increased capacity to fend off bacterial diseases [506].

Compared to males, the high-ranking spotted females are larger, more aggressive, and more territorial than the males, but they are hardly dictatorial bullies. They rule with the assistance of family and a core of supporters. A dominant female will share food and support allies and kin in conflicts [507].

These matriarchs share another trait with male cats of other species: infanticide. Together with their coalition partners, high-status females will kill the kittens of low-ranking moms [508]. Doing so provides a genetic advantage to the aggressor's motherline. In other felines, male cats kill kittens when the chance arises [509, 510, 511]. Infanticide removes a litter from a female, thus ending lactation. Statistically, the

---

[83] Although pseudo (fake) penises and scrota are rare in the animal kingdom, they occur in a few branches of the animal tree, and animals use them for different purposes. In hyenas, they reflect the status and strength of the females. Long-tailed macaques develop fatty psuedoscrota but not pseudopenises [3191]. In a group of small cave-dwelling "barklouse" (Sensitibillini), the genitals are reversed, with the female using her appendage to penetrate the male's pseudovagina to retrieve sperm [3285, 3284]. Several pollutants can override the expression of female traits and impose male characteristics on biological females creating inter-gendered individuals [3245, 3217, 3376]. In some species, the altered genitalia of intersexed females block the oviducts and lead to sterility or death.

kittens she nurses are not his. However, they can mate when she ovulates again and produce his kittens. Therefore, killing the kittens gives him a genetic advantage.

In spotted hyenas, the dominant female will kill kittens under the same hormonal imperative, and she receives a similar adaptive reward. More of her babies survive relative to the kits of other mothers. The alpha benefits from this differential survival. This genetic competition, with high-ranking females killing the offspring of the lower-ranking, has been observed in other social species, including our primate cousins, the chimpanzees [512] and capuchin monkeys [513].[84] Pregnant dominant meerkats will expel some female subordinates from the family to prevent them from killing their offspring [514].

The genitals of female hyenas are lower than in most cats. This shift gives the spotted female greater control over her mating choices. To succeed at getting sex, a male approaches her with a proper, subservient attitude. The position of his penis signals he knows his inferior place in hyena society.[85]

Variations from expected norms fascinate scientists. They can expose our biases or provide insight into why species behave as they do. No one knows what triggered the journey of the female hyenas toward increased aggression, but we have hints. Matriarchal animal societies have been known since the description of the pecking order of chickens in 1913 [515, 516]. However, our bias dismissed female aggression hierarchies as aberrations.[86] In at least seventy-six different mammalian societies, female leadership exists. Although uncommon, female-biased leadership, where females lead in most group decisions, happens among mammals as diverse as killer whales, bonobos, lions, elephants, and hyenas [517, 416]. This natural variation

---

[84] Infanticide by females is widespread among mammals. It is often found in social groups where there are synchronized bouts of breeding (seasons) and competition for resources for the young, including social position. Females target distantly or non-related offspring, but kin can be attacked if it benefits the mother's offspring [3167].

[85] If male hyenas respond to submission like the males of many species, their subservient position should increase their cortisol levels and decrease their T and A_4 levels. This hormonal regulation triggers the migration away from the natal unit. This move from his clan reduces his ability to form alliances and re-enforces his subservient position.

[86] It also blinds us to animal society with codominant genders, such as wolves, with alpha males and females in the hierarchy. Not every animal society organizes itself around gender lines.

makes biologists wonder why species with aggressive females contrast with patriarchal mammals. They challenge our assumptions about leadership and aggression and raise questions about cultural biases that might exclude females.

High-ranking females gain a survival advantage. If males compete for mating privileges, females compete for resources. Dominant females obtain food, status, and access to the safer core areas within a group [518]. A different plausible explanation holds that the transformation grew from selection pressure for aggressive cubs of both genders [519]. Such selection pressure would act as a pacemaker, setting the trend toward more profound female dominance. This transformation would give females greater control over mating and resources for offspring, yielding additional advantages in their social setting [416].

Yet another explanation reframes the role of aggression in matriarchal species. Dominant females are often not the most aggressive ones. Like bonobos, female hyenas form coalitions. Without supporters, a leader cannot maintain power. An alpha's ability to work with allies helps her retain status in a hyena clan [520] or bonobo troop. Brute bullying becomes maladaptive because it interferes with obtaining and maintaining powerful allies.

Hyena's power alliances resemble those of other matrilineal status systems far more than once believed [521]. At the bottom of the social ladder, allegiances provide protection and improved social standing [522]. For those at the top, alliances ease the maintenance of their rank.

A female hyena, if isolated from its pack when young, loses the social effect of her birth rank. Her androgen levels fall. Without someone like a mom to model supremacy, she takes longer to form coalitions and realize their full social potential [523, 524, 525]. Androgens in young female hyenas mold brain chemistry in different ways at different times. Both social background and previous learning affect how and when the hormones tweak brains already predisposed to violence and sociality by genetics [526].

The ratios and levels of testosterone, estrogen, and their precursors induce modifications during childhood. They also matter inside the womb and at birth. They participate in shaping the fetal circuitry of the brain [527]. It influences future levels of aggression, sexuality, and cognition. During development, testosterone activates androgen receptors in the brain. The male newborn's brain gets a whopping dose of estrogen as an enzyme (aromatase) converts fetal testosterone to estradiol (a form of estrogen). In rats, our frequent stand-in for humans in the lab, this estrogen exposure activates later male-style aggression and territoriality [528]. Together, these hormones outline the boundaries of what can become the personality of the brain's owner.

Androgens continue to sculpt minds during adolescence when the brain undergoes massive remodeling to meet the reproductive tasks ahead. During this development phase, exposure to artificial testosterones, including anabolic steroids, alters how the brain responds. It correlates with a heightened sensitivity to stimuli,[87] adjusted brain serotonin levels, and elevated aggression [529, 530].

Storytelling Finches and Habitats

> *"Oran a azu nwa."*
> *(It takes a village to raise a child.)* — Igbo proverb

Testosterone also provides mechanisms for *avoiding* conflicts by providing obvious messages about an animal's qualities. It helps a male signal his ability to withstand challenges from the fellows in his community. Tweaking hormonal levels, including testosterone, lets a society function. When a society functions, it allows its members to survive and reproduce.

---

[87] It increases the number of dendritic spines, the small bumps on dendrites (incoming connections) of neurons at "excitatory" connections, ones that increase the probability of a neuron firing.

Male house finches (*Carpodacus mexicanus*) signal social status through their plumage. Their color and behaviors tell other birds who can beat up on whom. Colorful males are well-fed and can defend territory better than drab males. Most birds build reds, yellows, and oranges in their feathers from the plant materials they consume. High testosterone levels let their bodies transform these raw building blocks (carotenoids) into bright pigments. Males incorporate them into their feathers.

To make red feathers, males must secure a hunting ground that gives them access to the necessary plants. Their colorful plumage tells females they have staked such a claim [531]. Females find the red sexy. Bright feathers mean that that particular male controls a territory loaded with the resources she wants for her young. It also signals a set of genes that would benefit her sons. Male levels of T remain high during the breeding season. It promotes the aggression needed to defend their territory and those resources needed for their little gene carriers—better known as babies.

However, as I have mentioned, testosterone does not cause ferocity. Testosterone is one factor. Habitat also plays a key role. When you compare breeding aggression among birds, hostility patterns do not map well onto testosterone levels among species. The type of nest they require predicts the concentration much better.

Many birds weave nests. Their territory's quality and the competition for secure nesting sites amid its bounty impact their willingness to defend their architectural miracles. Once a couple makes a nest, that site's value goes up, and they will protect it with vigor. Think of it as the difference between having a lease canceled before you move into an apartment and a home you've built with your own beak. If nesting sites are in short supply, bigger fights occur.

Cavity nesters, birds living in hollows, have difficulty finding nesting sites. Compared to related species that are not dependent on rare nooks for survival, both male and female cavity nesters aggressively defend their nest sites. However, neither gender has higher T-levels than their less-restricted, nest-building relatives [532].

Some animals have an alternative way of passing on their genes: if you can't breed on your own, assist your family or flock mates (alloparenting). Cooperative breeding styles do not arise from stressed-out subordinates. A balance among multiple hormones, not just testosterone, contributes to this supportive parenting in many social vertebrates [533].[88] Sometimes, support makes the best of a bad reproductive situation. In other cases, group parenting contributes more to genetic survival than striking out on one's own.

Humans fit this latter pattern. Our reproductive potential improves when we cooperate and gain the support of family and allies. Getting your mom to babysit the kids is among the caring strategies [534, 535]. Grandmothering supports reproduction among primates and across human cultures [536, 537, 535]. We also fit a male pattern; men reduce testosterone levels during early fatherhood by converting it to another hormone, estradiol [448, 538]. This change happens in both heterosexual and lesbian partners of new mothers [539], promoting parenting and partnering, both of which benefit the offspring.

In superb starlings, breeding males T peaks during the courtship phase [540]. As in us, the level plummets during the nurturing phase when they divvy up the care for the eggs and chicks. Both breeders and helpers[89] share a similar level of this hormone. In mammals and birds, nurturing fathering occurs when nature dials testosterone down a notch. Not every species reduces testosterone levels during fathering [541]. In those

---

[88] Oxytocin, vasopressin, testosterone, and prolactin all contribute to multiple areas of sexuality and bonding.

[89] Behaviorists call helpers "alloparents." Often, they are siblings or offspring unable to breed that year. Alloparents increase the fitness of their genes by contributing to the rearing of a related clutch, each containing a portion of their genetic code.

that do, from fish to mammals, the male body knows when and how to make this switch by converting testosterone into estrogens [542].

Low-status males don't always express lower testosterone than high-status males. Testosterone promotes strategic behaviors, not necessarily dominance-seeking ones. T can promote acquiescence in low-ranking males [543]. Remember, testosterone does not work in a vacuum. Subordinate males have high levels of another hormone, cortisol.[90]

Like testosterone and estrogen, cholesterol makes cortisol in the progesterone pathway rather than in the androgen loop. Any cholesterol shunted into the progesterone-cortisol path diverts from the testosterone one. Rises in corticosteroids correlate with high levels of stress. "Overdoses" of these hormones form a separate mechanism that reduces the ability of the picked-upon animal to triumph over its dominant neighbors [531]. Like testosterone, corticosteroids regulate our bodies and our behaviors. Also, like T, their precise results depend upon the physiology and social history of the body.

Corticosteroids: Stress to Success

> *"One ring to rule them all, one ring to find them, one ring to bring them all and in the darkness bind them."*
>
> — *The Fellowship of the Ring* [544, p. 50]

The body makes all steroid hormones from a common base molecule, cholesterol, and they all look like four ragged rings cut from chicken wire. Minor differences around the ring produce different effects for each "steroid" hormone. Enzymes alter the shapes of these molecules, shifting or adding atoms along the sides and bending them. Stick one bit of stuff on the side, and it becomes testosterone. Replace that stuff with

[90] In social species, high and low-ranking individuals have elevated levels of stress hormones. One is stressed because of his or her low rank. The dominant animal is stressed because of the challenges of and to their position.

other molecular adornments, and the steroid becomes progesterone, estrogen, cortisol, or other steroids, which can pick the molecular locks on our cells. Like keys with minor differences in their teeth, they fit different "receptors," the locks on the surfaces of the cells. These slight differences affect the body by unlocking diverse molecular doors.

Among plants and animals, there are hundreds of steroidal varieties. For example, the plant estrogens in bourbon, beer, and wines act like estrogen in our bodies, even at moderate drinking levels [545]. Steroids, like cortisol, estrogen, and testosterone, direct body fat to specific parts of the face and torso. Estrogen and progesterone help build the breasts. They also direct the fat deposits that soften the angles of the body to produce an archetypally feminine or juvenile appearance.

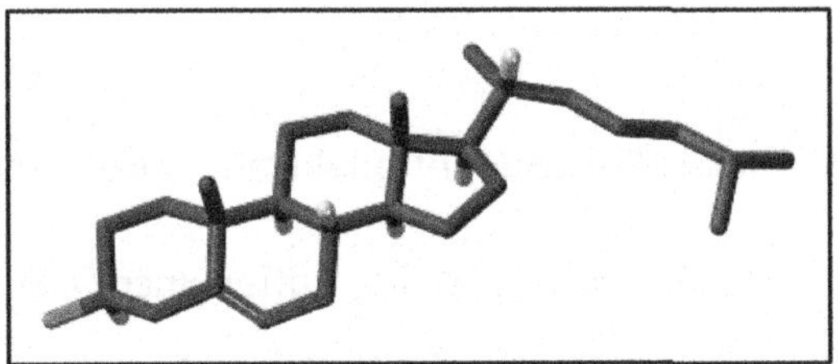

Figure 3-2. Cholesterol has a structure that looks a little bit like a do-it-yourself puzzle piece for building chicken wire. It is the base molecule for creating many different hormones. — (This skeletal diagram is copyright-free and courtesy of Jynto, posted to Wikimedia Commons, 2011)

Most cholesterol never leaves its original state, and it stands guard over our cell membranes. In both men and women, some cholesterol ventures forth to become hormones. Before reaching the testosterone-estrogen shunt, cholesterol receives its first modifications in the progesterone-cortisol (progestagen) passage. In vertebrates, that first passage makes a wide range of steroids. Pregnenolones and progesterones come first. The body converts these precursors into cortisol (C), the primary hormone of the adrenal cortex. Cortisol transforms into compounds that regulate mineral (mineralocorticoids), salt (aldosterone), and sugar levels (glucocorticoids). Mess with these corticoids, and you mess with the hormones that maintain health. Damaging the systems they regulate, including glucose and insulin metabolism, can lead to disorders such as insulin resistance and premature aging [89, 90, 91, 92, 93].

People with high cortisol levels gain weight more readily than people with lower levels [546]. C interacts with another hormone, leptin, that triggers hunger [547]. Our low-ranking Pleistocene ancestors would have used cortisol to improve the use of calories to which they had access. In early 21st-century North American society, this energetic advantage encourages weight gain and often leads to further bullying.

> *"Never thought you'd like fat little crybabies, Parvati."*
> — *Harry Potter and the Sorcerer's Stone*, Chapter 9, J. K. Rowling [2896]

Cortisol promotes insulin resistance and fat around the organs and belly fat [548]. It rounds the face (moon face), broadcasting a message of low status and signaling a reduced threat to higher-ranking individuals. It allows the stressed person to store fat efficiently, a good adaptation for low-status individuals with less access to resources. This low status interacts with one of the remaining socially accepted forms of prejudice and bullying: weight shaming [549, 550].

The prejudice against and bullying of people with "stress bellies" makes linked emotional traits harder to interpret. People with high waist-to-hip ratios often show reduced coping skills, suffer increased depression, and respond worse to chronic stress than people with lower ratios [551]. In young women, cortisol-induced fat reduces the hourglass appearance of the torso, a fertility (beauty) marker. Paradoxically, the increased body fat may serve an adaptive function: it harbors cholesterol storage, reducing the cholesterol available for immediate transformation in the cortisol pathway. The roots of such obesity issues stem from social stressors and lie in their higher C levels, not in the character of the overweight person.

A Multi-Function Problem?

Hormones do more than regulate our aggression and body shape. They also influence social bonding and reproductive behavior. Reproductive and non-reproductive sex and bonding have different functions [552]. One involves partnerships (including child rearing) and social bonding. The other, the passing on of genes.

As in other pair-bonding and harem species, women reap reproductive benefits by coupling with men who provide care, resources, or security for their young [553, 554, 555]. They also profit by pairing with those who can contribute heritable traits to make their offspring attractive [552]. Ideally, they hit the jackpot and get a "twofer" who combines both sets of features.

Women slightly prefer softer (low T) features during the less fertile phases of their menstrual cycles. Because of testosterone's association with a high sex drive and aggression, this fondness should reduce the chance of male hanky-panky and make for safer babies. When estrogen spikes and women are likely to ovulate, they find symmetric and testosterone-sculpted features alluring [556]. That shift could benefit her sons [557], but it is too subtle to evoke infidelity [558]. They may, however, sway single women's choices of short-term sexual partners [559]. Women in heterosexual relationships who rate their partners as unattractive report greater erotic attraction (but not more philandering) to other men close to ovulation than at different times during their menstrual cycle [560].[91] Single, unpartnered women and those interested in short-term mating follow the same pattern [561]. Translation: From a biological perspective, sensitive-looking guys should get sex from unattached ladies more often. High-testosterone ones should attract partners when it counts for the genes: the fertility phase.

Across species, males are not inclined to miss mating opportunities. Male humans have found ways to keep their mates outside the fertility period. More so than women, men try to retain mates through coercion, threats, and debasement [555]. Some men can synchronize their peak testosterone with their sexual partner's menstrual cycle. Most do not take this tactic [562, 563]; they may, however, still respond to scent.

There are brilliant ideas so bizarre, eloquent, and simple that one must sit back to marvel at the genius behind them. The "sniff tests" rank among them. Men or women (senders) wear T-shirts for several days;

---

[91] They also reported a higher rate of jealous (mate-guarding) behavior by their partners during ovulation. These data may not hold true for female-female partnerships, which show a different pattern of peak sexual interest [3372].

receivers sniff samples from those shirts and report their impressions. The swatches contain pheromones that broadcast information about the sender. The hormones that ripen an egg modify body odor, and men respond to it with a brief rise in testosterone and a cortisol drop [564].

Men also take a different approach, one we share with bonobos, chimpanzees, and even marmosets: social and nurturing skills that are antithetical to bullying. In many primate species, females prefer males who treat them well and who regulate and reduce violence within the social unit. They want ones that will protect their babies from roving males rather than killing their offspring [565]. Exposure to babies drops testosterone levels in human males. As a birth approaches, men and women share many hormonal experiences, including a rise in prolactin (milk-stimulating hormone) and cortisol. Following the birth, both men and women prepare for nurturing with reduced testosterone or estradiol. Even when fathers-to-be hold a cute, blanketed doll with large eyes and baby-like features, their circulating testosterone plummets, with an initial drop of about 33%. [566, 567].

Long Lives for Short-Lived Ghosts

"Monsters are real, and ghosts are real too. They live inside us, and sometimes, they win."

— Steven King, introduction to *The Shining* [568]

T and C levels are transient, changing rapidly. Alone, these hormones may influence immediate behavior during sex but are unlikely to be involved in long-term mate-guarding. Low-testosterone males express jealousy, too. The form it takes depends on the species. Crickets and songbirds emit a call to stave off other males, discouraging them from approaching females or entering their territory [569]. Elephant seals claim an exclusive breeding right to multiple females and will fight other males to retain that control. Other species, like the Parthenium (Mexican) beetle (a ladybird mimic) and dragonflies, will extend their copulation, preventing other males from mating with their belles.

Human males also use these T-based tactics, but they often follow a different path to maintaining their genetic advantage over rivals. Like our chimp and bonobo cousins, our species takes a bio-social approach that reduces the need for energy expenditure: we evolved as bands of semi-cooperative male relatives. For bonobos and chimps, females leave their birth troops and join a band of brothers and cousins. We share this innate behavior with our close relatives; about 70% of human cultures follow this same pattern. In most traditional societies, a baby's father would be the husband or his near relative (allowing for hanky-panky). Despite the "ick" factor this idea evokes in many of us, these shared genetics provide one scaffold upon which nature can build a less violent, nurturing male (more about this topic later). Laura Bohannan encapsulated this idea in her humorous essay, *Shakespeare in the Bush*, where she tells the story of *Hamlet* to a group of Tiv from West Africa. The cultural disconnect hinges on Hamlet's disgust over his mother's marriage to her brother-in-law and his father's cries for vengeance. Tiv society lauded the brother's actions because the king cared for the widow and her son, Hamlet [570]. For the Tiv, the marriage represented the proper response for a leader; in us, it evokes an uncomfortable feeling of cortisol-laden stress.

Hormones From T to C

The presence of a regular mating partner (wife) can reduce the costly testosterone production and cortisol levels correlated with stressful mate guarding [571] and promote within-group cooperation, a valuable trait for guarding the troop and turning aggression outward, against neighbors who may covet one's territory, females, and resources [572].

The corticosteroids (a.k.a. cortisol and corticoids) may be kissing cousins to the androgens and estrogens, but they are the daughters of progesterone. The adrenal glands produce corticosteroids in their outer layers (the renal *cortex*). These hormones let us cope with physical stressors. *Mineralo*corticoids regulate our electrolyte levels and blood pressure; they preserve the balance between sodium and potassium needed for life. *Gluco*corticoids regulate the metabolism of sugars, proteins, and fats. Cortisol (C) ranks among the most influential of the glucocorticosteroids. Adrenaline increases your heart rate and the fight-or-flight response, but it is virtually powerless without its corticosteroid partners that regulate its function. The corticosteroids fine-tune the brain's synapses to respond to immediate stress and trigger fear memories [573, 574].

> *"He that fights and runs away, May turn and fight another day; But he that is in battle slain, Will never rise to fight again."* — credited to Oliver Goldsmith

C and its derivatives redistribute the body's resources to meet emergencies, such as dominance struggles. It suppresses many functions not critical for immediate survival. It releases the energy (sugars) the body needs to respond to the adrenaline signal. It sends sugars to the brain to cope with emergencies and mediate emotions. Cortisol levels increase in response to fear, infection, injury, pain, and low status. As part of our behavioral safety system, it ups our response to predators and competitors. And during aggressive confrontations, C allows us to withdraw when losing.

Nature designed corticosteroids to keep the body on an even keel, releasing sugars and minerals when needed. But when traumas like bullying, rape, and shunning create epigenetic ghosts, they transform glucocorticoid regulation in ways that can haunt us through life [575, 576].

High glucocorticoid levels impair our immune function, dropping the number of white blood cells used to fight disease [93, 577, 91].[92] In high doses, they increase fat deposition and atrophy of muscles. Our low-ranking ancestors in the Pleistocene would have benefited from this effect. As subordinate animals, they would have had less access to resources, including food. Cortisol would have improved their use of the available calories and promoted the production of ghrelin. This stomach hormone stimulates appetite [578]. Together with another hormone, leptin, they facilitate the retention and development of fat reserves,

[92] Circulating lymphocytes and eosinophils.

plumping a woman's appearance and signaling she has enough calories to reproduce [579].[93] Today, that same effect subjects us to the social stigma of obesity. Adults who experienced bullying as children run an 80% higher risk for obesity than adults who do not report bullying [324]. Adults with other PTSDs have similar problems [580]. Chronically high corticosteroids relate to this morbidity, creating a positive feedback cycle that furthers weight gain.

In hierarchical social species, the lowest corticosteroid levels are typically found in members of intermediate rank. Combined with tweaked *serotonin* levels,[94] it reduces their aggression toward higher-ranking members. It makes them less of a threat to alphas and, therefore, less of a target. They still have higher social status than the social "losers" at the bottom. Their hormonal tweaking may help them interpret and send social signals, choose partners, and develop alliances [581].

Corticosteroids also regulate histamine. This hormone participates in our modulation of inflammation, immunity, and stress response. The tie between histamines and stress becomes evident when asthma flares, when our skin erupts in stress-related hives, or when we experience anti-histamines' sedation effects.

Histamines are among our oldest stress-associated molecules. Still, we consider them minor stress hormones in humans, often overlooking their role in this context. In fish and reptiles,[95] histamines are essential stress hormones [582, 583, 584]. In zebrafish, social bullying and dominance patterns regulate histamine, reinforcing status [585].

---

[93] Leptin is a hormone produced by fat cells. It favors weight gain by stimulating hunger, a critical adaptation for lower-ranking social animals confined to less favorable feeding sites, eating last, or having food taken from them.
[94] (5HT) is a hormone associated with regulating mood and novelty seeking.
[95] For those of you who keep up on classification, yes, I know that the concept of a "reptile" is now defunct—dead—gone. For the clarity of most readers, I use it in the somewhat traditional sense, meaning lizards, snakes, and their close kin (the lepidosaurs).

Histamine and cortisol both influence our sense of stress and interact to regulate our social responses. Cortisol accumulates in our hair, and samples from victimized children show more significant C deposits than their peer tormentors [586]. This higher level correlates with reduced reasoning skills, leading to riskier choices. In nature, this C-impaired judgment increases an organism's chance of premature death. Making a rival easier prey improves the survival chances of the higher-ranking animals.

Again, I caution: Specific hormones do not induce specific behaviors or feelings. Hormones interact with receptors, neurotransmitters, and other hormones to accomplish many goals. When we refer to corticosteroids as "stress hormones," we use a shorthand way of describing their impact on our physiology. Those impacts, in turn, affect our sense of stress and well-being. Feelings lay far downstream.

Hormones work by binding to receptor sites on cells. By using and modifying different receptors, the body can create a myriad of effects from the same hormone. Many asthma inhalers contain synthetic cortisol that decreases the inflammation in the tubes that lead to the lungs, thus reducing asthma symptoms. If taken for a long time, corticosteroids can trigger depression. In some people, corticoid inhalers bring forth panic, and in others, transient diabetes. And when steroids were prescribed for arthritis, an upsurge in insomnia and mood disorders happened among the elderly and others taking the drug [587, 588]. In effect, it mimes the depression and terror induced by sitting low on the social ladder.[96]

Feelings of inadequacy and being overwhelmed during periods of stress stand in sharp contrast to the sense of well-being provided by testosterone. Artificially increased C levels disturb sleep and increase appetite, leading to weight gain and shifted energy levels (either up or down) with an impaired ability to focus.

---

[96] From personal experience, I testify corticosteroid derivatives can induce a condition resembling ADHD, with much energy but an inability to focus. These side effects should not be surprising given that nature designed corticosteroids to help us through stress or low status when shifting attention and hypervigilance aids survival.

When stressed long enough and hard enough, the body fails to maintain its cortisol output.[97] High glucocorticoid levels contribute to developing PTSD and its recircuiting of the brain. Once inflicted, people with PTSD often drop their C levels. This change renders them less able to gate their adrenal response, resulting in hair-trigger reactions to adrenal bursts [589]. Half a century after the Holocaust, survivors with PTSD had significantly lower C in their urine than survivors without this syndrome [590].

Chronically high cortisol levels reduce the function of the hippocampus, a region of the brain critical to the formation and retrieval of memories [591, 592]. At least in part, it executes this transformation by stimulating the release of a different hormone, leptin [593, 594].[98] The paring back of connections in most brain regions would cause catastrophic meltdowns. However, the hippocampus can perform a trick that challenges the little gray cells in most of the brain: It can grow new neurons, even in adulthood. This genesis of new neurons contributes to our ability to adjust and recover from trauma, which is an integral part of our coping mechanism [595, 596]. This small, curled, seahorse-shaped region deep in the brain plays an integral role in stress management, buffering the impact of the hypothalamus, pituitary, and adrenal systems [597, 598, 599].[99]

Corticosteroids, Serotonin, and Bullying

The body ramps up its release of cortisol when physically or emotionally pressured. A low social rank produces stress. Those with a high position strive to maintain status and thus also feel the strain.

---

[97] Lowered C levels are found in survivors of physical and psychological stress. Long COVID survivors, for example, show a protracted drop in cortisol levels, accounting for many symptoms [3286].

[98] Reduced hippocampal connectivity is associated (correlated) with higher levels of leptin and insulin resistance, which help us gain weight [594]. Both increase with stress.

[99] Technical note: To further confuse things, the mechanism by which the hippocampus helps us overcome and rebound from stress remains unclear. One idea is that it facilitates our recovery by using another brain region, the ventral dentate gyrus (vDG), an area that helps regulate mood [3509].

Vertebrates release additional C when they occupy a subordinate or dominant role. The lowest levels are found in the middle of the social ladder.

Under short-term stress, a burst of cortisol gives us the edge we need to face sudden danger or challenges. It signals us to back away from a losing confrontation. When the C level remains high for too long, as they do when we undergo prolonged subjugation, abnormal feedback systems develop. This physiological response to suppression happens across vertebrates, from fish to birds and mammals [600, 585]. In humans, we find evidence of peer victimization in childhood in the cortisol levels of the hair in the late teens [601, 585]. The impact lasts through life.

After a dominance fight (e.g., a sporting competition), winners and losers have increased C levels. It remains elevated in the losers for much longer [602]. It doesn't matter if the challenge happens through a physical fight, chess match, or work task. Much of the data comes from studies of males. Based on limited data, this effect also holds true in females. Women's cortisol levels remain high in the losers of a contest longer than in the winners [502]. In high-testosterone, highly competitive women, testosterone allows social rank and dominance behaviors to blossom when cortisol levels drop [603], and in multiday chess competitions, the cortisol levels increase more in female winners than the losers [604], a pattern consistent with the elevated C levels seen in socially high-ranking males who must defend their position.

This pattern does not mean everyone responds to T or C the same way. People differ in their receptors, sensitivity to the two hormones, environments, and social training [605, 606].

Over time, the myriad of cortisol insults experienced by the low-status, subjugated person hypersensitizes the brain, intensifying its response to future losses [607].[100] The epigenetic regulations in DNA expression

---

[100] Among the insults is the damage to the HPA axis (hypothalamus, which regulates our mood; pituitary, which manages glandular functions; and adrenal, which governs fight or flight response).

happen throughout life,[101] and damage persists long after one removes the trigger. Long-term high cortisol generates long-term modifications in body function. Eventually, the brain signals the presence of threats when it ought to reassure us of safety. Paradoxically, some transformations relate to the body's failure to maintain freakishly high C levels, leaving adrenaline ungated to ride roughshod over the body, becoming a dictator of the brain and glandular regions that regulate emotions and commanding us to panic [573].

One need not remember suffering and neglect to suffer its consequences, and that distress can be social or physical. Trauma-induced modification in infancy can impact us throughout life. Baby rats who do not get enough tender loving care from their mother rat develop chronic high cortisol as adults.[102] This under-loving or neglectful scenario leads to elevated C and low serotonin. This combination yields the same responses in humans. The resultant epigenetic tweaks to DNA expression [608, 609] regulate this part of our hormonal lives.

Mice have endocrine responses that correspond to those in humans, and they respond to the same drug classes that work in us. Drugs that build serotonin levels also switch off the DNA that regulates and reduces the rise of cortisol in the brain. Depression and reduced cognitive function correlate with high C levels [610].[103]

Of course, a confounding factor must be teased out. Cortisol regulates energy use, and its level rises as our energy demands increase. An animal's work to gain resources or maintain status within a group influences

---

[101] Epigenetic tweaks happen when small molecules flip the on-off switches that control the codes in our cells. Embryos flip bits of DNA on or off to make livers, hearts, and big toes. Under stress, we make modifications that control our reactions to hormones.

[102] Recall that cholesterol is the precursor fed into the progesterone-cortisol and the androgen-estrogen pathways.

[103] Serotonin does not alleviate all depression and cognitive dysfunction. The study investigated one specific mechanism, the hypothalamic-pituitary-adrenal path. Imbalances here can spark multiple mood and thinking disorders [610].

its level [611, 612]. The stress and the energy expenditure needed to maintain social standing contribute to high cortisol levels in high-ranking organisms.

We can tease apart these two states—low-status cortisol and levels needed to maintain status—by looking at low-energy challenges, such as verbal altercations between competitive individuals.[104] Mehta, Jones, and Josephs [613] studied competitive males who entered their pets in dog-agility contests. The high-testosterone men whose dogs won matches had lowered C levels when they won (upped their status). Victors felt rewarded and wanted to participate again.

Cortisol increased in high-testosterone losers. Typically, they chose not to compete again. For them, that contest acted as a punisher. C did not vary in less competitive men (statistically lower testosterone). Women behaved the same way.

Cortisol increases with tensions as mild as the hassles of daily life [546]. It elevates with a lack of deep sleep and chronic stress. The body releases less cortisol during early sleep than before rising. Increased morning C prepares us for awakening and the day ahead. The lowest levels happen during our deepest sleep phase (i.e., delta or slow-wave sleep), about three to five hours after our head hits the pillow. Stressors that interfere with deep sleep, such as a disheartening day or the aftermath of violence, limit our ability to reduce our C levels and handle stress.

Corticosteroids let us maintain homeostasis [614]. They prevent insulin from dropping blood sugar too far [615]. It regulates acid secretion in the stomach and the passage of sodium through the small intestine [616] and adjusts the release of adrenaline [617]. In short, cortisol gets its regulatory finger into every hormonal pie in the body.

---

[104] Some studies suggest that cortisol is less related to rank than once believed. However, a proper neutrophil/lymphocyte ratio, which indicates a well-functioning immune system, improves with increased rank [327].

Cortisol levels rise as social status decreases from the norm. Mice who experience a social trouncing become depressed and show elevated C and low serotonin levels (similar to those found in under-loved mice). In humans, the risk of heart attacks also increases with social stress and low status [618]. The same relationships hold for depression [619] and work burnout [620]. Cortisol levels rise earlier in the mornings because stress increases with anticipation of the coming day [621]. These correlations do not prove that C or stress causes heart attacks; they show that events tend to happen together, just as we tend to eat more ice cream on hot summer days than in a cold winter blizzard.

For the victim of the bully, cortisol metabolism plays a pivotal role. The high-stress levels of victimhood can permanently alter the body's biochemistry, putting the victim in greater danger of depression and anxiety throughout life [622]. Both cortisol and epinephrine peak during dangerous and charged emotional events. They facilitate the formation of "flashbulb" memories that warn us about environments to avoid in the future [623, 624, 625]. As bullying occupies more of a person's life, he or she will dodge more and more situations. They will take fewer and fewer chances lest they upset the surrounding alphas.

Constant high C levels can permanently scar a part of the brain that lets us learn and make memories, the hippocampus. With chronic stress, the hippocampus atrophies and becomes less efficient in processing memories [591, 626, 593]. The bullied person suffers cognitive loads that impair and cause difficulty learning compared to a person who never or seldom experiences the torments of a bully [627, 628, 629].

Despite impaired memory and learning, victimized persons form *flashbulb* memories. Another brain region, the amygdala, joins the hippocampus to create traumatic recalls. This almond-shaped mass processes emotional content and boosts memory consolidation [630]. This enhancement explains why so many victims of Nazi atrocities, the killing fields of Pol Pot, the Bosnian genocide, the Rwandan genocide, and even schoolyard bullies can identify their tormentors decades later and at a single glance.

Flashbulb memories form faster, trigger more quickly, and vary less than most recollections, and they remain consistent longer. A 2021 study of the 2016 Brussels bombings, for example, showed consistent stories among the survivors and similar accounts told about the bombing over time [631]. Flashbulb images form during traumatic events and are often fragmented [632].

Just as rewarding chemicals, like dopamine, influence the detail of pleasant, rewarding memories, cortisol contributes to the speed of traumatic memory consolidation, its integration into the brain, and its choppy structure [633]. Far from being repressed, flashbulbs lurk near the surface, ready to prime the body and to keep us alert if we encounter that traumatic situation again. In nature, that quick response to danger can save one's life. In stratified animal societies, it can instigate reactions that lower your status in the social hierarchy.

Stress accomplishes these adjustments through epigenetics, altering what information gets copied from the DNA (transcription) and the pruning back (atrophy) of cells and their connections. The stress-induced glucocorticoid impacts stress and damages the hippocampus, an old part of the brain involved in memory and learning, shrinking this area and altering its function under chronic strain [634, 635, 636]. This correlates well with EEG studies demonstrating reduced working memory in low-status individuals compared to high-status ones [394].

A smaller hippocampus predisposes people to further stress-related damage [591]. A small hippocampus can result from chronic stress and/or predispose its owner to substantial stress-related damage. This relationship helps explain why dominant animals often harass lower-ranking ones "just because." The resulting cortisol bouts eat away at their confidence and ability to depose the higher-ranking despot. And, as is frequently true during development, increased cortisol in childhood can create long-term changes in the hippocampus, causing memory impairments and abnormal reactions to future stress [637].

A single episode of high cortisol (and possibly other glucocorticoids) can prime the adrenal gland for rapid firing. The resulting hair-trigger stress response can persist long after whatever initiated the fear has departed. The target may reduce their corticosteroid output and replace it with a fast adrenaline release [638, 639]. When this condition becomes chronic, we know it as post-traumatic stress disorder.

Chronic (long-term) exposure to C results in depression and anxiety associated with two additional classes of hormones, sleepy-time indole*amines* and wakey-wakey catechol*amines*.[105] Together, these "amine" pathways regulate sleep and wakefulness. They also handle our flight, fight, and freeze responses before, during, and following peril and conflict.

> *"I jes' hates to hab to weigh anything today, 'caze I 'members so well dat each day dat de slaves was given a certain number of pounds of cotton to pick. When weighing up time come and you didn't hab de number of pounds set aside, you may be sho' dat you was goin' to be whupped."*
> — Interview with Adeline Hodges, enslavement survivor, describing a memory trigger [3502]

One finds this ancient stress-response system throughout the vertebrates. Even fish behave as we do. In trout, cortisol triggers increased brain serotonin (an indoleamine) and catecholamines following a fight. In the dominant fish, cortisol rises in one brain region (the telencephalon or hindbrain) and dissipates within a day. In the subordinate fish, serotonergic activation happens in all three brain divisions and stays elevated longer. Subordinate animals also show a substantial increase in brain adrenaline-like[106] activity not seen in dominant fish [602].

Social defeat and subjugation have long-term consequences for many species, not just fish. In the male Syrian hamster, routed animals will avoid novel situations.[107] Following a loss, they avoid both the victorious opponent and non-threatening intruders. These generalized avoidance behaviors happen with reduced

---

[105] Indoleamines are hormones made from tryptophan, an amino acid. They induce a calming, sleepy-time feeling and include melatonin and serotonin. Catecholamines are made from the amino acid tyrosine, and they stimulate you to action. They include adrenaline and dopamine.

[106] catecholaminergic activation

[107] Here, gender matters. The aggression system triggers differently in males and females. These differences happen in a complex hormonal system that includes oxytocin and vasopressin, which have gender-specific effects [641].

serotonin activity. Activating serotonin-receptor sites in the brain can reverse some of this conditioned behavior, driving up aggression in male hamsters [640, 641]. In our ancient midbrain, the Ventral Tegmentum (VTA) also kicks in to reduce or amplify contextual fears triggered by specific circumstances [470, 642, 643] or to generalize those terrors, contributing to living nightmares such as PTSD [644].

High cortisol levels in lower-ranking animals serve a social purpose. By reducing cognitive function and increasing depression in subordinates, higher C reduces the threats to leadership, stabilizes group coherence, and reduces aggression [645, 646]. Such a system promotes genetic survival when functioning within a kinship group that shares a high percentage of genes.

> *"I tried to laugh about it, hiding the tears in my eyes, 'cause boys don't cry, boys don't cry."* — "Boys Don't Cry" by The Cure [3094]

Such induced states manifest themselves in physical and psychological pain. People who experience chronic physical, economic, temporal, and psychological distress seek to alleviate their discomfort. Their risk for substance abuse increases because we do not, as a society, provide sufficient relief from the corticoid-induced pain they suffer. Nor do we acknowledge the existence of their agony. American culture preaches that people should "suck it up" and deal with emotional anguish. Then, we wonder why substance abuse and depression surround us. We expect and demand that the high-cortisol individual act like the high-testosterone one. We dismiss him as a wimp or histrionic whiner when he does not.

Low social standing ranks high among the causes of chronic corticoid stress. Low status alters the pathway connecting the adrenal gland (source of the corticoids), the hypothalamus (a central emotional processing unit), and the pituitary gland (a "master switch" that controls the glands of the body).[108] Through this axis, elevated cortisol wreaks havoc.

---

[108] This pathway gets its own nickname: the HPA axis (hypothalamus, pituitary, adrenal axis).

A two-year investigation of grade-schoolers [647, 648] found that children with low socioeconomic status suffered from steadily increasing cortisol levels across their study period. They did not see the same pattern in children in higher brackets. Their social status correlated with their corticoid levels. The increased C levels also reflect a higher perception of personal threat and family dysfunction. These levels primed their brains for a higher chance of psychiatric disorders later in life, placing them at risk for depression and anxiety disorders in adulthood [622].

> *"Unemployment insurance is a prepaid vacation for freeloaders."*
> — attributed to 1966 California gubernatorial candidate, Ronald Reagan [3457]

Self-blame and self-bullying also harm us. Young men who suffer from chronic unemployment show a pattern similar to bullied people [649, 650]. A low social status history increases their chances of using substances such as alcohol, tobacco, and cannabis to alleviate the social shame of unemployment. However, the use of those substances further decreases their chance of employment. The attempts to mitigate stress establish a vicious feedback cycle they have trouble breaking.

Stress-related modifications are more than skin—or, in this case, gland deep. They transform how genes express themselves. We see the adverse effects of stress in vertebrates [651, 652, 653, 654] and insects [655, 656, 657, 658]. Children with a low social status during their first five years of life have genes for inflammation responses switched on at higher rates than experienced by high-ranking children [659]. This socioeconomic effect strengthens rather than weakens with time [660]. The regulatory shifts experienced by dominated persons allow them to cope with altered cortisol levels. In adults, this difference, along with stress-induced changes to the gut barrier [661], explains why people under chronic stress have elevated inflammation-related disorders like heart disease, respiratory diseases, and infections.

More insidious, women traumatized by early life stress experience added inflammation during pregnancies, contributing to stress and poor health in their children [662]. The stereotype of the low-ranking nerdy kid with the inhaler may have a grain of truth, but children without sources of encouragement, such as a nurturing family, experience negative gene expression adjustments more intensely than children with robust support systems [663]. Stress fosters physical and psychological problems in disenfranchised populations who lack power and status, with less damage in people comforted by reliable emotional support networks that provide safety and inclusion [664].

Socioeconomic status has no diabolical magical properties. Women, LGBTQA, and anyone identifiable as different can become the target of repeated bouts of degradation and violence. Any source of trauma will upset hormonal and receptor levels. Individuals with experiences predisposing them to power and status have the molecular shelters to weather these corticosteroid storms. What strikes the low-status person like a hurricane will affect the high-status person like spring rain.

CHAPTER 4:

Selection Pressure and the Bully Syndrome

> *"[My new mistress] had never had a slave under her control previously to myself, ... The fatal poison of irresponsible power was already in her hands, and soon commenced its infernal work. That cheerful eye, under the influence of slavery, soon became red with rage; that voice, made all of sweet accord, changed to one of harsh and horrid discord; and that angelic face gave place to that of a demon."*
>
> — *Narrative of the Life of Frederick Douglass, an American Slave*, by Fredrick Douglass, 1845, Chapter 6 [665],

Selection Pressure Favoring Bullies

> *"The murder rate in the United States is 0.5 per 10,000, and in Canada, it is 0.18 per 10,000."*
>
> — United Nations Office on Drugs and Crime [666]

We dislike the idea that bullies fare well. Our species evolved a tendency to see the world as a just place. We trust the evildoer will get a deserved punishment, especially when they direct malevolence against us [667]. Through this belief, we cope with stress and unjust surroundings [668, 669]. If we follow the rules and do good deeds, no harm should befall us [670, 671]. When no punishment seems forthcoming, or if the world becomes too evil, we blame the victim [345, 346, 672]. Doing so disrupts our lives less than acknowledging that bad things happen to good people.

Figure 4-1. Job on the dung heap. The biblical story of Job, who suffered without fault, cautions against our innate desire to blame others for their misfortunes. Hans Holbein II, Job and Eliphaz, woodcut, ca. 1525-26. (Public Doman dedication from the Rijksmuseum of the Netherlands.)

This pattern colors our world, from our beliefs about unemployment [673, 674] to fraud victimization [675]. Our concepts of deserved natural rewards and punishments sculpt our reactions from rape [676, 677] down to who deserves to get the last finger sandwich after a meeting. Without this coping mechanism, despondency descends, and we cease to function. We bristle at the notion that bullies often succeed better than saints.

Behaviors evolve for the same reasons as other features: they support adaptation to the environment. To a biologist, an "adaptation" is an inherited physical, social, or behavioral feature that allows an organism to survive in its environment and keep its genes in the gene pool, preferably producing more offspring than its neighbors who do not bear the same trait. By definition, one must pass an adaptation to future generations through DNA, epigenetic activation, the silencing of genes, or the promotion or suppression of gene products.[109] For a feature to be adaptive, it must promote its host's survival into its reproductive years and at least maintain its proportion of gene carriers (organisms) in a population. A really beneficial adaptation beats the odds and allows its host to reproduce more effectively in a specific environment.

Fitness comes in many forms. A mutation can change coloration, concealing the organism from predators or making it a stealthier hunter. It might make the organism more visible, advertising peril to its foes. An organism might run faster than its neighbor, play dead, or emit a foul stench. A beneficial gene could produce redder feathers or help a bird build a better nest. Adaptations let animals squirt a nasty-tasting

[109] If your mind feels like a shaken snow globe right about now, don't worry. Many molecular geneticists are still trying to wrap their heads around the ways gene expression manipulates us.

fluid, inject venom, or do imposing displays to scare predators or competitors.[110] Some mutations alter the neurotransmitters and receptors that control the sensitivity to social situations, leaving their bearers more cooperative or Machiavellian.

A specific behavior does not always correspond to a specific gene. Inheritance increases the *probability* that a particular behavior will occur; it does not guarantee it. Like physical adaptations, action patterns work through chemistry, physiology, and anatomy. The environment puts the final finish on behavioral traits, just as it regulates our height and weight.

> *"What are the odds that a cluster of frigging psycho-traits would follow one family line for over three hundred years? Over fifteen generations of genetics, there's zero chance, but... fifteen generations of privilege and power could produce a string of people like Richard Tarquin."* — *Silent Consent* by Circa24 [2905]

For example, an enzyme called MAOA[111] breaks down several small neurotransmitters for recycling. These include dopamine, adrenaline, and serotonin, a triad introduced in the last chapter. MAOA exists in several sizes, from long ones to shrimpy ones. The organism's DNA (genotype) codes for its form (phenotype). We have two versions of our personal genetic code (half from mom and half from dad). Therefore, we receive two of the five possible instructions for building shorter or longer MAOA. Our inheritance impacts our ability to withstand stress without becoming violent. Further complicating the instructions from our MAOA are the epigenetic and expression changes caused throughout our lifetime [678, 679, 680].

Tweaks to our proteins and lipids can alter our responses to environmental signals, increasing our chances of acting in a particular way. Humans have no innate desire to make money, but we crave the power,

---

[110] A praying mantis weighs about 0.14-0.18 oz. A 140-lb human comes in at 2,240 oz. Despite having 7.5 / 10,000 of that human weight, it confidently performs a threat display that makes most humans take an initial step away. That's what I call an adaptation!

[111] MAOA stands for "mono-amine oxidase A." It breaks down neurotransmitters formed from amino acids. The gene can have several sections that repeat. The ones with more repeats are associated with better mood control and a decreased tendency to react with violence. The shortest two alleles (forms of the gene) have the strongest association with violence. They do not act in a vacuum. The influence of the genes is tied to environmental switches.

resources, and status that help us get what we want. In modern civilization, cash lets us meet those intrinsic goals. As a young species in the Pleistocene, we scrounged for food. Now, we use foraging behaviors to chase money in the corporate world, hunt for cures for diseases, and rummage for information in print and on the internet.

Social animals with behaviors that promote power retention have an advantage [681]. Even in species with *relatively* egalitarian gender relationships, such as wolves and bonobos, those who maintain a higher rank in the troop tend to leave more offspring [455]. They can access or take resources that weaker troopmates dream of attaining. An experience of dominance shapes an organism's internal chemistry and brain architecture to buffer against stress, providing, on average, better health and a higher status [327, 682, 599]. Power rewards its holder with a nurturing internal sense of rank and denies self-confidence to those who experience subservience.

The logical mind says people who practice dominance should learn the skills needed to become influential leaders. They can call upon experiences and recognize the strategies of others in authority because they have used these tactics. One might assume they learned to maneuver the labyrinth of social power and should know how to emerge ahead of the pack. Unfortunately, being a leader differs from being a *good* leader.

Dominance has its dark side. It can corrupt. Like the monkeys who receive testosterone supplementation, people who receive a sudden jolt of power tend to become less sympathetic to individuals lower in the social hierarchy. After their ascension, many see themselves as worthy of inflated salaries and greater respect than the peons they left in their dust.

> *" 'Comfort women' is the term used for the sex slaves tortured and raped by Japanese soldiers in their occupied territories during World War II. The adult children of these women suffer from many of the same psychological disturbances as their victimized mothers with PTSD, including depression, hyperarousal, insomnia, and low self esteem."* — Lee on comfort women [3504]

As with the slave mistress described by Fredric Douglass (beginning of the chapter), whose demeanor changed from compassionate to cruel, a rise in power does not come with an automatic empathic remembrance of one's former state. Rather than lifting their fellows on the lower social rung, the bully syndrome often comes into play. Without vigilance, their increased status drives a newly dominant organism to secure its escalating rank by increasing the power gap. Initial empathy evolves into behaviors that subjugate the ones below them.

Hierarchical, stratified social structures are part of human and other animal social patterns [683]. Traditional societies have long varied in their power distribution. "Big men" systems recognize the skills or wisdom of those who can provide for or advise the tribe. Rather than procuring a secure spot at the top, these leaders constantly compete with others to secure resources for their followers. The prestige gained through prosocial behavior needed to maintain their position contributes to their sensual appeal to women and, presumably, their mating success [524, 684]. Other peoples recognize individuals for their roles, specific abilities, and talents, bestowing status in one area of expertise but less so in another [685, 683]. Some communities engage separate war leaders and peace leaders.

In Western societies, we have money. Most executives recognize the importance of cooperation and the positive impact of respect, and typically take responsibility for shared goals [686]. But corporate and executive bullying remains common. Those who use their office to crush their underlings sully all reputations.

In modern industrial (WEIRD) [112] societies, people exercise newfound despotism over co-workers, subordinates, post-office clerks, checkers, and others perceived as less influential or prestigious. Many engage in social distinctions that most do not consider violent or demeaning. Some cross the fuzzy, invisible threshold to a delusion of natural superiority, bullying, or microaggressions.

> *"Dehumanizing service workers looks to me to be mostly about insecurity."*
> — Finn Murphy [3181]

Corner offices, reserved parking spots, and executive washrooms are perks of power. These distinctions can slip from the deserved rewards of success and privileges of rank to supporting a belief in one's innate superiority. In the movie, *The Big Short*, when confronted with the immorality of a business position, a character challenges his accuser to compare bank accounts [687]. The audience catches the offensive implication: The man with the thicker wallet believes himself more intelligent and deserving than those with less.

Victims of corporate bullying report exhaustion and job dissatisfaction [688]. Their discontent leads to turnover rates that reduce corporate function but increase the power differential for bullying executives, further securing their jobs. Such abuses of power happen most often when people interact outside their family unit.[113] Domestic abusers disgust us because they act against their kin, a behavior lacking sound biological sense.

Human use and abuse of power took shape in the early days of our species. It built upon pre-existing behaviors supporting power structures in other primates, mammals, and animal life. These entrenched behaviors remain with us and can still aid the bully or the victim.

---

[112] WEIRD = Western, Educated, Industrialized, Rich, and Developed. Most published psychological and sociological studies have concentrated on this population. Although Westerners believe they represent all human behavior and people are the same everywhere, our generalist nature allows for learned variations in behavior and thought.

[113] Because of the intense competition for resources within a family, vicious family bullying happens. I will address the issue in a later chapter. On *average,* families cooperate more than strangers.

Harassment, Violence, and Sex: A Tale of Three Primates

> *"Repression is the only lasting philosophy."*
>
> — Charles Dickens, *A Tale of Two Cities*, Book 2, Chapter 9 [689, p. 138]

Many social organisms, including primates, resort to violence and submission as tools to achieve and maintain rank. Humans, and our genetic cousins, bonobos and chimpanzees, share many dominance and submissive traits with other social vertebrates. All three of our species also possess honed appeasement behaviors to dissipate aggression.

For a male chimp, bonobo, or human, his success at mating and reproduction correlates with social status. Bonobo males garner much of their social position from their mothers [690, 691].[114] Juvenile chimpanzees gain early experience of status from their mom. For both of our cousins, higher-ranking mothers gain more resources and provide the social environment and power to create a foundation for future leadership within the troop [692]. Paternity is unknown, so the father's social standing is presumed to have no *social* repercussions.

By watching their mothers, the young learn about aggression and Machiavellian politics and discover the necessity for cooperation and coalitions. As adults, chimpanzee stratification correlates with the alliances formed with others who experienced similar rank-shaping. With time, chimp males depend less upon their mother's position and support than male bonobos. Humans also learn the value of aggression and cooperation from family and society [693]. Teamwork can compensate for individual power differences, reducing the ability of a few alphas to monopolize resources [694].

---

[114] Bonobos and chimpanzee offspring have no way of knowing the identity of their fathers. Thus, their mother's rank matters. In humans, the social position of both parents plays a role.

Compared to our cousins, our species has flexibility in how we attain our social status and access to power, property, and prestige [695, pp. 37-40, 696, pp. 41-58]. Like the bonobo troops, the rank of our families matters to our position. Some human societies strictly ascribe status based upon the caste of one's family. Others offer paths to higher or lower standing based on abilities. Like chimpanzee societies, people create coalitions, banding together to demand concessions or overthrow leaders.[115] Humans experience many of the same status effects as other primates [460, 697].

Low social status suppresses the immune system, resulting in more disease and reproductive suppression. Subordinates work harder for calories and nutritional value and have more difficulty obtaining mates or sexual encounters. Reduced rank correlates with a higher number of dominance confrontations. These aggressions arise through bullying by higher-ranking individuals and defensive strategies designed to retain or gain status and resources [698].

Our species will use or ignore inequalities and the consequences of inequalities, which often emerge early in life, to encourage underperformance, violence, and maintenance of social position in underclasses [699, 263, 700, 701, 702]. This social-homeostatic mechanism stabilizes a society in favor of the privileged and upper echelons.[116] Animal communities use learned helplessness to stabilize their relationship with the environment [703]. When a social being gets humbled during a defeat, it becomes less of a threat to the hierarchies and subject to less of the costly aggression needed to maintain power [704].

---

[115] Humans behave like other primates, and males dominate most violent uprisings. However, females also participate, both as combatants and as supportive forces.

[116] These statements give a general overview of the impact of low social status and gender compared to high status. High-ranking primates also experience tension as they struggle to maintain their dominance. When threats to power escalate, the key markers for stress that correlate with ill health also increase [3451].

The Herod Effect

Figure 4-2. Massacre of the Innocents by Marc Antonio Raimondi (1560s) after Raphael. Etienne Dulane, printmaker. In the biblical account, King Herod ordered the slaughter of all the firstborn sons in Bethlehem because of a prophecy of the birth of a rival king. There is no objective (non-scriptural) evidence that this slaughter took place, However, Herod may have killed own firstborn son who had a claim on his throne. His younger son succeeded him. From a reproductive standpoint, eliminating a potential rival made good genetic sense. It extended Herod's reign and his chances for reproduction. The execution of such gene-carriers provides a benefit to males who practice infanticide, favoring the genes that support the behavior. — (Public domain, Cooper Hewitt, (Smithsonian Design Museum, 2017)

Throughout history, societies have killed the children of a deposed leader; they threatened the consolidation of power and the welfare of the victors and their families. Within families, rulers sometimes killed their offspring and put their resources into raising one or more grandsons to inherit the throne. These grandchildren threatened the ruler less but still functioned as effective gene and meme carriers. Other species also use strategic infanticide to conserve resources and gain advantages [705]. Males will use it to enlarge their reproductive advantage over their rivals at the expense of female reproduction. [706]. However, females use diverse strategies to reduce the threat to their young, such as multiple matings, e.g., bears and bonobos [707], and forming cooperative parenting groups that can defend against predators, including their own kind [708]. They may also spend more time with a dominant male during estrus (chimps and bonobos), align themselves with less infanticidal males, or just keep their distance by selecting territories [709, 710]. Females will also resort to the murder of the children of others to gain resources and social advantages for their own offspring. Spear-nosed bats, for example, will nurture and rescue a fallen baby of a relative or friend (close contact) but attack and kill offspring from different social groups [708].

Figure 4-3. Several religions use the theme of a child endangering a ruler. Here, the painter Raja Ravi Varma recounts the birth of Krishna, escaping the prison in which he was born. The slaughter of the offspring of a dominant male is a phenomenon also seen in the animal world — (1890, oil on canvas, Maharaja Fateh Singh Museum). Public Domain Source: [3249]

Chimps, bonobos, and human males critically differ from most social species. Although violence against children happens, we tolerate the offspring of other males more than many mammals, and the infanticide rates tend to be low among our kind [711, p. 410, 712]. In many social mammals and birds, a new dominant male injures or kills babies when he seizes power. Behaviorists call this phenomenon the "Herod effect," a reference to a biblical story of King Herod, who slaughtered all the infants in a city because he believed one threatened his throne. Infanticide exerts pressure on the social structure of a species, favoring females who protect their offspring and fetuses against the homicidal and feticidal intents of their troopmates [713, 714].

The previous dominant male would have sired the pre-existing babies at a regime change in a patriarchal species. By butchering cubs, a male lion or tomcat increases the chances the female will soon ovulate [715, 706]. The killings allow the newly dominant feline to reproduce more quickly than if he allows the cubs to live. For the less social house cat, males have no guarantee of paternity, and feral males regularly kill unattended kittens [511].

Primates behave likewise (e.g., [716, 717, 718, 719, 720]. Among the aye-aye, a type of lemur, males have been implicated in infanticide, reaping the same genetic advantages as the tom cats in their multi-male system [721]. This link between paternity uncertainty and the murder of infants holds in multi-male troops of primates, each male benefiting from the induced fertility [722].

Multi-male troops are not the only condition that favors the Herod effect. If they can locate the infant and force a female into a reproductive state by killing an offspring, then changes in leadership and stray males also pose threats [723]. When a band of isolated female mountain gorillas meets a strange male, they approach with caution. New males might perform violent displays of power that can include killing babies. This infanticide eliminates the offspring of a previous alpha male and hurries the female toward reproductive condition [724, 725]. To protect their offspring, females join the males of their troop to drive stray males from their territory [726, 727].

Chimpanzees and bonobos, who live in bands of closely related males and near-permanent female transfers, infrequently engage in Herod-like infanticide within their troops. Each male has a genetic stake in the other's offspring, with brothers sharing an average of 50% of their DNA, 25% with nephews, and 12.5% with first cousins. However, chimpanzees will attack the male infants of neighboring troops, consistent with eliminating genetic competition through the Herod effect [728]. Female infants, future sexual partners who will leave their natal bands, are attacked less frequently.[117] Females become breeding stock later. The son of another male is nothing but genetic competition. Females, who have a high genetic investment in the survival of their offspring, take protective countermeasures, including potentially costly ones. Some leave the safety of a troop with a defeated male; others trust female alliances [714].[118]

Among mountain and lowland gorillas, males also live in patrilocal groups that will kill unrelated offspring. Young males will break away and live solitary lives until they can form their own social cluster with a group

---

[117] The cadavers of the slaughtered infants are typically not cannibalized or only partially cannibalized, arguing against the aggressors hunting for food. Mothers frequently carry the corpses of their offspring following an attack [3247]. The lower reported rate of female infanticide also argues against a strict resource-competition pattern. These models, however, are not mutually exclusive, and we will return to this discussion in the chapter on warfare and genocide.

[118] Although there are reports of dominant females abusing the offspring of rivals, no one has reported infanticide in bonobos at this time [926]. Aggression toward younger males happens but becomes more intense as the young male approaches adulthood. Remember, we have far less data on wild bonobos than on chimps. Remember the old saying: Absence of evidence is not evidence of absence.

of wandering females. The progeny of these single-male troops endure a greater threat of infanticide by neighbors because the single male cannot defend them as well as a band of males. Stray males often target the offspring of immigrant females and their suckling infants [729, 725]. Their behavior reflects paternity uncertainty and the Herod effect. Females protect their future reproductive investment by abandoning a failing silverback to join a more powerful male in a different group near the start of his ascendency. In doing so, she sacrifices any current pregnancy for future gains [730].

Chimpanzees, bonobos, and several other primates show a variation on this theme, which could be called the "Herodias" effect [731]. High-ranking females interfere with the ability of lower-ranking genetic competitors to raise their infants [732, 733]. This spiteful strategy reduces the reproductive output of rivals and secures more resources for one's own offspring.

Bonobos can harm, kill, or kidnap the infant of a female with inferior status. The ranking female may keep it for hours or weeks [734]. Despite complaints from the mother, the dominant female retains the child. She might nurture and feed it with her own offspring or use it to replace a lost baby. Far from being a loving act, this abduction extends ongoing intimidation and violence toward the lower-ranking birth mother. This behavior differs from adoptive behaviors (alloparenting), where a female assumes responsibility and cares for the infant.[119] The high-status kidnapper mistreats the infant, often causing physical damage. In humans, young children separated from their parents (or other primary caregivers) also have shortened telomeres, segments at the ends of chromosomes that are associated with longevity [735]. When it weakens, she abandons it. The distressed mother can reclaim her child, but the damage will have diminished them both.

---

[119] Among the bonobo, females who have never given birth and who help rear other offspring suffer less aggression. Adoption, which could enhance a young female's standing, remains rare. In one known case, a four-year-old male was adopted by his older brother and survived two years [3459]. In two other instances, "permanent adoptions" (more than two months) of out-of-troop females were reported. In one, a natal offspring received more grooming than two adopted ones. The other was a postmenopausal female's adoption of a young female [3256]. Caring for the infant may reduce aggression toward the parent by falsely advertising fertility [763].

If the separation is short, the infant might endure traumatic memories and changes in behavior and brain function from the stress. If the kidnapping lasted days to weeks, the birth mother would have suffered decreased or ended milk production. That infant may die. If it lives, the child will never rise to a high rank [736, 737]. The alpha female has removed a potential threat to her children's dominance by decreasing the fitness of a rival. In these class-conscious societies, females benefit from killing or sabotaging the reproduction of other females.

Despite their own social competition, females have developed counterstrategies to defeat infanticidal males. Females need to reduce the chance that a male could murder their offspring. In some species, one or more females will attach themselves to dominant males who provide protection [738, 739]. Because they are mating partners, the offspring are likely his, and his protection extends to them. Other species confuse paternity by having sex with several males, leading to reduced aggression from the multiple partners.

In other species, a male accounts for the female's sexual behavior; his brain weighs the probability that an offspring is his child. Males of many species detect a female's fertility level through markers, such as sexual swellings, changes in scent, or changes in behavior. They watch the females, especially the fertile ones. Males use their observations; each male estimates his rates of sexual contact with a given female relative to her trysts with other males.[120]

In those species where a male can ensure some fidelity, violence toward offspring decreases with the perception of monogamy [740, 741]. In other species, the association of a female with a troop of related males decreases the genetic payoff for the spiteful murder of a child. Humans have used the former strategy for thousands of generations. Most women leave their village or band of origin to live with their husband's

---

[120] It is possible that some may monitor the social status of other contacts. His higher-ranking relatives would be potent gene carriers and protectors. More distant and weaker relatives, not so much.

family. In contrast, situations that reduce the chances for genetic procreation or promote male-male competition increase the payoff for infanticide [742].

If a troop consists of related males, the dynamics change. Both wild chimp and bonobo females, not males, leave the band of their birth and join another society [743, 744, 745], reducing the risks of inbreeding. The dominant male will father more offspring than other males. Because the males share a genetic heritage, all his male relatives share a stake in those offspring because they carry the genes held in common with their dominant brother. Each adult brother, son, uncle, and cousin may have a reduced chance of fathering a child, but they pass their genes by proxy through each other. All related males within the band of brothers have at least some of their DNA in those little gene carriers. As one's genetic distance from the dominant male increases, fewer of one's genes get passed to offspring. Thus, groups fracture as they expand, and the genetic distance from the dominant males becomes too great.

Within each troop, the sexual hierarchy mirrors the dominance hierarchy. Statistically, the dominant male will be the father of most of the offspring. Because early dominance often follows the lineage and status of the mother [692], his brothers are likely to comprise many of the beta males. Hence, the highest-ranking males of the troop will have the biggest stake in the survival of the young.

Humans, chimps, and bonobos (and probably our last common ancestor) possess male hierarchies.[121] The distribution of Y-chromosomes shows male chimps stay within their communities, leaving when a group splinters; they do not typically join neighboring troops. A given Y-chromosome remains within a community, seldom penetrating another society [746]. Within human communities, the lines and patterns of dominance

---

[121] When scientists look at a set of data, they must choose among the possible physical explanations for the phenomenon. A guiding principle in making that choice is *the rule of parsimony*. In the absence of data to the contrary, we choose the explanation that makes the fewest assumptions but which explains the relevant data. When we see the same pattern of behavior in humans, chimps, and bonobos, we accept the working hypothesis that our last common ancestor also shared the behavior. We remain open to the idea our hypothesis is wrong and stay willing to revise it in light of new evidence.

are not as absolute (linear) as in many other primates. The hierarchical and gender relations also vary across our three species.

Chimps have troops of 20 to over 150 members [747, 748, p. 167].[122] Dominant males form coalitions that subjugate females [749] even as individuals strive for good social relationships with the ladies. Males will protect new arrivals from established ones [750, 751]. Occasionally, subordinate males form coalitions to depose an alpha and his allies [752, 753]. Female support makes the coup a more manageable task [754].

Chimpanzee male cooperation extends beyond immediate kinship to an extended kinship group [755, 756, 757, 116]. These in-troop alliances allow the male chimps to keep peace within the group, engage in cooperative behaviors such as hunting, and seize resources from the neighbors. When small groups of males return from a hunt or patrol, the alliances and power structures allow the alphas to retain power, thus avoiding political upheavals. Despite these coalitions, males will become violent with each other when they vie for status, food, and sex.

Many assumptions about bonobos come from studies of a few bands in zoos. Human discord in their range has rendered non-captive bonobos understudied. Limited observations suggest they live in smaller troops than chimpanzees, seldom exceeding 100 members. These troops break into foraging units of about 27-51 members, a party size larger than the chimpanzees' 9-30 member parties [758, 759]. These again fuse to reform the larger group at night. This larger party size may explain the lower violence when foragers meet: the danger of serious injury may render the risk too high.

---

[122] Like humans, chimpanzees have cultural traditions and live in various habitats. They range from the rainforests at Tai, the evergreen forest of Kibale, and into the more open areas at Gombe. For these reasons, the size of chimpanzee troops varies, as does the specifics of chimpanzee behaviors and cultural traditions [747, 3393]. In contrast, bonobos have a limited rainforest ecology that has been less variable throughout their evolutionary history [958].

Bonobo males do not press their advantage in size and strength to dominate females [760].[123] For the male bonobo, bullying would be a foolish mistake. The genders are close in mass, and in a skirmish, males and females have roughly even odds of winning a fight. [758]. Unlike chimpanzees, female bonobos build short-term coalitions more readily than males. They do not typically defend each other from bullying. They do form alliances, often with high-ranking males, against pesky, lower-ranking, less desirable males who harass or threaten infants.

Like female chimpanzees, bonobo females move from their natal troops to join bands of related males. Despite their kinship, bonobo males fail to form the effective coalitions that characterize chimp males. This failure allows the less related female bonobos to have more influence within the community. Far from being a matriarchy, the social ranks of males and females overlap [761, 762, 763, 764]. Instead of forming alliances to gain and support their status, each male bonobo strives for personal power with the support of his mother [758, 765, p. 111]. This maternal attachment further hinders the male's ability to form male coalitions.

Far from the stereotype of the love monkey, female bonobos bully each other. For status-conscious bonobo females, rank matters. Females will express status by bullying. They will bite, hit, and harass subordinates, giving the higher-ranking females superior reproductive odds.

Much of the famous bonobo sexual behavior functions as homeostatic safety valves, reinforcing social rank. Unlike female chimps, female bonobos form temporary coalitions to gain access to a resource, such as food, or at a provocation from a lower-ranking male toward an infant. At least some of the sexual behavior among females stems from coalition building. No one has observed them defending an infant against a higher-

[123] On average, chimp females are 82.5% the size of males (or males are about 21% larger than females), a significant size advantage [760]. Adult bonobos differ by about 15%.

ranking female. Dominant females will align themselves with their sons [766] or with dominant males [765]. These behavior patterns should seem familiar to humans.

When female alliances form for protection, they can turn violent on an offending male. Their kicking, hitting, and biting can turn so brutal that they can sever the penis or scrotum [711, 767]. The attackers will gnaw flesh, causing severe wounds and permanently hindering the victim. The injuries can reduce his status and mating opportunities. These female alliances limit the ability of a male to bully them or attack their infants.

Within their male kinship group, the males vie for rank. The bonobo's preferred mode of competition appears to be physical displays of prowess rather than chimp-level violence [758, 455, 768] or the Machiavellian manipulation of relationships done by chimps [769].

Despite the high levels of bullying and violence among chimps and bonobos, males tolerate the other males of the troop. Male kinship groups result in less violent competition for mating and resources than would occur among unrelated males. Fighting one's brother or half-brother, someone who carries 50% or 25% of your genes, makes less reproductive sense than fighting a stranger. The family groups thus permit these societies to flourish.

Unlike the chimps and the highland gorillas (and many other social mammals), male humans and bonobos attack offspring less often. The aggression of bonobo and human males toward females and their offspring takes a different form. An unusual feature of bonobo and human female fertility cycles is how they reduce the genetic advantage of the Herod effect.

Both human and bonobo females have a hidden fertile period. During ovulation, female chimps display far more significant swelling between the vulva and the anus. This state signals her reproductive condition to the males of the troop. Human females show no overt sexual signs during ovulation. Female bonobos retain their swellings during much of their non-fertile periods.[124] Because male humans and bonobos have trouble telling which females are fertile, sex occurs throughout the female's reproductive cycle. This behavioral difference makes paternity more ambiguous than in chimpanzees. The kinship of the males in both bonobos and proto-human troops would ensure each male would have a genetic interest in each of the offspring.

The individual's reproductive stake must be reasonable for such a social structure to succeed. Based upon the size of our neocortex, Dunbar [770] estimated that protohuman troop sizes would have been no larger than 150 people. That is the highest number of stable relationships most people can follow. (Sorry, 151st Facebook Friend. You don't make the cut.) Estimates go as high as 290 people [771]. In a Dunbar band, males would share a significant percentage of common genes and a substantial stake in the survivorship of the young.[125]

Bonobo behaviors suggest both genders have an inkling about when fertility occurs. Males try to monopolize sex with ovulating females but do so with less success than chimps. Female bonobos will have sex at any time, but during their fertile phase, they spend more time near higher-ranking partners, a choice reflected in the skewing of fatherhoods toward these males [768].[126] Any given offspring probably came from them. By concealing their oestrus, females achieve another advantage: they hide ovulation from

---

[124] The presence of genital swelling makes it easier for a female to enter a new community [749], but females with swellings increase aggression among males [611].

[125] For the arithmetically inclined, here's a back-of-the-envelope calculation: Make a simplifying, low-end assumption that half of such a band is female, each with an average of two to three surviving offspring. We can now estimate the genetic investments of the males. Only one quarter (37 or 30) of the 150 people would be adult males. Fringe males should be no more than five generations removed from each other and share $1/2^5 = 1/32$ or more of the genes with any given offspring, even if none of their own mating opportunities resulted in offspring. The investment would double at each level of relationship closer to the core males. Suppose you use the higher estimate of 290 people. Then, the degree of relationship extends one more level of genetic distance to six generations, and a Kilworth band should have 1/64 of their variable genes in common with any given offspring for the most distantly related males.

[126] This partiality for high-ranking males during ovulation may be related to the *slight* preference heterosexual women *may* show for high-testosterone physiques (fertility markers) in males near their ovulation.

others, reducing competition from females who might otherwise disrupt their attempts to mate [772]; they may also use social swelling[127] to cement their relationships with the other females [773].

Women have fewer markers for fertility than female bonobos. Ancestors of human males would have had few clues about who fathered which offspring. However, if our ancestors followed the pattern of our chimpanzee-bonobo cousins, the child's genes would have come from the dominant male or his close family. The strategy of infanticide lost value among both the bonobos and the early humans as the paternity of an offspring became increasingly randomized among related males [774].

Mate-guarding would skew the genetic lottery toward the dominant males in chimps, bonobos, and probably our ancestral societies. If the high-ranking males were not the father, a related troop member probably was. Because motherlines and their status impact dominance in all three species,[128] the closest relatives to the alpha often share ranking positions in the troop and are likely candidates for the alternative father. Thus, during a single regime, the dominant males would have the best chances at the genetic lottery and the least reason to harm the child. No matter who fathered an infant, it still carried many of the same familial genes.

Each offspring represents a genetic lottery ticket. The males of chimp-bonobo and most traditional human societies increase their odds of winning the draw because the males of a troop are closely related. Even if a male does not father a given child, he wins the smaller payoff. His brothers or cousins transmit some of his genes [751]. Humans continue to follow the ancestral pattern of female migration, with about 70% of cultures practicing patrilocality—the movement of a woman into the home, the village, or the compound

---

[127] Bonobos do have a swelling phase, but it does not reliably mark ovulation [773].

[128] Maternal status provides social modeling and resource access. These advantages are found in many social species, not just primates [3496, 3148, 3474]. In primates, however, the maternal bond and impacts can last through the mother's lifespan [3464].

of her husband. About 10% practice matrilocality; the remaining are neolocal (new households) or have visiting marriages where neither party moves [775].[129]

Chimps, bonobos, and humans all have clusters of related males; the gene flow comes from females that join the band. In populations with little genetic exchange, the benefits of fighting are sensitive to small changes in the movement of genes or individuals. Your brothers are your gene carriers and your competitors [776]. One would expect rapid flips between self-sacrificing altruism and selfish aggression within these societies after minor shifts in population structures [777]. Your brother may harbor your genes, but you carry more of them; if it comes to you or your brother successfully wooing the lady, it behooves you to win [778]. The benefits from living with your brothers may thus not derive from reduced hostility toward each other but from reduced violence toward offspring.

Harassment has benefits for higher-ranking males in all three species. Chimps form coalitions not just to harass females but to bully low-ranking males and to protect themselves from higher-ranking rivals. Both female bonobos and humans form alliances that serve as units to guard higher-quality mates and limit the fertility of lower-ranking females.

[129] Foragers break this pattern; the more dependent a society is on gathered plant foods, the more likely it is to practice matrilocality and take advantage of the learned skills of the gatherers [1310].

Coalitions of Genes

> *"Calonice: What if [our husbands] grab us and haul us by force into the bedroom? ... And if they beat us?*
>
> *Lysistrata: Then you must submit—but do it grudgingly, don't cooperate. There's no enjoyment for them when they just force it in. Besides, there are other ways to make them suffer. They'll soon surrender. No husband ever had a happy life if he did not get on well with his wife."*
>
> — Lysistrata [779]

In contrast to the males, adult female bonobos, chimpanzees, and, in most cultures, humans share no set genetic relationship with the females in their troops beyond their daughters. Bonobos have more aggressive females than most apes [774, 780].[130] They will harass and murder males. Primatologists have also witnessed the abuse and murder of infants by rival females [737]. As in bonobos, human females have murdered or contrived to murder the offspring of rival women. Like the dominant female bonobo who kidnaps the progeny of a subservient mother, human stepparents often treat a birth child with more tenderness than a stepchild. This abuse benefits their gene carriers at the expense of the genetically unrelated offspring.

Such plotting is not a modern phenomenon. In one story from the Jewish Tanakh (סֵפֶר מְלָכִים, Christian Old Testament), King Solomon goads that a child be cut in half to appease a woman whose own child had died [781]. The mother objected, but the grieving woman saw the plan as fair. In the Hindu tradition, the demoness Putana tried to kill Lord Krishna by breastfeeding him with poisoned milk. In Greek mythology, Procne kills the child of her husband and imprisoned sister (Philomela) by having it cooked. She then feeds it to her philandering husband, Tereas. (Definitely, an overreaction to the affair.) The German folk tradition

---

130 This is an excellent time to recall that most of what we know about bonobos comes from artificial communities in zoos. Usually, these have too few males and may have abnormal social structures. Also, bonobos live in a different habitat than chimpanzees, and resource competition rather than mate competition is the likely catalyst for their aggression [3215].

recorded by the Grimm brothers tells how Hansel and Gretel's stepmother advocated for their deaths. Another unrelated female, a cannibalistic witch, further threatened them [782].

Figure 4-4. Procyone feeds her husband, Tereus, the body of his son after his affair with her sister. —(Artist: Antonio Tempesta, etching from 16th century edition of Ovid's Metamorphosis Book VI)

Of course, it is not only women who kill children. The Christian Bible tells the story of King Herod the Great, who massacred the infants to try to kill the prophesied Messiah. Although most biblical scholars dismiss the story as a literary device,[131] it could spring from another incident. Herod executed his three sons, Antipater, Alexander, and Aristobulus IV, who could have rivaled him for the throne. Ivan IV of Russia killed his son, Ivan, and Constantine the Great killed Crispus. Prince Sado of Korea died sealed in a rice chest by his father, King Yeongjo. The greatest threat to one's throne would be one's sons, so killing them in favor of a grandchild would be an excellent way to hang on to power. The ancient Varaha Purana (Hinduism) condemns to hell anyone who commits female infanticide for as long as the Indras exist. In the Srimad Bhagavata, Asvatthama is banished for infanticide. Moreover, the Holy Qur'an of Islam rejects the practice of infanticide for both males and females (Surah 6:151 and Surah 60:12). The widespread stories and condemnations suggest either a history of the behavior and a negative effect on societies or a demonization of those who are believed to practice it.

Humans, chimps, and bonobos can also nurture the children of others [783]. Culturally determined strategies of social (alloparental) care allow us to wean our offspring earlier [784, 785, 786].[132] Usually,

[131] The story only appears in Matthew, and no other source—biblical or other—claims it happened.

[132] Cooperative parenting is not confined to primates. It occurs in several social species, including dolphins, housecats, and bats [3153, 498, 708].

parenting by a non-birth parent goes well. Humans have multiple neurochemical safeguards that trigger nurturing reactions. Like penguins and other adoptive species, our physiology encourages the care of non-descendant offspring [787]. Our large kinship groups during the Pleistocene cold have catalyzed this easy triggering of nurturing responses. Alloparenting happens in other apes, but not to the same extent as in humans. Different human societies tweak our communal care to their environment and cultural history. The filters of our cultural norms and genetics shape these fitted practices [788].

Despite our typical lack of aggression toward offspring within social groups, our alloparenting response is not a perfect mirror of our parenting behaviors. Our unwillingness to confront the complexities of violence directed toward children hampers our ability to respond to it [789]. Stereotyped evil stepparents do exist. The "Cinderella Effect" refers to the statistically higher rate of child abuse by stepparents than by natal parents [790].[133]

Children who receive care from an unrelated adult suffer more abuse and neglect than they do from their biological parents. Child neglect accounts for 78% of reported child maltreatment in the United States and is higher in non-natal children [791, 792]. Estimates range from six times the fatal violence [793, 794] to one hundred times the likelihood of death [790].[134] Compared to murders by birth parents, those killed by a surrogate were a bit older and more likely to be male than female (71% vs. 51%). The killer is typically a man rather than a woman, 90% vs. 48% [795]. These findings fit the strategy set by our great-ape relatives: eliminate the genetic competition.

---

[133] Here, the term "step-parent" does not refer to the legal definition but to any adult figure who fills a guardian-like role in a child's life other than the natal mother and father. Boyfriends or girlfriends of a parent also abuse children at a higher rate than a natal parent.

[134] In one study of 219 fatal child abuse cases, 60% involved biological parents, and 29% involved surrogates (adoptive, foster, step-parents, and intimate partners of birth parents). Birth parents represent a far greater percentage of caregivers than surrogates, skewing the figures [795].

In one category, mothers appear to outstrip other caregivers: the killing of newborn infants. Often, these deaths accompany post-partum depression, have a self-sacrificing motive, and frequently include their own suicide. Fathers who killed their infants expressed jealousy and presumed infidelity, suffered from emotional regulation (personality) disorders, or had prior histories of violence [796].

Mothers may compete for status for their children, but they share a threat: males. The joint biological problem faced by all males is the uncertainty of paternity. For humans, the closer the offspring are in age, the more likely a man is to provide paternal care. Our species instinctively uses evidence of fidelity to one male (monandry), and the closeness of births serves as one marker for shared paternity [797]. The certainty of paternity correlates with reduced levels of bullying, neglect, and harassment.

Like their bonobo counterparts, men must contend with the behavior of women. Females in both species band together to protect their young from low-ranking males. No matter how much we want to believe otherwise, babies seldom look like their mothers or fathers. They carry no bar code linking them to their dads. To establish paternity, we use some of the same tools as our other social mammals, including our connections with the mother and the appearance (phenotype) of the child as it ages [798]. Our poor perception of paternity may be one reason mothers outstrip fathers in killing neonates. Lower-ranking males have fewer mating opportunities and are thus less likely to father offspring. In theories based upon sperm competition, these males would reap the highest benefits from aggression toward offspring.

For small, related groups of males, like our ancestors, the dangers involved in eliminating a potential offspring, who probably carried many of one's genes, outweighed the reproduction lost to competition from related males [799]. If the fluidity of a troop changes, if more unrelated males join a community, or if lower-ranking males experience a decline in mating, then this risk-reward ratio should shift. These changes would result in pressure for more aggressive male behavior toward each other and toward offspring [800].

CHAPTER 5:
Upon Reflection

"Better never means better for everyone. It always means worse for some."
— Commander Waterford, *The Handmaid's Tale*, Season 2, Episode 5: "Faithful" [801]

Dehumanization, violence, and status-seeking are not artifacts of human cultures. They are as intricately woven into our evolved behaviors as compassion and cooperation. We express our dark side in Machiavellian manipulations and sabotage, and honor killings. Although betrayal and aggression can provide short-term gains in our ongoing primate politics, cooperation and trust are essential for more than group survival. They are critical for maintaining status. Through friendly coalitions, we advance our social standing and forge protective groups of friends and family that support us and shelter us and our reputations from rivals intent on spite.

The Iago Syndrome

"I follow him to serve my turn upon him." — Iago, from Shakespeare's *Othello*

Any viable mating strategy—for both human and bonobo males—must contend with the hidden estrus. In both species, paternity remains obscure because the gentlemen cannot tell when a lady is fertile. Many powerful males could be the father and thus have a genetic investment in the genetic lottery known as offspring.

Our species shares an ancient history of living in troops of related males. This pattern extended the investment in junior to this broader circle of male kin; each child bears at least some familial genes. The combination reduces the genetic payoff for infanticide. We carry this history from our early days as a species deep within our behavioral programming, programming we extend to other aspects of our lives.

> *"Power concedes nothing without a demand. It never did and it never will."* – Fredrick Douglass

In this atmosphere of doubtful paternity, male bonobos and human males use Machiavellian tactics to increase their chances of mating relative to the next guy. They can achieve this end in several distinct ways. In many species, males can force sex upon females, an opportunistic or a last-ditch choice [802, 803]. Far from being a violent by-product of patriarchal social structures, forced sexual contact occurs from humans to insects [804, 805, 806, 807, 808]. Among humans, rape is an ugly global issue spanning cultures and a prime example of the objectification and degradation of another person's body, robbing them of autonomy and dignity. Yet men—including mainstream human-rights organizations dominated by men—minimize the trauma of rape and fail to recognize it as torture more than do women [809].

Violence, including sexual violence, is a natural outgrowth of our inherited psychology. That origin does not excuse sexual assaults nor make them acceptable. Despite Madison Avenue's brilliant confusion of the terms, "natural" does not equate with "good." Arsenic is a mineral in much spring water, but that doesn't make it a healthy food choice. Asbestos is a natural fiber, but I don't want it in my house. Rape may have biological motives, but it is no more acceptable than murder. It removes a fundamental right defended by animals as diverse as humans and insects: choice.

Human and bonobo females defend their sexual choices.[135] They will fight to maim, kill, or otherwise destroy their attackers [810, 811, 812]. They form coalitions to block attempted rapes, as do their kin.

---

[135] Often the sexual choices of women are culturally limited, but women will still defend their mating and family choices within that context.

Males, one would presume, should experience an approach-avoidance dilemma. Their primate programming would encourage them to mate guard to attain exclusive or preferred access to copulation. If their personal risks are not too high, they should also interfere with attempts by others to coerce sex.[136]

A male can impress females with his superiority—something hard to accomplish. To do so, he must actually *be* superior or lie convincingly. He can show his worth by defeating competitors (antagonistic dominance), as elephant seals, elk, and gorillas do. He can exhibit his physical features, like the peacock and red cardinal's feathers, the racks of horns sported by moose and deer, or the prolonged singing of chucking frogs, cicadas, or canaries. Or, he can bully. These competitive signals are tough to fake. Thus, maintaining or gaining access to an egg presents a predicament in species where males compete for females.

> *"According to her Doctor's Report, [Kamala Harris] suffers from 'urticaria,' defined as 'a rash of round, red welts on the skin that itch intensely, sometimes with dangerous swelling.' She also has 'allergic rhinitis and allergic conjunctivitis,' a very messy and dangerous situation. These are deeply serious conditions that clearly impact her functioning."* — Donald Trump, Truth Social, Oct 14, 2024 [3254]

Humans also use hard-to-fake (honest) signals, such as control of resources or complex vocabulary, to demonstrate intelligence (competitive dominance). Both men and women compete not just for power but for another reliable marker that can lead us to the top of the social pyramid: the prestige earned through skills [813, 814].[137] For our distant ancestors, these desirable, hard-to-fake features would have improved early woman's chances of accessing power and resources, thus bolstering their reproductive success. Indeed, we still value prestige-based leadership over bullying leadership and will often opt for female leaders perceived to exhibit this trait [815]. To combat the advantage of prestige, be it in a male or female opponent, competing leaders will resort to Iago's tactics,

[136] Low-status males are convicted of more rapes than high-status ones [3248]. It is unclear how much of this discrepancy comes from the greater use of this tactic, the higher number of low-status males, the ability of high-status males to avoid prosecution, and our perception of what constitutes "rape" [3484].

[137] Chimpanzees and bonobos share our proclivity for three major types of dominance: through aggression, competition, or bluffs [3207, 737].

instilling doubt, spreading distrust and creating an enemy from which only they can save us, encouraging their followers to inflict violence against those who refuse the collective identity [816].

To the Pleistocene mind, holding sway over resources implied that others allowed him to maintain control or that they could not dislodge him from those goods. That individual maintained their position because of their reputation for leadership qualities such as intelligence, skills, and confidence [817]. Status due to dominance or prestige had reproductive payoffs [818]. In closely related bands of males, competition and cooperation would form parallel roads to success [371]. Each male or coalition must impress females without the other troop members beating them to death.

Status-conscious species can access a third viable option: *The Iago Syndrome*. Males (and females) can gain status and power by squashing their rivals like bugs.[138] Given an opportunity, even low-ranking males can destroy their competition through lies, backstabbing, and innuendo. In *Othello*, ca 1599- 1602 [819], Iago is Othello's long-time friend and military comrade. Jealous of Othello's success and angered by the promotion of another man, Iago plots the downfall of both his friend and military rival. With no other realistic courses of action open to him, Iago proceeds with a campaign of intimidation, slander, and deceit.

---

[138] States also use the option of squashing their opponents. The ability to destroy another government with rumors or by undermining its credibility entails far less risk than an open assault. Also, I do not advocate for the unnecessary squashing of bugs. It's cruel, needless carnage.

Figure 5-1. Sic Semper Tyrannis (So Goes All Tyrants) is a cry against oppression attributed to multiple Roman sources. The 22nd US Colored Infantry used it as their motto. Their regimental battle flag shows a defender of slavery falling before his victim. —This photograph of the flag was taken by D. B. Bowser [3064].

Human males and females engage in direct and indirect violence designed to demean or destroy the prestige and reputation of outsiders or members of their social group. Like Iago, we will use these methods to denigrate rivals, even ones close to us. We face the same paradox as other social species: The welfare of a group and an individual do not always coincide. Society requires mutual trust to function, but the betrayal of trust can provide short-term rewards with access to power, resources, and mates, even if achieved at the expense of society, e.g., [820, 821, 822]. A person can gain status by crushing others. The more trust invested in a relationship, the more effective the betrayal.

Cooperation and stable hierarchies may benefit group survival, but cheating can provide the potential for upward mobility in the group. If a low-ranking organism can succeed, it can garner rewards measured in reproduction and resources. If discovered, the group can inflict devastating consequences on these social cheaters. In cooperative situations, superior abilities may nurture admiration and interdependence; competitive settings, however, engender malicious envy that can foster backstabbing and sabotage [823, 820, 821, 824]. Turning a cooperative situation into a competitive one—a common tactic of both schools and businesses—can raise the level of backstabbing, thus reducing rather than enhancing performance [825].

Behavioral violations, such as bluffing and within-group aggression of the social contract, should occur when the cost of the betrayal drops below the potential rewards [826, 827]. Those at the top of a dominance hierarchy may widen that gap with intermittent attacks on their underlings, thus upping the costs of advancement [828, 829]. Social stressors, from microaggressions to genocides, can serve this purpose [830]. One need not always trounce an opponent; shaming them can be as effective and less costly.

Just a quick reminder: Power relationships need not be linear, set in stone, or the same under all conditions. Social groups respond to their situations, and behavioral hierarchies can change. A teacher leads a classroom through prestige but readily surrenders his status to his doctor at the physician's office. Many societies have war leaders and peace leaders. The celebrity who cannot fix a toilet should surrender their prestige to the plumber.

Status signals and social cheating do not need to stem from conscious behavior. Paper wasps, a surprisingly intelligent species capable of recognizing individuals by sight [831], respond to dishonest messages. Weaklings who brag about their high status and fighting potential through behavioral communication or physical markers (more dark on their faces) are likely to have their stingers kicked by their nest mates. If a wasp's facial features are too high for their ability, it gets stuck in more dominance battles and finds itself on the receiving end of a whooping [832]. The honest, dominant wasps who won the contests showed a rise in "juvenile hormone," the insect equivalent of the androgens that rise in vertebrates after winning a battle. Other wasps eavesdrop on these conflicts and learn who they can challenge with little personal risk and who is a fraud [833, 834], and social bees also monitor the contests among their hive mates [835]. As with defeated humans, an unmasked pretender suffers a long-term loss of status within the colony [836].

Humans have more flexibility than wasps when we manipulate our images. To fake status, we lie, cheat, wear knock-off luxury goods, and just plain brag. When those below us on the social scale discover our ruse, they delight in our exposure, humiliation, and demotion. In the workings of our Pleistocene mind, our unearned status comes at their expense. Our brain assigns value to reputation [837, 838], and it does not like being fooled.

From wasps to primates, social interactions come with memories [839, 840]. Just as groups will move against an overthrown bully and his collaborators, they can turn on a discovered cheat. Tactical options among humans include aggressive behavior, peaceful negotiations, honest competition, and deceitful destruction. Somewhere in our heads, we determine which alternative provides the optimal strategic advantage for the lowest cost. Then, we use it for personal and political (coalition) power [841, 826, 777].

Humans do not stand alone. Other animals also act as Iago bullies, habitually lying, cheating, and misdirecting to foster their goals. Societies often reward Machiavellian background behaviors with increased rank and status. Individuals who can play the game, who manipulate the impressions of others, gain a commanding advantage over the politically naive. By turning the thoughts of others against our rivals, we elevate ourselves and our collectives. Backstabbing innuendo carries fewer hazards than physical confrontation. However, betrayals can backfire.

A Matter of Honor

> *"What were they doing other than trying to secure some Iraqi neighborhoods? What made people want to kill them for handing out candy and soccer balls, and delivering tankers of drinking water to them?"*
>
> — *The Good Soldiers*, p 151 [842]

> *"It is not man who is the enemy of the human species. It is the irrational; it is the spiritual when it is divorced from the material."* – Anne Rice, *Queen of the Dammed* [3222]

Most Westerners feel disgust and horror when they imagine honor killings in the Middle East. We recoil when reading about the murder of a family member for no reason other than causing damage to nothing but a reputation. Our minds picture wives killed by their husbands for sexual infidelity. We think about young women murdered because someone might interpret non-traditional behavior as loose morals. Because the Western press gives "honor killings" headline status, they loom large in our consciousness. We believe such vengeance to be commonplace and see it as something visited upon women. We imagine it as a foreign practice that does not fit our Western paradigm of behavioral norms. However, the drive for honor, for recognition, and for prestige within our societies is a fundamental primate drive, and we can find different tolerances for affronts to honor within society.

The Muslim world does not own honor killings [843]. Aggressive responses to affronts to prestige follow our reactions to Iago-style attacks that undercut our status and reputation. American cultural traditions, especially in the southern states, include violent responses to honor challenges [844]. The saga of Beowulf tells of a blood feud and a challenge to personal honor. In the Balkans, honor-based blood feuds among males outstripped those committed against women and may be executed by women in defense of their men [845, pp. 10,].

Cultural specifics, not religion, sanction or condemn domestic violence homicides [846, 847]. Though the specifics and excuses used for domestic murders differ, domestic homicides of women occur worldwide. In the United States, spouses or other domestic partners account for about half of the female victims of homicide. In approximately 90% of cases, American men who kill women, kill someone whom they know. [848, 849].

Killings to protect family or personal honor include the murders of LGBTQ people and those driven to suicide by families and communities [850]. In the United States, data specific to gay victims of violence was not analyzed until the 2017 National Crime Victimization Survey. 2021 was the first year the FBI released anti-gay hate crime statistics through the National Incident-Based Reporting System [851]. LGBTQ people are almost four times as likely to be victims of violent crimes, and men are about six times as likely to be assaulted by someone they know [852]. Between 2005 and 2014, the reported crime rates against the LGBT community surpassed those of hate crimes against Jewish citizens. Although Race/ Ethnicity/Ancestry form the most extensive set of hate crimes (54% combined), religion (13.3%) and sexual orientation/gender identity (19.2%) tease out as two groups leading all others in hate-crime victimization [853, 851]. LGBTQ people suffer victimization by people they know more than non-sexual and gender minorities. In some cultures, the honor killings of LGBT household members wipe away the family's shame with blood [854, 855, 856, 857, 858].

LGBT suicide victims tend to be younger than non-LGBT victims; men have more problems with their families than women, and transgendered victims report more childhood abuse [859]. The extent of gender-based violence to enforce heteronormative "masculine" roles (femicide) remains understudied [860, pp. 8,18, 19], as does the part of family and societal honor in the murder and bullying of gender and sexual minorities [850].

Until recently, much of the Middle East kept poor statistics on "honor" killings. The United Nations estimated that at the turn of this century, as many as 5,000 women per year were killed worldwide for dishonoring their families [861]. A 2013 study in Jordan found that 40% of adolescent boys and 20% of girls considered these deaths justifiable [862]. Contrary to the Western perception of honor killings, not every incident is concerned with sex. The reason may involve a woman seeking a divorce, or, in strict religious societies, it can happen for apostasy or marrying out of one's faith.

Despite their grip on our hearts, little reliable data can attest to the frequency or social acceptance of these attacks [863]. Decrying the lack of information, Fadia Faqir explored honor killings in the context of violence against women [864]. She estimated that worldwide, honor motivates about 5,000 annual murders. Although tragic, this number is paltry compared to the overall number of female homicides. Like the American statistics on LGBTQ violence, Faquir's estimate may mislead.

Faquir's figure probably underestimates the number of honor murders. Another study collected data from 520 autopsies at Khyber Medical College in Pakistan [865]. It accounted for women who died from nondirected violence (e.g., riots and traffic accidents) by removing them from the sample. They then examined the honor killing rate among women who were the intended targets. Most of these murders happened to women in their reproductive years,[139] between the ages of 13 and 40 [865], and most (63.4%) happened in rural districts. Interviews with family members implied that over 90% of these homicides had been honor killings.

The research team estimated that in Pakistan alone, about 1,000 women perish from honor killings annually. The Aurat Foundation, a women's advocacy group, estimated that in 2014, more than 700 women died in Pakistan from honor killings [866]. If we use the higher figure, 1,000 women, and estimate the population of Pakistan as 212 million [867] and women at approximately half that number (106 million), the upper rate of these honor killings is 0.09 per 10,000 women. Based on the data of Alam [865], roughly 90% of female homicide victims were honor killings. This estimate increases the murder rate for women to 0.1 deaths per 10,000. Compare this amount to an overall homicide rate of 0.7 per 10,000 in Pakistan [666]. This figure tells a different tale than the popular Western narrative. Overwhelmingly, murder finds male victims.

> *"The global economic impact of violence ran approximately $14.8 trillion dollars US in 2017, 12.4% of the global GDP, averaging $1,988 US dollars per person."* —Iqbal, Bardwell & Hammond, 2021 [3402].
> *"The combined costs of homicides and assault, are about 7.5 times higher than the cost due to war and terrorism."* — A. Hoeffler, 2017 [3129]

Killers have different reasons for extinguishing lives, but being male carries a higher risk of being murdered than being female. Male perpetrators account for 96% of the world's homicides and comprise 78.7% of its victims [666]. Despite the differences in the homicide rates of Pakistan, the United States, and Canada, all three countries fall close to this mean level of violence against men (76.7%, 77.8%, and 69.8%, respectively). If you have the questionable luck of being born a male, you

139 In chimpanzees, the reproductive years correspond to the peak period for mate guarding (jealous behavior) by males and their control over females.

are three times as likely to be killed by another person, usually another man. Homicide, like war, tends toward a gendercide of males.

Crime increased in the United States and Western Europe following the Second World War.[140] It peaked in the US from the late 1970s through the early 1990s. From the mid-1990s onward, homicides began to fall. Contrary to popular belief, the United States does not rank among the countries with the highest intentional homicide rates, but it has a high population. Only China, India, and the European Union had more people in 2022. America's large population and a constant barrage of homicide-related news stories and entertainment bias our perception, making us believe the US is more violent than it is. In 2022, El Salvador topped the list with 37.0 murders per 100K people. The average in the US is 4.96 per 100k [666]. As of March 2023, the US ranks 76th out of 197 countries.

Some groups bear the brunt of these figures. FBI statistics from 1975-2005 give the chance of becoming a homicide victim six times higher for Black than White Americans [868]. This span includes the 1980s when African American males carried a risk 8-10 times higher than Whites (4.7 vs. 29.6/100K). At its peak, homicide ranked fourth as the cause of death for African Americans, six to seven times that of European Americans before returning to the levels of the early 1970s. Consistent with our ape heritage, males constituted 77% of victims, with the highest death rate among 18 to 24-year-olds [868].

Affronts to honor play into many of these homicides, and challenges to honor activate more brain regions than technical failures or frustrations. In violent male offenders, personal insults reduce the active connections between the emotional limbic system and the prefrontal cortex that controls outbursts. This reduced communication between the emotional and logical brain. The change leaves the neural network

---

140 Crime rates probably also increased in the Eastern Block countries, but the figures remain blurry. Soviet officials classified and falsified crime statistics, and people could be convicted without an actual crime if it were for the good of the people [3069]. Russian homicide rates followed the same rise as the US in the mid-1980s but with different age profiles [3430].

more open to provocations in violent offenders than in men not involved in the justice system [869, 870]. Personal affronts can thus become flashpoints for violence as the normal anger-management pathways cease to function.

A woman's chance of being murdered remains far below that of a male. From 1976 to 2005, the FBI reported roughly four times as many male as female murder victims [868]. However, women die at twice the rate at the hands of their intimate partner and are four times as likely to be killed in a sex-related attack (81.2% vs. 18.8%) [871]. Males trump women as victims in other categories of murder. Honor killings that follow an affront to one's prestige contribute to this difference [872, 873].

> *"On August 24, 2025, Albert D. Sturgill was shot to death after flipping his middle finger toward an American flag and a war memorial."*
> *—Melissa Chan, NBC News* [3354]

Not all states partake equally in the bounty of American homicides. Mississippi, Louisiana, and Alabama often lead the nation with a yearly rate between 11 and 20 deaths per 100,000. Although other states have less consistent rankings, Illinois and Maryland regularly rank high with rates of about 0.80-0.88, ignoring the 2020 spike, when others seemed to scramble for the top spots. Except for Illinois, Ohio, and Michigan, the states with the highest homicide rates are in the American South and West. Most of New England and the American Northwest have homicide rates below 0.5 [874, 875].

Some blame the ubiquitous presence of guns in high-homicide areas, but guns could be present to protect oneself from pre-existing violence or a mistrust of the police [876]. These arguments draw on correlation, not causation, and do not explain why these regions seem more prone to gun ownership and lethal violence. Ethnic rifts do not explain the difference. The Secure Communities Program, which deports undocumented offenders, has not proven to decrease crime, while immigrant-tolerant legislation has reduced crime rates [877, 878, 879]. Although our xenophobic, primate nature assumes outsiders to be the cause of violence, sanctuary counties in the United States have decreased property crimes compared to non-sanctuary counties, possibly because of spillover effects on immigrant communities [880, 878, 881].

Regional divides in violence cross geographic borders, with Southerners tending to be more violent than Northerners even when one holds ethnicity constant. Socioeconomic arguments, although powerful [882], also fail to explain the difference. The socio-economic effect does not explain why the American South and West have more homicides per capita than the North and Northeast. Nor do they explain why Louisiana frequently leads the nation in homicides [883].

An understudied component of the American homicide rate is the US version of "honor killing." One's perception of respect factors into how people react to affronts in the northern and southern states [884, 885, 886]. Masculine honor ideology often dictates that men respond to an insult with aggression. Male students at the University of Michigan received a verbal affront to their honor before they met with a research team. (Someone bumped into them and called them an "asshole.") Students from the North paid little attention to the insult. Students with Southern roots had a peak in cortisol and free testosterone, a combination that primes the trigger for a potentially violent reaction [887]. This Southern value of personal honor and its links to white supremacy shine through in the correspondence preceding the American Civil War and contributed to the secessionist movement and a war that killed an estimated three-quarters of a million people [888, 889, 890, 891, 892, 893]. Likewise, personal honor played a role in the escalation of the Vietnam conflict, where President Lyndon Johnson feared being labeled as a traitor and coward if he de-escalated the war or withdrew [894], behavior consistent with the hubris syndrome (*see Chapter 9, The Pangloss Effect*).

As a part of the prestige path to status, honor matters. Men with different concepts of how to respond to affronts to their character also show differences in the Netherlands [895]. A Brazilian study of men incarcerated for homicide also found that affronts to honor and fairness played heavily into their crimes [896]. The impact of honor goes beyond verbal insults; it modulates our level of envy in unfair labor or financial exchanges [897]. As a surrogate for dominance, honor matters to everyone, but some cultures encourage hair-trigger reactions towards affronts. Southern women also hold to an honor code but are more likely to retaliate against an insult to their reputation by spreading rumors and shunning the offender [898].

> *"All people will speak of your infamy, and for one who has been honored, dishonor is worse than death."* – Lord Krishna on Arjuna's refusal to fight in battle, *Bhagavad Gita*, Chapter 2 [3304]

How we view honor shapes how we respond to an insult to our integrity. When primed to react by cultural conditioning, several regions of the brain become more active [899]. The angry brain increases activity in the limbic system (emotions). The hippocampus (spatial and memory skills) and the insula (our sense of worth as an individual) determine the magnitude of the threat. The anterior cingulate cortex's dorsal (back) part (emotional regulation and processing) decides how to respond.

> *"If a man comes into my hut and defecates on the floor, what do I do? Do I shut my eyes? No! I take a stick and break his head. That is what a man does."* — *Things Fall Apart*, Chinua Achebe [1997]

Consider one focal point of Western Scriptures, the rules of conduct known as "The Ten Commandments." Lying gets teased apart from "false witness," giving us wiggle room. Two acts that deal directly with assaults on honor against humans, adultery and theft, get mentioned twice, with a command against both committing the act and thinking of doing it. Depending on which version you ascribe to, three to four others deal with insults to the honor of God, whom they describe as jealous. In contrast, murder, which often results from affronts to honor, gets mentioned once.

Honor becomes a surrogate for status. In our brain, our innate sense of unfairness and drive for dominance conspire to prepare us for battle when we perceive disrespect in another person. To find killings based on honor, one need search no further than the United States.

Cooperation, Spite, and Bluffing on the Savannah

> *"Spite is a little word, but it represents as strange a jumble of feelings and compound of discords, as any polysyllable in the language."*
>
> — *The Life and Adventures of Nicholas Nickleby*, Charles Dickens [900, p. 162]

Humans are environmental generalists. We function in a wide range of habitats by altering our conduct. We take behaviors that evolved for one purpose and graft new functions onto them. Bullying exploits our flexibility as much as other endeavors. In addition to direct violence, people use biological processes to maintain power. Communication, eating, sex, and even habitat selection expand our influence and reduce the power and status of others.

Before a mammal fights, several systems engage. It activates an area of the brain that craves the battle (ventromedial prefrontal cortex) and primes the hormones to endure. Then, it assesses the costs and benefits of the match [841]. Flaws riddle our accounting mechanism, so the desire to fight often wins over a semi-logical assessment of our chance for success. When our social brains conspire to commit war, they weigh these same variables, and our sense of honor combines with the I-thou dichotomy to prevent our withdrawal from the conflict. Few dare instigate war knowing they will lose, but once involved, we'd rather see the battle through than admit defeat. Punishing our opponent becomes as dear to us as winning.

For years, ecologists recognized cooperative, mutualist, win-win behaviors in a multitude of organisms. Win-lose behaviors, such as competition, predation, and parasitism, have been widely studied. We even discerned behaviors where one organism benefited from the presence of another but gave its partner nothing in return (commensal relationships).

Malicious associations, where one partner harms another without a personal gain or at a cost to itself, made little sense. Thus, we presumed spite to be the domain of humans. As with every other time we've believed "humans are the only animal that..." we discovered little behavioral distance between us and others. Ground-nesting gulls and mantis shrimp spend much time punishing their neighbors in crowded breeding grounds. Chimps wage war on their nearby troops. Even bacteria behave spitefully [901]. Spite is a valuable tool that readily takes root in organisms [902], enhancing the spread of markers to keep one safe from the malice of kin.

Most of the time, our species does not go out of its way to harm others. We gain too many benefits from our high degree of cooperation. Malicious behaviors peak during resource depletion or increased competition [903]. The paradox of spite, the willingness to endure a loss of fitness to harm another, disappeared once we stepped back from the individual.

From the perspective of the gene, it doesn't matter if a spiteful act causes a specific gene carrier to suffer a drop in fitness as long as the competition suffers a more significant loss [904]. For the gene, that's a win. Even if a gene decreases the absolute fitness of an organism, it can be favored if it wipes out the competition [905]. The spiteful act leads to over-representing the "winner's" genes in the next generation as long as the next guy takes a heftier reproductive hit. Just as organisms will self-sacrifice to benefit their family, they will self-sacrifice to inflict costs upon non-kin [906]. Spite is the flip side of altruism. Unsurprisingly, organisms favor their recognizable kin with self-sacrifice and non-kin with increased malice [907, 908].

Selection favors genes, not individuals, and if a behavior renders enough support to family or hinders competitors, it can spread even at the expense of the individual. Among our bonobo and chimpanzee brethren, altruism and positive self-sacrifice spread their genes in a community of related males. Spiteful behaviors toward outsiders, including acts that reduce one's fitness (negative self-sacrifice), such as warfare, reduce the enemy's reproductive potential and let the spiteful troop gain control of the rival colony's resources. In humans, these same spiteful tendencies emerge in us-vs-them politics, racism, and discrimination [799].

Spiteful behavior is contagious, like laughter or yawning [909]. Its poison can spread through a population. Through spiteful propaganda, Nazis turned the populace against the Jews and other minorities, prioritizing torture and extermination. Under Pol Pot's regime in Cambodia, the Khmer Rouge turned on the Chinese, Cham, and Vietnamese, killing 1.5-2 million people. Among these were many intellectuals and tradesmen who could have aided in the rebuilding of the state.

Our large brains match the appropriate tools for violence to social situations, and we can fine-tune our malice. The warm feeling of vindication and joy we get at the harm others (Schadenfreude) varies in intensity and motive. We may desire to see outsiders suffer. At other times, we consciously impede them from helping someone else. And we can engage in pre-emptive strikes [910]. All these feelings, these tools, contribute to our "dark side."

Human bullies use the same physical and Machiavellian dominance tactics as other animals, e.g., fiddler crabs [911]. We hit, humiliate, disgust, or ignore victims to fine-tune the damage we inflict while minimizing the danger to ourselves. Individuals calculate their approaches based on a range of factors. We measure the differences in status and power, e.g. [841, 826, 912, pp. 141-202].

Our kind evaluates the probability of success and costs of a retreat, the presence or absence of onlookers, and the likelihood that others will join in the victim's torment, be it a child on the playground or a subjugated nation. We see our spite as justified, and from personal hubris to national pride, our biases cause us to overestimate how others will value our attack and punitive actions [913]. Our propensity for "following the leader" can leave us at the mercy of influential players who can instigate violence in their coalitions, reaping the benefits without a significant chance of paying the costs [914].

Although hard to believe, vindictive behavior has an upside: it helps control cheating and free-riding. Given half a chance, social organisms will take advantage of their neighbors. The behavior need not be voluntary. Ground-dwelling gulls, for example, cooperate to defend their brooding sites by flying off their nests to form a group defense. There are always a few stinkers who remain hunkered down on their eggs. They are not consciously deciding to scam their neighbors. The cheating gulls have an innate, genetically determined defense trigger that differs from most of the colony, and they fail to respond to the cry for community solidarity.

One might presume that cooperative individuals would be at a disadvantage compared to cheaters. They play a guessing game in each interaction, having no information about their partner. The less confident you feel about the trustworthiness of another, the more likely you are to make self-serving choices [915, 916]. However, social species develop interaction histories, and lingering memories of past relationships provide information about how group members behave. Defectors and those who always take find themselves without allies. In laboratory "fairness games," people will reduce their winnings to take away the winnings of an unfair competitor [917]; it is better to receive less than to have your opponent gain more.

Any genes favoring teamwork will be adaptive in a stressful environment where collaboration contributes to survival. Cooperation, a detrimental trait for a solitary animal like an octopus, could benefit a social one [918, 919]. Being treated fairly and treating others with dignity gives us a jolt of oxytocin that rivals the reward of spite. Maltreatment can drop oxytocin levels, leading to social disengagement. Occasionally, it raises it, pushing the victim to double down on their inclination to "tend-and-befriend," their abuser [920].

In social situations, vindictive revenge punishes cheating and freeriding. Social animals keep a running score of who does what to and for whom. Reciprocity may be a tit-for-tat relationship or function more like a balance weighed over time. Young human children tend to use the former; adults use the latter.

Like other innate activities underlying human societies, we grafted malice and appeasement onto behaviors that once supported the social organization of our early hominid forbearers. These Pleistocene and Pliocene ancestors would have lived in troops of related males, much like modern chimps and bonobos. Unlike these other chimps, we lived on the savannahs. This habitat difference would have made small troops easy pickings for predators.

> *"Vengeance and retribution require a long time; it is the rule." — A Tale of Two Cities,* Charles Dickens [689]

Many social organisms benefit from living in large, cooperative groupings [921]. Savanna mammals, such as zebras, gnus, antelopes, and baboons, depend upon the size of their troops to dilute the chance of becoming a snack. Plentiful food, large populations, and low predation favor single-species herds. When times are good and nature's buffet is bountiful, the center of the pack offers great protection. But when troop sizes plummet or times become lean, rivalry for food or water competes with the desire to avoid predators. Cooperation among the species climbs as the chance of becoming lion fodder increases [922]. Mixed herds derive benefits, both through the abilities each species brings and the joint properties that emerge through their cooperation [923, 924]. Their collaboration and information exchange reduce the need for vigilance and lowers stress levels.

Early hominids would have experienced similar selection pressures as they spent more time in the savannahs and less in the forest. The grassy habitat could have selected for larger troops with more pro-social relationships. With large, bonobo-sized troops,[141] our costs for war and competition would have been

---

[141] Bonobo troops average about 120 members; communities vary between 25-80 for chimps).

high, and the benefits of cooperation greater than for chimpanzees. And when predation peaked on the savannahs, learned cooperation rather than warfare between neighboring troops could have provided a survival advantage.

Compared to life in the rainforests, the grasslands posed new dangers from predation, increased by the reduced access to the haven of trees. Based on the skeletal remains of *Australopithecines*, a genus that includes human ancestors, we developed our vertical posture early in our evolution. [142] The erect, aggressive, bipedal dominance posture of apes became the proto-human's natural deportment [925]. Usurping the aggressive stance for prolonged upright movement benefited our ancestors in this more open visual field, allowing them to see over and navigate through tall grass while freeing their hands. This pre-existing behavior signaled the presence of larger animals (due to their increased height), providing a convincing bluff. Large troops of upright members would have proven a buffer against predation, attracting the attention of females and encouraging tolerance among males.

Like the bonobo and the chimp, our closest behavioral counterparts, proto-women probably had more flexible social patterns than men. Female bonobos and humans spend less time fighting and develop stronger social alliances than chimpanzee females [774, 926, 927]. We spend more of our lives in aggregations, contributing to our group size and decreasing the rewards for intergroup aggression. Female humans and bonobos have little trouble accessing groups of males, though high-ranking females can make a newcomer's life miserable for years. The higher social rank of females may have also contributed to another process that furthered our tolerant nature: the selection for intelligent, peaceful males.

---

[142] *Ardipithicus*, who lived in the Pliocene 4.4 million years ago, may already have moved to the savannahs, but debate still surrounds the nature of its habitat. Some argue it still lived in the rainforests. Its fossils and context remain insufficiently studied by independent researchers.

A Fox's Tale

> *"My unexpected finding is that inside every man there lurks ancient caretaking tendencies that render a man every bit as protective and nurturing as the most committed mother."*
>
> — Sarah Blaffer Hrdy [565]

From the 1930s through the mid-1960s, biological science in the USSR fell under the jaundiced eye of Trofim Lysenko,[143] who threatened geneticists with deportation, torture, and death. In 1959, Lyudmila Trut and Dmitri Belyaev braved this anti-science atmosphere to conduct a breeding experiment on the valuable but aggressive silver fox [928].

Dr. Trut selected the least aggressive and anxious males and females from the breeding stock on fur farms. Lyudmila and her colleagues would limit contact with the caged animals, assuring that their calmer natures stemmed from genetics and not training. They took copious notes on the ordinarily solitary animals, watching their reactions to humans and other foxes. She used their responses to humans when making choices for their yearly breeding. By the second generation, changes had started in the foxes. In 1978, she presented a fox as tame as any lap dog to the 14th International Congress of Genetics in Moscow [929].

During domestication, other changes took place. They began wagging their tails like dogs and developed multi-colored coats.[144] (Before you decide to adopt one of these adorable tame foxes, remember that Dr. Trut only bred the aggression out of them. They still have other wild traits and so may shred your designer drapes and scent mark.) Estrus shifted, with females entering the breeding condition earlier. The "elite" fox had higher blood serotonin levels, a condition associated with mood enhancement and reduced anxiety in

---

[143] Trofim Lysenko conducted fraudulent agricultural studies during the Stalinist era of the USSR, promising that plants could be bent to the will of the people. He rejected genetics as counter-revolutionary and killed seven million people through the starvation that swept the country in his wake. Contributing to his anger at geneticists was a humiliating public presentation where his early work was publicly criticized.

[144] Before you decide to adopt one of these adorable tame foxes, remember that they bred out the aggression. They still have other wild traits and may shred your interior decor and scent mark.

humans. Serotonin also regulates the hypothalamic-pituitary-adrenal system, which is critical to aggressive behavior. They had lower cortisol levels than their wild-type counterparts and an elevated level of melatonin in the pineal gland.

The fox also underwent anatomical changes common to many domesticated species. Floppy ears appeared, and their tails shortened, as did their snouts and legs, giving them a juvenile appearance. They underwent changes in their hippocampus (memory system) and oxytocin levels (affiliative hormone). The prefrontal cortex increased its connections with the hippocampus and cerebellum. And like dogs, the fox could interpret human social cues, such as pointing to hidden food. They liked to play into adulthood, and contrary to what they expected, the tame fox did not show a drop in brain volume [930]. The wide range of changes and the consistency with which these variations occur together, not just in foxes but in numerous species, indicate that changes to a few genes cause multiple effects (pleiotropy).

The copious (and often secreted) notes kept by the team prepared them for the return of rationality to Soviet and Russian science. Once freed from the restrictive insanity of the Lysenko era, they collaborated with geneticists in Western countries. Together, they identified genetic differences between the sweet lap-fox and its wild-type cousins.

Behavior is more than DNA deep. It also depends upon the expression (epigenetics) of the DNA. Because the brain's prefrontal cortex plays a significant role in behavior, they chose to study gene activity there first. Three percent of the fox's genes presented differently between the two strains (335 genes). The tames showed higher expressions in 280, including genes for memory and learning, immune system regulation, and dopamine and serotonin production; the others were lower. Forty-five genes corresponded to changes between the domestic dog and the wolf [931]. One gene, SorCs1, helps to move glutamate (an excitatory compound) across neural membranes and appears to be a key player in tame behavior.

The parable of the tame fox has implications for us. Humans and bonobos are less aggressive than chimps. Our heads are more rounded, and we have slender, less muscular builds. Humans may not have floppy ears, but we have smaller ones, as do bonobos. Where the fox developed shorter front and back legs, humans and bonobos have longer hind limbs but shorter forelimbs. Like the domesticated fox, bonobos have reduced muzzles, but humans have the shortest of all, with our faces severely flattened.

Humans show signs of "domestication," a collection of traits common (but not universal) to our few domesticated mammals. Bonobos, like humans, have benefited from pro-social behaviors in their environment [932]. Self-domestication could have happened through female choices or ones made by their families. Biased choices, favoring men with cooperative natures and abilities rather than aggression, create one of the pressures toward the hyper-social niche carved out by our two species [933]. Cooperation's benefits should be enhanced by the shared male genetics typical of hunter-gatherer bands, contributing to a feedback loop favoring greater cooperation [934].

The fox experiment has a flip side. The researchers also selected for aggression and quickly bred a more hostile variant. Although their behaviors were opposite, the brain's grey matter increased in both the aggressive and docile strains, and both shared many of the same anatomical changes [930]. These selected genomic and epigenomic modifications combine with yet another layer of complexity: micro-RNAs (miRNA), which change the expression (intensity) of the protein products in the strains [935]. Together, this triad yielded immediate results, with domesticated strains within a few generations.

Large, competitive societies have shifted away from the conditions that favored our self-domestication. We live increasingly in communities composed of unrelated males who share little genetic investment in each other's offspring. As trust and political control decline, such as with criminal gangs and hyper individualism, chimpanzee-like aggression toward others is likely to climb [936, 937].

Unlike the resource-rich habitat that favored cooperation among the bonobos and hunting and gathering ancestors, we have spent the last century exploiting the fossil fuel reserves and increasing competition for the precious remnants. In 2020, during the early days of the COVID-19 outbreak, resource hoarding extended to toilet paper and hand sanitizer. Per the resource-control hypothesis of mating choice, males stockpiled resources more than females [938].

The limited availability of high-paying jobs can further favor sociopaths who denigrate or sabotage the work of their colleagues, increasing the payoff for aggression and reducing the benefits reaped through skills. The "Cluster B" personality disorders, which include what we once called psychopathies, are now more common and have an earlier onset in the hyper-competitive atmosphere of the United States than in Europe [939]. The Iago syndrome should triumph as the pretext of skills by Machiavellian opponents magnifies the prestige of the aggressive rather than rewarding the competent. Coalitions of violent upsurges can be sparked by a few individuals in choice positions, often by people who will benefit from the violence or lose less than their pawns [914]. If changes in foxes can significantly shift aggression levels within one generation, how might our altered environments and mate choice impact us? The verdict remains out, but there are lots of ideas.

Predisposed?

> *"The day will come when, after harnessing the ether, the winds, the tides, gravitation, we shall harness for God the energies of love. And, on that day, for the second time in the history of the world, man will have discovered fire."*
>
> – Pierre Teilhard De Chardin [940]

As we evolved into our kind, we built upon behavioral traits our common ancestor shared with the chimps and bonobos. These commonalities include power politics, tool-making, prestige, and complex social structures and communication. Like other social species, we undergo hormonal disruptions and display more aggressive and appeasement behaviors when crowded or if resources become scarce [941, 942, 943]. Crowding creates hair triggers for group aggression, with coalitions forming around alphas to access resources or power [944, 914, 945]. Violence, torture, and murder take place in both chimp and human communities, e.g. [774], and we often save our worst punishments for "outsiders" of our species, the ones we have learned to perceive as subhuman.

In chimps and humans, resource competition and aggression can escalate to warfare among troops. These slaughters unite related males to defend their kin and reproductive partners [424] while providing reproductive opportunities for low-ranking, dominated males through forced sex and war rape [803].

To achieve warfare, cooperation must exist. Coalitions of cooperative male chimps patrol their territory, refraining from harming each other as they pick off stray males from other troops [946]. Female chimps partake in attacks less often and are likelier to chase away opponents rather than hunt them to maim or kill [947]. To achieve these ends, our cousins developed Napoleonic intelligence, the cognitive skills needed to join forces to defend against outsiders [948]. They formed cohesive groups, with each member willing to face danger for the good of the whole, thus preserving their genes in their brothers, possibly at the expense of their own lives [949].

The danger posed by outsiders could create a feedback cycle, driving chimps (and humans) toward coordinated in-group teamwork through the genocides of less cooperative troops [950, 951]. Both social birds and mammals show increased group cohesion in the face of threats, especially among those more likely to endure a fitness cost [952, 953]. Paradoxically, if the common ancestors of humans, chimps, and bonobos engaged in genocidal raids, it could have created conditions for disruptive selection. Chimps continued along the high-aggression path. Bonobos formed larger groups for protection, making warfare more costly while increasing the benefits of cooperation as a mechanism for advancement within the troops. Humans can follow either path, but our perceived peril and population size influence our movements toward war and peace [954, 949].

Hunter-gatherer societies collaborate within their bands but engage in more frequent inter-group conflicts than larger-scale societies [955]. Our technology makes our scaled-up direct confrontations between nuclear powers risky, mirroring the troop-size effect seen between chimpanzees and bonobos. Humans expand their alliances to surrounding groups, often related by genetics, language, or culture [937, 246]. Because most human groups are patrilocal, the males who join raiding coalitions are also typically bands of brothers, fathers, and cousins. We use learned behavior patterns to recognize who is "like us" and who can be eliminated, like pond scum. Both chimp and human coalitions will raid surrounding troops and commit "ethnic cleansing." These genocides remove the genetic competition and allow victors access to a loser's resources. The chimps' territoriality, mobility, male bonding, and band structure (fission-fusion) resemble those of traditional human foraging societies [956], and the cost, as measured in lives, impacts both.

Warfare is probably humankind's supreme expression of "spite." To a biologist, spite is any activity, planned or innate, that inflicts a loss of fitness, such as pain and suffering, on both sides, causing a lose-lose scenario. War has a simple goal: to impose a disproportionally greater loss of fitness upon the defined enemy than oneself.

Whoever initiates war does so with the belief that they will lose less than their opponent. Despite the casualties and pain, each side will fight with weapons and propaganda to inflict damage upon "them." The biological goal is to adjust the reproductive and resource outcomes in one's favor. Some will argue that human warfare is about ideology, not resources and reproductive fitness. However, we still choose our battles, initiating wars when we have a significant perceived advantage and can inflict disproportional damage to the "ideological" opponent.

As of publication, no one has recorded warfare among bonobos. However, the bonobo remains the least observed of the wild apes. We might not know about intergroup bonobo violence because of this paucity of data. Observed encounters between troops of bonobos have included screams and distance maintenance before approaching each other. When groups join, aggression happens more often between males from different communities than within a band or among females [957].

Bonobos move as troops; unlike chimps, small male patrol parties are unknown. When bonobos fracture into smaller units, these foraging parties include females [958, 774]. Because of their larger group size, battles would entail more significant injuries than among chimpanzees. The greater jeopardy incurred in attacks reduces the payoff and discourages the behavior [956].

Female bonobos often initiate contact, but we do not know if they do so with any troop or only with their birth community. Transfers of males into new bonobo communities can happen when there are few males and many females. However, when a male tries to join a troop, the natal males do not welcome him. Violent interactions decrease with time, but hostility remains [959]. Begrudgingly, they seem to trade limited access to the troop and its females to increase their group strength and chances of survival.

We remain generalists. Prosocial acts and kindness release bursts of oxytocin, a hormone that can foster warm, fuzzy feelings and a sense of bonding, increasing cooperation within a troop. This same internal drug promotes xenophobia and can predispose us toward violence and the dehumanization of outsiders.

Despite our focus on the violence of chimpanzee warfare, these encounters can be costly. Troops try to avoid each other, using auditory signals to avoid larger groups. These smaller groups prefer to stage attacks on stray individuals or smaller foraging parties,[145] with males killing or attacking the foreign males and

[145] Early reports—before our focus on chimpanzee violence—included instances of peaceful encounters and fluid troop membership in the Budongo forests [3444]. These data may reflect the impact of the environment (heavier rainforest than most troops), or our desire to find peace in our sister species.

absconding with females. Despite regular aggressive encounters among males, open, lethal attacks between large groups carry a cost and seldom arise. Adult females may attack females from another tribe, but before their first pregnancy, they may visit neighboring tribes, possibly scouting out their options [960]. Our focus on inter-troop violence can mask the often-tolerant relations among neighbors, just as a focus on war, violence, and bullying can obscure the often-good relationships among humans. Like humans, our close cousins have options.

Resources and Gendercide

In many animals, including bonobos, introducing a prized item, such as a favorite food, can trigger competition and aggression. Be it land for a human, a sunning rock for a lizard, a leaf vein for an aphid, or a watermelon for a chimp, an individual with a desirable resource will defend it against an intruder.

Jane van Goodall first documented chimp aggression when she tempted local chimps to her camp with bananas. At first, people believed that the bounty of delicacies caused the attack as factions fought to control the treats. Later, they found that a group unfamiliar with the resource victimized the chimps who held it [424]. The parsimonious explanation would be that chimps fight for territory in general, not just because of a specific resource.

Gaining exclusive access to a rare treasure can temporarily increase the possessor's status for as long as he holds the prize. She who controls the office's coffee cups, doles out the cherry kuchen at a meeting, or brings clean needles to a shooting gallery gains temporary status. He who has a large house, fancy phone, or suit worth more than my car garners prestige—until those objects lose their social value. The makers of trading cards and other collectibles use this principle, creating shortages by underproducing specific items needed to complete a set.

Status through possession goes beyond our chimp family. I spent an agreeable afternoon at a local zoo, watching as a gust of wind blew a young girl's hat into the gorilla enclosure. A female tried to retrieve it, but the silverback decided it was his. He sat, examining the lacey object with its ribbons for a good twenty minutes as the others approached, begging for the treasure. When he finished his study, he began tearing it apart, doling out parts of the bonnet to the others.

As with the silverback above, in chimps and humans, aggressive responses to attempted usurpations of needed resources generalize to other objects of interest, not just those necessary for survival [941]. When a limited or desirable resource suddenly becomes available, peaceful relationships can give way to desire and aggression. He who has that object of desire moves up the social ladder, at least while he maintains control and the item remains of interest. Both a paucity of resources and a sudden glut of goodies trigger aggression.

Our brains adjust slowly to the idea that this new wealth is no longer a benchmark. We continue to purchase a prestigious stock, even as its value falls. We collect all stuffed animals or sports cards in a product line even though, by design, few are genuinely scarce. We rush to be the first to board a plane even though the first and last people to enter will arrive at the same time. We even give our children names upper-class parents used a generation ago.

> *"And when the Lord your God gives it into your hand, you shall put all its males to the sword, but the women and the little ones, the livestock, and everything else in the city, all its spoil, you shall take as plunder for yourselves. And you shall enjoy the spoil of your enemies, which the Lord your God has given you."* — Deuteronomy, 20:13-14, *Holy Bible*, English Standard Version

We are programmed to desire what our cultures and primitive brains associate with group rank. Status makes men more desirable mates and attracts women. Women seek high-status spouses because social positions and resources can be passed to their offspring. Where

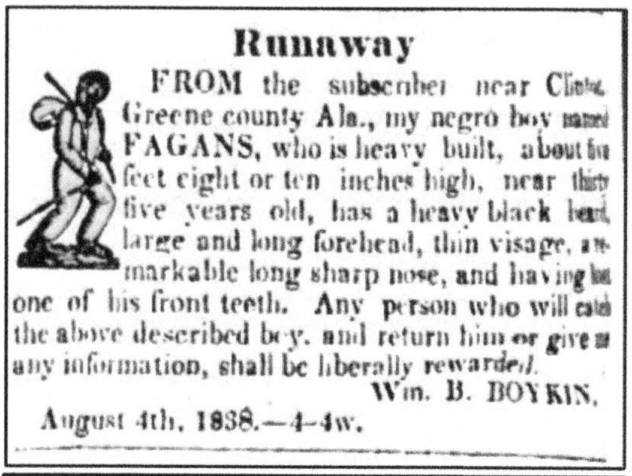

Runaway

FROM the subscriber near Clinton, Greene county Ala., my negro boy named FAGANS, who is heavy built, about five feet eight or ten inches high, near thirty five years old, has a heavy black beard, large and long forehead, thin visage, a remarkable long sharp nose, and having lost one of his front teeth. Any person who will catch the above described boy, and return him or give any information, shall be liberally rewarded.

Wm. B. BOYKIN.

August 4th, 1838.—4-4w.

Figure 5-2. Escaped Slave notice from *The Columbus Democrat,* a Columbus Mississippi Newspaper, published August 18, 1838. (This image comes from The Library of Congress Archive [3076]

prestige is not an option, we can express this competition through violence, bullying, and dominating others. On a larger scale, it becomes war and genocide.

Human history is replete with violence and degradation of our fellow humans and other organisms. Genocides and enslavements, arguably the worst forms of bullying violence, have occurred throughout history. People kill, torture, and humiliate because of ethnicity, politics, religion, or cultural differences. We seem capable of violence against anyone we define as not "us." Wars of attrition and genocide have raged between Persians and Greeks, Vietnamese and Cambodians, Hutu and Tutsi, Catholics and Protestants, and among rival bands of chimpanzees [754, 245]. The War of the Roses (1455-ca. 1487 CE), a civil war between the houses of Lancaster and York, raged for over 30 years and saw a scale of killing and brutality virtually unknown in prior British wars. It followed upon the heels of the protracted Hundred Years' War (1337-1453 CE) between the monarchies of England and France. In ancient times, as in our modern one, *gender*cide occurred regularly, with conquering peoples killing the males of the vanquished and turning their females into concubines, slaves, or wives.

In the United States, extrajudicial murders (lynchings) overwhelmingly targeted African-American males. Over 4,400 lynchings happened from 1883 to 1941; 4,000 were men, and 3,265 were Black [961]. These killings typically followed accusations ranging from petty theft and refusal to work to murder and sexual assault. Although crowds did lynch women, they more often targeted them with rape and sublethal violence. Then, as now, this gender-based violence, this inhumane objectification, instilled a sense of group vulnerability that contributed to the social power that favored males while rendering permanent physical, psychological, and social scars upon its victims [809].

Women were also used as surrogates for their male relatives. Lynchings continue to happen in over one hundred countries, with a steep increase in weak states, including Ghana, South Africa, and Haiti during the first decades of the 21st century [962, 963]. Accusations of sexual or religious misconduct often fuel the anger that spurs these murders [964].

At first glance, such violence appears maladaptive. Slaughtering and maiming each other harm both societies. The cost of wars and defenses has bankrupted nations. In the first nine months of the Russian war on Ukraine, the Russian Federation spent over 82 billion dollars, a quarter of its annual budget [965]. Adjusted for 2020 dollars, the United States spent over four trillion dollars fighting World War II, nearly 40% of its gross domestic product. America's 21st-century wars had cost the economy an estimated $1.5-1.7 trillion by 2014.

World War II's estimated governmental monetary cost across nations is about one quadrillion US dollars [966, 967]. The US financed the Second World War with domestic war bonds (63%) and a tax rate on the wealthy between 66%-94%; its current wars are funded through international loans and an upper tax rate of 35%. [968, 967]. These monetary figures do not include much of the losses endured by civilians or the inestimable loss of cultural heritage. Relative to their pre-war population, Poland and the USSR suffered the highest proportion of war casualties, with Poland, a focal point for the genocides by the Nazis, losing between 15%-20% of its population and the USSR losing nearly 15% [969, 970]. During the ongoing Russia-Ukraine war, Russia had sacrificed approximately 300,000 troops, and Ukraine had lost between 80,000 and 175,000 by August 2023 [971, 972]. However, the population of Russia is 144.2 million compared to Ukraine's 38 million, so the Ukrainians have suffered *relative* troop losses double those of the Russians, and this figure does not include the disproportional civilian losses.[146]

---

[146] The Russian Federation has capabilities that far outstrip those of Ukraine as of 2024, yet Ukrainian forces continued to fight despite losing as many troops in the first year and a half as the US lost during the two-decade-long Vietnam

War has eliminated entire peoples. Wars have deprived us of phenomenal talents, people who could have bettered humankind or worsened us. The bright mind can find itself a pack mule rather than a physicist, or a sex slave rather than a physician. Against intuition, these abusive behaviors, which destroy people and nations, can have adaptive value for genetic survival.

To understand war as an adaptation, we study our common ancestors and the behaviors we share with our war-faring cousins, the chimps. With them, we share our propensity for war and gendercide. Chimps defend territory and expand their influence by killing the males of neighboring troops. As with humans, males overwhelmingly initiate these attacks, and the victims are typically males of another community. Chimps aggress from positions of strength with a numerical advantage that can be eight-to-one over their targets [973]. This parallels human behavior, where attacks from positions of strength are the norm and where the benefits of aggression can far outweigh the costs.

> *"Imagine a room awash in gasoline. And there are two implacable enemies in that room. One of them has nine thousand matches. The other has seven thousand matches. Each of them is concerned about who's ahead, who's stronger. Well, that's the kind of situation we are actually in... What is necessary is to reduce the matches and to clean up the gasoline."*
> — Carl Sagan on the Nuclear Arms race [3239]

Violence does not always unduly damage its initiators or perpetrators; it can increase their relative social standing. The spiteful perpetrator needs only to improve his or her *relative* status or fitness through the conflict [907]. As long as the victim experiences a greater drop in fitness than the perp, the perp wins the battle. For example, Al Qaeda's attack on the World Trade Center on September 11, 2001, initially cost nineteen of their members' lives but killed nearly three thousand others. For the cost of some pilot training, it caused millions of dollars in damage and a reallocation of government resources to defend against future attacks. They banked on a limited response from the United States; they lost that wager.

---

conflict [3289]. The ability of Ukraine to resist the invasion comes, in part from another feature common among animals: A resource holder will fight harder and longer to maintain that resource than will a usurper.

> *"I do not believe you can segregate the notion of peace from the general conduct of international affairs... If we could only devote more resources to the study of arms control, we would make great breakthroughs."* — Henry Kissinger on the Nuclear Arms Race [3156]

The fear of terrorism has created other financial drains that could be earmarked to improve lives. Alongside nuclear threats, countries must contend with possible biological weapons. Biohazard research labs increased worldwide following the 2001 anthrax attacks on the United States. By the beginning of 2023, sixty-nine known biosafety level-4 labs had spread across twenty-seven countries working on a countermeasure for bioterrorism.[147] The number has roughly doubled since 2010 and nearly tripled since 2000 [974]. As with the nuclear threat, the potential for a biological war has locked brains designed for clubs and spears into a race to develop and defend against a threat that could destroy our species.

Vindictive, malicious behavior is not the only way to survive. A voluntary drop in rank to a more subordinate position can also benefit an organism. Sometimes, lowering its status can reduce its likelihood of victimization and increase its lifelong chance of procreation. Among fish, Emerald gobies have a clear dominance hierarchy. Lower-ranking emerald gobies have less access to food than higher-ranking ones. The disparity in the quality of their diet can exaggerate these differences in size. Subordinates will reduce their food intake if their large size can lead them into conflict with the dominant community members [975]. Programmed dieting increases the size differences, decreasing the chances of violence and expulsion from the group.

In cichlids—a family of freshwater fish— individuals of similar size vie for more than immediate power. They compete for resources, and most species are territorial. These fish will guard their feeding range more vigorously against similar-sized organisms than against those of markedly larger or smaller ones [976]. Land vertebrates display a comparable propensity to guard their resources more jealously against similar

---

[147] Level-4 labs contain the most dangerous pathogens, like Ebola, smallpox, and Lassa.

competitors. Even in human societies, businesses worry more about the competitor next door than the giants or the shrimps. People guard their spouses against sexual advances from more desirable mates. The United States, the Russian Federation, and the People's Republic of China worry more about each other than their less powerful neighbors; the Balkan states are more likely to fight each other over resources than to invade Greece, Turkey, Hungary, or Italy.

In the natural world, a behavior is destructive or maladaptive if it damages the chances of genetic survival. The individual who carries the genes must pass them to viable, fertile offspring who stand a prayer of reproducing. Those genes and internal genetic communities that pass this fitness test will enter the next generation. Because behaviors result from an interaction of our genetics with our environment, the genes supporting adaptive behaviors get passed to the next generation, just like one for an adaptive physical trait.

The morality of behavior, how good or how proper it may be, does not determine one's ability to deposit into this gene bank. We contribute genes that allow us to survive, encourage reproduction, and help us leave behind more babies than our neighbors. If an organism does not pass on its genes, for whatever reason, no matter how noble, the genes that promote genetically maladaptive but selfless behavior become lost over time. Similarly, genes that support manipulation and bullying become lost if no one wants to mate with the stinker. If teamwork enhances reproduction and survival, as often happens, nature favors cooperation and the genes that support it [977, 978, 979, 980]. If selfish behavior aids the genes, nature favors that.

The key to understanding this system comes from shifting our paradigm from the long-term survival of populations to the shorter-term, generation-to-generation survival of DNA. Even if a behavior causes long-term destruction in the future—like our propensity to engage in war mixed with our ability to build nuclear weapons—it will persist if it improves the short-term passing of its supporting genes through individuals in the present.

Genes that enhance behaviors destructive to a group can thrive if they are adaptive for the individuals who carry them. The genetic variants that make people prone to overindulgence in alcohol are maladaptive for society. Excessive drinking damages the long-term health of the individual. However, if these genes reduce barriers to reproduction during the peak childbearing years, they give their bearer an adaptive advantage. If collaboration benefits genetic procreation, genes promoting cooperation will thrive. If aggression encourages the passing of DNA to the next generation, genes promoting aggression will be favored.

As in the aggressive fox, genes, epigenetic changes, and regulators can predispose their carriers to higher levels of aggression and violence. A gene remains in the population because it promotes action patterns that benefit itself. Behaviors that profit genes do not necessarily need to help or harm the individual or society [981, 982]. They just need to provide a better survival advantage than other variants.

Even if a bullying behavior happens to be destructive to a species or troop, it can still be adaptive to individuals and their genes. Human populations, like populations of other organisms, grow until resources become limited. A village that slaughters its neighbors can access their land and resources. If these additional assets meet the offender's needs, the violent act will give them an edge for survival. If an aggressing community reduces the reproductive potential of a victimized group, it could leave a disproportional number of offspring in their wake. If the attack gives them a fierce reputation, they decrease the chance that another clan or village will attack them.

Like the chimpanzees that maim and kill related splinter troops, which vie for resources [424], humans attack their competitors for resources and power. Significant threats to these desires often include close but segregated kinship and linguistic groups (e.g., the Russian-Ukrainian war, the Balkan clashes, and the conflicts between Kinyarwanda-speaking Hutus and Tutsis).

Behavior that increases one's reproductive success or the reproductive success of one's children conveys an adaptive advantage. Males have the potential to make more offspring than females. The genetic advantages of sons generally exceed the benefits of raising daughters. Worldwide, the ratio of men and women is about equal. More boys are born, but men, especially young men, run a higher risk of dying [983, 984].

In social species, one does not need to kill opponents to out-compete them. One can reduce the reproductive potential of rivals by bullying them. Social stressors increase cortisol levels [985]. In social animals, high cortisol reduces fertility, and cortisol can cross the placenta and impact the fetus. For females, high cortisol levels can trigger miscarriages [986]. These reductions in offspring benefit the bully by reducing genetic competition.

In adults, stress can raise cortisol levels that trigger epigenetic changes leading to PTSD (post-traumatic stress disorder). It does the same thing across the life cycle [987]. In fetuses, stress triggers these changes at an earlier and more sensitive maturation phase. Normal corticosteroid levels contribute to neurological, muscular, and cardiac development. Low maternal cortisol can lead to poor lung development. Exposure to a raised cortisol (and other glucocorticoid) level during gestation alters the hypothalamic-pituitary-adrenal feedback system and can lead to offspring born with PTSD. These infants can suffer changes to gene expression that increase their risk of developing depression (especially in women) or anxiety disorders [988, 987]. Maternal stress also shortens the fetus' telomeres, a segment at the tips of the chromosomes that influences aging and lifespan [989].

In late pregnancy, high cortisol increases the probability of a stillbirth, a reduced fetal heart rate at birth, and alterations in the mitochondria that power the heart and skeletal muscles [990].[148] The PTSD of female genocide survivors also correlates with changed gene expression associated with faster aging and increased cancer risk; these epigenetic changes occur in their blood and the blood of their children [407], providing a potential marker for identifying trauma survivors.

The developing embryo responds to corticosteroid overload by downregulating its receptors, impairing neural development, and creating an easily triggered stress response. The fetus probably doesn't care if the traumatic maternal stressor is forced confinement, bullying, or a disaster; it responds to the cortisol [991, 992, 993]. Children of traumatized mothers start life with a neural system predisposition to PTSD and avoidant traits [994], and studies of puppies imply that young organisms undergo a similar molecular regulation [227].

Fetuses and offspring do not accept this signal without a fight. The body has safeguards and backup systems to offset some trauma. Regulatory genes, such as Sonic Hedgehog (and no, I am not making up that name), can counter some stress-induced damage [995]. Other systems of neurotransmitters, as well as enriched emotional and social environments, can provide buffers and allow the victim of stress to function within a normal range [996, 997].

From a biological perspective, it does not matter how sweet the bullied child may be or how much devil spawn is in the bully. It does not matter if he should have had his head nipped off before he was an hour

---

[148] Changes in transcripted proteins include cytochrome C, important in converting sugars to energy, and PDK4, associated with type 2 diabetes mellitus. In adult PTSD, abnormal neurons and vesicles that contain neurotransmitters [3140] and a paring back of neurons (apoptosis) in the hippocampus and amygdala accompanies the syndrome [3151]. Although I am unaware of a paper that shows this effect on fetuses, it is not a stretch to hypothesize that a similar response would impact embryonic brain development. In Rwandan genocide survivors, for example, the effects of maternal PTSD are seen in the gene expression patterns of their children, findings supported by animal studies [1446].

old.[149] Three factors ultimately matter: how many genes and epigenetic regulators fostering behaviors transfer to the next generation; how cultural norms become grafted onto that child from his or her social contacts;[150] and how many of one's genes get passed on *relative* to the genes of others.

By elevating themselves above the pack, even in childhood, the bullies, who dominate through subjugation, increase their later chances of reproduction. Aggression provides a path to high rank without the need to develop those pesky, hard-to-master prestigious skills. Social position includes access to high-ranking and fertile mates. Their childhood experiences of power often translate into adult status that, in turn, benefits their children. Most female primates prefer high-ranking males, but the path to dominance matters.

Primate offspring benefit from the mother's rank if she holds a dominant position. High-ranking parents can access resources and advantages not shared by the community. Their children mature with an expectation of status and the sense of entitlement that social animals naturally crave. Sons have a higher probability of close contact with high-ranking male models. Among our non-human cousins, sons and daughters learn their potential status from mom, and female behavior continues to play a role among humans.

Our tendency toward bullying gets checked because other organisms in a social group also crave power. Those not too downtrodden can sometimes rise, refuse their subordinate station, and overthrow a bully.[151] The principal check comes from other members of their pack, flock, or society. Once a mob completes a

---

[149] I credit this quip to Mr. Mark Dorroh, an old friend, musician, and night-club performer.

[150] Some cultures and cultural settings foster more aggression and competition than others. For example, North American hockey players display more aggression than players from other cultures [3378]. Competition aggression need not be through games or warfare; interactions that can be interpreted as competitive, such as driving, can also trigger an aggressive response [3371]. Although shame remains a constant across cultures, the perception of harm and the damage inflicted by bullying and shunning also differs among traditions, as does the induced shame [85, 3086]. For example, exclusion or shunning debilitates across cultures but plays a more significant role in collectivist societies, e.g., Korean *wang-ta* or social exclusion bullying [3503].

[151] The Arab Spring (الثورات العربية al-Thawrāt al-'Arabiyyah), which manifested in 2010, exemplifies this principle. In this uprising, as in many across time, there remained a danger of substituting one bullying dictator for another. Because revolutions typically include power struggles afterward, Machiavellian elements have a distinct advantage. Forcing a leadership change carries risk, be it for cause, frustration, or the sake of change.

coup, it has no guarantee that a new leader will behave better. Indeed, the traits driving a social animal toward a leadership position in an unstable society will often produce a similar result.

As tempting as it may be to stop the story of degradation there and blame our predisposition to violence, social species require teamwork to thrive. Alliances can kill or ostracize non-cooperative members who turn against or fail to support their collective, and societies will impose ostracism or banishment, effective death sentences. Social compassion and recognizing another's pain are critical skills in regulating social norms.

Mirror Mirror

> *"Sometimes you're the windshield, sometimes you're the bug."*
>
> — The Bug, by Mark Knopfler [998]

Specific brain cells fire when we perform an action or watch someone else do it—as if we saw ourselves doing it in a mirror. These internal simulations form the scaffold upon which experience, age, and learning build our sense of compassion [999, 1000]. Their presence in rodents [1001, 1002], primates, and songbirds [1003], and suspected in many other species [1004], suggests they evolved before our last common ancestor diverged.

In social species, each individual remains aware of some of what's happening around them, but not all of it. From schooling fish to humans, they depend on allies to provide collective information about the environment to enhance their foraging and avoid predation [1005]. Individuals who lack trust and default to a complex, time-consuming decision process whenever a shoal mate screams "shark" or "lion" lose this primary benefit of group life. They become easier prey for the predator. They may reap the short-term benefits of a specific feeding patch but lose the long-term benefits of sharing. The result is selection pressure favoring conformance to the wisdom of the crowd. Individuals trade their confidence in their own perception, weighing it against the pool of information provided by the majority. To this end, they engage their theory of mind, directing it at the group rather than an individual.

Specific parts of the brain take part in our internal deliberation. Be it the determination of beauty, economic decision or jury deliberation, group preferences activate the nucleus accumbens (a pleasure and social center) and the orbitofrontal complex (rational decision-making) fire, suggesting our ideas combine with social forces to assign values to concepts and decisions [1006, 1007, 1005, 1008]. These two regions assign much of our subjective value to something, with activity increasing with the value of the reward.

A third region, the anterior cingulate cortex (ACC), joins these to estimate the volatility of the situation and the likelihood of reward or punishment [1009]. Parts of the frontal lobe and parietal lobe[152] (top of the brain) also come into play to analyze what might have been—the "counterfactual" or "alternative realities" [1010]. Like a good book, these internal stories let us rehearse different responses, rendering them more readily available for the next time we encounter a similar challenge. Together, these regions transform the integration of social information from one of passive acceptance of information to an unconscious deliberation, with more weight given to trusted sources and less to those of enemies or to those who cry wolf [1005].

Networks of "mirror" neurons [1011, 1012] may be involved in tasks as diverse as learning language (copying the vocal movements), observational learning, imitation, and understanding the intent and feelings of others (theory of mind). They help us differentiate between intentional and unintentional harm, modulating our reactions [1013]. These neurons fire when we empathize, take the other person's perspective, or cooperate as a group, trusting the rewards outweigh the risks [1014]. They serve as a

---

[152] The lateral frontal polar cortex (lFPC), dorsomedial frontal cortex (DMFC), and posteromedial cortex (PMC) reinforce learning by helping you imagine alternative versions of scenarios [1010]. Understanding how things might have happened if you behaved differently may not, by itself, alter your behavior, but it is a step in that direction. Without picturing alternatives, change becomes less probable.

foundation for recognizing and learning that what hurts us hurts others and allows us to respond to their pain.[153] Even if we know we won't injure someone, we exhibit physiological aversion responses, such as vasoconstriction and the stimulation of the fight-or-flight (sympathetic) system, when asked to inflict simulated harm on someone we recognize as human[154] [1015, 1016]. And, counterintuitively, we often respond with more vigor against hate-motivated speech than actions when the outcomes are the same [1017]. This faster denunciation of ugly, spiteful words may have roots in the socially infectious nature of hate speech and the lower physical risk posed by the vocalizations.

Mirroring can spur us to action or warn us that our actions invite retaliation. Our internal simulations warn onlookers that the victimizer could turn on them. They can frighten resisters into submission or alert them to hide their activities. People with antisocial disorder and who lack empathy for others can also suffer from a deficit in mirror-neuron activity [1018, 1019].

Sociopaths with a faulty mirroring system learn that shooting or raping someone or shoving their head in a toilet is unacceptable without understanding why it is wrong. With fewer brain cells dedicated to empathy, sanctions against harming others make as much sense as a rule demanding everyone wear an orange tie. You might understand that everyone looks horrified at someone wearing a purple tie, but

> *"The average [identified] psychopath will be convicted of four violent crimes by the age of forty. And yet hardly anyone is funding research into the science. Schizophrenia, which causes much less crime, has a hundred times more research money devoted to it... Because schizophrenics are seen as victims, and psychopaths are seen as predators. The former we feel empathy for, the latter we lock up."* — Dr, Kent Kiehl, cited in "Suffering Souls," *The New Yorker* [3235]

---

[153] Associative learning and mirror neurons are not competing theories about how we learn to care about each other and respond to each other's feelings. Instead, they approach the problem at two levels of analysis: learning (developmental) and the mirror neurons upon which the acquisition is based (mechanistic). One could also analyze the benefits of mirror neurons and social learning to the organism, their evolutionary history, mechanisms of action, genetics, and epigenetics. All provide information needed to understand our emotional responses and actions toward others [3095].

[154] One must approach these studies carefully. They typically include members of the same recognized community or WEIRD populations. Traditional societies, which often do not recognize outsiders as humans, are unlikely to show the same responses or inhibitions to outsiders and ingroup members.

it makes no sense. If you slipped and wore a purple tie one day, the vehement reaction of others could take you by surprise.

Solitary bullies risk retaliation from coalitions of others, but when confronted, they are often taken aback by the anger of their opposition. With less integrated social cues, the animosity of others befuddles them. They see those who stand against them as bullies and themselves as innocent victims.

> *"The great concourse of Christians who witnessed the tragedy scraped for hours among [Hose's] ashes in the hope of finding a sufficient number of his bones to take to friends as souvenirs."* — On the lynching of Samuel Hose in Atlanta Georgia [1020]

We cannot know if Nazi leaders, whose actions killed millions and destroyed the lives of millions more, had deficits in their mirror neurons. Neither the concept nor the technology to test their function was available then. However, in 1945, all twenty-one defendants at the International Military Tribunal in Nuremberg pled not guilty to war crimes. Most were shocked when witnesses stepped forward to testify against them and at people's reactions to their atrocities. Many seemed to consider their behavior unremarkable [344], as did the many who posed for photos with pride or who brought their children to enjoy the lynchings in the United States [1020, 1021]. It seems unlikely that the trainloads of people involved in the lynchings or the thousands of participants in the Holocaust all suffered from defective mirror neurons. It is far more probable that social signals, including childhood indoctrination, skewed their perceptions, leading them to misperceive their victims as "non-human" or as "the enemy."

Mirror neurons are not the only deficit seen in sociopathy [1022, 1023]. Indeed, we seldom see a one-to-one mapping of a particular behavior onto a specific brain region; backup systems and alternative programming are valuable to our species. Our prefrontal cortex and temporal lobes[155] process images by

[155] The specific regions include the anterior dorsal prefrontal cortex (working memory, task planning, self-evaluation, inhibition, and choosing our responses to situations). A second area is the language-processing superior temporal sulcus, which correlates with our use of language to identify outgroups.

their social rank [1024]. People who lack empathy may also have modifications to their oxytocin system [1025]. Oxytocin promotes within-group social behavior and attachment but promotes defensiveness against outsiders, and a mammal's early history influences the activity of oxytocin receptors [227].[156] Deficits in the emotional core of our brain, the limbic and paralimbic systems, can also lead to callus behavior and interpersonal violence [1026, 1027, 1028, 1029, 1030, 1031]. Even changes in the amount of grey or white matter (neurons or connective tissue) in crucial brain regions can reduce emotional regulation and increase aggression [1032, 1033]. And when a subject in an MRI is shown the face of a hated person, specific regions of the brain that register strong emotions light up[157] in the mirroring, limbic, and paralimbic systems. At the same time, the right superior frontal gyrus, associated with impulse control and inhibition, drops in activity [1034].

> *"He is a perfect yogi who, by comparison to his own self, sees the true equality of all beings, in both their happiness and their distress."*
> — *Bhagavad Gita* (6.32)

The environment sculpts the oxytocin, limbic, and mirror neuron systems, producing the final product [1035]. A loving, supportive environment can mold someone with deficits in these overlapping systems into an emotionally functioning member of their culture; abusive environments and negative modeling push the other way [1036, 1037].

> *In a 2016 online survey, the Southern Poverty Law Center found over 2,500 "specific incidents of bigotry and harassment that can be directly traced to election rhetoric,"* — 2016 report from the Southern Poverty Law Center [3488]

Bombardments of hate speech or prejudicial images against a group or an individual can create neural biases against the target by dampening the feedback in these systems [1038, 1039]. For the victims, bullying with verbal attacks creates

---

[156] The neural regions with high oxytocin activity overlap but are not identical to those high in mirror neurons, suggesting the two systems interact to produce a functional pro-social individual.

[157] The medial frontal gyrus, right putamen, and premotor cortex show bilateral activation. In the right brain, the insula, premotor cortex, and frontal-medial gyrus grow more active with the degree of hate. Although sometimes called the "hate circuit," they also respond to images of people you love [1991] and with feelings of disgust [3517, 3092]. These suggest the circuit deals more with strong emotions than hatred per se. This alternative view is backed by its uncoupling during major depression [3260].

long-term psychological trauma on par with personal violations such as domestic abuse, physical assaults, and robberies [1040, 1041]. The social consequences of pervasive negative images can alter one's ability to gain employment or a fair trial [1042].

With the rise of online hate speech during these first decades of the 21st century, often on targeted social media platforms, the neurological repercussions of the ugly verbal assaults and their social implications, including terrorism and political radicalization, should not be dismissed [1043, 1044]. Hate, like spite, can addict and keep people returning for regular doses of this drug [1045, 1046]. The anonymity of the internet and the presentation of a group face should evoke less mirror neuron activation than accusations against someone standing before us. Given that our brains are pre-wired for caution around "others," the internet spread of hate, be it against a group or an individual, can spread like cancer.

We show less mirror neuron activity and empathic responses when watching people we have learned to view as outsiders. On MRIs, the pain matrix lights up less when we see out-group members being hurt, and our stress, as measured by changes in the electrical conductance of our skin, does not change as much [1047]. Fear and conflict monitoring by the limbic and paralimbic systems reflect our conditioned biases about people [1048] and make us more prone to strike out.

The Sins of the Father

> *"The sins of the fathers are visited upon the children."*
>
> — Biblically derived English idiom

Sometimes, being a bully seems like a pretty good deal. He or she gets to take what they want and treat a fellow organism like dirt. The bully often experiences an elevation in status, prestige, and physiological buffers against stress, even if their power happens through fear. If this portrayal were the complete story, social evolution should favor ever more violent behaviors. Individuals who practice bullying—the physical, verbal, or emotional degradation of others—should have more offspring than non-bullies and come to dominate the world. If they are to have their genes survive across the generations, low-ranking organisms should also gain more alleles and epigenetic markers that favor aggression. Only their social experience of low status should hold these latent bullies in check. In the wake of such pressure for ever-increasing dominance, the Machiavellian imperative should tear apart the threads of trust, kinship, and friendship holding society together. Cooperation should become increasingly tenuous, and alliances should collapse.

Instead, members of our species forge strong bonds, defying this prediction. Power struggles can turn violent, but most societies value cooperation. Discovering that their child has become a bully mortifies most parents. People sacrifice themselves for strangers. They donate to charities and wince when they see another suffer. Treating someone with grace and dignity lifts us higher than spiteful revenge. A force other than the quest for pure individual power must exist. The benefits of respect and cooperation must counterbalance the benefits of personal power.

Adult females nurture these brain regions along with their offspring. They provide safety and support for their young as they explore their world and interact with their age-mates [1049]. However, they are limited in what they can do to the offspring of a higher-ranking female. Primates will band together to seek protection, defend cultural norms, ostracize free-riders, and kill bullies [1050, 1051, 1052, 1053, 1054]. They will create social alliances with dominant animals, grooming and backing them in conflicts [1055]. These patrons defend their clients and help them acquire rank and core ranges [1056, 1057]. As with humans, subordinates hesitate to endanger their position by challenging more dominant ones without a supportive coalition backing them.

Allies can work together to accomplish tasks, as a defense against common enemies, to gain resources or as a defensive strategy [1058]. Male chimps excel at joining forces to enhance their status or to take food from females. With Machiavellian cunning, chimp alliances tend toward opportunism, with members willing to shift loyalties [754, 1059, 1060, 1061]. Like humans, chimps use microaggressions to instigate fights, with stronger individuals doing most of the needling [1062].

The weaker coalitions among the female chimps, which serve more for protection than for attaining high status, are more stable than male ones [1061]. Bonobo females will form and maintain alliances to protect resources and punish aggressive males, and female gorillas will form coalitions to protect their young from infanticide [1063]. These data, from both our sister species and data from gathering-hunting societies, counter the belief that women do not bond with each other like men [760, 1064].[158]

Considering the chimpanzee data and the female coalitions among bonobos, one must consider how reduced opportunities for frequent daily contact bears on our social relations and the levels of life satisfaction. Cities show higher levels of mental health issues; however, once the monetary deprivation of many city residents is accounted for, the denser social networks of cities seem to buffer against depression [1065].

In Western societies, women express their suffering from social isolation more than men. The loss of kinship ties in our mobile civilization *correlates* with the higher levels of depression in women relative to men in Western, non-collectivist cultures [1066], with the peak in gender's impact during the highly social teenage years [1067]. In contrast, Western men express more loneliness after the loss of a life-partner or their status role as the family breadwinner [1068, 1069]. From solitary confinement to a time-out in a classroom, social isolation stings with abandonment, as does bullying by exclusion.

[158] The repeated violence against intimate partners in rural areas, where women are more isolated, tends to be higher than in urban areas. This difference is often attributed to psychosocial causes, such as unemployment, poverty, learned attitudes, and reduced governmental attention and shelters [3103].

When crowding becomes problematic, we use many bonobo-chimpanzee distancing and grooming mechanisms to minimize stress. Like us, our cousins reduce physical contact with their neighbors (social grooming), bluff aggression (which could spark violence), and social activity. [1070, 1071, 1072]. The number of social connections does not increase as fast as the population. Using cell phone contacts as an estimate, the rate of contacts only grows by 12% when a city's population doubles [1073, 1074].

Chimps and bonobos band together to punish antisocial behaviors that, if left unchecked, would benefit the stinker. These "groupish" behaviors happen in both sexes. Female bonobos join forces to kill aggressive males; female gorillas can make it difficult for a bullying silverback to retain control, setting the stage for a coup d'etat by a contender. These coordinated assaults allow attackers to take out the offending party with less risk than a one-on-one attack [1075]. Humans also engage in targeted conspiratorial attacks where community members coalesce to punish an offender.

Groupish coalitions help explain class cohesion and alliances, interclass tensions over resource hoarding, and our spiteful delight at the social disgrace of high-ranking politicians and celebrities who abuse their status. Still, dominance hierarchies facilitate community function. Social animals have behaviors that indicate their position within a group. Dogs, lions, primates, chickens, and even social fish display social structures that support dominance hierarchies. True social insects, such as bees and wasps, have differences in rank and privilege within their societies [1076].

Figure 5-3. US anti-Japanese propaganda poster, WWII (ca 1941-45). It is a common tactic to dehumanize an enemy and portray them as vicious. Here, the Japanese soldier with feline features is depicted as abducting a naked white American woman. In the background, another soldier bayonets a kneeling man as others are hanged against a fiery backdrop.— [3290] This work is in the public domain in the United States because it is a work prepared by an officer or employee of the United States Government as part of that person's official duties under the terms of Title 17, Chapter 1, Section 105 of the US Code.

Within these social hierarchies, animals worry more about affronts delivered by troopmates of a similar status than with annoyances from lower social tiers. It takes little effort to put an inferior individual in their place. Peers pose an immediate threat to self and resources. Confrontations among similarly ranked group members chance reduced status, lost rank, and increased stress. Lost conflicts with peers often coincide with a corresponding loss of resources. Lower-ranking neighbors become more willing to challenge the loser. In contrast, winning a peer skirmish yields significant rewards, including a rise in rank, more access to provisions, and hormonal buffering against stress.

The effects of winning and losing reach beyond the individual contestants. Clashes affect the prestige of allies. The status of sports fans rises and falls with their team's fortunes. In some species, such as lions, a new male in control of a pride will kill the kittens sired by the previous male. Human power struggles often include threats to the families and friends of opponents. Politicians employ hate speech to denigrate and dehumanize their opposition, sanctioning violence and hate crimes [1077, 1078].

We destroy lower-status individuals for practice, profit, intimidation, or amusement. We attack our peers for our survival. Alliances may maintain power, but leaders beholden to specific coalitions must keep their supporters happy, even at the expense of other troop members [1079].

We find some of our most intense turf battles among organisms of similar status. Like trying to "fight city hall," the costs of making an open assault on the big guy are often too high to justify the slight chance for a pay-off. Animals, including humans, select their battles judiciously. When an underling attacks a higher-power group member, he or she usually lashes out in frustration, desperation, or misjudgment of the predicted results. For example, Louis Riel led a populist uprising against the Canadian government. He believed other First Nations would join the revolt initiated by his Métis people. The other nations did not enter the fray. They opted to watch and wait, letting the Métis take the brunt of the battle. The Cree leaders Big Bear and Poundmaker led similar but separate insurgencies; none coalesced into a successful cross-cultural rebellion.

People claim to love heroes who stand firm against a bully but seldom express support unless they believe he will win the battle. We stand back, weighing the possibility that the savior will win against the probability that he will become the next bully if he stands victorious. Once a dictator takes power, he needs different allies than when he fought the revolution. Early military supporters become his greatest threats, and often suffer from acute "lead poisoning" [1079].

Many assassins tweaked their actions to balance the threats of retaliation against their potential benefits. The cost of violence against an influential leader can include one's death and the extermination or dishonor of one's family and allies. Retaliation against kin is common. For example, following the failure of Operation Valkyrie, a plot to kill Hitler, more than three hundred of the conspirators' relatives were arrested under the Sippenhaft (blood guilt) laws. Clause von Stauffenberg, who planted the bomb, was executed the following day. His wife delivered their fifth child in the Ravensbrück concentration camp. The offspring of key conspirators, including Stauffenberg's four older children, were marked for transport to Buchenwald.

Their aunt, Alexandrine von Üxküll-Gyllenband, who was newly released from prison herself, found the transport truck and rescued them [1080].[159]

Benefits of rebellions and assassinations, such as Operation Valkyrie, include disseminating a message, inculcating fear, or increasing wealth, status, or power among peers or associates. These also apply to feigned and failed assassination[160] attempts. Hitler's survival of the 1944 July assassination attempt provided propaganda minister Joseph Goebbels with fodder for rallying support and portraying their leader as protected by God. Both Prime Minister Indira Gandhi (1984) and Princess Diana of Wales (1997) saw a surge in popularity after their deaths. Theodore Roosevelt's 1912, Ronald Reagan's 1981, and US Representative Gabrielle Giffords' 2011 survival of attacks rocketed their popularity. Critics have accused Belarus's Alexander Lukashenko of staging a botched execution (1994) to bolster his approval ratings and to consolidate power. Reported actions against the lives of Venezuelan Presidents Hugo Chávez (2002) and Nicolás Maduro (2018) also met with skepticism and claims that they or their political assistants may have staged the attacks to boost their popularity.

Assassination attempts may provide a short-term benefit but don't always yield long-term sympathy boosts. In 2018, mail bombs sent to former President Barack Obama, Hilary Clinton, and actor Robert De Niro caused a short-term drop in the approval levels of then-President Donald Trump, but left no long-term impact. Nor did the various assassination attempts aimed at President Trump have a lasting effect; his popularity may have risen briefly, but it had no long-term impact on the American polls leading into the 2024 presidential election. An attack on a leader triggers a primal group response, leading the troop to rally around the threatened alpha. As the threat disseminates, however, our loyalty dials downward, and we

---

[159] The rescue occurred during the fall of the Reich, and the invading Allied forces had detained the transport.

[160] There remains debate about what constitutes an "assassination." Like "bullying," the term segregates one class of violent behaviors from other forms. I use the word in the strict sense of politically motivated assassination of people in positions of power or the killing of people who threaten the consolidation or maintenance of power.

derive our ultimate response from the subsequent events and behavior. Thus, when partisans assassinated Benito Mussolini while he fled his war-torn country, he no longer functioned as the troop leader. Now perceived as a traitor and oppressor, Mussolini's death released a tidal wave of hatred and relief as power shifted toward the resistance.

A perceived common threat can unite a group against a perceived common enemy. The World Trade Center bombing served as a catalyst that brought together diverse elements of American society. Because the United States had superior military resources, the payoff for retaliation was high, and the danger to the collective was low. By attacking the big guy, the bombers created a significant threat. The United States considered itself a cohesive entity—a social surrogate for a kinship unit. The bombing heightened this unity by creating an external enemy that took precedence over internal squabbles. Before 9/11, most Americans had little animosity toward Islam; the attack changed that. Islamophobic attacks increased despite the Muslim-American community serving as the nation's primary source of information about the few plots that have surfaced since that day [1081]. These attitudes were promoted through hate speech and deliberate misinformation [1082, 1083, 1084, 1085, 1086, 1087, 1088].

According to the US Department of Homeland Security, terrorist attacks[161] on the United States fit no one profile. It includes a range of ideologies. The vast majority, 88%, of such attacks, as classified by the American government, were homegrown attacks [1089]. An analysis of 25 years of domestic terrorism by the Center for Strategic and International Studies found that most domestic terrorist incidents have come from right-wing zealots since 1994. In 2019, these accounted for 75% of the attacks; that figure rose to 90%

[161] The US DHS defines terrorism as the threatened or actual use of illegal force and violence by a non-state actor to attain a political, economic, religious, or social goal through fear, coercion, or intimidation [1089].

in 2020 [1090]. The perpetrators, however, see themselves as avengers, devoted warriors, or patriots defending ideals as they fend off "others" who would threaten their status and way of life.[162]

For high-status individuals, significantly lower-ranking subordinates are like mosquitoes. One can swat them, but there is little payoff in increased status. If their neural and hormonal systems amplify the pleasure of such minor victories, they will experience a physiological reward for striking a peer or swatting the annoying fly. If the attack has symbolic significance, the payoff can increase.

The bombing of the World Trade Center had such a high symbolic value for both the attackers and the attacked. Each side magnified the importance of the strike but through different lenses. People who saw the United States as a symbol of oppression may have seen Al Qaeda in a favorable light. However, those positive feelings did not translate into political power during the Arab Spring of 2010. Its promised destruction of the West magnified the appeal of Al Qaeda through a brief but impassioned lens. When an undertaking falls short of a pledge, a group loses credibility [1091, 1092]. Al Qaeda followed this pattern. The extreme violence of other revolutionary groups, such as the Islamic State, further alienated many in their target audience—those oppressed and disenfranchised by leadership in Syria and Iraq.

In contrast, the United States received sympathy and support following the bombings. Other Western countries recognized their own vulnerability in the attack on the US. This magnification resulted from our mirror neurons' ability to identify with the plight of others and to experience their pain or joy. Many countries joined with the stronger United States to attack Al Qaeda and the Taliban in Afghanistan. They placed themselves in danger because of treaties, loyalty traditions, and their stake in the outcome.

---

[162] Mass perpetrators do not fit a single profile. Smeuler [21] proposed a taxonomy of fourteen categories, including motivations from the self-important and manipulative masterminds to fanatics and true believers who work for their cause, avengers, status and profit seekers, the compromised, and the truly deranged and sadistic.

The retaliatory torture of prisoners, especially in Abu Ghraib prison, reduced the credibility of the United States, costing the support of allies. Both the United States and Al Qaeda represented themselves to their target audiences as the heroes in a battle between good and evil. No matter what finally happens in the Middle East, the cost of the attack has been tragic for people in the Western and Arab worlds.

Less powerful organisms or coalitions may challenge a higher-ranking bully despite the threat. At other times, subordinates make it impossible for the bully to maintain status. For example, in male-dominant primates, females may band together to keep a powerful male from displacing the current leader of a harem [754]. Among highland gorillas, a single dominant silver-back male can lay claim to a harem of females. Lower-ranking males live at the edges of the troop. If the females do not like the dominant male, they can alter their behaviors to make the community ungovernable by the silverback. As their protest grows, he must use more energy to maintain control. This drain on his reserves weakens him, paving the way for a different male to mount a challenge and take charge.

The new gorillas-in-charge may kill the offspring of the previous alpha, thus reducing his genetic contribution to the next generation. The hazard of infanticide decreases the pay-off for the ladies, making the coup a desperate attempt and preventing the casual use of regime change as a solution to troop problems.

By the way, do not presume the boss passively submits to the plotting harem. Silverbacks spend energy disrupting female coalitions and dominance hierarchies. By rendering them ineffective, males reduce resource competition among the cliques, yielding more widespread fertility and limiting their ability to form alliances against them [1093]. Both effects benefit the reproductive goals of the dominant male.

Occupied peoples have used similar combinations of low and high-key resistance against leaders they do not like. Small acts of disobedience combine well with violent attacks. During World War II, the Dutch underground coordinated attacks on the occupying Nazi government and produced an underground market that disrupted the economy. Combined with spiteful attacks, such as the destruction of the dykes, they made the Netherlands more trouble than it was worth. When combined, even small acts of sabotage pack a wallop. (My uncle, for example, served with a team that unloaded food bound from Amsterdam to Germany from the back of the train car as others loaded it.)

The Irish used similar tactics against the British. Enslaved African-Americans and Afro-Caribbeans combined low and high-key disobedience with sporadic armed rebellions. During the Indian Revolution, the passive resistance to British rule championed by Mohandas Karamchand (Mahatma) Gandhi contrasted with the armed resistance of other leaders, such as Barin Ghosh and Bagha Jatin of Kolkata. The combination proved effective.

Collaborative groups can repel rivals and defend joint territory more easily than individuals [1094, 1095]. Cooperation can bestow direct or indirect reproductive benefits [1096, 1097, 1098], delayed senescence [1099], improved feeding and resource access [1100, 1101], and social benefits, such as grooming and inheriting territory from non-kin [1102, 927].

Cooperative behavior, however, requires mutual tolerance [1103]. One road to such acceptance lies in our desire to retain control over our resources. Dominant males of many species employ other males (satellites) to help them keep their status and rank. Among gorillas, an aging silverback may allow a younger, submissive male access to females in exchange for his support. Males prefer weaker allies to reduce competition from those who could threaten their position or territory [1104]. Examples of species that use satellites include the wild relative of guinea pigs or cavies [1105], Mexican fruit bats [1106], wrasse fish [1107], and pied wagtails, a kind of bird [1108, 1109].

> *"And do good; indeed, Allāh loves the doers of good."* — *Quran*, 2:195

These assistants defend areas as "rent" in exchange for *temporary* permission to stay in the territory; if they are male, their rent may include limited access to local females [1110]. Because the territory holder chooses weaker, non-threatening partners, these alliances end at the landlord's discretion. Owners prefer vulnerable neighbors to stronger ones, and may weaken ones who become a threat. Satellites are allies; their interest in weakened neighbors coincides with their host's. In 1962, the US bordered on nuclear war when the Soviet Union moved missiles into nearby Cuba. In 2019-21, Ukraine pushed for full-fledged membership in NATO; in 2022, Russia invaded. Like the pied wagtails, neither superpower wanted a potent threat along its borders.

In some circumstances, ruling with an iron fist saves energy, but he who keeps power by force invites a revolt. Animals whose subservients grant alpha status expend less energy to retain their rule than the bully must spend to maintain their throne. A respected leader does not need to waste energy on threats of force unless his or her image becomes tarnished by the deceit of others. The despised ruler does. The dynamics and coupling of gentle power and coercion can be as complex in other animals as among humans [1111, 1112].

Based on modern hunters and gatherers and our ape cousins, our distant ancestors probably lived in small bands that broke into smaller parties for hunting and gathering. Like the bonobos and the chimps, these communities would have contained related resident males, migrant females, and their offspring. Political coalitions would have formed within our troops, as in chimps and bonobos [1113]. This pattern would have influenced our communities as we became human thousands of years ago.

In the late 19th and early 20th century, popular opinion held that these societies were cooperative and egalitarian, with no explicit authority. This view fed our desire for peaceful and supportive ancestors. The image of the sweet progenitors conflicted with the alternative hypothesis: We evolved from aggressive or "killer" apes. Each model simplified the complex behavior of primates into sound bites that fed preconceived notions of human nature.

Contrary to these early popularizations, studies of animal and human societies tell a different story. Contemporary hunters and gatherers have dominance hierarchies, though less extreme than those in agricultural communities [1114, 1115]. Their interdependence supports individual altruism and prevents individuals from gaining excess power [1116, 1117, 1118, 1119, 1120, 1113]. The members of these mobile human troops depend upon each other for mutual survival, and status within a group may rely more on reputation than aggression [1121]. That status could build on a skill for hunting, spirituality, or defense. Societies required cooperation, and prestige in one area would not guarantee recognition in another. The need for teamwork fights against the opposing force, a desire for dominance. In mathematical models, when two opposing forces put conflicting pressures on a social system, the resulting change from one stable state to another can happen quickly.[163]

Based upon common behaviors in all the great apes (including us), the first "moral" sanctions in Pleistocene humans may have been against bullying [1122, 1123, 1113, 754].[164] Even without verbalized social mandates, other apes act as if they understand the consequences of harming their troop mates or placing them at risk.

Far from being a thin, Hobbesian coating of learned behavior enforced only by social and religious codes, our ancient human admonitions against harming others build upon this biological sense of social hierarchy

---

[163] The opposing forces develop a mathematical system known as a "catastrophe." In catastrophes, systems remain steady across a large domain and suddenly shift. Once the shift is made, a new stable state can develop, or there may be a wobbly period as relationships settle into their new roles.

[164] Some form of the "golden rule"—treating each other as one wants to be treated—or the "silver rule"—not treating others as one would not want to be treated—is present in many, if not most, widespread religious traditions.

and morality that supported survival within a troop [1124, 1125]. Imposed sanctions would have had little negative consequence on the average cooperative community member's behavior. Cultural sanctions may have served to maintain social order and group recognition by imposing learned cultural norms, such as dietary taboos, or to warn individuals whose inborn behavior and moral compass lay outside the norm, the psychopaths.[165] The potential to raise one's standing by cheating can excite anyone, but that temptation would lure individuals with impaired consciences to action. They would know an act was "wrong" even if they did not feel it. High social consequences would serve as an external regulator and have adaptive value within a group. Harming a member of one's troop endangers oneself and one's genes. These original sanctions would have excluded outsiders. Often, as in racism, casteism, and warfare, they still do.

A group's ability to form coalitions directed against bullying by alphas would have increased as the ability for communication (language) increased.[166] Social selection pressure would have decreased the profit for overt aggression and amplified the reproductive payoffs for covert bullying and more generous, pro-social behavior [1126, 1127, 1128]. More subtle intimidation would have become a better strategy.

Women in that era would have selected *against* mates who abused them.[167] Females of most social species do not seek out bullies. They prefer males with high social standing who can control the resources their

---

[165] Sanctions would work against individuals with aggressive or sociopathic traits who did not have a high rank. Individuals who maintain status through those same traits would not impose corrective measures upon themselves, though, like the female bonobo enforcers, their clan members might. Nor would lower-ranking individuals have the power to impose sanctions upon their leaders. Leaders often reserve the most draconian measures to squelch their opponents, including family members. Frodo, the chimpanzee that reigned at Gombe for many years, overthrew his brother, Freud, and reigned using terror as a weapon. Once overthrown, his fate was not a kind one. One day, he awoke to find the others had abandoned him and refused to answer his calls. Once he was defeated by a coalition of other cooperative males, Frodo spent much of the next two years as an outcast. He eventually rejoined the troop and lived the rest of his life as a middle-ranking male, occasionally threatening others.

[166] Language is a two-way street. It can support the bully's ability to break coalitions or impose his will. Language also provides a cheap mechanism for social grooming.

[167] Recall that among the bonobos, our other closest genetic relative, females form coalitions to harass and punish bullying males. Male bonobos who treat the females well have more mating success than males who try to rule by violence.

children need. For this attraction to work in his genetic favor, a male must be willing to share his assets with the female and her offspring. In many societies, men walk a thin line between showing how well they can provide for their families and damaging their reputations by appearing to be braggarts. Do too well, and other community members might decide to knock you down a peg or two.

Like women today, Pleistocene women would have preferred males who could walk that line. Women then, as now, would have chosen unselfish men over males who hoarded resources for their exclusive use.[168] If we followed the bonobo pattern, early human females would have mated with generous males more often than stingy ones and preferred kind men to abusive ones. These preferences for generosity and kindness would create reproductive pressure against bullies despite their attempts to dominate. Roughly 4,000-5,000 generations have passed since our Pleistocene ancestors became recognizably human. Women could have begun the domestication and taming of our inner bullies across this time frame just by having babies with gentle partners. These choices built upon and shaped our primate brains, leading to our sense of morality among our families and community.

---

[168] I knew someone who could not grasp this concept. He gave his girlfriend an expensive designer purse for her birthday. Then he bragged that he found it at a resale store. She got angry. He thought she would be impressed by his frugality. She felt devalued and saw him as a skin-flint male who would hoard resources. Advice for my male readers: If you get that designer gift on clearance, keep that tidbit to yourself.

CHAPTER 6:

The Return of the Bully: Social Magnification

> *"For the hand that wields the sword and kills with it is not man's hand, but God's; and it is not man, but God who hangs, tortures, beheads, kills, and fights. All of these are God's works and judgments."*
>
> — Martin Luther, 1526, *Whether Soldiers, Too, Can Be Saved* [1129]

Humans who thrive in small bands are cognizant of their interdependence. Hunters and gatherers have an immediate stake in power distribution among families and clans. Leadership is not "winner-take-all," but split among those with needed talents and prestige in specific areas. She who makes better darts or pots becomes a mentor in making them. Skilled tacticians may engineer defense plans, but experienced hunters or gatherers decide where to hunt and when. No one excels at everything; societies will not long prosper with a single dictator making all decisions.

Small-band societies contain one or a few kinship groups. Families bear responsibility for preventing individuals or lines from accruing excessive influence. Instead of amassing dominion in the hands of a single dictator or oligarch, leaders share power to retain order [1130, 1131]. Individuals who attempt to bully the rest of the band soon find themselves without friends or support. This early social ecology, this negative response to tyrannical activity, still permeates our behavior. Humans continue to respond to community pressure and withhold support for the bullied, even when not actively participating in the abuse [1132, 1133], even though abusive leadership styles endanger the welfare of the community and engender distrust and resentment [1134]. However, like the would-be big man of the band society, bosses and bullies who rule through abuse ultimately find themselves without friends or support, facing potential retribution as their power fades. Even dynastic bullies ultimately find themselves on the wrong side of history, remembered as villains rather than saviors.

Feedback from their community and family keeps those who would be king in their proper place. Power and status spread throughout the population in these small units. This diffuse power structure makes it more difficult for one individual or subgroup to hoard power than in larger, less related societies. "Leaders" in our WEIRD culture often heap contempt on those they consider below them; in traditional bands, that attitude would render them pariahs who risk expulsion.[169] Our social structure has lost sight of the purpose of leadership for the primate group: the sustainability of the community and gene pool. Instead, it focuses on unsustainable, ever-increasing fiscal growth and restricting power to a few.

As with our chimp and bonobo brethren, the details of social constraints among early humans would have differed within and among groups. Some bands would have tolerated a greater concentration of power. Some would have displayed more group cohesion. Within the constraints of their environments, each band could vary in size, composition, or the use and distribution of resources. Each experienced pressure from within its ranks. Every group would have interacted with its neighbors as they roamed the African savannahs, searching for the resources needed to survive. Some of these interactions would have been peaceful (64% tolerance [1135], like the super-nesting groups reported in the lowland gorillas or the gatherings of related clans of people today). Other encounters would have created carnage.

Humans will defend their territories and resources from encroachment. If water, gas, or antelope exist on *our* land, then we perceive that resource as ours. In modern industrial societies, we defend our homes. We

---

[169] Personal note: I do not understand the attitude of a college administrator who denigrates their faculty, the executive that tries to micromanage their massage therapist, or the random bigshot who thumbs their nose at service workers. That massage therapist has your body in your hands, and if you did not need their skill to unplug your toilet, why would you bother calling the plumber?

create watch committees to protect our neighborhoods. On a larger scale, we form armies to guard our boundaries, king, or country.

> *"They are now catching humans like the dog-catcher used to catch dogs. They are already hunting for one week and have not yet enough. The imprisoned workers are locked in at the schoolhouse. They cannot even go out to perform their natural functions."* — Reichsminister for the Occupied Eastern Territories to Gauleiter Fritz Sauckel, on the enslavements in Russia, 21 December 1942 [3512]

Bands of chimps share this behavior, attacking and killing those who enter their territory [1136, 245], as do governments, gangs, and homeowners. In both humans and chimps, one band can engage in warfare against a rival to the point of genocide. The winner can access desired resources, such as food, water, shelter, energy, or fertile females. Many human cultures claim captive females as wives or concubines. We will enslave captured men, but as a species, we are at least as likely to kill adult males.

Gendercide (the butchering of neighboring males) and the sexual appropriation of females have followed us through history. The Bible (Judges: 20-21) describes a massacre where God commanded the other tribes of Israel to slay every male Benjamite.[170] The story holds that vengeance fell shy of killing everyone (Judges 21), and a small fraction of the men survived. They found wives for the remaining Benjamites by slaughtering the men and non-virgin females of Jabesh-Gilead, a town that refused to participate in the gendercide. They gave the virgin women to the remnants of Benjamin. Similarly, when the Athenians besieged and conquered neutral Melos, they killed the men, enslaved the women, and turned the land into an Athenian colony per Thucydides, [1137, pp. 362- 401].

---

[170] "Go and smite the inhabitants of Jabesh-Gilead with the edge of the sword, with the women and the children. And this is the thing that ye shall do, Ye shall utterly destroy every male, and every woman that hath lain by man. And they found among the inhabitants of Jabesh-Gilead four hundred young virgins that had known no man by lying with any male: and they brought them unto the camp to Shiloh, which is in the land of Canaan" (KJ, Judges 21: 1-4).

> *"Soldier: I can't do it. You can't just kill them, they're underage....The girl is a virgin. It's unprecedented to kill a virgin. It will bring bad luck to the city.*
> *Commander: Well, make sure she's not a virgin before you kill her. Now, get on with it."*
> — *I Claudius*, TV series. [3144]

The Romans took both men and women as slaves. (At a time before we used fossil fuels, why waste energy?) The chance of surviving an attack by the Romans improved if you were a woman. The English verb "to rape" comes from the Latin word "raptio," which refers to the kidnapping of women for sexual purposes. The story of the rape (abduction) of the Sabine women, a foundation myth of Rome, bears a substantial similarity to the kidnapping and sexual enslavement of the Jabesh-Gilead women. In addition to their labor, conquerors used women as wives, concubines, or sex slaves. Roman law banned the murder of virgin women, leading to the necessity of raping young women before killing them—just in case. Female captives who suffered enslavement understood sexual acts as part of their servitude.

The sexual enslavement of captured women was not limited to Rome and Europe. Neolithic archeological sites yield a disproportionately low number of young women among the bodies of slain males [1138, 1139]. DNA analyses following maternal and paternal lineages also support an ancient origin for "raptio" [1140].

When Bosnia-Herzegovina proclaimed its independence from Yugoslavia, the predominantly Serbian Yugoslavian army and Bosnian Serbs began the greatest massacre in Europe since WWII. By 1995, they killed over 100,000 people, 80% Bosnian Muslims. The objectified women were often raped, tortured, then killed or traded to others for provisions. To maximize pain and damage to the genitals, soldiers would also penetrate their victims with guns and broken glass [1141].

The primary purpose for the rapes was not the carnal gratification of the troops. The brutality of the acts stands in testament to their intent to humiliate and degrade the victims. In Bosnia, an estimated 12,000 to 50,000 women and girls were raped, 80% of whom endured repeated sexual assaults during imprisonment

[1142, 1143]. Many were kept in "rape camps" until they became pregnant or gave birth, testifying to the act's other purpose: rape as a reproductive strategy.

Although the preference for the killing or enslavement of enemy women varied, their abductions are not limited to European history or antiquity but are widely practiced atrocities now [1144, 1145]. The Tutsi genocide in Rwanda included similar gang rapes. At the end of WWII, the Soviet army committed 95-130,000 rapes, and the Japanese raped 20-80,000 women in Nanjing alone before their executions [1146, 1147]. The First Nations of the Northwest coast of the US and Canada used slaves captured during raids; most were female [1148, 1149]. Among the Yanomamo of South America, traditional warfare between groups often culminated in the rape of women or their abduction to use as wives.

In what is now South Dakota, at the Crow Creek massacre site (mid-14th century CE), the skeletons of young men outnumber the young women, but among the remains of people over forty, the reverse is true; fertile females were probably taken captive [1150]. The Piegans of the Blackfoot Confederacy of the North American Great Plains believed women should be spared and adopted into their villages to give birth [1151]. Aztec and Chaco Canyon archeological sites (American Southwest) of the 11th and 13th centuries have the remains of too many women, and contemporary remains at Mesa Verde have too few [1152]—strong evidence for "raptio." The European colonial period also saw widespread abductions and enslavement of aboriginal women [1153, 1154, 1155].

The period of the Rajahs of India oversaw the seizure of young girls as "gifts" into the early nineteenth century [1156, p. 12]. Following the occupation of Japan by the victorious Allied troops, many Japanese women committed suicide, afraid they would be gang-raped by the invasion forces [1157]. Korean women responded similarly to the sexual atrocities of the invading Japanese army [1158]. More recently, the abductions of females in Nigeria and neighboring Cameroon since 2013 by Boko Haram, e.g. [1159], rapes by Russian soldiers in Ukraine [1160, 1161], and reports of sexual assaults against civilians by both Israeli and Hamas forces[171] [1162, 1163, 1164, 1165] have become the contemporary face of this historical and too-human pattern of forced servitude.

In all of these cases, the differential abduction and sexual enslavement of women point to one outstanding biological feature: an increased ability of the male victors to sire more children at the expense of their neighbors. Even if bloody wars kill thousands of males, the victors produce more offspring than the vanquished. The leaders and their genes typically survive.

The Green-Beard Effect

> *"I stand against my brother, my brother and I stand against my cousins, and I, my brother and my cousins together stand against the world."*
>
> — Bedouin/Arab proverb

The manifest resource that sparks inter-group violence among humans need not be an obvious threat. It can be symbolic, such as cultural differences or the salvation of souls. Humans battle over points of faith, such as how one perceives or prays to a loving deity. The differences seem trivial to outsiders, like going to war over how to crack an egg.[172] These minor points can identify cultural links and perceived kinships that would otherwise remain obscure, leading to identity-based harassment and victimization [1166]. For the groups in conflict, cultural differences identify kin as readily as if they—and they alone—all had green beards [1167, 1168].

---

[171] As is common in such conflicts, both sides in the war deny they actively employed or encouraged rape and other sexual torture as weapons. Because sexual assaults inflame hatred, there is reason for conflicting parties to use it as part of information and misinformation propaganda campaigns against their rivals. Given the ubiquity of rape during war, its absence by either side would surprise me.

[172] Jonathan Swift satirized the causes of conflict in the first book of *Gulliver's Travels*. There, the Lilliputians wage war over how to break an egg [3458]. The battle of the eggs led to six rebellions and the death of one emperor. Swift uses this allegory to mock the wars fought among religious sects in England.

A "green-beard effect" refers to any marker identifying someone as family, even if you have never met them [1167, 1168]. These kinship markers promote altruism among kin and hostility toward outsiders [1169, 1170, 879]. One finds green beards across living things. Ants, bees, and wasps use odors (pheromones) to identify nestmates, determine rank, and defend against intruders [1171, 1172, 1173, 1174]. Squirrels listen to calls to identify kin [1175]. Yeasts that express the FLO1 gene "flock" together, providing mutual protection for those yeasts and the gene [1176]. Even the extracellular matrix of some bacteria in biofilms can be shared and lend protection with close genetic relatives or cooperative species [1177, 1178, 1179].[173]

> *"Marriage: A nepotistic diversion of resources."*
> — Richard Alexander
> [3510]

Humans also use green beards that interact with and direct our instinct for cooperative behavior; these markers shape our patterns of reciprocity, ethics, and violence. A critical green beard identifies from which vagina a person emerged. Except for artificial fertilization and implantation, the woman and her family can be assured the offspring is hers. A man has less assurance, but the odds are tipped in the husband's favor in marriages. Patrilocal societies increase the man's confidence in a genetic investment. In such hunter-gatherer societies, men typically share meat with their band, improving their genetic fitness through the survival of their and their relatives' children. When a maternal grandparent is available, she typically provides more care and resources than the paternal one [1180].

Western culture breaks from the traditional pattern. Our neolocal, genetically mixed troops reduce paternity certainty; the green beard becomes a mere tint around the sideburns and our collective's typical response changes. We often receive more alloparenting and attention from our maternal rather than

[173] Green beards are not the only social understanding that we share with bacteria. Some bacteria account for genetic cheats (free-riders) and adapt their strategies to combat unfair competition [3070, 3513]. The joint ability to recognize kin and punish free riders carries a substantial adaptive value.

paternal grandmothers. Contrary to what was once presumed, if the mother's community resides nearby, many patrilocal societies that involve the father's family in child-rearing may also include alloparenting through the motherline [1181, 1182]. As women move farther from their birth families, access to a parenting support network decreases [1183]. When social services place at-risk offspring within families, they suffer more neglect and abuse from paternal grandparents than maternal ones [1184, 1185, 1186].

As small family units melded into large ones, our biological responses to kinships remained. Humans still prefer our own offspring to those of others, and nepotism became problematic. Humans continue to aid their neighbors, even though they may no longer share the close genetic bonds that once held primate troops together. Generosity toward others, however, still proves beneficial. Individuals who support and defend others, behaviors primarily reserved for *within*-troop interactions in chimps, receive more reciprocal goodwill than those who earn reputations for selfishness [1187]. However, as populations grow and mobility increases, it becomes harder for our primate brains to monitor who has been generous to whom, making it easier for stinkers to defect from the social contract and reap the benefits of cooperation without the costs of aiding others [1188]. The cheaters benefit from anonymity and our inability to remember our thousands of social transactions. However, our societies use symbols to play the odds.

In small bands, family recognition was pretty straightforward. As our societies grew, humans developed learned social signals to extend our recognition [1189]. Symbolic extensions of biological recognition signals extend the genetic markers of relationships that glue bands together. Clans of families have developed symbols such as tartans (Gallic clans, Europe), Bògòlanfini (mud cloth, Mali, Africa), royal Kenta cloth (Akan people, Africa), and Janeu or sacred thread (Brahmans, India). Scarification patterns or circumcision distinguish groups in some regions; surnames suffice in others. In the early 20th century, facial scars from fencing Mensurs marked honor, status, and fraternal membership among German students. Among the Yakuza of Japan, some members identify their station by inserting small plastic pellets into their penises

[1190, 1191, 1192]. Some Russian soldiers have performed similar deeds (pearling) in solidarity with specific army units[174] [1193]. Imperial-period Romans knew a person by a three-part personal name (*tria nomina*), identifying him within his extended family. Among the traditional Dutch, children received their names in honor of family members. The firstborn children received the name of a grandparent. Other relatives received the recognition in an understood but more flexible order. (I thank the universe that this custom has faded, or my parents would have christened me "Wilma" or "Cornelia.")

As the number of generations from the founding ancestor increases, fewer genes unite a society. With time, clans break and fracture. Like the warring factions between broken troops of chimpanzees, they can become rivals rather than brothers. The more tenuous the familial ties, the weaker the loyalty. The more intense the competition for shared resources, the higher the tension and the tendency to fracture.

All animals, including us, interpret the world through their species-specific programming. Birds see trees as places to make nests. Cats do not understand closed doors because no such barriers exist in their natural world. Humans, who evolved in small kinship groups, must also adapt our small-group programming to survival within larger societies. Hence, we broadened our concept of the extended kinship group to include individuals we *perceive* to be more like ourselves.

In a species that is 99.98% identical, we have become adroit at finding minute differences between people. From what we can measure, humans place more symbolic meaning upon their world than any other species. These meanings extend to our definitions of our being. Clans can unite people either by genetic kinship or by *perceived* kinship. Even if the biological association is tenuous, a clan affiliation links a group back to a common ancestor or founder. In the Biblical story of Cain and Able, God places a "mark" upon Cain to

---

[174] Although not visible in day-to-day interactions, these hidden markers can demonstrate membership in a group upon demand. Infant circumcision once served a similar role.

protect him from the vengeance of related clans when he leaves Eden for the land of Nod.[175] The stain may refer to a tattoo or other permanent body art that marked its wearer as a member of a specific clan or moiety. The origin story of Judaism, a heritage stemming from Abraham, Isaac, and Jacob (See Figure 6-2), be it true or myth, ties the people to a common ancestor. Americans trace their democracy to a group of "Founding Fathers" who have taken on near-sacred status.

In India, caste identification can lead to within-caste loyalty and between-caste animosities. The perceived relationship continues even though the genetics among castes within a geographic area is often less than the genetic distance within a caste across the Indian subcontinent. Differences exist in the paternal Y-chromosome (haplotypes) among the castes; otherwise, the genetics across castes tend to be similar [1194, 1195, 1196].

In today's Western world, organizations such as the Masons and the Moose serve as surrogate clans. These brotherhoods have signs and symbols that function as their green beards, identifying lodge brothers. Agatha Christie's famous detective, Hercule Poirot, wore a blue forget-me-not on his lapel, symbolizing his Masonic affiliation in the post-World War II era. Although there is no firm evidence that German freemasons used this flower as a widespread symbol [1197, p. 100], the little flower has since become a cherished green beard, drawing attention to a Masonic affiliation.

Our perception of our green beards has changed over time. As societies grew, bands joined together to survive. At the end of the last ice age, about 10,000 years ago, some organized into agricultural

---

[175] Based upon their names, Cain (i.e. Old Arabic: *qyn or* metalsmith) and Abel (*ibil or* herdsman) could represent a foundation story that recalls a time when related nomads and settled villagers fought over resources.

communities.[176] These more complex societies included people with less in common and tenuous genetic affiliations. As familial connections weakened, people found new ways of recognizing "us" from "them."

Our green beards include language. Linguistic patterns permit us to distinguish members of related communities from outsiders. Native speakers of a language feel a kinship with each other that does not extend to non-native or non-fluent speakers. Before writing fossilized much of our speech, dialects developed readily as bands split apart. These green beards carried information about who was related and how long ago groups diverged. Even today, subtle variants can quickly identify a group member or an outsider. The pronunciation of the "ou" diphthong readily distinguishes a Canadian from an American. The term "bubbler" rather than "water fountain" identifies American roots in Wisconsin or Rhode Island. And, in the United States, the words "pop" and "soda" distinguish members of one's geographic in-group from the unwashed heathens.

> *"They're bringing drugs. They're bringing crime. They're rapists. And some, I assume, are good people."*
> — Donald Trump's campaign announcement speech, 2015
> [3223, 3415]

Our innate favoring of kin remains intact, but the size of modern communities makes it difficult to identify genetic kinships. Back in the Pleistocene, people would have recognized most of their band by sight.[177] Infants and toddlers learned the green-beard behaviors to act along familial or perceived kinship lines. Industrialized humans still favor our family, friends, and close associates in recognizable patterns of nepotism. We discriminate against those whom we conclude are not kin and

---

[176] Permanent agricultural and horticultural settlements are a relatively new innovation. Like other organisms, humans seek the easiest way to reach their goals. For us, hunting and gathering would be a more efficient lifestyle as long as game and vegetation were abundant. The amount of labor and organization needed to maintain agricultural settlements does not make sense unless hunting and gathering no longer support the population or the slight increase in birth rates gave the farmers a tactical advantage.

[177] As in modern children, an ancestral toddler would have survived by trusting their surrounding elders. Without their help, a child would have had little chance of self-defense against an attack by a powerful adult. The ability to discriminate between friend and foe develops with age.

stereotype them, simplifying our behavioral decisions and limiting our cooperative behavior with that group [1198].

We align ourselves with those in whom we perceive commonality, even when no biological kinship exists. Schoolchildren intervene to stop bullying on behalf of someone with whom they share attitudes more than for fellow students whom they perceive as different [1199]. In doing so, they support one of our great green-beard generators: societal memes.

People often harass, scapegoat, or bully individuals and groups whom they perceive as different in appearance or behavior [1200, 1201, 1202, 1203]. Defenders risk ostracism from the in-group but tend to have a higher status, bestowing self-efficacy and offsetting some of that threat [1204, 1205]. Nazi leadership encountered this effect; late in the regime, they discovered many of their most ardent opponents ranked among Germany's upper classes and elites [1080].

The courage of one person can make a difference.
Dr. Vittorio Sacerdoti: 45 lives saved;
Chiune Sugihara: 2,200-6,000 saved;
Oskar Schindler: over 1,000 saved;
Miep and Jan Gies, Victor Kugler, Johannes Kleiman, and Bep Voskuijl: Otto Frank and Anne Frank's diary preserved

Politicians weaponizing differences against a group can create humanitarian crises that bring the worst and best in our character to the surface. Most people within Germany and the occupied lands supported the law-and-order nationalist platform of the Nazis, if not their authoritarianism. Others formed resistance groups even though they were threatened with death if caught [1206, 1207]. Despite the risks to their lives and liberty, white abolitionists joined with African Americans as conductors, shepherds, and stationmasters along the Underground Railroad.

For an employer, prolonged corporate bullying leads to retaliation behaviors, including reduced work output and the purposeful wasting or damaging of company property [1200]. Corporations that gain infamy for bullying incur costs measured in productivity, recruitment, and damage to their reputation [1208]. Even when we recognize harassment as wrong, we often overlook or fail to notice it. We miss the signs, not by inattention, but because of the varied forms bullying takes. Temporal bullying through work scheduled to interfere with personal life, wage theft, denigration of accomplishments, refusals to promote meritorious workers, pay disparity, and assigning skilled workers to trivial jobs are often overlooked forms of aggression that devour the spirit.

Today, tribalism emerges in the surrogate kinship in schoolyard alliances and offices. In schools, homerooms become social units that compete with each other. Employees who identify with the workplace as a surrogate family become victims of bullying less often than those who do not adopt a robust corporate identity [1209]. Fans of books, films, or games relate to fictional characters or computer avatars as if they were part of their community. And phone snubbing—ignoring actual people in favor of a smartphone—has become annoyingly commonplace. (Sorry, I had to include my pet peeve here.) Although such interactions typically establish cultural norms and healthy in-groups of allies and friends, they can readily slip into bullying and exclusion.

The existence of insiders demands the presence of "others" outside of the social circle. In neighborhoods with more than one school, young children typically perceive theirs as the "good" one and the other as inferior and less worthy, and its pupils become the objects of scorn. They insult students from surrounding schools, calling them names. As they age, the institutions and types of degradation change, but the importance of affiliation remains. These young humans resemble the bonobos who, when encountering another tribe, may bellow and scream at each other, but, unlike chimps, they also try to minimize aggression by avoiding direct contact, with their encounters usually ending peacefully.

Debasing others extends beyond the school grounds. It can determine who gets to use facilities such as the neighborhood pool or the roller rink. Other venues can remain neutral, providing a place to meet and de-escalate quarrels. In high school, turf confrontations can become more violent. For our ancestors, territorial disputes determined who could hunt or collect berries in an area.

Today, the turf battles can decide who uses the skate park or dance club, who can date people living in specific areas, who can attend particular schools, and which schools get bullied. In 2013, the City of Chicago unleashed a firestorm of controversy when it announced plans to close forty-nine underused elementary schools and one high school due to declining enrollment. These schools, on average, had worse academic indicators, such as lower standardized test scores. Most of these institutions served African-American communities. The Public School CEO, Barbara Byrd-Bennett, chose this action to address a deficit of $600 million.

The city portrayed the move of 12,000 students as an attempt to save money and consolidate resources by placing children in better-performing schools. Many affected families had doubts about the motives, considering the move as financial and academic bullying. Most students transferred to academically weak schools no better than the ones they left, and the receiving schools had little time to prepare for the influx. The closings required many displaced students to cross street-gang territories to attend school [1210]. These boundaries could create repercussions for any student, not just rival gang members.

People formed camps, each attacking and maligning the other. Fiscal conservatives downplayed the danger and emphasized the potential savings to the beleaguered system. Student advocates downplayed fiscal savings and highlighted the risk [1211]. Both predicted the onset of Armageddon if the city failed to follow their advice.

By 2018, attendance neither improved nor declined, but students experienced short-term drops in their reading and long-term drops in their math scores [1212]. The Chicago school system continued to run a deficit of $600 million. Following the closings, Chicago experienced a correlated surge in teen violence in the affected school districts. In the transition, many school supplies were lost or destroyed. The city had to expand its 2009 "Safe Passage," adding $8 million to the program to pay 600 more workers to stand along the school routes to reduce the hazards to these students [1213]. Gang-related violence in Chicago spiked during 2013-14, but the comparisons to other cities, each trying to make their realm appear safer, renders the spike hard to interpret.

The story becomes more complex than either faction likes. Chicago closed 44 schools between 2002 and 2009 for poor performance or underuse [1214]. By December 2018, the city had shuttered 193 schools, the majority in predominantly African-American districts [1215, 1216]. The closures impacted over 70 thousand students and their families. Children were either transferred to pre-existing schools or, contrary to the stated goal of conserving funds and resources, new schools were built, often in different neighborhoods. By March 2023, Chicago operated 649 public schools [1217]; at the time of the 2013 school closures, 681 schools operated, a difference of 32 schools.

Few addressed the tax drain as a central issue. Chicago and its collar counties produce about 75% of the state's revenue, but for every tax dollar sent to the state, Chicago receives back only $0.58. The southernmost counties average a return of $2.88 [1218]. It is the only Illinois school system required to pay contributions to its district teachers' pension fund (others are state-backed). Instead, a pervasive myth holds that Chicago spends funds beyond its means, cannot budget, and that downstate counties support the lazy, dangerous minorities living in Chicago. In reality, the downstate counties, through their biases, have unwittingly engaged in a reckless pattern of gaslighting and financial abuse, bullying tax dollars from Chicago's captive population. Unfortunately, the United States Constitution makes it nearly impossible for Chicagoans to leave Illinois to join neighboring Indiana or Wisconsin. If feasible, Chicago would have attempted it long ago.

This diversion of funds from inner cities to wealthier districts reflects the infant abuse patterns seen among our chimp and bonobo cousins. By reducing the resources for the young of potential rivals, disrupting their youth, and creating cortisol-enhancing conditions, we reduce their chances for status and the probability they will compete with our offspring. By working against the perceived "other," we clear the path for our children's social and financial success.

Such discrimination and prejudgment spring from deep in our psyche. Our brains still perceive the people surrounding us as members of our Pleistocene family band. Our ancestors' strongest rivals would have been the neighboring troops, and we retain hard-wired programming to see nearby out-groups as the evil "them." However, like the brothers in the old proverb, we will unite and fight alongside "them" if a more significant threat arises from ones less related. We fight for our families and friends. We'll funnel services to our neighborhoods and villages at the expense of others. We will struggle to succor those who share our green beards: language, culture, and religion.

As our ancestors moved from small bands and clans to organized villages, cities, and states, they had to overcome problems of understanding similar to those of the schoolyard. A neighbor may no longer be a brother or cousin, but human instincts lag behind the changes. Despite losing track of our extended genetic kinships, we recognize each other through common language patterns and complex belief systems. We still feed the friends of our children when they visit. We no longer provide for them because of familial bonds. Our hardwiring, imperceptible, innate, and subconscious programming, with which we come equipped, speaks to us from our past. The Pleistocene brain whispers from beyond awareness, "*Those children carry our genes.*" After all, they are in the middle of our troop and pose no threat. Thankfully, that brain does not yet comprehend the reality of city life: the person in the next house or apartment is not kin. If we did, civilization might collapse.

The Move to the Cities

"Beware if a naked person offers you clothes." — African proverb

Lee [1219, 1220, 1118] and Sahlins [1221, 1222] challenged the idea that agricultural communities opened doors to leisure unknown to hunter-gatherer societies. When Lee examined the time budgeted for work, he found hunter-gatherers required far fewer work hours and enjoyed more leisure time than we do. Gatherers, it seemed, needed to work 20 hours or less per week (4-5 h/day) to survive. Modern industrial cultures average more than twice that. Sedentary, agrarian people also suffer higher rates of viral and helminthic infections and more infant mortality [1223]. That information tells us about our environment. Our ancestors probably moved to large, organized, agricultural societies not because it provided a more prosperous life but because we were in trouble.

Once established, the agricultural and sedentary lifestyle provided a sufficient adaptive advantage to keep it going. Women no longer needed to carry their babies as their troops moved, freeing their energy. Sedentary peoples have higher reproductive fitness than nomads. Their higher reproductive rates provided another benefit beyond producing gene carriers: the children offered security. The higher population densities allowed the sedentary food producers to drive out and displace their hunter-gatherer neighbors.

Organisms, in general, do not work more than they need to. Energy is a precious commodity, and selection pressures program animals to maximize their nutritional intake while expending as little energy as possible to get it [1224]. Organisms will follow the path of immediate least resistance whenever possible. Humans like pull-tab cans and twist-top bottles because they eliminate our fights with can openers. When trying to open a tasty snail for dinner, a crow will drop it to the ground. They fly only high enough to minimize the energy expended (a combination of the height of the flight and the number of drops needed to break the shell [1225]. Viruses that prey on bacteria use an inbred optimal foraging strategy in host selection [1226].

Caribou choose their habitats according to available food, the energy expended in getting it, and avoiding biting insects [1227]. Ground squirrels select food that provides an optimal diet while putting forth as little energy as possible [1228], and the sparrows that devour my grass seed leave once they've consumed the bulk of the feast.

Humans follow the same inbred pattern. We often choose fast, greasy food over tasty sit-down meals. We are programmed to like high-calorie grease and minimize our hunting times. We can learn to devalue the slips of paper we hand to the cashiers because we tend to discount the past effort (work hours) we invest in the meal, but corporations count on us not factoring those hours into our calculations when making purchases from fast food to plastic flamingos.

Animals, including humans, want the highest return for their efforts. They move into low-quality habitats when denied access to preferred niches. In the wild, better conditions contribute to reproductive fitness. When the prime real estate degrades and no longer provides more value than a lower-quality niche, the colonization of the suboptimal property begins [1229, 1230, 1231, 1232].

The movement of humans from a gathering-and-hunting work commitment of 10 to 20 hours per week to the intense labor of agrarian societies had disturbing implications. Clans that could no longer sustain themselves as hunters and gatherers became sedentary horticulturalists and farmers. They faced the unpleasant prospect of working much harder for the uncertain hope of future food. To accomplish this goal, they submitted to another unpleasant situation: They formed broader coalitions with people they usually would mistrust.

Anthropologists and biologists agree on at least one reason we got together in larger groups: survival. In some areas, a rapidly changing environment at the end of the last ice age favored survival and reproduction in larger communities over smaller ones [1223]. We could not have accomplished this feat without our natural primate affinity for the community living we inherited from our earliest common primate ancestor. Those same forebearers also infused in us a desire for status and the intelligence to use other humans as allies to gain and hold it.

Larger states intensified many inborn primate behaviors. Social amplification occurs when individuals begin a chain of events resulting in large-scale changes in group behavior. For example, in both beluga whales and humans, the presence of young increases play behavior in adults. This increased play can, in turn, improve the health of the group [1233].

The amplification stemming from individual connections creates divergence among societies. Different habitats, gender ratios, or group sizes can yield radically different behaviors. For example, Hamadryads and olive baboons are close relatives. Habitat differences have favored aggressive Hamadryad males who control harems of about ten females through their fertility cycles [1234], the only primate to show this behavior other than gorillas. When females become a limited resource and the competition becomes fierce, mayhem increases as males vie for sexual rights, overriding their powerful social adaptations designed to keep the peace [1235]. In an infamous 1925 case, the London Zoo had a gender ratio of 15 males per female, resulting in a massacre of both genders as the males battled each other and trapped the females in their quest for mating rights [1236].

Olive males, in contrast, cannot achieve this same control, but they attempt to dominate the females during oestrus. The resulting difference in the pattern of bully-like behavior toward the females creates a different social structure among female and male troop members. Male olives engage in more multi-male/multi-female interactions, and the males will defend females and offspring [1237]. Although the social structure of this distant relative is more like our own, primatologists have had difficulty dislodging the false norm of the violent, dominant male from popular consciousness.

Habitat differences and colony size influence behaviors across various animals, not just primates. Food availability helps regulate colony size in Magellanic penguins [1238]. Nest size alters how hard a wasp will work [1239] and the ant's ability to recruit allies against an attack [1240, 1241].

Among humans, male cooperation declines and freeriding increases when the group size rises above about eight members [1242], and most groups have key members with more interconnections. At least among teenage boys, cooperative and aggressive interactions rise with the number of links. Increased group size reduces the frequency of contacts, encouraging smaller clusters to form across the community's increased territorial size. More space and connections dilute the interactions among pairs in core contacts. This reduction makes developing the within-group trust necessary for cooperation harder and should tip the scales toward more aggressive, bullying behavior [1243]. Remember, however, that bullies tend to be popular, and smaller communities can provide enclaves for degrading aggression. In grade schools, smaller class sizes show higher affiliation than larger ones, but they also generate higher bullying rates [1244, 1245, pp. T-10].

In human societies, the potential for anonymity has grown with the size of our communities.[178] Bullies can amass power in small enclaves within the obscurity of a large group. They develop coalitions among larger groups, just as dominant chimpanzees do. Like Iago, they spread rumors to undercut or destroy rivals. They also use other organisms to enhance their advantage.

[178] The internet has, to some degree, reversed part of this trend. Today, many sets of eyes pierce the codes that safeguard privacy. Despite these quasi-professional information thieves, cyber-bullies typically hide behind the anonymity shield of the web while making their attacks widespread. Authors have written volumes on cyberbullying and fraud, but they differ little in their composition from the flesh-and-blood bullying that has existed across time.

Satellites

Satellite (noun): a smaller object that orbits a larger or more powerful one.

A need for mating or a resource range is a common cause of aggression. Territorial organisms—and territorial societies— need legroom. Dominant animals can access the best locations and defend their resources and territories against all comers. They may also bully their neighbors and infringe on their space. An aphid heroically kick-boxes with intruders who attempt to usurp her spot on a leaf. Bees and wasps guard the entry of their hive against members of other colonies. Squirrels chase other squirrels from their foraging grounds, and troops of chimps and humans will drive away or kill other troops that wander into their territory.

When times are plentiful and the habitat is forgiving, the dominant organisms have little trouble defending their space. The less able individuals might not score the best territories, but they know the competition will whoop them if they get out of line. In good times, the reduced access to food, mates, and other resources does not threaten their lives. Fighting would pose a risk and reap few rewards.

The territorial size of many lizards depends on a limited resource. Among the potential limiting resources, sunning rocks are the easiest to manipulate in the wild. As cold-blooded creatures, lizards need the sunning sites to warm themselves, and their territory often centers on this resource. When kind researchers add high-quality sunning stones to a crowded colony, territorial sizes decrease [1246]. When they remove some of this nice furniture, the lizards compete for more space and access to the remaining rocks. The rocks do not cause the territory to expand or to contract. The lizards weigh the effort needed to maintain the limiting resource against their need for it. When the human thieves returned the missing rocks, peace returned to the colony, and the lizards decreased their territorial holdings.

In black iguanas, males defend their slice of the territory against all neighbors except for one: A colony can contain a tyrant who can trespass on the domains of others without worrying about the consequences. The high-ranking lizard meets opposition with an open-mouth threat that intimidates rivals and sends them skedaddling to hide in crevices. In Evan's study [1246], the tyrant exercised his right of position over the limited resource but never challenged other males over nearby food and water. Those resources were plentiful, and the energy needed to defend the reserve exceeded the benefits gained by hoarding. Spaces with sunning rocks, however, were in short supply, and rank had its privileges.

Some species have a more intense problem to solve. Their competitors can move more easily, have bigger brains, or have more energy to spare. Organisms consider more than one factor, weighing the costs of food, water, and security. [179] Territorial males can be overwhelmed by trespassers desperate for survival. In many species, such as *Rana clamitans*, the "green frogs" [1247], devious males hang around the edges of a territory, keeping a low profile in the hope of gaining the real estate later—even in the brief absence of the owner.

In other species, such as kribensis or rainbow cichlids, male satellites will defend an area for a dominant male [682]. This service increases their chance of usurping territory or of sneaking sex in the present. Times, however, do change. When starvation, thirst, or unsuccessful mating threaten an animal, it grows desperate, and attacks on the neighbor's resources increase. Coalition leaders must spend more effort retaining loyalty as more and more low-ranking underlings attempt invasions to gain what they need to survive.

---

[179] Abraham Maslow [3189] proposed that humans make such motivational distinctions in their day-to-day lives. According to his hierarchy of needs, a person must first meet their physiological and safety needs, a task stymied by bullying. Only then can they focus on love, self-esteem, and reaching their potential.

Every so often, species with similar problems converge upon similar solutions. Many territorial individuals allow one or a few weaker individuals to join them. Many species prefer relatives, often siblings or offspring. By choosing a relative, the owner gains another chance to pass on some of his genes.

As satellites, these individuals never gain full access to the territory. The seat of power remains under the control of the dominant landowner, but the territory owner allows satellites to guard his or her property in exchange for a lesser share of the spoils. The owner seldom admits potentially challenging neighbors as satellites. He benefits if threatening rivals lose relative status. By surrounding himself with males of lower or intermediate ability, a dominant male can gain their support and still maintain the ability to kick them out when he no longer needs them. The satellites win by improving their chances of mating, even if they never have the same opportunities as the owner.

Satellites have a tacit agreement with their patron. Should they violate that agreement, the owner can turn on them. The territorial male may not grant mating privileges; however, if the satellite defends a remote region, he may gain access to females on the sly. Should the client defect to another neighbor, he would invite an attack.

The satellites hold a subservient position, but in some species, their service positions them for a potential future usurpation should the territorial owner weaken [1248]. Although the territorial owner chooses a weakling, the desperate weakling sometimes wins. The Cold War relationships among the United States, the Soviet Union, the People's Republic of China, and their "satellite" states provide a human example of this same kind of territorial understanding. Although a union of any two major powers could have weakened the third, none of the three superpowers trusted the other enough to work as its satellite. And when satellite states defect, as Ukraine did with its move toward NATO, the reaction of the former patron can turn genocidal. At a more local level, lower-ranking friends who ingratiate themselves with high-status males can serve as approachable "wingmen" and improve their own chances of scoring a date.

When we look at our closest cousins, the bonobos and the chimps, their tribes vary in the details of their social structures but consistently do not like new males. If a community allows a male to join, he holds a peripheral (satellite) status. In emergencies, the male troop members provide him with less support, and he is more likely to be targeted for harassment than members of the band of brothers. Because he cannot access the protected center of the troop, predators have opportunities to eat him. Natal males will steal choice foods or locations from him, expect a disproportionate amount of work from him, and interfere with his ability to mate. In short, they act as bullying energy thieves. In their society, like ours, status confers privileges, including reproductive benefits [756, 1249]; a marginalized male has few chances to mate.

The changing ice-age climate near the end of the Pleistocene was a critical time for humans.[180] Then, we were a young species, had little genetic variability, and had become recognizably modern humans (*Homo sapiens*) only about 85 thousand years earlier. The ice age provided an ecological version of a financial depression and resource redistribution. Territorial incursions must have increased as we scrounged for food. The water cycle slowed, rainfalls decreased, and the lush rainforests of Mother Africa yielded to drier savannahs. As people retreated to the mountains in what is now Ethiopia [1250, 1251], the Sahara became lush with vegetation. These competitive conditions would have magnified the benefits of collaboration, enhancing the ability of the more cooperative band to survive, both as attackers and as the attacked.

Communities will attack other tribes to gain or protect resources [1252]. They attack, fight, and sometimes kill intruders who wander into what the troop considers its territory. Chimps with overlapping domains can

---

[180] The last ice age happened in the Tarantian era: 115 to 11 thousand years ago). Although the *H. heidleberensis* migrated from Africa earlier, *H. sapiens* left our mother continent between 60 and 90 thousand years ago.

go to war to gain exclusive access to those assets, even if their neighbors were once part of the same troop.[181]

Ancient humans would have behaved similarly [1253]. Pressure for resources would have forced together bands that, in better times, would have hunted and gathered apart. They would encounter each other with a modicum of distrust and possibly aggression. If the bands included close relatives, they would have recognized each other by green beards. Their dress, looks, family tales, rituals, and language would have identified them as kin, just as these things can signal group affiliation today. By joining forces, combined groups could defend precious resources against outsiders.

As with birds, lizards, and humans today, our ancestors' relationships and boundaries would have shifted with levels of resource competition. Perhaps the willingness to trade resources among surrogate territories quelled hostilities where exchanged benefits outweighed the risks of contact.

Those human communities who joined forces might have had trouble separating again. Like the lizards with their sunning rocks, nearby niches would have filled with other joined bands. The social setting changed, but the mind did not. Family groups and clans would still have vied for social power in the new environment. Satellites, once invaluable aids to territorial defense, could no longer be dislodged.

In these new conditions, hierarchies developed to maintain order and power. Societies survived if they made an adaptive transition to new power structures appropriate for their new social environments. The clans and proto-settlements that experienced too much internal instability plunged into chaos created by

---

[181] Like humans, chimpanzees prefer asymmetric warfare. Parties of chimps will attack smaller parties or lone members from other communities. People have made many claims about bonobos being less prone to violence and fighting. Because the home range of the bonobo is in the Congo, little fieldwork has been conducted on wild populations. The limited field studies have shown that they tend to form larger parties than chimps and seldom travel alone. When bonobo troops meet, hostility and tension reign. The paucity of isolated victims can explain the lack of reported murders. Remember, it took many years of field observation before we observed chimpanzee warfare. When first encountered, chimp warfare was dismissed as a by-product of human interference.

power vacuums. Such societies would have been easy pickings for their neighbors. Excluded satellites would have little stake in defending a chaotic society. The same disenfranchised satellites in a neighboring organized society would have a significant stake in an attack: the potential access to more resources.

We can use these insights to interpret some of the legends and foundation stories that have come to us from thousands of years ago. Livy, the Roman historian [1254], describes how the Latins, under the instigation of Romulus and the direction of the Senate, approached their Sabine neighbors and asked for women as brides. The Sabine men refused (a very primate thing to do). The Romans then organized a festival to draw in the Sabines. On a signal from Romulus, the Romans attacked the outsiders, seizing their women as wives and concubines. (Romulus then told the women why they should be flattered to become the property of the new marauders. History did not record the women's responses, but my mirror neurons give me a pretty good idea of what they were thinking.)

A similar story in the Biblical book of Judges describes eleven tribes of Israel turning on the twelfth tribe, Benjamin, in a civil war. It describes the incident that started the genocide: the gang rape and murder of a concubine of Levi and the refusal of retribution for the murder, a sex-based atrocity.[182,183] The oath taken by the other tribes not to provide wives for the surviving Benjamites reinforces the sexual aspect of the genocide. This denial of resources would have extinguished the line except for the slaughter of another group. The other tribes gave those women as wives to the Benjamites (Judges 20 [1255]). Because writers recorded the events almost 600 years after the described incidents took place [1256, p. 153], many biblical

---

[182] Even if one assumes elaboration over time, the story shows a structure similar to modern propaganda: the demonization of the opponent. Similar accusations were made during the Roman civil war of the late Republic era. Supporters of Octavius (Augustus) decried the marriage of Mark Antony to Cleopatra. Rather than characterizing her as Pharaoh, she was portrayed in Rome as a promiscuous and treacherous woman—the whore of Egypt. Our popular image of Cleopatra as a sensuous temptress descends from this Roman perversion of her. Upon defeat, Antony killed himself at the battle of Actium. Rather than facing the humiliation of enslavement, Cleopatra also killed herself [3145].

[183] Helen of Troy notwithstanding, I doubt any nation waged war over one woman. The outrages described by The Bible and Homer have a ring of propaganda, justifying other less noble motives for genocide.

scholars question the account's veracity. (Ancient historians were often more intent on a good storyline and a politically astute message than accuracy.)

A kernel of truth may lurk behind the story. It might have its origins in a tale of retribution in which the followers of one faction (represented by Saul, a Benjamite) were killed by the supporters of another (David of Judah).[184] Scriptures state that while he worked for King Saul, David had to flee the royal court, indicating he did not feel safe there [1256, p. 157]. During the civil wars of an early Israelite monarchy, this account would have made good political sense. It would have mustered support for a strong kingdom under Judah. The story also explains the differences in the power structures in the early histories versus the later ones.

This perspective of the civil-war story has all the elements of resource and power competition. The ruling Benjamites, who had choice property, fell to a coalition of related neighbors (the satellites) who would benefit from the fall. All it took for Benjamin to fall was a rise in the power of one satellite group, Judah, and the switching of allegiances.

If human nature were the same then as it is now, the halt of the attempted genocide in the story of the Benjamites would not have come from pangs of conscience but from the desire to counterbalance a strong Judah. When American President George H. W. Bush did not dislodge Saddam Hussein after the first Gulf War, he kept a second strong power in the region as a counterbalance to a potentially hostile Iran. The story mirrors the aftermath of the abduction of the Sabine women. War erupted between Rome and its neighbors over the assault. They conquered the neighbors (typically a prelude to enslavement), and Romans colonized

---

[184] Multiple artifacts refer to the "House of David," but the stories of David and his relationship with Saul have probably been embellished and distorted. Most of the sources are religious rather than historical accounts. There is no definitive proof for the existence of either Saul or David beyond the sacred texts. They may be real or historical figures may have inspired their descriptions.

the defeated towns. The stories of the Benjamite genocide and the Sabine abduction are riddled with more war, sex, power, shifting resources, and violence than a Tarantino movie.

> *"You know, sometimes I get awfully tired of trying to save the white man's soul"*
> — Thurgood Marshall [3331]

A family, clan, or another social unit of animals can maintain power over their satellite populations by bullying the competition, especially the close ones. Defecting satellites risk losing the protection of their patron and attacks by others who will scrounge their resources. If satellites can gain strength, they can weaken and overthrow the powerful. For at least this reason, coalitions of satellites threaten dominant organisms.

> *"And besides all this, if any of them ever appear too handsome or distinguished-looking for slaves, they impose death as the penalty, and their masters also are fined for not checking them in their growth and fine appearances."* — Athenaeus [1257]

Bullying, harassing, and setting satellites against each other reduce their threat. Many societies waged terror campaigns against their most dangerous satellites for these reasons. Slaves and slave rebellions met with brutal reprisals. The Romans punished the Spartacus rebellion by crucifying one rebel on each mile of the Appian Way. Spartans would systematically humiliate and kill Helots (serfs). Their levels of cruelty probably did not rise above those reported by the rival Athenians, but the Spartans did practice a particularly pernicious form of enslavement. The Spartans assigned them the most degrading task, forced them into humiliating acts, forced public drunkenness, and imposed mandatory yearly scourging. These acts instilled learned helplessness into the slaves and internalized their inferior status. Because self-confident Helots posed a threat, such would be executed [1257]. Each autumn, the Spartans declare a ritual war against their serfs, requiring the "*kryptia*"[185] who were finishing their military training to terrorize and kill helots without

---

[185] Scholars debate the role of the kryptia. They attended a boarding school (*agoge*) and may have trained for stealth military service (like modern special forces) or acted as an internal security force for the Spartan elite. Others consider it to have been a right of passage. Because the Greek word "kryptos (κρυπτός)" means "hidden" or "concealed," I tend to favor the former interpretation.

repercussion. As they culled the herds, they targeted uppity-looking slaves and took food from the fields. This practice removed the greatest threats among the helots and reduced their overall risk.

> *"Injustice anywhere is a threat to justice everywhere"* — Dr. Martin Luther King Jr, Letter from Birmingham Jail [3158]

More recently, the United States sold and separated enslaved families, imposed scourgings and lynchings, and forbade education for the enslaved population. They squelched revolts, such as the Stono revolt of 1739, with executions and tighter strangleholds. Following the Nat Turner Rebellion (1831), the militia beheaded anyone believed to have participated in the uprising and, echoing the Spartacus rebellion, mounted their heads atop poles at frequented crossroads. One road in Southampton, Virginia bore the name "Blackhead Signpost Road until 2021 [1258]. It commemorated where the severed head of an enslaved man was left to rot following the Turner Slave Rebellion.

With peaked fear and rage, mobs of enraged Whites slaughtered hundreds of free and enslaved Blacks across the American South for months. They doubled down on slavery, and the modes of control became increasingly oppressive. On November 5, the Southampton County Court sentenced Turner to hang. The wheels of injustice turned quickly, and his execution occurred on November 11. Turner was skinned, and surgeons dissected his body before he was rendered into grease. The tanned skin was sliced into bits and distributed as souvenirs. A physician purchased his bones as a medical specimen, but the head disappeared.[186]

---

[186] Nat Turner's skull may have resurfaced in 2002. Richard G. Hatcher, a former mayor of Gary, Indiana, received the head as a donation for a civil rights museum. In 2016, he returned it to two of Turner's descendants. They placed it in the care of the Smithsonian Museum pending DNA analysis. As of this printing (2025), the DNA test results have not been definitive, but an injury to it is suggestive. In a 21st-century twist to public desecrations of bodies, after shooting 17-year-old Trayvon Martin, George Zimmerman retweeted pictures of Martin's corpse after receiving it from an admirer who called him a "one-man army" [3315, 3316]. Zimmerman states he did not see the image and retweeted it by accident [3093].

With the abolishment of legal chattel slavery, states substituted restrictive Black Codes to limit the rights of the freed and an illicit peonage (debt) slavery system that lasted 75 years beyond the Emancipation Proclamation. These took the form of chain gangs, sharecropping, and convict leasing. The state would lose the debt records, leaving the prisoners the virtual property of the state or those who purchased their labor [1259]. As with the Spartan helots, African Americans experienced and continue to experience terror through arson, lynchings, and rapes.

The cruel subjugation and dehumanization of enslaved people serve two essential purposes. It provides a permanent underclass that elevates the lowest-ranking societal members, thus reducing tension and promoting stability within the free community [1260]. By preventing uprisings, enslavers also prevented significant interruptions to a principal source of energy: slave labor. The Romans crucified 6,000 rebels along the Appian Way after defeating Spartacus. This wanton destruction of human life shocks modern sensibilities, but Nikiforuk [1261] argues that Rome sent a different message; it terrorized the remaining slaves and reassured the public that the state would not tolerate more energy interruptions. Until we learned how to tap fossil fuel, enslaved labor provided much of the power that supported civilizations. This parasitic benefit to the enslaver complements its injury to the subjugated peoples. Enslavers gained improved access to energy and resources, providing them a reproductive advantage while suppressing the conquered population and confiscating their land.

Europeans do not hold a monopoly on the cruel subjugation of enslaved peoples. It happens worldwide. The despair of slavery has occurred in East Asia, from Vietnam to Korea and Japan [1262]. The Maori of New Zealand enslaved the Moriori of the Chatham Islands in a genocide that approached 95% [1263]. Zanzibar once housed great markets that sold luxury goods, including ivory and Africans. East African nations, such as the Yao and Makua, would enslave captives, then use them on clove plantations or sell them, with an estimated 50,000 people sold annually. Despite being banned by the Sultan in 1873, the lack of enforcement allowed the practice to continue until 1909. In 1981, the North-West African country of Mauritania became the last country to abolish maternal descent-based slavery; they did not criminalize it until 2007. The government denies its continued existence but bans the study of enslavement and its demographic analysis [1264].

The Native American First Nations of the Southeast served as safe harbors, slaveholders, and slave catchers [1265, 1266]. Many First Nations practiced enslavement, with the Nations of the Northwest Coast practicing hereditary enslavement and intercommunity slave trade, impacting up to one-third of the population. The Maya enslaved thieves, and debt bondage for gambling was common in North America. Throughout the Americas, people practiced slave raiding and followed the typical gendercide pattern: men were the most likely to be killed, and women and children were enslaved or adopted. Among the Aztecs, the children of traitors were enslaved for five generations [1267, 1268, 1269].

> *"You have as many enemies as you have slaves."* — Seneca (Roman Proverb)

The Reuters Foundation [1270] and the International Labor Organization [1271] estimate that 40-50 million people survive some form of enslavement today, the highest number in history, though the lowest recorded per capita level. The desire for slave energy grows where resources and fossil fuels cannot provide the labor we take for granted [1261]. Estimates of the profitability of enslavement, the ultimate form of bullying and degradation, vary. Conservative approximations of $28 billion annually use a narrow definition of human trafficking [1272]. Most estimates place the annual revenue generated by enslavement near $150 billion [1273, 1274]. Despite the disparity, all place enslavement among the most lucrative illegal businesses, below drug trafficking ($426-$652 billion) and counterfeiting ($923 billion- $1.1 trillion). At $150.2 billion, only illegal logging ($52-$157 billion) ties in third place with the profits made on human misery. If one adds the illicit harvesting and trafficking of human organs to other forms of enslavement, its profitability increases by $840 million to $1.7 billion [1273], leaving logging in its dust.

Rulers and high-ranking animals stay wary of all threats to power, not just those posed by their enslaved. Partnerships in any subdominant group pose a threat. Silverback gorillas and male chimps interfere with female alliances. After a military coup, a sensible dictator will kill the leaders who supported him; they pose his most significant threat, and the backing he requires to maintain power differs from the help he needed to gain control. Human social hierarchies include bullying tactics and general interference with any alliances threatening to upset the status quo and the current distribution of benefits.

Torture, such as whippings, contributed to the control of the enslaved population, who, as you might guess, were not always passive victims. Torture included elements of degradation to break people's confidence. It suppressed communication through fear of reprisal and forced its victims to swallow their rage. The separation of families during both the Shoah and African-American enslavement reduced the social cohesion of the enslaved communities. This isolation severed kinship ties that could serve as a spark for rebellion. Often, the easiest path to control is not by direct interference but by setting such backstabbing and distrust into motion.

Another Magnifying Force: Women

> *"When, say, a female lemur or bonobo dominated a male, or a female langur left her group to solicit strange males, a woman fieldworker might be more likely [than a male fieldworker] to follow, watch, and wonder than to dismiss such behavior as a fluke."*
>
> — S. Hrdy, *The Woman that Never Evolved* [1275]

Another social change contributed to the development of solid hierarchies in early human settlements. Chimp and bonobo females without small children can switch troops when conditions worsen [1276, 744, 1277]. Female humans would have lost much freedom of movement in early male-dominated settlements. The WEIRD Western world takes this condition for granted. However, this change would have led to profound social changes as we moved to larger bands. Permanent settlements amplified the impact of the more stationary females.

Our closest relatives have less social continuity across generations. Chimp and bonobo networks pale compared to human ones [1278, p. 30]. Our coalitions extend far beyond the borders of family and kin. Chimpanzee tribal structures follow fission and fusion patterns, breaking into smaller groups with their society reuniting. Males partner for specific activities, such as patrolling, feeding, or power alliances. Males seldom stray from their birth territory. When males do leave, it is often due to pressures from other males [1279]. Humans express staunch, devoted male cooperation beyond kinships.

The most substantial ties in these groups are those between mothers and dependent offspring, and these family units serve as links for the community [1280]. Male chimps seldom move to the nearby troops even if they have close genetic ties. If they make such a move, they do not fare well. The new band may kill him; if he manages to join, he usually becomes a satellite male, exiled to the edge of the troop and bullied into doing more work for fewer resources.

Females range farther from the natal troop to join a new community. Proto-humans would have shared this fluid female movement among bands of related males [1278, p. 28]. Even among modern humans, about 70% of cultures show patrilocality [1281, 751, 1282]. Many cultures accept migrant women who marry into their society but remain steadfast against granting membership to other men.

Proto-human female mobility gave males access to different females over time [1283, 1284, 1278, pp. 27-28]. Like the bonobos, who share our hidden estrous and gender similarities, human females probably developed alliances with each other and within their maternal family units. The homebody nature of the males ensured that virtually every male in the troop shared a genetic interest in the offspring.

> *"If a man's wife bears him children and his maidservant bears him children, and the father during his lifetime says to the children which the maidservant bore him, 'My children,' and reckons them with the children of his wife, after the father dies the children of the wife and the children of the maidservant shall divide the goods of the father's estates equally. The son of the wife shall have the right of choice of the division."* – Code of Hammurabi, Law 170 [3137]

Despite the Western desire to see our species as monogamous, it doesn't jibe with our evolutionary history or cultural variety. Jewish scriptures describe ancient Hebrew society as polygynous, and today, Orthodox and Conservative men may marry without a Get (religious divorce), but women may not. Culturally sanctioned polygynous marriages occur among several West and Central African, Middle Eastern, and Polynesian cultures. Formal unions may bestow the status of wife or the lower position of concubine. The Code of Hammurabi (Babylon, ca 1700 BCE) provided for the children of concubines as well as those of wives. Besides recognized unions, polygyny can happen outside marriage with mistresses, multiple casual partners, and serial monogamous relationships.

In polygynous cultures, the homes of powerful men often house many women. In ancient religious texts, male leaders from Krishna to Saul express prowess and strength through their number of wives or concubines and in their treatment of those women. Men derive a clear reproductive benefit from larger harems, producing more little gene carriers and blocking other males' access to eggs. Women benefit by accessing high-status or high-quality partners who might otherwise be unavailable. In some traditional societies, multiple wives cooperate with farm work, cooking, and childcare. The household economic unit benefits not just from the labor of several wives but also from that of their offspring. In societies without social safety nets, the children (typically the boys) will care for elderly household members in their old age. Children are the social security system, and families with multiple wives have more chances to produce surviving progeny.

> *"I'm just one of the Master's robes. He can wear it or he can take it off."* — *Raise the Red Lantern*, Su & Zhen, 1991 [3242]

In large polygynous households, women develop hierarchies, often spurred by favors received from the husband or the female head of the harem or household—usually the mother or another blood relative of the patriarch. Women in a harem may need to vie for positions to ensure their sons' inheritance and prevent their deaths as threats to succession. Wives also form female and mixed coalitions to rear offspring, deflect violence, and advance agendas. Ramesses III's wife, Tiye, conspired with court officials to kill him and have her son mount the throne. Queen Jeong of Joseon (Korea) used her skills at court to instill strife and dispose of rivals and their sons, position her husband as heir to the throne, and convince him to name her son (Sejong the Great) his successor. The imperial harems of the Medieval Safavid Empire, located in what is now Iran, included wives who conspired with eunuchs to secure the throne for their sons.

Lekking and Behavioral Magnification

> *"Well, you can tell by the way I use my walk, I'm a woman's man, no time to talk."*
>
> — "Stayin' Alive" by the Bee Gees [1285]

As a species, we engage in many mating behaviors. These behavioral patterns include "lekking,"[187] a form of competitive mating. In a lek, one gender competes for the attention of potential mates. Leks differ from other sexual competitions because the competing gender offers no resources beyond their sexy appearance and genes. Leks are competitions, and the fussy gender evaluates the contestants. The rivals must be within the chooser's sensory range. For a peacock, the males compete with their beautiful tails. Not every species

[187] Lekking is not the only pattern of mating behavior in which we indulge. Animal behaviorists classify humans as "generalists," animals that can switch behaviors. An organism can be a generalist in its diet, habitat, or sociality. Humans tend to be generalists at many different kinds of behavior. Our reproductive programming includes resource and territorial choices, sneaky behaviors, variations on our number of mates, and even a subtle focus on fertility and intelligence.

shares this bias for vision. Songs drive crickets wild with desire, and carnivores, from pet dogs to grizzly bears, respond to pheromone signals. Even humans use music and colognes to attract mates.

Like peacocks flaunting their tails, human males display their value to women with elaborate sexual signals of mate quality. The shape and efficacy of a flirtation or lekking tactic vary among contexts, genders, and cultures [1286]. The Wodaabe, who live along the southern shore of the Sahara in Niger and Chad, prize male beauty, and men adorn their bodies to impress young, marriageable women [1287]. In modern industrialized societies, power suits, expensive watches, fancy cars, and the latest electronic gizmos all function to advertise status and finances [1288]. Sports, music, and NASCAR driving also flaunt skills that offer no resources. In our hypercompetitive, industrialized societies, we lek with unneeded high-status symbols, monopolizing them so others cannot gain their advantage. Okipa (Sun Dance) scars once enhanced male mate quality among several peoples of the North American plains, including the Lakota and Mandan, but these marks bear no meaning for most non-indigenous people. Talking about your top position in a corporation might benefit a man in the hypercompetitive United States, but traditional Irish women would be repelled by someone so full of himself. And face it, verbal lekking does no good unless you know the language.

Humans extend their lekking skills beyond mate procurement. Within organizations, people lek to gain pay raises and promotions [1289]. In group therapy, participants lek to impress others with the severity of their problems and their work toward recovery [1290]. Large companies spend thousands of dollars designing elaborate reports on glossy paper to court stockholders when they could place the same information on recycled newsprint. Corporate lekking skills attract the love and admiration of a different sort: investment dollars.

We have generalized our specific lekking patterns to many aspects of our lives. When competing with people from other cultures, contrasting lekking practices threaten our ability to get the mate, position, or attention. Changes in lekking and flirtation place us at a disadvantage. It confuses the signals, making one's quality as a mate less apparent. Combine this handicap with our desire for status and our preferences for rare, quirky twists, and you have a green-bearded recipe for xenophobia, the dislike or fear of people who differ from you [1291].[188] Xenophobia and the scapegoating of outsiders spike during economic downturns or political instability when increased competition ratchets up the tension [1292]. With its accompanying violence and degradation, xenophobia bolsters the power difference between the insiders and the perceived threat, relegating newcomers to a lower social status.

Proto-humans would also have displayed their reproductive and social value. Then, as now, language skills and cleverness would have appealed to females. Men with superior problem-solving skills stood a better chance of living and mating. Women who mated with survivors had offspring with a better chance of survival and reproduction. The valued traits would pass to offspring at a disproportionately high rate. Our big brain became our peacock's tail. Women prefer intelligent men. In cultures where women have a choice, they avoid men who are less clever or successful than themselves. This sexual selection explains how the enormous human brain developed so quickly, with reproductive choices ever ratcheting upward the importance of greater problem-solving skills.

Members of leks do not play fair. Less desirable males have a reduced chance of mating and don't like it. A quick and dirty way to gain status is to eliminate the competition. If you get a predator to eat a desirable competitor, you win. If you are a peacock and yank tail feathers from a sexier bird, you win. If, like Iago, you

---

188 Other factors recognized as contributing to xenophobia include gene guarding (by excluding those who do not share green beards); disease avoidance (e.g., anti-Asian sentiments during the first days of Covid-19; spikes in homophobia during the early HIV pandemic); and resource and mate guarding (e.g., claiming the oilfields during the Russian invasion of Ukraine).

slander your rival and trick potential mates into seeing your opposition in a jaundiced light, you decrease their odds of breeding, and you win. If you are a privileged group member, you can grumble about reverse discrimination or demand access to programs, draining them of funds.

Both male and female humans spread rumors about our competition to reduce their chances of advancement. Look at any political campaign's lack of virtue and fair gamesmanship [1293]. The supporters of the candidates hurl accusations. Often, the actual contender remains above the fray, looking noble and superior. In business, competitors portray their company in the best possible light while highlighting their opponent's shortcomings.

> *"O Shahrazad, I pardoned thee before the coming of these children, for that I found thee chaste, pure, ingenuous and pious! Allah bless thee and thy father and thy mother and thy root and thy branch!"*
> — The King spares Shahrazad, *1001 Nights*, Burton, 1886 [3508]

Men also make sexual choices beyond the apparent signs of fecundity. Men prefer intelligent companions. When heterosexual women lek, they enhance their fertility markers, the initial attractants. WEIRD women display status with designer clothing and jewelry. The Dai women in southern China gather to demonstrate their spinning skills while men court them with instruments. Of course, flirtation increases around competition and, like male lekking behaviors, can take many forms. 19th-century Victorian noblewomen used their fans to attract a suitor's attention and to communicate with him.

Female primates have a pecking order like the males do. A high social standing does not entitle her to more males. However, high-status women are more likely to bond with a socially dominant male during estrus. Neither gender seeks to mate down the understood social ladder.[189] Both men and women consistently rank higher-status people as more attractive.

---

[189] We have too few studies on non-binary and transgendered people to make statements about mating patterns and preferences in these groups [3381].

Multiple mates do not convey as significant a benefit to a female. She gains her advantage by securing the highest-ranking males with the best genes and resources for her young. That aim is in direct opposition to the sexual goal of the male. Traditionally, males have desired exclusive sex, and females wanted complete access to attention and resources, i.e., goals that benefit their genes [1294]. Where women complain about men treating them as sex objects, men complain about women treating them as success objects [1295].

Like the lekking males, human females try to take out the competition. For long-term relationships, human males prefer intelligent females. In both grade school and high school, girls who are different, bright, or unusually beautiful become targets for bullying by other girls. They engage in less physical fighting than boys but are likely to shun or spread rumors about each other. Smart girls often find themselves isolated, especially in primary school. They become branded as a geek or a nerd. Beautiful girls can band together but fear betrayal and character assassination. When women physically attack another woman, they often focus on disfigurement—of the face or reputation—rather than death or maiming.

In traditional band societies, people would have known each other and their networks of kinships. As bands grew, clan affiliations and clan-based marriages kept track of these interrelationships. The ban on bullying, that ancient golden rule, would have kept the worst abuses from becoming customary.

As societies grew, all bets were off. Competition among unrelated individuals no longer had the same natural obstacles. The new larger social milieu of unrelated or distantly related families magnified our tendency to cooperate and compete. Larger societies have more individuals with fewer shared genes (= more interactions outside the family). Primates find these conditions ripe with possibilities for bullying and spite.

For males, the chances of finding potential sexual partners increased if they prevented the few high-ranking males from hoarding the women. For females, the choices of available sexual partners increased under the same conditions. The lekking competition and paternity uncertainty magnified our tendency for bullying and cheating in this new setting.

PART II

Schadenfreude: Or the Lie of Silent Assertion

Their Pain, Our Gain

"It is what we all do—we that know. I am speaking of the lie of silent assertion; we can tell it without saying a word, and we all do it—we that know. In the magnitude of its territorial spread, it is one of the most majestic lies that the civilizations make it their sacred and anxious care to guard and watch and propagate. For instance. It would not be possible for a humane and intelligent person to invent a rational excuse for slavery; yet you will remember that in the early days of the emancipation agitation in the North the agitators got but small help or countenance from any one. Argue and plead and pray as they might, they could not break the universal stillness that reigned, from pulpit and press all the way down to the bottom of society—the clammy stillness created and maintained by the lie of silent assertion—the silent assertion that there wasn't anything going on in which humane and intelligent people were interested."

— from *My First Lie and How I Got Out of It* by Mark Twain [1296, pp. 159-170]

Figure 7-1. Coin of Emperor Magnentius (350-353 CE). This reverse side shows the emperor in armor, holding a standard and an olive branch (peace). He rests his foot on the back of an enslaved captive, a significant energy source for the empire. — (Image released to Wikimedia Commons by the Staatliche Museen zu Berlin without restrictions.)

CHAPTER 7:

Experiments in Bullying

"Better to reign in Hell, than serve in Heav'n." — Book 1, Line 263, *Paradise Lost*, [1297]

Humans are a social species, and social species bully. We inflict physical or emotional violence on members of our communal unit. Bullying degrades the victims, transforming them into something less than members of the group, something less than human. In this way, we reign over them, preventing them from achieving a position where they could gain power over us. Even if it detracts from our group's success, the destruction of another can increase our fitness if it reduces the competition's fruitfulness. The fundamental primate desires for power, control, prestige, and sex have shaped human biology.

Our willingness to appease authority figures includes an innate caution against challenging an accepted leader or extending our personal dominion beyond our proscribed limits. Among our closest cousins, the bonobos and the chimpanzees, individuals who win dominance competitions usually gain the group's support. A bonobo will not initiate an attack on a higher-ranking member without a coalition of allies [1298]. In chimps and bonobos, victims of an attack show more fear when a coalition forms than from a single attacker. Murderous chainsaw killers and other homicidal maniacs aside, humans exhibit the same pattern. Coalitions reinforce social ranks, and high-ranking members need group support to maintain or capture power.

From the horrors of genocide to our tolerance of the bullying boss, our species listens to and follows leaders we perceive as legitimate. In the late Pliocene, when our *Australopithecus* ancestors differed little from other apes, this tendency kept us alive. Our elders, the survivors of our troop, taught us how to survive. As humankind gained its complex language skills in the Pleistocene (~2.6 MYBP-11,000), it exploited the collective knowledge of a band by following the survivors and listening to those with honed skills and memories of past conflicts [1299, 1300]. In our modern age, with its weaker social bonds and deadlier weapons, this same *unquestioning* tendency to obey threatens our survival.

The obedience of subordinates helped early human communities function, which tied intimately with their ecology [1301]. We are the ultimate generalist organism. We have no set geographic or dietary niche. We engage in a multitude of sexual patterns, and people choose partners for varied reasons. We live in huts, trees, and underground. We eat almost anything—plant, animal, fungus, bacteria, or, in a pinch, Uncle Fred. We use flexible dominance hierarchies that can be diffuse or linear. In environmental and financial crises, we rally around a leader, increase our xenophobia, and nudge toward more authoritarian preferences [1302, 1303, 1304, 1305]. As our global competition for our declining natural resources increased in the first decades of the 21st century, this tolerance of, if not preference for, more authoritarian governments appears to have accelerated. By 2012, the level of global democracy had peaked and then fallen back to 1989 levels [1306].[190]

Most of us resist cultural changes. (I'm still acclimating to soft drinks served with lime rather than lemon.) Many traditions promoted survival in our past, improved group cohesion, and stabilized societal power structures. Still, our ability to adjust our behavior and acquire skills throughout our lifespan has let us adapt to shifting environments. Through these abilities, we have become the most invasive mammalian species on the planet. Our habit of occupying almost any habitat requires ongoing learning. It demands a joint use of our acquired knowledge and communicating what has and has not worked in the past. That collective

---

[190] In this analysis, the United States scored only in the top 10-20% of democracies, but it's status has slipped since 2011 with a significant trend toward greater autocracy [1306].

use of skills changed us. Supremacy through power and strength remains with us, evidenced by bullies and authoritarian dictators. However, our focus on *prestige* challenges Machiavellian manipulations and force as a road to recognition [1307]. Our preference for social skills over bullying and our kind's disdain for subjugation have created a feedback cycle, driving us toward more egalitarian foraging societies than our chimpanzee cousins [1113].

Until recently, we ignored prestige in other primate societies, presuming their dominance hierarchies and mating potential reflected strength. Anthropologists have begun questioning that assumption [1308]. In many non-human primates, females express mating choices that do not follow physical power, and in chimp societies, it's the womenfolk who determine who becomes the reigning alpha male. Frequently, they prefer less aggressive males who will guard their young or ones with superior social abilities, a prestige-like selection. These skilled males form stronger alliances than Brute-Strength Charlies [115], creating selection pressure against bullies. In short, other primates also differentiate between the most aggressive and the dominant.

We derive dominance from skills, intelligence, and culturally approved lekking behaviors. We flash our wealth to signal prestige, not brute force [1309]. We gain status and goodwill through empathy and generosity, not by hitting our supporters over the head with sticks. We also squash our rivals like bugs or target their offspring, often assassinating their characters rather than their bodies. We gain and hold resources through family ties and revolutionary coalitions. And from secrets among siblings to the proprietary information of corporations, hoarding knowledge remains an essential avenue to power.

Language acquisition ratcheted up our primate intelligence. Learning about our environment took less effort once we could access our collective knowledge and share information through words. With speech, we could warn each other not to eat the yew berries and explain that the seeds are toxic. We didn't need to learn through experience, observation, or trial and error. We could tell each other what did and did not work. We learned to muddy communication with verbal bluffs and lies.

Speech increased women's capacity to convey information about productive foraging sites and their patterns. This ability-sharing increased their contribution and survival value to their communities, contributing to greater egalitarianism in equatorial foraging cultures. In most traditional modern gathering and hunting societies, women gathered most plant materials, which comprise the bulk of their diet and probably did so in ancient ones. Equatorial foragers thus break the residency pattern, with many matrilocal cultures that take advantage of women's local knowledge [1310].

Language also endowed us with a powerful path toward prestige: verbal manipulation. Oral communication allowed the selective and judicious exchange of information. As needed, someone shared an experience with another band member or described encounters with outsiders. This strategy sparked a group memory that was easier to maintain than encyclopedic knowledge within each brain. This interdependence helped our collectives function. It enabled our ancestors to outcompete, fight, or kill competing groups of protohumans, much as chimps wipe out their competition today [960, 711]. During the Pleistocene, multiple *Homo* species[191] lived or evolved, and four species of *Australopithecus* lasted into the era. By 1.5 MYBP, we lost the last known *Australopithecus*; only one human genus, *Homo*, survived. By 30,000 BP, only modern humans, *Homo sapiens*, endured.

We may never know if we slaughtered or outcompeted these closest relatives or if they went extinct for other reasons. However, many organisms prefer to attack competing species that occupy a similar niche

---

[191] *Australopithicus robustus, A. africanus, A. aethiopicus. and A. boisei* arise during the Pleioscene and survived into the early Pleistocene (~2.6 MYBP-11,000). The earliest known members of the *Homo* lineage, *H. gautengensis and Homo habilis* come on the scene around 2.8 -1.4 MYBP. Not everyone agrees on the boundaries among lineages or which were separate species. We're pretty sure that the three most recent members of the *Homo* branch all originate with *H. erectus*. We know less about the other offshoots of our tree. As with other sciences, human paleontology is not set in stone. Science updates and evolves as new data emerge.

(intraguild predation). Ladybird beetle larvae, best known for eating aphids, attack younger (weaker) ladybird larvae of the same species and the offspring of other ladybird species [1311, 1312]. Brown widow spiders will hunt the related black widows [1313]. The ciliate *Tetrahymena vorax* can transform into a behemoth to prey upon rival *T. pyriformis*. Curly-tailed lizards vie with brown anoles and will eat them if their shared pickings become lean [1314]. Nematodes turn on their competitors when hungry [1315], and intraguild predation and rivalry occur among mammalian carnivores [1316]. Closer to us on the Tree of Life, chimpanzees will mount attacks on competing gorillas [1317] and confiscate prey from baboons [1318], a competitor whom they occasionally attack [1319, 1320].[192]

> *"I don't know [what weapons might be used in World War III], but in the war after the next war, sure as hell, they'll be using spears."* — attributed to an unnamed Army lieutenant, 1946 [3287]

Humans, bonobos, and chimps recognize their group members and know whom they can trust. Like our primate cousins, humans remain intolerant of strangers [1321] with easily triggered xenophobia [1305]. Today, we fashion more horrific weapons, but our human warfare looks strikingly like that of the chimps [960]. Like us, chimpanzees and other species who engage in group-on-group violence do so when a substantial power differential permits one group to attack another with a high probability of success [246].

Like many other social animals, ape societies organize along a power and prestige hierarchy.[193] We put faith in powerful, self-assured authorities and take collective actions on their behalf. We accept the decisions of our superiors as reasonable and justified when he or she displays confidence. Before the first Iraq war, American President George H. W. Bush's ability to portray certainty of victory increased the national levels

---

[192] Researchers have rarely seen predation on baboons by chimps. As baboons have become more common in chimpanzee territory, the level of predation has increased, but chimps still prefer hunting other species. This predation on a guild member may be nothing more than opportunistic hunting.

[193] Not all social animals use a hierarchical society. Bees and wasps, for example, have a caste system, but the jobs performed are set by the conditions under which they are raised. Queens perform a reproductive function for the hive but do not rule it.

of optimism; statistically, such positivity increases aggression and risk-taking by troops [1322]. The faith conveyed by President Vladimir Putin about the outcome of the Ukraine Invasion rallied the Russian forces in the first days of the Russia-Ukraine war; the expressed confidence of President Volodymyr Zelenskyy and the Ukrainian government fortified the Ukrainian people, helping them resist the invasion.

In social systems, being an alpha isn't about bullying; it's about getting your way more often than not. Gaining the trust, respect, and cooperation of your fellows is a good way to achieve status. As a high-ranking individual makes choices, it bears responsibility for the welfare of its troop, and wise alpha males and females understand that a contented troop fosters their retention of power.[194] A male chimp who leads his community into danger or to feeding patches of questionable quality is not likely to maintain the confidence of others and may become a target for a coup. A dominant male who pushes the females beyond their endurance will lose control.

Far from being passive possessions in chimpanzee troops, females contribute to many of the decisions, including when to rest, stop to forage, or nest for the night. Bearing the primary responsibility and care for the offspring, female chimps can let their intentions be known by refusing to continue. In essence, they stage sit-down strikes, often initiated by the alpha female.

An alpha, be it male or female, human or chimp, who rules via terror and intimidation without reciprocity or who does not aid their troop, invites defection and encourages their community to transfer their loyalty to other high-ranking members [1323, 1188]. An American Marine tradition holds that leaders eat last—after serving the enlisted men [1324]. In many primate societies, a male will defend babies born after he joined the troop. In Hanuman langurs, with one male per troop, the resident male protects the infants born on his watch [1325]. In multi-male communities, female primates typically show a mating preference for males who forge alliances, demonstrate superior social skills, and secure the troop and her offspring. These same abilities promote the attainment and retention of power [115]. Should the community lose confidence in a high-ranking member, defections to an opposing coalition skyrocket.

Interdependence with our leaders also increases expressed trust in them and our willingness to do their bidding [1326]. For a chimp, this hierarchical behavior includes accepting an arbitration of fights by the alpha or his allies and the willingness of the dominant animal to protect the troop. In nature, one-sided kleptocracies don't work. Male chimps hunt for meat; sharing it strengthens social bonds, aids related community members, and, like me with my dogs, stops persistent begging [1327].

We understand power at this primal, biological level. The standard ploys of authoritarian regimes include creating an enemy only the leader can protect against. The Nazis portrayed the communists and Jews as threats only they could eliminate. North Korean leadership warns its people about imminent invasions by South Korea and the United States. Before the invasion of Ukraine, Russian propaganda portrayed Ukrainians as war-mongering Nazis bent on genocide. During the first Iraq war, the American press was replete with stories about weapons of mass destruction and Iraqi soldiers killing Kuwaiti babies.

> *"The foreign policy aim of ants can be summed up as follows: restless aggression, territorial conquest, and genocidal annihilation of neighboring colonies whenever possible. If ants had nuclear weapons, they would probably end the world in a week."* — Bert Hölldobler, Edward O. Wilson [3126]

Among male mountain gorillas, an alpha gets access to the best food and other resources, but the ultimate payoff for being the silverback is access to sex. He has privileged access to females as they enter reproductive readiness. He exerts dominance through a complex of protection, affection, and aggression expressed through infanticide, bullying females, enlisting their aid, and attacking other males. To ensure their genetic survival, male orangutans who lack attractive facial flanges resort to rape [1328, 1329, 1330].

In chimpanzees, rank provides earlier and more frequent access to sex—and they are not above coercion to enhance their reproductive chances or their dominance [1331, 1332]. Chimp societies will wage war against other groups of chimps to access females and other resources. In human conflicts, the spoils of war have often included women. In theory, this access to females by victorious males provides triumphant male primates more chances to pass their genes to the next generation.

> *"The real crisis isn't buried in Epstein's black book, a flight log, or some list. It's in plain sight. The deeper, more pressing emergency is one we've been refusing to face for decades: the widespread, systemic epidemic of child sexual abuse in America."* — *Epstein, the Headlines, and the Children We Ignore,* Kathryn Robb [3220]

Unfortunately, we have poor data on how widespread rape is among human males. We do not know if a majority of men are capable of rape or if serial offenders commit most of these hate crimes. We know we differentiate among rapes. As with murder, kin selection comes into play. Our tolerance of the deed depends on the target. A kinship group tolerates forced copulation by one of its members more if he directs it outward, away from the populace. Hence, many men who would never consider sexual assault acceptable in their community will commit the atrocity during a war or genocide aimed outward against another group [1333, 1334, 1335]. They will rape "outsiders"—disenfranchised or despised people viewed as outside or fringe to a culture and beyond moral consideration [1336, 1337, 1338]. This distinction and hesitance to attack related women have exceptions: marital and date rape are far too common, even though we consider the behavior abhorrent. We share this reluctance with our close primate relatives.

Forced copulation among our non-human primate relatives happens, but it is comparatively rare. The only species in which we regularly observe it is the orangutan. Gorilla males use their size to threaten other males and could use it to bully much smaller females; instead, they appear to have an inborn inhibition against attacking the ladies of the troop. Go figure. Who would have thought that intimidating and alienating females would be an ineffective way to maintain control of a harem or to ensure a regular supply of willing sex? The females revolt, and the bullying male finds himself displaced. Bonobo females also form coalitions to correct such behavior in sexually bullying males and have been reported to have killed them, and even among the more sexually aggressive chimpanzees, attempted forced copulation may have limited success [1339].

Among humans, rape is far too common, with an estimated one-in-six American women self-reporting the trauma of rape within their lifetimes [1340]. Our faceless society provides avenues for rape that traditional societies lack, as does the adulation and praise we often give bullies in general. (Think of all the books published on how to be a bullying "alpha" to get what you want. Spoiler alert: In the long term, it doesn't work as well as coalition building.) Unlike other primates, women in our society typically live with one adult male rather than in a communal setting, providing further opportunities for sexual bullying not generally available in other primates.

Traditional human hunting and gathering societies would have experienced similar pressure, with bullying and sexually aggressive males experiencing loss of face, physical violence, or even death or expulsion from the social unit. Once alone, the bully's prospects for survival drop to near nil. As the old adage of anthropology says, a lone primate is a dead primate. So, yes, forced copulation can be a natural, if repugnant, reproductive strategy, but in a traditional society, the social cost of such a within-group action would exert selection pressure against the behavior, minimizing any adaptive advantage. Rather than raising the perpetrator's status, it would render him a cursed anathema, decreasing his chances to form coalitions, attract mates, or even survive.

Bonobos (and possibly lowland gorillas) break the pattern of male-male competitive violence common in other non-human apes.[195] Females can dominate males, often through coalitions. Anthropologists measure

---

[195] Lowland gorillas form another possible exception. We know little of this animal, which was only recently recognized as distinct from the more thoroughly studied mountain gorilla. Lowland gorillas have varied interactions with neighboring troops, from aggression to various degrees of tolerance. Some reports describe peaceful encounters with

status by recording deferential feeding and submission gestures [1341, 1342]. Higher-ranking females can access prime mating partners and gain control of other resources [1343, 762]. Human dominance behaviors fall somewhere between those of the chimp and bonobo. We have dominant males, but like our other generalized behaviors, dominance is not absolute. Both bonobo and human women make choices for their communities. Female bonobos can decide where to feed or nest. Similarly, in many hunting-gathering societies, women (the gatherers) decided where to forage for the plant foods that comprised most of the diet and physical culture [1344].

Humans and apes compete for resources beyond sex, such as food, grooming, and aggression appeasement [1343, 1341]. Our two closest relatives, the chimp and the bonobo, respond differently to stress from such competition. High-ranking members of both species show reduced stress hormones (cortisol) before sharing food. In both apes, cortisol spikes in a lower-ranking animal if a dominant male gets more food, a reaction akin to our sense of equity and fairness.

Differences in their hormone levels demonstrate that the two species experience stress differently. For the bonobo, the incident of social inequity included a higher spike in cortisol levels than in chimps, indicating bonobos perceived the encounter as a more potent stressor. To the same unfair treatment, the chimp showed increased testosterone levels, a response associated with perceived dominance testing [1345].

No one likes being on the bottom rungs of society, but for a social animal, low positions trump the slow death from living alone. For the hope of inclusion, our species will sacrifice much personal autonomy, including our professed morality.

---

neighboring troops and sharing common nests. Recent shifts in highland gorillas to multi-male troops correlate with increased killings due to within-group coalitionary conflicts [726].

The Nuremberg Defense and The Eichmann Experiments

"Befehl ist Befehl." (An order is an order.) — Old German military saying

Different forms of mass violence, from warfare to bullying, give its participants a tactical advantage. In a group, we share dangers and rewards we cannot access alone. The risk of harm to any individual decreases in a mob attack, so we benefit from a higher degree of safety than when attacking alone. Other mobbing species, from wasps to ravens to pack-hunting carnivores and warring chimps, exploit the power of a group when staging attacks [1346]. Apes, hyenas, lions, and wolves draw upon their hierarchies to coordinate attacks. Mixed herds of mammals, such as wildebeest, gazelles, and zebras, will take advantage of and respond to their different species-specific sensitivities to signs of danger. Organized societies can divide the labor (and responsibility) of an attack as effortlessly as we distribute the work of farming or gathering. Calorie for calorie, an attack can be more cost-effective than a peaceful assembly or an honest negotiation.[196] Despite their inherent inequalities, in groups, we draw upon each other's strengths, and healthy hierarchies promote social bonding, cooperation, group efficiency, and safety [1347, 1348, 1349, 1350].

The 1956 signage at Wounded Knee was titled "Site of the Wounded Knee Tragedy;" in 1973, the title of the new sign proclaimed it a "battle." Both signs describe the carnage. Only the later sign uses the word "massacre."

Social species can organize with flexible or linear hierarchies for power, control, or job allocation. Wasps will attack nestmates who sport a coloration that boasts of too high a rank. In chickens, the pecking order determines whom one can peck and who can do the pecking. Vigorous puppies push runts away to gain more milk for themselves. Adult dogs and wolves know their place and function in a pack. Black eagle parents stand idly by as one of

[196] Attacks may look costly, but one must consider the damage inflicted on the competing party and the relative change in status. Negotiation often reduces the relative distance between parties. Attacks typically increase the comparative resource (wealth) or status of the aggressor.

their two offspring kills the second one; when they were young, they, too, killed their nestmate. In these cases, survival depends upon one's ability to endure peer pressure and violence.

> *"Although no federal law directly addresses bullying, in some cases, bullying overlaps with discriminatory harassment when it is based on race, national origin, color, sex (including sexual orientation and gender identity), age, disability, or religion. Federally-funded schools (including colleges and universities) have an obligation to resolve harassment on these bases."*
> — US Federal laws on bullying [3300]

Human societies also organize through hierarchies, and we have an inborn craving to belong and a tendency to yield to authorities when we perceive them as legitimate. We can challenge an unfair or unjust leader but often defer to the one in command. If that authority bullies other people, we usually say nothing. In the Nazi-occupied lands of the 1930s and 1940s, a relatively small number of dictatorial leaders manipulated a substantial portion of the population. Under the pressure of the Nazi governance, people committed genocide or turned a blind eye as it happened. In Azerbaijan and Turkey, the same thing happened during the mass murder of Armenians. In Rwanda, a few authorities orchestrated the mass murder of the Tutsi people. The Cultural Revolution in China saw the systematic elimination and harassment of those associated with its imperial past. The government of the United States systematically committed genocides against the First Nations of the land. In Russia, Stalin systematically starved Ukrainians [1351]. It took a handful of people to turn a high percentage of civilized societies into butchers. For each one who merely "followed orders," many more people stood idly by or cheered as the massacres took place.

We read about atrocities and wonder how anyone could partake in such events. We shake our fists in moral outrage and cry for the victims. We think of the Shoah, the massacre of seventeen million human beings, and position ourselves as morally superior to the perpetrators. Our mirror neurons fire, making us confident we would take a stand and refuse to participate, no matter the cost.

Perhaps we should not sing our own praises too hastily or presume an advanced conscience. A primate drive to conform to social pressures and dominance remains with us, even when those forces demand we demean others or commit violence. We will bully others when our community approves. Social coercion can push a schoolroom of children to torment and humiliate a peer. If the class believes they have the support of the alpha—in this case, the teacher, monitor, or principal—the group receives tacit permission to proceed and amplify the torture. When political authorities express racist, ageist, or sexist remarks, they give voiced consent for their followers to take action against those groups.

In contrast, an alpha's condemnation of a behavior inoculates against it. When a leader steps forward and denounces the abuse, the bullies tend to reduce the severity of or cease the attacks. Our mirror neurons and primate deference to authority allow our alphas to reduce bullying by modeling consideration and compassion [1352]. The inoculation declines when the intervener is removed. Bullying may continue away from the school or workplace. Still, compassionate leadership provides a haven from torment, at least for a while.

In the wake of the Second World War's atrocities, people craved reassurance. They wanted to know what kind of person could have followed the commands to enact a Holocaust.[197] From 1963 to 1974, what kind of person would, like Adolf Eichmann and many other Nazis, just follow orders to torture? Stanley Milgram and his Yale University team explored who could obey inhumane orders to torment or kill in what have since been deemed the "Eichmann experiments" [170, 1353, pp. 56-57].

---

[197] The idea that someone bears no responsibility for following orders given to them by an authority (the "Nuremberg Defense") goes back centuries. During the Nuremberg trials that followed the collapse of the Nazi government, many people, most notably Eichmann, pleaded "Befehl ist Befehl" (An order is an order). Over time, this defense has had inconsistent results. During the Vietnam War, American conscientious objectors reversed this reasoning, refusing to participate in an action they believed to be criminal.

Milgram's crew asked subjects to participate in a teaching-learning experiment. The researchers would hook a "learner" to a device that doled out electric shocks when the "teacher" threw a lever. If the student made a mistake, the teacher administered an electrical shock by pressing a lever on a panel. With each error, the instructor advanced one toggle, subjecting the student to a greater voltage. The labels beneath the first switches on the control console read "slight." Before the experiment began, the research team subjected the teacher to a low-level shock so he would experience how it felt. The toggle descriptions progressed in a graded series from slight to extreme intensity. The ones at the end of the series bore the warning "danger: severe shock." The last three switches beyond the danger zone bore the label "x."

The subject believed he had randomly received the part of the instructor, but Milgram's team had determined his role from the beginning. The researchers showed the teacher how the student was restrained and how the electrical leads were attached. Once they escorted the teacher to his post in an adjoining room, others released the student and played recorded, scripted responses for each administered shock.

They escorted the instructor to a different room, where he could not see the student but could hear him (the recording) through an intercom. Each time the teacher threw a switch, it elicited grunts or complaints (the recorded responses). These ranged from a simple "ouch" to pleading and frantic cries from the student who screamed about having a heart attack [1353, pp. 56-57]. The research team scripted every phrase and reaction used in the experiment, from the researcher's responses to the student's cries.

Miller's Eichmann experiments did not employ threats or coercion. When someone refused to continue giving shocks, the experimenter made a straightforward, non-threatening statement, such as "It is important for you to continue." The pressure to conform came from a researcher who reassured the teacher that the shocks inflicted pain but would not cause permanent damage. This kind of prompt is known as a "demand characteristic." A better term might be "social expectation." Milgram's team never made demands of his subjects, but they responded to his prompts with obedience.

Before these Eichmann tests, both psychologists and laypeople predicted few would participate beyond a mild shock—probably no more than 0.1% of subjects. Surely, no one other than a sadist or a psychopath would proceed into the range of strong or severe shocks.

Despite their compliance, the participants expressed distress. They swore, perspired, cursed the experimenters, and chain-smoked. The teachers complained and expressed various levels of hesitancy and resistance but submitted to authority, administering shocks they believed could kill the student. Milgram expected obedience from a minority of psychopaths. Instead, 63% to 65% of the participants complied with obedience up to the lethal 450-volt setting. Far from being mere automatons, the subjects complained, cried, bargained, and begged to stop, but most conformed to the request [170, pp. 375-377, 1354, p. 11–12]. Our primate deference to hierarchical power prevailed, leaving the subjects feeling disempowered, defenseless, and unable to say "no." With the blessing of the alpha, they exercised their power and tortured the student with minimal prompting.

For any women who may feel smug, Milgram's team found no difference between genders in their willingness to torture another [1353]. Nor has our innate capacity for cruelty or compassion changed since the original Eichmann experiments [1355]. We continue to delude ourselves, believing in our moral superiority over the next guy. We evaluate others in the light of Milgram's experiments but see ourselves as the exception to the rule [1356].

Our willingness to harm another increases under the orders of an alpha. Our mirror neurons and the empathic regions of our brain, including the insula, become less active [1357], and we feel less guilt about our behavior [1358, 1359]. Post-experiment debriefing interviews revealed that the subjects saw themselves as sharing a goal with the researcher. This perspective shifted their self-image from one of violence to a sense of virtue [1360]. Such cognitive rescripting extends beyond obedience and influences our willingness to accept falsehoods and emotional venom perpetrated by our leaders.

We share our propensity for behavioral and emotional contagion with other social animals, including wasps [1361, 1362], zebrafish [1363], songbirds [1364, 1365], elephants [1366, 1367], rodents [1357, 1368], and other primates [1369, 1370, 1371, 1372, 1373, 1374, 1375, 1376]. In humans, rumors, xenophobia, and discrimination follow this pattern, but so can playfulness and compassion, particularly if modeled from above [1377]. Milgram's subjects also complied with orders to stop the experiment and to *stop* shocking their victims (experiment 12 [1353]). Our willingness to obey prosocial directives provides a mechanism to mitigate behaviors. By expressing forceful public disapproval of rape, discrimination, and violence, leaders can change the world for the better, and betas can shame alphas. By modeling cooperation and pro-social behavior instead of conflict and hatred, a true pathfinder decreases the cost of standing for justice against the mob. The choices made by moral individuals provide a kernel upon which moral unity can develop.

> *" 'If you push the button,' Mr. Steward told him, 'somewhere in the world someone you don't know will die. In return, you will receive a payment of $50,000.' "* — "Button, Button" by Richard Matheson (first published in *Playboy*, 1970) [3426]

Although Milgram receives credit for these archetypal experiments on obedience, others paved the way for him, and others followed. Landis [1378] studied how emotions impact facial expressions, and he also performed a more shocking experiment, which would never be approved as ethical by today's standards. He asked people to perform feats that evoked strong emotions, including intense adverse reactions like disgust.

In one study, Landis instructed them to decapitate live rats. As in the Milgram experiment, people protested and complained. Landis bullied his participants into compliance, often through name-calling. Although the subjects had to restrain and execute the animal, they complied [1379]. People abdicated their will and morality to Landis as they caused the gory death of the fluffball while holding their victim in their hands.

Milgram found much variation among his participants, ranging from a few who refused to give any shock to those who progressed to the maximum value [1353]. His team uncovered four factors that reduced our willingness to commit violence against another. If the victim was in the same room or if the teacher made physical contact with the victim, compliance dropped to 30%. Today, we recognize the relative ease of pushing a button to launch a bomb that will kill hundreds or thousands of people compared to the relative difficulty of killing a single person who stands before us.

In the following years, people looked for conditions that could mitigate the violence done on command. The closeness of authority also shapes our willingness to commit an atrocity [1380]. Obedience to repugnant orders drops when the boss is away. When Milgram's teachers received instructions over the telephone, they did not elicit as much compliance as when an alpha remained in the room. This result, however, contrasts with a study done in a strongly hierarchical hospital setting. After receiving telephoned orders from a doctor to give patients a lethal medication overdose, nurses complied [1381].

People obey orders to harm less frequently as the perceived difference in rank between them and the alpha diminishes [1382, 1383]. Since the 1966 study of obedient nurses, our respect for nursing skills has surged. By 1977, Rank & Jacobson [1384] found that as the prestige of the nursing profession rose, obedience to questionable orders from doctors plummeted. The mere perception of legitimacy creates a sense of obligation that can override personal moral standards [1385, 1386, 1387]; a reduced difference in relative status can, in part, compensate for this effect.

Milgram also discovered that obedience dropped if a substitute power figure presented as having a lower rank. In one case, the instructor moved to rescue the student when he thought the orders came from illegitimate hands. 80% of teachers refused to obey a questionable substitute; 20% still followed orders.

One nagging problem remained for Milgram: the status of Yale University. He worried Yale lent an air of legitimacy to the study. The subjects' faith in Yale's morality could bias their results. Milgram removed the name of Yale, repeating the experiments at a shady-looking clinic in Bridgeport, Connecticut. The compliance rate dropped—slightly. From the initial 63-65% compliance at Yale, the compliance rate dropped to 48% at the Ye Olde Creepy Clinic. Half the population obeyed commands to cause distress and potential death to their fellow human.

For our social species, it doesn't take much to establish dominance and obedience; even a dubious celebrity status, such as a television game-show host, suffices to elicit confidence and cruel behavior in contestants, with 81% of the participants willing to administer the lethal shocks [1388]. Those who self-identify as conscientious and agreeable (conforming) also tend to deliver more shocks, as do those with authoritarian leanings [1389, 1390]. People who identify as non-conformists were more likely to stop earlier or refuse to administer the punishment [1391]. However, this effect may relate to how the participants direct their empathy toward the perceived legitimate authority or toward the victim [1392]. Thus, the capacity for compassion may not vary between these groups. Instead, empathy proved to be a two-way street. People can empathize with the victim or the experimenter. In these studies, most identify with the alpha, but personal histories can tip the scales. For example, in a Rwandan study (that modified the method), participants whose families suffered during the genocide were likelier to disobey orders to harm another [1393].

The perception of legitimacy helps explain why some nations, such as Denmark, Sweden, and Bulgaria, refused to participate in the destruction of European Jews [1394]; they did not see Nazi rule as legitimate. Countries with histories of pronounced anti-Semitism, such as Poland and Czechoslovakia, complied more readily because the orders fit a pre-existing paradigm. Compliance in Milgram-esque studies decreases when subjects identify with the student [1392]. The less social distance, the more alike they were, the less likely they were to administer a shock. This relationship jives with our species' focus on culturally learned green beards and kinship markers.

In the United States, for example, White subjects are less likely to shock students whom they perceive as belonging to the same ascribed race [1395] and more likely to shock those perceived as "others." Prisons create feedback systems that accentuate an us-them mentality by dehumanizing prisoners and increasing the acceptability of maltreatment [1396]. Extreme examples of this process include the Stalinist prisons in Poland, Abu Ghraib Prison under both Saddam Hussein and the occupying Coalition forces, and the Nazi concentration camp system.

Prejudices trigger responses in the amygdala, the ancient part of our brains that respond to fear and hate; these emotions increase our willingness to harm others [1397], and that willingness rises when we encounter someone we perceive as different from ourselves and consider them a potential threat [1398, 1399, 1400]. Two clusters of cells inside the left anterior insula correlate with authoritarian and social dominance orientations, stable personality traits that favor greater obedience to authority [1401]. Other persistent personality features, such as a willingness to adopt routines to avoid uncertainty, the desire to avoid unpleasant emotions (like confrontation), and cultural indoctrination, contribute to one's obedience or disobedience to commit a Holocaust [1402]. The activation of these areas in authoritarian personalities provides one physiological reason why people at the bottom of a social pyramid accept the rules, opinions, and orders of those at the top [1403, 1404]. Prejudice, hatred, and obedience to dictates from on high all impact activity in the insula (social processing and empathy), amygdala (emotional regulation, stress, and social processing), and anterior cingulate cortex (emotional regulation, social awareness, discord detection, and pain and reward processing).

Today, ethical considerations prevent us from replicating Milgram's exact work [1405] despite no objective evidence of long-term harm to the participants [1406]. The unexpected emotional suffering of the subjects in Milgram's first experiments leads most researchers to consider a direct replication unethical. However, from 1968 to the present, dozens of researchers have replicated the Eichmann protocols in modified forms and across numerous cultures, including volunteers who gave mutual shocks [1407]. Each well-designed experiment failed to disprove Milgram's results and lent further credence to those initial findings: When prompted, most of us will torture another human being [1408].

In 2009, Burger [1355] published his "150 volt-solution". In this abbreviated version, he used a weaker shock as his maximum; he studied the willingness to shock subjects rather than the willingness to bring the severity of the shock to a lethal point. He then extrapolated the higher levels of obedience based on Milgram's results. Despite over half a century of education about the Holocaust, despite our exposure to stories about blind obedience, Burger saw no significant reduction in obedience rates. Burger found the compliance rates in 2006 were only slightly lower than those found by Milgram 45 years earlier. It did not matter if participants saw someone else refuse instructions to obey, nor did gender matter. Women reported more discomfort with the procedure than men but complied at the same rate as Burger's male volunteers. Others have substituted psychological or sensory torture, each attaining similar results. Most people will obey.

Circumstances may not call upon us to refuse to enact blatant physical cruelty, but we may ignore the imposed sufferings around us. One classic study placed 40 theology students in a quandary where they had to choose between rendering aid to someone in distress or ignoring them. The situation reflected the biblical story of the Good Samaritan. Some of the students were rushing to give a sermon about helping; the control group hurried to talk about an unrelated subject. Their presentation topic made no difference to their likelihood of rendering aid. What mattered was how rushed they were and *how* they internalized their religious values [1409].

As the joint responsibility we bear for personal intervention decreases, we feel less influential and become disempowered to make decisions. As this personal agency drops, we grow less likely to take responsibility or become involved [1410, 1411, 378]. For example, when passing a woman in distress (actually a confederate of the researcher), people stop to intervene 70% of the time when alone. The presence of one other person drops that rate to 40% [1412]. The reduced probability of targeted help from others provides insight into why rural residents seem more inclined to render roadside aid to a stranger than city dwellers.

> *"My armed group killed 7 people. Sorry, it was actually rather 7 children and their mother, making it 8 in total."* — from an interview with a Rwandan genocide perpetrator [3091]

This same effect happens with obedience. If we can shift the blame, we become more inclined to harm others [1413]. When provided with a choice, participants in Eichmann experiments will shock less and take longer to shock than when ordered or coaxed [1407]. This change correlates with activity in the regions of the brain that generate empathy, including the insula, TPJ, ACC, and striatum. By shifting the blame for an atrocity, the commands of authority reduce activity in the brain regions rich in mirror neurons that let us empathize with the victim's pain and those that produce guilt [1359].

Bolstering the Milgram experiment, social psychologists have repeatedly demonstrated this "bystander effect." And, if we can rationalize that the danger isn't real, if the threat is remote, as on the internet, or if the victims suffer verbal or psychological dehumanization rather than a physical threat, we can rationalize our failure to intervene even more.

Sticks and Stones

> *"To separate them from others of similar age and qualifications solely because of their race generates a feeling of inferiority as to their status in the community that may affect their hearts and minds in a way unlikely ever to be undone. The effect of this separation on their educational opportunities was well stated by a finding in the Kansas case by a court which nevertheless felt compelled to rule against the Negro plaintiffs."*
>
> — U. S. Chief Justice Earl Warren [1414]

Not all violence involves guns or knives. Most of the time, the torturer does not use electric shocks or burning coals. Instead, he or she uses words. More often, bullies manipulate and torment us through a psychosocial approach that alters our physiology and our ability to regulate it [1415]. Hate speech, the constant inference of inferiority, or the constant imposition of an inferior status can destroy the self-image of an individual or a group. They learn they are worthless, and suffering defenselessness affects events in their lives.

Figure 7-2.

Following the First World War (pre-Nazi period), many emergency currencies included debasing code words and images that built upon a history of European antisemitism. They encouraged dehumanization and violence against Jews. These four examples above are typical.

Upper Left: A notgeld (Tostedt, 1921) portraying the lynching of two Jews. The casual portrayal suggests such violence was acceptable. A 1939 Nazi publication makes direct reference to this note [3353].

Upper Right: The dunking of the Jew (Brakel, 1921 series) depicts a Medieval dunking of a Jew and uses antisemitic code words ("money lender/price gouger," and "profiteer") to call for the persecution of Jews in revenge for crimes against Germany.

Lower Left: (Sternberg, 1921). The culmination of a notgeld series depicts the 1492 burning of 27 Jews accused of stabbing a host by puncturing it with nails and knives till it bled.

Lower Right: Shaving David (Beverungen, 1921 series) plays on a German expression (to shave someone over the spoon = cheat them) and depicts an image of payback. (Text: "This time, you have been shaved over the spoon.")

Dehumanization distorts hormone production and processing. Cortisol rises, opening us to stress. Chronic stress increases our levels of adrenaline and its sister, noradrenaline, hypersensitizing our emotional center (the amygdala) and triggering the release of more cortisol, which harms memory formation. Although stress impairs memory formation, the brain treats those memories that do form as critical for survival and resists altering them. The brain can extend that fear to create a more generalized emotional paralysis [1416]. Serotonin levels drop, plummeting our mood, and in our prefrontal cortex, its reduced quantity impairs our ability to think and problem-solve.

With the aid of these hormones, chronic powerlessness rewires bits of the brain. The prefrontal cortex is the center of decision-making and problem-solving, two functions whose impairment virtually defines helplessness [1417]. It also regulates our willingness to conform [1418]. The repeated emotional trauma of imposed impotence alters the structure of the amygdala and hypersensitizes it, leaving it more open to fear and stress [1419, 36]. Repeated ordeals impair the hippocampus, a memory processing unit, leaving us less able to access positive solutions to problems. Activity also drops in the nucleus accumbens, a region that anticipates rewards [1419] and one functionally tied to the amygdala by major information highways of nerve tracts [1420].

This learned helplessness removes one's sense of efficacy [1421, 1422]. When a person, animal, or group experiences the constant stress of being at the mercy of powerful others or of fate [351], they become increasingly unwilling to take chances and suffer from continuous hopelessness [262]. Even the denial of promised compensation for a job or task creates an inequity that alters mood and fosters a sense of vulnerability [1423]. The bullying of employees costs corporations billions of dollars in lost ideas and productivity each year. Employees cease contributing, realizing they gain no recognition for constructive input—or worse, they are denigrated for their efforts, often in front of their colleagues. However, professional and managerial bullying can cement or advance the bully's position by eliminating his competition, improving his relative power and status, and providing scapegoats for his failures at the expense of the corporation.

A perception of powerlessness creates a feedback loop. We value traits and achievements we recognize as under our control. We give ourselves little credit when we hold an external person or force responsible for an accomplishment. When caught in the biological, psychological, and social spiral of learned helplessness, we devalue more and more of our abilities and worth [1424]. The sense of worthlessness intensifies hopelessness and self-blame, further devaluing our self-worth [1425].

> *"[We] admitted we were powerless over alcohol—that our lives have become unmanageable"* — First Step, Alcoholics Anonymous [3265]
> *"I have a life-threatening problem that once had me."* — First Affirmation, Women for Sobriety [3157]

Physically and emotionally battered women show the core features of this lost control [1426, 1427]. Disenfranchised status contributes to the internalization of powerlessness. Children who endure adverse childhood experiences, who struggle with name-calling, verbal abuse, rejection, or exclusion by their peers, also show long-term learned helplessness that extends across their lifespans [1428, 1429]. Indeed, social exclusion shifts the locus of control from internal to external [1430, 1431], making them more subject to this imposed inefficacy.

> *"As I said, I may have used poor judgment. Perhaps I should have taken angry or even militant steps, both when I was in the agency, or after I left it. But I must confess to the world that the course that I took seemed the better as well as the easier approach."* — from Anita Hill's Opening Statement to the Senate Judiciary Committee about sexual harassment by Clarence Thomas, 11 October, 1991. [3387, p. 26]

Childhood bullying imposes some of its most devastating transformations by removing the sense of safety at school [1432, 1433, 1434]. It immerses the target in a psychologically hostile environment throughout the school day and often through all peer interactions. Learning becomes dangerous as the bullies pounce on errors and successes [1428]. The student has no practical alternative to their self-deliverance into the punishing school environment day after day.[198] In theory, adults subjected to workplace bullying can quit and seek employment elsewhere. However, fear, doubts about prospects, and learned powerlessness can keep subjugated underlings working in a demeaning, dehumanizing environment with a bond stronger than iron chains.

Lest you think this is an excellent way to retain employees, remember how behavioral modeling works. When the first employee rebels, others become more likely to follow. When Rosa Parks refused to

[198] A parallel but more extreme situation existed in some slave-holding areas before the American Civil War. Slave owners could require a slave to visit a specialist for punishment. The enslaved person had to comply and report to that person or place for whipping or other torture. The rationale behind the practice was that household members knew each other and, therefore, would not be severe enough to administer the chastisement.

surrender her bus seat to a White man in 1955, the woman sparked the Montgomery Bus Boycott. Anita Faye Hill's testimony about sexual harassment by Circuit Judge Clarence Thomas led to thousands of women coming forward about harassment in the workplace and a movement that made workplace gender violence less acceptable, and in multiple countries, regional populations, shocked by the vehemence of Nazi anti-Semitism, led to resistance to the deportation of Jews and forced a local moderation of the anti-Semitic laws by the Nazi governments [1435].

Abusive treatment and loss of core efficacy correlate; the inverse also holds true. An internalized sense of control provides the confidence to impact one's life. The empowered person is more likely to see fate as contingent upon his or her behavior [351]. The more predictable the circumstances leading to a sense of control in one's life, the more confidence one has in exercising power within those parameters. When the boundaries for success become broad and unambiguous, a person internalizes failures less. Conversely, as the limitations caused by exploitation become erratic, a socially savvy abuser can move the criterion for success to increase or destroy the confidence of his or her target, manipulating the sense of failure.

Figure 7-3. The Jolly Darkie Target Game, by McLoughlin Bros., ca 1890. This company also made a racist puzzle titled "Chopped- Up N-s" (ca 1875). The company went out of business in 1921. Racist objects like these give insight into the ubiquity of dehumanization and other attitudes from the not-so-distant past. [3291].

When a person of influence prompts the action, people will attack another's self-efficacy with words as well as violence [1436, 1437, 1389, 1390]. Upon the request of a power figure, people inflict hateful and demeaning speech against others. In one experiment, the subjects were instructed to inflict stress-inducing words on someone taking a test for a job interview [1426]. The willingness to engage in demeaning speech did not differ from the willingness to inflict physical pain in Milgram's classic studies. Even with a remote authority and a nearby victim, thirty percent of people escalated, increasing their hostile speech [1390].

Humans use bullying techniques to improve their relative status by reducing the efficacy of the competition. Job dissatisfaction and poor working conditions create situations where peers turn against each other. People direct their dissatisfaction toward their co-workers, insulating their powerful bosses. Researchers looked at this effect on nurses, a low-status job compared to doctors, within the strict medical hierarchy at an American hospital [1381, 1438]. Peer bullying amplified job dissatisfaction by reducing their sense of internal control and power. This peer-on-peer verbal violence decreased morale and reinforced the internal pecking order.

The results are not culture or WEIRD-specific. Researchers repeated this study in Chinese [1439] and Spanish [1440] hospitals and obtained similar results. During the early days of the COVID-19 pandemic, supervisors prone to demeaning and disruptive behaviors often doubled down on their bullying. These tactics increase error rates, reducing job satisfaction and jeopardizing patient safety [1441].

The results are not specific to nursing. Stress, loss of efficacy, and burn-out rates skyrocket wherever supervisors degrade and minimize the value of their workers, students, or volunteers, or when they fail to control peer-on-peer bullying [1442, 1443, 1444]. It happens to athletes and parents, people receiving pay, volunteers, and people who once loved their work. Burnout has little to do with employment status. It has its roots in a lack of respect and recognition, in a reduced sense of accomplishment and purpose, and in exhaustion and alienation [1445].

In the early stages, the body releases more adrenaline and the cortisol family of stress hormones. It prepares to cope with fight-freeze-or-flight options by increasing blood sugar and pressure levels. The physiology is maladaptive in a long-term setting with no escape, such as an office or school. Nature designed this system for short-term responses, not long-term adaptations to hellish environments. In time, the stress overloads the hypothalamic-pituitary-adrenal (HPA) axis that regulates these hormones. The body adapts to the increased adrenaline levels and becomes desensitized to it. These responses contribute to the primary symptoms of burnout, including depression, fatigue, and difficulty concentrating. Early life traumas create long-lived changes in this pathway, making the victim more sensitive to later trauma and health problems [1446, 1447].

> *"Give the lady what she wants"* — Motto of the Marshall Field Department Stores, 1852-2006

Volunteers have notoriously high burnout and no-show rates. Ones working with drug addicts ranked among the first recognized cases of burnout [1448]. Typically, these unpaid aids hold the lowest status within organizations and lack social support. Often, they have no clearly defined role, perform monotonous tasks, and have skill sets other workers seldom recognize [1449]. Lacking integration within the organization, they suffer depersonalization and satellite status [1450]. Supervisors often assign them monotonous tasks, and they receive token recognition or credit for their contributions. Organizations that wish to retain their volunteers, workers, and consumers should remember respect, honor, and chances for accomplishment are core retention strategies that apply to all [1451]. For salaried workers, adequate, fair compensation and a reasonable differential between their salaries and salary increases with those of the CEOs is a critical component of worker respect, as is the principle of equal pay for work of equal value. Recognition matters.

Because I Can

> *"Each prisoner has a 'contact officer' who monitors and helps advance progress toward return to the world outside—a practice introduced to help officers avoid the damage experienced by performing purely punitive functions: stress, hypertension, alcoholism, suicide, and other job-related hazards that today plague American corrections officers, who have an average life expectancy of 59."*
>
> — Doran Larson, *The Atlantic*, 2013 [1452]

Like many social animals, people exercise power over others for a simple, circular reason: they can. Prisons overflow with good people who entered the criminal justice profession to serve society and support others. In traditional correctional facilities, the power differential between guards and prisoners combines with their different roles. Though starting with the best motives, guards can fall prey to the prison norms and begin to see their charges as less than human. Accepting the role of a prisoner or guard changes a player's perceptions and behaviors in this game of power.

In a disturbing study, Philip Zimbardo [1453] randomly assigned college students to play the roles of inmates and guards in a mock prison. Once chosen, the guards met with Zimbardo and his team. They prepared the guards for their experience and, as in Stanley Milgram's obedience study, prompted them to fulfill their roles, encouraging behaviors that would escalate the violence faster [1454, 1455].

Like Milgram's "teachers," the "guards" knew what they must do to fulfill their part in the study [1456]. More than their role conditioned the guards. They also identified with the "wardens" (researchers) as leaders and conformed to their expectations [1457]. The "guards" received authority symbols: uniforms, whistles, truncheons, and reflective sunglasses that obscured their eyes. The others were locked in barren cells and wore pseudo-prison garb. Their long shirts and lack of pants heighten the power differential. Upon arrival at the facility, they were blindfolded, strip-searched by the guards, and assigned identification numbers.

Self-identification with their roles happened fast. About one-third of the student guards (high status) took less than one day to morally disengage from the "prisoners" and to turn sadistic. They developed a pejorative, dehumanized view of the prisoners, even though they had arbitrarily received their position of power. The mock guards devised humiliating and callous punishments for the "detainees." Their prisoners showed emotional and behavioral changes congruent with being bullied. These student volunteers became apathetic. They cried and developed depression. On the second day, an inmate rebellion happened; harsher punishments followed. Ultimately, the prisoners acquiesced to the sham authority and their vulnerable position.

Far from remaining an impassionate observer, Zimbardo and his team fell into the role of wardens [1453, 1458, 1459]. Rather than releasing prisoners who wanted to end their participation, the wardens began to interpret their desire to leave as attempted prison breaks. They started to minimize or cover up the abuses. The repercussions of their leadership roles altered these mock wardens' perceptions of their identities and social situations. Like the "guards, they began treating the experimental prison as if it were real.

Zimbardo intended to run the prison experiment for two weeks. It broke down after six days. Some students who played the role of guards had become increasingly cruel and sadistic, demeaning and debasing the "prisoners" in their care, while others showed signs of acute traumatic stress [1460]. Those students who served as prisoners became dehumanized and consumed by mounting hatred for the guards. In those six days, *"human values were suspended, self-concepts were challenged, and the ugliest, most base, pathological side of human nature surfaced"* [1458, p. 3]. None of the participants expected the prison experiment to have such an incredible effect so quickly or the random circumstances of each group to have such power [1458, 1460].

Due to its unexpected fallout, no one has attempted to replicate the study because of ethical considerations. However, naturally occurring events relate directly to it. In the United States, conflicts between police and disenfranchised communities, most notably African-American and Latinx communities, partially replicate these conditions. The animosity on both sides mirrors the results obtained in the Stanford study. Suspicions and disdain for the other side intensify with each negative encounter.

> *"I have become convinced that we do not have a race problem in policing. Rather, we have a race problem in society that is reflected in policing."*
> — Seth Stoughton, former police officer and law professor at the University of South Carolina. [3201]

The American history of extrajudicial murders (lynching) has left a legacy of suspicion. In the 21st century, mobs seldom attack jailhouses and pull out prisoners to torture and kill. Social norms no longer accept such direct execution of "justice." However, extrajudicial killings still happen. Trayvon Martin died at the hands of a neighborhood watch coordinator on Feb 26, 2012. The shooter was suspicious of a Black teen in his area; he called the police, confronted Martin, and shot him following an altercation. Trayvon's shooter claimed self-defense and was acquitted.

An off-duty Chicago detective shot Rekia Boyd in Chicago on March 21 of that same year while she partied in Douglass Park with friends. The shooter claimed one of the people had a gun; no gun was ever recovered. He was charged with involuntary manslaughter. Judge Dennis J. Porter directed the verdict. Because the shooting was intentional, it could not be involuntary [1461]. The prosecution could not change the charges to first-degree murder because it would constitute double jeopardy.

Eric Garner died in a prohibited chokehold on July 17, 2014, in Staten Island after being accused of selling individual cigarettes from packs that lacked tax stamps. On August 9, 2014, the shooting of Michael Brown, after he was accused of "robbing a convenience store" in Ferguson, Missouri, sparked nationwide protests. Mr. Brown came to the attention of the police for stealing a box of cigars. George Floyd (May 25, 2019) was accused of making a purchase with a counterfeit $20 bill. Seventeen minutes after the first squad car arrived, Mr. Floyd lay unconscious, pinned to the ground with a knee on the back of his neck. The paramedics pronounced him dead [1462]. Character assassinations followed that claimed Mr. Floyd had a history of armed robbery. He did not, but he had a record of minor drug possession (less than half a gram of cocaine) and petty theft, neither of which warranted the death penalty.

In these last three sample cases, the police responded to a report of a minor, non-violent crime: selling single cigarettes without a tax stamp, stealing a box of cigars, and passing a forged $20 bill. None of these should carry the death penalty. Yet, identifying with the police, community, or victim (the us-them divide) and the precipitating incident shape and distort our impressions of shootings [1463, 1464]. Those who believe race influenced the decision to fire a gun are more likely to see the shooting as unreasonable [368]. Two-thirds of Americans believe that the police will use force against a Black American faster than against a White one; 79% consider the problem serious [1465]. About one-quarter of people killed in police shootings are African Americans, approximately double their representation in the population. In 2022, Black victims comprised 34% of police shootings where the victims turn out to be unarmed, about 22 per year [1466]. Although the figure seems low relative to the total American population, the disproportionate numbers provide strong evidence that race factors into enough police decisions to warrant attention, and it is implausible that the police are the only decision-makers who experience this behavioral bias.

In the wake of the George Floyd killing, Whites and Blacks agree that police violence is a serious problem; they disagree on how serious it is. 83% of Blacks rate the danger “extremely serious,” and 14% moderately so. 39% of Whites, who are impacted less by police violence and more likely to identify with the police, consider the violence to be “extremely serious,” and 35% rate it as “moderately serious.” 26% of White people and 21% of the overall population do not consider the problem severe [1465].

> *" 'You're missing work, you're getting arrested, you're having to find a bondsman, because you can't use your property [for a bond],' Jackson said. 'They said mine [her bail] was going to be $2,500. Who's going to have $2,500 when you don't have $20 to pay your trash bill?' "* — on debt Peonage in Vally, Alabama, D. Fatherree, Southern Poverty Law Center [1467]

The Zimbardo studies and kin selection provide insights into the long-standing situation. A positive feedback cycle of suspicion exists between the African-American community and law enforcement, with each negative encounter spiraling the tension upward. Through much of the 20th century, many police departments openly suppressed African-American civil rights and liberties. Many states enforced peonage laws that repackaged and reinstituted enslavement under a different label [1259]. As of 2023, there are reports that Valley, Alabama, still jails and fines debtors for overdue municipal bills despite its prohibition by state and federal laws [1467]. This history creates the first self-identified "kinship" group, African Americans. The "brotherhood" of the police forms the second family group. Like anyone, they will protect themselves when answering calls in potentially dangerous situations. White officers, however, elevate their level of force more often than Black ones and use it more often in Black neighborhoods [1468].

> *"The degree of civilization in a society can be judged by entering its prisons."* — attributed to Fyodor Dostoyevsky

Enter alpha modeling. The expectations of commissioners and department heads (the alphas) can inoculate against violence. A clear message that the department will stand behind justified but not unjustified force, such as mandating the use of body cameras during confrontations, reduces the incidence of complaints and reports of brutality [1469].[199] The Milgram Obedience study demonstrates that we remain cognizant of our primate social structure. Group expectations and leadership from respected authorities can mitigate or enhance violence. Violence by the police, though the exception to the rule, happens often enough and is reported widely enough to reduce people's faith in them, making the

---

[199] This study occurred in Las Vegas (Force: 10% Black, 56% White). In Washington, DC (Force: 50% Black, 35% White), the use of force increased. The rules for using body cams also differed, making it unclear if the disparity in ethnic diversity between the civilian communities and police or the body cameras had an impact.

population less compliant with their orders and increasing the animosity towards them. Unlike the Scandinavian prison model, where the prison guards eat with and prepare prisoners for reintegration (noted at the beginning of this section), the traditional American model feeds an adversarial system that is challenging to break.

In light of these data, we can look through history and better understand the replete cycles of cruelty. Fredric Douglass spent his early years in slavery. He recounted the arrival of a kind northern woman as the new mistress of the plantation and her quick transformation into a "demon" [665, pp. 57-58]. White America's frequent animosity toward the Black Lives Matter movement across the past decade reflects Zimbardo's studies. It is mirrored in the stories of Nazi-era guards who openly wept upon arriving at the concentration camp but soon partook in the sadism. We see it in the indoctrination into some cults, and we see it in the phenomenon of burnout in health care and social workers who too often go from caring about their patients to seeing them as whiners and complainers. The attitude leaks through to their clients, who see these professionals as lacking regard and of little benefit. And, in a more hopeful vein, we can glimpse the force of social role-expectation in General von Hannecken and other Nazi officials and SS troops stationed in Denmark who adopted the local abhorrence of Jewish deportations and adopted Danish attitudes. These lost their zeal for Nazi-imposed anti-Semitism and even hindered the transports [1435].

The depression resulting from burnout and learned helplessness happens at the most basic levels of anatomy and physiology. It alters our body cells and the composition of our microbiome [254, 1470, 1471]. These tiny microbial symbionts live throughout us; with over 100 trillion bacteria alone, they outnumber our body cells [1472]. 95% of those cells reside in our digestive tract, an area prone to stress-related damage [661]. The microbiota plays a significant role in neuro- and hormonal regulation [1473]. The strength of these multitudes led us to conceptualize a "brain-gut" signaling pathway that steers our reward pathways. They shape mood, behavior, and even our proneness to addiction [1474, 1475, 1476]. They interact with our environment and genetics to regulate our responses to stress and bullying [1477, 1478]. People with different gut flora learn to recognize and react to threats at different rates [1479].[200]

This microbial behavioral system, like the microbiota, includes every system in the body. Gut bacteria affect stress disorders, depression, and impaired thinking [1480, 1481, 1482, 1483, 1484]. These symbionts influence our ability to form and recall memories (the hippocampus), make rational decisions (prefrontal cortex), and incorporate new learning. Our flora manipulates our social behaviors [1485, 1486] and, by extension, our acceptance and status within our primate community. And when these changes happen early in life, the gut biota can disrupt brain development, function, and behavior, reducing or enhancing the potential for status in later life [1487] and our propensity toward violence or victimization [1488].

Mitochondria are the intracellular symbionts that allow brain cells to burn sugars and fats. In rats, favorable social conditions, such as frequent grooming and high social rank, ups the level of mitochondrial DNA immune-system cells as compared to socially stressed individuals [1489]. Under social stress, the brain mitochondria malfunction, decreasing the energy needed for survival [1490, 1491]. In these altered power stations, cells store "memories" of past metabolic strains by altering their genomic expression [1492, 1493]. Mitochondrial changes link to social deficits, reduced sociability, and increased aggression [1494]. Their overlapping yet gender-specific expression in the amygdala influences how we process post-traumatic stress and why different genders express it differently [1495]. In the HPA axis, activity in this organelle impacts stress management and hormonal regulation [1493]. Stress affects neural mitochondria throughout the life cycle. In rat models, traumatized mothers will even pass the expression for mitochondrial damage to their pups [1496].

---

[200] Technical note: not all bacteria altered our threat reactions. Eight genera contributed to most people's responses. Active in most people were *Butyricicoccus, Agathobacter, Alistipes, Phocea*, and *Veillonella*. Three additional genera stood out in people with extreme reactions: a *Faecalibacterium* taxon, Tyzzerela_3, and Bacteroides. [1479].

Reduced metabolic function depletes our ability to regenerate and retain brain tissue. Rats who cannot escape shocks, no matter what they do, lose neurons in the hippocampus [1497]. In humans, this region shrinks and undergoes structural changes under traumatic stress, depression, and burnout syndrome [1498, 1499, 1500, 1501]. Helplessness decreases neurogenesis in this region, which generates more neurons than in other brain areas [1502]. This part of the brain processes short-term memories to produce and recall long-term ones [1503, 1504]. The hippocampus also promotes the brain's association between emotional trauma and specific triggers and regulates a principal hormonal system, our old friend, the HPA axis [1505]. Traumatic stress, such as social defeats or chastisements, also enlarges the fear network of the amygdala as the atrophying prefrontal cortex and hippocampus become less capable of regulating its output [607, 1506].

For years, people assumed an inability to escape punishment was a crucial component of learned helplessness; however, a perception of reduced capacity relative to others or a lack of social support may suffice to instill the same social paralysis [1507]. Punishers can evoke a sense of powerlessness *even when a subject can flee*. Social defeats decrease aggression and increase fear in Syrian hamsters despite escaping the chastising episode [640, 1508, 1509]. Repeated defeats in social encounters similarly overwhelm rats and hamsters, even if the animal obtains temporary relief through a retreat [1509, 1508, 1490], and social rank remains a crucial factor in submissiveness to others [1510].

The responses to medications further demonstrate the physical nature of the phenomenon. Some drugs instill helplessness [1511],[201] while others can mitigate or block the effects [640, 1512].

---

[201] *Technical note:* In animal models, learned helplessness can be modulated by giving an animal desipramine, GABA, and Norepinephrine. Each compound influences specific brain regions, including the hippocampus [3448, 3198]. Glutamatergic dysfunctions, which can happen in the parts of the brain that process pleasurable and punishing stimuli, are the current focus of much research on mood disorders and vulnerability. Overactivity in the LHb (lateral habenula) alters the ability of the brain to regulate Depression reduces GABA/glutamate signals, and the treatment of depression up-regulates GABA in the LHb [3233]. This regulatory function of GABA makes it critical in learned helplessness.

Social stressors that induce a sense of powerlessness can be physical or verbal. They can spring from the vitriol of a single person or arise from an organization, a place of employment, or a church. Wai [1513] described his native Nigeria as suffering from a collective form of acquired helplessness, depicting the people as feeling powerless and apathetic toward the corruption and injustice around them; they had learned that no matter what they did, they could not improve their lives or their country. As a society, Americans have a pervasive sense that their government is corrupt, that elections are rigged, and that their vote does not count. These instilled beliefs fit well with a hypothesis of learned helplessness. This same effect explains why Americans vote in such low numbers at elections despite the relative ease and safety of voting, and why so many people, worldwide, willingly follow autocrats who promise to provide them with a modicum of control.

Although we minimize the damage from violence on our young by dismissing the childhood offender as "nothing but a bully," the evidence contradicts that minimizing statement. Peer-on-peer violence is at least as devastating to children as adults [1514]. Children in all age groups experience similar trauma associated with peer and sibling victimization. Relative to adults, children have little control over their own lives. This reduced level of autonomy makes this trauma worse rather than less damaging, leading to long-term depression and anxiety [1515]. Indeed, being bullied as a child provides a background for easier manipulation and subjugation later in life. The still-developing neural net, especially in small children, makes their damage longer-term [622, 1516]. Because these experiences shape the physiological underpinnings of future learning, these social interactions form the core of understanding others and how we relate to the world [1517, 1518].

Neurotensin, which functions throughout the central nervous system as a regulator of two reproductive hormones, LH and prolactin, also regulates dopamine pathways. It is also involved in the epigenetics and physiology of depression and learned helplessness [2496]. Desipramine alters activity in the locus coeruleus, a region that regulates how we translate thoughts into actions [1511].

Experiences also regulate the behaviors of individuals. Our backgrounds can ease or intensify our traumas, sense of powerlessness, and sense of autonomy. By reducing the psychological preparedness for success and social dominance, these experiences can function as social homeostatic mechanisms. These formative histories transcend the concrete day-to-day experience of success or failure. They include lessons from our culture, literature, and religion.

Religious and Political Inoculation, Religious and Political Bullies

> *"I don't even know if God exists, but I do know that if God is your emotional role model, very few human relationships will match up to it."*
>
> — *Oranges Are Not the Only Fruit*, J. Winterson [1519]

Based on cultural experiences, we make assumptions about how the universe functions. Religion and social organization fit within this framework. Human beliefs in non-physical causes are ubiquitous, though a belief in a single, transcendental, supreme being is not. Many contemporary band-level, tribal, and small communities lack the big-man structure familiar to larger agricultural societies. In parallel, they seldom develop a faith dominated by an overlord deity.

People call upon the perspectives of nature and embed them with symbolic meaning in many ways [1520, 1521]. Religions invoke various incorporeal agents, from deities and ghosts to nature spirits and ancestors, and spirituality often deals with the relationships between people and their surrounding world. Far from a source of unending solace, these beliefs can threaten one's safety with terrifying warnings about witches or damnation. Ancestors or nature may guide one and render aid in times of need. Rituals and beliefs vary, forming effective communal green beards. Religious functions can include healing, understanding prey and procuring food, identifying kinships, and recognizing the sources of evil.

In contemporary hunting and gathering societies, "shamanism" is pervasive [1522]. However, lumping all shamanistic religions into one batch is like lumping all theistic and priest-centered faiths together. From the worship of Odin to Marduk to Christ, they share similarities but also display differences [1523]. The shaman's connection to the spiritual world provides the knowledge the band could not otherwise access.

Though often dismissed as superstition by WEIRD societies, shamans play a critical role in their communities. They are reservoirs of tribal symbolism, serving to interpret the relationships and intentions of members and outsiders. Their skills serve as ecological mediators, bolstering sustainability. Many island religions encode ecological wisdom, including sanctioned and tabooed use of lands and resources. They can play an environmental role by supporting sustainable use of the environment across generations [1524]. Some shamans take their prey's perspective by calling upon their mirror neurons, helping them understand it better. Shamanistic elements can be seen in the deer dance of the Hiaki (Sonora, Mexico), in which the performer plays the hunted deer; at a deeper level, it recounts a massacre of their people in 1900 [1525].

A shaman's contact with the spirits or ancestors can identify dangers to a person or community. They help unite a people in preparation for war and provide the confidence and psychological support needed for battle. The Shawnee prophet Tenskwatawa, for example, warned against adopting the ways of the White settlers and rallied the troops at the 1811 Battle of Tippecanoe (Indiana). Although discredited, his warnings were sound, and the strategy may have worked under traditional warfare conditions.

Vision quests through altered consciousness continue into agricultural cultures (e.g., drugs, fasting, mortification of the flesh), and even today, few desire to enter battles believing their God stands against them. WEIRD societies may not denounce someone as a witch. However, under the cloak of religion or politics, sermons can still incite violence against an idea, a person or a group as the cause of misfortune.

Some experience religion as a revelation by a divine being. For others, it is a spandrel, an ornamentation that just happened as we achieved greater consciousness and cognition. Supernatural beliefs may be part of our hard wiring, helping us avoid acting upon simple correlations. When we do not understand something, we search for causations, using supernatural causes to bridge gaps in our understanding [1526].

However, a feature as universal as faith in supernatural powers suggests that an intense selection pressure favored the development of faith, and it may well have played a role in our species' survival [1527, 1523, 1528]. Our tendency toward religion and religion-like behavior has adaptive physiological underpinnings [1529, pp. 86, 312]. Faith in an All-Father like Oden (Norse), Marduk (Babylonia), or the genderless Olodumare (Yoruba) could unite societies despite weak genetic links. Faith in a totipotent political leader can serve the same purpose, sometimes combined with a conviction that they rule through the Gods' favor. Medieval European kings and queens claimed a divine right to the throne; the populace accepted the pharaohs as sacred beings; the Nazis banned secular education and portrayed Hitler as having a messianic mission to save the world, inculcating religion-like zeal into its followers [1530]. As with religious traditions, political saviors can be benign or malignant, indoctrinating tolerance and cooperation or inculcating hate. Extreme ideology, political or religious, correlates with mistrust and a need to control others [1531].

Granted, that last example is political, but the politics of agricultural societies can don the garb of agrarian religions. Both systems relegate authority and individual wills to either spiritual or corporeal leaders. Both establish social orders and control that reflect their values and assure their continuance. Both can rule from fear, power, or inspiration. Both can unite people through modeling and expectations or divide the world into an us-verses-them confrontation. Both can be tools for change or stability, and both can derive power through transcendence, love, or hate.

Shared religious tradition helped our ancestors solve the problems of cooperation and maintaining pro-social behavior as our communities expanded [1532, 1533]. The "golden rule," to treat others as you would be treated, is widespread throughout the world's cultures, suggesting an ancient origin; it is part of our fundamental psyche, directing prosocial behavior toward those surrounding us. The rule made good biological sense when our kind roamed the planet in extended family bands. Our ability to generalize the rule facilitated our expansion into communities of unrelated members without slitting each other's throats.

We do not know when the ability to engage in religion emerged in our ancestors. Still, we can presume it developed on top of our primate brain and was a refinement of our other primate behaviors. Multiple brain regions activate during religious experiences, and different experiences employ different areas, including parts of our social brain. Spiritual experiences stimulate the prefrontal cortex regions necessary for self-representation, goal-driven behavior, and social conformity. Religious faith and *internalized* morality (independent of faith) stimulate our older, intuitive brain more intensely than external, secular belief [1534]. This intense experience can both increase our susceptibility to faith-based fraud and our rage when we expose it.

The precuneus (perception, memory integration, mental imagery, pain responses) goes into overdrive along with the anterior insula (subjective feelings/ emotional stimuli including disgust, fear, and happiness). We increase our energy levels in the ACC (anterior cingulate cortex), which monitors inconsistencies and errors, does cost-benefit analyses, and scrutinizes the motivation of others. The reward center of the ventral striatum activates more, possibly assigning value to the stimuli and assigning a reward to belief or the process reinforcing it, be it a social protest, church service, or ritual dance [1535]. Then, the regions of the cortex associated with emotional regulation, self-image, and the resolution of cognitive conflict flare [1534]. This internal grounding is independent of belief in any deity; instead, it finds its roots in internalized ethics and applies equally to people of faith, atheists, and agnostics [1536, 1537, 1538].

The social brain can provide a sense that gods or ancestors watch over you, understand your motivations, and know what you do. It can fire your mirror neurons to create empathy with your fellow humans, group, or deity [1539]. These internalized mechanisms can convince you that God approves of *your* desires. Rites can activate or suppress activity in the caudate nucleus (CN), a region associated with memory, emotion, reward, and romance. Silent prayer, for example, increases oxygen use (brain activity) in the CN's reward system [1540]. Combined with alterations to the ACC, amygdala, and frontal cortex, a targeted ritual can heighten our sense of danger, tweak awe, or prepare us for battle [1541].

This adaptive basis for religion (or secular institutions) stands independent of any supernatural origin. Science can provide a physical, evolutionary basis for a phenomenon. It does not and cannot speak to the spiritual. None of these elements excludes or negates the others. There need not be winners and losers in a war between the spiritual and physical. Physical causes and spiritual beliefs are not a battleground but different, non-competitive models of how things work.

Religious beliefs and disbeliefs provide a signaling system that can replace more costly rituals or exchanges. An organized system of trust and behavioral standards, whether secular or religious, improves the chance that people will emulate each other [1542, 1543]. A political or religious movement fails unless it reinforces philosophies and behaviors that work and exposes variants that do not. In a generalist species like ours, imitation reduces the complexity of social interaction by limiting choices. Selection pressure would further aid the survival of a clan as a whole—and hence its clan members—by discouraging beliefs and activities that are not innately unhealthy but which damage their long-term survival or cohesion. In other words, social sanctions amplify behaviors.

> *"Witnessing the aftermath of battle is a great shock. But it is the way people at war systematically pervert and distort 'civilized' values that so affronts the rational mind. Calm, loving friends become soldiers and killers, religion sounds the battle cry, and the 'us versus them' mentality shows no mercy."* — R. Paul Shaw & Yuwa Wong, preface to *The Seeds of War*, 1988 [3449]

Cultural practices, including religious dictates and beliefs, change over time. For most of our existence as humans, we moved over limited territories. We exchanged mates with nearby troops. Cultural norms and dictates spread slowly across communities. The more green beards we shared, the more likely we were to have close kinship ties with the neighbors. The more another group followed our cultural practices, the more likely we would be to bestow nepotism on a societal scale [1544, 1545] and inoculate against competing ideas and groups.

Religion's variations and internal contradictions also serve as green beards [1546]. Such signals expose outsiders and affirm kinships by providing a mechanism for group recognition and organization [1547, 1548, 1549, 1550, 1551, 1552]. As our territories spread, group membership and kinship became more tenuous. Patterns allowed outlying group members to reaffirm their commitment to the core identity. As we expanded beyond our initial community size, belief systems provided a cheap way of recognizing kin and organizing the population [1553]. Costly rituals, such as pilgrimages to the home territory or circumcisions, signal loyalty to the core group and the continued friendly intent toward the extended community [1554]. Today, calling upon such robust core identities may provide buffers against social stressors, including perceived ethnic and racial discrimination [1555].[202]

---

[202] The buffering effect of group membership can also contribute to the allure of some criminal organizations, which offer an enhanced group identity and a rejection of external labels.

Figure 7-4. This 1926 image shows the Ku Klux Klan driving Irish Catholics from the United States, along with the evils they were believed to propagate. The Ku Klux Klan considers itself a religious organization. — from *Klansmen: Guardians of Liberty*, Ama White, 1925; Branford Clark, Illustrator, p. 21 [3270].

Many precepts and organizations in religious systems are complex and often withheld from outsiders. Although learning a doctrine imposes a cost on the individuals, paying that cost (investment in society) engenders mutual faith within the community [1556]. In other words, membership has fitness value.

Although my dog contends that everyone wants to be her friend, the world does not function that way. An ability to distinguish between community members and potentially hostile outsiders bestows another fitness benefit. Religion, like grammar, can serve in this protective role. Outsiders can learn and integrate the signals with beliefs, but the process takes time and is often tricky [1557]. Facets of religious culture depend upon a system laden with internal contradictions and no physical reality [1550].[203] These features make it difficult to reconcile the concepts and learn them. Integrating these ideas and practices into daily life makes it easier to recognize strangers. Precepts and expected behaviors could clash with their earlier training, increasing the difficulty of integration.

In addition to dictates with immediate and apparent cohesive value to the group, such as not killing each other, beliefs often impose additional restraints. These behavioral and cognitive nuances provide mechanisms of social control and conflict reduction [1558, 1559]. They can signal commitment to the group [1560]. They can support an adaptive distribution of labor within the group (e.g., priesthoods and caste

---

[203] A Biblical example would be the foundation story of Abraham preparing to sacrifice his only son to God. At the last minute, God stops him (Genesis 22:1-18). Typically, this passage is taken to represent a condemnation of human sacrifice. Later in Judges 11:31-39, the general Jephthah must sacrifice his daughter in a pledge to God.

systems). Such commandments include abstinence from beef (caste Hindus), feeding pork to the ancestors (traditional Highland New Guinea), abstaining from meat during a particular season (traditional Catholic), or refraining from work on a designated day (traditional Judaism). These examples provide no *immediate* survival benefits but have distinct cohesion value. These added dictates strengthen public conformity, cohesion, and identity. These features increase the chances that the group, and hence the genes of its members, survive.

Religion provides additional solidarity by creating bonds among people with tenuous ties. It brings members of a society into shared settings, increasing the sense of community and mutual empathy, and it may have played a crucial role in the formation of the first non-kin societies. Some religions threaten apostates with damnation or expulsion if they fail to follow their tenets. Many houses of worship provide for needy parishioners despite lacking biological kinship. Among the Hopi, a man's bloodline or clan membership does not determine his membership in a particular religious (Katsina) society, serving as a unifying force that increases group cohesion across clan distinctions.

The dictates of religion and culture help members recognize each other [1551, 1561]. Upon this foundation, we once built stabilizing pseudo-kinship structures. The ingrained behavioral green beards identify who were family through shared cultural traits. By extension, it tells us who merits the benefits of our often-taxed supply of generosity and who will sacrifice for us [1556, 1562, 1563, 1564]. Society often imposes arbitrary commandments extending beyond the actions that cause harm or bestow direct benefits. Such social (kinship) selection favors individuals who conform to the group norms and creates barriers against infiltration by potential rivals.

> *"The tomato and the nightshade are of the same family... The one is nutritious and the other poisonous... The assumption that mass movements have many traits in common does not imply that all movements are equally beneficent or poisonous."*
> — Erick Hoffer, *The True Believer*, Preface [3128]

Some argue religion provides a moral compass for societies; the picture is more complex. Personality and temperament matter. The motivation for religious behavior factors into the mix, as does the type of morality. People whose belief system forms an internal scaffold for pro-social values behave differently from those who use religion mechanistically as a means to achieve an end. Belief systems, be they social, political, or religious, can reduce some prejudices while inculcating others. The repeated messages of leaders function as a social priming agent to promote love or hate, enhancing benevolent or malignant values [1565]. A powerful sermon or political speech can shame a negative behavior or whip listeners into a homicidal frenzy.

You can find some ideals spread across an array of cultures. Be nice to each other. Don't kill or steal from your community. Such values promote group stability and survival. These fundamental traits became the foundation for our value systems. Like other primates, we apply these niceties to our families and societies but don't always extend the same courtesies to outsiders.

Religion plays no apparent long-term role in the internalization of such ideals. Pro-social behavior, such as responding to an appeal for charity, increases right after visiting a house of worship or reading sacred texts about giving. At other times, this "Sunday Effect" is not as apparent [1566]. Internalized spiritual and moral qualities correlated with giving [1567], and the social involvement of a church may provide a brief oxytocin boost that promotes altruism. Muslims who read verses from the Holy Quran that exalt prosocial values temporarily boost their trust in others; this effect does not happen after reading ones that glorify Allah [1568]. Pro-social religions slightly suppress antisocial behavior [1569], possibly through the Sunday Effect and by providing an explicit roadmap for psychopaths who lack an internal moral compass. [204] Unfortunately, externally imposed religiosity has little chance of inculcating the internalized values associated with pro-social behavior.

Other moralities are religion-specific. Eating pork, combining dairy and meat, or mixing wool and cotton is immoral in the Orthodox Jewish faith. (In the mid-1970s, one Kosher grill in Chicago used an early vegetable-based, vegan cheese to create a mock cheeseburger for their menu. The taboo against mixing meat and dairy was so strong that people vomited in the bathroom.) For traditional Roman Catholics, eating meat on the Friday before Easter (Good Friday) is wrong; for a practicing Latter Day Saint or Muslim, drinking alcohol is abhorrent, and for Brahmin Hindus, killing cows is comparable to murder.

Religion also has a dark side. Specific traditions can tacitly encourage or sanctify dehumanization and attacks on people. Slogans and sermons claiming "God Hates Fags" have nudged people to attack members of the LGBTQ community, and microaggressions have become more common [1570]. At the same time, the mainstream American Episcopal Church preaches tolerance and acceptance, as do the Community of Christ (LDS) and sects within Lutheran, Presbyterian, and Methodist faiths. Some caste Hindus appeal to religion to excuse attacks against Dalits (Children of God or scheduled castes) by claiming the victim killed a cow;[205] others call for affirmative action, increased tolerance, and inclusion [1571, 1572]. When political or religious beliefs support internalized negative concepts about an identifiable group, the chance of prejudice and violence against them increases, regardless of internal or external orientation [1573]. This concordance

---

204 Studies about religion's influence on morality are heavily contaminated by the researchers' beliefs about its malignant, benevolent, or neutral nature. A jumble of confounding factors muddle the study of religions, including who is drawn to a faith, the intrinsic or extrinsic motivations for participation, and how codified beliefs interact with personalities [3164]. Studies often failed to distinguish between traditions, many of which rank their values differently, and the influence of the beliefs and prejudices of their leaders.

205 As in Christian countries, politics, and religion often mix. "Religious" oppression of the Dalits often combines with Hindu nationalism. Kancha Ilaiah Shephard, an Indian political theorist, warned that the Hinduvata movement is nothing but "Brahaminism" [3204, 3231].

with belief systems is not unique to religion, but the power of faith provides a sturdy scaffold upon which hatred or tolerance can build [1574].

> *"Even a hasty perusal of this factual document convinces one as to the President's Jewish ancestry. From the viewpoint of eugenics, it explains his natural bent toward radicalism"*
> — US Antisemitic political flyer
> ca 1935-39 [1576]

Weaponized religion has become a fixture on America's evening news reports. When the United States was attacked on September 11, 2001, the Al Qaeda ideologues attacked the protected "people of the book" in the name of Islam. They forgot that when you worship hatred, you violate the first tenets of the faith: Only God is worthy of worship. Jerry Falwell, a conservative pastor and televangelist, claimed the "US got what it deserved" because of its tolerance of feminists, gays, liberals, and abortionists, and that the American Civil Liberties Union had removed God's protection [1575]. (He forgot to mention the role of the kitchen sink.) American President Obama was repeatedly labeled a "secret Muslim" despite attending a Christian Church. Franklin Roosevelt was portrayed as Jewish when world-Jewish conspiracy theories were favored cudgels [1576].

The World is replete with headlines of people who justify murders based on religious or moral convictions. They range from killing individuals who disagree with them (e.g., Alan Berg, talk-show host murdered by Wotanist David Lane in 1984), to the 2014 antisemitic mass murders at the Village Shalom and Jewish Community Center in Overland Park Kansas by Frazier G. Miller, and the 2016 Pulse nightclub slaughter in Orlando, Florida). Words from the pulpit, the mimbar, or the political stage can radicalize a congregation and instigate violence; through these, leaders can also reduce tension and appeal to our better natures [1577].

Religions impact group security and xenophobia by skewing our perception of human behavior. People often contend that outsiders must be morally inferior because they are not subject to the same dictates. In reality, most humans possess an innate moral sense independent of a formalized religion or belief.[206] We absorb morality from the people around us and the culture in which we are immersed [1578]. Different groups place different emphases on care, fairness, liberty, loyalty, authority, and sanctity [1579, 1580, 1581, 1582]. In various mixtures, these foundations underlie human morality. No community or faith has a monopoly on virtue. Still, codifying common moral elements increases the sense that one's group is the only trustworthy one. This perception extends beyond physical interdependency to transmitting cultural information and trustworthiness [1550, 1583].

Religion can assure followers that perpetrators will receive a just punishment in this life or the next. It can also absolve a perpetrator of blame. Many sects treat non-believers and apostates as social cheats and castigate or exclude them accordingly [1584, 1585]. Some have imposed penalties, such as higher taxes or bans from specific professions. If society fails to enforce the norms, some doctrines can provide comfort, assuring the believers that the cheats will receive chastisement in another life, spiritual or physical, while the faithful receive rewards for their conformity.[207]

People engage in religion for different reasons [1586]. For some, religion and faith form their core beliefs and become a positive end in their own right. For them, faith, belief, and good works are predominantly internalized. People with this intrinsic spiritual value have more extensive pro-social behaviors than those who manipulate religion as a means to an end. Those who live in accord with their conscience as part of their core values experience a more positive sense of well-being and control and greater resilience than those who use religion as a social tool. Those with intrinsic values have more positive attitudes toward

---

[206] People who lack a moral compass are known as sociopaths, psychopaths, or antisocial personalities. People with the disorder will understand that a behavior is wrong because they have learned it from their society and family. However, they do not feel it is wrong, i.e., they do not experience guilt like a typical person.

[207] Not all religions accept the doctrine of an afterlife or reincarnation.

giving to charities and other pro-social judgments and behaviors than those with ulterior motives [1587, 1588, 1589, 1590, 1591, 1592].

> *"Terror is nothing but prompt, severe, inflexible justice; it is therefore an emanation of virtue. It is less a special principle than a consequence of the general principle of democracy applied to our country's most pressing needs."*
> — Robespierre, 1794, on the French Reign of Terror [3440]

For those whose religious life functions as a means to an end, donations can serve as a form of status-driven lekking [1593, 1594]. People with an external orientation typically pay less attention to behavior and ethical attitudes [1587, 1588, 1592, 1595], including the moral implications of consumption [1596, 1597] and tax avoidance [1598]. They use their "faith" to increase their standing within the social group or rationalize otherwise unacceptable behaviors such as hate, prejudice, and discrimination [1599]. An extrinsic religious orientation uses membership in a dominant in-group for ends such as protection, power, comfort, revenge, and status [1599, 1600, 1601]. In some countries, including the United States, this effect is confounded by authoritarians who self-identify as "religious" and use that identity as a cudgel to further political agendas [1602, 1603, 1604]. High levels of externally expressed religiosity can stem from xenophobia and the mistrust of others and become a cudgel to restrict and control outsiders through preaching and moralizing [1605].

In 1985, an American study found that internal orientation correlates with reduced racial prejudice, but externalized, goal-oriented religious behaviors exacerbate it [1606]. Several studies have replicated these results; however, the gap has shrunk as social acceptance and tolerance for overt racism declined [1607, 1603, 1608]. Internalized faith does not automatically reduce prejudice and discrimination. If religious indoctrination supports discrimination, it can increase it in both intrinsic and extrinsically oriented people [1609, 1610, 1611, 1612].

The third kind of orientation, questing [1613], focuses on the search for answers to the problems and questions raised by life [1614]. It remains the least understood of the alignments. By definition, seekers tend to be open to experiences and are willing to change as they pursue truth and answers.

People are a mixture of all three orientations, but one often dominates. Some studies suggest that of the three emphases, a dominant quest orientation aligns most closely with charity, compassion, and understanding toward others [1615, 1616]. However, threatening statements and stances can still trigger activity in the regions of the brain designed to recognize and process aggression [396].[208] In a Christian sample, WEIRD questers were less willing to assist identified fundamentalists with tasks than people whose religious stances remained unknown [1617]. They directed their intolerance more toward normative contraventions than at the person, and they showed less prejudice when the task was unrelated to the violation [1618].

Religion bestows a mechanism for social integration and a way to assert spiritual power. When culture and circumstance reduce the physical control of one's life, it can provide an immunizing effect against learned helplessness [1619, 1620]. Internalized religious beliefs and motivation, where faith is part of one's core identity, have an inoculating effect against anxiety, stress, and depression.

People with an internalized morality or spirituality have a more significant buffer against stress than those who use religion for extrinsic purposes [1621, 1622].[209] One sees this effect across diverse communities [1623]. It occurs in both Muslims and Christians. Studies replicated these findings in Europe, North America,

---

[208] The primary threat-processing regions are the amygdala, insula, temporal lobe, and supramarginal gyrus.

[209] NB: These studies are correlative. It could be that coping problems stem from a pre-existing personality or learned coping strategies that reduce coping skills and increase the tendency toward externalized rather than internalized religious or spiritual experiences.

and the Middle East. People with an extrinsic focus, who use religion as an instrument to gain social status or to impress others, do not reap these benefits [1624, 1625, 1626, 1621].

Faith and mental health do not follow a linear pattern. People with strong beliefs, be they theists or atheists, experience less anxiety and depression than people with lesser confidence levels [1627]. The strength of our convictions, not religion itself, seems to be the inoculating agent. Similarly, no link exists between faith and morality or the motivation for good works [1628, 1629, 1630, 1631, 1632]. In contrast, an extrinsic religious orientation correlates with a higher expression of prejudice and intolerance of others [1633, 1634, 1635, 1636, 1637].

Religious orientation can shift our locus of control and sense of powerlessness. People who feel constrained by their sect experience less empowerment than those who find it a source of encouragement [1513, 1638, 1639].[210] No great surprise. An external locus of control interacts with an extrinsic religious orientation to create a substantial negative fallout. Intrinsic spirituality supports an internalized sense of control, which reduces stress hormones, such as cortisol [1640].

> *"Hamleigh, you must take care of Philip and his friends. I cannot be seen to be involved, but you must make it clear that you are doing this to protect the church."*
> — Bishop Waleran Bigod, *Pillars of the Earth*, Ken Follett [3122]

The people with an integrated, internalized faith, those who practice religion or engage in other social causes because they believe in goodness, kindness, and other moral values, also receive a higher buffer against stress, even without a religious label. Religious affiliation can feed into this pattern of self-efficacy, but it does not, on its own, create it. It can instill social tolerance and trust or undermine it [1641]. Those who exploit the trappings of religion because it elevates them socially are the same puppet masters who manipulate people in other aspects of life. They practice

[210] N.B.: Wai also found that the locus of control mattered less in Nigerian subjects than in the European and American ones of earlier studies. His findings suggest that culture influences the strength of this factor. He researched in Jos, a city with recent religious conflicts between Muslims and Christians. Christians, who held more power, showed more resilience to learned helplessness than Muslims.

religion to meet their own goals and agendas, not those of a deity, using scriptures and rituals as weapons rather than as an embrace. Unfortunately, in religion, as in other aspects of life, these same Machiavellians often rise to the top of the social heap.

Because our spiritual beliefs are fundamental to our core being, they become dangerous when distorted and provide an avenue for ethnic and religious violence. Behind a shield of pronounced righteousness, they bend the identity of a people. Unscrupulous leaders can bring together true believers to commit microaggressions or genocide as they destroy the credibility of true piety. Their followers, who execute heinous acts in the name of their community or God, will presume absolution. In contrast, effective leaders, be they religious or secular, model pro-social behaviors that increase cooperation within their group [1350].

A Natural Experiment: Genocide

> *"Traditional peoples tend to view other people as split into three types: known individuals who are friends, other known individuals who are enemies, and unknown strangers who must be considered as likely enemies."*
>
> — Jared Diamond, *The World Until Yesterday* [1642]

The concept of kin preference and Milgram's principles provide explanations for atrocities. They explain the formation of the Khmer Rouge's obedient and murderous child soldiers [1643] and the Janjaweed (militias) who displaced 2.5 million and killed 400,000 people in Darfur, Sudan. They shed light on the sadistic personnel of Nazi concentration camps and the Magdalene Laundries run by the Catholic Church in Ireland.[211] They provide insight into genocides.

---

[211] Magdalene "Laundries" or Magdalene "Asylums" were initially created as halfway homes to reform "fallen" women, i.e., prostitutes. Quickly they became workhouses for any young women who showed signs of acting out sexually or being too high-spirited. Often, women would effectively become slaves, trapped in the institution for all or for most of their lives [2089, 2088].

> *"Keep them yelling their devotion, add a touch of hate at Rome. You will rise to a greater power; we will win ourselves a home!"* — Simon Zelot in *Jesus Christ Superstar*, A. Webber, 1970 [3391]

"Genocide" is a modern concept [1644]. In earlier centuries, eliminating populations who posed potential threats was considered sound practice. Killing—or better yet, enslaving—a potentially rival population made sense. Genocides consolidate power and eliminate resource competitors [1645, 1646]. In an era without fossil fuels, the enslavement of one's rivals parasitized their labor, which drove the economy upward.[212] It eliminated threats, and conquests brought land, energy (slave labor), and wealth. Civilians usually accepted the legions involved in a massacre. The public understood their military combatants protected them. The Roman genocide of Carthaginians consolidated the Roman dominion of the Mediterranean. The Catholics of France inferred the blessing and approval of God in the 1572 slaughter of the Protestant Huguenots by assassins and mobs of Catholics. At Sơn Mỹ Vietnam, members of the United States Army obeyed orders to butcher 347-504 civilians. Through the mid-19th century, many Canadian, American, and Mexican communities offered body-part bounties payable for the scalps or ears of Native Americans as evidence of the kill [1647, 1648, 1649, p. x, 1650, 1651, 1652, 1653]. To achieve these successful holocausts, soldiers had to obey directives to slaughter.

For most of our history, people accepted weapons of mass destruction, civilian deaths, and genocide as acceptable practices. During the First World War, all sides used poison gases against their enemies. Chlorine and Bromine gases transformed into acid in the lungs and the eyes as they ate through the skin, melting them. At the Second Battle of Ypres, Belgium, the German army became the first combatant to use chlorine gas on a mass scale. French, French Colonial, and Canadian troops became the first victims of these large-scale assaults. Two thousand deaths occurred during that first brief gas attack [1654, p. 81]. Britain launched attacks with "Red Star" (chlorine gas). The French developed the deadlier but slower-acting phosgene gas.

---

212 In a world without fossil fuels, the enslavement of one's rivals provided a source of energy, their labor, that could drive the economy and improve the living standards for the victors.

Soon, American, German, French, and British troops would use a combination of chlorine and phosgene (White Star) against each other. With few exceptions, troops obeyed the order to gas the enemy.

In the 1994 Rwanda genocide, members of the Hutu population killed an estimated 500,000 to 800,000 Tutsi despite significant intermarriage between the groups [1655]. Many Hutus who refused to participate also lost their lives as much of the world watched [1656]. The initiators of the extermination cited the assassination of President Habyarimana to restore stability after the "events" (i.e., the last genocide of Tutsi). This coup by assassination provided a unifying rallying point for participants.

To many people, the cries of "civil defense" sounded legitimate in the wake of the social unrest. The broadcast media aired propaganda from the government, including accusations of the Tutsi militia poised to attack the Hutu majority [1657]. These broadcasts rallied people to arms, activated the threat circuits of the brain, and instilled ethnic hatred.

The importance of "kinship" or "in-group" relations came to the surface quickly. Perceptions of ethnicity trumped years of cultural integration, intermarriages, and a shared language. The propaganda campaign spread a loathing of the Tutsi among the Hutu [1658]. Where support for the coup was reliable (the perceived in-group), people needed less prompting to kill Tutsi. In regions where Tutsi and Hutu lived side by side, where there was greater recognition of Tutsi humanity, people were more hesitant and refused to attack. Only when the government exerted physical force, including the destruction of property and threat of death, did many acquiesce to the genocide.

> *"If all Jews were like him, there would be no Jewish Question"*
> — allegedly said by Adolf Hitler referring to his childhood physician and his "Edeljude" (noble Jew), Dr. Eduard Bloch, March, 1938 [3236, 3390]

The Rwandan government used the same foot-in-the-door technique as Milgram. They incited small aggressions against Tutsi to desensitize resistant Hutus to violence. Once the process started, the slaughter intensified. Attacks on property escalated to attacks on people labeled as "terrorists." Assaults spread to anyone Tutsi. And like Milgram's teachers who remained in the room with their "students," even the staunchest supporters of the genocide would occasionally save the lives of individual Tutsi known to them.

Colonel Bagosora moved to seize power after the presidential assassination of President Habyarimana. He maneuvered key supporters into positions of influence. Their offices provided the trappings of legitimacy that Milgram found compelling in manipulating his subjects. They blamed the murder on the Tutsi, though no one had traced the blame to a specific person or group. The combination of strategically placed officials and public accusations broadcast through partisan media (RTLM) provided the recognition of legitimacy needed to recruit support. The leaders of the genocide passed their orders through a chain of command to the police and neighborhoods, providing the direct personal contact with an authority figure that had proven effective for Milgram.

The state compartmentalized the genocide [1659]. It tasked some authorities with driving the Tutsi from their homes. Others obstructed the legal procedures lest the Tutsi, deemed responsible for the assassination, escape conviction. As the hatred built, the government entreated citizens to detain "suspects" and surrender them for questioning. Others, especially militia members, committed much of the killing. This division of labor reduced the perception of direct responsibility for the murders and, like Milgram's increasing electric shocks, provided comfort levels for compliance.

Obedience increases when subordinates depend upon their leaders. The government coaxed compliance in a depressed economy with offers of cash, food, land, and jobs in exchange for eliminating traitors. Those who participated or turned a blind eye profited by acquiring Tutsi possessions and positions for their families. Most did not kill; instead, they partook in compartmentalized actions that delivered the Tutsi to their executioners.

Milgram found less blind obedience when he placed the torturer close to his victim. This finding dovetails with human evolutionary history. In early human communities, people would have had their closest contacts and allegiances with their kin. Knowing someone makes brutalizing them more challenging (but not impossible). However, financial distress can amplify prejudice and discrimination against non-green-beards. Knowledge does not preclude the execution of one's neighbor, but contact with a person or a group inoculates against cruelty, though compliance with torture remains high.

By the end of the approximately 100-day campaign against the Tutsi, three-quarters of a million (minimum estimate) to over one million people had died. The Tutsi comprised over 90% of the murder victims. Most of the remaining 10% were Hutu who participated in the resistance or Tutsi sympathizers.

Dissent increased as the regime lost its military power and its international standing. Most Hutus never approved of the genocide and readily switched their support to the resistance fighters. Hutu factions began quarreling among themselves. The people at the core of the attacks, who lost their license to kill Tutsi, started attacking other Hutus. Prime Minister Jean Kambanda, who led the genocide, became the first person to plead guilty to the crime under international law [1660, p. 1]. Many other prisoners at Arusha, the United Nations tribunal and detention facility for such prisoners, continue to blame their unjust fate on their Tutsi victims, claiming them to be masters of deceit who had tricked the world.

The 1994 genocide was not the first campaign against Tutsi or Hutu. Despite their similarities, Belgian and German colonial authorities had emphasized minor differences between these groups. A genocide of Tutsi happened in 1963 as the Belgian colonial government withdrew from Rwanda [1660, p. 9]. In 1975, another genocide happened in neighboring Burundi. There, Tutsi perpetrators slaughtered 200,000 Hutus. What had begun as class warfare and resentments in the 1960s had become an ongoing blood feud. Blood-driven vendettas are another long-standing feature of human intergroup conflict, seen in diverse places such as the traditional inter-village warfare of Papua New Guinea, the Kikuyu of Central Kenya, and the Beowulf legend of England.

We may never know how many perpetrators engaged in the mass butchery of the Rwandan genocide. The figures range from a few thousand to three million. The massive span rises from estimates of the murders per person and how many people participated in each killing. After the government fell, the new Rwandan government held thousands of detainees on allegations of genocide, but those figures included people identified for political motivations. Between 1996 and 2002, over 7000 people came to trial charged with genocide [1661].

Straus [1662] estimated 175 to 210 thousand people cooperated to the point of direct execution. He used the numbers of perpetrators held in detention, court records, and accounts of accomplices to estimate the sizes of the attacking groups (the cells). At these figures, 14% to 17% of the adult male population would have been involved directly in the slaughter. Note that this figure falls below the 30% compliance Milgram saw at the presumed lethal edge of his study. Neither he nor his successors could model this direct murder scenario; we know even less about the levels of sub-lethal compliance for this or any genocide.

Hitler seized power in 1933 when the government building, the Reichstag, caught fire (February 27). The Nazi Party scapegoated the communists. Jewish persecution flared after they desensitized people to the elimination of communists.[213] By 1934, the Nazi government had built eight large camps within Germany. By 1944, the country housed at least 13 primary and 44 satellite camps. The occupied territories held far more of these facilities. Ninety million people were under the direct political control of the Greater German Reich, and only approximately 55,000 people (0.06% of the population) served as concentration camp guards. Even when considering the additional personnel needed to run the railroads, load the cattle cars,

---

[213] The first concentration camps opened in March of 1933, soon after Hitler's appointment as chancellor. Most of the early prisoners were members of the Communist party and Hitler's political enemies.

and build the gas chambers (often slave labor), the figures remained below Milgram's level of active compliance.

Lest we believe leaders of atrocities are all prejudiced, abnormally evil figures, consider the following. Sporadic reports of the physical abuse of children by members of the Irish clergy in institutional schools circulated for years before sufficient outrage against the Church built in Ireland. Children suffered sleep deprivation, soiled sheets, beatings for soiling clothing with menstrual blood, hunger, and gang rapes at the hands of hundreds of brothers, priests, and sisters [1663, 1664, 1665].

This system of concentration camps for children continued for over 70 years while such involuntary confinement waned for other groups [1666]. Because of how the state paid institutional schools, these places made money by cramming as many children as possible into their facilities, including ones who had families and should not have been there. The Magdalene laundries, also run by the Catholic Church in Ireland, held young women whose only sin may have been that they were a bit hard to handle. Inside these asylums, these girls became women. Frequently, they never left again. Like Zimbardo's guards, the staff abused their power. The last of the Irish Magdalene laundries closed in 1996. We do not know how many women spent their lives in involuntary menial servitude in these religious slave camps.

The Nazi regime tried to desensitize the German population to genocide by introducing it as a humane way of ending the suffering of the mentally challenged and eliminating "terrorist" threats. The Rwandan government deadened its people with claims of national defense and an end to terrorism. The church portrayed the women of the laundries as wanton harlots.

Genocides encapsulate the paradigms produced by Milgram, Zimbardo, and their intellectual descendants. Our primate nature primes us to surrender our will and please higher-status animals. Like genocidal chimps, we expect coherence to group norms and cooperation. As we dehumanize our victims, we create an outgroup, reducing our sense of kinship. Violence against them becomes manageable. The rational mind (orbitofrontal and prefrontal cortex), which inhibits the emotional amygdala, ceases its regulatory functions, releasing our primitive brain. It equates pre-emptive attacks with survival, and we become more compliant with murder and deaden to violence.

Most of us believe we would never obey a Hitler or a Pol Pot. Armed with our knowledge of the Eichmann-type obedience studies, our courage and better nature would drive us toward righteousness. We would resist. This comforting fallacy, the "better than average" effect, applies across a range of self-comparisons [1667]. Most people believe they are more intelligent than average, have more reasoned political opinions, and have more common sense, hampering our reasoned self-assessments [1668]. We also extend these judgments to our social circles and to groups we hold in a favorable light. This bias grows particularly strong when we compare our morality against others [1669, 1670, 1671]. Poles, for example, believe they would stop administering shocks sooner than Russians or Germans but not as quickly as the French or English—reviled and lauded ethnicities [1672].

The damage done to the decimated population exceeds the death toll. Traumatized survivors of a holocaust live, work, and raise their offspring with fewer resources and a reduced social status relative to the population that perpetrated the atrocity. Hypersensitized adrenal and cortisol responses haunt many for life through depression and stress disorders. These precarious adrenal or corticosteroid systems reinforce their reduced status. Many overcome the barriers, but others never recover. The murders shatter cultural traditions and extended families, the critical support systems for the offspring.

The damage diminishes the next generation's threat to the dominant population. It reduces their ability to compete through epigenetic changes [1673], long-term maternal stress, and changes to the HPA hormonal axis and its receptors [1674, 1675, 1676, 1677, 1678]. Early-life stress can even reduce one's ability to feel pleasure throughout life [120, 121, 122], thus reducing the reward value for achievements and rank.

Transmitted epigenetic changes leave offspring vulnerable to maladaptive risk-taking or avoidance behaviors [1679, 1680] as well as PTSD and other stress disorders [1681, 1682, 1683, 1684]. Fathers, as well as mothers, convey epigenetic changes [1685], and a maternal or paternal origin of stress-impacted DNA can create different effects on an offspring's mental health behavior [1679, 1686]. Combined with the too often transmitted societal prejudices, these benefits to the victor's next generation could make genocide a costly but effective strategy for the genes.

Genocide, enslavement, and just plain bullying may also predispose us towards passive if not active, microaggressions and revenge against the perpetrators. When Rwandan genocide survivors or their children were hooked up to an encephalograph and given a choice to aid either a former perpetrator or a member of their own group, they chose former perpetrators and their offspring less often. When they did select a former tormentor, their reaction time lengthened, indicating more internal discord about their decision. In contrast, those who committed genocide favored the former victims, as did their children. [1687]. Slave revolts have a history of bloody revenge. As for me, if someone were to enslave me or commit genocide against my family, I suspect that if I had the chance, the only power on this Earth that could stop me from exacting revenge would be my primate drive to obey.

Resisting and Resistance

In our rush to condemn our species as sheep who readily obey monsters, we forget that both known and unsung heroes have withstood the drive to conform. Franz Jägerstätter refused to fight for Nazi Germany; they martyred him for his resistance. Six people hid Anne Frank (and seven other people) from the Nazis, using their own resources to provide them with food, clothing, and a few niceties; they protected them for two years before being captured; only two of the helpers were detained and sent to forced labor in Amersfoort.

Pliny the Younger (first-century CE) opposed Christianity as a load of divisive, subversive, superstitious claptrap. Yet, he opposed the harsh punishments for practitioners and sought to treat them fairly. His appeals risked alienating the emperor and, hence, his reputation and his life. In 331 CE, Emperor Galerius issued an Edict of Toleration, ending the persecution of Christians. Although it may not sound like much of a risk for an emperor, remember that those who opposed the wishes of the military often encountered death through untimely "accidents." (In 238 CE, they slaughtered their way through six of them.)

Although punishable by death, Oung Chanthol, a member of the Khmer Rouge, secretly helped Cham Muslims escape, providing them with false documents, food, and information to avoid arrest. Svetozar and Slobodanka Vidović sheltered their Muslim neighbors during the Bosnian Genocide, Serbian journalist Nebojsa Pajkic exposed the death camps at Omarska and Trnopolje, and Yugoslav General Rifat Kadić was imprisoned for his opposition to the campaign of "ethnic cleansing." Paul Rusesabagina, manager of the Hôtel des Mille Collines in Kigali, sheltered over 1,200 Tutsi from government militias during the 1994 genocide (about as many as Oscar Schindler did during the Shoah). Zura Karuhimbi, a traditional healer, saved more than one hundred Tutsi refugees by hiding them in her home and threatening to curse any militiamen who entered. Abdul Aziz al-Hilu fought against the Sudanese forces and Janjaweed militia to protect people during the Darfur genocide. Kurdish communities in Iraq sheltered displaced Yazidis, a religious minority, from ISIS in northern Iraq, as the village of Le Chambon, France, resisted the Vichy government to save thousands of refugees.

For each name we know, many more struggled against genocides and bullying autocrats. We lack data on the rates of resistance. In the Eichmann studies of Stanley Milgram, only 3 participants—who believed they were torturing people—threatened to report it to the authorities and have the study terminated. Only 2-3% resisted administering the lower shock levels despite the demands and begging of the learner, but from 300 volts onward, people began to say no. (The "learner" had started demanding his release at 150 volts.) Only three participants threatened to have the study shut down; they did so at the 330-volt level, and all three of them continued to the maximum voltage. None attempted to free the student.

As shocking as this is (pun intended), challenges and refusal did occur. Far from the clear-cut obedience vs. disobedience dichotomy, most subjects showed opposition to orders with their nervous laughter, smoking, hesitancy, crying, and attempts to convince the researcher to stop the experiment [1688, 1689]. Although considered a form of resistance, this distress of the torturer would hold cold comfort for his victim. At 300 volts, active resistance, the refusal to continue, increased.

Disobedience also intensifies when at least one other person initiates resistance. When two others refuse to comply, noncompliance is induced 90% of the time [1690], suggesting that adherence to social norms may dominate our conformity rather than strict adherence to authority. Social approval releases the rewarding hormone oxytocin even when confronting social tensions [1691]. In times of social stress, such reinforcement could offset the distress caused by increased cortisol levels, increasing the need for social acceptance.

The desire to conform, be it to refuse or to obey, better fits with the experiences of known members of Nazi execution squads, many of whom complied with orders to murder to meet the expectations of their comrades, even when their commanding officers expressed doubt about the orders to kill or told them they could be assigned other duties without penalty [1692]. This jives with reports from Rwandan and Cambodian genocide perpetrators who said that without an external military force, they would have continued to kill [1693]. A restored social order alone would not have stopped them; the shared culture of genocide, the conformity to the established group norm, was too strong.

Most people find harming others distressing and repugnant. Pressure to inflict pain changes activity in the frontal lobe. On an electroencephalograph (EEG), the resulting conflict appears as an increase in theta waves, which are associated with conflict monitoring, cognitive control, memory retrieval, and emotional engagement [1694]. This change indicates increased coordination with other brain regions involved in rational control [1695]. The jump in theta activity tells us that the brain is analyzing and resolving discrepancies, such as a gulf between information and behavior or morality and social demands. Receiving an order to inflict pain, even on a virtual subject, reduces this theta activity, making it easier to commit the act [1696]. A field study of Rwandans born after the 1994 genocide demonstrated this effect: people who showed more resistance to delivering a shock to another person also exhibited higher theta activity [1393].

Although obedience clearly played a factor, Milgram may have undervalued the social bonding within the dyads he created. The grouping of the subject and experimenter inherent in the Milgram paradigm reinforced the power of obedience. Compliance dropped in the trials if the authority figure was isolated from the subject. The interactions of the duet could have forged a sense of personal bonding [1697], strengthening the power of the orders and further reducing the cognitive conflict and activity in the frontal cortex and the ACC.

Caspar [1698] found that most people motivated by personal identification aligned themselves with the source of power, not with the victim. Those who connected with the experimenter were more likely to follow orders to torture. However, it is unclear if they bonded with the power figure or the “cause” he or she represented. In contrast, identification with the victim had a much lower impact on the probability of disobedience. This result fits with our primate instinct to favor powerful coalitions over weaker ones within our social groups. Getting along with the powers-that-be can provide protection and elevated power and status within a community.

Laurent Bègue found that people who scored high in conscientiousness and agreeableness, both community-oriented personality traits, administered higher electrical shocks than those who scored lower. In other words, those willing to buck the trend and be disagreeable to the crowd were more likely to disobey [1699].

Resisters in such Eichmann studies typically cite their morality as their reason for noncompliance. Yet, people regularly override their moral compass and participate in actions they consider immoral. A primary reason given for obedience is identification with the cause, with some variation on doing it for science or data acquisition. This corresponds well with the reason given for protecting people from genocide or other persecutions [1700, 1701]. The more responsible they felt, the more freedom from coercion they felt, the more likely they were to intervene. However, in Caspar's 2021 study [1698], just as many people claimed they obeyed because they respected authority and believed that one should follow their orders. This could explain why many of us will ignore our own ethical convictions when the immoral choice aligns with the influence of secular law or religion. Throughout our childhood and for much of our evolutionary history, conformity and obedience aided survival.

So When Does Bullying Backfire?

> *"Die Gedanken sind frei."*
> *("Thoughts are free.")* — Von Fallersleben [1702]

Our intrinsic desire to please our leaders facilitated our survival for much of our human history. As children, conforming to rules handed down by our parents keeps us safe; we are built to be credulous and trusting of our elders. Just look at our willingness to believe our parents' tall tales about supernatural beings like

Figure 7-5. Gestapo photo of Hans Scholl [3355]. Hans, his sister Sophi Scholl, Alexander Schmorell, Willi Graf, and Christoph Probst founded a German student resistance movement, The White Rose, that denounced the Nazis as criminals. The Nazis executed all the founders. Hans continued to suffer another indignity after the end of the Nazi era: his founding role in the White Rose was portrayed as secondary to his sister's because he was homosexual [3229, 3453].

Santa Claus, a figure that violates every single bit of our experience [1703]. As we age, we see parents' control as less legitimate. This change of perspective confers adaptive value. It allows us to transition from the dependent child to the nurturing parent who instructs the next generation [1704]. We mature into a different set of relationships and expectations within our society. Traditionally, these transitions happened within kinship hierarchies.

Kinship groups that operated as interdependent units survived better than those that behaved as collections of independent individuals. Members of kinship groups with better survival rates also passed on their genes differentially. Those sets of genes included the ones that facilitated cooperative personalities. The inborn chemistry that makes us want to conform and obey flourished under kin selection pressures that favored pro-social behavior. It also sets the stage for genocide and the elimination of competing or potentially hostile neighbors [1705].

Sometimes, obedience fails. Someone defies the bully and says, "No, I will not participate in this torture or degradation anymore." In 2011-2012, the Middle East saw the fall of existing dictators in country after country as people used a new tool, electronic communication, to reorganize the power of groups. The resistance fighters in Nazi-occupied countries fought against the oppressors at high personal costs. The traditional German folksong (ca 16th century), *Die Gedanken sind frei* (Thoughts are Free, [1702], became a veritable anthem of resistance for political prisoners.[214] The White Rose of Munich sacrificed their lives in opposition to the National Socialists. After the Gestapo executed six of their leaders, other members continued the struggle, smuggling their manifesto from Germany to Scandinavia and the United Kingdom. The revolutionaries of France, Russia, Iran, the United States, and a myriad of other countries each chose to defy the powers that ruled them.

Why take the risk? Milgram's Eichmann experiments demonstrated that obedience was the norm but not absolute. A resistant minority existed. One good predictor of future defiance is early defiance. The sooner someone says "no," the more likely they are to keep saying "no" [1706]. Milgram and those following him identified several factors that allow someone to resist compliance with a bully.

Obedience or Conformity?

Acquiescence to the experimenter's will is the most common explanation given for conformity in the Eichmann experiments. However, the post-experiment interviews done with Milgram's participants suggest other factors may be in play, such as identification with science (the cause) rather than the person giving the orders [1707].

Most participants (72%) reported trust in their leader (the experimenter); they did not believe the "learner" was being harmed. However, this does not match well with the distress expressed while administering the shocks. Rather than constructing self-accusatory reasons for their actions, most subjects unconsciously produced after-the-fact, face-saving explanations for their compliance with torture [1708]. This explanation

[214] In one famous act of courage, the prisoners of concentration camps were ordered to produce something beautiful, a performance to celebrate Hitler's birthday. Hans Litten, a prominent and *early* critic of the Nazi government, presented this poetic text to the guards. Although the song's origin is unknown, it is believed to have originated in the 16th century during a different peasant rebellion.

helps explain the rationalization of non-prosocial behavior during the COVID-19 pandemic. Then, many ascribed their refusal to wear masks, isolate, or social distance to a defense of personal freedom and rejection of social control rather than bending to the will of a leader. However, much of the hateful overt narrative framed mask-wearers as compliant sheep who surrendered their wills to the state.[215] The symbolism of an imposed burqa, combined with common anti-Islamic sentiments, transformed the mask into an image of oppression [1709]. This, combined with calls for individual freedom, overrode empathy for at-risk people [1710].

When complying, we may remain largely unaware of our complicity. People who obey orders to harm another often show more theta brain-wave activity in their mid-frontal brain than those who obey immoral orders less frequently [1393]. On an EEG, this brain wave indicates increased conflict monitoring and error processing, suggesting that some Eichmann subjects are more likely to think about the immoral order than others.

There is another interpretation of Milgram's results: conformity to the group rather than conformity to the leader. In Milgram's experiments, the "teacher" stayed or communicated with the researcher, not with the "students." The subject may have conformed because the leader set a small-group norm, not because of a need to please the authority. The primate drive to conform with peer behavior is powerful. Humans will ignore the smell of smoke and the clang of a fire alarm if no one around them seems concerned about the threat [378, 1412]. Even babies converge on group norms, crying when they hear the cries of other babies [1711, 1712].[216]. Western parents know the social pressure of the schoolyard to possess the latest must-

---

[215] This hostile response to the mask and distancing mandates may seem like a self-centered response rooted in the hyper-individualistic culture of the United States. However, it can also be viewed as conformity to a sub-group norm that championed personal choice. Most middle-class Chinese reflected their more community-centered, cooperative culture. They considered distancing and mask mandates fair and considered themselves the beneficiaries of the mandates rather than the targets [3446].

have toy, and teens maintain a narrow range of hairstyles, clothing, and music options to avoid scornful labels from their peers.

The primate drive to conform with peer behavior is powerful. We may not aid someone in need if those surrounding us show no such inclination, just as the actions of others can prompt us to help [1713, 1714, 1715]. People are more willing to assist another person in a dangerous situation if at least one other person moves to render aid [1716].

Both group conformity and obedience to leadership involve group cooperation and contribute to the Eichmann results. Remember, our groupthink consists of an offstage analysis that assigns weight and importance to signals we receive from our friends, family, and surrounding culture. A small but powerful pleasure center in our brain, the ventral striatum, especially one subsection, the nucleus accumbens, plays a pivotal role in processing emotions and motivations [1006, 1717, 1718]. Together with our default behavior-center (posterior cingulate cortex), our social and introspective insula, and our cognitive prefrontal cortex, it evaluates the potential rewards for our actions [1719, 1720, 1721].

When we conform with our peers, this reward center, the ventral striatum, activates. Spoiler alert: Cooperation and conformity make us happier than disagreement and conflict. We typically ascribe higher reliability value to dominant leaders and to people we trust, and trusted leaders have less trouble retaining power and group loyalty [1350]. We also look at the numbers, and significant opposition can undo the leadership effect. Neighborhood engagement, for example, can stigmatize the perpetrator of domestic violence and increase the probability of the victim stepping forward for help, thus reducing the prevalence of spousal abuse [1722]. For autocrats and bullies, eliminating the opposition silences the voices that could threaten their control.

"The First Thing We Do, Let's Kill All the Lawyers"

Shakespeare did not affront lawyers in this often-quoted statement [1723]. He acknowledged their status. Lawyers know the law and can topple a revolution or de-legitimize a government's commands in the people's eyes. The perception of illegitimate claims of mastery inoculates against following orders [1394, 1724].

Maintaining a perception of legitimacy is crucial to gaining the cooperation of the governed [1387]. To become a successful dictator, one must possess at least a veneer of legitimate command. If the leader cannot back commands with action or power, if the status is questionable, or if the paradigm of control does not fit a population's beliefs, then the bully's influence decreases.

In Scandinavia, most notably in Denmark, the anti-Semitism of the Nazis made them less legitimate than they might have otherwise been perceived. This rejection of Nazi authority contributed to the inoculation of the Scandinavians against the madness. In annexed countries whose dominant anti-Jewish traditions followed different rationalizations and paths, the extreme measures imposed by the Nazi government instilled an unexpected resistance to their Operation Final Solution [1435]. What the Nazi hierarchy imagined as a unifying force across Europe became an unanticipated source of contention. Many countries with a history of anti-Semitism still denounced many Nazi dictates while accepting some, but not all, of the anti-Semitic aspects of their doctrine.[217] As the leaders and populations in these nations renounced the imposition of anti-Semitism, their modeling sometimes inspired local National Socialist officials who often rejected the more extreme stance of their home government and the orders emanating from Berlin [1435].

---

[217] Western Europe showed greater resistance to Nazi-style antisemitism than Eastern Europe. Many occupied countries also demonstrated kin-selection, with greater resistance to the deportation and maltreatment of their own Jewish population [1435].

The perceived distance from the authority figure and the length of association impact obedience. Compliance dropped when Milgram left the room and phoned in his orders to torture. In an experiment in which doctors telephoned lethal orders to nurses, they, too, complied [1381]. The standing of the institutions (a university and a hospital) enhanced the legitimacy of both commanding authorities. The long-standing relationships between the nurses and doctors in the hospitals contrasted with the transient ones between Milgram and his "teachers." The group norm and hospital culture of obedience also led to more compliance.

Weak hierarchies have less acquiescence than strong ones. Military discipline uses solid chains of command to ensure soldiers obey orders that can either save their lives or get them killed. Trained soldiers have stood side by side in a phalanx in mutual defense and sacrificed themselves in trenches and on beaches, knowing they were destined to die. This counterintuitive behavior does not occur spontaneously. A history of interaction between soldiers and their superiors combine with a commander's perceived legitimacy and the desire to conform to the expectations of their platoon mates and retain honor in their eyes. These factors build shared confidence between the troops and their commanders.

Studies of Reserve Police Battalion 101, an "Order Police" from Hamburg, suggest that conformity to group norms influenced their compliance with the genocide of the Jews and Poles at least as much as the tendency to obey orders. As in Milgram's study, when officers told the soldiers they could opt for alternative duties if they felt uncomfortable with mass executions, only a handful refused to participate in direct involvement. Despite their disgust and guilt, many who complied did so not to please their commander but to avoid ostracism from their troopmates. Fear of breaking the group norm and being labeled a coward held a stronger sway than their commanding officer's permission to refuse [1692]. The same peer pressure that goads children into tormenting a victim in schoolyards and contributes to gang and intergroup violence drove these men to mass murder. The same desire to please our community, when channeled into prosocial norms, prevents most antisocial violence and has the potential to reduce violence by making hatred and bigotry unacceptable [1722].[218]

Similarly, people from collectivist cultural backgrounds, such as Chinese and Korean, share more common goals and obey slightly more than people from individualist cultures, like Western Europeans and North Americans [1725, 1726, 1727]. Responses to bullying also differ; collectivist victims ascribe less blame to the bully and more guilt to victims, including themselves, for not adopting group customs [1728]. Collectivist societies tend toward more homogeneity and value diversity less than those emphasizing individualism [1729]. Their social solidarity and pressure to conform let them address mutual challenges, such as economic slumps and pandemics, more effectively than individualistic WEIRD cultures [1727, 1730].

People who can handle unpleasant emotional stress also gain resistance to blind obedience. One tragedy of schoolyard bullying lies in its ability to produce chronic anxiety at an impressionable age. Victims of psychological or physical bullying tend to avoid attracting attention, fearing incurred wrath. This fear can last through life, making chronic victims more compliant as they desperately avoid unpleasant or challenging dominance encounters.

This fear is widespread. Approximately 25% of children expressed fear of bullies as their most pressing concern [265]. Low estimates indicate that 40% experience at least one bout of victimization [1731]. As many as 36% or more suffer post-traumatic symptoms extending into adulthood [1731, 312]. Adult intimidation and workplace social stress carry a high price tag in induced diabetes, immune, gastrointestinal, and musculoskeletal system disorders, and mental health disorders, including depression and burnout [1732]. In a 2007 study of trainee doctors in Tamil Nadu, India, 64% of the men and 48% of the women said the medical and paramedical staff had harassed them [63]. In Pakistan, 80% of psychiatrists in

---

[218] The question thus becomes: Why does it sometimes fail so spectacularly that humankind endures yet another holocaust?

training reported having been bullied within the past year [58], and burnout from workplace bullying contributes to teachers leaving the field and teacher shortages [1733, 1734, 1735, 1736, 1737]. During the 2020-21 school year, elementary school teachers related threats of violence and physical attacks at rates near 8% and 7%, respectively. High school teachers reported 4% vs 1%, with women bearing the brunt of the physical abuse [1738]. With both adults and children, most bullying incidents are believed to go unreported, and many victims lack faith that administrators will address the problem.

Finally, groups can reinforce the authority of a leader, including bullies, or they can reduce compliance [1690]. Collective acquiescence to the leader increases our programmed reticence to go against an established leader and to follow our moral values. Dissent within the ranks opens the leader's legitimacy to question; group compliance decreases when individuals openly oppose an alpha. Once the first few opponents defect, it becomes harder for a dictator or ruling coalition to maintain power.

Evil People or Evil Situations?

> *"I returned, and saw under the sun, that the race is not to the swift, nor the battle to the strong, neither yet bread to the wise, nor yet riches to men of understanding, nor yet favour to men of skill; but time and chance happeneth to them all.*
>
> — Ecclesiastes (Book of Kohelet) 9:11, *The Bible* [1739]

We react to the behavior of others differently from how we respond to our own. Most people see themselves as basically good people. When we do something against our "character," we ascribe our shortcomings to the situation in which we find ourselves. Of course, we did "that," but look at the circumstances—or if I hadn't cooperated, worse would have happened. Personal failures fall beyond our control. We extend this internalized leniency to family and friends; we take a jaundiced view of the motives of outsiders.

> *"Until the last quarter of the twentieth century, racially explicit policies of federal, state, and local governments defined where whites and African Americans should live."*
> — R. Rothstein, 2018 [1743]

For someone outside our social circle, we ascribe their behavior to their character. If I lack work, I blame a disability, my employer, or the economy. If a stranger is without employment, he or she is a lazy freeloader. We find this I-thou bias worldwide. In a Zimbabwean study, students ascribed success to intrinsic characteristics of people with whom they identified while attributing their failures to circumstances. They reversed the causes and effects for people from other groups [1740]. Women recognize *institutional* sexism more easily than men [1741]. American Whites are more likely to consider racism solely as discrimination by individuals than are African Americans, failing to recognize the systemic roles of institutions [1742, 1743, 1259, 1744], and cis-gendered people are more likely to ignore discrimination against the trans-gendered and nonbinary [1745]. In short, whatever the "-ism" is, those harmed by the inequity develop a greater mindfulness of it than those who reap a benefit from imposing it.

People excluded from privilege recognize inequity and institutional prejudice and discrimination with more ease than the advantaged; those benefitting from social status are likely to dismiss the other's experience with pejorative use of terms such as "wokeness," "political correctness," or "reverse discrimination" [1746, 1747, 1748, 1749]. In the United States, civil rights activists use the slogan "Black Lives Matter" to remind people that we often overlook the extrajudicial murders of African Americans (lynchings). The phrase evokes wrath in others, who twist it to imply only Black lives matter and suggest reverse discrimination [1750]. The term "All Lives Matter" attracted slightly more White than Black support [1751]. This fits the general principle: we identify with and provide more support to "us" than to "them," and, like it or not, tailoring the image toward the empowered can be powerful.

This attribution difference lets us retain the illusion that we, as individuals and families, are superior to others, deserving more rewards and less punishment than the other guy. This attitude became evident in the Zimbardo prisoner-guard experiment, though both parties were intellectually aware that their roles had been determined by chance alone [1752]. This attribution interplays with our internalized premise that the world is fair and the less fortunate deserve their fates.

Beliefs about the world or a situation being fair or biased originate in two brain regions—and yes, they are among the usual suspects: the insula and the ACC (anterior cingulate cortex). These respond to social breaches and unfairness. Fairness perceptions also activate areas at the border of the temporal and parietal lobes, the rTPJ, that lets us shift our attention and alerts us to the incorrect beliefs of others.[219] When watched in an MRI, all three regions increase their activity in response to a norm violation; when prompted with a just-world scenario, only the usual suspects, the insula and ACC, light up [1753, 1754]. These areas process the contradictions while we suppress our rTPJ, which could alert us to the thoughts of others.

Faith in universal justice links to negative attitudes toward affirmative action among the lucky and enfranchised [1755]. After listening to a live broadcast of America's Vietnam conscription lottery, the young men less likely to serve considered the system more fair than those who received low draft numbers [1756]. On average, CEOs see themselves as more worthy of success than their subordinates, tenured faculty consider themselves more capable and intelligent than those outside the tenure track, and those who live in the developed world see their state as proof of their and their country's innate superiority. Similarly, we tend to perceive our own social group as morally superior and more deserving than others. An Israeli study showed a film of a staged taxi accident. Both Arabs and Jews attributed less blame to the driver if they thought he was a member of their own group [1757].

---

[219] The rTPJ = right temporal, parietal junction.

Few admit these feelings, even to themselves. These prejudices seldom enter our conscious thoughts, but we recognize them when we scrutinize the behavior of our kind. They reveal themselves in nepotism and our differential standards for trust and protection among family, friends, allies, and others.

> *"She told [captain] Turner that her son's death had been unnecessary—that it was caused by the lack of organization and discipline among the crew."* — words of a woman to Captain Turner of the Lusitania [1654, p. 297]

For most of our evolutionary history, this instinctive "I-thou" dichotomy, this biased attribution of intent, helped us predict the actions of troop members and strangers. By tuning their responses to the social environment, our ancestors increased their probability of staying alive long enough to reproduce. They benefited from defining fundamental differences between "us" and "them," building trust among kin and suspicion of less related, competing primates.

> *"Here, also, are the veterans of King Philip's War, who burned villages and slaughtered young and old, with pious fierceness, while the godly souls throughout the land were helping them with prayer."* — from: *At the Indian Killer's Grave* by Robert Lowell, 1946 [3168]

We make assumptions about the behavior of the target group. The people who lynched African Americans or who took pride in being "Indian Killers" believed their victims deserved neither respect nor human status.[220] Our species makes assumptions about the behaviors of the target groups. The horror of violence perpetrated on the victim diminishes through those simple assumptions, justified or not. By starting with different axiomatic beliefs, people reach different conclusions about the morality of torture, murder, and war. Humans can draw different conclusions about using violence against groups, individuals, or animals.

---

220 "NHI" is a nasty and unofficial bit of American slang: No Human Involved. It refers to crimes against society's throw-aways, such as prostitutes, drug addicts, the homeless, gang members, or sex offenders.

Figure 7-6. Ota Benga (ca 1885-1916) at the Bronx Zoo, ca 1915-16. Photographer unknown, ca 1906. [3250]. The maltreatment of the Bambuti and other Ituri forest people has deep roots, as does world collaboration with their oppression. The St. Louis World's Fair (1904) displayed Mr. Benga as part of its ethnology exhibit. He became a featured attraction in the Bronx Zoo's monkey house in 1906, forty years after the "official" end of American slavery in 1865. He drew huge crowds until African American protests forced the exhibit's closure 20 days later. After years of insisting that Mr. Benga served as a zoo employee, officials apologized for the display 114 years later—in July 2020 [3071, 3211]. His attempt to return to Africa was thwarted by the outbreak of World War I. Mr. Benga took his life in 1916.

Today, we experience more non-kin interactions than ever before. We fight societal disputes with guns and bombs, not spears and axes. Our technology has amplified the potential for mass violence. Despite our enhanced ability to slaughter, the millions that died during the twentieth-century wars can pale when compared to tribal warfare. Although genocides or gendercides may target specific groups, overall civilian deaths typically fall below 1%, and combatant deaths range near 5%. The estimated 70-85 million World War II death toll represented 60-66% of the European Jewish population but only about 3% of the overall population. This figure includes the firebombing of Dresden, the deaths in the concentration camps, and the atomic bombs dropped on Nagasaki and Hiroshima. In tribal warfare, the overall casualties average about 15% (range 10%-60%), and genocide of one's neighbors can be the stated objective [1758]. We underestimate the carnage of traditional wars with spears and arrows because of the small bands involved. Relative to their population size, these clashes have been Holocausts.

Remember, for much tribal warfare, genocide has been the point. When we compare modern wars to tribal ones, we compare conflicts with different goals. The lower civilian kill rates in industrialized wars include people not targeted per se. Groups earmarked for genocide fare on par with the higher losses seen in band warfare. In Bosnia and Herzegovina (BH), the 1991 census estimated the population to be 4.4 million; the next census (2013) estimated the population at 3.5 million, a 20% population drop. Before that holocaust, their crude total death rate was similar to that in the United States (0.86%) and the Netherlands (0.9%).

> *"A plague o' both your houses!"* — William Shakespeare, *Romeo and Juliet*, Act 3, Scene 1, (1597)

Between 1992 and 1993, 100,000 people died in a genocide, 80% of them Bosnian Muslims. Adult male Bosniaks suffered disproportional casualties that were triple the overall mortality rate. In the Srebrenica massacre alone, nearly 8,000 men and boys were slaughtered. Following the traditional gendercide pattern, the death rate for women stayed below the death rate for men [1759]. With these figures, the kin-selection advantage of gendercide becomes apparent.

Following the genocide, the male death rate in Bosnia dropped below Serbia's and approached that of the United States by 1996. The survival rate for Serbian women remained relatively constant. In contrast, Bosnian women had an enhanced chance of survival, with their mortality rate dropping to near the pre-war level [1760].

In these modern genocides, the carnage is proportionate with that in tribal wars. 2002-2003, operation *Effacer le Tableau (Clean Slate)* wiped out an estimated 40% of the Eastern Congo's Bambuti civilians. The massacres included reports of enslavement and cannibalism [1761, 1762]. When the Nazis came to power in 1933, there were 9.5 million European Jews; by 1945, 66% had been killed. The Rwandan Genocide killed between 8% and 10% of the country's population but slaughtered 77% of the Rwandan Tutsi population [1763, 1659].

Paradoxically, our violent side may have driven our cooperative one [954, 951]. Like bonobos and unlike chimps, our ability to cooperate and live in larger bands and foraging groups increased the cost of raiding and warfare [1764, 1765]. We harbor a propensity to nurture blood feuds for generations, with each side blaming the violent disposition of the other while overlooking its own role. This inclination can fuel local hatred, making self-sacrificing behaviors worth the costs. By surrendering the autonomy of revenge to a state, we reduce the peril of violent individual disputes and provide ritualized mechanisms for justice and reconciliation [1766, 1767].

However, the adult offspring of genocide survivors are not necessarily inclined to forgive and forget. Indeed, the children of traumatized mothers can be less favorable to reconciliation efforts than their parents, even after controlling for the PTSD of both generations [1768]. Evidence of humanization by a former tormentor can contribute to conciliatory attitudes. It does not significantly enhance the victim's trust or render them open to negotiations or future interactions [1769]. Unless relatedness is high (kin selection), the cooperation of a subordinated population is likely to have strings attached and to be accompanied by resistance and low-level aggression [1770]. Memories of attacks, enslavements, and betrayals can span generations, influencing multiple facets of our lives, e.g., [1771, 1772, 1773].

When blood feuds rise to the group or state level, our inner chimpanzee surfaces. We join together in troops to hunt down and kill "the others" who threaten us or our assets. In the ongoing Israel-Gaza conflict, for example, both sides consider the contested land and resources rightfully theirs. With the boundaries for an acceptable compromise miles apart, with each group believing itself to be the wronged party, the prospects for bloody revenge are great.

The cry of "never again" that rose in the wake of the Shoah could not stop the holocausts of Rwanda or Cambodia. During the Bosnian "ethnic cleansing," many Westerners resisted labeling it as a "genocide," even as Serbian forces slaughtered thousands. The hesitance of the West to intervene on behalf of the targeted Bosnian Muslim deepened the mistrust of the West in Muslim countries and deepened the animosity toward it.

Our biases can make the recognition of an attempted genocide difficult, but forgetting past abuses can render it near impossible. Governments that commit holocausts—either directly or through the support of a perpetrating entity—will seek to erase their role. Many pass memory laws designed to suppress the history of the atrocities, forbidding the dissemination of information or revising the context to favor the executioners [1774, 1775]. The USSR denied the existence of the Ukrainian famine (the Holodomor) in 1933, the executions of kulaks (dekulakization), and the dehumanization and internment of political opponents. Turkish policy denies the mass slaughter of Armenians. The Nazis portrayed their extermination of the Jews as resettlements. Some American states have introduced bills to limit the ability of schools to discuss the enslavement of Africans or the post-Civil War Jim Crow laws and peonage slavery. As of August 2024, approximately 96 such bills have been introduced; most of them have failed or remain in committee, e.g., SC HB4325; AR HB1218; WI SB411 [1776]. And the historical marker at the Little Bighorn National Monument in South Dakota continued to refer to it as a battle rather than a massacre until the 1990s. Although the results are mixed, with sufficient political support, the acknowledgment of tragedies can reduce group retribution and continuing blood feuds if it encourages dialogue (the reintegration of opposing sides into one) and includes a framework for sufficient fairness and justice to provide a sense of vindication without fueling resentments.

We may have some safeguards against violence, but industrialized societies have driven up the cost of war. We have entered an age where our traditional, instinctive views of outsiders threaten to destroy us with the push of one nuclear button. Even at the precipice of disaster, politicians vie for power using zealous propaganda to create the perception of internal and external threats only they can control. Nuclear and biological warfare are not childhood games where we can call a "do-over" if we get it wrong. Once our drives for vengeance and spite take hold, no victor will stand with their spear poised, kicking the cadavers of their enemy into the muck. If we survive, civilizations will collapse, and our depleted world will lack the resources to rebuild.

CHAPTER 8:

Vengeance and Spite

Multiplying Mistrust

> *"With every day, and from both sides of my intelligence, the moral and the intellectual, I thus drew steadily nearer to the truth, by whose partial discovery I have been doomed to such a dreadful shipwreck: that man is not truly one, but truly two."*
>
> — R. L. Stevenson, *The Strange Case of Dr. Jekyll and Mr. Hyde*, chapter 10, [2]

Humans engage in violence, cooperation, and trust, balancing them to survive. We err on the side of caution, presuming danger rather than safety. If incorrect, we can revisit the innocuous situation; if we assume a malignant condition is benign, that error could kill us. This bias easily tips the scales toward intergroup conflict and our darker angels.

> *"The best thing would be if there were not a single Muslim left on this Earth. Then we would have reached our final goal."*
> — ascribed to a leader of a Danish-Swedish anti-immigration group, Stram Kurs (Hard Line). Cited by Omar Suleiman [3241].

A few leaders professing blame and violence can undo the efforts that countless others spent healing rifts. In his quantitative analysis of violence, Lewis Richardson [1777] argued that we inch closer to homicides, feuds, and warfare with each hostile action; peaceful gestures do not readily negate ill will [1778]. Personal rifts, like national ones, can escalate with posturing and memories of mistrust and threats. The history of tension between American police forces and African-American communities illustrates this concept. In India, the perceived lack of government action during the Godhra massacres fuelled ethnic hatred between Muslims and Hindus

[18].[221] One minister who threatened to burn copies of the Quran on September 11, 2010 [1779, 1780, 1781, 1782] provoked rage toward Americans in the Middle East. Word spread that no Qurans were burned,[222] but a handful of mullahs urged revenge and generalized the actions of that small group as if they reflected those of all Americans [1783]. The minister made good on his threat on March 20, 2011. The anger it evoked resulted in riots and further eroded trust between communities, making overtures of peace more challenging to achieve.

Most comments by Arabic-speaking YouTubers had nothing to do with the affront (33.8%); about 20% of commenters directed their anger toward the minister, the US, or the American forces. Only 2.5% called for war against the US [1784]. Most Americans were horrified at the desecration, and the minister soon fell from the headlines. Unfortunately, Quran burning continued in other parts of the world, often as part of anti-immigration or anti-Islam movements.

Humans fall prey to a reasoning error in which we see ourselves as superior. We see the mistakes of everyone but ourselves, and so telling others how to live their lives becomes tempting. Our families, clans, and nations magnify this human tendency, and we assume that the rest of the world should be like us. Only through our specific interpretation of the world can peace, harmony, and freedom ever exist on Earth—again, at least for everyone who counts in our unacknowledged worldview.

Some people find their assurance and solidarity under a flag or a tribal tattoo. To others, religious affiliations are critical to distinguishing friends from foes. Signaling membership has roots deep in our ancient psyche.

---

[221] The Godhra massacres started in 2002 when 59 Hindu pilgrims were killed in a train fire. Although no one could trace the cause of the fire, many Hindus blamed Muslims and retaliated with three days of violence in the far-western state of Gujarat, killing over 790 Muslims and 254 Hindus, and destroying 230 mosques and 274 shrines [3358]. Many more went missing, were raped, or were injured, and some sources have considered it a pogrom: a riot designed to kill or expel a minority population.

[222] I refuse to give the names of the ministers and politicians here, not to protect them but to deny them the publicity upon which their hatred thrives.

At one time, the imposition of tattoos or scars, the complexities of ritual, the understanding of mythologies, the reactions to inconsistencies, and the parallel non-physical reality of faith served as reliable markers to identify who was related and to brand strangers amid our kinship group.

Our species readily develops distorted beliefs about outsiders. Communities are composed of people, most of whom see themselves as superior to those who surround them. People instinctively extend this sense of self-enhancement to their companions (pseudo-kinships). Our individual and group sense of moral superiority showcases this self-enhancement effect [168], as does our propensity to believe that we will prevail in a fight or war.

We combine our distorted perceptions with learned prejudices and biases. Thus, White supremacists believe that non-Whites are inferior, that they want special privileges that are not available to Whites, or that there is a conspiracy to dominate, replace, or enslave people of European descent. Many Qutbists believe non-Muslims, especially Westerners, are uncouth sexual animals and have a deep-seated hatred for Islam and Muslims. The average American thinks of active White-Supremacists and Qutbists as hate-filled, but many "typical" Americans (of all races) hold less-extreme beliefs about some group. They may love family and country but express hostility to people they perceive as outsiders.

Humans do not experience these perceptual differences because some have inborn evil traits. We learn our prejudices across our lives, building them upon a biological scaffold that warns us to err with caution. Discrimination involves multiple brain regions. Biases about outsiders light up our amygdala, medial prefrontal cortex, and insula (a.k.a. the usual suspects). Intense dislikes bring more parts online, including those involved in threat detection [1785, 1786]. In contrast, becoming a victim of discrimination impacts the regions of the dorsolateral and dorsomedial prefrontal cortex, parietal lobe, ventral tegmentum, and hippocampus associated with increased depression, anxiety, and PTSD [1787, 1788].

The oxytocin that wraps us in a warm-fuzzy and protective feeling with our friends and family also increases our hostility toward strangers, fine-tuning our ethnocentrism and intolerance of others, and sculpting our empathy and revenge [1789, 1790, 1791]. That same boost we experience with our peers intensifies hostility toward those who violate our sense of social norms. However, pairing people who score high on xenophobic checklists with charitable and accepting peers takes advantage of the group norm and drives them toward reduced prejudiced responses [1792]. Activity in the facial recognition region of the brain (fusiform gyrus) increases in people with high xenophobic responses to visually identifiable groups [1793]. Our sense of danger—the feeling that we should retreat—increases activity in our left anterior brain (approach-withdrawal conflicts). It activates parts of the parietal region, providing sensory interpretation and rousing the body for passive avoidance by activating the right hemisphere or active avoidance by activating the left [1794, 1795, 1796].

The parietal region is chock full of mirror neurons that anticipate the actions of others and let us understand their objectives. Most studies associate mirror neurons with empathy and social perception. Their role in empathic behavior and social understanding may also elicit stereotypes. They bias gender perception based on stereotyped expectations [1797]. Because xenophobic fears are based on stereotypes, cultural misunderstandings, or past negative experiences that activate these neurons, they *could* lead to biased conclusions about the intent of strangers, including dangerous conspiracy theories used to cope with abrupt life changes. During the early days of the COVID-19 pandemic, unproven, xenophobic beliefs fueled violence against Asian Americans [1798], much like AIDS directed blame and vengeance toward gay males in the 1980s, and the Muslim community suffered misdirected retaliation following the bombings of September 11, 2001. Each event triggered cultural upsets, and each had an identifiable scapegoat population.

> *"Gloria: Ma, how do you feel about the new neighbors? I mean, you never told us how you feel about Black people.*
> *Edith: Well, you sure gotta hand it to 'em. I mean, two years ago they was nothin' but servants and janitors. Now they're teachers and doctors and lawyers. They've come a long way on TV."*
> – Lear, Nicholl, & Joseph, *All in the Family*, Season 1, Episode 8 [3351]

Another brain region, the temporal-parietal junction (TPj), regulates the social context of our behavior. This neural hub integrates language with social processing and memory [1799]. The TPj participates in our "theory of mind," our interpretation of the motivations, thoughts, and feelings of others. The TPj becomes more active when hearing phrases that violate our preconceived norms about a group [1800]. It drops its activity when exposed to hate speech, often coherent with unspoken social norms, reducing our empathy toward that group [1801].

The N400 effect is a normal EEG (electroencephalograph) response where the needle deflects around 400 milliseconds after receiving a stimulus inconsistent with our beliefs or experiences, or when we hear an unexpected concept [1802]. Sentences such as "I swam in the book" or "I read the pool" would elicit this response. (A friend telling you they just received thirty dollars to share, so they are giving you five dollars would also cause an N400 deflection.)

People also show a more robust N400 response for phrases and images that challenge ethnic or gender stereotyping than those congruent with stereotypes [1803, 1804, 1800, 1805, 1806]. Our brain processes the conflicting information as if it were fact-checking for an error [1807]. Positive contact with outgroups can reduce bias [1808]. Even films presenting positive images that challenge internalized negative prejudices temporarily suppress the N400 response [1803, 1809]. Contrary to the assumption that media images of successful, likable African Americans (or other disenfranchised groups) will lead to the misperception that discrimination has ended (Enlightened Racism), these portrayals increase awareness of the discrimination around us. That positive effect vanishes when we are reminded that the subjects are atypical [1810]. That result may reflect the perceived threat to the social status of the dominant caste.

The condemnation of negative images by trusted authorities can counter their impact [1811, 1812]. Something different happens when we view "media literacy" videos. If perceived as preachy, these videos prime negative prejudices, raise hostility in their aftermath, and suppress positive bias (positive stereotyping). They activate bigoted outlooks and make the viewers more skeptical about the content, so we discount it. [1813]. Given the low production value, stilted situations, and tacky, over-the-top acting in many educational videos, the observer's brain also spots the artificial cadence of the actor's voices. It registers the message as false [1814]. When content is concordant with our views, we overlook those signals. Such eusonance helps explain the online political echo chambers that often overlook facts or culminate in conspiracy theories [1815]. It also explains how so many adults of their day could have considered as realistic the hokey, melodramatic drug-exploitation film *Reefer Madness* [1816] and the anti-drug fictional book *Go Ask Alice* [1817] while the desired teenage target audience considered them an affront to their intelligence. More recently, polarizing echo chambers have led otherwise rational people to believe absurd claims such as space lasers caused the California wildfires [1818], that the British royal family ranks among a conspiracy of shape-shifting space lizards trying to control the planet [1819, 1820], and that Haitian immigrants are eating people's pets in Ohio [1821, 1822].

Overt, self-righteous tones countering preconceptions entrench the targeted beliefs.[223] The brain shifts, tweaking activity in multiple regions,[224] responding as if the statement were "false." The anterior cingulate cortex responds to the perceived social breach and monitors for conflicts and errors. It processes the moralistic tone as a lie and closes itself to the perceived propaganda. The amygdala kicks in, warning us of danger [1823]. The left TPj (temporo-parietal junction) increases activity, making us consciously aware of

[223] Rejecting the message and ceasing to listen is not unique to social prejudice. We respond the same way when someone attacks other long-held beliefs. Our brains tell us the other person is lying or deluded.

[224] I provided only the highlights in the main text. Multiple brain regions change their activity. Detecting lies calls upon social knowledge, self-protection, and our theory of mind. For more technical information beyond the text, see: [3183, 3399, 3150, 3398].

the other person's words and increasing our ability to filter them [1824]. The rational and social prefrontal cortex registers dissonance between its learned patterns and the new idea. We respond to the input as if it came from an untrustworthy stranger. If the brain concludes the incoming data is manipulative and false, these barriers prevent dangerous "misinformation" from entering.

In a genocidal world where contacts also bestowed benefits, we would gain a survival advantage from this combination of heightened physical awareness and sense of danger, rapidly accessed models (prejudices) about strangers, and sharpened facial (kin) recognition. Consequently, strangers can trigger our decision-making center in the right dorsolateral prefrontal cortex [1825]. This executive region holds a trial, weighing the evidence against the others with the aid of the prosecutor, our right temporoparietal junction (rTPJ), which is essential to our theory of mind. Together, these regions hold sway over our sense of us vs. them [1826]. The livestream of data reaches our motor cortex, and we make survival plans as we categorize the new people into a primitive paradigm of good "us" and bad "them."

We view ourselves as principled people who do bad things because of circumstances. However, our perception of "them" contains less forgiveness and understanding. We believe their strays from righteousness happen because they are innately wicked. When we fail, we blame circumstances; when others fail, it is not because of inequity but because we succeed on merit [1827]. Hence, we ascribe fundamentally different reasons for why *we* do things and why *they* do things [1828]. These inborn tendencies lead us to snap judgments about safe and unsafe interactions. Communities disagree about who constitutes a threat, the bad "them," and how the good "us" should respond.

This thinking error does not depend on age. Children also consider themselves morally superior to their peers and classify the world into in-groups and out-groups. They believe themselves more likely to intervene to stop bullying than other children. Like adults, they worry about what others think. If they sense their mates approve of the bullying, children defend less and join the bully more often than if they detect disapproval [1132].

When different groups interact, problems can arise. The fundamental attribution error leads each conflicting party to see the other as morally inferior. They also presume creative, intellectual, and rational superiority. Such beliefs about our families and cultures are essential to our well-being. They alert us to dangers and damage our ability to listen to each other and resolve conflicts [1829].

Much misattribution error stems from the I-Thou, good-bad way we think about other people. It blinds us to the complexities of situations; it allows us to ignore any part of a story that does not fit our narratives. When we dismiss people like Adolf Hitler and Osama bin Laden as evil incarnate, we free ourselves from the need to probe and understand the circumstances that led to their atrocities. Without that information, we have little hope of stopping similar events. Osama failed to recognize the complexity and diversity of American opinion and society when he engaged in anti-American prejudice.[225] Others will follow his path unless we understand what led to his hatred. We fall prey to the standard fare of dehumanizing metaphors that bombard us during conflicts and the prelude to oppression and genocide. It takes more effort to travel the more difficult path of looking beyond the images toward a critical analysis [1830]. Someone who chooses understanding rather than explicit condemnation and says "no" to a lynch mob can also find themselves labeled as traitors to their community. They can also spark an army of resistance.

This groupthink, this natural shortcut in our cognition, allows us to simplify and focus when we have only partial information. It enables our species to respond quickly to other individuals and groups. For our ancestors in the Pleistocene era (~2.6 MYBP-11,000), rapid decisions about other humans based upon

---

[225] A thought on this note: how many of us would gain the title "evil" if we were remembered for the worst thing we did in our lives?

minimal information would have given our ancestors an advantage over clans who pondered an outsider's intentions at length.

For most of our evolutionary history, our cautious ascription of positive motives to our troop or clan and negative attributes to outsiders contributed to our survival. In our twenty-first-century, multicultural world, it has become a double-edged blade prepared to disembowel us as we hover our spear-wielding Pleistocene hands over nuclear triggers.

Today, most live in crowded cities and towns with thousands or millions of others. Presuming all outsiders constitute threats no longer serves its original purpose. A prudent paranoia keeps us safe from potentially violent neighbors, but assuming all strangers are potential threats tears apart our spirit with fear and hate.

> *"For generations, the institutions of slavery and involuntary servitude were defined by the race, color, and ancestry of those held in bondage. Slavery and involuntary servitude were enforced, both prior to and after the adoption of the 13th amendment to the Constitution of the United States, through widespread public and private violence directed at persons because of their race, color, or ancestry, or perceived race, color, or ancestry. Accordingly, eliminating racially motivated violence is an important means of eliminating, to the extent possible, the badges, incidents, and relics of slavery and involuntary servitude."* — Article 7 of the Matthew Shepard-James Byrd Jr. Hate Crimes Prevention Act, Senate Bill 909, 111th Congress, 2009 [3293]

Hatred and its resulting dehumanizing violence vented against a person or a group can yield unforeseen consequences. Brutal attacks can awaken a collective conscience. Matthew Shepard (1976-1998) was brutally tortured and left to die alone, crucified on a remote rural fence. His murderers killed him because he was gay.[226] His torturers used Matthew's sexual orientation in their initial rationalization for their crime. That same year, three murderers tied James Byrd Jr. (1949-1998) behind

[226] The prosecution alleged McKinney and Henderson pretended to be gay to lure Matthew from a local bar. McKinney's lawyer mounted a "gay-panic" defense. Some argued that McKinney and Henderson targeted him only because they thought he had money to steal. Even if the latter proved true, the defense used Mr. Shepard's homosexuality as a ploy, hoping the jury would consider homophobia acceptable grounds for a heinous crime.

a truck and dragged him for two miles across a stony pavement, leaving a trail of body parts [1831]. He died after his body struck a culvert, tearing off his right arm and head. Later, people found his limbs along the roadway. His body was barely recognizable as human. His murderers slaughtered him because he was Black. These horrific crimes led to the bipartisan passage of the Matthew Shepard and James Byrd Jr. Hate Crimes Prevention Act (SB.909) by the US Congress on October 22, 2009.[227] American President Obama signed the bill into law six days later.[228] (Not everyone agreed. Ninety-five US representatives who voted against this 2009 hate-crime protection act still held office in 2017 [1832].)

> *Republican Representative George Henry White of North Carolina introduced antilynching legislation to the US Congress in 1900. It failed. It failed over 200 times. In 2005, the US Senate expressed remorse for its failure to act, but efforts to pass it stalled again in 2018 and 2020. Finally, On March 8, 2022, over a hundred and twenty years after Henry White's first proposal, the US Congress passed the Emmitt Till Antilynching Act that imposes criminal penalties for hate crimes that result in death or serious bodily injury or ones that include kidnapping, aggravated sexual abuse or an attempt to kill.*
> — US Congress. H.R.55—117th Congress.422 Yeas; 3 Nays; no abstentions. Republican representatives A. Clyde (Georgia), T. Massie (Kentucky), and C. Roy (Texas) voted nay. [3216]

The brutality of these crimes and their resulting publicity caught the country's conscience. Untold millions suffer from dehumanizing bullying and other violence. Worldwide, people are humiliated, tortured, and killed for political, ethnic, religious, gender, or social reasons. Communities hesitate to think poorly about or to act against one of their own. Societies often overlook the cruelty or support the tyrants who perpetuate it.

In part, bullies commit violence because they can get away with it.[229] An imbalance of power in the schoolyard, workplace, or political arena allows one party to attack another with impunity [1833, 1834,

---

[227] Despite the professed bipartisan support for the bill, congress did not pass it for another ten years after the murders. To see the text of the bill, visit: https://www.congress.gov/bill/111th-congress/senate-bill/909/text.

[228] In the wake of Mr. Shepard's murder, his parents, Judy and Dennis Shepard, organized the Matthew Shepard Foundation to "Erase Hate." They campaigned for the Matthew Shepard and James Byrd Jr. Hate Crimes Prevention Act and have provided hate-crimes prevention training to police and prosecutors.

[229] McKinney and Henderson, the two young men who killed Mathew Shepard, were barely adults themselves. The loss of these other lives to a moment of hatred also impacted the people of Laramie. No one wanted to believe that

1835]. Others suffer more subtle forms of bullying. Sometimes, the tormentors leave bruises on the body. They always leave bruises on the soul.

There must be reasons for our acts of cruelty. Something in our nature makes us condemn and lash out against fellow humans, often at the slightest excuse. We sacrifice ourselves for families and allies but also persecute our most vulnerable and sabotage our most capable.

Violence and prejudice seem to permeate our nature. We mirror many territorial social mammals with a higher violence rate than solitary species. Like our sister species, the chimpanzee, we engage in murderous coalitions. (Recall that World War II mortalities hovered around 3% of the population.) We kill outsiders at a higher rate than most other primates studied [1834, 1833]. Archaeological data suggests about 2% of prehistoric agriculturalists died at the hands of their fellow humans. This rate is about halfway between the *cumulative* violence rates (not just wars) of our closest relatives, the chimpanzee and the bonobo [1836].[230, 231]

Denying the violent side of our being has failed. Efforts to thwart the violence with civil sanctions have failed. For thousands of years, religions have preached the core rule of kindness and respect for others. For thousands of years, this social control has also failed. To circumvent the sanctions, we rationalize and encourage the belief that outsiders are less than human. Different cultures and religions have shared this

---

the perpetrators were locals from their community. The authorities soon had them in custody. The people of Laramie then had to grapple with the ongoing horror that two of their own had beaten and crucified Mr. Shepard on a fence. To them, McKinney and Henderson were Aaron and Russell.

[230] Contrary to our image of bonobos as love monkeys, they do engage in violence. Bonobos cause severe, maiming violence in zoos. Many subordinate males disappear from communities between seasons of field observation. Dominant males and females regularly bully low-ranking males, and they must subsist at the edges of the troop. No one knows how many become the victims of predation, how many bonobos die from wounds inflicted by other bonobos, and how many are murdered by their troop-mates.

[231] This study [1836] remains controversial because it does not cover all forms of brutality, such as infanticide. However, the study suggests that cultural and environmental components of violence can have a massive impact on the rates of violence.

excuse for violence[232] despite having the same golden rule at their core.[233] This professed standard has not stopped vicious attacks. At one end of the spectrum is violence and degradation to the individual; at the far end, enslavement, genocide, and warfare. Our kind detests the horror of murder but enjoys a strange satisfaction from trouncing a rival.

Vengeance

> *"For our part, we want the [shipping] traffic—the more, the better; and if some of it gets into trouble, the better still."*
>
> — Winston Churchill on the shipping of goods to England under German threat, cited in *Dead Wake: The Last Crossing of the Lusitania* by Erik Larson [1654, p. 190]

One gets a unique, pleasing blend of vindication and sadistic satisfaction when we see someone we consider innately evil get their comeuppance. From ancient theatre to modern movies, people have reveled in the villain getting a taste of his own sadism. We feel contentment. The outcome reassures us that the world has at least a modicum of justice. Although the physical universe is entirely neutral, these images reassure us that Fortuna will turn against the wicked and that society will avenge evil deeds.

We accept and understand the concept of vengeance. Our species has rejoiced in it and sung its praises from the ancient tales of blood feuds, such as *Beowulf* with Grendel's mother, to the modern morality tales in movies. Taking a life, at least for members of one's society or caste, demands satisfaction. Genesis (9:6), Exodus (21:23), and Deuteronomy (19:21) explicitly forbade the acceptance of blood money to compensate for the loss of innocent life. It demanded the killer's life be forfeited.[234] The Samurai class of Japan upheld

---

[232] Recall the illusion of moral superiority [168]. People often assume that the golden rule is unique to their belief system. We ignore the widespread nature of this same value among beliefs.

[233] Although the wording differs among religions, the golden rule holds that we *should* treat others as we wish to be treated. It is also phrased as the silver rule: Do not treat others in a way you would not wish to be treated.

[234] To be fair, remember the Bible would have people killed for working on the Sabbath, eating pork, or mixing cotton and wool into a blend. To my modern mind, only the last of these makes any sense.

family honor through vengeance killings (敵討ち). In cases too frequent to recount, a perceived offense against one nation has led to violent and bloody retribution.

Blood feuds can cross generations. They have descended into murderous rampages involving families, clans, or tribes. Trans-generational rivalries in Western cultures include the Campbells and the MacDonalds, a dispute dating from the Massacre of Glencoe in 1692. The American feud between the Hatfields and the McCoys spanned 1878-1891. The Donnelly–Biddulph community feud (1857–1880) terrorized central Ontario near London. In modern times, blood feuds among street gangs, from the Morans and the Capones of gangland Chicago to the Crips and the Bloods in Los Angeles, have caused horrific violence and sorrow for survivors and those caught in the crossfire. The deadliest grudges occur near the top of the food chain, where politics and quests for family power mix. Examples of national blood feuds include the War of the Roses (1455-1487), fought between the thrones of Lancaster and York; the Israeli and Palestinian conflict, in which both parties feel aggrieved and claim rightful ownership of the same land; and the Rwandan genocide in which the Hutus massacred the Tutsi.

Before the agricultural states regulated conduct by law, blood vendettas were long recognized as a legitimate recourse against a wrong. They remain a recognized form of justice in many traditional societies. Lee [1118] reported that among !Kung San of the Kalahari, 15 of the 22 homicides on which he gathered data had stemmed from blood feuds.

The earliest known codes of law, such as the code of Hammurabi (ca 1700 BCE), attempted to seek fair compensation for violence within their contemporary understandings of castes [1837]. These codified rules for vengeance across society promised a modicum of formalized retribution and reduced the need for inter-familial warfare. Blood feuds must have been frequent in ancient times. The Biblical Book of Numbers (35:6) declares that six cities should be refuge sites for those who spilled blood.

Even in a modern industrialized society, we demand vengeance. Sweden, which has a relatively enlightened penal system aimed at reform, has contended with a population that increasingly calls for harsher, more punitive sentencing [1838]. We want to see punishment imposed upon a person who harmed us. We bristle when we believe someone has broken an understood social contract that regulates society. Our desire for vengeance is not for a simple, just solution or comparative punishment. Instead, we aim for a disproportional infliction of harm, ostensibly to serve a dual purpose. The sentence imposes a significant handicap upon the offender to remove the benefits brought by multiple offenses. Secondly, we hope severe punishment will deter evil "others" from breaking the same contract.

Two additional reasons haunt our desire for disproportional punishment. As part of our I-thou dichotomy, we weigh wrongs with our thumb on the scale. We consider the harms perpetrated against us more significant than the wrongs we inflict. Therefore, we inflict disproportionate vengeance on the perpetrator to achieve vindication.

The other unstated reason for unbalanced punishment should haunt us. Vengeance has roots and tendrils in our reward centers [1839]. To the revenged, the infliction of a "just" castigation feels good, and we like to prolong agreeable sensations. Because the perception of justice differs between perpetrators and victims, the gap between those definitions can fuel further cycles of violence and spite.

Spite: Cutting Down the Tallest Poppies

> *"Tarquinius was in his garden when he received the messenger: he walked along in silence, striking off the heads of the tallest poppies with his stick, and dismissed the man without an answer. On this hint, Sextus put to death or, by means of false charges, banished such of the Gabines as were able to oppose him: by distributing their fortunes, he purchased partisans among the lowest class; and acquiring the uncontested rule brought the city to submit to his father."*
>
> — Titus Livius, *The Beginning of Rome and its Earliest Tribes* [1840, p. 375]

We engage in spite when we lash out at another being to the point of causing us some personal harm. After a person injures us, we see the behavior as revenge. If someone directs that malevolence at an "innocent" party, we call it bullying. Although we understand, often tolerate, or secretly applaud vengeance, we disdain the person who turns against an individual perceived as innocent. Instead of gaining sympathy, openly spiteful people who harm or torment a perceived innocent who falls within our circle of moral concern receive no such adulation. We applaud their downfall.

Between these extremes of righteous vengeance and petty sadism fall most acts of malice. Spite punishes someone else. Hamilton [1168, 1841] defined spite as one of the four payoff patterns for social interaction: mutualism (win-win), altruism (generous), selfishness (done for gain), neutral (no winner or loser), and spitefulness (lose-lose). These patterns mirror the interactions found in nature: mutualist (win-win), commensal (win-neutral), predatory (win-lose), neutral (no winner or loser), and spiteful (lose-lose).

Neutralism occurs when no benefit or harm comes from a relationship. Most experience it with the many tiny bacteria that grow on our kitchen counters. Most are harmless, and we live together quite peaceably. We cross the boundary into spite by taking steps to kill the peaceful co-habitants. The people with whom we ride on the train or drive on the highway have a neutral relationship with us, too. In contrast, we fight to the death with predatory cohabitants, such as bed bugs and mosquitoes.

Commensalism happens when one organism benefits significantly with negligible harm or benefit to the other party. The small mites inhabiting our eyelashes and hair follicles fit this category. They benefit from us, but we receive no serious harm or significant benefit from this association.

In altruism, one organism sacrifices itself or a resource for another individual. Self-sacrifice may seem maladaptive, posing a "Darwinian Puzzle." However, in social species, such as humans, most altruistic behaviors occur among relatives, thus providing an indirect method for our genes to reproduce through our family. Because many social species live within kinship groups, genes benefit from the survival of their carriers. We extend this courtesy to others who share our valued green beards and whom we treat as relatives, creating a network of pseudo-kinships. Some organisms extend their altruism to other species. Both humans and canines, for example, depend upon their communities for survival. Both assume risks for the benefit of their packs. These shared social values helped dogs and humans form a bond of affection across our evolutionary divide.

The perception of altruism and benevolence improves the chance that a primate leader will retain control. Defying social norms for personal gain reduces their air of legitimacy; violating standards for the benefit of another community member raises the trespasser's prestige and enhances it [1842, 1843]. Those who risk their status or resources for others, the "rebels *with* a cause," gain support during power struggles [1844].

Spitefulness happens when an organism takes steps to hurt another group or individual who has done no harm or who has provided it with a benefit. Such malice assumes many forms. One can deprive others of resources just to keep them from having them. One can eject a beneficial partner from one's territory to harm them or to derive a sense of dominance and superiority. One can imperil another organism or keep it from gaining safety. One can harm another without risking damage to oneself or accept personal harm to take down another.

Like self-sacrifice, spite can come with a cost to the individual. To a biologist, this price makes the behavior interesting. Like altruism, spite poses a "Darwinian Puzzle." Nature typically does not equip us with physiology or behavior that works against our welfare. Natural selection should eliminate such behaviors. Genes that encourage individuals to harm themselves should reduce fitness in their hosts. Therefore, both apparent altruistic and spiteful behaviors must confer benefits that are not immediately obvious [1845]. The genes supporting self-harming behaviors should become less frequent if it does not provide an advantage. Without a high pay-off, lineages engaging in spiteful behaviors should go extinct.[235] Malicious behavior, likewise, must help the procreation of genes by giving their carrier (the organism) a relative advantage that exceeds the cost of the malice.

The puzzles disappear if one reframes the goal as one of *relative* harm. If the spiteful animal can inflict severe damage on another being but sustain only minor harm to itself, it gains a *comparative* advantage over the target. Such asymmetry is the guiding principle of spite.

In social species, malicious behaviors set boundaries for community members and punish infractions of understood norms. If you offer me a quarter of your sandwich and I think it is an unfair offer, I can commit a vindictive act and destroy the entire sandwich. My "goal" would be to deprive you. If I were to destroy your property or deliberately sabotage a situation that benefited you, it would count as spite. When a bonobo, chimp, human, or other primate harasses another community member because they can, it is an act of spite.

---

[235] Technical note: There are ways to circumvent this global statement. Genes that promote spite could be linked to beneficial genes that outweigh the disadvantages. The messages made by the genes could also perform other functions, making both spite and altruism spandrels (by-products) of different processes. The widespread appearance of these behaviors across the animal kingdom suggests these arguments are less likely than the one presented in the text.

In social mammals, *perceived* injustices can trigger vindictive behavior. Perceptions form the foundations for interactions with any being. Our analyses of observations provide the only insights into what another thinks or feels. We cannot know their internal motivation, but we form conclusions about fairness based on what we see.

The desire to be treated justly has roots inside our heads. We get a boost of oxytocin when another treats us well. Remember, our brain is a series of exquisite, interlinked circuits created by a miniaturist. Its many regions specialize in functions. Our sense of fairness starts behind the middle of the forehead in the ventromedial prefrontal cortex (VMPC). This area modulates emotional reactions, defines our desires, and shapes our conceptions of fair treatment and our responses to abuse. Lesions in the VMPC create abnormally strong emotional reactions to perceived unfair treatment [1846, 1847, 1848].

Emotional responses from the VMPC and amygdala can create physical upset (somatic symptoms) and drive much of our learning, pro-social behavior, and aggressive outbursts [1849, 1850]. Some forms of depression, impulse control, and psychopathic behaviors also link to abnormalities of the VMPC [1851, 1852, 1853, 1854]. This region interacts with the emotional amygdala to construct parts of the gut feelings that guide our moral intuition [1855, 1856, 1857].

This same region weighs the long-term consequences of behaviors against their immediate payoffs [1858, 1859, 1860, 1861]. We can thank our VMPCs that more people do not succumb to the impulse to seriously harm someone who took the last donut from the break room. Spite can give immense gratification without regard for its possible long-term costs. When we act spitefully, we repress the inhibitory function of this region, essentially saying, "The devil with the consequences, vengeance feels good now." This region can work with other parts of the prefrontal cortex to remember which behaviors and events previously yielded a reward, such as prior fair treatment or other acts of spite. The VMPC can also conspire with other brain parts to mute or extinguish responses that elicit punishments or no longer receive rewards [1862, 1863, pp. 1-5, 148-166, 1864].

Through small acts of spite, we receive satisfying hormonal rewards for our dominance or aggression. Our brain chemistry does not distinguish among the sources of the rewarding cortisol or testosterone. It recognizes the hormones, not necessarily the difference between triggers. Acts aimed at "justice" and acts of spite trigger similar chemicals. Given the many reasons that a member of a social species seeks vengeance, justice, or status, it makes adaptive sense to interrelate these mechanisms and to keep their lines of distinction blurry.

Cooperative social organisms will engage in spite following perceived inequitable treatment, though not all social animals consider the same things just. Like humans, other primates keenly perceive the fairness or unfairness of a situation when they are the victim. Within their species, this indignation increases with the social distance between the individuals [1865, 1866, 1867, 1868]. Chimpanzees and other primates who receive an inferior reward from a human compared to a prize received by another will often refuse the treat. Chimps consider a benefactor's effort when judging fairness [1865]; organ-grinder monkeys (brown capuchins) take unequal pay as an affront without regard to effort. When a capuchin sees another monkey getting a better reward for the same task, they often reject their prize and throw it back at the human. As in humans, if that delicacy had once sufficed, it becomes unacceptable if another monkey gets something better [1869, 1870, 1871].

One can turn the tables on the spite experiment and ask not *how* a non-human primate responded to spite but *if* he or she would initiate a spiteful encounter without provocation [1872]. Chimps were given a chance to get food for themselves or for another chimp. The chimps did not care if their companion in the next room received food as long as they got theirs.[236] The snubbed chimp, however, was cognizant of the slight. Bonobos also ignored the desires of other adult troop members, acting for their own good [1873]. Humans may be aware of local snubs and fairness within our social circles, but as long as we benefit, we seldom think about the working conditions of laborers who teach our children or make our socks.

Humans respond to fair treatment, much like the chimps and capuchin monkeys. We reject unfair ultimatums to the point of self-harm. Humans get angrier and are more likely to rebuff a consciously or purposefully inequitable offer. Like the capuchin monkey, we bristle when we receive unfair treatment [1874]. Part-time workers who receive fewer benefits for doing the same jobs as full-time employees show less loyalty to the company and become less willing to meet calls for higher productivity.

> *"If a monkey hoarded more bananas than it could eat while most of the monkeys starved, scientists would study that monkey to figure out what the heck was wrong with it. When humans do it, we put their picture on the cover of Forbes."*
> — ascribed to Nathalie Robin Justice [3130]

Spiteful behavior increases both when resources are scarce and in the presence of unfair treatment [1875]. A business manager whom workers perceive as unjust, domineering, or dictatorial can expect resistance through psychological, economic, or physical sabotage, even at the expense of the company providing their jobs. Although demeaning employees appears to reap short-term benefits for the company or perpetrator, it can also result in long-term loss of productivity. Unfair treatment of a group drives them toward increased cooperation against a shared oppressor [1876]. In a world where fewer people and nations hoard more of our dwindling resources, we can expect less trust and more acts of spite and sabotage directed at the resource holders.

---

[236] In these chimpanzee spite experiments, the chimps came from the same social troop. Based upon the violence often perpetrated on outsiders, one would not rule out the possibility of spiteful behaviors if the experiment were repeated with strangers.

Power-to-take experiments model employee-labor relations and taxation. In both cases, profit flows from the work of the laborer (taxpayer), but the company (government) takes a share. Thus, worker and taxpayer discontent flows from the same sense of being ill-used. Both are variants of a dislike of "freeloaders." In these experimental scenarios, one person receives the right to take profits that another player earns. If the receiver takes a percentage deemed fair by the earner, he or she will continue to generate income. However, if the taker begins to demand too high a share, the earner will sabotage his or her earnings to spite the receiver [1877].

In a similar experiment, the ultimatum game, one player receives a small sum of money to split with a second player, the receiver. An offer perceived as unfair (differentially benefiting the offering party) creates an N400 shift on an EEG, a needle jump consistent with an unexpected (and, in this case, unacceptable) response [1802, 1878]. (This is the same deflection that happens when we receive information that contradicts our beliefs, prejudices, and stereotypes.)

Receivers can either accept or reject an offer. Although it varies across studied cultures, people often agree to an offer of 30-50% of the prize, with 50% being the most commonly accepted minimum [1879, 1880, 1881, 1882, 1883]. If the other player refuses the division, he loses the money for both parties. If the receiver considers the division unfair, he becomes more likely to reject the offer, punishing the selfish splitter, though the rejection also causes a loss for him.

> *"In the US, the federal minimum wage was $7.25 per hour. US Senators can take home up to $608 per hour for the hours they work in D.C., presuming 10-hour work days."* — A. Bolton [3308]

The receiver evaluates the proposal, factoring in the size of each share and his inferences about the splitter [1884]. In most cases (about 76% of the time), the proposers offered a fair split [1885]. This fairness is slightly higher than the 72% of reasonable offers chimps make in ultimatum games [1886].

We show a bias toward ingroup members, accepting lower proposals from them [1887, 1888]. Cultural norms lead to differences in offers and acceptance levels, and stereotyping others interplays with these biases, leading to unfair behaviors as cultures interact [1889]. Gender stereotyping also biases players in their proposals and willingness to accept an offer [1890, 1891]. Consistent with the concepts of sexual-competition behavior, men offer women more significant shares than they do other men. In contrast, women discriminate far less based on gender [1892], propose more generous allocations, and accept lower ones [1893].

WEIRD splitters make higher offers to thin partners than fat ones, and receivers accept a lower offer from a svelte splitter [1894].[237] We also take a smaller allotment from people we deem attractive [1895, 1896]. These biases affect employment, salaries, and advancement [1897]. Racism, casteism, and sexism may carry a professed stigma, but short, fat, and ugly remain socially acceptable grounds for discrimination [638, 1898].

Cheated parties tend to act vindictively in both power-to-take and ultimatum games. People with impaired consciences offer lower shares and respond with more spite than those with functional ones [114, 1899, 1900, 1901]. Although retaliation entails an additional cost for the injured party, the comparative price of vengeance decreases as the perception of injustice increases.

The relative punishment for being unfair goes up as the swindler claims more of the reward [1846]. By sabotaging a greedy partner, the earner prevents the receiver from increasing his or her relative power. In contrast, fair treatment builds trust and reciprocity across encounters, benefiting both parties in social species as diverse as apes [1902], bats [1903], reindeer [1904], and birds [1905, 1906, 1907]. In humans,

---

[237] The value placed on the reward also matters. Smokers were less willing to reject an unfair allotment of cigarettes (addictive reward) than cash [3468], and break time from a dull task could work as well as a positive reward such as cash [3123].

we give long-lasting reciprocal relationships a special name: friendships, and the betrayal of friendships—or even rejecting the offer of friendship—can trigger hurt, rage, and disgust.

Anger based on perceptions of unfairness can lead to dehumanization and behavioral spirals during negotiations. Opposing parties become more entrenched and secretive [1908, 1909].[238] As group members vie for power, they risk forming opposing coalitions. This hazard increases if the rival group members perceive their competitors as unfair or dictators. The strategies, sometimes conscious and sometimes on autopilot, unfold like a chess game. "Players" cannot follow a straight line toward their own goals; they must consider the objectives and strategies of others, many with incompatible objectives. The contenders may find that some former competitors transform into valuable coalition members. They need to squash others like bugs.

Whenever we put faith in another person, we preconceive fairness. We would not trust him if we did not believe the other would respond reasonably. Social animals, including humans, build histories with community members. Through memories, we make informed decisions about another's probable actions. New encounters are a gamble. That new person might not be honorable. We balance the benefits of placing our confidence in someone with the potential costs of betrayal. Our inclination to trust may link to our disposition toward risk, but the empirical data are equivocal. Different data support and contradict this hypothesis [1910, 1911, 1912].

People care about more than the payoff; they also worry about how the outcome came to be. We behave as if the price of a betrayal exceeds our actual losses [1913, 1914]. When we cheat, bully, demean, or betray others, we assume this social cost. Histories and expectations of disloyalty reduce the probability of future faith [1915]. When the memory of a cheater—or an implanted perception of deceitfulness—taints a

---

[238] An example of this behavior happens during labor-management negotiations. Negotiations often stall due to parties becoming more entrenched in old issues, not just on new points or through new arguments.

community, demographic, or business, confidence becomes harder in future encounters, e.g., [1916, 1917]. We see this history in the low trust levels women and disenfranchised people display when interacting with males from an empowered group. It also shows in the difficulties of reconciliation movements following aggressions and genocides.

We weigh the social conditions when encountering outsiders. We place more faith in someone if we believe our goals coincide. Our VMPC, the same area that responds to fair treatment and extinguishes unrewarding behaviors, shapes our confidence or skepticism in conjunction with the ACC (social awareness and rewards). When someone's interests do not coincide with ours, they have reason to lie. A false perception of shared goals forms the foundation of many confidence schemes. This difference also contributes to the animosity between labor and management. In these situations, we rely on input from the VMPC and ACC to trigger the amygdala, giving us that gut-level sense that someone will betray us.

During initial encounters, we harbor doubts. Suspicion relates to our innate desire to retain a position of power and our understanding that the other has the same goal. We find ways to increase the relative pay-off necessary to engender confidence and to counterbalance our loss of control and the risky nature of first contacts. This premium compensates for the increased threat [1914]. We offset the danger by seeking vouchers from third parties. In traditional tribal communities, encounters could include seeking common ground through common kinships. In modern societies, we perform internet searches for reliable, skilled tradesmen, or we ask friends about their experiences. The process takes more time than automatic acceptance, but it can save your roof or your life.

Why Spite?

> *"Oh, no. No, a man washes his hands before or after tending to his needs. It tells you a lot about a man."*
>
> — Strickland in *The Shape of Water* [1918]

On the surface, spiteful behavior, like unquestioning trust, has a negative payoff. Spite creates an enemy. At its worst, it harms us physically, emotionally, or financially.

Until recently, few people believed species other than humans engaged in spite. It was dismissed as a weird human behavior. However, whenever someone thought that he or she found a pattern unique to humans, someone else found it in other animals. No one should feel surprised to find that other creatures can act spitefully [1919]. We find spite among social insects [1920, 1921]. Prairie dog mothers will leave their own litters to kill the offspring of nearby (competing) females [1922], and female stickleback fish will target nests where other females have laid their eggs [1923]. Even members of a bacterial clone will sacrifice themselves for their colony, releasing toxins to kill encroaching competitors [902]. Sexually immature male stump-tailed macaques will risk injury to interfere with the mating of other males [1924].

In humans, "cheating" by refusing to pay a fair share of donations to a mutual fund increases as the payoff for selfishness rises [1925]. Cheaters don't just reduce the size of the commons. They also increase their *relative* profit at the expense of other players. The selfish competitor decreases the financial standing of others while increasing his own.

One community near me had a parking lot maintained by local merchants for over thirty years. Cooperation in the venture was voluntary and non-binding, but most entrepreneurs pitched in.[239] As new businesses entered the community, some refused to contribute to the maintenance fund, pocketing what should have

---

[239] Both adults and children hold to non-binding agreements based on public declarations more often than if they never intend to honor them [3311].

been their share. There were no effective means of preventing the free-riding customers from using the shared resource, encouraging the non-cooperative behavior of the cheaters [1926]. As costs rose, this strained the contributing businesses while the penny-pinchers benefited from their "free" parking lot. The cheaters enjoyed an increased profit margin—one of our society's surrogates for rank—until the contributors sacrificed the lot rather than tolerate the free-riders.

Rank does have its privileges in social organisms. Higher-status individuals get better access to resources and mates. Their offspring start life with benefits derived from the position of their immediate kin. Survival rate increases with higher status. Dominant animals can tell the others in their troop, pack, or clan what to do. High rank provides the males of most social species with preferential reproductive access. Social species, like primates, need mechanisms to counter spiteful behavior. If there were to be no offset to the benefits of spite by high-ranking members, one would expect bullying to increase in each generation and social organization to break down.

Because unfair behaviors provoke spiteful responses, the menace of reciprocal spite favors the evolution of "fairness" and rewards an ongoing practice of fair treatment within a social group [1927, 1054]. However, pressure for "fairness" must overcome the inherent benefits derived from supporting one's family by decimating others. Reciprocal spite inflicted by lower-ranking group members would suffice. The parking-lot cheaters mentioned above banked on the importance of that lot to the local stores. They underestimated the power of spite. The merchants who had long maintained that facility ended their funding. Soon, the community lot faded into history. Most of the original contributors, with whom people did business for years, survived. The merchants willingly sacrificed the community lot to punish those who benefited without contributing. The apparent "winning" strategy, not sharing the cost, became the downfall of many of the new entrepreneurs. (Who'd have thought it: When others experience spite, they tend to hold a grudge [1928].)

The real world abounds with such inequity, and the spiteful payoff hypothesis explains much of the perceptual differences seen among economic classes. High-ranking individuals depend upon the cooperation of small coalitions of supporters. Alphas need not keep everyone happy to maintain power; they must only keep critical allies supplied with power and status. The remaining population needs only to receive enough to prevent revolt.

Paradoxically, the privileged pose the most significant threat to rulers despite sharing the green beards of wealth or status. Following a coup, dictators often purge the upper echelons of those who helped them attain power. Those former allies understand how to overthrow the leader and can drive another rebellion. Whether the despot rules in government, business, or on the playground doesn't matter. When striving for power, a tyrant needs one set of supporters; he requires a different coalition to retain it [1079].

The dominant parties in industrialized societies have money, rank, and privilege. The privileged are as prone to the fundamental attribution error as anyone else. They see their accomplishments not as a result of any social advantage but as the result of their hard work and industry. They have money because they earned it. To them, fundamental character flaws, such as laziness, spendthrift behavior, or poor decision-making, explain why others have less. The select ones bristle and rebel against paying into general welfare funds to benefit others they perceive as “free riders” who profit from a resource without paying for it. The other end of the financial spectrum sees inequitable compensation and barriers. They focus on impoverished living conditions, poor school funding, and reduced employment and transportation opportunities. They bristle at providing benefits or tax breaks to the rich and perceive those who receive remuneration far above their contributions as social parasites. It does not matter if a worldview is correct or deluded; it still shapes the thinking and behavior of its adherents.

The perception of free-riding raises spiteful behaviors. People will punish spongers, even if the punishment carries no gains or imposes a significant cost to themselves [1584, 1929]. The greater the deviation from the perceived norm, the greater the impact on the community, the more severe the punishment it wishes to administer. Students who depend on hard work to earn grades typically call for harsher penalties for cheating on exams or AI-generated papers than their professors. Copying a test may anger a teacher; it enrages fellow classmates. The punishers will also call for more significant sanctions if they lack a conscience or the community leaders push for harsher chastisements [1690].

Our attribution errors warp our perceptions of the value of our relative contributions to society or commerce. We see ourselves as over-contributors and others as cheaters who take advantage of *our* work or the effort of our social circle. Consequently, we cannot agree upon who free rides because we use different scales to measure our financial and societal support and the give and take of others.

We experience frustration and anger over doing what we consider a disproportionate amount of work. These emotions color our decisions about our future social contributions and cooperation. They increase our willingness to punish the perceived deadbeats, leading to a potential spiral of revenge. The value of law rests in its ability to mediate these disputes, dispensing unbiased justice for both parties [1929].

Because of our differences in perception, both parties seldom see third-party mediation as unbiased. For example, the views of Arab States and Israel differ dramatically over the ability of the United States to address distinct grievances. The acceptability of third-party mediation depends upon a mutual perception of impartiality, if not absolute justice. Third-party mediation in labor-management disputes results in more mutual satisfaction when both parties cooperate in the choice of arbitrator. As with individuals, the sense of being in control of one's own actions and fate matters. Our perceived personal freedom of action plus a close temporal connection to positive outcomes of our actions increases our feeling of satisfaction [1930,

1931]. When perceptions of fairness diverge too far, the legitimacy of that authority collapses. Both sides feel cheated, and the mutual perceptions of freeriding return or become entrenched. Unsurprisingly, freeriding, the tendency of some group members to benefit from goods held in common by a group (= public goods), builds resentment in the group members who see themselves as doing the work or paying the bills.

Societies can maintain public goods with minimal cheating if repeated social interactions reward cooperation and punish swindling [1932, 1933]. However, our industrialized societies ask us to make altruistic commitments in which the cooperators continue supporting the commons even when the scammers remain unknown. Consequently, we search for someone to blame and often settle upon classes of people. In the United States, the words "welfare" and "entitlement" have developed pejorative undertones of financial and social cheating. This view has widely replaced the positive connotation of temporary aid for those devastated by circumstances. As the number of people in need has increased, these lifelines for the most vulnerable have become targets for vitriol. As America's demographics shift toward a higher proportion of the elderly, the young increasingly see the elderly as free riders. Many young adults see pre-paid programs, such as social security and pension funds, as handouts and financial burdens that benefit the current retirees but which will be bankrupt when the younger ones reach maturity. The generated frustration and anger contribute to the increasing hostility and dehumanization of the old.

Many upper and middle-class taxpayers cry for the abolition of much of the social safety net. These demands stand despite data indicating that the further downtrodden someone becomes, the harder it is for them to regain their social footing. Between 2010 and 2024, the United States elected several "tax-reform" or "Tea-Party" candidates. They focused on cutting taxes and government spending, often directing their attention toward reducing entitlements, ending collective bargaining rights for public employees, and breaking public sector unions. Attacks on these buffers often play to understood racist, casteist fears that specific sectors, especially African Americans and Latinx, are sucking the lifeblood from Whites.

This tactic has extended to include ageism with mounting attacks on pension plans and attempts to deprive long-term employees of their earned and agreed-upon compensation—a.k.a., pensions and health benefits. Several have argued that the states cannot afford the negotiated pension plans—the delayed payments promised to workers in exchange for their labor. They say little about the disproportionate distribution of pension dollars between high-paid and low-paid employees nor the systematic looting of pension monies by public officials to pay other bills without having to raise taxes during their terms.

Societal beliefs about poverty also reflect this biased attribution. Many workers have few reserves and a significant stake in a functioning safety net. Most people have no current requirement for support and barely recognize a potential need. They presume they thrive because of their hard work and thus will never need the safeguards of such programs. However, in 2022, approximately 12.6% of 42 million Americans, about 12.6% of the population, participated in the Supplemental Nutrition Assistance Program (SNAP or Food Stamps).

Contrary to stereotypes, the highest rates are not concentrated in the large cities of the north but in the South and Southwestern states, with New Mexico leading the nation at 24.3% participation [1934]. Two-thirds are children, the elderly, or people with disabilities. The remainder is overwhelmingly workers who cannot survive on their wages or suffer temporary joblessness. In 2019, one-third of recipients had earned income. 74% of the unemployed who were capable of working found employment within a year of receiving SNAP. The remaining quarter is often saddled with caregiving activities for children, the elderly, or disabled family members [1935]; nationally, 65% of recipients are families with children [1936]. In 2019, the average gross monthly income for a household on SNAP was $872, and their net income was $398. The most common source of income was Social Security benefits for the disabled and elderly [1937].

Members of vulnerable groups see the upper economic classes as cold, uncaring, and selfish. They picture them as lacking compassion and unwilling to pay their fair share for a social fail-safe that protects everyone. They have difficulty taking the perspective of the people in the upper financial strata who feel put upon to pay a disproportionate amount for services they are never likely to use. Across this gulf, both sides consider themselves the aggrieved party. Both see the other through the jaundiced lens of the fundamental attribution error. Political parties exploit and reinforce these views to gain or retain their dominance and privileges. In the United States, Democrats and Republicans use taxes to reward their coalitions and punish the opposition. Both parties redistribute wealth to their core supporters from their opponents and from the people whose support is unnecessary to retain power [1079].

The inherent asymmetric nature of power in modern societies makes enforcing cooperative behaviors, such as contributions to social insurance funds, hard or impossible to accomplish through informal injunctions [1938]. Believing they are the aggrieved party, each social caste is unlikely to enforce sanctions on itself. This disconnect expands as the divide between the upper and lower classes grows more pronounced. As camps dehumanize each other, it becomes easier to engage in spite as the "I-thou" schism leaves each believing the outsiders act from intentional malice.

Both sides in this fiscal conflict desire to see the other punished as parasites. Despite the long-standing belief that subjugation will eliminate the problem of uppity underlings, the data says otherwise. Social pressure may reduce free-riding but cannot eliminate it [1939]. Increased spiteful pressure also backfires because each side sees itself as victimized and the other as the free rider. Therefore, increased punishment increases spiteful resistance. This disconnect may serve the agendas of politicians, but it sows discord.

The desire to punish is spiteful. It takes energy to inflict retribution. It can also bring physical or financial losses, but the cost-reward ratio of castigation can tip to the reward side in humans and other animals [1940, 1941, 1942, 1943]. The benefits of punishment lie in its ability to stop a behavior that harms the welfare of others. High-ranking individuals and coalitions chastise to maintain dominance and to discourage freeloaders and other social cheats. Subordinate members can defect and engage in vindictive acts toward those with power, especially if their position or rewards appear illegitimate. Humans express such rebellious behavior through reduced cooperation, even when it involves a loss of personal welfare [1944].

Freeriding can include access to ephemeral goods, such as reputations. Pretenders can damage others by cashing in on their good name and prestige. When we abandoned band societies, our interactions with kin became more limited, and our interdependence on others became more profound. The importance of our reputation and honor as a social shortcut to trust blossomed [1945]. Because we are biased toward judging outsiders as "bad," freeriding by a member of an identifiable community damages that group's reputation and impedes future acceptance. Lawyers, doctors, police, politicians, and used-car sellers are among the more evident victims of the slime-balls that lurk amongst them.

As the *perception* of freeriding increases, the willingness to continue our support of the free-rider decreases. Even young school and preschool children recognize and resent mooching. They reduce their team efforts (become more free-riding themselves) when others behave non-cooperatively [1946, 1947, 1948]. This pattern persists throughout our lives, and we get vindictive pleasure from the retaliation.

In a social species, individuals can flaunt resources, markings, jobs, or allies. A country can brag about its power by committing an act of war or losing territory to gain a greater goal. At a more primal reproductive level, spite enforces fairness in a population through the repetition of punishment [1927].

Vindictive punishments can affect a disproportional benefit relative to their investment. Spite can inflict a cost on the perpetrator and advertise the fitness of the inflictor. An act that strikingly harms the doer promotes reproductive fitness by proclaiming, "Look at me. I am so strong and powerful, I can lose all these [fill in the blank] and remain on top!" For a male, superiority can advertise status and his power to endure a costly humiliation or apology and retain dominance [1949].

If the payoff is high enough, the spiteful organism reaps a genetic benefit. It does not need to leave behind more of its offspring; it just needs to create a situation where the competition produces fewer gene carriers. Because of the link between non-kinship and the perception of free-riding, native populations often treat new immigrants with hostility, presuming exploitation [1950].

Primates show the same general range of maliciousness and cooperation as humans. Like us, chimps respond to unfair intentions and distinguish them from unfair outcomes [1951, 1952]. Like humans, chimps will sacrifice to watch a punishment inflicted on another if they consider it justified [1953]. Other social mammals and birds also express spite. Elephants have attacked villagers who killed juvenile elephants, and have destroyed human cropland following eviction from their grazing grounds [1954]. And among the birds, tightly packed, ground-dwelling gulls spend much of their nesting time punishing their neighbors with pecks and jabs.

In North Dakota, white-tail deer, or as farmers call them, rats with antlers, are infamous for urinating on haybales, imposing a health cost to other herds that might stop for a nibble [1955]. The urine renders the hay undesirable to the farmer and other deer. Lest you leave with the idea that mammals are the only cruds out there, note that we have observed similar spiteful germ warfare in bumble bees [1956]. Bees infected with the gut parasite *Crithidia bombi* spend more contact time with bees from colonies susceptible to this disease than with immune bees. This behavior harms non-kin at little direct cost to the bio-warrior's home hive [1956, 1957].

In social species outside the vertebrates, we find animals with fewer behavioral options but who are still programmed for spite. Like the gulls, colonial mantis shrimp sneak behind their neighbors to strike their tails. Corals and anemones engage in clonal "warfare," with the polyps at the colony's edge sacrificing themselves so their genetically identical "twins" can expand their territory.

The larvae of the little parasitoid wasp, *Copidosoma floridanum,* lay eggs in caterpillars and use them as living nurseries. In wasps, a fertilized egg becomes female, and an unfertilized one becomes male. In this species, eggs can subdivide to produce multiple embryos (cloning). A single *Copidosoma* lays several eggs. The species isn't particularly fussy about prior property rights, so several wasps might colonize the same caterpillar. When repeated parasitism happens, competition erupts. Two or more larval broods will fight for the limited resources inside their unwitting host. Some members of the clones will change developmental tracks to become assassins. These soldier wasps are a nasty lot that nature designed to kill their own kind.

Copidosoma soldiers sacrifice their ability to reproduce, a loss for the clone. These sterile workers kill other larvae in the host. The soldiers usually turn most of their fury against unrelated larvae from different broods [1958]. They identify their siblings from chemical cues (pheromones). The closer the genetic relationship to the soldier wasp, the less spiteful behavior the siblings receive [1959, 1960]. Male soldiers are more likely to kill females, and female soldiers are more likely to kill males. This distinction serves as a further safeguard against killing their clonemates. When resource competition is high, they do kill their sisters and brothers.

For this wasp, spiteful behavior eliminates the chance for direct procreation for the little sterile soldier. By differentially killing the competing larvae, however, it increases the likelihood that some of its clones endure. It thus gives its genes a chance of survival by proxy in its kin.

Kin selection happens when behaviors favor the passing of genes through close relatives. If kin selection is to occur, there must be a mechanism for recognizing one's relatives, the infamous green beards from earlier chapters.

Not all species developed green-beard technology. When present, green beards allow organisms to estimate their genetic similarity. The closer the *perceived* relationship, the more likely one is to risk harm to defend or to support that individual. Spite, in theory, should follow the opposite path. More spiteful behaviors should happen when the perceived genetic relationships are low and the cost of the spite does not outweigh the benefits of the behavior (fewer offspring from the competition). Altruism and spite involve sacrifice. They are intimately linked and challenging to distinguish. Often, an altruistic act for one party is spiteful toward another, and it is through the evolved function, the adaptive design of the action, that we understand it [1961].

Relationships provide an imperfect guide towards whom one acts altruistically or spitefully. Among close kin, the cost of helping can still become too high. When self-sacrifice works against the reproduction of the genes, green beards be damned! Selfish and spiteful behavior against kin prevails when it favors the perpetrator's endurance [1962]. Even bacteria will attack related strains to ensure survival or gain access to their genetic resources [1963, 1964, 1965].

Many species reduce their brood size, with either the parents or their offspring serving as the executioners. If parents cannot provide enough food for everyone in a litter or nest, they or the offspring correct that problem. Many birds of prey engage in siblicide; in some species, one's ex-sibling also becomes dinner

[1966, 1967, 1968]. Among hawks, owls, and eagles, the first hatchling controls the resources and has differential access to food, often starving their brother or sister. In other species, like the lesser spotted and black eagles, the first to hatch will use its size advantage to attack its younger sibling [1969, 1970, 1971].

In some species, this Cainism is obligatory, with the healthier, larger nestling killing the weaker one. Other species have more behavioral flexibility, letting the little one live if conditions are good [1972]. The cliff-dwelling Kittiwake gulls use spiteful, punishing pecks or push their weaker nestlings to their deaths [1973, 1974]. In many species, the parents do nothing to stop these aggressions; odds are, they, too, murdered their siblings. Besides, they cannot provide enough food for everyone. By surrendering one child to death, they improve the survival chances of the others. If they were to rear an extra chick, it could decrease the brood's health, reducing their children's mating choices later in life.

Research on vertebrate siblicide has focused on birds [1975]. The offspring stay in or near the nest, and their pecks and pushes are easy to understand, making them a tempting target for study. However, mammals have their share of lethal sibling encounters. Cainism happens in about half of all spotted hyena litters of same-sex twins [1976]. Pigs and lynx engage in siblicide. Mother lynx will intervene in fights, but like most siblicidal birds, hyenas and pigs let the offspring contend for survival [1977]. Young wild dogs (*Lycaon pictus*) engage in fights leading to sibling death [1978], and domestic dog pups will push weaker whelps from the nipples to hoard the food. The disadvantages of runts can begin in the uterus, with a poor connection of the placenta to the mom, followed by less access to her colostrum and milk.

In humans, women conceive more multiple births than are born, reabsorbing one or more embryos during the first trimester [1979]. Dubbed the "vanishing twin syndrome," this natural process provides a healthier, more nutrient-rich environment for the remaining offspring. Twins suffer more complications and risks than singleton births, including lower birth rates and more early developmental delays. Mothers deliver twins two to three weeks earlier than solitary fetuses. Offspring born after a twin vanishes (reabsorption) are similar to singleton births, giving the surviving offspring a safer start to life. The causes of these disappearing twins remain unclear. Some may not be viable. The mother's body may reject some as incompatible or recognize her inability to support a multiple pregnancy. Or, the competition for resources among the embryos may play out like the pup's search for the nipple or a real-life version of *Highlander* [1980]: In the end, only one can survive.

Fig wasps engage in spiteful behavior against close kin. Males compete to copulate with a limited number of females. They can be as selfish about this competition with their kin as they are with strangers. Males favor personal procreation over that of another wasp. Brother wasps are not programmed to recognize each other as such. They typically act *as if* they have no green beards. Perhaps the payoff for cooperation, leaving one's genes behind through brothers, fails to outweigh the gamble of making an error and letting a stranger copulate with a potential mate.

In theory, brothers should pass on half of each other's genes. If they can recognize each other, they should cooperate. The presence of competition rather than cooperation begs for an explanation. These tiny wasps are not social. They might lack green beards and not have the ability to spot kin. They might recognize their siblings, but without the same fidelity. With an iffy identification, the wasp risks making a mating error—letting a stranger rather than a clone copulate with a potential mate.

Relatives can compete fiercely for resources. (Just ask heirs who have warred over trivial things, like who gets mom's pots and pans.) Although it appears to be a Darwinian Puzzle, this "reverse kin-selection" reduces inbreeding and enhances the genetic variation in their offspring [1981]. Nature favors negative kin selection when competition from one's close kin presents a greater danger than the benefits derived from having them carry one's genes.

Among mongooses, too many low-ranking packmates lead to social tension. Individuals can voluntarily leave an overcrowded pack, or both sexes attack subordinate members, driving them out. Although mongooses have a cooperative breeding system with communal nursing care, they recognize and avoid mating with their siblings [1982]. High-ranking males in the banded mongoose preferentially evict their closest female relatives from the clan. This reduces the chances of their offspring inbreeding with close cousins [1983]. Because mongooses avoid mating with their sisters, expelling unrelated females would reduce their potential breeding partners and benefit unrelated males. When too many fertile females occupy an area, evictions lessen the competition for resources, and driving away siblings can spread your genes as your sisters become founders in new colonies.

If you think these spiteful evictions by mongooses have no relevance to us, remember that chimps, bonobos, and humans all follow patrilocal patterns, with young females leaving their natal troops as they come of age. In our society, where adolescence remains a time of dependency rather than the onset of adulthood, girls tend to engage in more relational spite than boys, which could reflect an ancient wanderlust. Boys tend toward more physical attacks. An Israeli study found that the opposite-sex parent is usually the primary aggressor toward adolescents. Overall, sons receive more verbal and corporal abuse because both parents typically contribute, but daughters report more paternal verbal and physical aggression [1984]. In contrast, boys engage more in classic physical violence. In children with conduct disorder and a lack of empathy, this gender difference becomes exaggerated, with CD girls engaging in more hurtful and manipulative relationships [1985].

The changes happen at the point when traditional societies would have seen the youngsters as ready for adulthood. Dealing with these unwelcome behaviors decreases parental satisfaction [1986]. It is not a stretch to conjecture that, like the mongooses, these conflicts may prepare the family for the emotional changes needed for the offspring to form new core (reproductive) units within or across bands. Despite the distancing of direct relatives by evicting one gender (typically female) from a family band, the seemingly spiteful act encourages outbreeding and fosters positive relationships with the neighbors.

> *"To Sundries got to Replace in kind those which were taken from people in the Hospital to Convey the Smallpox to the Indians Viz: 2 blankets, 1 silk handkerchief, & 1 linnen do."* — 1763 invoice submitted by Levy, Trent and Company and certified as "used for the above mentioned" by Captain Ecuyer [1989]

Spite can also limit the spread of parasites among kin but increase it among non-kin [1987]. When the strategy works, biological attacks provide a significant advantage to the attacker and potentially spiral out of control. Human biological warfare is a prime example. Bio-weapons are used against "them" to subjugate, control, or eliminate "them." In the Middle Ages, siege troops catapulted rotting corpses over city walls to spread infection. Following the rebellion of Pontiac (1763-1766), Captain Simeon Ecuyer gave two blankets and a handkerchief from smallpox victims to diplomatic representatives of the Delaware Nation during the siege of Fort Pitt. [1988, p. 73]. Though debate remains about their intent or knowledge of how smallpox spread, correspondence suggests they hoped for an outbreak [1989]. The American Civil War had widespread allegations of selling clothing from yellow fever and smallpox victims to opposing troops [1990]. Today, despite knowing the danger, the United States and Russia have retained their stockpiles of the smallpox virus in a world without immunity.[240]

During the Second World War, Allied and Axis forces stockpiled biological weapons. Japanese troops poisoned over 1000 wells in Chinese villages with cholera. The Japanese released typhoid germs into a river to infect Soviet soldiers during the Nomonhan (Battles of Khalkhyn Gol) and exposed targets to plague-infested rats and bombed them with containers of plague-infested fleas[241] [1991, pp. 75, 128, 129-130,

---

[240] As of 2021, the official stores of smallpox are in the CDC in Atlanta, Georgia, USA, and VECTOR in Koltsovo, Novosibirsk Region, Russian Federation [3079]. Both countries keep the virus for vaccine research. Each suspects the other of weaponizing the virus.

[241] The plague program did not start as a weapon of genocide. Ishii's original intent was to develop preventative vaccines for use on the home front [1991, pp. 17, 75, 127].

1990, 1992, pp. 118-126]. Humans have used such insects and insect-borne diseases against their neighbors throughout recorded history [1992].

Humans also engage in small acts of biological warfare outside our consciousness. Sick males cannot mate as readily as healthy human males. Therefore, an infected male would benefit from spreading diseases to his reproductive competitors. In humans, men cough more violently than females, dispersing respiratory pathogens further. Men also shed their skin cells and associated bacteria more readily than women. These simple, unintentional differences pass infections to rival males [1993].

Among social animals, healthier ones, especially males, are better able to engage in sex. A sick one benefits if it also makes the ones around it ill. By spreading the joy of parasites, the ailing animal levels the playing field and re-establishes its rank relative to the other players in the herd [1993].

Elephants also use germ warfare against non-kin. As a herd moves across Africa, it defecates more near watering holes and salt licks than elsewhere. This defecation reduces the availability of clean resources for both the spiteful and competing herds, reinforcing the nomadic niche of the elephants [1994, 1957]. When the next clan of pachyderms arrives at the infected pond, they have few options besides using the polluted resource. Because the pooping pack is unlikely to return to that area soon, their behavior carries only a slight cost to themselves. However, their victims of biological warfare get sick with nematodes, bacteria, or whatever other fecal parasite was in the stools of the past herds. The spiteful poopers thus increased their relative fitness.

The unkind defecation pattern of the elephants has a more substantial fallout. The spiteful poopers already have their infections and have survived their parasites. Many parasitic worms cannot reinfect immediately after release, giving the malicious herd an extra layer of protection. The targets of their spite, other packs, may or may not have an immune system prepared to protect them from these diseases. Relatives in different communities are more likely than non-relatives to have similar immune systems to the poopers and thus some protection from the onslaught of these tiny invaders [1994, 593]. This natural resistance provides an additional edge in the reproduction game.

Since the turn of the 21st century, spite has become a more acceptable field of study in evolutionary biology. Gardner and West [1959] summarized the increasing evidence for a natural history of spite in everything from primates to microbes. Some species of bacteria will self-destruct on making contact with members of a different clone. This self-sacrifice releases toxins that destroy their rivals at the cost of their own little lives. By killing themselves, they benefit their genetically identical siblings. The survivors can scavenge resources, such as iron, from the dead bacteria [1995].

Bacteria also develop supportive networks. If you've ever left your dishwater sitting in the sink for too long, you may have seen a film of these critters form across its surface. As disgusting as we find such biofilms, those microbes joined forces to colonize that niche: the surface water that contains yummy nutrients from your plates [1179]. Sometimes, they share green beards with other species, phenotypic markers that let them work together. Where teamwork is adaptive, it happens. During deprivation, the rules change, and the stinkers may cheat their partners [1996]. In other cases, self-destructive behavior in bacteria, like self-destructive behavior in animals, carries a lower cost, or it is easier to achieve than the evolution of cooperation.

The self-sacrifice of the bacterium has a disturbing flipside. Because it kills itself to protect the resources of its relatives, its behavior is malevolent to strangers and an act of organismal altruism toward kin. Malicious behaviors that reduce the fitness of a rival simultaneously qualify as selfless behavior toward a kinship group. Any trait that reduces the outsider's reproductive potential increases the relative fitness of the family [908]. Similarly, any behavior that gives one's kin a selection benefit necessarily puts another group at a disadvantage and can thus qualify as spiteful.

So, Who Are Kin?

> *"When we gather together in the moonlit village ground, it is not because of the moon. Every man can see it in his own compound. We come together because it is good for kinsmen to do so."*
>
> — *Things Fall Apart*, by Achebe [1997]

Despite our perceptions, humans have few differences; we are an incredibly uniform species. We share about 99.9% of our coding DNA. All mammals tested share about the same number (25,000-30,000) of coding sites (genes), and so far, every gene tested does about the same thing across species, be it a human or a mouse [1998]. Plants, animals, and yeast have many hereditary sequences in common with us. Our coding DNA is about 60% identical to fruit flies [1999] and 23% to brewer's yeast [2000]. Many mutual sequences code for proteins that build the structural framework of the cells, direct cell growth, and regulate cell division events. All multicellular organisms share genes that tell cells to stick together rather than to fission apart, like amoeba. Every animal above the level of a sponge shares DNA codes for muscles, skin, tissues that interconnect, and nervous tissue that responds to stimuli. The similarities among animals are so pronounced that if one takes the genes that code for an element of the human eye and puts them in a fruit fly, the recipient will use that gene to form a normal insect eye. Even a human and a poppy contain compounds that convey evidence of an ancient common ancestor.

Of course, one can compare genes in different ways. The most basic comparison level is to look at how many *functional* genes we have in common. Approximately 99% of the active coding sequences in a mouse have a *similar* coding match in the human genome, and 96% of these arrangements lie within corresponding regions of mouse and human chromosomes. Both humans and rodents share about 44% of their genes with roundworms,[242] with a 50%-70% match in amino acid sequences in shared proteins [2001].

Beyond these rough, back-of-the-envelope similarity genomic measures stands the proteome: the collection of proteins made by the genome. Proteins formed by the 44% shared DNA with the nematode create a 50% to 70% amino acid sequence-match in our shared proteins [2001], with different levels of similarities among those proteins. Some are more conserved over time than others. In the proteins coded by the DNA of mice and men, we rank at about 78.5%. By comparison, rats and mice are about 97% identical in their coded proteins [2002]. Chimps and bonobos are about 99% similar, and both are about 95% to 98.8% similar to us. That tiny bit of genetic material makes a huge difference.

As a species, we are incredibly inbred. Humans share more than 99.9% of DNA in common.[243] Every human reading this chapter has nearly identical DNA to Einstein, Desmond Tutu, William Shakespeare, Sun Tzu, Mother Theresa, Oscar Wilde, and Adolph Hitler. Everyone on this planet is a member of a young species linked by descent from a common ancestor. Depending on ones definition of "human," that descent goes back as far as 99,000 to 580,000 years [2003, 2004, 2005, 2006, 2007, 2008].

Human traditions have foundation stories that trace their history to a single ancestor. Many men who self-report as members of the Jewish priesthood (the Cohanim) share a distinct form of one particular gene on their Y chromosome, the Cohen modal haplotype. By religious tradition, they trace to a common male ancestor, Aaron, the brother of Moses, who would have left a marker in all his male lineages about 3300 years ago [2009].

---

[242] Roundworms, or nematodes, are one of the more successful phyla of animals. The clade includes animals in virtually all common niches, including many common parasites.

[243] We may be 99% identical in our DNA but remember: DNA does not tell the whole story. The most similar DNA deals with fundamental adaptations, such as mitosis and cell division. Then different species dice and splice the messages made from the DNA to create a proteome that differs more than the DNA itself suggests.

Any living human whose ancestry goes at least in part to Europe or the Middle East has a good chance of being related to Charlemagne and Alexander the Great. Members of the African diaspora can include among their distant relatives Amanirenas, the Kushite queen who defeated the Romans, and the nameless people who built Great Zimbabwe. Similarly, any given Native American stands an excellent chance of a relationship with the builders of the great pyramids in Mexico, the founders of Cahokia, the mound builders of the Mississippi Valley, or the ancestors who drew the petroglyphs outside Petersburg, Ontario. Everyone is likely linked to a common ancestor in our family trees by 1400 BCE [2010].

Despite our 99.9% similarity, few consider all humankind as close to them as their brothers and sisters. Understanding biological kinship should be straightforward. The people with whom we share an unusually high percentage of our genes are our kin. With these people, we share the little genetic tweaks that make us look and behave more alike than we do with other random humans. Our closest kin are brothers and sisters, mothers and fathers, aunts, uncles, and cousins. The farther the genetic link, the less we think of someone as family. We share about half of our distinguishing genes with each parent and about the same with brothers, sisters, and children.[244] On average, our children pass half their DNA to our grandchildren. Because that passage is random, our grandchildren share about one-quarter of their genes with us. Our great-grandchildren will have an eighth of our DNA, and each subsequent generation removed will be about half as related to us as the generation before.

On this matrix, we graft our investment strategy. We care for organisms that share more of our genes. We invest less time and effort into organisms we perceive as non-gene carriers because they do not transmit our precious hereditary load into the future. Despite their identical genetic distance from us, we invest more time in caring for our children than for our parents. Our children represent the future of our genes;

---

[244] By half of our genes, I mean those specific genes, those alleles not held in common by mankind.

our parents are their past. We should be willing to help siblings raise our nieces and nephews but be less inclined to sacrifice for a neighbor's child who is not a blood relative.

On average, we respond to our *perception* of genetic similarity with nepotism. We favor our children over those of the neighbors. Most nesting birds will hunker down on their nests to nurture their babies; most do not willingly incubate an unrelated bird's egg. Social species often watch the offspring of group members, many of whom are relatives. The working caste of social bees and wasps will raise their sisters; because of a quirk of their genetics, these sisters are more closely related to them than to the queen.

Heredity channels our thinking about resources. Parents and offspring differ on this point. The debate will continue because one generation is 100% related to itself and only 50% related to its parents and offspring. Parents want to maximize their gene carriers' potential. For many in the upper and middle classes, that involves saving money for college educations, housing, and emergencies.[245] Children want to spend that money on enrichments for themselves, such as their food, vacations, and trips to amusement parks.

Despite the competition, organisms express more selfishness and spite toward outsiders than family members. They display more mutualistic and altruistic behaviors toward family members than toward outsiders. Indeed, the larvae of the humble ladybird beetle eat younger (presumably unrelated) ladybird larvae with gusto, eliminating the competition while also getting a satisfied tummy [2011, 2012, 1311].

Our most significant rivals are our kind. We compete with others of our species for food, space, and mates far more than with other species. To ensure genetic survival, we strive to maximize our gene deposits in the next generation *relative* to the genes left behind by our neighbors. We benefit most when we do well in the

---

[245] As an alternative reproductive strategy, families may discourage college or trade school educations, encouraging their older children to contribute to the family income as soon as they graduate. This viable alternative may be favored when there are fewer financial resources to spend on education.

breeding season and our neighbors strike out. For that reason, both altruism and its wicked stepsister, spite, make sense.

Sometimes, the gene-recognition mechanism goes awry. Other organisms crack the behavioral codes and have us raise their offspring as our own. The cowbird and the cuckoo get little songbirds to adopt their young. We have bred furry little lap dogs, such as miniature poodles, to resemble our offspring in size and the pitch of their bark. And when our companion animals die, we grieve for them as we do our children.

At other times, parents adopt offspring as shields. Ducks and geese often have large gaggles of babies trailing behind them. Many of these ducklings and goslings abandoned their natal parents to join the larger gaggle. The foster parents don't object; the babies are reasonably self-sufficient, and it takes little energy to raise them. If a predator finds the gaggle, its large size dilutes the chance it will eat one of their offspring [2013]. The babies prefer to live in big flocks for the same reason.

Penguins adopt for a different reason: They lay one egg per season, and each baby is precious. Nature hardwired penguins to care for this valuable genetic load. If the baby of good providers dies, another chick may move from lower-quality parents to the grieving ones. The birds have robust programming to care for this changeling because the risk from offspring rejection is high. Penguins will battle over contested ownership of a chick. No matter who wins, that baby will find a loving, nurturing home.

Although most humans no longer live as hunters and gatherers, our bodies have kept their old adaptations. We still crave fats, a taste preference that kept us alive in our active migratory lifestyle but now clogs our arteries. We enjoy sitting and relaxing, a luxury that would have minimized our calorie expenditure in the old ways. From Kinshasa to Chicago and Beijing, we have moved from the plains of Pleistocene Africa into modern cities, but our physiology has barely changed. Our genes and their environmental interactions determine much of our social nature. Our ways of doing things have lagged behind our changing environment. In our natal state, in small bands of hunters and gatherers, daily interactions would have been with family members. We knew relationships through family, clan, and tribal affiliation. We knew our kin, and we knew who had blood feuds with them.

As we moved into the towns and cities, interactions became less confident. We still carried our innate nepotism, but we now had trouble recognizing who wore the green beards of our clan. Today, through our social programming, we make the adaptive mistakes upon which civilization depends. We embrace our coworkers and neighbors and help them, treating them like extended family. Multiple brain regions still interpret continued physical proximity as an indicator of kinship despite consciously knowing the correlation no longer exists. Their familiar faces activate our anterior temporal lobe, our amygdala signals safety, and the hippocampus retrieves a memory of friends or foes. Most people will assist their office mates rather than sabotage them, and we still feed the playmate of our child, albeit he is no longer likely our direct relative.

On the surface, helping our direct office competition or feeding our children's playmates looks maladaptive. This behavior, however, serves a greater purpose. Being friendly and supportive to coworkers and local children increases group and neighborhood cohesion. It informs others that you are a valuable community member, not a social leech. Ultimately, the benefits of cohesion and perception outweigh any slight disadvantage incurred from sharing a sandwich or a bowl of curry.

We still look for green beards to identify our clans. Consciously, we know members of the same army unit are not genetic relatives, but we recognize them as such. The American Marines boast that they leave no man behind. Combat personnel are interdependent and will stake their lives for each other as if they were blood kin. We identify our social kin through national flags, sports colors, and gangland symbols. We form life-long bonds with friends in college. Despite a lack of genetic bonds, we may call someone "aunt" or "uncle" from childhood. For social relatives, our closeness and love are as intense as biological ones.

Humans shift loyalties with their perceptions. Many LGBT people have suffered friends and families turning against them after revealing their orientation. In the United States, Americans get trained from childhood to consider skin color as a social marker. Other cultures recognize facial features, height, scars, or tattoos as caste distinguishers. Recognition of differences, by itself, does not cause problems. Social problems arise from how we deal with or fail to address those prejudices [2014]. Biases, however, shift.

> *"I go to bed on September 10th white, wake up September 11th—I'm an Arab."* – Dean Obeidallah, The Axis of Evil Comedy Tour, 2007 [3206]

Our preconceptions depend upon who we consider "other." Our fluidity lets us change identifications rapidly. When Americans see photos of people with similar skin tones, they see color as a uniting element. If one puts the same people in uniforms, their perception shifts, and they identify the people with that collective. The group sense created by the attire shifts our attention from a physical feature to the more narrowly focused sport, medicine, or profession associated with those uniforms.

Many people consider their businesses as second families. Street gangs will fight and die for their organization that serves a kinship role [2015]. In Europe, team violence can erupt over soccer matches. Football hooliganism ranges from malicious taunting to murderous riots in the name of surrogate (team) families. The United States and Canada have also had intimidation and violence erupt among fans surrounding major sporting events, and the police prepare for violence after championship games.

In our modern world, allegiances shift as we redefine surrogate kinship groups. Employers who fail to treat employees as valued contributors exacerbate the I-thou dichotomy between labor and management. They invite spite and rebellion as surely as a bullying silverback gorilla does. They can expect reduced work and quality from their employees, who will seek loopholes to circumvent the demands of spiteful employers. Workers who fail to acknowledge the power of the silverback in respected employers hazard more than unemployment. If most laborers see management's authority as legitimate, the freeloader or malcontent risks the alienation of peers. When labor and management understand and appreciate their interdependence, they treat each other with respect. When they see each other as competitors, they stifle each other with spiteful acts, undeterred by their own expense.

As we increase the gulf between the rulers and the ruled, we become more aware of our side as wronged and their side as abusive. We set ourselves on a downward spiral of suspicion and conflicts over power. In this setting, we continue to behave as both spiteful and altruistic animals. None of these behavioral associations should surprise us. Spite, altruism, power, violence, kinship, and green beards are inextricably linked, but we can break the spiral of venom—if we can find a way to break the circles of warped attribution and mistrust.

Part III:

The Golden Rule

"That which you hate to be done to you, do not do to another."

— As stated in a late-period Egyptian papyrus (c. 664 BC – 323 BCE),
A Late Period Hieratic Wisdom Text [2016, p. 95]

"Never impose on others what you would not choose for yourself."

— Confucius, *The Analects* XV.24 [2017]

CHAPTER 9: Rationalizations and Provocations

> *"What the hell are you newspaper men doing here? We're just killing a few negroes that we've waited too damn long about leaving for the buzzards. That's not news."*
>
> — Raper [2018, 2019]

We retain primeval wiring to make snap judgments based on minimal information. Our evolutionary history hurries us down the spiraling roads of prejudice, hatred, or desire for vengeance rather than guiding us toward dialogue, understanding, and regard. In the end, respect might yield more benefits, but our ancestral programming responds to short-term pleasures rather than long-term advantages.

The hormones and neural networks of our past make the choice of delayed reward and respect less immediately gratifying than a good punch. Nature gave us a strong push that reinforces these primordial urges. Among the chimps, higher-ranking males make more babies. They make more babies if they lack self-control, delayed gratification, and planning [2020]. This triad makes them less predictable and tame—excellent traits for passing on one's genes, but dangerous when mixed with our penchant for mistrust and nuclear weapons.

Vindications of Violences

> *"Caedite eos. Novit enim Dominus qui sunt eius." (Kill them all; God will know who belongs to him.)*
>
> — attributed to Arnaud Amalric ordering the 1209 massacre of the Cathars at Beziers, Albigensian Crusade

Almost everyone recalls being bullied. Wellnigh no one remembers or admits to bullying. Often, those who engage in physical or emotional violence against another do not appreciate the extent of their harm, rationalizing their actions. The victim may represent a threat to their well-being. At other times, *it* ceases to be human. The aggressor may intend to harm or lash out to fend off a potential danger where none exists.

Our brains dismiss the injuries we inflict on others. Such harm can increase our position relative to those around us by reducing their status or increasing our rank. Like a wasp who stings her hive mate or the elephant who contaminates a watering hole, we remain oblivious to the biological, adaptive reasons why we lash out. Instead, our brains construct explanations for our behavior.

Humans concoct pretexts to perpetrate violence. These arguments can reflect an unabashed hatred for another person or group. The murderers of Matthew Shepard, the gay American tortured and then crucified on a remote ranch fence, apparently believed they would suffer no consequences for killing a gay man. Matthew became an object upon which they could vent their loathing. People need scant justifications when society sanctions hatred, such as honor killings or attacks on LGBTQ Americans.

Although we strike out consciously, we often camouflage the motives for an assault, hiding them even from ourselves. When a person, group, or nation victimizes a rival for love, status, or resources, they seldom strike without a pretext. Instead of acknowledging the selfish motive for our attack, a human will seek a palatable, manifest cause—a face-saving guise for the cruelty. Then, we believe this after-the-fact construction was our driving force all along.

Expo Facto

> *"Any excuse for a real war will do, as long as it is seen as necessary to protect the tribe. The remembrance of past horrors has no effect."*
>
> — *The Social Conquest of Earth* by E.O. Wilson [2021, p. 63]

Humans take pride in our large brains, believing they will guide us with sound judgment. However, we often offer retrospective models for our behavior while clinging to faith in our ability to act upon logic. Seldom do the rational centers of the frontal lobes rule. More often, the members of the large, lumbering committee of conflicting brain centers push their individual agendas, striving to have their voices heard, urging one to down that extra-large pizza with double cheese on a Friday night. We like to believe our analytic prefrontal cortex makes the final decisions. But our rational mind draws conclusions from the communications it receives from its co-workers, and often, the amygdala and insula write their memos in large, bold, angry fonts. Even if the prefrontal cortex were the brain's CEO, the right and left sides would function as semi-independent co-chairs.

"Split-brain syndrome" happens when the left and right hemispheres fail to communicate [2022, 2023, 2024, 2025]. It allows us to study which side of the brain exercises dominion over different functions. The left hemisphere has primary control of language, including the meaning of words. If you project a picture of a sock or a foot so only the non-linguistic right hemisphere receives the image, a split-brain owner will likely look at her feet. If the two sides of the brain cannot communicate, the more verbal left hemisphere does not know that the right side saw the picture. When asked why she looked at the ground, she will construct an answer with the linguistically fluent left side of the brain. It finds a motive for the behavior, such as a sudden concern she has mismatched her socks or her shoes hurt. The left brain attributes a cause to her actions despite not knowing what the right side saw. Although the subject concocted the excuse after the fact, she *believes* the intent stems from the described motive [2026, 2027].

Although strokes can disconnect the hemispheres [2028, 2029], most studied cases arise from treating severe epilepsy by cutting the primary connection between the right and left hemispheres (corpus callosum). Split-brain syndrome establishes that we develop explanations following an action yet believe our justifications came first. These conditions show that our species will search for a pretense to account for its behavior. Many of Anna Freud's ego-defense mechanisms show this same lateralization in brain function [2030].

We rationalize not just our actions but our emotions. People who unknowingly received a dose of adrenaline blamed their mood change on their environment [2031]. [246] If a female confederate were present, men assumed they were sexually aroused. Otherwise, they explained their symptoms as anger at the experimenter, panic about participating, or the temperature in the room.

We have an imperative to make life into a package of acceptable behaviors. We create motives for looking at our feet or slicing another person with a machete. We seek external causes when shifting adrenaline levels, oxytocin, or cortisol play ping-pong with our feelings. As with our successes, we have little trouble taking credit for our good moods, but we project our failure and anger into the environment, where it often lands on others.

---

[246] Hecht performed this study in the 1960s; it would never make it through an ethics review today.

The Return of the Mechanical Man

> *"The genial Nurembergers around them became transformed and taunted and insulted the woman. The troopers at her sides abruptly lifted her to full height, revealing the placard hung around her neck... As best Martha could garner, the placard said, 'I HAVE OFFERED MYSELF TO A JEW.'"*
>
> — *The Garden of Beasts* by Larson [2032]

By definition, psychopaths are manipulative people who lack empathy and guilt.[247] They are callous and express little concern for social norms. They also show multiple disconnects in the brain that impede routine and emotional feedback. A disconnect separates the tempering right hemispheric dominance in empathy, guilt, and fear from the impulsivity, stimulation-seeking, aggression, and risk-taking of the more verbal left hemisphere [2033]. Some people who suffer from psychopathy are also afflicted with narcissism (excessive self-love and a need for constant affirmation), and the tendency to lie and manipulate others for self-gain (Machiavellianism). When these occur in combination, we say the person suffers from the "Dark Triad."[248] When intentional sadism enters, they have the "Dark Tetrad."

In psychopathy, the anterior temporal lobe and ventral frontal region don't always connect right, creating a rift between the object-function understanding and language processing of the ATL and the functions of the VFL's language expression and executive controls (e.g., planning, self-control, and maintaining attention). They don't hear that little inner voice saying something's wrong. They may understand that society condemns an action, but they don't feel it.

---

[247] In this book, I follow the custom of equating sociopathy and psychopathy. Most psychologists use the terms interchangeably, but some consider them different degrees of a continuum of antisocial behavior [3246].

[248] Dark empaths—people who show many dark-triad traits but retain a sense of empathy and understand, if not appreciate, the perspectives of others—also exist [3323]. Other combinations of these traits, such as the dark tetrad and the vulnerable dark triad [3320, 3321], suggest that, like cancer, these disorders form a cluster of diseases rather than a single pathology.

Our old friend, the medial temporal lobe (more emotional processing and strong memories), may suffer impaired connections with the medial frontal lobe (self-awareness and emotional content) and the cingulate cortex (follow-through on plans, emotional regulation, and decision-making for rewards) [2034]. Making matters worse, psychopaths have reduced activity in the prefrontal cortex, the area associated with planning, decision-making, and impulse control—the triad that restrains impulsive sex and baby-making [2020]. With these offline, they may never consider the personal consequences of their actions, let alone how they wreak havoc on others. For males, the triad reduces sexual inhibition, a procreative advantage that can result in more offspring.

These biological underpinnings of psychopathy result in a combination of an altered perception of fear and reduced inhibition of violent response [2035, 2036, 2037]. People with these disorders typically experience blunted and delayed positive emotional responses and empathy [2038, 2039]. Their ability to ignore societal norms (emotional distancing and low guilt) and use aggression as a finely-honed tool allows psychopaths to pursue their goals at the expense of others.

Although most of us do not qualify as psychopaths, we have much of the same underlying hardwiring in our brains. The "normal" person differs in only a few critical connections among a quadrillion.[249] These fundamental differences in the wiring and learned "software" lead to a distinct difference: The psychopath does not care about social norms; the non-psychopathic person does. When we lash out, we feel compelled to have a good reason—or at least a good rationalization—popping from our functional right and left hemispheres. We use our talent for justification to concoct creative excuses for bad behavior, including the perpetration of bullying and other violence.

---

[249] In a volume about the size of a large grain of sand (1 cubic millimeter), a mouse brain has about 100,000 neurons and a billion synapses connecting them. We're a bit more complex. We estimate that humans have 10^{15} synapses among an array of neurons rivaling the number of stars in the Milky Way at a potential 100 billion [3107, 3084, 3134]. In case you're interested, a fruit fly has about 25,000 neurons in its itty-bitty pinpoint of a brain [3280, 310].

Sometimes, our grounds stem from open, unabashed hatred and dehumanization of another person or group. The murderers and tormentors, who have beaten, humiliated, and killed gays, African Americans, Dalits, Aboriginal peoples, and other disenfranchised groups, presume they will suffer no consequences for torturing and murdering someone with little social standing. They may not believe we hold a shared humanity. When societies demean and dehumanize a sector, the empowered gain the freedom to vent rage upon these victims. Repetition of hateful, degrading messages desensitizes the amygdala, insula, and frontal cortex to violence against a recognized sector. It activates the reward centers for executing spiteful or violent acts against them. The perpetrators might imagine their actions benefit their social group or humankind, the real people, to maintain or improve their honor and rank. Many who commit genocide deem they earned a reward for taking out the trash rather than a punishment for mass murder. When Ilse Koch, the Butcher Widow, left dogs to guard an apple core among starving children, she did it to torment and because it activated her reward center. It felt good.

When society sanctions vitriol and revenge, as in honor killings or active discrimination, people need little encouragement to do heinous things. The reward centers activate in the ventral striatum, making subsequent acts easier and potentially pleasurable. Society forgives the action; it sees the victim as having earned violence and degradation. When society condones the actions of bullies, the boundary between the uncaring psychopath and the rationalized violence of the average person becomes a blurred distinction.

Most of the time, we search for a "need" to justify violent behaviors. We see ourselves as good people and feel uncomfortable escaping that mold. Our thoughts (cognitions) become discordant with negative behaviors that conflict with our self-image.

Humans take the path of least resistance toward resolving this inner conflict. Making a choice, be it free or imposed, shapes future decisions [2040]. We redefine ourselves and the nature of an act as good or bad.

We analyze the situation, hunting for a ruse to dismiss or resolve the conflict between behavior and belief. Humans achieve these goals in an old-fashioned way: We change our neurological function to meet our needs.

As we explore ways to absolve our changes and realign our attitudes with our actions, specific brain regions glow in an MRI scan. Two particular areas, the dorsal anterior cingulate cortex (DACC) and the anterior insula (AI), become more active. Like the corpus callosum responsible for split-brain syndrome, the DACC transfers information between the two hemispheres. It lets us know when we've made an error, helps us weigh choices, and evaluates social interactions, including shunning. In the brain of someone who lacks a conscience, the cingulate cortex, including the DACC, and insula, is pruned back.

Some psychopathies correlate with thinner cingulate cortices and anterior insulae [2041]. Because the DACC and AI contribute to both our sense of self and right and wrong, pruning this region contributes to an inability to feel guilty about harming others. Guilt requires that the rational cortex, with its vast intellectual knowledge base, communicate with the ACC and AI. These areas serve as conduits to our emotional amygdala. People who suffer from a missing conscience do not understand guilt because they do not feel it. They do not feel guilty because they cannot make the physical connections needed for that sensation. To a psychopath, feeling guilt can be as foreign as humans feeling pain at the tips of their tails.

Not everyone with damage to their ACC and AI becomes a habitual, sadistic, bullying offender. Most people with antisocial disorder never cross the threshold into criminal behavior. Some become bullies; others will use their skills in business. Others appear as slightly aggressive or overly protective of family members. Most compensate through coping strategies and never recognize their condition. They learn to avoid the pitfalls without a functional conscience guiding them.

The thinner cingulate can also deprive a sociopath of the usual physiological warning signs of guilt. Because of this modification, they often deceive lie detector tests. The ACC helps regulate blood pressure and rational cognitive functions such as reward anticipation and empathy [2042, 2043]. More generally, the ACC evaluates social situations and provides feedback about social errors. This part of the brain plans. It goes nuts when it knows we have made an error.

The anterior insula becomes more active as adolescence approaches; the change in the AI contributes to the strong sense of identity, justice, and purpose that most teenagers experience. By stimulating activity in the AI and ACC, we alter our perception of right and wrong to match our needs [2044]. In other words, it provides us with a sense of fairness [2045, 2046, 2047]. With the appropriate wiring, they shape the pacifist monk or an outraged potential bomber.

We respond in horror upon hearing stories of elder abuse in nursing homes or patients in psychiatric hospitals. We assume the worst of the victimizers. When healthcare workers bully or dehumanize patients, they violate their positions of trust and power. Frustration or self-protection can also drive their actions. Many workers suffer maltreatment by staff members or patients, not necessarily by the one they abused. One Australian study reported over 67% of healthcare workers experience workplace violence in the form of verbal abuse, and 33% of the workers had received threats from clients.[250] 12% reported physical assaults by clients, and another 10% reported other bullying by clients [2048, pp. 29-30]. The same study reported physical assaults in American psychiatric facilities as high as 41%.

When we look at the workers' behavior in this context, the perception of the violent acts shifts from aggression to personal preservation. By redefining oneself as the victim of brutality, the dissonance between the role of caregiver and self-defender lessens. Unfortunately, the redefinition can create an

---

[250] Note the similarity of this figure to the percentage of people experiencing other forms of bullying.

ethical quagmire, yielding an increased risk of assaulting a weaker or wrong party and inappropriate violence.

But We are Special!

> *"ACTION: Well, he acts like he don't wanna belong.*
> *BABY JOHN: Who wouldn't wanna belong to the Jets!"*
>
> — "The Jet Song," Stephen Sondheim & Leonard Bernstein [2049]

Humans use cognitive tools to shape their perceptions about the ethics of dehumanization and violence. These justifications vary with personal experience and cultural definitions of acceptable behavior. The concept of proportional or appropriate punishment varies across cultures, as do the measures of the severity of misdeeds. Cultures also demand different retribution for different categories of people.

Many traditional human cultures have defined outsiders as non-people. Self-labeled tribal names often translate as "the people," "the true people," or some variation on a theme that implies superiority over others. Other self-imposed labels include green beards such as tribal markings, diet, language, or customs. They can consist of foundation myths (kinship links).

Ancient city-states and nations also tied their inhabitants together through foundation mythologies.[251] Romans considered themselves descendants of Romulus. Jews trace their ancestry to Abraham. "Europe" (Greek for "broad face" or "broad eye") refers to a foundation myth involving Zeus and a beautiful woman; it originally referred to parts of Greece. Many cultures tie their Gods and heroes to the origins of villages and states. Athena is the patron of Athens; Ireland refers to the Celtic goddess Ériu. Origin stories provide

---

[251] The term "mythology" refers to the stories of a people. It does not address the veracity or falsehood of the stories. In most cases, people within a group understand the mythology of their own people. The degree of acceptance and interpretation varies.

roadmaps to understanding proper behavior and etiquette and exerting their claims to land and other resources. The accounts provide a unified understanding that can be virtually impenetrable to outsiders.

Traditional societies tolerated greater violence against outsiders than they accepted within their communities. The Romans restricted crucifixion to non-Romans. Aztec priests had few moral qualms about sacrificing non-Aztecs to the Gods. The code of Hammurabi called for different levels of rights and retribution for offenses by or against high-ranking patricians (*amelu*), landless freemen (*muskinu*), and slaves (*ardu*) [2050]. In recent years, many Americans tolerated the restriction of civil rights and liberties, such as writs of habeas corpus, for perceived outsiders or the suspension of the posse comitatus act (18 U.S.C. § 1385, June 18, 1878 [2051]) to monitor or police aliens and marginalized insiders such as African, Muslim, and LGBT Americans.[252]

Within the last century, WEIRD societies have accepted a radical idea: All members of our species should be free from undue harm and treated with respect. We extended civil liberties to more people, condemned the degradation of the disabled, and worked toward less barbaric prisons. Our brains and behaviors have not kept pace with our intellect. We vindicate not extending respect and tolerance to people outside our circle of moral caring and justify our refusal to safeguard the environment for the next generation [2040]. Generations yet to be conceived have no rights.

We have difficulty overcoming our natural classification of others along an I-Thou spectrum that places our needs and the needs of kin ahead of strangers. Nature programmed our species to question the traditions and behaviors of others. We often replace the I-Thou relationship with something more akin to an I-It mode of action [2052]. Throughout history, as we moved further from the genetic "I," we incurred more significant

---

[252] Under US Law, the federal government limits the ability of local governments to call upon the national military to enforce the law. Although the military can be invoked, a state must go through specific channels in the federal bureaucracy. The act was weakened in 2006, but those changes were repealed, and it was returned to its original form in 2008.

risks and received lower rewards for cooperating with "them." Although most people now live in societies with little true kinship among their members, we still respond to the old rules that pit "us" against "them."

Even though the best interest of humankind resides in limiting conflict, conflict continues to dominate in part because of our loyalties to the related "us." Wars, revolutions, interethnic violence, and individual conflicts can increase suddenly [2053]. A familiar pattern of violence and peace lies in resource allocation and possession. The interplay of resources and the security needs of communities can shift each faction's perception of the others. With these shifting relationships among and within groups, individual thought can yield to the rule of dictators and self-direction to the actions of mobs. In Rwanda, for example, a few leaders shifted a nation's beliefs and behaviors, creating a social climate conducive to the genocide against the ethnic Tutsis.

Positive learning about the other party creates sudden shifts in thinking, causing the system to become more resilient to such relational changes [2053]. Education about cultural and ethnic diversity leads to greater understanding and less acceptance of violence. People familiar with members of sexual, religious, or ethnic minorities harbor fewer negative attitudes toward them than people who do not know a group member.[253]

We have difficulty treating out-groups as deserving of respect because of our evolutionary history. Nature equipped us to recognize its familial lines; thus, we derive a genetic benefit by assisting those we perceive as kin (altruism). We hesitate to aid competing lineages and may actively inhibit their access to power and reproduction (spite). Kinship groups whose normative behaviors promoted genetic survival could pass those genes to the next generation through the members who lived to reproduce. If altruistic behaviors

[253] Most of us know members of minorities, but many do not realize it. If I were a student at certain religious institutions, for example, I would be much more in the closet if I were Gay, Mormon, Jewish, Muslim, or Catholic because of the attitudes condoned by those institutions.

favored the genetic family over outsiders, we would expect modern humans to retain an inborn (genetic) tendency toward familial and group nepotism and for distrust or spiteful behavior toward outsiders. Today, we are desperately seeking ways to counter any of these tendencies because we have entered an era in which they have become maladaptive for the good of humankind.

Society benefits if we understand the aspects of cultures that promote or legitimize violence in direct or indirect forms [1830, 2054]. That knowledge can inoculate us against its acceptance. We know we victimize groups described in dehumanizing terms and ignore violence against them [2055]. That bit of information increases our capacity to intervene against such portrayals by governments, the media, or schoolyard cliques.[254]

Knowledge equips us with one critical component to counter our innate behaviors and substitute pro-social actions. For example, we tend to ignore someone needing assistance more often in a high-traffic area than in a low-traffic area. When more people are responsible for helping, each person becomes less likely to deliver relief [378]. We feel free from our social responsibility. Consequently, a person in need is more likely to receive assistance on a lonely country road than on a city highway. Once we learn about this tendency, our forebrains can incorporate the information. At that point, we have empowered ourselves to choose our actions.

Despite the potential inoculation of knowledge, we face another powerful barrier: the desire to conform. Our species will deny the evidence of their eyes to comply with a societal norm. When students are told to match the length of a line to one of three other lines, they have no problem choosing an obvious, accurate

[254] As our brain responds to the conflicting information, our EEG would show a robust N400 response, as discussed in the last section.

reply. If a peer group first identifies a wrong answer as correct, 37% of people will conform with the group and select the same incorrect answer [2056].

In an automated version of the same experiment, researchers seated subjects in adjacent booths. They could see what they thought were the other's perceptions. In reality, the experimenter fed them pre-packaged responses. After a few preliminary trials, subjects waited to answer until they knew what the others said. Then, they usually conformed their answers to match the group norm [2057]. This conformity effect extended beyond questions about measurements. People accept ideas they would describe as repugnant in the absence of social pressure [2058, 2059].

Although more challenging than instilling hate, we can reverse negative pressure by portraying bullying and violent behaviors as unacceptable to the community. Sometimes, one person can alter a norm. Although difficult to maintain, buddy systems for at-risk people, such as neurodivergent students, can model acceptance and disability awareness, destroying the perception that bullying is acceptable [2060, 2061, 2062]. Literature can create a basis for understanding. Judicious use of pro-social stories and experiences can shift behaviors away from identification with and support of the bully. This shift in perception moves people toward a positive image of victims and advances positive moral engagement in their defense [2063, 2064, 2065, 2066]. Such pro-social attitudes involve ongoing efforts. The anti-bullying river flows upstream against our drive for dominance and status.

We construct ethical narratives to justify our engagement in or abstinence from violence. These arrays often happen after our action or refrainment, but we believe them to be our motivation. Rationalizations permeate government propaganda, newspapers, and playground arguments as we choose sides and decide who is human and who crawls beneath our moral concerns.

Dehumanization

> *Dene/Diné = people; Deutsch/Dutch = the people; French/ Frank = noble people; Hopi = well-mannered/peaceful people; Inuit = people; !Kung = people; Lenape = the people; Magyar (Hungarian) = the people or the messenger; Saxon = warriors with knives; Swede = kinsmen; Ute (nuuchi-u) = people*
>
> — Online Etymological Dictionary [2067]

Anthropologists agree on one fact about human cultures: Each has a sense of superiority over others. Our species uses culture to grapple with problems and threats from the environment. In every region, that area's cultures help the inhabitants survive. No wonder each group thinks it has the answers.

> *"When an object is lied about, all objects, not just the one involved, are distorted."* — Max Picard, 1955 [3197]

Socially dominant classes assume stigmatized people suffer less pain [2068]. For example, Black Americans are systematically denied the same pain management as White Americans because of such prejudices [2069, 2070]. We ascribe more compassion, love, curiosity, and emotional subtlety to our ingroup than to outsiders defined by xenophobia, racism, and casteism [2071]. And when individuals rate their intelligence relative to others, they usually rank themselves above average.

> *"In Springfield, they're eating the dogs, the people that came in, they're eating the cats... they're eating the pets of the people that live there."* — Donald Trump's accusation about Haitian Immigrants in Ohio, 10 September, 2024 [3301, 3089]. This recycled accusation has been hurled against many immigrant groups, including the Chinese Vietnamese, Irish, Italians, and Latin Americans. Such claims typically include racist and xenophobic elements [3494, 3360].

All humans share this innate belief in personal or cultural superiority. That belief supports the imposition of our ways upon others or degrades them [2066]. Once dehumanized, victims no longer appear worthy of dignity and consideration; we exclude them from moral standing. Violence against them ceases to carry horror and outrage as it would against an equal. Dehumanization reduces another to no more than a tool and is a violation of their personhood and dignity. It gives tacit authorization for horrendous acts and absolves perpetrators of moral responsibility for rape, torture, or other abuses of power [2059, 2072, 2073, 208]. For the victimizer, reducing the prey to vermin obscures the true nature of the violence, freeing his power to torment. Once unrestrained, one can abuse a prisoner, employee, or schoolyard scapegoat without fear of retribution. Dehumanization also reduces the physician's tension and distractions, letting them serve the dying or operate on a gall bladder attached to a liver rather than a person with a beating heart [2074].

The stigma of sexual humiliation magnifies dehumanization across cultures and is often associated with a less-than-human status [2075, 2076, 2077]. Once imposed, it gives tacit societal approval for further violence. Deep-fake porn, which superimposes the faces of unknowing victims onto the bodies of actors, has become a degrading reality haunting people of all ages. Once victimized, the dehumanizing image can forever torment and upend a person's life [2078]. The deep-fake porn humiliates and degrades, undercutting the person's self-efficacy and social voice. Once associated with an obscene image, the damage to one's reputation and life can be difficult or impossible to overcome. [2079, 2080]. Both real and generated images can combine with sextortion to exert control over victims and further traumatize them [386, 387, 2081, 2082].

> *"On the morning of January 9, 1997, as she walked from a friend's apartment on the second floor to her grandmother's apartment on the sixth floor of the Cabrini Green housing project in Chicago, a little girl who would become known as 'Girl X' was raped, strangled, had gang signs scrawled on her body, roach spray was sprayed down her throat, and she was left unconscious in a stairwell."*
> — The Girl X assault, Chicago, 1997 [3068]

Sexualized dehumanization has implications beyond the aggressive eroticization of female bodies through pornography.[255] It plays into learned cultural norms that regulate and define our value to ourselves and others. Sexual objectification impacts beliefs, attitudes, and actions. It facilitates a sense of ownership, denies human autonomy, and ignores the victim's wishes and feelings [2083]. In both males and females, even a brief experimental exposure to pornographic content decreases their view of women's competence, morality, and humanity [2084]. Sexualization

[255] One should remember that correlation does not mean causation. There is likely a positive feedback cycle between pornography and sexual dehumanization rather than a simple one-way cause and effect.

contributes to physical and mental health issues, including stress disorders and depression [2085, 2086]. Its imposed fear restricts women's movements and opportunities. In a person or group who has endured a history of sexualization, a gaze that a typical man might ignore or respond to as a mild threat transforms into a trigger for body shame and anxiety in women [2087].

> *"November 2, 2000, Marietta Georgia: Up to 25 males — ages 12 to 25 — raped and molested the girl for 12 hours after luring her off of a bicycle and into an apartment building, police said Wednesday"* — Disabled Ga. Girl Gang-Raped, Chakraborty [3087].

By defining women or any group by their genitals, we dehumanize and tacitly approve of harassment and violence against them. From drunken fraternity parties in the United States to Cambodian *bauk*, societies have given a pass to young men who commit gang rapes, even if they do not overtly approve. Despite official condemnations, society often treats gang rapes as bonding opportunities, harmless recreation for young men, or a way to make money [2077]. The victimized women, in contrast, suffer shame and castigation. Sexualized dehumanization imposes fear and control on the sexualized groups and reasserts the primacy of the entitled.

> *"In the cold shower, the beatings continued. Under an icy flow of water, wearing heavy cotton undergarments, the children would be hit with a belt or a cane, depending on which nun was doing the beating that day."* — description of punishment for wetting the bed given by a Magdalene survivor [2090]

Cultures use sexualization to control women and disenfranchised minorities and as a mechanism for moral exclusion. From the 18th to the late 20th century, sexual accusations against women in Ireland could lead to incarceration and a life of servitude within the Magdalene Laundries [2088, 2089].[256] Survivors of this enslavement reported emotional, verbal, and physical torture, including electric shocks to the hands [2090]. Branding

---

[256] According to the *Irish Post*, the Irish state has apologized, acknowledged the wrong, and paid £50 million in compensation to survivors. This gives the survivors about €11,500 of €100,000 each for their years of suffering. As of 2013, the congregations of sisters who operated the laundries have refused to compensate survivors, arguing they were not the abusers [3435]. As of 2019, about half the money, €25 million, had been paid to 700 women. In 2017, the ombudsmen extended compensation rights to women who resided in other institutions but had to work in the laundries. As of February 2019, nine of the 87 additional Magdalene women who applied for compensation had received it [3258]. Only one received the maximum award.

them as harlots makes ignoring their claims and denying justice easier despite the public recognition of the atrocities. In these facilities, the state and church also imprisoned pregnant, mentally disabled, and homeless women.

Dehumanization through sexualization undermines the perception of rape as a severe and demeaning crime [2091]. Men tend to exonerate perpetrators and blame victims more than women do, consistent with the correlation between sexual dominance and the acceptance of rape myths [2092]. Hostile attitudes increase the likelihood of objectifying a target group, and hostile sexism is linked to greater acceptance of acquaintance rape, a condition too many still consider an acceptable excuse for the violation [2093, 2094].

When men fall under blame for their actions, communities often scapegoat mothers for poor parenting [2095]. Society casts women into the unattainable role of the idealized mother. Far from the stereotype of a woman who always smiles, bakes cookies, and makes all problems disappear, mothers, like other humans, are fallible beings and can do only a human job. Mother-blaming has been pervasive throughout history, and moms continue to garner culpability for the actions of their adult children [2096, 2097, 2098, 2099].

WEIRD societies have far less affinity for blaming the father for the actions of their children [2100]. Americans are likelier to blame high-school and grade-school teachers [2101]. As of 2018, 89% of grade-school and 64% of secondary-school teachers are female. Like mothers, they are desexualized and expected to self-sacrifice for their students [2102, 2103]. During the 19th and early 20th centuries, many American school systems contracted female teachers under "marriage bars" that terminated their employment if they took a spouse. These bans reinforced the asexual nature of the teacher. They also enhanced married women's financial dependency on their husbands by limiting their commercial opportunities.

> *"The boy paid his sister a dollar to tell police that Cordie had tried to rape her, but a grand jury refused to indict and Cordie was released and abducted that same day by county officials, who took him to a cedar tree and hanged him."*
> — G. King, Devil in the Grove [3331]

Note that marriage bars were not limited to teachers. Until they were found unconstitutional in the 1970s, many professions required women to resign if they married. At its peak, 751 local school boards imposed bans against married women, and half of all female office workers were impacted. These bullying bans forced women from the labor force, affecting their ability to get raises, promotions, and tenure [2104, 2105]. Those who remained contended with the low salaries paid to women workers and the feminization of poverty, an issue exacerbated for women of color who also faced discrimination based on skin tone and a dehumanizing stereotype of hypersexuality.

In the United States, African-American women have endured a hyper-sexualized fetishization aimed at reducing them to a set of sexual attributes. The imposed image adds an additional layer of challenge both in academia and the workplace while providing a tacit pretense for rape and other violence while diminishing their chance for just retribution [2106, 2107].[257] Latinas experience similar stereotyping, exacerbated by the images of the hot-Latina lovers that much media, including Latinx media, has considered acceptable [2108, 2109, 2110]. In the late 19th and early 20th centuries, Asian-American men endured the same stereotype, and Asian women continue to bear the demure "Lotus Flower" and predatory "Dragon Lady" fetishizations [2111, 2112, 2113] that contributed to the 2021 Atlanta massacre of massage therapists [2114].

---

[257] Try googling "Black girls" and "White girls." As of June 2025. The images I got included many sexualized Black women. The images retrieved of "White girls" include erotic images but are far more diverse. I ran this search after reading *Algorithms of Oppression* by Safiya U. Noble. More positive images of black women now appear than when Ms. Noble publicized the dehumanization [3208, 3438], but the search continues to return sexualized images of adults (search run on September 12,2025). The AI results incorporate the learned prejudices of their users, a problem shared with other AIs that learn from biases, such as hiring algorithms that shape the candidate pool and often incorporate past hiring decisions and downgrade the qualifications or abilities of female and minority candidates [3072]. (NB: Since I first ran these searches in 2023, the returns for "white girls" have also become more sexualized, and now includes images from the *White Chicks* movie and *Two Broke Girls* (TV series).)

Consistent with the primate pattern of sexual hoarding by powerful primates, human males of entitled groups sexualize disenfranchised males by diminishing their physical attractiveness or casting them into the role of bestial predators.

Like the sexual aggression and gendered racism aimed toward Black women [2115], Black male bodies have also been fetishized and burdened with false sexual accusations. D. W. Griffith's film, *The Birth of a Nation*, originally titled *The Clansman*, played on White fears of carnal, depraved Black men. White mobs lynched African-American men for many reasons but often defended their actions through allegations of sexual offenses [2116, 2117, 2118]. These ritualized, extrajudicial murders often involved hanging or burning alive after torture and sexual mutilations, including castration. Many of the accusations were after the fact, and when lynching was questioned, racists would point to these accusations as if they were the norm, arguing for the protection of White honor and purity. Portraying the lynching victims as sexual predators discounted the ongoing, ritualized acts of terrorism used to control Black Americans with fear after the fall of legal chattel slavery.[258] The accusations reinforced racist hierarchies and turned Whites who opposed and spoke out against the lynchings into social pariahs.

> *"Before the torch was applied to the pyer, the negro was deprived of his ears, fingers and genital parts of his body."* – Article from the *Springfield (Massachusetts) Weekly Republican*, April 28, 1899 cited in [3112]

Given the minimal time our species spends engaging in sex compared to other activities, dehumanization through sexualization makes little logical sense. Yet, LGBTQ people are defined by that minuscule slice of life. The sexualization of the community contributes to its marginalization and feeds the brutalization it often endures. Lesbian, bisexual, and transgender women understand the humiliation of fetishism, erotic objectification, and being treated

[258] During the enslavement era, the enslaved were the legal property of owners who could inflict any punishment, including death. Still, many things motivated extrajudicial killings (lynchings). The Turner rebellion, for example, resulted in hundreds of executions by terrified and enraged White mobs.

as an exotic sex toy available for conquest [2119, 2120, 2121, 2122, 2123]. LGBTQ people also share the experience of media sexual stereotyping with women of color [2124, 2125, 2126, 2127]. Bisexual women bear the double stereotype of being hot-blooded by the straight community and unable to admit they're gay by the LGBTQ one [2128], while gay men bear a feminized stereotype that reduces their social standing in our heteronormative, male-dominated society [2129, 2130].

> *"Abuses have not been restricted to Abu Ghraib. Numerous people held in the US Air Bases in Bagram and Kandahar in Afghanistan say they were subjected to torture or other cruel, inhuman or degrading treatment in US custody, and the administration has failed to comply with the Geneva Conventions with regard to the Guantánamo detainees."* – Amnesty International Press Release, 7 May, 2004 [3518]

Sexual dehumanization has other tendrils used to strangle humanity. At Abu Ghraib Prison, guards stacked naked men into piles and had women lead them on leashes like dogs in sadistic degradation. The prisoners were sexually abused, raped, and coerced into engaging in non-consensual sexual acts and simulated sex acts while being photographed. Photos showed penetration objects, including a truncheon, wire, and a phosphorescent tube [2131]. Men were forced to masturbate while female soldiers watched. They were ridden like animals, forced to get food from toilets, and had female soldiers fondle them [2123, 2119, 2120, 2121, 2122]. The torture cast the men into sexualized, feminized roles using an archetypical masculine power for this abuse. The prisoners were stripped of personal honor and the ability to exact vengeance. When integrity is challenged and satisfaction is denied, violence often erupts.

The Abu Ghraib tortures grew from the anger Americans experienced after the September 11, 2001 attacks. Then, the US demanded retribution, and the sacrifice of a few of the "others" provided an outlet for the involved soldiers and contractors.[259] Despite most Americans condemning the torture, there was no

[259] Some argue that torture does not work. They are wrong. Torture may not extract *accurate* information, but it will elicit a confession. People will say anything to end their suffering. The physical and psychological damage can limit recall and the ability to share information later. It may be a terrible way to get reliable data, but like it does in lynching, it can provide cathartic relief and propaganda opportunities for the torturers.

significant outcry or demand for the perpetrators' hides. Instead, the sexualized hell left people with a moral blindness, now known as the "Abu Ghraib Effect" [2132]. When viewed in light of the Stanford Prison Experiments, the degradation of the prisoners should not surprise anyone.

The sexualization of prisoners at Abu Ghraib did more than humiliate and dehumanize them. People will torture a sexually objectified victim more readily than a non-sexualized one [2075, 2068, 2133]. Sexual objectification desensitizes perpetrators and inoculates them against the violence, quickly blinding them to the humanity of their victims and freeing them to engage in other tortures, including electric shocks, sleep deprivation, beatings, and prolonged standing and kneeling.

Self-imposed sexualization also happens among the empowered. Cultures can impose behavioral and dominance ideals, leading members to measure their worth in carnal prowess or conquests. Failure to meet societal standards can result in shaming and loss of status. Young Cambodian men, for example, may gain peer-group status by engaging in a *bauk* (gang rape); refusal to cooperate can be viewed as a deviation from the presumed social standard. The objectification of women as an expected norm can restore a sense of power and control and provide a mechanism for attaining the expected conquests. Hostile attitudes toward women correlate with rape and other physical abuse [2093, 2092, 2068, 2134]. In intercourse, at least, downtrodden men still claim dominance.

The "illusory superiority" or the above-average effect leads us to overestimate our positive qualities and ignore our negative ones [2135, 2136]. We use our sense of primacy to uphold our enslavement of other people and the colonization of different cultures. Aristotle [2137] argued that natural authority dictates the right to hold a higher status, and the superiority of these elites obliges them to guide the less capable. He intended to explain why Greek citizens should control free non-citizens and slaves. He never clarified why birth or membership in a cultural group should transmit these skills, nor did he need to do so. The implications were evident to his audience of Greek citizens. Unfortunately, this attitude has remained with us and permeates modern industrial cultures. Many of society's elites continue to perceive their position as earned, even if they have inherited all or much of their status through family or membership in a dominant social group.

We sometimes receive tacit or manifest societal approval for violence. When society rewards violence and degradation, we learn it pays to harm others. Successful aggression can transform a social animal like an ape or a hyena from a docile individual into an active aggressor after a few successful fights [2138, 2139, 704].

Like other social animals, humans increase aggression after successful bouts [2140, 2141, 2142, 2143, 2144, 2145, 2146]. Victors and losers feel the emotions of dominance and submission: pride and shame [830]. A horror of violence can give way to bullying in the wrong circumstances and rewards. The news is replete with stories of celebrities who bully coworkers or express disdain for the lesser beings surrounding them. As obnoxious as this behavior is, we can trace its roots to the privileges and power ceded by primates to the leaders of their troops.

Not everyone exposed to the circumstances of fame becomes an insufferable bully. Many successful people enjoy the privileges of success and power without actively demeaning the lower-ranking humans around them. The strong social ties of our species favor leaders who are givers and actively aid others. Trustworthy leaders enhance cooperation within social systems [1350]. Such pro-social behaviors help them to develop allies among the ranks of the troops. In a society of multiple lineages, such as our own, beneficent behaviors can backfire if the public perceives them as nepotism.

Once the blatant flaws in the concept of natural superiority become apparent, we have a better chance of avoiding the obvious pitfall. However, the trap can take a different guise—a belief in our moral primacy. Our faith in our ethical stances again appears benevolent. The attribution error leads us to see ourselves as motivated by a sense of right and wrong and to see others as driven by egoism and personal gain [2147].

> *"The visual evidence and the verbal testimony of starvation, cruelty and bestiality were so overpowering as to leave me a bit sick. In one room, where there were piled up twenty or thirty naked men, killed by starvation, George Patton would not even enter. He said that he would get sick if he did so. I made the visit [to Ohrdruf concentration camp] deliberately, in order to be in a position to give first-hand evidence of these things if ever, in the future, there develops a tendency to charge these allegations merely to 'propaganda'."*
>
> — Dwight D. Eisenhower [3131]

We rely on our convictions about our moral, intellectual, and rational supremacy to defend the subjugation of a group or excuse the imposition of our will on others. We assume it is for their good. We will accept that the leaders of an opposing group have subjected them to an intense campaign of propaganda. We see outsiders as lacking access to accurate information or under the malevolent control of a different government, members of a criminal gang, or another [false] religion. From this perspective, we make a logical jump: because their leaders have influenced them, that community has had its free will removed. This handicap renders them unable to exercise judgment. Therefore, they do not know they suffer or that their behavior is wrong; the presumed propaganda campaign deprived them of the words to express their frustration or unhappiness with their current state.

We condescendingly see other cultures as children, unable to make informed decisions about their lives. This narrative allows an outside power to support wars of liberation against groups that might not want it. Various religious groups have used the same argument to justify "conversion" programs aimed at turning LGBTQ people into heteronormative heterosexuals.

We take the idea of verbalization one step further. We presume people must be able to verbally express a desire for something before they can have thoughts that differ from ours. This arrogance arises from a widespread reasoning error in our species. We confuse what is *necessary* for something to happen with what is *sufficient*. The ability to verbally describe misery *suffices* to show one cares about one's state. Just because something, such as language, is "sufficient" to confirm suffering does not mean that language is *necessary* to experience it.[260] Young children often cannot verbalize what is wrong in their world, nor can animals [2148, 2149]. Just because they lack the vocabulary to explain their victimization does not mean the bullying is not real. An inability to express displeasure or satisfaction with their current state does not negate their experience.

Sign Here: Financial Bullying

> *"No promises were ever made to give Jackie anything. Every dollar a kid earns before he is 21 belongs to his parents. Jackie will not get a cent of his earnings."*
>
> — attributed to the mother of Jackie Coogan, a silent movie child star [2150]

Communities use social contracts to regulate everything from marriage to selling paper. Within cultures, members agree upon most aspects of customary rights and privileges of rank, and they tie these rights to social obligations for the benefit of the group [2151, 2152, 2153]. Societies often ostracize or kill people who become a perceived threat by challenging the established social order. These punishments carry little risk of consequence, and they reflect the behavior in other primates, where resource allocation is influenced by dominance hierarchies, social alliances, and competition over available resources.

---

[260] An example of this happened to me recently. I wanted to make an analogy with a familiar party favor, a small, curled paper tube into which one blows air. It uncurls and then re-curls when the air pressure is again reduced. What is that thing? I know it, even though I have no word for it.

Group consent and social contracts work better in theory than in practice. Unequal bargaining power often leads to coercion. Many countries have laws to protect young people from contractual exploitation. Most governments acknowledge the inability of children to enter into informed contracts. They hold that children lack the competence to enter binding agreements because of their emotional immaturity and limited experience, knowledge base, and cognitive development. Courts may appoint independent legal counsel to interpret the children's wishes and protect their rights, and ombudsmen may investigate complaints.

A prime example of financial bullying directed toward children happens to child performers. Jackie Coogan, better known to later generations as Uncle Fester on the original *Addams Family* television series, began as a child superstar, working with Charles Chaplin and other greats of the silent film era. In several films, Coogan received top billing over equally famous adults. As a child actor, Mr. Coogan earned millions of dollars. When he reached adulthood, he discovered his parents had raided his earnings and spent the bulk of it. His stepfather had even confiscated a diamond-studded pocket watch given to him [2150, p. 172].[261] Instead of entering adulthood as a multimillionaire, he relied on the kindness of friends for survival, especially his close friend and co-star Charles Chaplin. He had company. Other child actors, such as Shirley Temple, were also exploited. Under the old common law that regulated the earnings of children, the parents, not the children, had all rights to the income of a minor.

In 1939, California passed the *California Child Actor's Bill*, better known as the Jackie Coogan Act. The state effectively defined a percentage of the child's earnings as belonging to the child. Revisions to the bill have also included guarantees of education, vacation time, and working hours. The bill provided some legal protection of their earnings and legal standing for the children when they reached the age of majority.

---

[261] Recall the Cinderella effect. Stepparents can be wonderful, but statistically, they commit more acts of abuse and neglect than biological parents. As with the Herod effect, evolutionary biologists tie this phenomenon to the lack of a genetic connection.

It barely scratched the itch. Egregious financial exploitation of Hollywood child performers could still take place. In California and many other states, only 15% of the child's earnings had to be placed in trust for the minor. For every dollar a child earns, the bill protects only fifteen cents. Canada mandates 25% [2154]. The remaining earnings remained available for use at the discretion of the parents or guardians, and yet, that protected 15% can still encounter risks. Shirley Temple discovered that no one had deposited her 15% of her multi-million-dollar earnings into her Coogan Trust. Despite having earned millions, little money remained that she could call her own. In 1989, Gary Coleman (*Diff'rent Strokes*) sued his parents for misappropriating the millions of dollars he earned as a television star.

This manipulation of loopholes goes beyond the film industry. Dominique Moceanu (Olympic gymnast) sued her parents for her money [2155]. United Press reported that singer LeAnn Rimes brought suit against her father [2156]. And, in 2012, parents of child actors brought suit via the Coogan Act against a bank holding a trust, suing it for withdrawing funds in the form of fees. The bank lost upon appeal [2157]. Although no lawsuit documents confirm that Christopher Robin Milne (of *Winnie-the-Pooh* fame) underwent financial abuse per se, his father's writings, in which Christopher featured, primarily benefited his father. His parents profited from his identity and the popularity of his father's children's novels, making him a target of bullying by his schoolmates. The emotional impact on him contributed to his foregoing of royalties.

Today, technology has democratized the parental exploitation of children. Despite the Children's Online Privacy Protection Act (COPPA), postings of private information can be done with parental consent [666, 2158]. With the advent of blogs, vlogs, and other social media, more parents have begun exploiting their children and their images, often for personal ego gratification [2159].

"Sharents" have posted mortifying photos and pulled cruel stunts on their children in their quest for more followers and the dollars it can bring. What may be hilarious to a parent can be mortifying to the child. In a

Spanish and Czech study, 3.5% of parents acknowledged posting a photograph of their naked child or infant [2160]. In an American survey, 56% of parents admitted to posting embarrassing information about their children, and 27% shared an inappropriate photo [2161]. By adolescence, most people disapprove of their parents' sharenting; it embarrasses them and damages their own impression management and autonomy [2162].

Children's personal privacy has decreased. It has degenerated from having their schoolbags and rooms rifled through by concerned parents to the non-consensual posting of their images across the web [2163]. Assuming control of their child's image and broadcasting intimate moments in their lives to a worldwide audience effectively robs them of agency and opens them to future abuse [2164].

Where an in-school, workplace, or neighborhood bully humiliates and tortures, the victim has respite away from that site. With online bullying, the images and words follow them everywhere, even if they change schools or move. They lose their right to anonymity and the freedom to be forgotten [2165, 2166], exacerbating internalized depression and anxiety and enhancing externalized symptoms such as aggression [2167, 2168, 2169].

In addition to cyberbullying and harassment by peers, sharenting opens children to identity and information theft, sexting, deep-fake manipulation, and the exploitation of current and old family posts by bullies [2170]. An Australian study found pedophile sites obtained 50% of their images from social media postings [2171].

Before the internet era, parents might embarrass their children by showing their date a naked photo of them as a baby in a bassinet. My parents were particularly fond of a home video of me dancing alone at a family party at age eight. Then they wondered why I never danced at family parties again. Today, "sharenting," the habitual posting or oversharing of children's images by parents, can involve humiliating instances that will haunt their offspring for years.

Sharenting influencers can reach millions of dollars in income per year by exploiting their offspring for profit [2172]. In 2025, only a handful of states, including Illinois, Utah, California, Maryland, and Washington, require that a parent protect a portion of the income for use by the children whose privacy is violated. The use of children in online content is not yet considered a "job" in most states, and the Coogan law does not apply to most of these victims [2173, 2174]. However, California has expanded the protection of the California Child Actors' Bill, mandating that 15% of their earning be set aside until they reach adulthood [2175].

Old people also experience age discrimination and financial bullying. The elderly fall prey to financial abuse through constant demands upon resources, excessive charges by trustees and care facilities, or theft by their children that parents may hesitate to report. Despite evidence of capacity, adult heirs regularly challenge wills and contracts entered by senior relatives.

Although such financial exploitation may seem like an evolutionary mismatch within our modern economy, this monetary abuse does not surprise biologists. From the perspective of genetics, an offspring is one hundred percent related to herself but has only fifty percent of each parent's DNA. Consequently, each offspring has a more substantial stake in gaining access to parental resources than acknowledging their parents' rights or siblings' share of any inheritance. Our pro-social cultural overlays and understanding of our society's norms reduce the severity of the legacy conflicts over estates, pots, and pans. Where the contributions of the elderly outweigh their costs, expect to find care and respect. Where they become burdensome, expect bed-bug-ridden nursing homes and dehumanizing treatment as they await the release of death.

A similar rule holds for the financial abuse of children by their parents. In day-to-day decisions about spending family funds, parents determine what percentage of family resources to invest in themselves and each child. Children, in contrast, do best when all the resources come to them [2176]. This fundamental bias plagues our associations with other family members and more so with non-kin. Contractual law, at its best, balances the needs of all parties fairly. At its worst, it becomes a framework for legalized exploitation.

Governments seldom form based upon viable, even-handed social contracts [2177]. Such pacts work better for small, confined groups like bands of gatherers or pirate ships [2178] than for more extensive assemblies. With larger groups, contracts are often biased, bestowing privilege upon one sector and enforcing the subjugation of another.

> *"Another day older and deeper in debt. St. Peter, don't you call me 'cause I can't go, I owe my soul to the company store."* — "Sixteen Tons" by Merle Travis [3255]

Following the end of legal slavery in the Southern United States, sharecropping became a norm. This system reframed slavery into a debt bondage that included Blacks and poor Whites. One-quarter of the population owned most of the land (predominantly cotton farms). Tenant farmers, most of whom were sharecroppers, cultivated 73% of the farms. The sharecroppers received little pay, and the landlords controlled the prices of goods their tenants needed. Landowners made loans at a high interest rate, creating a system that enforced perpetual financial debt. In cities, company stores and towns did the same, effectively placing their employees in servitude.

In 1938 Nazi Germany, one lopsided social contract held that Jews should not fight back against subjugation by Nazis. On November 7, 1938, Herschel Grynszpan tried to avenge the deportation of his father and the oppression of Jews. He planned to shoot the German ambassador to Poland. By mistake, he hit the ambassador's third secretary, Ernst vom Rath.

This one man's breach of the rules led Joseph Goebbels to arrange for "spontaneous" demonstrations. The riots became known as "Kristallnacht" because Nazi stormtroopers broke the windows of thousands of Jewish-owned shops. Looting ran rampant. The violence took the lives of at least 26 people. Rioters raped women. Fire engulfed 119 synagogues. The arson and vandalism left untold numbers injured or homeless.

Grynszpan broke a critical social norm: Jews should not fight back. On the pretext of that incident, the government tacitly approved the riots. The victims faced fines because they had damaged property. Afterward, Goering devised a solution to prevent Jewish victims from accessing funds from their insurance companies; he declared the contracts void.

The Nazis redefined the damage to Jewish properties as damage to the state. Accordingly, the state seized the insurance monies, leaving the victims without payment [2179]. Insurance companies voided the contracts on pension, life, and health insurance policies held by Jews [2180]. The Reich prosecuted only one set of rioters after the carnage; they indicted men who raped Jewish women. The government would not punish the vandals, but they did charge the rapists who had broken the ethnic purity laws imposed by the Nazis.

When Franklin Roosevelt addressed the "forgotten man at the bottom of the economic pyramid," he and his advisors referred to the landowners, not the sharecroppers who remained at their mercy [2181]. Landlords dominated committees designed to enforce the New Deal contracts and investigate complaints [2182]. Benefits seldom trickled down to the tenant farmer.

Dysfunctional trickle-down ideology continues to plague workers today. Since the 1980s, universities have increasingly depended upon non-tenure track[262] faculty whom they pay less for higher workloads, can fire

[262] Contrary to what most people assume, tenure does not mean a person cannot be fired. It means they cannot be fired without due cause and a hearing. Non-tenured faculty do not have this right and can be dismissed far more easily.

with little to no warning, and disenfranchise from university governance [2183]. They lack the power and autonomy of tenured faculty and can be disciplined or dismissed at the pleasure of their chairs. Their classification as "temporary" employees makes it harder to obtain research grants. Often, they are denied research opportunities, decreasing their value and limiting their chances to qualify for tenure-track posts elsewhere. Departmental policies, including the assignment of classes for which the non-tenured person is minimally qualified, micromanaging other courses, higher class loads, and minimal support contribute to lower student ratings for non-tenured faculty, reducing the likelihood they will ever be considered for a tenured position [2184].

The combined stressors contribute to job dissatisfaction, higher cortisol levels, and the health problems associated with perpetual stress and a degraded self-image and sense of worth. Although they hold most university-level teaching positions, non-tenured faculty are widely excluded from university governance (no or abridged voting rights). Though the tenured minority pays lip service to equality, they seldom fight for equal pay for work of equal value or the extension of their privileges, even to the long-term non-tenured.

Still, the myth of equal, dependable, shared governance through a shared social contract prevails. Both individuals and governments continue to use the ruse of violated social agreements or understood norms to support violence against a person or group. (I have little doubt that Matthew Shepard's murderers believed society would protect them because Matthew had violated the norm of being straight). Offered excuses gain momentum when one group needs the resources of another. Seldom do the laborers who do the bulk of the work get treated as equals [2185].

Water, Water Everywhere

"Water, water, everywhere, Nor any drop to drink."

— Samuel Taylor Coleridge, *The Rime of the Ancient Mariner*, (1834) [2186]

Around the world, the ongoing communal violence against native peoples and ethnic minorities exemplifies the problems of forging fair social contracts in the face of kin selection and unequal power. Coalitions have eaten away at communities by unilaterally rewriting the terms of agreements. Face-saving pretexts, plausible to the aggressor, usually precede the seizure of territory or other resources.

The expansionist phases of nations wreak havoc on others. Rome entered into treaties with their neighbors only to break them when the agreements no longer suited its needs. In the mid-20th century, the military of Imperial Japan massacred and subjugated people in Korea and Manchuria. In search of resources, European countries seized land in Africa and North America and took Africans as slaves. Americans have backed brutal dictatorships in pursuit of oil. Aztecs enslaved their neighbors as they built their empire. Like warring, genocidal chimpanzee tribes, to amass power and resources, human coalitions subjugate and destroy potential competition. Like chimps, we take wealth from weaker communities, including land.

> *"Every person who, without the consent of the Superintendent General expressed in writing, receives, obtains, solicits or requests from any Indian any payment or contribution or promise of any payment or contribution for the purpose of raising a fund or providing money for the prosecution of any claim which the Tribe or Band of Indians to which such Indian belongs, or of which he is a Member, has or is represented to have for the recovery of any claim or money for the said Tribe or Band, shall be guilty of an offense and liable upon summary conviction for each such offence to a penalty not exceeding two hundred dollars and not less than fifty dollars or to imprisonment for a term not exceeding two months."* — 1867, Chapter 98, Section 141 of the Canadian Indian Act [3485, 3187, 3132, p. 301]

Through the lens of history and with our current twenty-first-century mindset, the behaviors of the conquerors seem appalling. If we could look through the eyes of those conquerors, we would see a different picture. Romans had confidence in their natural dominance and right to rule. The existence of their lifestyle and culture depended upon resources gained through conflicts. Because of their confidence in their superiority, they knew they had a right to survive. Many cultures share this conviction and do not consider their victims fully human. During the Rwandan genocide, the Tutsi were characterized as "cockroaches." Misogynists often refer to women as "bitches" or "cunts." The Nazis spoke about concentration camp inmates as "blode Schweine" (stupid pigs). Animalistic and mechanistic descriptions, like the leashing of the Abu Ghraib prisoners, reduce the perpetrator's burden of guilt. They set a stage on which one can commit genocide—or seize the labor, land, food, or water—of a brutish enemy portrayed as little more than animals incapable of managing their affairs [2187]. The abrogation of treaty rights and even enslavement cease to evoke horror as the enslaved become no more than machines that happen to talk.

Such annulments are not limited to the distant past. In the 1950s, the word "termination" circulated in political circles. It referred to the dissolution of all treaties with Native Americans and the intent to sell the remaining tribal lands to bidders and cronies. The Taos Pueblo of New Mexico challenged the thinking and introduced a series of protests and lawsuits, successfully reclaiming the Sacred Blue Lake. Other nations followed suit, and many reclaimed specific treaty rights [2188]. But, as John Philpot Curran warned, eternal vigilance is the price of liberty; without it, our inner chimp rises.

> *"The condition upon which God hath given liberty to man is eternal vigilance; which condition if he break, servitude is at once the consequence of his crime and the punishment of his guilt."*
> — John Philpot Curran [32]

In 2004, a bipartisan bill assaulted Hopi and Navajo water rights. The Arizona Water Settlements Act limited the amount of water the Arizona tribes could claim to 67,300 acre-feet for all future settlements. On 14 February 2012, former United States Senator Jon Llewellyn Kyle cosponsored proposed legislation to change the status of water rights of the Navajo and Hopi peoples within Arizona [Senate Bill 2109-45 [2189]. Representative Ben Quayle of Arizona sponsored an almost identical bill, H.R. 4067. This bill went to a committee that same day. The bills, known as the *Navajo-Hopi Little Colorado River Water Rights Settlement Act of 2012*, would have ended many sovereign water rights for the two minority nations. Sometimes justice prevails; both bills died in committee.

Had they passed, the bills would have effectively ended the water rights negotiated during the reservation settlements of the 19th century. On the surface, the proposals sounded fair. In exchange for limiting their water rights, the state of Arizona would reallocate water to the Navajo Nation: "*certain Central Arizona Project (CAP) water pursuant to a delivery contract that effects [sic] the delivery of that water through its diversion and use from the San Juan River.*"[263] The Navajo-Gallup water project would deliver these waters to and through the Navajo Nation. The Navajo and Hopi nations would enter into leases and exchanges of water within Arizona. Certain waters would be retained for use in future settlements of the claims by the Navajo and Hopi nations.

Aside from the assault on the historical and cultural ties to the Little Colorado River, another problem lurked within the wording of the proposed changes to the water rights. No feasibility study had yet been done

[263] The San Juan River is a large tributary of the Colorado River and a drainage area for the Four Corners region where Utah, Colorado, Arizona, and New Mexico come together. In the 20th century, the waters of the San Juan were polluted by drilling for fossil fuel and uranium mining. It flows through the Navajo nation after joining the Animas River. It has substantial annual flow variation, and the droughts of the 21st century have reduced the yearly peak flow from approximately 30,000 to less than 10,000 cubic feet per second [3288].

(noted in the bills). The Hopi and Navajo would have needed to forfeit their legal rights to water from a river they had used across time. In exchange, they would become dependent on a pipeline others would control and which might never be built.

The bill would have ended agricultural and ecological self-sufficiency in both minority nations. Even before the 2012 proposal, they had been excluded from the 1922 Colorado River Compact and have fought for their water rights through the courts. The Hopi and the Navajo would no longer have retained the right to bank their water or recharge their aquifers.

The bill limited their ability to bring suit against companies that polluted the sacred river's waters, their reservoirs, and the water table. It would have effectively ended their input into the water quality control of neighboring industries [2190], including the strip-mining coal industry at Black Mesa that continued until 2019. The Hopi tribal council signed a contract with Peabody Energy with the aid of legal counsel that Peabody may have influenced. The company earned millions for its stockholders while depleting the water table and polluting the river.

The Hopi constitution should have prevented the exploitation of mineral rights without the people's approval. Tribal councils were imposed on Hopi in 1936, removing a significant obstacle to negotiating for mineral rights. Washington ignored tribal rule and authorized the council to make the leasing decision [2191]. In the Hopi worldview, no one owned the rights to the land, so they could not be ceded.

Although neither of the 2012 bills passed, both nations felt the insult. Their proposed congressional bill sent a clear message. The United States could take away the remaining rights of minority nations within its borders. No contractual treaty rights should be considered safe.

Enforcing the Norm

"But I do feel you're in the wrong... I feel it intuitively."

— from *Rhinoceros* by Eugène Ionesco [2192]

Societies have long used brutal, humiliating punishments against people who challenge social conventions. Tsarist Russia would execute or exile dissidents to Siberia. The Soviet Union used executions, "corrective labor camps" (Gulags), and isolated colonies to suppress nonconformists. Between 1937 and 1938, the Soviet security police arrested nearly 1.6 million people. Almost 87% of these arrests were on political grounds. The Soviet government had an 85% conviction rate [2193, p. 1023]. Between 1940 and 1954, the Soviet government executed over three-quarters of a million people for counterrevolutionary activity or other political charges.

The United States did not have a formalized system of Gulags and internal exile colonies; it did have peonage, prisons, and chain gangs that, despite the 13th amendment to the US Constitution, effectively reinstituted slavery under a different name [2194, 2195]. The conditions endured by the prisoners were horrific, and the American public seemed unwilling to recognize the treatment.[264] By 1930, the American state of Georgia alone had over 8,000 convicts working on chain gangs over whom they held virtual property rights [2196]. Three-quarters of these prisoners were African Americans. A semi-novel, *Georgia Nigger* [2197] used fictional characters to bring the horrors of chain gangs to public attention, much as *Uncle Tom's Cabin* [2198] did for the horrors of chattel slavery almost a century earlier. Photographs of mistreated convicts at the end of the book emphasize the reality of the stories told in the novel. Spivak [2199] simultaneously published a pamphlet that gave the same information in a nonfictional format.

---

[264] As Mark Twain noted, "the lie of silent assertion" [1296, pp. 159-170].

Just as governments use prisons and labor camps to control their populations, religious institutions have used violence to enforce religious norms and to stop new beliefs and practices that threaten the stability of local societies. In the West, the best-known example of this intimidation comes from the inquisitions against heresy by the Roman Catholic Church [2200]. The Church imprisoned and executed heretics like the Manicheans for years. However, it did not turn the Inquisition into a systematic process of carnage and oppression until Pope Lucius III declared the major heretical sects anathema in 1184. He took this step during ongoing political disputes between the papacy and Emperor Frederick Barbarossa in the north.

The first expansive inquisition was the episcopal inquisition of 1184. Pope Lucius III established it to eliminate the threat posed by the Catharist sect in France. Under Pope Gregory IX, the Vatican commenced a Papal Inquisition in 1231. The Vatican ostensibly designed it to combat the spread of heresies that had attracted many followers. The Papal Inquisition wielded Papal power with an iron fist, especially against Fredrick II in their continued dispute. It also established the social doctrine of *perpetua servitus iudaeorum*—the perpetual servitude of the Jewish People [2201]. That doctrine remained a subjugating force upon the European Jews until the late 19th century.

But, They Beat Me First!

In addition to governments and religious institutions, individuals and mobs will also gather and execute their punishments on individuals who do not conform to the group's standards. One British propaganda poster from the American Revolution depicted the tarring and feathering of a customs commissioner [2202]. American mobs reserved this brutal punishment for people suspected of being British informants or sympathizers.

This particularly heinous practice did not end with the Revolutionary War. It continued, and its victims included members of religious minorities. In 1851, Father John Bapst, a Jesuit Priest, was tarred and feathered for opposing Protestant religious indoctrination of Catholic children in public schools [2203]. In 1832, Joseph Smith, founder of the Church of Jesus Christ of Latter-Day Saints (a.k.a. LDS or Mormons), also suffered this punishment in Hiram, Ohio [2204, pp. 178-180].[265]

Through the 19th and mid-20th centuries, lynch mobs enforced norms. People could be dragged from their homes or jails and killed by mobs outraged at the actual or alleged behavior of the victims. Lynching (extrajudicial murder) usually involved hangings or shootings, but it often involved prolonged torture or burning at the stake.

> *"Everybody granted that if Tom were white and free it would be unquestionably right to punish him—it would be no loss to anybody; but to shut up a valuable slave for life—that was quite another matter. As soon as the Governor understood the case, he pardoned Tom at once, and the creditors sold him down the river."*
> — Conclusion of *Pudd'nhead Wilson*, Mark Twain, 1894 [3356]

Before 1850, when slaves were a valuable commodity, angry masses perpetrated many of their lynchings against white abolitionists. Live slaves could be sold, so killing one constituted the destruction of property. With rare regional exceptions, by the 1880s, mobs used lynching to impose a doctrine of white supremacy and intimidate African Americans and, in the American southwest, Latinx.

There are no reliable records of how many people were lynched before 1882. That year, the *Chicago Tribune* began recording these events through their appearances in local papers [2018]. Many accounts reported the gruesome details [2205]. From 1882 to 1968, mob violence in the United States resulted in the lynchings

265 The mob might have suspected a carnal relationship between Smith and a local woman, Nancy Marinda Johnson, who had been a guest in Smith's home. Marinda later became Smith's 10th plural wife [2204, p. 179]. Others claim that the incident had a financial motive, with the punishment instigated by people who believed they had been cheated by the prophet [2204, pp. 178-179]. Many non-Mormons despised early Mormons. Men expressed more anti-LDS sentiments than women. The LDS's successful conversion of women and the early LDS practice of polygamy contributed significantly to this animosity.

of approximately five thousand people. At least 73% of the victims of this period were African Americans, though African Americans constituted about 12% of the population.

Vehemence, such as lynching, is neither confined to the United States nor the past. In many parts of the world, the empowered justify violence against disenfranchised groups and women who step away from their confined roles and existing social norms, e.g., [2206]. In parts of India, groups dedicated to enforcing the old caste system have killed Dalits because they worked for a better life or dared to love, defend themselves from, or take vengeance upon someone from a higher caste. In some regions, honor killings and humiliations continue to subjugate people who challenge caste barriers or sexual norms. In 2012, for example, a Dalit woman was beaten and paraded naked through her Mulgaon village in Maharashtra because an upper-caste girl eloped with her son [2207].

All of these examples of degradation have been championed based on violations of social norms by the victim. The perpetrators viewed the victims as forfeiting their rights because they could not or would not abide by the set rules of their societies [2149].

If a person or group does not agree to a social contract or norms, they gather no innate societal protection. Society, which can also serve as an aggressor, determines right or wrong and the punishment imposed upon the offending target. In the case of Phoolan Devi, labeled a gang leader and bandit queen, the upper castes in India declared her a brigand and a killer. However, *before* her outlaw actions, these same groups did not stop the seizure of her family lands, her forced marriage, nor the repeated rapes by police and bandits that led directly to her life of vengeance [2208, p. 245–251, 2209, 2210]. Rationalizations, such as that given by the assassin of Devi, that it was to avenge the murder of 22 members of his caste, provide social support for continued blood-feud violence.[266]

We are more likely to attack another person (or group) who has harmed us if we believe the harm is intentional [2211, 2212]. We create explanations to stereotype the victim into the role of the aggressor, reinforcing our faith in our superior moral stance by insisting that the target has refused to comply with *our reasonable* social requirement. These excuses can legitimize anything from public humiliation to a civil war [2213]. Here, the attacker chooses the moral standard and defines the social values based on his or her own beliefs or best interest. The target of the aggression automatically assumes a morally inferior position. Tautologically, the aggressor thus deserves superior or special rights because they possess exceptional moral qualities.

The Pangloss Effect

> *"If this is the best of all possible worlds, what then are the others?"*
>
> — *Candide* by Voltaire [2214]

In the comic satire *Candide*, Voltaire mocks the then-popular belief that ours must be the best of all possible worlds, an idea championed by the German philosopher Gottfried Leibniz [2215]. If God is all good, he reasoned, then God must have chosen the optimal circumstances for us. Evil exists because it provides opportunities for courage; it brings out the best in humanity. Voltaire satirized this view in his character Pangloss, the excessively optimistic philosopher in his book, who finds a positive side to the harshest of trials [2214].

---

[266] Devi, a member of the Mallah (a fisher caste), served 11 years in prison for her alleged part in a "massacre" of twenty Thakur men. She became a folk hero to lower castes and served as a member of parliament. Sher Singh Rana assassinated her in 2001. Revenge motivated the killings by both Rana and Devi.

Although displayed differently across cultures and genders, a sense of pride is a cultural universal [2216]. We take pride in personal achievements, such as weight loss. (I was 25 pounds down when I wrote this and up ten reviewing it), and in the successes of a group, region, or ideal with which we identify, e.g., [2217, 2218]. That pride can be mobilized as motivation to excel for a company or neighborhood [2219] or, as we have seen in earlier chapters, to commit genocide, e.g., [2220].

When we tacitly act as if our country, religion, school, or baseball team exceeds all others, we become de facto fanatics. By aligning ourselves with a defined, select group, we place ourselves in a position of superiority. As true believers, we unknowingly substitute pride in membership for self-confidence and self-fulfillment [2221]. Arrogant, hubristic group pride can also lead to dehumanization and peaked violence against outsiders, be it between countries, schools, or sports teams [2222].

In Greek mythology, the Goddess Nemesis blinded those who committed hubris, bringing down the arrogant and the powerful.

Our sense of pride originates in the usual suspects involved in our social self and our understanding of others; thus, it ties to our sense of self-worth and our position in our social power hierarchy. The anterior cingulate cortex (ACC) compares our performance with our internalized expectations. The prefrontal cortex (PFC) provides our self-referencing framework. The dorsolateral and medial PFC process information about social hierarchies and our position in them. The amygdala modulates the intensity of our sense of achievement, and our nucleus accumbens (NA) of the ventral striatum (VS) provides the dopamine reward when we or our group succeed. Once we process the feelings, the insula lets us feel the states generated by other regions and prompts us to adopt dominant postures.

Hubris, the exaggerated sense of self or group importance, pride, and honor can be mild or pathological. Hubris syndrome (or hubris incompetence) happens when a person with power grows enamored of themselves, their abilities, and their potential to succeed [2223, 2224]. The syndrome, typically recognized in those who have achieved a modicum of status, leads to contempt for others. The condition can occur in political leaders, administrators, or anyone with an exaggerated sense of self-importance and a taste of power [21]. Unlike narcissism, hubristic disorder involves being isolated from feedback and accountability because of one's position.

Both hubristic and narcissistic disorders lead to bullying of perceived inferiors. Unique to the hubrist is an excessive identification with their organization or nation, believing its interest and theirs are one; a belief that history/God will vindicate their actions; restless and impulsive actions spurred by their sense of power; and disregarding practicality for their broader vision [2225]. As with other personality disturbances, having mild versions of the traits can improve functioning in some fields and harm it in others. Minor obsessive-compulsive characteristics make for a good housekeeper; obsessive-compulsive disorder means he or she could spend the day cleaning one doorknob. Hubristic *traits* can help a leader function without second-guessing each decision; the *disorder* impairs functioning.

Anyone can fall victim to hubris, or narcissistic arrogance. It replaces healthy, authentic pride in our accomplishments. The same hormonal and brain areas remain involved but may slip into hyperdrive or dampen their input. Changes in our levels of indoleamine (sleepy-time hormones) and catecholamines (wakey-wakey hormones) in the striatum and limbic system can alter our risk-taking behavior by making us more rigid or impulsive [2226, 2227, 2228]. The ACC malfunctions and ceases monitoring our performance against standards, resulting in a more rigid, self-enhancing system of beliefs. The PFC becomes overactive and exaggerates the importance of oneself or one's reference group.

In response, the reward circuits of the NA develop a hair trigger for self-referential input. In contrast, the amygdala becomes less sensitive to social feedback, and we disregard the accomplishments and feelings of others. Still, it can be recruited to provide the emotional bump to retain our position should the dominance hierarchy become volatile [2229]. This defensive response ties to our comprehension of our social relationships and the motives of others.

Unquestioning, absolute certainty about our status, a belief that we cannot fall—that we are entitled not to fall— blinds us to understanding. Attacks upon our unquestioned superiority threaten our worldview. The logical course of action is to become defensive. We lash out against the offending person or group who threatens our position or who expresses an opposing view or a counterexample to challenge our idealized world.

We desire this best of all worlds for ourselves and our own and expect that our leaders will provide it. When threatened, our emotional brain kicks into gear, and our expectations become demands. However, we share this limited world with many who want the same resource access. If we take a utilitarian approach and hold that all lives have value, then the best courses of action follow paths leading to the best balance between harm and benefit for the greatest number [635]. Conflict minimization models [2053] usually have such a utilitarian bias. Unfortunately, our brains differentially weigh those values, and it often pits our feelings and survival instinct against a more reasoned assessment of fairness.

As with anything else in the brain, specific regions become involved when drawing conclusions about an action's morality. Moral utilitarian decisions depend on a functional, reasoning ventromedial prefrontal cortex (vMPC) and its connections to the emotional amygdala [1569]. When these two regions malfunction, they impair the conscience.

Utilitarian decisions require that we weigh our emotions against our reason. One group loses to benefit another. Theoretically, we consider the harm and benefits to both before reaching our verdict. In reality, everyone has a thumb on the scale. People with impaired consciences have less trouble making utilitarian decisions because they put less weight on emotional arguments. However, that decision is biased in their favor. They are more likely to endorse choices that injure one to benefit another [2230, 2231] and to undervalue or dismiss the emotional consequences of accidental harm [2232, 2233, 2234].

On the surface, minimizing social conflict would seem to be a good thing. However, our species weighs the pros and cons of actions on our biased scales. We under-assess the personal benefits we receive from a deed and overrate the blessings the same act bestows upon others. Our minds exaggerate harm to ourselves while minimizing the wounds inflicted on others. In other words, we err on the side of genetic caution. We systematically underestimate our gains in resources and fitness while overestimating those of our competitors. This is the cracked cornerstone of our utilitarian decisions.

One does not gain a reproductive advantage by doing as well as one's neighbor. Anyone without this inclination would be less inclined to attack their neighbors for resources. They would be less likely to fight to gain resources. They would be more likely to become targets of others of their species who do not share this proclivity for fairness at their own disadvantage.

We can reduce but not eliminate our innate bias when we reflect on emotionally charged decisions, mulling them over during the day and sleeping on them—preferably for several days—and then deciding. Sleep processes our experiences [2235]. Rapid eye movement sleep (REM) suppresses our adrenergic activity (fight-or-flight instincts) and downregulates regions of the vmPFC, dPFC, and amygdala critical to our sense of morality. This is one reason why we can imagine things in our dreams that we would never consider doing. It's not some Freudian wish fulfillment; just a whispered exploration of possibilities with critical parts of the brain tuned out. REM lets us explore consequences, even if that journey occurs in the bizarre world of dreams. People who spend one night or several days processing Trolley-Type dilemmas[267] make fewer

---

[267] Trolley Problems are moral dilemmas involving killing one person to save others. Philippa Foot introduced the original version, along with similar problems. [3491].

utilitarian decisions and become less willing to perform a harmful action aimed at the greater good [2236].[268]

Violence often erupts over resources. Rome waged wars to access the wood they needed to fuel their cities. In the modern era, nations have battled over oil, farmland, and coal. The required resource, however, does not need to be something physical, such as food, water, or fuel. Inadequate space can spur aggression in humans and other animals. When held in cramped, crowded conditions, rats and deer, both social animals, experience more stress and aggression than when they enjoy more personal space [2237].

Humans in crowded urban areas also show higher stress levels [2238]. We respond with more aggression and experience more fear with increased crowding [2239]. Our accustomed levels of contact and our perception of others as kin or non-kin modulate this impact.

We can warp the utilitarian ideal of optimum good for the greatest number to uphold violence against any person or society. If an assassination or surgical military strike can prevent the deaths of more people, we justify those few deaths. The schoolyard bully endorses his humiliation of a peer because others will join in the fun. Fraternities and sports teams will promote misogynistic behavior, from jokes to rape, on the pretext that it encourages social bonding among the group members [2240]. People with variations in behavior, be they autistic, LGBT, or a religious minority, can be bullied or otherwise tortured based on the specious argument that they disrupt the societal thread. Through one-sided utilitarian reasoning, minorities can be enslaved for the greater good, and animals can be killed or tortured to benefit our valued species. Morality collapses without the emotional component, learned prohibitions, and a sense of compassion.

---

[268] We do not know the proportional impact of sleep and conscious daytime processing. The architecture of sleep and the amount of REM experienced correlate with the changes, but both could be downstream from other daytime experiences.

Reverence for Life

"Non-combatant, n. A dead Quaker." — *The Devil's Dictionary* [2241]

Few people advocate for bullying or other violence because it makes us feel good or serves our group's utilitarian purpose. Instead, we cloak the arguments that commit us to violence with more skill. Today, people across many traditions see life as inherently sacred and its wanton destruction as inherently evil [2242, pp. 32-37, 2243]. On the surface, we contend humans should have a right to freedom from harm by our own species. We consider indiscriminate attacks profane.

As idealistic as this argument sounds, it can bolster violence if it stops a greater tragedy. The United States government inspired support for the war against Saddam Hussein and Iraq by highlighting the atrocities of the Iraqi regime. Americans responded to the argument that we should topple one man to free a nation, believing Hussein and his government had no reverence for life. The proposed war of liberation would reduce the harm imposed upon the Iraqi people. Many Americans saw President Hussein as a malevolent dictator and widely believed his removal would alleviate Iraqi suffering. However, America fell victim to the hubris trap. Rather than greeting the Americans as emancipators, Iraqis saw the invasion as a painful loss of dignity and freedom. Rather than cheering their liberation, they saw the Americans as invaders who bombed them back to the Stone Age. No one wishes to lose honor, and we will fight to retain it.

We also make an exception to the sanctity rule for self-protection. Most of us stop short of the extreme position that would cast self-protection as immoral. Few would allow themselves to be harmed or killed just to remain non-violent. To do so is to fight against an evolutionary history of adaptation for self-defense. This wiggle room can provide a rationalization for pre-emptive violence.

Like humans, other species are not always prudent judges of what constitutes a danger. Pain and annoyance, for example, make us more irritable and prone to lash out at a convenient target [2244, 2245].

We share an innate sense that something must be blamed for suffering. Many species, from crayfish to snakes to mammals, escalate their aggression towards cagemates as their pain level rises [2246]. Even a pissed-off pigeon may torment his neighbors, unleashing a reign of terror. With the proper conditioning, spite can become its own reward [486, 487].

Like humans, these animals were not choosy about their targets. They attacked whoever seemed close at hand as if they blamed the other organism for their pain. It did not matter if the perceived aggressor was their species or another. If it could not escape the negative stimulus, it attacked what it perceived as the possible source of its misery. From an evolutionary standpoint, the reasoning error makes sense. The ground is less likely the source of your pain than a nearby organism.

Humans make a similar cognitive leap and blame apparent targets for their problems. The more blame we ascribe to a source, the greater our anger, fear, and retaliatory aggression [2247, 2248, 2249, 2250]. Whenever possible, we prefer to blame outsiders. When Australian TV personality Steve Irwin (*The Crocodile Hunter*) died from the sting of a ray in 2006, many people went forth to hunt and kill innocent rays. Stingray carcasses littered Queensland, Australia's beaches, ostensibly killed by angry fans [2251]. In April 2011, hundreds of angry Afghan citizens attacked United Nations personnel after hearing a Florida minister burned a copy of the Quran [2252]. In 2021, a mob of enraged Americans attacked the US Congress after their presidential candidate lost the election. We tend to commit violence against a perceived threat, even when a moment of reflection would tell us our focus does not deserve punishment. Humans will target someone or something when enraged. We accept our anger, believing that the world is just, so someone must bear culpability.

The noxious stimulus for humans does not need to threaten life, culture, or religion. It could be any offensive element or idea in the environment. Domestic violence elevates as temperatures become hotter

[2253, 2254], as does social unrest such as riots [2255, 2256]. Heat, cigarette smoke, and air pollution also positively correlate with aggression [2257, 2258]. As with other animals, we lash out when in pain.

As discomfort increases, we lash out with less provocation. Even socially acceptable aggressions, such as hitting a baseball with a stick, increase in frequency on hotter days [2259, 2260]. When a close friend suffered the end stages of AIDS, he was in excruciating agony. He regularly lashed out at his caregivers with emotional abuse that, under other circumstances or a less forgiving team, would have left him isolated.

Humans seek ways to alleviate such tension. Cultures provide outlets for our frustrations and peaceful ways of settling disputes. We crave a safety valve, a safe place like team sports, where we can express anger. If misused, such outlets can intensify violence rather than dampen it.

Release Valves or Pressure Cookers?

> *"I'd never seen men hold each other. I thought the only things they were allowed to do was shake hands or fight."*
>
> — *The Rubyfruit Jungle*, Rita Mae Brown [2261, p. 132]

Laughing or swearing in an uncomfortable situation can reduce tension. Through signaling, appropriate laughter may provide a bit of a status uplift and diffuse tension. This mechanism works well if participants hold a similar status and the cost of conflict is high. If one participant has a significantly higher rank, he or she can use it to bully the subordinate, embarrassing them for their discomfort by attacking their diffusion response. In this classic scenario, the bully doubles down on the aggression, attacking the behavior meant to diffuse the tension.[269]

---

[269] On a personal note, I recall a bullying seventh-grade teacher lecturing us about the Holocaust. One uncomfortable boy was squirming and gave a nervous giggle. The jerk pounced on him, demanded to know if he thought the deaths of millions were funny, then asked him to stand in front of the class as he forced the pre-adolescent boy to look at pictures of the Shoah from a book. Until I began writing this paragraph, I had never questioned why this particularly sadistic teacher had that book with him or if he intended to victimize someone that day.

The style of the laugh signals one's position to others [2262]. Both swearing and authentic laughter increase as dominance rises. We know status increases laughter, but we do not know if laughter increases status. A momentary increase in rank could provide the emotional boost needed to contend with a stressful moment. Laughter can remove some severity from a situation and allow a subordinate individual to meet the need by temporarily raising their social rank [2263].

Humans depend upon the diffusion of aggression to survive social conflicts. We have a menu of methods to dissipate potential violence before it happens, and we function well when we choose a mechanism that works for both parties in a possible confrontation [2264]. Some forms of emotional release, such as a good cry, *may* work under some conditions [2265] but be humiliating in others. However, serious questions surround the efficacy of cathartic aggression.

Psychologists once believed verbal venting reduced the pressure that drove someone to violence. Unfortunately, the data suggest otherwise. It may act more as a rehearsal than an outlet. Rather than reducing violence, aggression increases when subjects express hostility [2266, 2267]. Far from releasing tensions that lead to violence, cathartic acts can release hormonal rewards, such as testosterone and dopamine, reinforcing violent thoughts. It can also reduce the cortisol levels associated with low status, situational helplessness, and the inability to respond. The act of expressing hostility creates a foothold toward greater aggression. Verbal retaliation, even without the original antagonist, feels good, and the likelihood of further retaliation increases. After all, military training films are designed to heighten combat readiness, not decrease it.

Warfare and political conflict mirror the rise in violence after sporting events. During wars, murder rates at home, away from battlefields, also increase [2268, 2269]. Peacekeeping actions may reduce military conflict by filling power vacuums but can also increase street violence and homicide rates [2270]. This latter effect can result from the sudden increase of unemployed soldiers or the insult to national pride during an occupation. In contrast, programs like the *Cure Violence* model, which focus on attitudinal changes rather than violence suppression or catharsis, have had a modicum of success [2271].

Violent sports tap into our sense of membership and affiliation [1032]. For some, the competitions serve as cathartic events, allowing fans to vent their anger by proxy [2272]; others respond with heightened aggression.

We can see anti-cathartic stimulation in hooliganism following sporting events. Fans of aggressive sports who have had the opportunity to vent their anger by proxy through their teams are more, not less, likely to commit violence [2273]. The intensity of this response correlates with increased free testosterone levels in these fans following the match, especially in the blood of the fans of the winning team [2274]. In violent sports such as hockey, soccer, and football, the violent outbursts of players create an additional social reward: statistically, those players make more goals [2275, 2276].

Cathartic aggression, far from being a uniformly beneficial way of decreasing violence, increases anger. Most people like television shows, video games, and movies with some action. Violent movies are safe and often much more satisfying than non-violent ones. Fictional violence or dehumanization creates moments of tension that stories without conflict would lack. Movies with avenged violence feel cathartic. However, mentally rehearsing violent scenarios feeds our tendency toward violence. It does not quell it [2277].

Even passive catharsis, obtaining emotional release through a violent video or game, can yield increased hostility levels through modeling [2278, 2279, 2280, 2281, 2282, 2283]. Many people find violent video games more fun than nonviolent ones. Often, we try to use them as release valves. People who believe in catharsis like competitive and aggressive games more when angry than calm. Practicing violence in games increases the probability of future violence; it does not decrease it [2284, 2285, 2286, 2287].

Entertainment modeling *proactive*, violent behavior increases three core elements that lead to aggression and social difficulties [2288, 2289]. It ups our level of anger and hostility (aggressive affect). It makes us think more antagonistically (aggressive cognition). It stimulates us, making us more alert (arousal). These traits predispose us to belligerent acts. They also are features exaggerated in dominant animals and ones that activate the ACC (our internalized standards and modeling), PFC (self-referencing and hierarchical information), the NA (motivational processing), and the amygdala (emotional responses).

After playing video games involving violence, anxiety goes up [2290]. Participating in fake violence makes us more suspicious of the intentions of others. We learn to distrust by seeing attacks modeled repeatedly. Rather than decreasing our hostile intentions, we become more likely to aggress because we start perceiving hostility in others [2285].

Proactive violence in entertainment also reduces the probability that someone will help another person. When someone rehearses aggression through entertainment, it primes them to be less prosocial.[270] It can lead us to reduce our altruistic behavior as it promotes hostile, competitive behavior [2291, 2292].

Fortunately, one can also model behavior that reduces violence. If modeling violence serves as a rehearsal for violence, modeling and practicing restraint should also modify our behavior. Our generalist brain can learn from empathy and cooperation in games, creating behavioral changes in as little as two weeks [2293].

Rituals also provide a path to violence or peace. Rituals are stylized, repeated, formalized behaviors. These societal traditions carry feelings and influence personal or interpersonal actions [2294]. A symbolic slap in the face could once initiate a duel in Europe; in parts of the Arab world, showing the bottom of one's foot or shoe constitutes an affront to honor. These ritual slights are not leveled against the body but against dignity. Far from an extreme institutional or government-only phenomenon, the subjugation of others to

---

[270] For a review of this topic, see Whitaker and Bushman, [2286].

the status of an animal or a mechanism occurs in daily routines, often so ingrained that we do it without conscious recognition [2295, 2296].

The mocking of a person's name, hate speech, and other identity-based derogations also target personal honor and humiliate their mark. The Nazis effectively combined name-calling with the denial of individual identities, identifying all Jews as either Israel or Sarah. This practice escalated from what many dismissed as harmless teasing, such as calling male Jews "Itzig" during the pre-Nazi era, to the social exclusion and debasement of the Shoah [2297]. We also encounter it in the use of infantilizing diminutives of the names of nursing home residents.

Material violations, such as the subjugation of others into poverty, not through lack of work or productivity but through cheating, wage theft, and the use of unethical, even if sometimes legal means, to hoard resources, also violate human dignity.[271] Inflicted poverty creates a state of dependency and deprives the sufferer of their most basic needs [2298]. Absolute poverty violates human dignity, including being forced to eat the trash that others discard and excluding them from society and its rituals of inclusion [345].

Humans love rituals. They let us slide into the regular patterns that our species craves. We sleep better if we have personal bedtime routines. We use conventions to treat medical conditions. We also use them to regulate power, submission, and violence.

[271] This human behavior has parallels in other animals. Pigeons trained to associate different lighting conditions with the effort needed to gain a food reward can be taught to peck targets that force neighboring pigeons to work harder for their food [486, 487]. They become bullying boss-pigeons who resemble corporate executives who aim to squeeze more labor from their employees for less compensation.

Rituals can also control us when another has the privilege of aggression or retribution, and they can reinforce group cohesion. Though lacking a formal structure, lynchings in the postbellum American South were rites that reinforced White privilege and power [2299].[272] Fitting with the Southern culture of honor killings, although 27% of the victims were accused of raping a White woman, the perception around lynching held that it was the principal reason [2019]. These ritualized murders symbolized honor and supremacy through the debasing and humiliation of the victim, provided a festive air, and bolstered community bonds for the White mobs [2300, 2301, 2302]. Be it the enslaved by masters, natives by colonialists, Jews by Nazis, women by men, homosexuals by heterosexuals, or individuals by bullies, each degrading act reinforced the moral exclusion and dehumanization process, creating a vicious circle (positive-feedback loop) that encouraged more assaults on person and dignity. As the degradation accelerates, the targets become objects of fear, disgust, and contempt [345]; the power of these feelings can lead to murders and genocides as the perpetrators resort to pre-emptive self-defense against the perceived inhuman enemy of their own creation.

> *"Three shining-eyed, healthy, clean children, headed for school, approached me... The eldest, a ruddy-cheeked girl of nine or ten, asked if I was going to the place where the 'niggers' had been killed... Animatedly, almost joyously as though the memory were of Christmas morning or the circus, she told me, her slightly younger companions interjecting a word here and there or nodding vigorous assent, of 'the fun we had burning the nigger.'"* — Rope and Faggot: A Biography of Judge Lynch by W. White, [3268, p. 3]

During the Nazi era, the government used indoctrinating practices, such as the mandatory "Siech Heil" or "Heil Hitler" greeting, party rallies, and frequent Hitler Youth activities. Churches were expected to display Nazi banners, and religious events, such as Christmas, were often stripped of Christian significance and used to propagate pseudo-religious beliefs in Aryanism and the power of Hitler. During the post-war De-Nazification period, which followed the defeat of the Nazis, the victors banned these sacraments, including the display of Nazi emblems and literature. They instituted their own rites, such as the highly ritualized war-

[272] Among children, group or joint bullying classmates serves a similar social function, with friendships often developing between those who bully the same victim [2331].

crimes trials. They removed the former leaders and changed the images on the currencies in former Nazi territories. Repetition of information about the atrocities committed by the Nazi government spread through the media daily. The rituals of De-Nazification changed attitudes.

The proper ceremony for the right situation can delay, direct, diffuse, reduce, or remove aggression. Ritualized play-fighting reduces anxiety and bonds people in stressful situations such as combat zones [2303]. We use rituals to get psyched for warfare and others to soothe social tensions and bond with groups. Conventions of morality, as simple as using polite words, can diffuse otherwise tense situations [2294].

The use of ritual simplifies our lives much the way that repeats in music allow our minds to wander. Even subtle patterns of behavior can elevate or reduce our sense of self-discipline. Accepted formalities surrounding shock, anger, and loss can diminish the calls for retribution after a tragedy [2304]. Conversely, ceremonies such as rallies, protests, and marches can bring together a society, or they can organize and prompt aggression [2305].

Rituals work directly on our sense of self-control [2306]. Participation in group rites increases the sense of community, a cozy, oxytocin-laced feeling that decreases internal aggression. They work as rehearsals for action or restraint and strengthen the neural pathways linked to those responses.

Since the 1960s, children have spent much of their free time-consuming media with violent content, from cartoons to super-bowl commercials and video games [2307]. We accept that a child's brain acts like a sponge, drinking in experiences from the schoolyard and home to the shopping mall and hamburger stand. We acknowledge that small children do not always distinguish between fiction and reality. We know the adult brain has formed the bulk of its core connections by the end of adolescence. For the child's welfare and the sake of society, we limit their access to dehumanizing, sexually explicit materials. Yet, we continue rationalizing the marketing of violent entertainment to this exact demographic.

CHAPTER 10:

What About the Bully? What About the Victim?

"One should wish to be both [loved and feared], but, because it is difficult to unite them in one person, it is much safer to be feared than loved, when, of the two, either must be dispensed with."

— *The Prince* by Machiavelli, Chapter XVII [2308]

Rehearsing

"The play's the thing wherein I'll catch the conscience of the King."

— *Hamlet* by Shakespeare [2309]

Few self-identify with tormentors. To be honest, most of us believe we were the bullied ones. Even people with power see themselves as victims. When workers demand what they perceive as fair compensation for their labor, bosses feel bullied by their employees. When the disenfranchised demand respect and an end to discrimination, empowered groups complain about being put upon. Our willingness to follow subtle suggestions from an authority exacerbates this situation; without thinking, we discriminate against specific subgroups as part of our pattern of obedience [2310],[273] often never realizing we're doing it.

The powerful are not necessarily cold or unsympathetic people. They fall prey to the same attribution error that haunts us all. This inherent I-thou dichotomy stems from the kin selection that has shaped our

[273] Racism, by itself, does not fully explain patterns of employment discrimination in the United States [2310]. Employment discrimination also follows the pattern predicted by Milgram. Racist and caste-based rationalizations from presumed authorities established conditions that encouraged other employees to participate in discrimination, often without questioning their actions.

worldview. We perceive the flawed behaviors of others as products of their inner personality and our faults as natural responses to the circumstances surrounding us.

Not surprisingly, the bully, violent boss, and bloody dictator do not make sympathetic characters. A standard, gratifying comic trope serves bullies with a deserved comeuppance. Their just humiliation satisfies us. Comedy and drama, be it in theater, literature, or graphic novels, provide the psychological and physiochemical contentment that only richly deserved revenge can yield. Through the magic of theatre, we indulge in joyful fantasies of vengeful retribution without the jeopardy of aggression and without inflicting actual physical or psychological pain on another. Far from dissipating our demons through catharsis, the resulting hormonal lift, complete with a dopamine surge, gives us a short-term boost of courage, indignation, and perceived status. Our insula, that kernel of tissue that regulates emotions and blends them with our cognitions, responds to the comedy (or drama), heightening our awareness of our physiological responses. With the anterior cingulate cortex (ACC), our prefrontal cortex (PFC) processes the dramatized social situation, and we begin to use our theory of mind to analyze it (see Chapter 8: Multiplying Mistrust).

> *"You will always be fond of me. I represent to you all the sins you never had the courage to commit."* — O. Wild, *The Picture of Dorian Gray* [3267, p. 58]

In the PFC, these internal rehearsals build self-confidence, reshape our attitudes, and motivate us to move against the bullies in our lives. Our imaginations can envision murder and humiliation with no intent to follow through. In mentally healthy people, the planning centers of our PFC reject illegal, socially condemned, and dangerous options up front. It listens to and evaluates the fear levels from the amygdala and weighs the potential costs of prepared behaviors against the possible benefits and our normal disinclination to harm others.

Throughout our evolutionary history, fear memories have kept us alive. We had to remember the lion at the watering hole, what enemies could kill us, and who beat us up when we said their mothers ate cave bears. Fear memories thus form quickly and tend to be long-lasting. The hippocampus remembers previous encounters. A limited "working through" of a brain-changing trauma can reframe experiences and reduce long-term neural, hormonal, and emotional damage from stressful and life-threatening events. However, far from being cathartic, experiencing ongoing, repeated recollections of traumatic memories (encouraged by some counselors in the past) strengthens the psychological fallout of the trauma. The PFC and already damaged hippocampus feed that information to the amygdala, which urges us to err on the side of caution. Repeated visualizations of positive outcomes can reduce the impact of past defeats. Comedy violates our expectations, and when bullies fall prey to their own cruelty, the unexpected result activates our temporal-parietal junction that integrates language with social processing and memory, reducing our inclination to see them as alphas.[274]

Most of us lose little sleep over what happens to the schoolyard bully, the self-entitled jerk who degrades a homeless person, the sadistic genocidal guard, or the boss who throws his weight around. We all fall into the fundamental attribution error, assuming these people automatically browbeat others because of internal character flaws rather than surrounding circumstances. Even people who know the research fall into the instinctive trap. We forget power begets power, and *anyone* can fall prey or become a predator.

What Happens to the Habitual Bully in Life?

> *"He wanted to kick someone in the kneecaps with his heavy work boot. So he had been caught with a few credit cards he had lifted from the wallets of his crew members... Nobody was liable for more than fifty bucks or so from purchases made on a stolen card anyway."*
>
> — *Ill Wind* by Anderson & Beason [2311, pp. 4-5]

Psychiatrists worry when frequent bullying behavior in children happens in concert with other psychiatric symptoms. The little tyrants who have suffered torment themselves have a higher risk for a poor psychiatric

[274] Recall the N400 effect that happens when we experience discordant or unexpected information.

outcome than if they had either marker alone [2312]. For this unfortunate subgroup, being a bully predicts later psychological and psychiatric problems, even without other childhood psychiatric disorders [2313].

Conditions that lead to a decreased ability to control mood and restrain behavior (externalizing disorders) increase the threat of becoming a bully [2314, 2315], but they do not alone cause it.[275] They complicate behaviors stemming from inward emotions and hostility, turning anger and depression outward toward the world. Examples of externally directed antisocial disorders include impulsivity disorders and childhood conduct and defiant disorders [2316]. Among adolescent girls admitted to psychiatric hospitals, both the tyrant and target have an increased chance of self-mutilation and suicide [2315].

Habitual childhood bullies risk remaining cruel through life, progressing to antisocial disorders as an adult [2317, 2318, 2319], and expressing more varied delinquent and increasingly violent behaviors [2320, 2321, 2322]. As identified childhood bullies progress to adulthood, they are more likely than non-bullies to abuse substances and commit crimes [2323].

When diagnosed with psychiatric problems, they share many mental health issues with their marks. They share high levels of depression and anxiety [2324, 2325]. The conferred dominance increases self-esteem in many pure (non-victimized) offenders, particularly females. Others, especially those who also suffer abuse, can share a reduced self-efficacy with pure victims [2326, 2327, 2328]. However, by adulthood, childhood bullies, victims, and bully-victims report more loneliness than people who do not report extensive experiences in these roles [2329]. The research on what happens to the aggressors leads to conflicting conclusions, possibly because we try to lump all tormentors into a single, overarching category. In doing so, we lump disparate things together like apples and schnauzers.

---

[275] Psychologists consider adults to have more fixed personalities. Only in adulthood can one receive a diagnosis of antisocial, histrionic, or borderline personality disorder.

What Traits Do Bullies Display?

> *"Years later, I learned an English word for the creature that Assef was, a word for which a good Farsi equivalent does not exist: 'sociopath'."*
>
> — *The Kite Runner* by Hosseini [2330, p. 38]

Bullying takes different forms, and so does the nature of bullies. Adolescent oppressors show a range of personalities and styles of torment.[276] Some habitual offenders prefer to initiate their harassment (proactive aggression). Others join an ongoing incident or support another's intimidation (reactive aggression). Ones with high callousness and unemotional traits typically enjoy both forms of entertainment. They initiate persecution and join ongoing attacks more often than abusers who lack these traits, benefiting from the bonding that can develop between the abusers of a shared victim [2331], [2299, 2332]. When cold-hearted and detached behaviors start early in life, the person has a poor prognosis. He or she suffers a high risk of displaying a personality disorder as an adult.

By tradition, psychologists treat adults as having more fixed dispositions than children. The most common diagnostic progressions for the early-onset callous predator are the diagnoses of antisocial or borderline personality disorder [2333, 2334]. These labels can follow the adult through life, and the courts often treat him or her as a sociopath.

Those who do not show uncaring and callous behaviors in other situations combine proactive and reactive bullying less often than the heartless ones. Abusers with marginal consciences will more readily initiate a tormenting session or join ongoing hostility. Those who show a conscience as a child have a better chance of escaping the cycle than the unfeeling ones. Usually, they have a lower genetic contribution to their browbeating behavior, as do those who start their denigration careers later.

---

[276] Psychologists define "personality" as a pattern of responses, attitudes, and behaviors that persist over a wide range of situations.

Bully-Victims

"Terror made me cruel." — *Wuthering Heights* by Emily Brontë [2335]

> *"Our mothers all are junkies, our fathers all are drunks. Golly Moses, natcherly we're punks."* — Gee Officer Krupke song, *West Side Story*, Leonard Bernstein, 1961 lyrics by Steven Sondheim [3170]

We ascribe our success to hard work and superior internal talent; we blame circumstances for our faults. We credit alphas with exceptional skills and earned rewards. However, we extend our just universe model downward to attribute the failures of others to their lack of motivation and innate personality flaws. Western culture pays lip service to a counterbalance, proclaiming the environment shapes everything, marking it as the singular reason for our behavior turning sour. We become fat because food surrounds us. We become monsters or saviors because of how our mothers potty-trained us. We become credulous because of Ouija boards, dust mites, or a lack of greens in our diet. Openly challenging this social taboo by suggesting that any genetic or physiological predispositions to violence exist invites personal attacks and character assassination. Still, we often defend the concept of environmental causes, but we never quite internalize it; we still ascribe malice committed against us to the nature of others, not to their nurture. In truth, nature is seldom so simple.

Enter genetics. Animal behavior, including human behavior, springs from the interaction of inheritance and the environment. Our genetics and epigenetics, not just our surroundings, predispose us to our roles. Some tendencies are innate and easily provoked. Others have a robust environmental influence. The two intertwine, making them challenging to tease apart. On average, bullies have histories of harsher punishments than non-bullies [2336]. It remains unclear if violent or demeaning reprimands caused them to become tormentors or if pre-existing behavior problems evoked drastic disciplinary measures.

Genetics contributes up to 65 - 70% of the variation in bullying [2337, 2338, 2339] and up to 60 -73% of the variation in victims. Predisposing factors include genetic and epigenetic traits that make someone a potential target for tormentors, such as autism, learning disabilities, and sexual orientation. Anxiety disorder alone has a heritability of 30-40% [2340],[277] and exposure to stress, including the humiliating social stress of being bullied, heightens the impact of pre-existing susceptibility [2341].[278] Even our ability to cooperate with others or stand against a bully may include genetic or epigenetic components [979].

Some bullies become victimized, and some victims become bullies [2342]. Surprisingly, the ones who suffer oppression and abuse tend to be more callous than those who never experienced it [2343].[279] The callous-unemotional trait is, in turn, associated with higher levels of direct and aggressive behavior [2344]. Again, there remains a question of correlation and causation. We do not know why bullies who experience victimhood tend toward greater insensitivity. They may have started with less compassion or their ongoing maltreatment could have numbed them.

Maltreatment through harassment, violence, and dehumanization prunes the neurons in the anterior insula and cingulate cortex. In bully-victims (BVs), this interacts with their aggression. This trimming decreases emotional responses and empathy, worsening behavior in those predisposed to harassing and intimidating others. These losses increase the chances of expressing antisocial disorder by adulthood. Psychopathic offenders with a loss of emotion report more abuse during childhood than similar offenders without these emotional deficits [2345]. And, like pure bullies, BVs who start their careers early and torment more often

---

[277] These figures represent the percentage within people with anxiety disorders that genetics explains. It does NOT mean the portion of the population with these disorders. About 31% of the population suffers from an anxiety disorder at some point, with about 19% experiencing one in any given year [3486, 3298]. The figures in the text are the percentage within that number.

[278] Although the text focuses on events that create negative epigenetic changes, life events, such as nurturing, can buffer these effects [3471].

[279] Bully-victims can either be bullied first, bully first, or experience both simultaneously in different aspects of their lives. Less is known about retaliatory bullying; unless specified, the articles in this chapter refer to people who started as targets or had simultaneous bullying.

are more likely to garner criminal records later [2346]. Keep in mind that self-reports of bullying provide biased views of their cause and development, courtesy of the fundamental attribution error.

Bully-victims dominate our modern folklore. We want to believe some controllable force prompts some people—especially children—to turn bad. With a contemporary twist, we still crave evidence that the devil made us do it and that we can exorcise or prevent the possession. In truth, 50-70% of school oppressors have not suffered prior abuse [2347]. People who serve in both roles comprise a fraction of people in each category, not the majority [2348, 2349]. About 10-20% of targets and 30-50% of bullies are both [2350, 2351, 2352, 2347]. BVs tend to have damaged self-images like pure victims. Still, like pure bullies, they can attain social status through aggression [2347, 2353, 2354]. However, respect and prestige do not necessarily accompany such status.

Think of bullies as out-of-control, aggressive, wannabe alpha primates who use a social shortcut to power. As in our chimpanzee cousins, the cost of rebelling against them is high, but so is the ultimate price paid by the bully. Heavily intimidated BVs stand little chance of rising to the level of a pure tormentor. Bully-victims typically gain an intermediate rank on the school, social, or other pecking order. And as in the chimps, when power begins to slip from the bully or the bully-victim, the lower ranks have cause to exact revenge.

The prognosis for BVs is poor before their aggression begins. They have suffered torments at the hands of another. They may have genetic characteristics or physical stigmas that single them out. As a group, bully-victims may include people who have genes or social experiences that make it hard to modulate emotions [2355, 2356, 2357]. These people show more severe psychological problems than pure bullies or pure victims. They also suffer from more depression and anxiety than pure bullies [2324, 2342]. Pre-existing outwardly directed disorders haunt a larger proportion of them than pure bullies. The frequency of attacks increases in people who cannot regulate their emotional outbursts. Other risk factors, such as attention deficit disorder, occur more often in BVs than in pure victims or bullies [2342].

Like pure bullies, BVs can obtain a critical payoff for oppression and tyranny through improved social status, but they typically do not achieve as high a social rank. Their status as sub-human targets might impair their rise, but another factor may play a role. The attacked can face social pressure to retaliate and harm the tyrants in proportion to the damage inflicted [2358]. Bully-victims, with their already lower self-esteem and intermediate status, may be more susceptible to condoned retaliatory attacks.

Bully-victims suffer higher levels of depression and experience more anxiety disorders than pure bullies or victims. Both flow from factors lurking in the background. The genetic and epigenetic profiles that regulate belligerence can interact with the environment to increase our chance of expressing anxiety,[280] mood, and personality disorders [2337, 2359, 2360]. Some genes increase the risk of becoming an aggressor or a sacrificial lamb. When these combine, they enhance the odds of becoming a BV. The environment tweaks the expression of genes, fine-tuning their output as needed through epigenetic modifications, e.g., oxytocin [2361, 2359].

[280] Anxiety disorders, including social anxiety, are the most common mental health problems [3432, 3483]. *Technical note:* Genome-wide studies *implicate* multiple genes, some overlapping and others unique to specific anxiety syndromes such as panic disorder (PKP1), anxiety disorders (LOC15225), generalized anxiety disorder (ESR1), and anxiety-stress-related disorders (PDE4B). These genes are nearest the studied chunks of DNA and are not necessarily linked as causal [2360]. The ESRI and PDE4B associated with generalized and stress-related anxiety are also found in rodents.

Hormones 102: Hormones and their Genes

> *"Worst case of testosterone poisoning I've ever seen."*
>
> — Lt. Commander Ivanova, *Babylon 5*, Season 1, Episode 15 [2362]

Earlier chapters noted the impacts of hormones, their genes, and their metabolism. Testosterone modifies aggressive behavior, and a sudden rise increases the likelihood that an animal will harass lower-ranking troop members. Progesterone, a hormone structurally similar to testosterone, also escalates aggression and territoriality.

However, testosterone and progesterone do not function in a vacuum. They work within a biological and environmental mosaic. Social stress changes the balance of hormone production, receptors, and recyclers (breakdown enzymes). Even glucose (sugar) metabolism is altered as stress interferes with the production of cortisol and corticoid receptors on cells [227, 2363]. Hormones and brain functions compose the primary bodily portion of this sophisticated, living design.

For example, high serotonin (5-HT) levels increase our thinking flexibility and reduce depression. Selective serotonin reuptake inhibitors (SSRIs) are the most prescribed prescription antidepressants. They prevent the brain from recycling the hormone (reabsorption) and letting it accumulate outside the neurons. How we produce and process serotonin influences our inclination toward boredom and impulsive behavior. Serotonin levels have been implicated in male-male aggression in humans and mice [2364, 2365, 2366]. The amount of serotonin made, the amount deposited in the synapse, and the rate of reuptake by the presynaptic neuron all play a role.

No single gene creates a bully or a victim; however, many genes regulate the propensity. Variations in the DNA code for serotonin transporter (5-HT) genes affect the rate of reuptake [2367, 2368]. This protein sits on the cell membrane and moves serotonin from the gap between neurons (synapses) into the neuron before that space (presynaptic neuron).[281] This reduces the amount of 5-HT acting on the connecting cell. Bullies with low serotonin would be more likely than others to attack—if only to relieve the tedium of life. A low level increases aggression and suicidality [2369]; it also reduces our ability to plan, control impulses, and regulate our behavior [2370]. Targets with serotonin transporter variants that reduce 5-HT concentrations are more prone to depression [2368]. Serotonin's effect on the prefrontal cortex accounts for much of this damage. This area makes rational plans and decisions. Low serotonin inhibits its ability to sort through and regulate its responses to inputs from other brain regions. Someone with low free serotonin (outside the cells) is thus more likely to act without first considering the consequences. This serotonergic system combines with dopamine and its allies (the dopaminergic system) to regulate both our self-defensive aggression (reactions to threats) and our more cold-blooded proactive attacks against others [2371, 2372].

Although genetic predispositions can render us more susceptible to damage from bullies [2373, 2374], the brain does not need a genetic predisposition to undergo damage from social rejection and bullying. Harassment can reduce serotonin levels by altering how much 5-HT transporter the body makes. Bullied twins have higher methylation (production change) on at least one transporter (SERT), upsetting the regulation of serotonin and blunting cortisol levels under stress [2375].

Among the hormones, oxytocin (OXT), a.k.a. the "love drug," probably has the best public-relations agent. Once considered a female reproductive hormone, we now know oxytocin gets its little binding sites into a wide array of functions in many brain parts. High levels of oxytocin encourage more than orgasms. It provides a sense of empathy and generosity. When the OXT level is high, it reduces risk-taking and fosters

---

[281] The brain transmits messages among neurons (brain cells) across "spaces" called synapses (a fancy word for "joining together" derived from Greek). A neuron that sends a signal is a presynaptic neuron. One that receives the message is a postsynaptic neuron.

trust and group tolerance. Oxytocin increases facial recognition, improving social bonding. Falling OXT levels can warn us that something is wrong in social encounters, urging us to protect ourselves and leave [2376]. Dropping its level amounts to love withdrawal. Empathy for others drops. We become less trusting, less tolerant, and more selfish. Anything that reduces our oxytocin levels or interferes with the brain's ability to receive input from that hormone fosters these unpleasant states.

Few things in biology, including oxytocin, are as simple as they first seem. Despite its reputation as a love drug, OXT levels yield more troubling results. This hormone does not encourage universal trust and social affiliation; it does so within a social group. When hyped up on oxytocin, we exaggerate our distinction between "us" and "them," increasing the I-Thou dichotomy in our thinking [2377]. It promotes bonding within the in-group but increases ethnocentrism, contributing to intergroup hatred and the likelihood we will discriminate and aggress against out-groups [1789, 2378]. For example, women show increasing xenophobia during their critical first trimester of pregnancy when the oxytocin levels first begin to rise [2379].

In rodents, we can explore the impact of OXT under controlled conditions we cannot achieve with humans. Oxytocin helps female mice survive social defeats [2380]. Shifts in this hormone result in social avoidance, increased behavioral flexibility, and reduced aggression. In contrast, male mice show less change in OXT after a social trouncing. Instead, a similar hormone, vasopressin, comes into play. Following defeats, vasopressin urges males toward retaliation. It also reduces their range of behavior, making conquered males less flexible and more stuck in their ways and intent on revenge than defeated females.

Remember, no single gene can explain bullies and victims. By itself, low oxytocin or abnormal OXT metabolism does not produce a bully. Nor does it create a victim. It does urge someone to respond differently to social cues. People with autistic disorders show some modifications of the OXT receptors

[2381], a finding supported in animal models [2382]. These variants would contribute to the altered interpersonal styles of the spectrum. Far from being tyrants, people with autism are likely to become targets for abuse. Little evidence exists about the ramifications of low OXT on aggressive personalities.

In contrast to the publicity received by oxytocin and MAOA, Neuropeptide-Y (NPY),[282] a tiny hero that adds to our molecular resilience, remains in the shadows, ignored by the paparazzi.[283] The body builds its NPY levels after social defeats. Like oxytocin and vasopressin, NPY is a highly conserved molecule with close structural cousins in animals as diverse as mammals and fruit flies [2383, 2384, 2385]. It comes in variants that tweak our sensitivity to anxiety and stress through its interplay with other hormones and activity in the hypothalamus, hippocampus, and amygdala [2386]. Through the hippocampus, NPY can reduce or extinguish contextual (association) fears and triggers associated with combat, dehumanizing events, rapes, and other tortures [2387].

NPY is active in multiple brain regions besides the usual suspects. It influences the septum (the connection between the emotional and rational self), caudate-putamen (speech, reward, cognition), and locus coeruleus (see footnote) [2386, 2388, 2387].[284] Neuropeptide-Y impacts both mood and the vascular system. Like vasopressin, neuropeptide-Y constricts blood vessels. Together with ghrelin (the hunger hormone), NPY makes us hungry, increasing our craving for high-carbohydrate foods, alcohol, and fats

---

[282] In flies, the hormone is known as NPF (neuropeptide-F). NPF and NPY show significant overlaps in their activities and effects on the invertebrate and vertebrate bodies [2384]. These functional similarities persist despite a divergence that predates the Cambrian explosion (1100-900 MYBP).

[283] The study of NPY goes beyond the theoretical. It offers hope to survivors of war, rape, and bullying as a potential treatment for PTSD [3482, 3507, 3436]. It also holds promise for obesity disorders, which often happen jointly with PTSD [2398].

[284] The locus coeruleus (= the heavenly blue body) is a part of the brain stem that sticks its finger into multiple functions. It is involved in our sense of arousal and behavioral expression, helping us to do adaptive, practical actions while suppressing maladaptive ones. It helps us interpret cues from within our body and sends information to our organs. It is intimately involved in stress and panic responses [3182]. When we do or say something stupid, like bolting from hiding when a dinosaur is looking for us (a classic in the movies), the locus coeruleus plays a role.

[2389, 2390, 2391, 2392], but protects against the stress-induced acids that can eat away at our stomachs [2393].

Like NPY, ghrelin protects against stress by tempering our responses to rewards. During our early evolutionary history, dominant troop members would have controlled access to food during hard times, creating psychological and physiological stress for the subordinate. Even for a dominant animal, a few extra pounds could enhance status and survival should food become scarce. Ghrelin levels, like NPY, rise with stress [2394]. For lower-ranking individuals, a hearty appetite driven by ghrelin and NPY would bestow a survival benefit. Scarfing down a large cheese pizza may clog our arteries today; however, in our last low-ranking common ancestor with chimps, a constant readiness to eat would have yielded a reproductive advantage. In the WEIRD world today, the stigma of obesity creates a spiraling feedback system between stress and bullying [2395].

Ghrelin activates a third hormone, Agouti-related protein (AGRP), which joins this crew to encourage eating to build our energy reserves. AGRP and NPY increase food consumption, our willingness to expend energy to get food, and our interest in food odors [2396, 2397]. When you follow that intriguing smell of pizza or popcorn down a hallway at work or give into the temptation to open the oven door to smell the brownies, AGRP and NPY are manipulating you. Elevated AGRP levels promote premature aging, providing yet another way oppression kills and disrupts families long after the tormenting event.

You may wonder why I wasted two paragraphs on a sidebar about eating disorders. Obesity correlates with PTSD. Oxytocin, Neuropeptide-Y, and cortisol all regulate eating [2398]. The "feeding" hormones also impact sleep. NPY also promotes sleep and reduces anxiety by regulating noradrenaline in the brain [2399, 2400]. Noradrenaline is our brain's version of adrenaline. It's a neurotransmitter in the same chemical family (catecholamines) and is structurally very close. Like its better-known brother, noradrenaline is an excitatory molecule that prepares us for action. Ghrelin levels rise during sleep deprivation (a stressor), and restorative slow-wave deep sleep (a.k.a., non-REM/ non-dream, quiet sleep) may regulate its production [2401].

NPY activity is also linked to reduced fear, including social anxiety [2402, 2403, 2404, 2405], aids decision-making in approach-avoidance conflicts [2400, 2406],[285] and regulates activity in the fear circuit of the amygdala [2407]. Paradoxically, high neuropeptide-Y levels are reported in people who die by suicide, possibly due to reduced fear of death [2408]. During chronic tension, our NPY levels elevate, letting us cope and providing a good excuse for carb-rich double-helpings of rocky road ice cream.[286] People with low levels run a greater risk of major depression and post-traumatic stress disorder [449]. This paradox explains our stoicism in coping during prolonged social strains, such as the deaths of loved ones and bullying, but emotional collapse afterward.

In contrast, a plethora of information points toward another hormone, dopamine. This hormone regulates our desires and pleasures. As a close cousin of adrenaline, it peaks in emotionally tense or frightening situations and guides our attention. It moves us toward things that can please us and away from punishing experiences. Dominant animals have higher levels of dopamine than subordinate ones. Faulty regulation of this hormone also links to childhood behavioral disorders and adult antisocial disorder.

The most persuasive evidence follows one specific gene that makes an enzyme called MAOA. This enzyme regulates the levels of small compounds called monoamines. These include dopamine, noradrenaline, and serotonin. These neurotransmitters and hormones regulate our moods and emotions. The MAOA gene was

---

[285] Multiple (not single) electroconvulsive shock therapy (ECST) has some benefits for alleviating major depression. ECST elevates neuropeptide-Y levels after several treatments [3424, 3272].

[286] The concept of comfort food is widespread across world cultures, but each culture has its preferred version. Commonly, they are high in carbohydrates and/or fats. Familiar comfort dishes in the United States and Canada include macaroni and cheese, ice cream, and grilled cheese sandwiches. Italians prefer pasta with tomato sauce. The Japanese often rely on hot ramen, while people in India might crave biryani, dal, or sweets.

among the first human genes found to influence aggression [2409]. It codes for an enzyme, monoamine oxidase A (hence MAOA), and the gene has received the dubious nickname "the warrior gene." Tinkering with MAOA metabolism—including modulating interacting hormones—alters aggression across multiple vertebrates, including puffer fish, sticklebacks [2410, 2411], chickens [2412, 2413], pigs [2414], and cattle [2415, 2416, 2417], and is suspected in dogs [2418].

Male mice with low MAOA attack their cage-mates more than normal mice. They are more predatory than typical mice, assertively defending territory and showing greater aggression after isolation. The dopamine secretion in the nigrostriatum (a reward/ motivation center) increases NPY, improving our resilience and letting us weather social stress [2419]. In contrast to OXT, male mice with high MAOA activity display low levels of aggression and more anxious behaviors than mice with typical levels. They also practice more peaceful, friendly behaviors with their fellow mice [2420]. We cannot do caged-behavior studies on humans, but the similarities across vertebrates imply we behave likewise.

In humans, the gene for this enzyme sits on the X-chromosome, just like the gene for color blindness or hemophilia. Men get one copy, and women get two. The Y-chromosome does not carry it. Because men get only one version of the gene, its action cannot hide behind the products of its twin. As with hemophilia, MAOA has standard and abnormal versions (alleles). Each genetic variant produces a slightly different MAOA enzyme. This protein regulates several neurotransmitters and hormones in the brain, including serotonin and dopamine [2409].

> *"Secondary school pupils were left battered and bruised after an unofficial 'Kick a Ginger Kid Day.' Police are thought to be investigating incidents at the Wingfield Academy in Rotherham, Yorkshire, where at least six red-headed students were attacked on their way to classes on the same day."* — *The Telegraph's* report of violence against redheads. *[3505]*
>
> *"It can come in anything as little as being called 'ginger' or the mortifying highs of a drunk asking about the colour of your pubes — in front of fellow commuters... But it's not just strangers. One colleague continued to tell me I 'had no soul' because of the colour of my hair. Yep, those occasions were just a laugh-a-minute"* — Amy Nelmes [3212]

Women with high-activity MAOA enzymes suffer more depression than those who metabolize their monoamine hormones less efficiently. Low-activity gene copies on both X-chromosomes may also lift self-confidence and happiness [2421]. Men with one copy of these low-activity alleles experience a different result. If their solitary copy is a lower-activity MAOA variant, it predisposes them to become more territorial, more aggressive, and have a higher risk of developing antisocial disorder.

Low-activity MAOA genes start their impact early. During development, embryos with these variants remove too many cells from the corticolimbic circuit (a.k.a. the usual suspects) as it sculpts the fetal brain, trimming away cells [2422]. The low MAOA-enzyme levels result in smaller limbic areas, including the amygdala. The limbic system regulates our emotions. The rostral (above the nose) portion of the prefrontal cortex is reduced; this area regulates executive function and fear learning [2423], and the pruned medial prefrontal cortex leaves us with less plasticity and resilience in the face of social defeat [2424, 2425].

These areas' anatomy and development lay a foundation for a predisposition toward aggression starting in the uterus, but the environment also shapes it [2422]. These neurological modifications mirror social information showing that violent children become worse when they experience abuse themselves. For children who already suffer from low MAOA activity, physical punishment makes the situation worse, not better [2426, 2427, 2428, 2429]. These data explain why some maltreated children become bullies and psychopaths while others live peaceful lives [2430, 2431, 2432].

Abnormal MAOA enzyme levels or activity do more than change the size of brain regions. They alter how we process data flowing from our emotional limbic area to information-processing parts in the brain. This corticolimbic circuit runs from the amygdala (emotions) through the rostral cingulate (error detection and social monitoring) to the medial prefrontal cortex (executive decision-making). Together, these regions act as social scoreboards. They track who is responsible for what. Together with the cingulate cortex, also part of the limbic system, they let us bond with others and manage social behaviors.

Despite its links to antisocial traits, these "abnormal" genes may confer an adaptive advantage on their hosts. Most men with the gene do not become criminals. Instead, they express the trait far below where it causes them harm. They display dominance, territoriality, and aggression through more vigorous defenses of their homes, associates, and families. They are forceful in business dealings, making them competitive negotiators and valuable as CEOs [2433]. The distinctions they draw between "us" and "them" may be starker. These traits provide a competitive advantage—unless they cross the fuzzy, indistinct boundary into a "disorder."

Humankind loves the ideas of peace, the evil "other" disappearing, and life returning to an idyllic state that never existed. This paradise cannot occur—if for no reason other than our inborn behavioral variations. Societies need both their hawks and their doves. Cooperative peacemakers forgive more readily and ascribe slights and offences to an accident more than those with low-activity MAOA variants [2434]. Doves glue us together, but hawks quickly invade a community composed entirely of doves, perceiving them more like pigeons. Among the peace-loving doves, cheaters who fail to keep the social contracts also thrive. Although female hawks with low MAOA activity contribute more to the public good when others fail to contribute [2435], male hawks are more likely to seek retribution.[287]

When provoked, people with the low MAOA variants (MAOA-L) get angry more readily and have more trouble regulating it [2436]. These males will spitefully punish perceived free-riding more readily than others. In studies, MAOA-L people will pay to punish free riders by feeding them hot sauce [2437]. Their increased aggression against thieves and free-riders suggests that people with the warrior gene may also

---

[287] True to the difference in how MAOA variants are expressed in women, women behave differently under its influence. Low MAOA females are less likely to become free-riders than low MAOA males.

control the level of free-riding in society by increasing its cost. And, when they perceive political oppression or instability, the low-activity MAOA variant hawks, not the peaceful doves, are more likely to revolt with acts of anger directed against the state [2438].

Reframing Aggression

"Kingship is not infectious." — Bantu Proverb

In WEIRD countries, social norms prevent many overt patterns of verbal abuse; we prefer to express deniable hostility and intent through nonverbal channels. Subtle variations in our vocabulary, voice tone, and grammatical construction can carry an unwelcoming message without overt threats. We degrade others by comparing them to despised animals, a common verbal violence that often flies under the radar [2295]. Less overtly, we bias the adjectives used to describe comparable communities. We represent men who have had sexual escapades as "experienced" but describe similar women as "sluts" or "easy." Young people who forget an appointment are "overworked," but old people have "senior moments." In the United States, if a White Christian attacks a crowd, the media portrays him or her as "disturbed" or "a murderer." If a Muslim does the same thing, we use the word "terrorist" and look for conspirators. Such biased descriptions reinforce the neural networks that shape our opinions.

Sports are a powerful, image-forming medium in the United States, with athletes often becoming role models. It also provides a microcosm for understanding ethnic and gender-based prejudice. Often unwittingly, sports coverage reinforces stereotypes of disenfranchised groups through its choice of vocabulary and chosen metaphors [2439, 2440]. Successful White athletes are described as intelligent more often than Black ones; Black athletes are more likely to be characterized as naturals or with animal metaphors. Print media also presented a higher ratio of African-American athletes in strength-based sports than Whites, reinforcing stereotypes of race-based athleticism and the black brute [2441]. This bias is not confined to the United States. Consistent with our species' predisposition to differentiate outsiders, we display widespread, culture-specific ethnic preconceptions in sports and other reporting [2442].

The language used during the Olympic Games showed fewer references to dominance, strength, and conflict for female than male athletes, reinforcing gender stereotypes [2443, 2444]. These gender differences are compounded by highlighting the female competitors' personal lives and appearances [2445, 2446]. Sports receiving airtime also differed, with the bulk of 1996 Olympic coverage of women dedicated to the individual rather than team sports, with women's gymnastics getting the lioness's share of the airtime compared to men (M:F, 16%:84%). For men, the most broadcast events were Basketball (M:F, 74:26) and volleyball (75:25). Men were described as strong players 41% more often than women, and the sportscasters frequently included qualifiers, following the praise with a "but" or "only to" statement to highlight a weakness in the women. Men's teams were typically called the "US" or "American" teams, but the female teams had the gender qualifier added [2444, 2446]. By the 2004 Olympic Games, verbal coverage had improved on American television; however, the analysis of one sport, women's volleyball, showed that over 20% of the camera angles focused on the women's breasts rather than their athleticism. Another 17% were butt shots [2447].

Sports studies may seem trivial, but they encapsulate attitudes that leak through despite attempts at media self-censoring. These differential comparisons increase the perceived distance between "us" and "them." This exaggeration makes committing violence easier. If soccer fans of River Plate and Boca Juniors (Buenos Aires), Honduras and El Salvador (Central America), or Manchester United and Liverpool (England) can clash over a kicked ball, imagine how associated, repeated, distancing imagery directs our thoughts.

We often endorse aggression with subtle facial expressions that shift from friendly to hostile in less than a few millimeters. If an approving visage can reinforce one's status as a trusted group member, a furrowed brow can instill doubt. With these gestures, peer groups and leaders can sanction or end bullying without saying a word. The violence foisted upon children by adults ranges from widely accepted disciplinary practices to brutal dehumanization, physical and emotional violence, and forced service of child soldiers in times of war [2448].

Tyrants love child soldiers for more than their obedience. During early to late adolescence, an area of the brain, the insula, goes into overdrive. Adult identities and an adult sense of justice form. Accessing these two developing forces creates dedicated warriors willing to kill and die for a cause. Below the level of national warfare, community violence takes its toll on the body and mind, interfering with education and instilling stress, mood, and cognitive disorders [2449].

In American schools, gay, sexual minority, and transgendered youth experience verbal and physical abuse at a far higher rate than heterosexual and cisgendered students who conform to gender norms [2450, 2451, 2452, 2453]. Students, faculty, and administration may focus on definitions of "bullying" and "violence" to dismiss unacceptable, dehumanizing treatment, such as a "Kick a slut in the head day," as deserved, just plain fun, or a harmless prank [2454].

This social ostracism and dehumanization of those who fail to conform to gender norms remain an accepted practice in many schools and communities [2454]. It happens in many countries, including Canada [2455], China [2456], and the UK [2457, 2458]. As of 2023, the bullying of gay and lesbian youth still receives social support in the United States, and faculty often remain silent and unsupportive of these foci of violence and microaggressions [2459]. Peers reinforce group norms by harassing students who are different with taunts of being "gay" or "queer" [2460]. This accusation moves them from the safety of the in-group to the status of a despised outsider. In 2015, 80% of LGBTQ youth reported victimization and had double the rate of suicidal ideation [2461]. Self-serving, browbeating politicians have since joined the fray, enhancing their careers by condoning the social denigration of LGBTQ youth.

> *"Bullying on the basis of actual or perceived race, color, religion, sex, national origin, ancestry, physical appearance, socioeconomic status, academic status, pregnancy, parenting status, homelessness, age, marital status, physical or mental disability, military status, sexual orientation, gender-related identity or expression, unfavorable discharge from military service, association with a person or group with one or more of the aforementioned actual or perceived characteristics, or any other distinguishing characteristic is prohibited in all school districts and non-public, non-sectarian elementary and secondary schools."* – Illinois House Bill HB5290, 97th General Assembly; lost on third reading, May 29, 2012 [3090]

In May 2012, the Senate of the State of Illinois defeated anti-bullying legislation that would have protected many school children from harassment, partially because it included protections for people with minority sexual orientations [2462, 2463]. Opponents defeated the broad-based anti-bullying bill, demanding that teachers and students be exempted from any anti-bullying components that violate their religious beliefs. The senators who favored the opt-out measures failed to state what religions sanction degradation, torture,

> *"[Propaganda posters] have to be sharp, attractive, and to the point, as well as exploiting ancient prejudices, showing a good knowledge of how people think and using the techniques that sway opinion. They have to be understandable in the blink of an eye from a distance of at least six feet and if they don't appeal directly to the emotions they should immediately be torn up and thrown in the bin."*
> — Tony Husband. *Propaganda Art* [3124]

or dehumanization. Instead, they acted on an understood social norm that allows empowered members to harass and intimidate people who differ from them.

Social consent to dehumanization has occurred before in numerous (if not all) genocides. The Nazi government in Germany and the occupied territories demeaned people they considered subhuman and unworthy of respect. These degradations took many forms, including propaganda on posters, film, music, and currencies. The arts by Jews, Slavs, and others not designated as the "master race" (Herrenrasse) were portrayed as "*Entartete*" (degenerate), while the works of select "Aryan" artists received praise. Propaganda films, such as *Jud Suss* [2464] and *Der ewige Jude* [2465], vilified Jews and instilled imaginary dangers into the minds of the dominant population.[288] Such targeted propaganda hacks our primitive survival instincts, stimulating fear regions like the amygdala to sow discord within communities and foster hatred [2466].

Until recently, American movies adhered to a so-called morality code that did not allow open or positive depictions of LGBTQ characters but readily joked and inculcated negative clichés such as the effeminate or "nelly" or "nance," or as psychopathic killers while giving tacit approval to homophobic prejudice [2467, 2468]. In short, propaganda reinforces stereotypes, dehumanizes its targets, and aims its message at our emotional limbic system.

---

[288] Not all the arts of Nazi Germany were aimed propaganda. Despite state control and censorship, most films produced during the Nazi era differed little from those of other film industries, such as Hollywood. It is easy to condemn the artists who lived under these regimes as collaborators. It is more difficult but wise to remember that most photographers, composers, and film artists were not involved in the Shoah, many were oblivious to the atrocities committed by their government, few had options for survival beyond what the government presented, and many opposed the antisemitic works.

Figure 10-1. Anti-Nazi propaganda poster depicting Poland as a handsome prince slaying the Nazi dragon. Dehumanization is used on all sides during war. —Lithograph poster by Arthur Szyk, 1943, produced for the US Government as part of official duties. Ⓟ [3240]

Under the Nazis, books for children included pejorative images of Jews. Some teachers would threaten to punish children who refused to memorize anti-Semitic poems in school. Popular music encompassed pieces such as the "*Horst Wessellied*," the anthem of the Nazi Party [2469][289] which paid homage to a Nazi youth they claimed was murdered by communists, and the Hitler Youth sang songs that urged putting Jews against the wall for execution. [290] As discussed in Chapter 7, the Rwandan government also effectively sanctioned dehumanizing propaganda against its Tutsi minority [2055], just as the Nazi government approved it against its targeted groups. The media used songs that extolled Hutus and debased Tutsis to create and reinforce an ethnic hatred of Tutsis and camaraderie among the Hutus.

One famous composer, Simon Bikindi, sang such songs. During his war-crimes trial, two pieces, (*Twasezereye = We Say Goodbye to the Feudal Regime* [2470] and *Nanga Abahutu = I Hate These Hutu* [2471] aimed at pro-Tutsi Hutus) were found to "manipulate the history of Rwanda and to extol Hutu solidarity" [2472, pp. . 48-68]. Although Bikindi denied they rallied people to genocide, the court interpreted his lyrics as calling for the murder of Tutsi, whom he compared to snakes [2472, pp. 6.1 p 34- 39, 1.2.3

[289] Although Wessel received credit for the song, its melody and lyrics bear an uncanny resemblance to the hymn *How Great Thou Art* [3073], which in turn, was based on an old Swedish folk melody. Wessel receives credit here for simplicity and to avoid tarnishing the original composers.

[290] Many people objected to the singing of anti-Semitic songs and demanded that Berlin ban their use. The cries fell on bureaucratic ears that stymied efforts, although those ears were not always deaf to the problem.

p128].[291] Hutus who refused to condemn Tutsi or who protected them risked violence, and the dehumanizing influence of this music increased their danger [2472, pp. 56, 268].

The music of Simon Bikindi and the Nazi composers lies at the extreme end of such abuse. They use overt ethnic stereotyping to incite hatred while building solidarity toward genocide. However, many accepted kinds of music exaggerate I-thou dichotomies.[292] Nationalist movements use lyrics to extol the values of involvement within the nation. When we sing songs that celebrate our membership within a society, we instinctively place ourselves in a superior position, maligning the value of other countries.

Despite their double-edged nature, patriotic songs and anthems remain stirring and lauded reminders of group membership. The United States, a country composed of people from many ethnic backgrounds, indulges in ritual nationalism and jingoism to the point that more unified countries consider vulgar. Numerous upbeat songs remind people of the advantages of American citizenship and the social expectation to act as a cohesive unit in times of trouble. In the early 20th century, one composer, George M. Cohen, wrote dozens of these songs that remain popular today.[293] Music expressing such cohesion can help society weather hard times or attacks.

> *"Well, try that in a small town, See how far ya make it down the road. Around here, we take care of our own." — Try That in A Small Town* [2473]

If a culture's folk songs and language elevate unity and pride, they can also manipulate people to wage war. The Nazi government in Germany identified *Sprachinseln* (speech islands) outside their national boundaries; they used the guise of this green beard and promise of reuniting the German people as a pretext for invading neighboring countries. Jason Aldea's hit song, "Try That in a Small

[291] In his defense, Bikindi said that his songs were not meant to advocate hatred, but more akin to smacking a small child to stop his misbehavior [2472, p. 59].
[292] Consider music that denigrates and dehumanizes women as another example.
[293] Songs by G. M. Cohen that remain popular with the American public include "For the Johnny Q. Public," "Over There," "Revolutionary Rag," "The Yankee Doodle Boy," and "You're a Grand Old Flag."

Town," claims to laud small-town values, but many have interpreted it as endorsing vigilantism, lynchings, and even racism [2473]. And when Bela Bartok, the Hungarian composer, found regions where Hungarian folk melodies persisted in Romania and Slovakia [2474], he urged the army to reunite these isolates with Hungary. In each case, the empowered population rewarded conformity to the national norm and increased aggression against an outside group.

We also sanction, even prefer, belligerence in our leaders—as long as they direct that aggression against others. We pay lip service to our dislike for aggressive and violent people but reward hawkish behavior. In democracies, voters prefer moderately antagonistic political candidates to doves [2475]. Right-wing populist politicians often exploit this silent preference with threats of punitive military action against unpopular groups or nations [2476]. Dictators, presidents, prime ministers, and corporate leaders who appear weak can become tempting targets for attack from others. Americans seldom vote presidents out of office during warfare if they present as strong alphas. Hitler's proactive rejection of the Treaty of Versailles and his war on crime and terror made him attractive to the beaten and ailing German populace following World War I. Even aggressive facial expressions in the British Parliament receive rewards of increased status within a political party [2477].

The defiance of Ayatollah Khomeini against Shah Pahlavi of Iran made him a popular leader, even among those who disagreed with his other stances. The hatred of the Shah ran so deep that despite their disagreements with Imam Khomeini, many still worked for him to spite the Shah. During that same period in Iran, a militant group took American embassy personnel as hostages. In the United States, President James Carter lost status with the American people when he refused to take military action against Iran. Many saw this refusal as a weakness. Soon afterward, Carter lost the presidential election to Ronald Regan despite the captives remaining alive. Ultimately, all the hostages survived, and only eight US military lives were lost.[294] Further illustrating differences in perspective, the Americans know the incident as the Iranian Hostage Crisis. The Iranians refer to it as the Conquest of the American Spies.

What of the Victims?

> *"Where there is charity and wisdom, there is neither fear nor ignorance. Where there is patience and humility, there is neither anger nor worry."*
>
> — St. Francis of Assisi [2478]

> *"There is nothing in the world as painful as the feeling you are not liked."* — Sei Shonagon, Japanese poet and lady-in-waiting to Empress Teishi, ca 1000 CE. Known only by her nickname at court, her true name remains unknown [3226]

A bullied person experiences a lower quality of life, including lower quality of health during and after harassing episodes [2479, 2480]. People who suffer bouts of persecution become isolated from peers who would rather avoid the losing side. Social rejection alters how our cannabinoid (CB1) receptors process social and physical pain [2481, 2482]. Casualties of shunning will attempt to ingratiate themselves, mimicking the behavior of the surrounding in-group in a subconscious attempt to gain membership [2483, 2484]. For a social organism like humans, isolation itself further damages the recipients; malevolents can use rejection and isolation as effective methods of terror [111, 2485], an imposition inflicted more often on females than on males [2486]. Those shunned and isolated by coworkers need more time to heal from their emotional wounds than people with strong peer support [2487]. In contrast, those who ostracize others report positive effects, including an increased sense of personal control [2488].

The ostracism need not be intentional to sting. Getting "phubbed" can lead to depression and jealousy [2489]. Phubbing, ignoring someone in favor of a social media phone feed, sends a signal to the snubbed

---

[294] Eight servicemen died in a helicopter crash during the one rescue attempt on April 24, 1980. President Carter's failure to pursue a rescue contributed to an image of weakness and his defeat by Ronald Reagan. The October Surprise conspiracy theory holds that the Reagan campaign conspired with Iran to retain the hostages until after the election. That conspiracy has not been proven.

person that they lack importance. The forgotten person is often someone close to the phubber, possibly because of increased opportunities to phub. The phubber may manifest this feeling, derogating someone if they interrupt their affair with the electronic feed [2490]. Unfortunately, we have no data on how many retaliate by tossing the phone in a piranha tank.

Childhood survivors of dehumanizing aggression experience a higher level of psychological and psychiatric problems later in life than similar children who do not have comparable torment in their background [2480, 2313]. Pure victims usually do not become oppressors. Instead, they typically show more aptitude for caring about others than pure bullies, bully-victims, and non-victims [2343]. Pure survivors also have more social problems, maladaptive behavior, and generalized defensive aggression [2491, 2492]. Because of their coping skills, they also have a potential for integration and adaptation into later, positive social contexts [2493].

Only about 25% of survivors become tormentors. The stereotype of the victim-turned-bully arises because of a logical fallacy in our thinking. Histories of oppression and dehumanization raise the probability of developing antisocial behavior *in specific subpopulations*. It does not increase it for everyone, but our brains do not like to make this distinction. Many who fill both roles have genetic or developmental propensities toward aggression [2337].

Those prey capable of becoming competent, judicious predators can stop their downward spiral in social rank. However, to transform from target to bully, the survivor must overcome the fallout of oppression on their neurochemistry, the hypothalamic-pituitary-adrenal axis, the autonomic nervous system, and the immune system. The survivors who overcome these barriers to become aggressors show high levels of callous behavior and high levels of delinquency by adolescence [2494, 2495]. In other words, bully-victims tend to be different kinds of bullies and may differ physiologically.

Evil Twins: Not so Identical

> *"Sometimes the greatest scientific breakthroughs happen because someone ignores the prevailing pessimism."*
>
> — *The Epigenetics Revolution* [2496, p. 29]

Some of us have a better chance at this negative salvation than others. Various alleles exist for a gene that controls our reaction to a specific neurotransmitter, dopamine. This neurotransmitter and its receptors control our natural proclivities for aggression. If someone inherits one kind of transmitter, then he or she tends to be a pacifist. With a less-active version of the same gene, that person would lean toward violence. These genes interact with the environment to place boundaries on how aggressive a person is likely to become. One form of the gene provides a naturally even temper, and another variety gives it a hair trigger.

We hate to believe a genetic or epigenetic fluke could predispose someone to bullying or subjugation, or that traumas experienced by a distant ancestor could contribute to depression or self-deprecation in a current generation. Our core beliefs hold that we have free will. We believe we choose to be good or bad, but the data does not agree. If it were strictly an environmental cause, one would expect the relationship between bullying and other factors, such as overall mental health, to vary globally. Cross-national studies consistently link bullying and psychosocial adjustment [2497].

Within their social context, the families of bullies and victims across cultures tend to be less functional [2498]. On average, bully-victims' families use an authoritarian[295] parenting style with high demands and

---

[295] *Authoritarian* approaches are top-down systems in which the leader's words are virtually law. Autocrats typically have high expectations of their underlings, but they provide little guidance and are not responsive to their needs. Mistakes receive punishment, but success receives little reward. In contrast, *authoritative* leaders place high demands on their subordinates but make explicit expectations, respond to feedback, and engage in more nurturing behaviors. They provide rewards for good jobs and positive behaviors. Authoritarian leadership (or parenting) has high demands but little direction or compassion. Authoritative ones have precise, reasonable, expected claims and reasonable, understood consequences and give subordinates guidance and support. Authoritative leaders provide flexibility and encourage more initiative than autocratic leaders. The other two common leadership and parenting styles are neglectful and abusive.

little nurturing [2499, 2500]. However, these data do not differentiate between cause, effect, and correlation. Without other data, we cannot know if family behavior caused the problems or if the aggression and degradation impacted the families. An underlying genetic or developmental problem could cause family issues, predispositions to stress could interact with events, or the family environment and bullying could vary because of other co-existing factors.

Twin and adoption studies are the classic methods of comparing the influence of the environment and inheritance. Our genes strongly influence adult antisocial behavior and other personality disorders. Studies attribute between 32% to 60% of antisocial conduct to genetics [2501, 2502, 2503], with cross-study analysis focused on a 51% genetic contribution [2502, 2504, 2505].[296] We also know that an early onset of callous antisocial behaviors (before adolescence) correlates with an adult diagnosis of antisocial behavior.[297] A traceable life-long tendency toward aggressive dominance or submissive deference suggests a genetic or epigenetic origin for the early onset of these traits.

One does not inherit a gene for antisocial behavior per se. Instead, antisocial behavior correlates with genes and gene regulation that mediate structural elements of the personality, such as neural networks, neurotransmitters, receptors, and hormones. For example, we have preferred levels of stimulation, sensation seeking, and impulsivity. Body chemistry sets these levels as we interact with the environment. How we process hormones such as serotonin can predict our ability to think before we bully [2506], our propensity toward depression [2507], and anxiety disorders [2508]. Our theory of mind and how the brain interprets social interactions contribute to the trigger pressure for aggression, as does how the body regulates emotions [2509, 2510, 2511].

---

[296] Polygenic risk scores (PRS) examine the total number of genetic variants contributing to a condition. The use of PRS for antisocial behavior has not shown a consistent, additive relationship with the disorder. I have excluded PRS studies from the discussion. However, using PRS may enhance the early detection of genetic risk factors [3475].

[297] Psychologists will not diagnose someone as antisocial before age 18.

In a study of 1,116 families with 10-year-old twins, the genetic component accounted for 73% of the variation in victimization and 61% in bullying. In other words, the environment accounted for less than 30% of the variation in victimization and less than 40% in bullying behavior [2337]. The genes influencing our likelihood of becoming a predator or their prey overlapped but were not identical.

Body-environment interaction accounts for the remaining variation in other ways. Both high and low social standing leave their imprint. Abuse, malnutrition, and stress produce specific changes in targets, reducing their status and chances of advancement. Dominance, power, and access to resources strengthen the positions of those organisms that experience them. Specific hormones peak or drop in response to these environmental burdens, and brain areas become rewired. These alterations underlie the behavioral modifications prompted by the environment.

Changes in an organism caused by altering the activity of genes are known as epigenetic features. The most well-understood imprinting method is methylation [2512, 2513]. Cells attach a small chemical lock, a methyl group, to a particular site near a gene. That lock prevents the cell from accessing the information at that site. Epigenetic changes alter organisms, and some modifications cross generations to impact our descendants [2514, 2515, 2516]. As we age, some genes get turned on, and others get turned off. Some changes, like the loss of the human tail on an embryo and our growing interest in sex during adolescence, happen at a programmed point of development.

Social pressures cause other genome-wide epigenetic changes [2517, 2518]. These changes are far from short-term. Changes can induce chronically high cortisol levels or a tendency to secrete excess adrenaline after bullying and other traumatic stress [2519, 2520]. Neglect and shunning lead to epigenetic damage [2518, 2521], and the confluence of ongoing macro and micro-aggressions, such as regular encounters with discrimination and ethnic disparity, can lead to premature aging [2522, 2523, 2524] and the transgenerational epigenetic harm [2525, 2526].

Discrimination and other psychosocial abuses suffered in childhood and before birth can create lifelong changes in inflammation and cortisol (stress) metabolism [2527, 2528, 2529, 2530], reductions in grey matter [2531], and blunt our ability to monitor errors, reducing self-esteem and making one a target for further humiliation [2532]. Epigenetic changes in metabolism and brain structure increase the chance of suicide during adulthood [2533, 2534, 2535, 2536]. The adult brains recovered from people who ended their lives show pronounced silencing of normally active genes in the hippocampus if they were abused as children [2537, 2534].

The brain consists of more than neurons. Once treated as sidekicks to the neurons, we now recognize glial cells as partners that are valuable to our physiology and mental health. They, too, can incur damage through social and physical stress [2538]. During our embryonic stage, they guide the development of neuronal networks [2539] and differentiate into subtypes (astrocytes, oligodendrocytes, NG2s, and microglia) that partner with the neurons and keep them safe. At this early stage of development, genes respond strongly to environmental insults, and maternal stress can translate into changes in the glial network of the embryo [2540].

These specialized cells hold neurons together, nurture them, keep them safe, and regulate the flow of information among them. Graceful, star-like astrocytes protect the neurons and clear away excess neurotransmitters. They promote the formation of the junctions between neurons (synapses) and regulate the flow of materials from the blood to the brain, acting as gatekeepers. Microglia are immune-system cells that move through the brain, sampling the neural environment; they support and care for a sick or injured brain by directing the immune system toward the damage. NG2-glia regulate body weight and transform into oligodendrocytes. Oligodendrocytes provide structural support and create elegant, insulating covers for the neurons, preventing short circuits in their electrical system. Other glial cells secrete and circulate the cerebrospinal fluid that bathes the central nervous system, provide scaffolding, attack invaders, clear debris, or respond to cellular chemistry. Glutamate is a neurotransmitter that stimulates a postsynaptic membrane, urging it to fire and send its signal/message to the next cell. Glial cells regulate our mood by adjusting the amount of glutamate in the synapses. They also prevent glutamate from leaking and affecting other synapses [2541].

Astrocytes store and dole out the glucose that fuels the neurons. As the most abundant glial cell, star cells provide the basic architecture of the blood-brain barrier. Astrocytes regulate the ions and neurotransmitters the brain uses to send and receive information. Early life adversities, including neglect, shunning, and physical, emotional, or sexual abuse, damage the valuable system of glial cells [2542]. Epigenetic damage to this critical system impairs function in affected brain regions.

On autopsy, people with histories of childhood abuse show distinct changes in the appearance of their astrocytes [2543, 2544]. These altered astrocytes interact with other stress-mediated epigenetic modifications, such as reductions in the expression of GR. This gene regulates the chemical docking ports (receptors) for glucocorticoids, a keystone in our response to stress [2521] and changes in serotonin metabolism [2507].

NG2-glial cells regulate hunger and temper brain inflammation [2545]. Stress damages the NG2 system, resulting in depression [2546]. Unlike other glial cells, NG2s are known to connect with neurons through synapses [2547]. These connections can send messages to neurons that aggravate stress-based anxiety [2548].

Some NG2 cells transform into oligodendrocytes (OLs). These modified cells make the connective tissue that covers the neurons, the myelin sheaths. This secreted wrap coats the part of the neuron that sends information to the next cell (axons). These myelinations build during our first years of life. Adults who experience stress and abuse as youngsters have epigenetic changes in these cells and their lining [2549]. The insulation decreases in their anterior cingulate, prefrontal cortex, and amygdala compared to those without a history of abuse [2550, 2551, 2552]. The ACC integrates emotions with reason, contributes to empathy and impulse control, and helps us comprehend gratitude and reciprocity. Another region also joins the fray, the lateral habenula, an area involved in aggression, learning through unpleasant experiences (aversion), and our response to traumatic experiences [2552, 2553, 2554].

As protection against immediate stress in infancy and childhood, too many NG2s may differentiate prematurely, reducing the pool available to make cells that support and insulate axons (oligodendrocytes) later in life [2555], acting as the electricians of the brain, keeping the signals fast and efficient. Healthy, functional OLs and NG2s are vital to many learning tasks, including information processing and retrieval. Early life stress alters the protein signature in the NG2s of the hippocampus, the area most intimately associated with memory and recall [2556]. Oligodendrocytes also play a role in our habituation and sensitization to stimuli,[298] long-term memory, and recalling paired things, like particular circumstances and places with bullying events (associative learning). Memory consolidation also depends on the input of the OLs [2557]. Their memory and recall roles are core elements in our responses to tribulations.

Of course, the body acts as a unit, and body systems and cells interact. Physical and mental stress can cause immune system dysfunctions. Physical stress releases a compound (leukotriene B4) that speeds mitochondria reproduction in the CD4 cells. These hyped-up cells produce and release a compound related

---

[298] Associative learning happens when we pair a stimulus with another event, like the telephone ringing with a pleasant voice at the other end of the line or associating being bullied at work or school.

to caffeine (xanthine) that changes the functioning of the oligodendrocytes in the emotional amygdala, leading to depression, anxiety, and social disorders [2558].

Not every epigenetic effect happens during our lives as independent organisms. The body also silences some genes because they originate in either the sperm or the egg [2559, 2560, 2561]. Developing embryos seem to know which gene copy comes from mom or dad. It can treat the same gene differently if it comes from the egg or the sperm. We use proteins from the unsilenced copy of the gene. Once, we thought that such parental biases were rare. Since the turn of the century, we have found them to be far more common than we thought [2562, 2560]. We express many of our genes as though they were a single gene from one parent. The body turns off the other parental gene in one or more body parts [2563].

Social dominance, a feature bullies exhibit relative to their quarries, shows epigenetic features, with the impact of some genes depending on their origin [2564]. Grb10 was the first such parent-specific gene found to influence social behaviors [2565]. The copy that comes from our mother regulates cell growth. The father's copy functions as a "social strategist" gene, contributing to patterns of competition, cooperation, and resource use. Mice unable to use this paternal gene (because some meddling researcher repressed it) tended toward aggression and often won social dominance encounters with other mice. They were also more likely to attack cage mates.[299] The same gene, if silenced in the mother, just made the mice large and fat. Mice also expressed the two copies of this same gene in different parts of the body. The brain expresses the Grb10 gene inherited from dad, and the rest of the body manifests the one inherited from mom.

Remember, Grb10 has company. The body uses epigenetic mechanisms to raise and lower its level of oxytocin, dopamine, serotonin, and all other hormones and neurotransmitters that contribute to our ability

---

[299] NB: This same gene, Grb10, influenced the development of leukemia in mice. The interaction of genes is complex. No one yet knows how or if the function of this gene correlates with these two functions: status and blood cancer.

to compete and our propensity toward victimization. Our genes contain the first hurdle we must overcome. After that, our physical and social environments contribute heavily to our gene regulation.

Making the Most of a Bad Situation

> *"Without followers, evil cannot spread."*
>
> — Mr. Spock, *Star Trek*: "And the Children Shall Lead," Chomsky [2566]

Not everyone is involved in repeated acts of violence, though virtually everyone experiences and inflicts aggression at some point. People who witness violence must cope with its images and the knowledge of their reactions to it.

Each party—aggressor, quarry, and witness—develops methods of making the most of their situation. They learn to spot and respond to their potential allies, victims, and tormentors. They know how to minimize damage, stay clear of the line of fire, or ally themselves with the winning side. Like disciples of Sun Tzu's *Art of War* [2567], they learn when to act and when to refrain. The cognitions, behavioral patterns, and rank expectations instill identities in these parties by manipulating the chemistry and architecture of their minds.

> *"You win by knowing when to fight and when not to fight."* — Sun Tzu, *The Art of War* [2567]

> *"Even your own shadow leaves you when you are in darkness"* — Ibn Tamiyyah, 14th-century Sunni theologian

Like the guards and prisoners in the Stanford Prison Experiment, tyrants and their marks develop behaviors consistent with social expectations of that role if they occupy that part regularly. They learn to manage atypical responses when they arise. Predators test the limits of their powers, and their prey discovers its boundaries. Early childhood experiences with torment leave their mark on the bullies and the bullied, triggering changes in mental health and teaching each to construct specific paradigms for dealing with the world [2568, 2569, 2570]. They develop and retain behavioral scripts, remembering how past encounters proceeded. They can then string together a series of events to comprehend the proper behavioral sequences. Neither the hunted nor their oppressors can easily overcome these teachings. Life has ingrained physiology and epigenetics that support its ongoing psychology.

The learned sequence of transitions, the patterns of cause-and-effect, have written changes on the epigenome that reinforce their role. As victims learn how one action slides into another, they learn new ways to become invisible and elude bullies, even as their sense of personal effectiveness erodes. Although they often fail to recognize the extent of their violent behavior, predators develop the eye of a hunter, honing techniques to find game to provide needed gratification.

The wounded learn something will destroy their happiness if they feel too comfortable or secure. The bully knows the sequence of triggers and cues that will optimize damage to his or her target. The aggressors learn to rationalize their behavior, often recruiting others to support their acts and blame the prey.

People develop cognitive models that guide their responses based on their perceived experiences of power, victimization, or both. These guides predict what will and will not work to keep them safe and give them status. They learn to evaluate the potential risk and reward for each sequence of events they encounter [2571]. In each situation, the superior temporal gyrus and inferior parietal lobe generate a theory about the minds of others. Internal models, developed through the prefrontal cortex, amygdala, hippocampus, insula, and ACC, guide us as we do mental calculations to determine if a fight is worth the effort. In extreme cases, the person is cognizant of the process. However, we often make these decisions on autopilot, never thinking about the humiliation or the power in the background.

Our cognitive models take much into account, weighing experiences. Our genetic, epigenetic, and social history combine to determine how our brain responds to danger. Some of our experiences will convince us to take the safer, low-risk roads. Other occasions will prod us toward dominance. These experiences go beyond our actual battles to rehearsals for status.

Part IV: Positive Violence

"A Denver boy was disciplined for throwing a nonexistent grenade. [He, age] 7, was pretending to save the world from evil forces on the playground and pretended to detonate an imaginary grenade. But his school has a no-tolerance policy, meaning no weapons, real or imaginary. So Evans got suspended."

— KATU News, Portland, OR, Feb 5, 2013 [2572]

CHAPTER 11:

Playfights are Not Bullying

Helicopter Bullying and the Abdication of Blame

> *"A Texas school required parents wanting to help with the second-grade holiday party to have a background check first. Schools auctioned off the right to cut the carpool line and drop a child directly in front of the building — a spot that in other settings is known as handicapped parking."*
>
> — *Time* magazine [2573]

In recent years, America has made childhood illegal. As a country, the United States has become so obsessed with treating everyone equally that, as a group, it has abandoned the idea that a history of good behavior should earn trust. To calm their fears for their children, some authoritarian parents ignore their childhood experiences and exert excessive control over their children's lives.

Extreme brooding behavior provides children with few chances to practice life skills independently of their parents [2574]. Helicopter parents will hover as their children play. They organize their children's time, leaving them little time for the unstructured play that prepares them for adulthood. Hovering can continue into the college years, leaving their offspring depressed, socially inept, and unprepared for adulthood [2575, 2576, 2577].

The healthy concern parents have for their children's safety has thus become pathological. Society has bombarded us with images of missing children through many media streams. Placing these horrific images in our sight has led parents to believe that child abduction and murder by strangers are commonplace. Although rare, no one wants to venture opposition, lest they be blamed for one of these cases. Because of our instilled fears, we have begun to deny children the right to play [2578]. Children need supervision; the issue is knowing how to balance control and autonomy. We choose to err on the side of immediate safety, not of potential developmental effects. We observe our children incessantly, ignoring the psychological links between being watched and being threatened. Our species seems to have difficulty finding the middle ground between overprotection and neglect, between being a helicopter parent and having a latch-key kid.

Like other mammals, humans exercise increasing autonomy as they age [2579]. (Just ask any parent if their five-year-old still wants to hold their hand in public—they'll shoot you a look as though you just arrived from Pluto; then, they'll expound on the story about their latest parking lot or bus standoff.) Children, as they grow, develop more independence from their parents, have an increasing sense of self, and spend more time with peers. However, in a country with increasing legal liability and unrealistic expectations of parents, those who advocate for children to enjoy even slight autonomy can find themselves pilloried in the press [2580, 2581, 2582].

Zero-Tolerance: An Impact of Fear

> *"Zero-tolerance policies forbid persons in positions of authority from exercising discretion or changing punishments to fit the circumstances subjectively; they are required to impose a pre-determined punishment regardless of individual culpability, extenuating circumstances, or history."*
>
> — Wikipedia [2583]

> *"The assistant principal told me [my daughter] had a stick and she was using it to shoot and kill other students."* — *A child's mother citing her child's assistant principle.* [2590]

Parents and other adults often fail to distinguish between bullying and healthy play. As a consequence, many schools have taken zero tolerance to absurd extremes, fearing being blamed if something goes wrong. By harsh punishments and invoking arrests, these authorities become the de facto bullies, traumatizing and stigmatizing the children [2584].

In one notorious (but not isolated) example of extreme enforcement, a well-behaved student received a suspension from school for five days (pending an expulsion hearing) for having a butter knife in her lunch [2585]. In a similar case in Florida, an 11-year-old honors student received a suspension for the same offense with a threat of criminal charges [2586, 2587, 2588]. The child used a children's "animal friend" butter knife, which the administration apparently considered on par with switchblades, bowies, and daggers, to cut a peach to share with a hungry classmate. The school placed the incident on her permanent record, and the police passed the case to state prosecutors with a threat of criminal charges. The shaming caused by expulsion hearings and the disruption of parental work schedules compound the damage in such cases.

In another incident, a five-year-old was suspended for pretending a small stick was a gun, and she was guarding the king and queen. This was not a long branch that could hurt someone by swinging it; it was small, and she pretended it was a handgun. In a rare show of solidarity between liberal and conservative-leaning media, reporters at the Huffington Post and Fox News found common ground, and both condemned the overreaction. [2589, 2590].

The inflexibility of these school administrations crossed the line into bullying and abuse. They wielded their disproportionate power to dehumanize their victims by lumping all students together as threats, regardless of context and intent. Under the guises of discipline and school safety, their actions resulted in emotional harm and humiliating social abuse. Although exclusionary discipline (suspension and expulsion) has its place in a hierarchy of punishments, it is not a primary intervention. It should not be used as a cudgel to brow-beat children for minor or imagined offenses, and where it is used, administrations should track biases.

Such automatic behavior from adults follows the path of least resistance, not what is best for the child. Asking everyone to behave by strict, invariant rules is far less risky than facing accusations of bias or mistaking an actual threat for healthy youth behavior. When one teenager sells amphetamines to his peers, our society follows the easy road and bans all teens from carrying their medications. This punitive and institutional control fosters prolonged dependency in exchange for temporary security against perceived threats.

Erring on the side of immediate caution is built into our primate psyche. Our ancestors survived by presuming a lion was hungry, not by believing he had just finished a hearty helping of antelope burgers. The *Ardipithicus* or *Australopithecus* who trusted outsiders could have met the fate of the stray chimp, beaten and bitten to death by a rival patrol party. In our diverse culture, educators like myself seldom share familial or community ties with the youth they encounter.

The children may not pose a significant menace, but irate elders do. Bullying parents strive for the fantasy of a danger-free zone for their offspring, demanding any hint of danger be removed, no matter how remote the risk. They might not tear the teacher or administrator to shreds with their teeth, but they can bankrupt them with litigation or cost them their jobs. Schools balance the threats from allegations of bias and neglect. Zero-tolerance policies give administrations professional cover.

Black students receive a disproportionately high number of suspensions, expulsions, and school-based arrests — often at about twice the rate of their peers. This suggests an enforcement bias. African Americans are also the only tracked group that is disproportionately transferred to alternative schools.

In the American public school system, suspensions begin in preschools. African-American toddlers accounted for about 18% of preschoolers in 2017-18, but 43% of the suspensions and 38% of the expulsions. For Whites, the ratio was 43% of enrollment, 37% of suspensions, and 37% of expulsions. Latinx students constituted 29% of registrations, 11% of suspensions, and 17% of expulsions [2591].[300] A 2017 cross-borough report on Student Safety Act Reporting in New York City followed this trend, with 59% of the reports filed against Black students, 30.9% against Latinx, but only 4.3% against white students [2584]. New York City is approximately 31% white, and whites comprise about 15% of the public school population.

Despite their professed even-handedness, school zero-tolerance policies (ZTPs) disproportionately harm Black, Indigenous, and other people of color, often setting the stage for future imprisonment [2592].[301] ZTPs in American schools skyrocketed during the 1980s' war on drugs and peaked in the early 2000s. Rigid one-strike policies were popular with voters (and hence with politicians) who believed they would reduce crime. However, administrations implemented them differently in Black, Latino, and White schools [2593], with Blacks less likely to receive mediation and mentoring prior to expulsion [2594].

Harsh, unbending policies without wiggle-room stigmatized and criminalized daily behavior. Exclusionary discipline peaked between 2004 and 2006 before the Obama administration urged reforms to the ZPG policies. Still, between 2011 and 2012, nearly half of California's expulsions were for "willful defiance," a category so broad that one could be charged for not picking up a piece of paper or removing a hat. And all the examples cited in the opening paragraphs of this section happened in 2016 or later.

Expulsion reduces a student's chances for success, and the stigmatization creates social stressors that could trigger the epigenetic changes discussed in the last chapter. ZTPs have resulted in schools that function as feeder systems for prisons in a "school-to-prison pipeline" that can criminalize even minor misbehavior, such as refusing to work or belching [2593]. Strict ZTPs result in early contact with the penal system, which

---

[300] This same report noted schools disciplined 50% of White students for bullying based on race, color, national origin, disability status, or English proficiency. The discipline imposed for bullying protected classes in Latinx and Black students amounted to 19% and 20%, respectively. These figures are more in line with their percentage of the population, slightly higher for Whites (3%) and Blacks (5%) and low for Latinx (-9%).

[301] A practical, and potentially more effective approach, pioneered in Spain, is the Zero Violence Brave Club, aimed at changing peer culture to recognize those who intervene to stop bullying as brave [3441].

has little to no positive influence on youth behavior [2595, 2596, 2597, 2598]. They increase the probability that children from low-income and marginalized communities will have criminal records, have their families haunted by the Department of Children and Family Services, or be placed in foster care.

School closings during the early COVID-19 pandemic provided an environment to assess the negative impact of American schools, their biased application of punishments, and their repercussions on disenfranchised students. The pandemic affected everyone, but it especially limited the educational opportunities for the economically disenfranchised. However, with the advent of near-universal home education, schools no longer called the police to refer students to the juvenile justice system. The number of teens in detention dropped by 24% in the first month of the lockdown and by 27% in July [2599, 2600]. It remains to be seen if the reduced contact with the justice system can offset the loss of classroom time. In the meantime, adults must confront the demeaning and dangerous aspects of biased and unequal enforcement of zero-tolerance policies and recognize that not every disruption is prison-worthy. Sometimes, it's not criminal. Sometimes, it's just play.

Playfighting

Our society has become so fearful of differential treatment among students that we shy away from reasonable distinctions toward absurd conformity. We have even pathologized playfighting, a behavior necessary for healthy development across a wide range of social animals. There is no evidence that rough-and-tumble playfighting (RTP) among agemates increases aggression [2601, 2602]. Zero. Zilch. Instead, it provides an innocuous environment for children to test their limits and learn about interpersonal cues. It teaches them communication skills and the social limits of cooperation and competition [2603].

Roughhousing may not happen in every species, but it is widespread among social ones from primates to rats, birds, and dolphins. It is such an attention-demanding, energy-sucking activity that it must have a high payoff to survive [2604]. In social species, play, including RTP, supports cognitive development, intent recognition, and societal adjustment [2605, 2606, 2607]. Playfights often involve multiple partners, letting children practice group skills in complex situations. Boys participating in RTP have better social skills and are more popular than other boys [2602]. In dogs, playfights develop interpersonal skills, producing more outgoing and open individuals, two traits linked to higher social rank [2608]. In adults, the experience of RTP can contribute to tolerance and cooperation [2609].

In humans, we cannot test whether popularity, social skills, or horseplay came first. The critical play experiments done in rats show that competent RTP contributes to other skills. Adult rats play less than juveniles, so pairing a youngster with a non-playing adult can reduce play experiences but keep different interpersonal practices, such as cuddling and grooming, relatively normal. One can also cage a juvenile with a cagemate, genetically or chemically altered to change or eliminate specific forms of play, such as RTP, then later transfer them to a colony [2610, 2611]. From such studies, we know that the RTP during pup-hood leads to popularity and enhanced social skills as an adult, not the other way around [2612].

Play, especially RTP, exercises the neurons of the cortical, limbic, and connecting striatal areas [2613, 2614]. Play and the number of playmates impact the orbitofrontal and medial prefrontal cortices, influencing social behavior in both rats and primates [2615, 2616, 2617]. Play has a pivotal role in pruning the connections among nerve cells, reducing noise, and letting critical neurons thrive [2618].

Play primes the PFC's responses to social signals, rendering them more in tune with social experiences throughout life [2619], and RTP moves its executive functions toward a mature state [2620]. Rats deprived of play or contact with others have reduced cognitive, emotional, and memory function, plus poor impulse control and social skills. When introduced to other rats, they become targeted and bullied outcasts.

RTP between parent and child teaches self-regulation, improves social resiliency, and improves executive function, talents linked to the prefrontal cortex [2621]. Instead of increasing aggression, RTP with fathers strengthens their bond and may decrease aggressive behavior [2622, 2623]. RTP develops motor skills and teaches self-regulation, an executive function of the prefrontal cortex. The behavior tweaks the stress hormones in the developing brain, preparing it to recognize dangers and appropriate risk-taking later in life [2624]. Where isolation and shunning diminish monoamines (dopamine, noradrenaline, and serotonin) in the ventral tegmentum (reward, learning, memory formation) and locus coeruleus (noradrenaline regulation, alertness, stress responses), socially rewarding behaviors, such as roughhousing and other RTP, enhance them [469, 470, 2625].

Most studies on human parental play concentrate on mothers, but RTP research has focused on fathers. Less is known about RTP with mothers, the forms these activities take, and their long-term consequences. For mothers and fathers, playfulness, including RTP, correlates with higher positivity, with the child interpreting events with more optimism [2626].

Although children know the difference between childhood play and childhood bullying, many adults seem unwilling to make the simple distinction. One is degrading, violent, and dehumanizing. The other is uplifting, voluntary, and joyful. Bullying is imposed through power. Playfighting involves taking turns and surrendering control, feeling assured that the game will remain fun [2627]. Bullying reduces competence in the victim. Roughhousing builds it in both parties.

Children need unstructured play, including voluntary rough-and-tumble activities. Without this roughhousing, children fail to learn normal adult social behavior [2628, 2629]. Adults too often assume something is wrong with children who engage in such activities. We reason these "aggressive" children may have too much anger and need counseling or a cathartic release structured by therapists [2630]. Perhaps, we think, roughhouse play signals hostility or is a risk factor for perpetrating violent crimes as adults.

Balderdash. Such nonsense is a gross generalization and an incorrect cognitive leap. We jumped from recognizing the infective nature of violence and the influence of the environment, e.g., cartoon violence [2631] to accepting the demonization of ordinary rough-and-tumble play that includes ritualized elements of aggressive behavior. Far from fostering aggression, pretend play, including playfights, aligns with positive, pro-social behaviors rather than with problematic activities [2632].

Why Rough Play?

> *" 'You may go, then,' said Odin, 'to live among the flowers, and play with the wild bees and summer insects. You must, however, find something to do, or you will get to be mischievous like the dwarfs'."*
>
> — Advice to the light fairies, in *The Dwarfs and the Fairies* by Keary & Keary [2633]

We have become so preoccupied with physical violence that in many American communities, a child risks expulsion for participating in normal playfighting. Healthy, child-directed activities include roughhousing for both boys and girls. Without it, our brains do not develop properly.[302]

Several brain areas will not mature correctly when adults ban childhood RTP. These include the prefrontal cortex, which is necessary for intellectual development and social competence. We share our reliance upon such horseplay for social development and maintenance of social structures with our close cousins, the chimpanzee and the bonobo [2634, 2635, 2605]. Roughhousing occurs in both adults and juveniles. It teaches us about competition and cooperation, and it builds social competence. RTP reinforces accurate

---

[302] In the past, other people have targeted forms of play, especially make-believe. There, confusion occurred between normal pretend behavior and lying or delusions. When I wrote my first short story as a child, my Irish grandmother scolded me for lying and made me go to confession. I didn't write again for years.

communication among troop members. Through these ritualized tournaments, we have fun as we learn about personal boundaries.

A second area of the brain, the *corpus striatum* (a.k.a. neostriatum, striatum, or striate nucleus), also depends upon this play for its proper maturation. The striatum organizes the sequence of events in playfighting and signals that the activity is fun [2636]. The striatum lies below the massive roof-like neocortex that covers much of the brain. Among its tasks, the striatum connects the neocortex with another set of brain control centers, the basal ganglia. Throughout life, the striatum undergoes epigenetic changes in response to stress. These predispose us to future bouts of depression and reduced cognitive function [2637, 2638]. Our mock battles build confidence and inoculate us against future stress, protecting these brain regions.

The striatum mediates planning and links the thinking and movement centers of the forebrain to output channels. In the striatum, the motivational centers in the basal ganglia join forces with the planning and movement centers of the forebrain. Together, they send messages to the output channels. If an incoming signal from a behavior releases dopamine into the striatum, it feels good and gets reinforced.[303] If the striatum finds the stimulus aversive, it punishes the action, and we repeat it less often.

The striatum also interacts with the amygdala. This almond-shaped region knows what it likes, and it wants what it wants. The amygdala considers pleasure deprivation tantamount to an attack. It creates tension in the brain, kicking up a fuss till it gets its way. The amygdala processes the background reward value of items, situations, or actions. It evaluates the *potential* for reward, while the striatum evaluates the *actual* one [2639]. Like the striatum, the amygdala has regions rich in dopamine receptors critical to socialization.

---

[303] You may be thinking, "Wait, I thought the nucleus accumbens was the major reward center." You're right. The NA is the brain's number one reward center, and it is part of the striatum.

These areas depend on early-life experience for proper development and the transition to peer interactions in adolescence [2640].

This disconnect between anticipation and reality explains cravings and why a double-cheese pizza is never as good as expected. In chemical or behavioral dependencies, the combined signals from these areas anticipate a greater reward than the one that happens when we finally get our way. It's why a child who has badgered you for a particular toy will play with it for half an hour and then forget about it forever afterward. Discord between these areas frustrates us. However, when these areas cooperate, reward values skyrocket.

The striatum is loaded with rewarding dopamine receptors and connects the rational neocortex to more instinctual nuclei. Dopamine may not motivate all goal-directed and reward reactions, but it contributes to many. The brain registers behaviors as pleasurable or adverse by assessing dopamine levels and translating that information into the relative reward value received for doing things from listening to music to sex [2641, 2642, 2643, 2644, 2645, 2646]. We don't even need to "do." We get tinges of pleasure when we observe or expect fun behavior [2647, 2648].

By segregating dopaminergic functions into different regions of the forebrain and midbrain, we create a sliding scale for reward values. The dopaminergic neurons respond to rewards and interpret cues that predict them. Based on this evaluation, they adjust the signal strength to fine-tune feelings [2649, 2650, 2651], helping us anticipate how rewarding experiences will affect us and others [2652].

The striatum also coordinates sequences in strings of behaviors. This ability arises from its role as a mediator between the basal ganglia and prefrontal cortex. The striatum assists the prefrontal cortex to make sense of others' intentions and fine-tune our interpretations. During RTP, the striatum accesses and rearranges behaviors, such as attack, predation, and submission, into patterns that will not trigger defenses in playmates [2653, 2654]. [304]

Horseplay stimulates the striatum, which explains why playfights in young rats show a different sequence of events than aggressive fights [2636, 2655]. Rats engaging in good-natured battles use play defenses that shorten scrimmages between playmates. Once they gain a tactical advantage over their partner, the striatum switches strategies, allowing their playmate his or her turn. A functional striatum also mediates other changes in the game plan, such as the commencement of grooming or playful sexual mounting.

By itself, dopamine cannot explain all the rewards of boisterous games. When young rats receive Naloxone—a drug that blocks the brain's opioid receptors—rowdiness drops and so does the stress-buffering effect provided by play [2656, 2657]. Naloxone blocks opiate receptors in the brain, interfering with a different pleasure pathway. This information suggests our internal opiates, the endorphins, also contribute to the joy of RTP.

Although it may seem like a no-brainer that pleasure-producing opiates reinforce play, scientists remain a questioning lot. They explore other possible explanations for why an opiate-deprived rat does not play right. When RTP occurs on Naloxone, the animals pin their partners less often [2658]. In contrast, morphine-treated rats increased their rate of pinning. The marmoset, a distantly related mammal, has a similar pattern [2659]. Giving morphine to a mammal elevates the level of overall opiates. These differences create abnormal playfellows for other rats or other marmosets. That difference alters the desirability of a rat as a play partner, thus decreasing their overall frequency of play, disrupting social dynamics [2660, 2661].

---

[304] We do the same during loving sexual foreplay, taking turns and organizing playful acts, rendering them unthreatening.

The complexity and variation in horseplay enlist many neural pathways. Scopolamine is a drug related to Jimson weed; both drugs hijack the brain and dim the light on reason by disrupting how we metabolize yet another neurotransmitter, acetylcholine (ACh). Scopolamine initially blocks RTP, but participation quickly returns to normal [2662]. Withdrawing scopolamine causes a rebound effect, with exaggerated play behaviors, with playfighting returning in full force [2663].

Like most transmitters, acetylcholine has multiple functions, including roles in arousal and rewards. Scopolamine blocks ACh action in the parasympathetic nervous system. This pathway dominates when the body relaxes and feels no threat. It helps regulate the secretion of exocrine glands, smooth muscle function, and the function of heart muscles.

The drugs pilocarpine and arecoline block the action of the same ACh receptors and also suppress pinning during play bouts [2664]. These data suggest the parasympathetic (rest and relaxation) pathway does not start playfighting per se. Instead, it regulates how play unfolds and how the animal switches from one move to the next.

Even at moderate levels, amphetamines, which push the brain's gas pedal, also inhibit aggressive play behavior [2665, 2666], but they do so using a different mechanism. These drugs ramp up the "wakey-wakey" hormones and transmitters known as catecholamines. Catecholamines boost alertness and are key players in our fight-flight-or-freeze response. They rise as morning approaches, then drop low as we drift into deep sleep.

When a vertebrate takes an amphetamine, catecholamines surge in the brain, activating the wake-up system. Two players influenced by amphetamines are norepinephrine (a brain-based cousin of adrenaline) and dopamine (a compound involved in rewards). Amphetamines move these transmitters from the cytoplasm of neurons into the synapses and inhibit the attack phase of RTP [2667]. At higher doses, the drug affects both attack and defensive behaviors. Because amphetamines inhibit the physical banter necessary for normal neural development, there are implications for amphetamine use, its impact on socialization, and the interpretation of aggressive behaviors in children treated with these drugs.

Playfighting is more complicated than first meets the eye. It provides far more than an outlet for pent-up energy in the schoolyard. Different species combine different elements of behavior into their good-natured tussles [2668]. Horseplay blends predatory, aggressive, and dominance behaviors. The blend of skills and patterns changes across the life cycle [2669, 2670]. These data suggest roughhousing does more than help the young hone these skills for adulthood [2671]. Instead, rough-and-tumble play develops normal competent pathways in the social areas of our brains.

In hamsters, these tussles increase the density of fibers that release corticotropin-releasing hormone (CRH) in the lateral septum (LH), one of the brain's social switchboards and emotional tuners. The LH is another region involved in social behaviors, mood, and motivation. This change in the LH facilitates playfighting, letting the hamster access energy reserves and reinforcing play with this chemical reward [2672].

Corticotropin and serotonin rise during acute stress. So, on the surface, one might expect that this increase in CRH would react like an alarm signal because it releases the corticosteroids involved in stress production. Enjoyable playfighting, like other positive social experiences, increases the levels of these hormones and increases the number of CHR receptors in juvenile animals [2672]. Think of this exposure as a kind of controlled stress rehearsal, preparing the body for real threats [2673, 2674]. It increases one type of corticotropin receptor (CRHR2), making it harder to trigger future threat responses [2675, 2676].[305] Through epigenetic modifications, well-regulated CRH makes the brain more resilient to tension, reprogramming the

---

[305] A receptor is a cellular attachment for a hormone, and there can be different kinds of docking ports for the same hormone in different parts of the brain. In contrast to CRHR2 style receptors, higher densities of CRHR1 are associated with higher levels of stress-related disorders, such as PTSD [3498, 3269].

brain to provide a broader range of behavioral options than fight or flight [2677]. By allowing a social animal (including humans) to adapt to low-level stress in a supportive situation, play fighting creates a positive-feedback loop—a situation where success breeds confidence, which breeds more success.

Another answer to the conundrum lies in the primary functions of corticosteroids and their interaction with dopamine and adrenaline. During rough-and-tumble play, corticosteroids' first mode of action is not in the brain but in those organs that regulate our power stores. They modulate the activity of the brain, liver, pancreas, and fat cells so we can access energy to engage in physical activity. In stress, the levels of corticosteroids remain forcibly elevated for prolonged intervals, establishing a feedback loop between the adrenal cortex and the brain. In playful aggression, the levels of corticosteroids and their feedback never rise to the level necessary to trigger the stress response.

During a bout, the striatum also floods with dopamine, a transmitter associated with reward interpretation. Individuals subjected to violence do not receive this dopamine reward with the sugar high and corticoid activation. Under the influence of this chemistry, the brain interprets high adrenaline levels, stimulated by both play and actual fighting, in entirely different ways. The emotions triggered by the rise in adrenaline depend on the perceived conditions, recognizing stress, pain, and sexual stimulation.

Again, dopamine is a critical modulator, helping the critter determine if the fight is fun (rewarding) or not. Depleted dopamine levels result in premature defense behaviors, cutting short the play session. In addition, animals may switch to behaviors like mounting or grooming due to depleted dopamine [2636].

Instead of predisposing a mammal to stress in later life, the development of the striatum, with its mix of dopamine and CRH activity, inoculates the individual against stress. The entire mechanism for this is not yet known (as of June 2025. We understand that an increased number of these CRH-producing cells dilutes future bouts of stress; it spreads stress hormones across a larger target. When combined with the CRH

> *"Some boys take a beautiful girl and hide her away from the rest o' the world. I wanna be the one to walk in the sun. Girls just want to have fun."*
> — Cyndi Lauper, 1983 [3173]

function, dopamine creates a different chemical environment in the brain. The nucleus accumbens is a part of the ventral striatum that interprets reward and success. This region increases activity when someone sees stressful images [2678]. This response bolsters the idea that the architecture of the striatum interprets the meaning of social encounters and regulates the internal and external response to those encounters.

An increase in CRH does not always signal a stress response. The release of CRH during play creates stronger striatal connections among the neocortex (interprets meaning), the basal ganglia (input-output of behavior), and the trigeminal system (control of facial movements) [2679, 2680, 2681]. CHR release also links to the sense of smell [2682], which provides critical social cues in many species. As these links strengthen, the animal has better intuition about the intent of others, thus reducing its overall stress response. It can better modulate its level of aggression and submission. These skills lead to different physiological changes that aid in mate selection and successful reproduction (i.e., biological fitness). Western prairie voles, for example, undergo a hormonal switch that reduces playfighting and promotes monogamous attachments at their first sexual encounter [2683, 2684]. These voles live in dangerous open areas. Long-term commitments with a single spouse lets them live and reproduce longer than if they had to travel between sexual partners.

Cooperative interactions, such as RTP, aid in brain maturation, foster the ability to form social ties, and develop empathy [2685, 2686]. A finely tuned system of hormones and neurotransmitters tweaks the involved brain regions, including the amygdala. Striatal activity is crucial in these functions, signaling when rough-and-tumble actions are fun and when they become a threat. Along with the connected posterior cingulate cortex (the PCC), it prompts us to change strategies to meet situational needs, joining forces with the behavioral output centers. By directing our responses and cooperative interactions, these regions promote the formation of social ties and make social evaluations. These decisions impact our status in life by affecting our fiscal choices, perceptions about fair treatment, and prospects for inclusion [2687].

Who Plays Rough and How?

> *"Rock-a-by baby, on the treetop, when the wind blows, the cradle will rock. When the bough breaks, the cradle will fall, and down will come baby, cradle and all."*
>
> — Traditional nursery rhyme, 17th century, origin uncertain

Social vertebrates engage in roughhousing, arguably the most common form of juvenile play in social mammals [2688, 2689, 2690]. Aggressive players include mice, rats, birds, dogs, cats, pigs, goats, and primates.[306] The elements vary, but all parties have a good time in playfights. Indeed, some species crave it. This mutual enjoyment distinguishes fake attacks from bullying. Through them, young animals learn social limits, how to signal submission, and how to protect themselves in a serious fight [2691, 2654]. Where real attacks and discrimination dysregulate the HPA axis and leave victims open for health problems and depression [1446, 1447], mock battles have the opposite effect. They reduce the rate of real aggression both in childhood [2692] and adulthood [2693, 2612, 2694]. These early-life experiences enhance adult sociality and temperament as well as skills such as manual dexterity, behavioral flexibility, innovation, and territorial defense [1364, 1365, 2695, 2696]. When the behavioral effects spread across a social species, it can reduce conflict and enhance cooperation across a community [2609].

In a social species, an inadvertent attack has far-reaching social consequences, and animals have evolved elegant ways to signal their intent. Misreading a play gesture as a threat could harm a family member or

[306] There is conflicting interpretations of jumping play in poison dart frogs. Hurme [3384] suggests it might be territorial defense or courtship play, but this view is not widely accepted. The presence of RTP in both birds and mammals could have arisen from convergent evolution or trace back to our distant common ancestors.

another valuable social ally. Conversely, if one mistakes an actual attack for play, the results could prove deadly.

Communication takes many forms, including facial movements, behavioral sequences, body posture, and vocal cues [2697, p. 266], and context matters. Coyotes and other canids, for example, respond differently to attack signals if a play-solicitation precedes them than if they stand alone. Canids also exaggerate their happy, cooperative movements during play to assure their playmates that they just want to have fun [2698, 2699]. They reinforce their intentions through self-handicapping and switching roles so a playmate has a chance to chase or attack. These actions signal companions and allow the friendly frolickers to hone their defensive skills [2700]. Humans extend this signaling to good-natured verbal tussles, using cues, such as laughter and smiling to reassure us that the teasing remains tongue-in-cheek. Changes in tone and body language to communicate when the play has crossed an acceptable line [2701]. Our canine companions also signal non-contact with ritualized play-biting with a relaxed, open-mouth homologous to our open-mouth smile [2702].

As in much biology and psychology, the study of playfighting has focused on the rat. Like human children, young rats love RTP. A casual observer can distinguish between actual rat aggression and a fair wrestling match. Like most other animals, rats take turns when they tussle. They target different parts of the body and send social signals indicating they harbor no intention of harming their partner.[307]

Even within species, not every creature enjoys the same games, but there are trends. Roughhousing in rats and dogs tends toward cooperation. In contrast, cattle compete during RTP, episodically interrupting the play to reassure their friends of their intent [2703, 2704]. House cats and lions enjoy play that simulates predatory behaviors, but different individuals prefer different games [2705, 2706]. Genetics influences

---

[307] Differences in attack points, the sequencing of attacks, and greater variation in play attacks is common across a range of species, including dogs, humans, other primates, and pigs [3363].

some recreation components; life experiences shape others. Patterns among partners build upon genetic predispositions and affect different strains of rats differently across their lives [2707].

The nature of rough-and-tumble changes across the lifespan in mammals [2708]. In rats, the probability of an actual attack increases as the brain matures, so the time spent in aggressive play decreases as they approach puberty. Attacks become rougher, the defense techniques become more retaliatory, and they engage in less playful nuzzling of the nape of the neck than when they were young [2661]. They become more inclined to attack the haunches and legs of an opponent—sites used in actual attacks. These changes happen even in animals reared in isolation.

Still, the cerebral cortex needs roughhousing to mature correctly. Rats with damage to this area do not make age-appropriate adjustments to their rough-and-tumble play behaviors [2709]. If an adult rat has had part of its frontal cortex removed early in life, it will continue to engage in and exaggerate juvenile roughhousing. These results suggest an animal needs a properly functioning cortex to switch from youthful play to functional adult behavior [2710].

In social species, patterns of interactions, including play behavior, create social bonds and networks of friendships that can contribute to support, status, and power within the group [2711, 2712, 2713]. Social species, including humans, favor specific playmates and prefer different mates for particular activities, including play [2714]. Belding's ground squirrels (*Spermophilus beldingi*), for example, develop preferred playmates, i.e., friendships, and spend a disproportional amount of time with that mate [2715].

In contrast, golden hamsters do not engage in much social behavior beyond childhood, and male and non-ovulating female adults show aggressive behaviors [2716, 2613]. These non-social rodents don't navigate the same complex web of peer behaviors as rats and humans, so golden hamsters have evolved a different pattern of play. The young do not prefer one littermate over another. Their playtime remains consistent despite the level of maternal care a pup receives [2717]. The number of siblings influences the distribution of play because more siblings provide more partners; it does not significantly alter the frequency or duration of RTP [2718]. Some differences exist in between-gender interactions with mixed-gender tussles, including more contact time and less pinning than playfights within genders [2719]. However, the presence of different sexes or the number of siblings has little influence on hamster tussles.

Despite these differences, rats and hamsters share two significant commonalities. As they attack and pin each other during RTP, they take turns [2719]. First, one partner attacks, then he or she gives up the advantage and lets his or her mate attack in return. Indeed, they share this commonality with many mammals and birds. Humans who share their lives with two or more dogs can observe this give-and-take in their canine companions, and we admonish our children to "play nice" and "take turns."

In the social animals studied to date, playfighting is fair fighting. The aggressor attacks and then retreats without mounting a defense. In an actual fight, the aggressor will press the advantage once he or she has the opponent pinned. In a playfight, the aggressor invites the "victim" to reciprocate the attack to continue the play. Many species have a roughly 50:50 (tit-for-tat) rule. Although this rule is not universal, the fighting remains fair. Dogs, for example, do not evenly distribute the chances for attacks and pursuits. Still, the dominant players allow their friends an opportunity to play the aggressor role under certain conditions. The subordinate dog can take his or her turn at chasing and tackling but not at robust indicators of dominance, such as mounts, muzzle bites, or muzzle licks [2720].

The structure of a playful attack also differs from that of a real one across species [2668]. Animals bite softly with no intention of harming their mate. They may nuzzle or lick the attack zone rather than play-bite it. They aim their strikes at different parts of the body than in an actual fight. When a rat pounces on its play partner, it does not attack its haunches as it would during aggression. Instead, it pushes its nose into the back of the neck and nuzzles it in the same spot a mother rat uses to carry it back to the nest. When the "attacker" has finished nuzzling, it does not press its advantage. Instead, it turns away and expects the opponent to reciprocate.

The defending rat will block the attack by jumping away and landing belly-side up. In a violent confrontation, this position would render it vulnerable. In a playfight, the belly-up posture prevents its mate from gaining access to the nape of the neck. This defense of the nape is an integral part of the play, and rats quickly tire of a partner who does not "fight back" with the proper play defense [2661].

A third aspect of play aggression is its ambiguity [2700, 2721]. Games aren't fun if you can predict every move. Although there are rules and elaborate signaling to distinguish between play and fighting [2668], sometimes things get out of control. Even in a game, someone can get hurt. The target must signal to the attacker that he or she has gone too far. The attacker must reassure the recipient that no matter what has happened, the intent remains playful and without malice. All parties must show their continued willingness and desire to play.

Humans share these norms with other mammals, e.g., dogs, wolves, and coyotes [2722]; sea otters [2723]; and rats [2612]. We detest social partners who always play to win or who do not take turns. Playfights, from the playground to the bedroom, cease to be fun when one partner presses an advantage over another. We also do not find aggressive play fun if the other partner does not play back. The share-and-share-alike pattern of the RTP extends to other social activities, including social decision-making, such as taking turns to decide what to watch on television or do on a date. No one wishes to work with a domineering office mate, and bosses who never yield an advantage to their employees have trouble forming coalitions when they need them.

As a social species, we learn the signals that differentiate playfights from actual ones. In Egyptian baboons, *Papio anubis*, the time the players spend in physical contact with each other is higher during aggressive play behavior than in real aggression. These primates also have fewer prolonged chases during play than during violent encounters. As with our play, getting caught is part of the fun. Like rats and humans, these baboons attack different parts of the body during play behavior than during actual violent attacks [2724]. Most of the time, baboons, humans, and other social species can distinguish between the patterns and sequences of play and hostile behavior because social species possess an inborn user's manual.

How to Roughhouse: A User's Manual for Humans and Rodents

> *"Everybody was kung-fu fighting. Those kicks were fast as lightning. In fact it was a little bit frightening. But they did it with expert timing."*
>
> — "Kung-fu Fighting" by Carl Douglas [2725]

Children playfight. Adults playfight. With the possible exception of some neurodivergent individuals, both children and adults need rough-and-tumble play in their lives. Children know the difference. They know when someone bullies them, and most know when a victim is not enjoying an interaction. As a society, we must re-learn these differences between bullying and mutual, consensual play. To do so, we must be willing to recognize the features of play and distinguish them from the telltale signs of aggression. Play should be a right of both adults and children. To grant this right, we should accept that when two young boys smile and laugh as they pretend to be Kung-fu fighters, they intend to have fun, not kill each other.

RTP borrows features from sex and aggression, but they serve different purposes [2726, 2654]. RTP teaches us to recognize interpersonal signals and expands our social options. Through it, we learn to distinguish pretend and friendly banter from violent or humiliating aggression. Rough and tumble play during childhood aids in the social and emotional maturation of the brain [2727, 2726]. Both genders crave this play. Both genders *need* this play, but boys need more of it than girls. Patterns in other species show that this gender difference traces back to ancestral gender roles that we inherited from the early species of humans.

RTP takes many forms. Sometimes, it involves attacks. It can include stalking and pouncing. It can involve pinning. When physical attacks occur, they do not intend to injure. In addition to fake attacks, playfights can include nuzzling, tickling, or grooming [2728]. Vocal signals, facial signals, and postures, like upward wrists in primates, smiling faces in humans and dolphins, and play bows in dogs, reassure the participants of their mutual intent to have fun and commitment to the game [2698, 2722, 2729]. Games end when one party indicates it has had enough, and the other recognizes the signal. An animal that breaks the rules, or does not respect the wishes of others to stop, soon becomes unable to find willing playmates.

Roughhousing takes many forms, but sharing roles is vital to this play. One participant may chase and catch another, but then it will relinquish its advantage and become the quarry for its partner. Sharing roles is one of the most evident signs that play, not bullying, is occurring.

During playfights, animals expose vulnerable areas to their mates. No adult rat will flip on its back and expose its belly in a territorial or mating dispute. They readily roll over during play. It signals trust and their interpretation of the tussle as whimsical. They take this chance for two reasons. First, a fair partner has no interest in harming them. Second, by flipping, they also protect the targets of the play attack. Making the game a challenge provides much of the fun.

Teams

> *"Ther dukys, and erlys, lorde and barone, Were take and slayne, and that wel sone, And som were ledde in to Lundone With joye, and merthe, and grete renone; Deo gratias" (Their dukes and earls, lords and barons were quickly captured and slain. Some were brought to London with joy and mirth to garner great renown [for our king]. Thank God!)*
>
> — The Agincourt Song, 15th century, [2730]

Our species loves playfighting, not only as individuals but as organized teams. Although chimpanzees, bonobos, and humans all engage in cooperative behaviors and enjoy rough-and-tumble group play, they do not share our adoration of team play. All three species have similar patterns of reassuring behaviors and calls, but our cousins are not known to organize themselves into play teams. Yet among humans, collaborative or competitive team games occur across cultures, with men participating in more team and combat sports than women [2731, 2732]. Our superior ability to form alliances, especially the playful leks of mating competition and power alliances [2733], might transfer to recreational activities. Perhaps it takes words to form play coalitions, or the smaller sizes of ape communities reduce the probability. Maybe language or our ability to reach beyond one's home village to form alliances plays a role. To date, we do not know why we appear to stand alone, not just among our triad but among most species. Most likely, we are blind to similar but less developed team behaviors in others. [308]

Human males have a predilection for contest competition and within-gender aggression [2734, 2735, 2736]. Although overlapping with women's abilities, men, on average, can run faster, have superior aerobic capacity, and gain muscle mass more readily than women. These features help with more than chasing down prey. They also benefit men in within-gender contests, resource competition, and between-gender attraction [2737, 2738, 2739, 2740, 955].

---

[308] Whenever scientists think they have found a feature that appears unique to humans, some animal, usually a chimp or bonobo, has proven us wrong.

> *"This is my fight song, Take back my life song, Prove I'm alright song, My power's turned on, Starting right now, I'll be strong."* — "Fight Song" by Rachel Platten [3194]

These features also aid in bride thefts, the abduction of women from other villages. Although culturally condemned by WEIRD societies, bride theft (raptio) was once familiar and accepted, e.g., Homer's *Iliad* and Ovid's *Metamorphosis* [2741]; Medieval England [2742]. It remains a way of procuring mates in several societies [2743, 2744, 2745, 2746, 2747]. Beyond sports, men engage in more direct confrontations and contests than women. Men distinguish between outgroup and ingroup members more readily and engage in more coalition violence.

Although female athletic teams have made headway in some countries, men still engage in more physical, competitive team sports than women [2731]. The socialization of women away from sports has decreased, and the acceptance of women athletes has improved [2748], but both men and women still prefer watching men compete. These cross-cultural, near-universal patterns hint that such preferences may have biological roots.

Despite its monetary significance and cross-cultural ubiquity, the evolutionary psychology behind team sports has received less attention than individual playfighting. There are parallels between team sports, organized raids, and warfare [2749]. Similarities to genuine combat include community (team) green beards, such as team jerseys, team caps, and team tattoos.

Fans sing fight-songs, much like the "Agincourt Song," written to commemorate the 15th-century battle of Agincourt. Home games ring with favorite chants, such as Manchester United fans singing "We Shall Not Be Moved" and the "Glory Glory Man United" song, the "Aggie War Hymn" of Texas A & M University, the relatively new "Go Leafs Go" of the Toronto Maple Leafs, or the less confrontational "Take Me Out to the Ball Game" sung during the 7th inning of a home game of the Chicago Cubs. Unlike the Agincourt battle, which claimed the lives of 5,000 to 11,000 men, sporting events take their toll on the players, whose injuries can be severe but typically fall short of death. Occasionally, the bruised or inflated egos of the fans result in crowds erupting in hooliganism that spreads the (typically sub-lethal) mayhem further.

Fans usually engage in good-natured provocations among sides, which reach their peak with contact sports like lacrosse and hockey [2750]. These similarities to battle are no more doomed to awaken primal, destructive urges than playfighting makes us yearn to destroy individual opponents.

Many organized sports hone skills practical in non-industrial warfare. They require strategy and the subjugation of personal vainglory to their group needs. Champion athletes garner status like victorious human warriors, CEOs, and alpha chimpanzees.

Watching competitive sports lets us evaluate the physical prowess of males, serving as a surrogate for sexual selection both within genders and between them. Sports help people choose potential allies without costly fights [2751]. In the absence of violent male-male competitions, sports serve as a surrogate, enabling men to evaluate each other and impress women [2752]. Social bonding through ritualized activities, such as golf outings among managers, allows us to assess social behavior, deference to authority, strategies, strength, and coordination. In past centuries, these traits would have contributed to survival and forging strong alliances. We still hone our ability to identify potential allies by watching violent, competitive team sports like football and hockey.

Traditional measures of physical adaptations have roots in genetics. Variants of a gene for collagen-I chain (COL1A1) can make a person susceptible to brittle bone diseases or reduce the healing time following an injury to a tendon [2753]. Polymorphisms (variations) in the gene ACTN3 contribute to muscle fibers' protein mixture and composition. It also improves the adaptation to exercise, exercise recovery, and sporting injury outcomes [2754, 2755]. Activity in CDH13 (cadherin 13) defends the heart from oxidative stress, protects it from arteriosclerosis, and may play a role in protecting against cancer [2756]; it may also impact GABA function in the hippocampus and contribute to ADHD [2757, 2758]. Men and women with a single gender assignment have different ratios of muscle fiber subtypes and activation capacity [2759, 2760]. The interactions of dozens of gene products determine the *limits* of strength and the speed of the reaction. In cultures that depend upon physical competence, they contribute to potential mate quality.

In many civilizations, families exercise dominion over the marital choices of their offspring, especially daughters. A potential suitor's physical aptitude, ability to work within a community structure, and societal status weigh in their decisions about mate choice. Sports intrigue men more than they do women, and men rearing daughters show a greater interest in watching sports than men without them. This finding supports the hypothesis that sports can serve, at least in part, as an ancient method of spousal evaluation by male kin [2761]. Recall that women usually leave their homes to live in the village or clan of their spouse. By selecting socially fit spouses, fathers reduce their daughters' jeopardy from the traditional bouts of sexual abductions ("raptio") commonplace for much of our tenure on this planet. The marriage also could solidify a powerful alliance between his community and that of the spouse [2762].

When women play team sports, they do so differently. On average, women on teams cooperate more, often sacrificing individual scoring for team success. Play patterns differ in men's and women's basketball [2763]. Women experience more ligament injuries than men [2764], possibly because they tend to use their muscles differently [2765]. Since 1983, men have outperformed women in sports at a nearly consistent rate of 5.5% (swimming) to about 18% (long jumps), primarily due to differences in musculature and physiology [2766].

Playfighting and Gender

"What are little boys made of? Snips and snails and puppy-dog tails."

— Traditional 19th-century nursery rhyme

Social species like roughhousing, but not all social species play these games the same way. Some have differences between the RTP of males and females, with more significant similarities among animals in which adults make their living doing roughly the same thing [2767]. Both genders of cats and dogs playfight the same way [2768], as do marmosets [2769] and grasshopper mice [2770]. Other species show differences in how males and females organize their roughhousing. Primates (both apes and monkeys), ungulates, pinnipeds, and most rodents show gender-specific differences in their horseplay [2771, 2768, 2772].

In juvenile rats, play is contagious [2660]. The playfulness of their friends is the primary factor that regulates the frequency of play. Furthermore, bold, confident rats initiate more play than timid ones. This self-assurance and high-play initiation provide clues about an individual's social status when it matures. For males, the ability to engage in roughhousing seems critical to the establishment of dominance, self-confidence, and social functioning. We do not have evidence of a similar correlation in female rats.

Differences among species and genders reinforce the notion that playfighting is not a single, homogenous activity. Instead, it has distinct components, each with its own regulatory mechanisms [2668]. Different species have evolved and adapted different mixes to promote their survival advantage [2773]. Time budgets[309] for rough-and-tumble play differ among species. For example, deer mice jump and nuzzle for 70% of their playtime, while rats do so for about 40%. Voles do more face-and-defend behaviors during play than rats. Like humans, other species will change their recreation across their life spans. Some also show different play tactics between genders.

---

[309] A time budget is how much time an organism allots to an activity.

In human children, RTP accounts for about 10% of all play across cultures when not restricted by well-meaning adult intervention [2774, 2775]. Like other social animals, we do not playfight for a needed resource (though a symbolic one can be present, such as in the game "king-of-the-hill"). Like other social animals, we switch roles often and do not attack to inflict injuries. We reciprocate, taking turns during the fight. We use ritualized signals to initiate play and continue to signal our friendly intent throughout the spirited tussles. In short, our species behaves like other social animals but with its own set of special moves, sequences, and combinations.[310] Problems do not arise from the normal give-and-take of roughhousing. They happen when one or more partners get it wrong.

Getting it Wrong: Crossed Signals and Crossed Purposes

> *"Speak roughly to your little boy, and beat him when he sneezes: He only does it to annoy, because he knows it teases."*
>
> — *Alice in Wonderland* [2776]

Bullying and violent behavior differ from playfighting. Bullies enjoy inflicting pain and humiliation on others. Children with conduct disorders (CD) rank among the most vicious childhood bullies. People with the early-onset form can be wired differently than others for empathy and aggression [2777, 2778, 2779]. These differences in hardware and programming make it difficult for them to distinguish between play and torment.

Much of the research on CD has focused on boys, but there is reason to believe the disorder manifests differently in males and females. Physical aggression is more common in healthy boys than in girls and is a gender-biased benchmark for conduct disorder. Both males and females with CD show callus behavior

[310] Conditions known to promote abnormal social integration include fetal alcohol syndrome, which also *correlates* with reduced and abnormal playfighting in rats [2679]. This physiological correlation across species again points to a normal physiological, developmental need for boisterous fun.

[2780]. But, where boys tend to compete for physical supremacy and express CD with physical violence, females compete for social and relational status. Girls with CD often communicate aggression by sabotaging the interpersonal relationships of their peers [2781, 2782].

When shown images of someone suffering from inflicted pain, the MRIs of children with conduct disorder show different brain activity patterns than those of typical youth. In both groups, the perception of others in pain activates the "pain matrix"[311] that encodes the sensory and emotional aspects of pain. Like mirror neurons, it lets us empathize with the suffering of others. When researchers presented images of intentionally inflicted pain to conduct-disordered and typical youth, they responded differently [2777].

Typical brains add information. More areas come online when presented with images of pain. Regions dealing with planning complex cognitive behaviors increase activity.[312] Centers for personality, self-expression, and decision-making fire up. The medial prefrontal cortex, which dictates correct social behavior, glows on the MRI. Regions for cognition and decision-making (lateral orbitofrontal cortex) demand more calories as they assess the situation. The pictures trigger activity in the temporal-parietal junction, which lets us distinguish between ourselves and others, as do the sectors that allow us to regulate our emotions [2678].

In other words, when evaluating pain, the average person engages brain regions that plan and decide right or wrong. We use our pain matrix and mirror neurons to superimpose our personality, self-image, and

---

[311] These brain regions include the anterior cingulate cortex (motivation, conflict monitoring, emotional modulation, and anticipation region); the insula (a cortical area involved in self-awareness, purpose, negative emotional experiences, empathy, and compassion); the somatosensory cortex (perception of touch and personal pain); the supplementary motor area (planning movements and learning sequences of actions); and periaqueductal (central) gray region (modulation of pain and defensive behavior). In short, they all recognized pain and prepared to deal with it [2777].

[312] The increased planning activity of the prefrontal cortex may help us decide our course of action following a violent or dehumanizing episode. It helps us determine if we should flee, render aid, or ignore the event.

experiences on the victims in the images. We feel empathy and concern. We understand the pain of the other person.

Tellingly, people with conduct disorder show a range of responses and impaired function. On average, their MRIs show *fewer* rather than more active regions when exposed to intentional violence. Intentional violence suppresses circuits experienced by a healthy brain. However, the insula, which regulates self-awareness, risk-reward, and negative emotional experiences, often becomes more active. The insula also adjusts empathy and compassion. Images of intentional violence can also stimulate parts of the primary motor regions (precentral gyrus), the region that comes online when we first consider a voluntary movement.

CD youths also show less connecting (striatal) activity between the emotional amygdala and the rational frontal cortex [2777, 2783],[313] and behavior-disordered children have difficulty regulating their emotions [2784, 2785]. Both males and females have blunted stress responses [2786]. In other words, like the adult psychopath, they intellectually *know* a behavior to be wrong but do not feel it in their gut.

Studies of adults with antisocial disorder have shown an additional difference from typical non-psychopathic adults. When most adults see images of someone in pain, it activates their pain matrix and mirror neurons. The mirror neuron system (MNS) lets us put ourselves in another person's position, and the pain matrix allows us to understand pain. Through them, we comprehend when another person suffers. Normal subjects empathize when shown a picture of inflicted pain; they know the experience hurts. Psychopaths show less MNS activity [2787, 1018]. Without the activity in the mirror neurons, they cannot empathize effectively, and they mirror the behaviors of others less [1019]. People who score high on scales

---

[313] The striatum was one of the regions with connections strengthened by playfights. One cannot know from these data how or if these bits of information fit together.

measuring psychopathic traits are also less prone to the contagious yawning most people experience [2788].

A lack of compassion, by itself, does not make one a sociopath. Instead, it can contribute to sociopathy. Many people have trouble with empathy but do not lean toward harming others. Autism, for example, does not lead people to induce mayhem or inflict pain on others. Because they differ from the norm, autistics are far more likely to be victims than aggressors. [314] In contrast, children with conduct disorder and adults with antisocial disorder run a higher risk of becoming bullies.

In short, habitual bullies with antisocial or conduct disorder get it wrong. They see the pain of others, but their brains do not make the same empathic connections as a healthy brain. Psychopaths do not make the standard distinctions and extrapolations that link our experiences to the feelings of others. The psychopath does not fully comprehend nor fully care that something he or she finds enjoyable would cause harm to another person or animal. The feelings of that other do not complete the information circuit in the conduct-disordered or adult antisocial brain. Either that information never makes it to the brain region that triggers empathy, or that region does not register compassion.

Bullies are not the only ones who get it wrong. Occasionally, everyone falls under the spell of social pressure. Occasionally, everyone misses the cues to stop. Sometimes, we get so involved in the game that we stop paying attention to social signals. Sometimes, we listen to an admired figure without thinking or fall under the spell of dehumanizing propaganda that dials down our mirroring system. When these happen, even well-wired people function as the occasional bully, often without realizing they have entered the role.

---

[314] Some autisms, for example, *might* include reduced mirror neuron activity; here, the data remains inconclusive [3389, 3479]. Autistics differ from people with conduct disorder in another relevant feature. Autistics feel pain and discomfort acutely; children with conduct disorder tend to have a higher-than-normal tolerance for pain [3314, 3411]. We do not know how, or if, this difference plays a role in their abnormal empathy levels or if the same mechanisms underlie the divergent features.

Many species, including our best friends, join us in this social error. In dogs, littermates often join in playfights. When one dog has and maintains an advantage over another (i.e., switches from RTP to bullying), the other littermates differentially join on the attacker's side [2789]. For that bout of play, reciprocity ceases, and they stop switching sides. Instead, they adopt a power imbalance rather than play. These action patterns foreshadow the essential cooperative behaviors to come. For an adult canid, the ability to identify and follow the leader can mean survival.

Humans also join groups that cheer for a winning team and pounce on the losers. O'Connell, Pepler & Craig [109] found that human children behave much like puppies. Children reinforce the schoolyard bully 54% of the time by passively watching. In 21% of cases, children follow the bully's lead and mimic his or her behavior. Only one-quarter of the time did children actively try to stop the bullying. Older boys (grades 4-6) were more prone to joining the bullying actively than younger ones, and girls intervened for the victim more than boys [109].[315]

Like dogs, we are social animals, and our children follow the lead of a bully for the same reasons. We share the innate tendencies to follow leaders and social norms. For small children, these skills kept them alive in traditional communities. They have less practice at environmental interpretation than adults, and their brains do not develop adult reasoning capacities until well past adolescence.

In our formative Pleistocene era, we would have lived in extended family or kinship groups. Many, if not all, adult males in a band had a vested genetic interest in the children. A troop of proto-humans would have cooperated to keep their children safe. In turn, children who trusted and accepted adult authority had an increased chance of survival. These survivors became our behavioral progenitors, who passed on their tendency to obey adults. Since the Pleistocene, most humans have dramatically changed their social

[315] Scientists are so quick to swoop down upon someone's flawed hypothesis that it's often said that they are the only species who eat their own wounded alive.

environment. Few of us still live in small kinship bands of hunters and gatherers, yet the dictate remains strong. We follow the alphas despite the use of "no" by two-year-olds and workarounds by older offspring.

In a world where our most dangerous mammalian predator is our own kind, we depend on our complex social groups. Today, most of us interact more with non-family members than with kin. The children's trust in adults can backfire. Although we understand we must shelter our children from the dangers of modernity, we do not extend our protection to all children, even though rationally, we know all small children are vulnerable.

For some, the primitive brain does not meet the tasks set for us by modern ethics any more than it always bends to cultural taboos on physical, sexual, and emotional abuse. Our ancient system of checks and balances can run amok. Two patterns of sexual harm remain consistent across cultures. Females suffer more sexual abuse than males, and men commit more sexual abuse than women. Different surveys vary in their estimates of the different kinds of child maltreatment and their frequencies, e.g., [2790, 1514, 2791, 2792, 2793].

The tendency of young children to obey and follow the lead of adults and older children plays into this problem. Their pattern makes them easy targets. As with the bonobos, the probability of maltreatment varies with the perception of kinship. On average, abuse is more common and severe when perpetrated by those who do not perceive themselves as the child's relative.

Parents whose own children compete with another child tend to be more vindictive and vicious. Although it seems a stereotype, the old fairy stories of wicked stepparents contain a kernel of truth. Per the predictions of the parental investment theory, stepparents abuse offspring more often than natal parents or adoptive parents [2794]. To this end, our literature and traditions may provide warnings and prevention.

CHAPTER 12:

Justice, the Villains of Fiction, and Our Escapes from Terror

> *"If morality represents how people would like the world to work, then economics shows how it actually works."*
>
> — *Freakonomics*, Stephen J. Dubner and Steven Levitt, [2795]

Is the World Just?

> *"John calls you Light; the Books of Kings calls you Lord; Exodus calls you Providence; Leviticus, Sanctity; Esdras, Justice; the creation calls you God; man calls you Father; but Solomon calls you Compassion, and that is the most beautiful of all your names."*
>
> — Victor Hugo, *Les Misérables*, volume 1: Fantine [2796]

In our daily lives, we struggle to instill positive social values in our children and to keep our values paramount in our own lives. We teach that honest labor and hard work will reap rewards. We believe our children stand little chance of success without such values. [316]

[316] We assume that the "just world" hypothesis, a cognitively complex idea, is limited to humans. However, other social species display behavior that hints at expectations of fairness from their social mates. Primates may reject what they see as an unjust partitioning of rewards for tasks [1869]. Dogs will cease cooperating with packmates if they perceive inequity in their treatment [3227].

Figure 12-1. Bernard Picart (1775). Susanna's Accusers Stoned. The apocryphal tale follows the beloved literary pattern of an innocent triumphing over evil. The elders who destroyed Susanna's reputation and life with their false accusations forfeit their lives through the same punishment Susanna would have suffered had she been found guilty. The story illustrates our desire for and our inherent belief in a just and sane world. [3196]

To cope with the world, we cling to these core beliefs of justice and fairness despite frequent evidence of a contrary and unjust world. We create personal narratives that inject cause-and-effect patterns into our lives and the surrounding world, even without a genuine causal relationship [2797]. However, as we make more upward comparisons, as we compare our fates to those society views as our social superiors, we become less confident in the justice of fortune [2798].

To maintain our belief in a fair world, we minimize the dissonance between what we see and how we want the world to be. We draw upon coping strategies. To bolster our views, we evoke the social norm of belief in divine justice and old chestnuts such as "What goes around, comes around" and "Live by the sword, die by the sword" [2799]. We seek mechanisms to reconcile three core and often conflicting experiences: our morality of pro-social behavior, our expectation of justice and its meaning, with the clear evidence that injustice often prevails.

We use cognitions to reduce this dissonance. Religion assures us that, in the end, goodness and perseverance will be rewarded in either this world or the next. People who believe in a deity that metes out retribution tend to have less compassion for the suffering, not more [2800]. A belief in divine justice contributes to a mindset that *we* are good, but *those* who suffer are evil and deserve their fates [2801, 2802]. Our separation of the world into an I-thou dichotomy allows us to see ourselves as innocent victims while we hold others responsible for their misfortunes, e.g., rape: [2803, 2804]. Traumas such as rape, illness, and neighborhood gang violence are held against the victims, soiling them and sometimes blaming them for the social and economic costs of their victimization [2805, 2806, 2807].

Although the just world assumption holds in both collectivist and individualist cultures [2808], cultural and historical context modulate how it influences our thinking and willingness to denigrate the scapegoat [2809]. Prejudice, low compassion, and devaluation of victims increase our conviction that they got their just deserts. They presume more of the claims are false than those who empathize with the accusers.

In contrast, people who strongly identify with victims presume a lower rate of false accusations. Supporters of the #MeToo Movement are less likely to see a survivor as overreacting or exaggerating their trauma [2810]. Those with less empathy for the accusers are more likely to disbelieve them, blame them more, or focus on how they should have felt more empowered, chiding them for not taking control or for wallowing in their victimization [2811, 2812].

Personal experiences of victimization shake our just-world foundation [2813, 2814], as can awareness of inequity and deprivation [2815, 2816, 2817, 2818]. External religiosity and an authoritarian personality, in contrast, reinforce victim-blaming [2819, 2820].

When we can alleviate someone's suffering, we blame them less, and the probability of our aid increases [2821, 2822] while increasing our perceived pool of positive karma [2823]. However, when successful intervention seems impossible, we protect our sense of self by distancing ourselves. The less confident we feel in our ability to reverse their pain, the more likely we are to blame and dehumanize the victims. This moral disengagement distances people from suffering and eases their path to victim-blaming [2824].

To the privileged, the disempowerment and subjugation of the oppressed can itself seem fair, and they tend to express stronger just-world beliefs than the oppressed [2815, 2816, 2817]. People who hold a rigid fair-world belief are more likely to render aid if they see the suffering as temporary and they have the power to effect a change [2825, 2826]. If the voice of a trusted authority says we should empathize with and support a victim, we tend to mute our criticisms of their character and lend a loving hand [2827]. If that same voice tells us to denounce them, we do.

> *"I'm collecting money for an AIDS charity. Did you know there are lots of innocent children out there who do not deserve it?"* — Paraphrased from a telescammer who called me in 1990, just before I started screaming.

If the victim looks attractive or has a high status, we empathize more than if he or she is of low birth. We aid the attractive, high-status person more and blame him less for his fate [2826, 2828]. People judge good-looking children and adults more favorably than less appealing ones, attributing more positive qualities to them [2829, 2830]. These biases extend beyond our evaluation of individuals. Understanding this bias, political campaigns and major corporations spend millions of dollars on beautiful, glossy presentations to create images of corporate beauty and morality.

In rape cases, women do less victim-blaming than men, and people with felony convictions tend to be more compassionate jurors than those who have had no contact with the criminal justice system [2831]. When we see ourselves in fellow or potential victims, we blame others less [2832, 2833, 2831]. However, we also spend more time avoiding contact with them [2833]. By evading interaction, we have fewer opportunities to see our reflection in their eyes. Avoidance reduces the psychological damage to our sense of security and self-worth.

As our experiences of justice and injustice broaden, as we become aware of the inequity and injustice around us, it becomes more difficult to indulge in the comfort of the socially normative just-world belief [2834, 2813]. Direct experience with injustice decreases our faith in fairness [2820]. We seek new evidence to bolster our beliefs and support our worldview. We become proficient in coping strategies, those psychological safeguards that maintain our convictions.

When confronted with a threat, be it physical or psychological, our physiology provides us with three options: fight, flight, or freeze. Unlike deer, who often freeze to avoid predation, our species responds to most immediate, apparent threats to life with fight or flight. In contrast, we often choose the third option to cope with stealthy, insidious dangers, such as the COVID-19 pandemic or economic collapses. About half of Americans favor two freeze-related strategies—denial and repression—as their psychological coping styles [2835]. These related forms of emotional defense are virtual twins. Denial means refusing to accept the world as it is; repression involves ignoring or forgetting the danger together with the feelings and psychological discomfort that surround it. They help people strive when the world seems doomed. Many cope with death, rape, cancer, a pandemic such as COVID-19, or social upheavals by ignoring it or denying its severity or impact on them.

Despite the bad reputation these receive in popular psychology, functional, psychological defenses, including appropriate denial and repression, form cornerstones upon which we build our mental health [2836]. As with other coping styles, the freeze options can drift from a survival strategy to maladaptive behavior. People who habitually cope through repression deny having troubles. Consequently, they place greater responsibility for victimization onto the sufferer than those with diverse, non-repressive ways of making sense of life do [2837].

In the long term, freeze options reject uncomfortable realities, creating unresolved tension if they fail to manage the threat. This tension creates a psychological climate ripe for the emergence of scapegoating, a maladaptive fight-response in which we project our negative emotions and frustrations onto others. Blaming others as the external cause for our personal or collective woes lets us create a physical enemy we can attack, a situation more concrete than a nebulous social, economic, or political downturn over which we lack control [2838, 2839, 2840].

Scapegoating an identifiable population, such as Jews, transgender people, or refugees, provides a release valve that lets others shift their focus from a failing repressive style to an object of blame and contempt

[2841, 2842, 2843]. As our world and lives become less stable, this pattern, this flipping of the psychological switch between denial and scapegoating, can be weaponized by organizations, political parties, and governments to manipulate us into collective stigmatizing, bullying, and even genocide [2844, 2845, 2846]. These stigmatizations can co-exist with strategic false (manipulative) kindness, such as humanitarian policies designed to incite resentment as our environment ceases to jive with our distorted perceptions [2847].

When our worldview and reality become dissonant, we seek ways to restore order through our practiced defenses. Often, we bolster our belief in a just deity or an authoritarian political savior who will mete out vengeance and assure final fairness [2801, 2848, pp. 247-266, 2849, 2850, 2851, 2852]. Our preference for authoritarian figures and spiritual certainty in times of need mirrors the tendency of chimpanzees to opt for more aggressive troop leaders when resources become scarce or when the family unit becomes endangered by other troops [372, 1113]., and for ones that focus more on social integration and cooperation during times of plenty [246, 1113].

Figure 12-2. The Antisaccharrites by J. Gillray, (1873 reprint of 1792 print). The artist ridicules the upper classes who forewent sugar, ostensibly to protest slavery. Gillray suspected their true motive was the high cost of sugar, not justice. The Antisaccharrite boycott did not extend to other luxury goods produced by enslaved labor. Here, he mocks the hypocrisy of King George, who exploits a moral stance for personal gain. The princesses are not amused. [3113]

Although their forms of justice vary, many religions foster a belief in a supernatural force that ensures an ultimately just world [2853, 2854]. Not all religious people experience the same force of faith in their personal defense systems. The impact of religion on our attitudes melds with the role it plays in our lives. People whose principal focus is on the externals and status provided by religion (religiosity) express more faith in divine justice. They use more just-world strategies, including demonizing perpetrators and victims, and demanding conformity [2855]. Those who cling to this firm belief in the righteousness of life tend to have more conservative attitudes and are less likely to act altruistically or render aid [2856, 2857]. People with an intrinsic religious experience typically believe in benevolent justice with more kindness and less brimstone. They look for positive results in unjust situations. They often favor victim restitution and place less blame on the perpetrators [2855]. Religion thus becomes a force that can tap into our motivation and self-awareness centers (e.g., the nucleus accumbens and the insula and its associated temporal lobe) and push us either towards tolerance and kindness or towards violence and the degradation of others.

The Biblical and Quranic story of Job/Ayub (אִיּוֹב)/(أيوب) illustrates an intrinsic belief in postponed justice. Job suffers for no apparent reason other than a contest between God and Satan (literally "God's adversary"). The parable tells of a holy man, Job, who did nothing to deserve his agony. People scorned him, believing he must have committed a heinous sin to receive such punishment. The story eventually bolsters a sense that the karmic balance of payments will ultimately be fair. It ends with the restoration of Job's health and a doubling of his wealth. With its happy ending and comforting resolution, the parable fails to explain why bad things happen to good people. Instead, it assures us that, in the end, pious, socially accepted behaviors will receive their recompense in our lives, through benefits to our children, or in the afterlife.

The successful tend to express faith that their intrinsic value, rather than luck, has determined their blessings. The more someone ascribes to this view, the more likely he is to blame the needy or the traumatized rather than understand them as victims of circumstance. The belief that fortune or work determines fiscal and social success links to how we want society to use its limited resources. Our faith in a just or unjust world affects how we see social programs (safety nets) and what we are willing to spend on these programs as a percentage of our gross domestic product (GDP). Belief in a functional meritocracy and its links to economic advancement impact our reactions to the disadvantaged and our levels of compassion [2858, 2859, 2860, 2861]. Americans and Canadians share a strong belief that work, rather than luck, decides one's status, and they spend less than 10% of their GDPs on social programs. In Denmark and the Netherlands, more people believe luck plays a critical role, and they spend more than 15% of their GDP on support programs [2862, 2863].

Our belief that bad things only happen to bad people is so entrenched that we fall prey to it even when we know that the misfortune occurred by chance alone. During the Vietnam War era in the United States, a random drawing of birth dates (the draft lottery) determined the likelihood of being drafted. Men whose draft numbers gave them a low probability of conscription showed greater self-esteem than individuals with a higher chance of being drafted [2864]. Furthermore, the men with a strong faith in a just world tended to like others better if they had a decreased chance of getting drafted. At a gut level, those with little chance of being drafted felt everyone deserved their fates, be it safety or submission to the draft and war.

We also form biased memories of justice as we write the narratives of our lives. We linger over instances that illustrate just outcomes. When we live our lives perched on the edge of sudden and unpredictable changes, these reassuring patterns instill faith in our social order. One set of forces, our belief in justice, pushes us upward, driving us with a hope that we will benefit from our labor and receive fair compensation for our suffering. An opposing force, the evidence of social and economic injustice around us, pushes downward on this same curve and tells us that the cheater and the bully often win. Our perception of justice can abruptly change when our world tilts too far toward one of these axes.

Slime-Balls and Bullies in Literature

Literature can increase our experience of fairness. Like religion, a good story can reinforce our sense of justice. We begin with the morality fables we learn as children and continue to devour tales throughout our lives [2865, 2866]. Stories teach us about social norms and that people are not always what they appear to be [2867]. They serve as agents that condition us to social norms and that blind obedience is not always synonymous with moral behavior [2868].

> *"A sad autumnal gloom enclasps my heart, and drives all sleep away! In person I approach the autumn screen to snuff the weeping wick."*
> — Cao Xueqin, *The Dream of the Red Chamber* (p. 366) [2869]

Through our imaginations, we share and explore philosophies, love, and power e.g., *The Dream of the Red Chamber* [2869]. We consume fiction that makes us feel good (happy endings or Cinderella stories), feeds our superior self-image (indignant observers of injustice), and shows we can overcome extreme hardships (misery memoirs). We read the sci-fi and fantasy genre, which promises to transport us to a better world or give insight into survival [2870, 2871]. The structure of television plots habitually has the good guys conquering the bad. Villains may torture the innocent, but at least in fiction, justice or salvation prevails, often with a satisfying dose of spite and vindication.

Although fiction can reinforce unrealistic narratives such as the just-world belief [2872], the range of fiction lets people explore new ways and possible life outcomes, leading to a dynamic, multidimensional view of the world [2873]. As complexity adds stability to an ecosystem, the multifaceted worldview acquired through diverse literature adds resilience to the reader. People who consume more of these narratives experience an increased expectation of fairness [2874], but they also have the tools and empowerment to believe their thoughts and actions matter. The more intricate and ambiguous the situations and characters, the more significant the cumulative impact of literature on the reader. Life-long consumers of literary fiction with rich plots and character development see the people around them with greater depth, express more empathy, take more perspectives, and become more critical of society's failings than habitual non-readers or readers of pulp genres. [2875, 2876]. They become more empowered generalists.

Storytelling is in our blood. Stories, be they fiction or accepted as true, can confuse outsiders and serve as green beards for tribal recognition. We tell origin narratives about our tribes and religious stories about our Gods. We recount the deeds of our ancestors and the time Aunt Daisy used her hat to empty the free candy bowl at the restaurant. We share family histories and broad cultural narratives. Some transcend cultural barriers, but most call upon the cultural experiences of the reader and lose meaning if translated.[317]

> *"He hath disgraced me, and hindered me half a million; laughed at my losses, mocked at my gains, scorned my nation, thwarted my bargains, cooled my friends, heated mine enemies; and what's his reason? I am a Jew."* — Shakespeare, *The Merchant of Venice*, Act III, Scene 1 [1619]

Although the form and plots differ, reflecting social realities, every culture participates in the ritual. Whether recounted by an elder around a campfire, or woven into the play of children, or watched by families on high-def television, we crave stories. This human adaptation transmits our legacy, letting us explore behaviors that have benefited our band and reject scenarios that have led to prior tragedies. They teach group norms, serve as community green beards, and facilitate cooperation [2877, 2878]. They give insights into others by expanding our experiences beyond our possible boundaries [2879], and they can buttress our prejudices and perceptions of others. When filtered through our cultural biases, the portrayal of Shylock in Shakespeare's *The Merchant of Venice* can reinforce antisemitism. But Shylock's soliloquy about his victimization and the humanity of Jews provides a nuanced view, and in many interpretations, it cautions against bullying people beyond their endurance.

Our enjoyment of literature links with our desire for fair outcomes. In our hearts, we trust that we live our lives as one of these narratives. They rescue us from a chaotic world, plunging us into one with

---

[317] Because I expect this book to have a primarily Anglo North American audience, I will pull most of my examples from English, Canadian, and American literature. I will refer to non-WEIRD literature sparingly.

consequences from good and bad behavior [2880, 2881, p. 95]. Our favored genres reflect worlds that dispense justice before the end of the story [2882]. We like characters who fit our sense of morality. Our heroes and victims must receive final vindication, even if it happens after death. Still, complex villains with few redeeming features enthrall us, and likable rogues must prove themselves unworthy if they are to remain scoundrels.

Few condone murder, but we accept extreme punishment in fictional media [2882, 2883]. Our antiheroes must present enough humanity to overcome the terror they instill. The multidimensional Tom Ripley [2884] and Joe Goldberg [2885] struggle to reconcile self-preservation and what they have done but do manage to justify their murders. Hannibal Lecter [2886] functions as a public service serial killer, as does Dexter Morgan [2887], both slaying monsters for whom we shed few tears.

We pass moral judgments on fictional personas. Indeed, our investments in fictional characters can exceed our expectations of the people surrounding us. We crave principals who fill the roles of good guys and villains, not morally lukewarm ones. We want justice to prevail [2888, 2871, 2889, 2890]. Through our robust fair-world link, we often perceive innocent victims, even fictional ones, as deserving of punishment.

When simple blame fails, we seek a just and acceptable meaning for their plight [2834]. Successful writers overcome this disquieting tendency to blame the victim by making them unambiguously innocent or their tormentors base and heartless. Charles Dickens used this technique to create sympathetic, too-good-to-be-real children like Oliver Twist, Little (Amy) Dorrit, and Esther Summerson of *Bleak House* [2891]. Although I personally want to slap some sense into *Madam Butterfly* [2892] and *Miss Saigon* [2242] (honestly, he's not coming back), in most people, the images invoke sympathy for the long-suffering and loyal woman who finds herself misused by a cad.

Figure 12-3. In Victor Hugo's *Les Miserables* [3125], the image of the long-suffering Cosette pulls at our heartstrings. Abused by the Thénardiers, she struggles to survive. Her character stands as a beacon of hope and victory over oppression. Émile Bayard, Hugo's illustrator, rendered this image, "Cosette balayant", in charcoal, gouache, pastel in 1862.

In Victor Hugo's *Les Misérables* [2796], the ever-sacrificing Fantine and Jean Valjean contrast with the rigid Javair and morally bankrupt Thénardiers. While Javair's inflexibility condemns weakness, we forgive and justify the occasional moral lapses of our heroes as long as they accept the consequences, provide a contrast with the antagonist, or move the story toward a desired outcome [2893, 2894]. In *The Hunger Games*, Katniss risks her life for her sister, contrasting with President Snow's ready assassination of rivals [2895]. From the first book of the *Harry Potter* series, Harry's desire to avoid the ethical menace of Slytherin contrasts with Voldemort's ambitions [2896]. The Squid Games [2897] humanizes the morally flawed gambling addict, Seong Gi-hun, by focusing on his kindness, contrasting it with the ruthlessness of other players, and portraying the trauma of his post-game guilt.

Breaking Bad [2898] pits morally questionable decisions against each other. We relate to the motives, if not the methods, of its protagonist, Walter White. We understand why he turns to drug dealing in the plight imposed by an unjust healthcare institution that threatens his family's welfare. We continue to root for him as he endures the consequences of his choices. Following the knight slaying the dragon archetype, we accept the desperation that drives Walter's immorality because of the institutional corruption he must conquer. The mixed emotions created by the complex characters of Katniss, Harry, and Walter move us to reflect on our values, recognizing them as complex [2899].

Not surprisingly, the violent bully, the self-absorbed, demeaning boss, and the bloody dictator do not make sympathetic characters. They do, however, make excellent foils. In standard comic formulas, bullies get their comeuppance near the end of a novel or skit. Their treatment of other fictional characters creates a scaffold upon which we construct our disdain [2900]. In fiction, we enjoy exacting tolls far exceeding what we would accept in life [2882, 2883, 2901], and as our personal stress increases, so does our appreciation of over-retribution in fiction [2902]. The bully's gruesome death, dehumanization, or humiliation satisfies; it feels justified [2894]. The punishment meted out to fictional villains stimulates at least nine brain networks, including the insula, prefrontal cortex, and cingulate cortex, to provide vicarious satisfaction at their downfall [2903]. Through the magic of these media, we indulge in their defeat and punishment, exacting vicarious vengeance, indulging both our desires for retribution and our fantasies of vindictive power without actual aggression. For a brief moment, our status rises above them as we bond with the justice-dispensing alpha.

In the safety of comedy, we explore and strike against oppression without feeling its full impact. We can mock stereotypes and racism to denigrate socially inappropriate behaviors. We revel in the poetic justice of a tyrant getting comeuppance, degraded as they have harmed others. Through humor, we experience the psychological and physiological satisfaction of exacting retribution on a bully who richly deserves it.

Be it in a comedy or a sci-fi drama such as *Star Wars* [2904], *Silent Consent* [2905], or *The Hunger Games* [2895], we love to witness an empathetic character administer vindictive justice to an unjust higher-ranking person or group [2906]. This heightened satisfaction relates directly to our sense of kinship and shared victory.

We prefer detective novels in which the hero exposes, catches, and punishes the perpetrator. We feel the least satisfaction when the victim is flawless and the murderer does not suffer. For example, the eighty-two *Perry Mason* murder mysteries of Erle Stanley Gardner featured despicable killers and, often, dislikeable victims. This formula ascribes fault to the victim and perpetrator, protecting our just-world ideology [2907, 2908].

> *"He had it comin' He had it comin'. He only had himself to blame. If you'd have been there. If you'd have seen it, I betcha you would have done the same."* — The Cell Block Tango from the musical *Chicago* [3281]

In this genre, the demise of the bullies-turned-victim gives readers their first dopamine rush and a sense of pleasant vindication. Their death restores the feeling of control we lost when we identified with the initial victims [2834]. I got a colossal endorphin and dopamine rush from *Murder on the Orient Express* [2909]. Even Poirot had to admit the victim had it coming.

Fun mysteries include ambiguities that would frustrate us in real life. Several characters will have understandable motives for delivering the blow. The text exercises our theory of mind and empathic skills as we delve into the suspects' heads. The death of the flawed victim, this bit of plenary justice, promotes satisfaction. We likewise excuse the immoral actions of our fictional good guys because we accept their intent and know their universe will be fair [2910, 2888, 2871].

From Conan Doyle's Sherlock Holmes to Grafton's Kinsey Millhone series, murder mysteries involve more than justice for the victim. They often seek redemption for a falsely indicted suspect. The reader knows the accused is innocent, and should the legal system fail, the fair world of the novel will prevail.[318]

We get our second dopamine and cortisol rush at the ah-ha moment when the killer is exposed, saving the accused from death and humiliation, e.g., *The Case of the Velvet Claw*, [2911]. Many episodes of fan-favorites such as *Monk*, *Law and Order*, *Blue Bloods*, *Murdoch Mysteries*, *Father Brown*, and *The Blacklist*

---

[318] *Murder on the Orient Express* and *Curtain* [2909], two of the Hercule Poirot mysteries by Agatha Christie, exemplify a variation on this theme. In both of these stories, individuals must correct the failure of the legal system to mete out justice, creating ethical dilemmas for Poirot. This mirrors the groundswell of sympathy for Luigi Mangione, who shot the CEO of UnitedHealthCare. The brutal greed practiced by the private US health insurance system and a prosecution widely seen as vindictive sparked an outpouring of support [3317].

illustrate this point. The Kinsey Millhone mysteries (Sue Grafton) include similar calls for justice by the falsely accused [2912] or from the unavenged dead, e.g., [2913, 2914]. Robin Cook (medical mysteries, e.g., *Coma* and *Viral* [2915, 2916]), John Grisham (legal dramas, e.g., *A Time to Kill* [2917]), and Circa24 (dystopian thrillers, e.g., *Silent Consent* and *Endured by the Gods* [2905, 2918]) elicit outrage and satisfaction with these same techniques.

The killer's willingness to allow another's conviction removes our sympathy for her plight. Their betrayal of the troop overrides our delight that they slew the bully and our desire to give them a ticker tape parade. If the story portrays the victim as innocent, this subsequent behavior cements our view of the murderer. Our mirror neurons fire with compassion for those we perceive as actual victims, not self-serving slimeballs. Increased testosterone and cortisol mix to fuel the rage sparked by adrenaline as it spikes our cardiovascular output. A satisfying dopamine surge through the nucleus accumbens joins with the amygdala to celebrate when the real culprit gets his or her comeuppance. The hippocampus processes the story arc, passing it to long-term memory, reinforcing the norm of belief in a just world. That faith promotes coping skills and an overall sense of well-being as it decreases distress and fear [2919, 2808].

Before they end, crime novels present us with a clear-cut victim and perpetrator. They provide us with a safe distance from the victims and the temporary injustices in the stories. Unlike reality, in which we often allay our fears by blaming the victim [2805, 2834], the simplified fictional world of theatre and literature allows a clear sense of vindication. The works gain credibility as the principle, vested with solving the crime and ensuring virtue, overcomes internal and external difficulties that threaten to prevent the triumph of justice [2920].

The ever-present "motive" for the crime reassures us that no one will have a reason to harm us if we remain moral and upright. It bolsters our belief in fairness, countering the contrary evidence that mounts across our lives. In our minds, we can find and destroy the source of evil without inflicting physical or psychological pain. Life is more complex, and evil can triumph over good. The prosocial punishments that promote cooperation by chastising the bad guys enthrall, but they survive alongside antisocial ones that target cooperative community members. Some communities, especially those with weak traditions of group cooperation, impose penalties on cooperators to negate or minimize the benefits of teamwork and mutual support [2921, 2922]. (You know the target: the coworker who clocks in before sunrise, hoards employee-of-the-month awards, and makes the rest of us look like we're on an eight-hour coffee break.) Antisocial punishment, including revenge against a hero, adds spice to a tale, but it doesn't provide the satisfaction of watching the hero take down the bully.

These sagas can pit heroes against Herculean tasks. Our empathy with the central character assures that we, too, would willingly sacrifice ourselves for a just cause or overcome mind-boggling odds. Never mind that most who can donate blood or register as marrow donors never do so. Literary fantasies let us bolster our pro-social image and may reduce societal violence by expanding our circle of empathy. Of course, other tales can foster hate and shrink our sphere by excluding others.[319]

Through storytelling, we rehearse future actions and prepare ourselves to be less submissive in our daily lives, perhaps by exploiting neural, hormonal, and receptor mechanisms similar to those that bestow benefits through playfighting [2673, 2674]. Quite possibly, by transforming our thinking, stories prepare us to perform that selfless act that saves another, or at least allows us to assume their viewpoint.[320] The

---

[319] Stories filter through our cognitive and emotional machinery. Hence, they can feed our humanity or depravity. The Christian account of the crucifixion is meant to engage our empathy for the suffering innocent. In *A Clockwork Orange* [3074], Alex DeLarge becomes enthralled by the Bible, empathizing not with Christ but with the Roman soldiers inflicting torture on the way to the crucifixion.

[320] Historic literacy rates and book production strongly correlate with drops in *within*-societal violence, but correlation is not causation. In-group homicide rates also dropped with the concentration of power into state hands and reduced regicide. As the upper classes (alphas) reduced reliance on violence, the lower-ranking humans followed suit. With few exceptions, drops in executions and non-state homicides have followed similar downward-sloping curves. An emphasis on the outward green beards of the non-violent movement, such as court manners that emphasized self-

principal character, be it a lawyer, detective, priest, or little old lady, serves as a vehicle for justice and empowers the reader with a sense of control. Yet, sometimes, stories must break the standard formula of right versus wrong to achieve a satisfying outcome.

Murder on the Orient Express [2909] exemplifies greater moral ambiguity than the typical detective novel. It remains one of the most beloved novels by Agatha Christie because it calls to our human desire for revenge. *MOE* holds a disturbing mirror to our world, reminding us that the boundaries between victim and offender, and justice and injustice, can blur. The story compels its principal detective, Hercule Poirot, to break with his innate concept of integrity. He is driven to seek an even-handed world through an impartial formal system of laws. What happens on the train forces Poirot to re-examine his paradigm and acknowledge that the search for fairness extends beyond the dictates of the law.

Poirot learns that allowing humans to demand redress for wrongs can sometimes deliver closure and satisfaction more efficiently than the courts. Conan Doyle's character, Sherlock Holmes, faces similar ethical dilemmas in *The Blue Carbuncle* [2923] and *The Speckled Band* [2924]. Unlike Poirot, the character of Holmes has few problems with people who mete out deserved retribution. Sherlock's goal is justice and salvation, not just impartiality and punishment under the law. Both Christie's and Doyle's stories explored our tolerance for this ambiguity in literature. In life, we would define their fictional justice as lynchings.

Fantasy and science fiction have long histories as vehicles for social commentary. Both of these genres present universes removed from our daily realities. There, we can explore social issues and reassure ourselves that justice, in the end, will prevail. *The Last Colony* [2925] reflects on the balancing of moral obligations owed to others, to one's community, and to a corrupt society. From *Star Wars* [2904] to *The Lord of the Rings* [544] to *Little Fuzzy* [2926] and *Avatar* [2927], our heroes fight overwhelming odds to

---

constraint and literacy, and the use of shame to enforce the trend, also grew during the downward spiral, as did government regulations that circumscribed daily life [3101, 3100].

bring righteousness, peace, and tranquility to their worlds. The form retains a just-world theme, but the focus shifts to the personal strength of the central or heroic figures. He or she still achieves a fair and satisfying ending because of the innate worthiness of his or her character, even if it follows a salvation arc.

What holds for the mystery and sci-fi genres transfers to other fantasies. Rewarding literature explores scenarios, solutions, and consequences. In Vika Swarup's *Q & A* (= *Slumdog Millionaire*), caste bigotry threatens to deprive young Ram of his prize money [2928]. In *The Kite Runner*, Amir seeks redemption for a boyhood betrayal [2330]. Jack London's *The Call of the Wild* immerses us in the perspective of a pampered pet plunged into brutal slavery as a sled dog [2929]. Stories pump our hormones in situations, spurring our confidence and teaching us to empathize with subjugation as we rehearse the role of the compassionate alpha. Be they happy endings or introspective spiritual salvations, satisfying fiction provides us with the closure we lack in life.

Literature also prepares us for another reality: Evil does not always present itself as the genocidal maniac intent on destruction or the tough-guy schoolyard bully restraining you and drooling into your mouth. It also appears in the guise of saints intent on self-aggrandizement as they destroy others with a smile. Literature overflows with beaming heroes presenting themselves as friends and saviors. Although *David Copperfield's* polite and "humble" Uriah Heep slithers into people's lives on the pretext of friendship, he aims to climb the social ladder with their destruction [2930]. In the final television episodes of *Buffy the Vampire Slayer*, Caleb, the evil priest, hides behind his cassock as he serves The First Evil [2931]. Nurse Ratched feigned care for her psychiatric patients while sabotaging their recovery in *One Flew Over the Cuckoo's Nest* [2932]. In the *Harry Potter* Series by J. K. Rowling, the soft-spoken and beaming Dolores Umbridge may outdo the over-the-top evil Lord Voldemort in their race for our disdain as her malevolence oozes from behind her grin. In *Mean Girls* [2933], the beautiful, sunny, sociopathic clique-queen, Regina

George, manipulates fears and anxieties, making her inevitable downfall sweet—until the audience recognizes its own disturbing complicity in the bullying.

Although everyone indulges in impression management, each of these characters symbolizes the bullying technique of false kindness, a power-dependent mechanism used to disarm a weaker person [2934]. The boss who remains calm and smiling, who keeps an even, sweet tone as she berates you, imposes her will as she transforms the natural behavioral option, gaslighting the victim and observers into believing that their anger evidences an immature or unhinged personality.

One can see fraudulent kindness at the social level as well as in individuals, and participants may not be aware of their hypocrisy. Among the many wonderful humanitarian and relief policies and efforts organized by host governments and Nongovernmental Organizations lurk others designed for internal and foreign propaganda purposes rather than to aid people [2847, 2846, 2935]. During the first year of the COVID-19 pandemic, for example, people lauded the efforts of nurses. People would stand outside and bang pots to show their support, and companies cashed in on positive images of nurses as heroes. However, none of that translated into calls for safer working conditions or better pay [2936]. Many meetings now begin by acknowledging that we stand on land stolen from the First Nations of the Americas. Such declarations may make the participants feel better, but they seldom translate into meaningful engagement, policy changes, or reparations for the descendants of the victims.

Our primate tendency to identify with and submit to an accepted authority enables this strategy. With a dulcet tone, the alpha can block all responses beyond a smile and compliance. The subjugated, rather than the intimidator, becomes the troubled one. The power differential weaponizes the acts of kindness to deliver the malignant intent. For disenfranchised people, the repeated experience of fake kindness rises to another symbolic aggression and justice denied, dropped against a lifelong training to be good and not make waves [2937]. Such aggressions must include the social niceties that support group and individual repression [2938, 2939]. In literature, feigned beneficence and compassion enrage us as they rehearse us for these experiences in the social contexts of life.

Satisfaction and Happy Endings

> *"It is the fate of all authors or chroniclers to create imaginary friends, and lose them in the course of art."*
>
> — Charles Dickens, *The Pickwick Papers*, 1836 [2940]

Humans are social animals. We act within our social context. So far, we have focused on the benefits of literary victims and bullies for readers. Individuals consume stories, but their behaviors resonate through their communities. When a book lights a prosocial spark or thought, that power can spread.

In life, we often avoid the bullied; their status can carry a stigma. The safety of a tale permits us to engage with a fictional offering. When we participate, we expect an emotional resolution by the end [2941]. High moral engagement with wounded protagonists reduces tolerance for literary bullying and instills more positive attitudes toward victims [2942, 2060].

Our empathy for beloved characters lacks the discomfort caused when someone within our circle of family and friends—our "in-group"—bears mistreatment [2943]. It lets us endure and contemplate their stress. The fictional setting promises redress and recompense in our own lives as we see the justice meted out within the tale. When a story reaches a positive resolution, we feel good. A small oxytocin rush thrills us when we experience fair treatment. We ride the dopamine high of dominance and revel in relief in a sweet release as our stress hormones drop.

Reading shapes more than our worldview; it impacts our mood. When a story reassures us that the world is fair, it taps into our attribution biases. Books can soften our I-thou dichotomy, inviting us to empathize with the pain and joy experienced by innocents and the falsely accused. With fiction, we step outside ourselves in ways otherwise impossible.

Those positive changes magnify across society. Lifted spirits decrease our propensity to blame the victims. It helps us cope with information about others' misfortunes without holding the injured party responsible for their suffering [2944]. A depressed mood should increase our tendency to see the world through a jaundiced lens, yet paradoxically, some of our most tear-jerking literature can have the most positive impact.

Give Me a Hankie, Please!

> *"Never be afraid to cry. It will free your mind of sorrow."*
>
> — Hopi proverb

We also enjoy poignant, emotional plots [2945] in which a weak protagonist fails despite trying, e.g., *Tess of the D'Urbervilles*, [2946], *The Mayor of Casterbridge*, [2947]; *The Great Gatsby* [2948]. We shake our fists at their unhappy endings, raging against the injustice in the world. We gain a moral high ground through compassion for the central suffering character, and our comparisons of those surrounding her endow us with a sense of virtue and superiority.

Although the weaknesses and ethical failings of the protagonist allow us to accept their final tragic fate, we expect a resolution. In *Casterbridge* and *Tess*, we obtain our moral closure as the principals (Henchard and Tess, respectively) accept their fates and punishments and gain redemption. Henchard requests the anonymity of an unconsecrated grave. Tess, whose complicity in her complex fate is more tenuous, receives absolution from her beloved (if undeserving) Angel in a poignant condemnation of gender-biased standards and expectations.

The treatment of storybook bullies and victims holds yet another fascination for us. It reduces the despair caused by random pain and suffering. With rare exceptions, literary narratives tie together the foremost threads to form the plot [2949, 2920, 2950]. Literature obscures the randomness of life's inequality and its paucity of happy endings.

> *"A reinforcement is, by definition, something that is rewarding to the organism; that is to say, it makes the brain direct the body to repeat the behavior in order to get another positive reward."* — Michael Shermer on dopamine [3230, p. 117]

Although we enjoy novels with twists and turns, we expect reasons for those circuitous paths. Our brains thrive on multiple suspects and intricate possibilities. Indeed, our enjoyment climbs with a moderate level of head-scratching as we try to figure out a "who-done-it," but we bristle at narratives that become too complex or encumbered [2951]. We want to understand what is happening. For a plot to hold together, it must provide order, structure, and the ability to follow the principal characters.

We crave consistency in our world, and literature is a haven in which we look for it. Even when a story remains open-ended or has a non-linear framework, we desire scenarios that not only tie together the threads of the verbal tapestry but also relate the end of the story to its beginning [2952, p. 23, 2953, p. 206, 2954]. (As Chekhov's Principle puts it, if there's a gun in the first act, it must be used by the third one.) This interconnectivity reassures us that things happen for a reason. As the story proceeds, we gain rewards through little bursts of dopamine. We reinterpret the earlier segments of the story and reshape our understanding of characters, including the victims and bullies, as the revelations become clear. For example, as Katherine Paterson's *Bridge to Terabithia* unfolds, it provides insights into the story's bully and heroes, letting us accept both as flawed, fallible humans.

Some literature ends with tragic results or unresolved issues, but most classical formulae bring closure for the protagonist. If the antagonist dies, the circumstances make it clear he has received punishment for his misdeeds. When the hero dies, he has achieved his quest or salvation arc, punished the guilty, or passed the mantle to a successor. In *Les Misérables*, Jean Valjean dies, but not before reconciling with Cosette and Marius [2796]. Hamlet dies as he exacts vengeance and achieves justice for his father (Shakespeare, ca. 1600-1602 [2309]).

In both of these stories, as in the tragedies of Thomas Hardy,[321] the reader leaves the tale satisfied, knowing the principals are poised to receive a reward in an assumed afterlife. In *Oedipus Rex*, the king dies as the price of redemption for his kingdom, suffering for his failings and unintentional violation of social mores [2955].[322] We leave the theatre, exit a game, or close the book, reassured of the world's structure and meaning. We reinforce our programmed belief that good people receive rewards and the crud of the earth will be punished [2805, 2956].

As literature strengthens the just-world phenomenon, it shapes our image of and response to bullies, and it can significantly impact teens. In adolescence, the brain changes its sensitivity to dopamine, and circuits of the nucleus accumbens, striatum, and insula become more active.[323] These functional units alter thinking and give insight into the interconnections and feelings of others. They also give teens the sense that no generation has ever thought like they do or cared as profoundly about social issues and that the answers to everything reside in them. Adolescents who score high in their sense of justice also have lower tolerance levels for bullying and a stronger conviction that the thugs should be punished [2957, 2919, 2958, 2959]. Such menacing violates their emerging self-reflection and sense of justice.

---

[321] Ok, I admit it. *Jude the Obscure* doesn't have a clear salvation arc, but most of his books do have tragic, subtle ones based on esoteric themes, such as love, dignity, atonement, and the earned release from suffering.

[322] In the ancient Greek worldview, it didn't matter that Oedipus didn't know his wife was his mother. The social dictate stood if the violation was intentional or not.

[323] These brain regions accentuate decision making, habits, rewards, pain, self-awareness, and our ability to empathize with others.

But I Still Like Darth Vader More Than Anakin!

"Power is the ultimate aphrodisiac."

— attributed to Henry Kissinger, US Secretary of State, October 28, 1973 [2960]

Despite our demands for the triumph of justice, fictional bad guys attract us. Well-constructed villains enthrall. We even revise morally ambiguous or corrupt characters like the Chicago gangsters of the prohibition era and the Japanese Yakuza [2961], transforming them into folk heroes. Many Indians treat Malangi, a bandit in British-occupied Punjab, as a Robinhood figure. The English highwayman Dick Turpin has experienced a similar upgrade. In *I Claudius*, Robert Graves [2962] elevates the Roman emperor, described by some as a brutal butcher, to a savior of Rome. Nineteenth-century Gothic romances had sexy cads as standard characters. In the late 20th and early 21st centuries, we experienced a flock of sexy vampires (and other supernaturals) with many of the same traits as their gothic counterparts: flawed but with complexity and depth, hardly two-dimensional [2963, p. 78, 2964].

Through these characters, we practice something deep and essential to our survival in a social world: We exercise our theory of mind [2965]. Without condemnation or danger, we can shift our morality and adopt the warped mental role of Joe Goldberg, the murderous stalker in the *You* series, or of Thomas Harris's social-service psychopath, Hannibal Lecter. By using our theory of mind, this temporary shift in morality engages us with shattered sanities, bullies, and sociopaths in a safe context. Through this typically flawed insight, we can broaden our perspective and understanding of others, a valuable survival skill.

The Anaheim Mighty Ducks aside, we have christened many a sports team with names laden with violence, such as the Tampa Bay Buccaneers, Nashville Predators, Minnesota Vikings, Golden State Warriors, and Pittsburgh Pirates. And lest you believe this to be an American obsession, consider the British and Korean soccer teams, the Arsenal (a.k.a. the Gunners) and the Horang (tigers). The list continues with Glasgow's rugby team, the Warriors; Turkey's Besiktas (Black Eagles); Malaysia's Johor Darul Ta'zim (Southern Tigers);

South Korea's Ulsan Hyundai Horang (Ulsan Hyundai Tigers); Brazil's Clube de Regatas do Flamengo (nicknamed the "Vultures"); the Russian CSKA (Central Sports Club of the Army); and Finland's Kärpät (ermines). (Don't laugh at that last one. Ermines are aggressive little devils.)

Our enjoyment of such media goes beyond a questionable cathartic release. These stories demand we disengage from the character's morality and indulge in different aspects of their experience [2966, 49]. The writer must make the villain look less like a bully and more like a well-rounded human, albeit with a fatal character flaw. We experience this same effect when we approach crafted images of public figures, and critical literary experience can create a base for recognizing manipulation.

> *"I have a dream that my four little children will one day live in a nation where they will not be judged by the color of their skin but by the content of their character."*
> —I Have a Dream speech, Martin Luther King [3159]

As in life, we shift our perspective and readily forgive the moral failings of likable and interesting characters, albeit to a greater degree [2967, 2968]. In the original *Star Wars* [2904], the malevolent Darth Vader exerted a commanding presence. Vader, one of the most potent movie villains of the 20th century, has an iconic reputation as a symbol of evil. Even before the series revealed his final fate, the character had a forceful attraction. Large audience segments preferred the evil Darth Vader to the less compelling movie heroes. They applauded and cheered him, not necessarily his immoral deeds. (When *Star Wars* first ran in theatres, I knew several men who emulated him by wearing all black and women who considered him incredibly sexy.)

As in life, we use specific criteria to determine whether we like a literary figure. These preferences let us rationalize our attraction and connect to them [2967, 2910, 2888]. Despite our deeply held illusion that we judge people by the content of their characters, humans treat short, fat, and ugly people with disdain, e.g., weight: [2969, 2970, 2971, 2972]; height discrimination: [2973, 2974, 2975]; and unattractiveness: [2976,

639]. We also project other negative stereotypes onto people with these features, such as laziness or selfishness [2977, 2978]. Tall, slender, and handsome men experience the opposite reaction.[324]

In *Star Wars*, George Lucas instilled Vader, and later Kylo Ren, with qualities that, despite our protests, instill empathy for them and let us forgive their faults. From the first episode, we know that Vader overcame severe obstacles that placed him in his life-support suit, yet he attained a position of power. Vader's costume thus serves as a peacock tail—advertising an impediment the character has the fortitude to conquer. Vader plays the plot's most prominent alpha character. He is tall, has a deep voice, and wears armor that gives the impression of a huge muscle mass. He is a self-confident risk-taker, a dominant figure who makes rapid decisions and shows reduced fear. All of these are perceived as high-testosterone features. Though the audiences recognize the evil of the empire, they accept that Vader uses his power for his cause rather than self-aggrandizement.[325] Because we find so many reasons to like him despite his character, we rationalize his behavior as the result of control by the emperor and plead for his ultimate salvation (a redemption arc).

When we forgive the evil of our antiheroes, we do so with the same zeal that we take in punishing unsympathetic villains. Villains of fiction can remain evil incarnate; they need not even accept an offer of salvation. When we hate villains, we want them to be as surreal as our heroes, and we want them to suffer. We use them as surrogates for the bullies who torment us in life, be they political figures, office bullies, or playground tough guys from our past. In fiction, we can watch them suffer the punishment they deserve, the chastisement without limits that society can no longer impose upon their living counterparts.

---

[324] The impact of height on success in women is not as clear, with papers pointing both to a height premium for education and success, and others that suggest the relationship is reversed.

[325] Of course, the true alpha is Obiwan, who conquers through his sacrifice, and brings about the outcome he desires.

Gothic villains, sexy vampires, and Vader appeal to both men and women. Despite their malice, these classic images depict financial and social power, status, physical prowess, and access to material resources. For our Pleistocene foremothers, these traits provided survival and reproductive advantages for their offspring. A tragic backstory, such as Loki from the Marvel universe, or rich emotional complexity can let women explore and practice recognizing the "dark triad" of psychopathy, narcissism, and Machiavellianism [2979], which occur in some short-term male mating strategies [2980, 2981] without the pitfalls of encountering them. The power, rebellious spirit, and keen wit of an antihero can create an attraction for both men and women [2982, 2983].

However, we remain a generalist species. It takes more than power and status to attract a mate. Women are predisposed toward un-fakeable predictors of Pleistocene reproductive success. Power is joined by social skills (status), bilateral symmetry, generosity, kindness, a deep voice, and intelligence. From these additional factors, symmetry, social skills, deep voices, and high language skills frequently appear in sexy villains. (Kindness and generosity are antithetical to the role of a sexy literary bully, though one could build a case for Walter White's intelligent, persistent, socially manipulative *Breaking Bad* character as self-sacrificing for his family's financial welfare [2898].)

We associate the baritone delivery of many favored movie villains with dominance [2984, 2985, 2986, 2987], sexual attraction and reproductive success [2988, 2989, 2984, 2990, 2991]. It signals high testosterone, large size, and aggression toward females and rival males [2992]. A smooth, deep male voice also indicates pathogen resistance [2993]. Beautiful vocalizations, like attractive faces, enthrall us, leaving us more vulnerable to social and fiscal compromises whether we come from the same gender or not [2994, 2995], as do voices that sound like members of our kinship group—speech with our regional accent [2996].

Intelligence has a high heritability as an adult psychological trait [2997]. Language skills, a feature we use to size up a person's problem-solving potential, can include up to a 76% correspondence in twins [2998].[326] The interaction of genes with the physical and social environments shapes the gifts bestowed by the genes [2999]. The subtle epigenetic changes from these interfaces let us exercise our intellect, helping it reach its potential or suppressing it [3000, 3001]. Intelligence exerts a powerful sexual attraction on both genders [3002, 3003, 3004], and most of us have met someone we considered gorgeous until they opened their mouth and spoke. (I know I have.)

Both men and women use vocal cues for social and sexual evaluations, but research has focused more on women's appraisement of men than the reverse. Women can and do estimate intelligence from the size and style of a man's vocabulary and use the marker as an indicator of mental fitness [3005, 3006]. Males respond by using more uncommon words during courtship or even imaginary courtship encounters [3007, 3008]. Poetry and lyrics, expressions of this wordy courtship, exist in most cultures and societies. The oldest known love song is ascribed to Shu-Sin, king of Mesopotamia, dating back to about 1960 BCE. Be it the Medieval troubadour songs, R&B masterpieces such as *Hello* [3009] and *All of Me* [3010], or modern rap like *Beautiful* [3011], the wordplay fills the same role in humans as the tail does in the strutting peacock; our facility with words is sexy and announces our fitness as a partner.

Many bilateral organisms, including us, consider high bilateral symmetry sexy [3012, 3013, 3014]. Intelligence, health, and strength correlate positively with innate body symmetry, suggesting that strongly bilateral features also mark a good breeding partner [3015, 3016, 3017]. Facial balance increases our initial attraction to men and women [3018, 3019]. In humans, this attraction also translates into positive social

---

[326] This correspondence may be genetic, but it could also be due to the similarity of environment. The mother's vocabulary and use of complex language also shows a strong relationship to the vocabulary growth of her offspring, even in adoption studies [2999].

perceptions, including trust, approachability, and assumption of morality [3020, 3021, 3022]. Actors cast as highly sexualized villains typically have strong bilateral symmetry. For example, the helmet and costume of Darth Vader maximized this adaptive fitness indicator.

Our amygdala, left striatum, and medial frontal cortex respond to physical features, providing our first impressions of people [3023, 3024].[327] The striatum and the insula push us toward our biases [3025]. We struggle to ignore someone's appearance because these "beauty" centers overlap with threat processing. Moral behavior may make us see someone as more attractive, but a beautiful face also makes us perceive them as more trustworthy and honorable [3020, 3026, 3027].

The arts warn against melding good and beautiful. Everyone manipulates their appearance to make positive first impressions. There is no known biological link between physical beauty and sociopathy.[328] Yet, we often perceive those with dark personalities as attractive [3028]. Many people who score high on the "dark triad" (Machiavellianism, narcissism, and psychopathy) pay particular attention to their appearances [3029]. In addition to grooming, clothing, and other adjustable physical features, their expressed emotions impact their projected image and provide the social bait to reel in targets for exploitation [3030, 3031]. Through initial impression management, they leave us with a positive social image, paving the way to excuse their malignant behaviors later. Through the attractive villains in the arts, we experience the deception from the safety of our armchairs, letting us dissociate beauty and morality. Literary scenarios allow us to rehearse, helping us recognize the traps and resist the handsome devils when they troll us in life.

---

[327] For people with blindness, the first impression of beauty comes through other senses. Attractive voices, odors, and textures impact the sighted, but I am unaware of any studies that have explored the concept or neurology of beauty among the blind.

[328] Some correlative data shows a link between antisocial behavior and symmetry [3319]. However, more agreeable people may have less symmetry because of more negative experiences, lower social standing, or more dominance traits in the less pro-social individuals. High cortisol levels can impact a person's projection of "beauty" through its impact on symmetry, facial features, and weight [3362].

Physical appearance measures physical health, providing information about potential mates, rivals, and leaders [3032, 3033]. Both men and women key into beauty quickly and prefer attractive, eye-catching people over less alluring ones. Recognizing healthy skin, youthful appearance, hair texture, and symmetrical facial features requires little attention, and these impressions sway future evaluations. In traditional societies, the learned markers distinguish the safe "us" from the potentially dangerous "them." People with anomalous, unfamiliar features experience more hostility, less empathy, and less positive social contact than faces with familiar features. In essence, we extend our just-world paradigm to appearance [3034, 3035]. Attractive features provide social and pecuniary benefits dubbed the "beauty premium" [3036, 2994],[329] and teachers with appealing features receive higher student and faculty ratings than those deemed plain or ugly [3037, 3038, 3039], a bias that can interact with other learned prejudices [3040, 2996, 3041, 3042].

Far from being a WEIRD obsession, our equation of beauty and goodness crosses cultures, and many have words confounding an attractive appearance and virtue. The Norse goddess Freya encompassed both beauty and love; Balder was just and beautiful. The English words "angelic," "charming," and "beauteous" muddy the ideas. The ancient Greek word "kalos" could mean either concept, as can the Arabic word 'jamil,'(جميل), and the Persian "noor." Yorùbá included a cultural proverb, "Ìwà lẹwà," or "character is beauty." In Korean, "areumdapda" (아름답다") can convey both physical and moral beauty. The Japanese word "utsukushii" (美しい) usually denotes beautiful, but it can express virtue, and bitoku (美徳) roughly translates as "beautiful virtue." The Classical Chinese character (美， měi) can express both concepts, and in ancient Egyptian, the polyphonic hieroglyph "nfr," found in the names Nefertiti and Neferhotep, can

[329] Although more good-looking people are found in jobs where appearance is an asset, the beauty premium is found within occupations, not just among them.

mean "good" as well as "beautiful" and "pleasing." True to our concept of beauty, Nfr (𓄤) is also a highly symmetric symbol.[330]

Symmetric features give important clues about the developmental history of the potential mate and his or her health [3043, 3044]. Diseases, ill health, and dominance fights take their toll on the body, marking it for life. High levels of symmetry function as an honest (hard to fake) indicator of fitness. Recent exposure to a dangerous pathogen may increase our preferences for symmetry as we search for mates who remained healthy during the outbreak [3045, 3046],[331, 332] possibly due to social cues we receive from others [3047]. This preference for a right-left balance helps us decide on the Goldilocks mixture of physical hardiness and parental qualities in prospective mates [3048].

The attractive *fictional* villain poses no threat. The combination of power, status, symmetry, and intelligence can thus override the negative cues that would otherwise warn us of trouble. We perceive literature as a safe environment in which nothing bad will happen. From its safe confines, we indulge in otherwise risky behaviors with these evil, malevolent characters. Sexual fantasies and stylistic modeling may occur, but with an understanding that the anti-hero's specific traits attract us, not their antisocial behavior.

Through its array of characters, literature provides escapes from our fears and opportunities to confront them. Books and theater reinforce our surface conviction that we live in a safe and just universe. We can forgive the sexy fictional villain as he fulfills our core power and social dominance fantasies or lets us cast

---

[330] Of the 40 triliteral signs (combining three consonant sounds) listed in Wikipedia, 15 are symmetric. Of the 27 uniliteral symbols, 9 are symmetric, and one of those (blue = 𓊪, 𓊪) has asymmetric equivalents (𓊪, 𓊪). Most biliterals contain multiple characters, but again, most of those composed of single hieroglyphs are asymmetric [3297].

[331] The link between disease adversity and a preference for symmetry remains a matter of debate. Several studies failed to find this link [3253].

[332] During the early Covid-19 pandemic, the perception of masked faces shifted. In a Japanese study, the wearer's attractiveness increased relative to those who wore preventative masks before the pandemic [3407]. It is unclear if the general evaluation of healthiness caused the shift or if the perception of caring for others (pro-social behavior) and courtesy altered the perception.

ourselves as the alpha and hero as we confront him. We identify with the sympathetic victim whose circumstances remind us that sometimes society rages beyond our control. As we revel in the destruction of bullies, we receive that bit of vindication we often miss in real life. And the hero's triumph reinforces our core faith in satisfying endings. That belief raises us from our beds to live each day with less fear despite the assaults from the reality around us, a reality likely to worsen as our resources dwindle.

CHAPTER 13:

The Passenger Pigeon

"If this earth should ever be destroyed, it will be by desire, by the lust of pleasure and self-gratification, by greed of the green frog skin, by people who are mindful of their own self, forgetting about the wants of others."

— John Fire Lame Deer, Lakota Prophet [3049]

Violence: Future-Perfect

For most of our tenure on Earth, we lived year-by-year on what the sun provided. We exploited the plants, animals, rocks, and fungi, gathering what we needed. Later, we grew crops and learned to cache food for hard times. We made shelters we could transport or found temporary lodgings around us. We limited our physical possessions to what we needed because we carried our civilization on our backs.

Later, we learned to exploit the dead. At some point, our ancestors discovered the calories preserved in ancient plants: peat, coal, and oil. By burning and releasing these sequestered calories of the dead, we could heat our homes and power machines. Mechanized labor replaced human labor. We grew more crops on more land and captured still more calories. We used the energy to dig deeper into the Earth to access more resources, such as metals, potash, and even buried water supplies. Ocean vessels that had once whisked us to new lands on the power of the wind gave way to steamers; then, flight became possible. Goods traveled around the planet, and travel became more commonplace.

Intellectually, we knew the source of these riches was not infinite, but we failed to grasp the implications of our dependence on depleting resources. In less than 100 years of intensive use, within just a few generations, humans squandered over half the energy stored in the once-rich oil fields of Saudi Arabia. Much of it we have burned in engines used to move metal boxes filled with people and goods. With massive amounts of oil, we haul cheap, disposable plastic toys from China to America. We ship yogurt from France and water from New Zealand around the world. We create geographic disconnects between suppliers and consumers and manipulate markets on unprecedented scales.

We used part of that fossil energy to mine for water to irrigate the deserts of the Middle East and the grasslands of North America. Now, both the massive Oglala aquifer of the Midwestern United States and the Sahara aquifer of North Africa stand in danger of depletion before the middle of this century. Like the oil we have exploited without thought, the fresh water in these underground stores is effectively non-renewable.

Oil and coal fuelled a global economy that yielded incomes and lifestyles far beyond anything previously imaginable. To secure profits, we harnessed this cheap energy to shift the sites of production around the planet and to replace expensive labor with cheaper workers. This strategy reduced the proportion of worldwide poverty and decreased the power differential between men and women. It also exaggerated the gap between the richest and the poorest among us.

Fossil fuels, nuclear power, and hydroelectric plants have assumed the work once achieved through human sweat. On May 14, 1957, American Admiral Hyman Rickover gave a speech in which he noted the "slave equivalence units" needed for the tasks of modern life in the mid-20th century [3050]. When Rickover wrote, the labor-saving devices around a typical American home (from electric lighting and central heating to cooking and communication) then required an average energy equivalent to 33 slaves. That figure did not include the labor units needed for transportation and work (2,244 more men). Rickover's statement highlighted our dependence upon a continued flow of fuel to prevent the collapse of our civilization. Despite the improved efficiency of internal combustion engines, despite advances like LED bulbs, our power consumption has increased, as has our thirst for it.

Our energy use now stands far above the enslaved labor available to even the wealthiest nobles throughout pre-industrialized human history. The craving for quaffing energy levels is not unique to the United States; it holds for all industrialized nations. It also serves as a harbinger of what will happen to our descendants should we waste precious oil reserves.

Over the past two generations, we have experienced a paradigm shift in our understanding of the environment. As we passed the halfway point for easily accessible oil, we began to awaken to the damage it inflicts on our world. To further our pleasure, we engage in activities that have increased the carbon dioxide levels in our atmosphere far beyond any level seen by our species. Within the roughly 200 years since the industrial revolution started, we raised atmospheric carbon dioxide levels by 50% to a level not seen since the mid-Pliocene, over 3 million years ago [3051]. The bulk of this increase has occurred since the 1970s [3052, 3053, 3054].

We see the damage oil, industrial waste, and farm chemicals have inflicted on our society. We now fathom the dangers these compounds pose to us and our descendants [3055, 3056, 3057, 3058, 3059]. We know that the environment we shall leave to our children and the future is neither as clean nor as healthy as the one into which we were born. However, like the gradient of vegetarians who do not see shrimp, fish, or Uncle Fred as animals, we, as a society, do not see future generations of humans as worthy of respect or a legacy of conserved resources. We retain the ancient programming to protect our children from immediate harm. Unfortunately, nature wrote that program for the good of individual gene carriers, not for the good of our species or planet. We still sacrifice the generations yet to come to protect our immediate children. Our programming does not include that future failsafe, and we have never managed to produce a stable cultural override to protect our distant descendants. We have little incentive to change our ways. Like the obedient subjects in the Eichmann obedience experiments, we obey societal demands to consume. Humanity bears the costs of the depleted resources, and the unseen billions who have not yet been born will shoulder much of that burden.

Our Legacy

> *"The Essence of all religions is the same—service to humanity."*
>
> — Swami Vivekananda

We continue to use our biological and psychological legacy to rationalize our continued and rapid use of resources. Like the Pleistocene humans, we still function on innate programming that looks for the biggest benefits for ourselves and our kin as we put forth the least effort. Just as we continue to crave the fatty foods that kept us alive as hunters and gatherers, we crave energy and the status bestowed through material goods and labor-saving technologies. Those foods once saved our lives, but their abundance now leads us to clogged arteries. Human material culture once protected us. As long as it still gives our children the advantages of status, we may be unable to change our paradigm of consumption.

I have never seen a passenger pigeon or an elephant bird. The children of Lebanon read about the great cedar forests that once covered the area, but past generations cut the bulk of those timberlands for lumber far before any of us took our first breath. The ancient Romans plundered much of the vast forests of Italy, reducing the trees to fuel. Domains that once supported hunter-gatherers and migratory peoples have fallen prey to the plow. Our collective ancestors extinguished those resources long ago. Management for quick profits has since rendered much of that land infertile.

When we take those precious gallons of oil for our use, we cannot repay future generations. It is a theft of energy from our children and our grandchildren. As each generation ignores the consequences of its actions on the future, it condemns its descendants to a life without the resources upon which we built our society.

Some of these losses have fallen short of extinction. Some of the forestlands of Italy have returned. Small patches of cedars still grace Lebanon [3060]. Although they once roamed North America in family herds, I will never see the woolly mammoth; extinction is forever. [333]

At some point, one of two things will happen with our oil reserves. In the first scenario, the non-renewable resource will become so depleted that future generations will lose their capacity to capture and use this fossilized sunlight. Like the mammoth, our energy reserves will have gone extinct. With time, our kind will regress to living once again on the short-term energy supplied by the sun, and our populations will fall to a level that the depleted reserves can sustain on our now polluted planet. There will be no cars, no phones, no travel, no advanced medicine—the people of that time will look back at us in rage.

Alternatively, our species could find new sources and continue its addiction to fossil fuels. We would continue polluting the air and raise our global temperature beyond the Pliocene levels.[334] Our already bloated energy budget and population could continue to grow. Even if we were to find clean energy sources, we would increase our pressure on the land and water until some other resource becomes too diminished to sustain us on a polluted and depleted planet. Then, the population will still fall, and in an even more decimated world, our children will curse us.

---

[333] Despite some recent attempts to "de-extinct" a few species, a true replication based only upon the organism's genome, its epigenome, and its microbiome remain beyond our reach.

[334] The Pliocene epoch spans the time between 5.3-2.6 million years ago. It is when our kind first diverged from the other chimpanzees. Toward the end of this epoch, the first known member of our Genus, *Homo gautengensis,* arises about 2.8-1.4 million years before the present.

> *"Bad men need nothing more to compass their ends, than that good men should look on and do nothing."*
> —John Stewart Mill, 1867 [3322]

Either way, our descendants face a future of wars, poverty, and population collapse. Our environmental apocalypse condemns them to the resurgence of enslavement. We might postpone their fate with new technologies and cleaner energy sources. However, such stopgaps do not break our fundamental thirst for over-consumption. At some point, our descendants will demand to know why we took away the energy that should have been their birthright. As their children cry with hunger, they will want to know why we used that energy to drain the aquifers and over-farm the land. How could we *knowingly* do such violence to them? If we look at our habits, when we study all those times when we could have conserved but did not, we must recognize ourselves as bullies who inflict degrading, dehumanizing disregard on coming generations. We can equate our depletion of their resources to a future siege, an embargo of goods from the past that will be at least as violent to our descendants as the sieges of castles and towns in the medieval world. The only difference will be that those ancient sieges ended. We will have left our future generations without the resources to rebuild.

A QUICK AND DIRTY INTRODUCTION TO THE KEY TERMS

Here are some key terms demystified. These will simplify the more technical chapters without needing to scratch your head 'til it bleeds.

Above-Average Effect: See *Superiority Bias.*

Accumbens: See Nucleus Accumbens

Androstenedone (A_4): One of Testosterone's big brothers. Where free testosterone does not stay in circulation long, this hormone is more stable in the blood.

Amygdala: An impatient, almond-shaped region deep in the brain that regulates and experiences emotions and wants what it wants when it wants it. Think of Veruca Salt in *Charlie and the Chocolate Factory* [3061]. It is part of the limbic system.

Antisocial Personality Disorder (ASPD): A recognized spectrum disorder highlighted by a disregard for the rights of others. Symptoms include a lack of remorse, a tendency to ignore social rules and conventions, and shallow or restricted emotional responses. Along with borderline and narcissistic personality disorders, ASPD has largely replaced the older term "psychopathic" disorder (See also Spectrum Disorders, Psychopathy, Narcissistic Personality Disorder, and Borderline Personality Disorder).

Attribution Bias: See Fundamental Attribution Error.

Borderline Personality Disorder (BD): A recognized spectrum disorder distinguished by unstable moods, self-image, and relationships. People with BPD often fear abandonment and can have rapid shifts in how they view themselves and others, frequently discounting personal histories in favor of the immediate situation. Contrary to stereotypes, many people with this condition can manage their symptoms and live stable lives. Along with antisocial and narcissistic personality disorders, ASPD has largely replaced the older term "psychopathic" disorder (See also Spectrum Disorders, Psychopathy, Narcissistic Personality Disorder, and Antisocial Personality Disorder).

Brain: More like a system of miniature organs than a single one. Each region is responsible for specific tasks, and many tasks involve multiple areas.

Bullying (working definition): Dehumanization, debasement, objectification, social exclusion, or humiliation directed against a person in any form, often combined with some form of violence or coercion (e.g., emotional, social, physical, sexual, temporal, economic).

Caudate nucleus: The CN plays a key role in associative learning, the process of developing connections between events or stimuli. Together with the putamen, this region forms the dorsal striatum. Both are involved in motor skills and reward processing.

Cerebellum: A large region near the back of the brain best known for its role in motor control. In humans, the cerebellum contributes to language processing and directs our attention to relevant events around us. It also assists in the regulation of our fear and pleasure responses.

Cerebral Cortex: The largest brain region in mammals, covering the other parts of the brain. It is composed of many subregions, each with distinct functions in motor, sensory, or association activities. These include executive function (thinking), movement, language, and sensory processing. When most people picture a brain, they envision the cerebral cortex.

Cognitive Bias: Systematic deviation from rational judgment/ objective reasoning, often caused by mental shortcuts such as stereotyping, emotions, or social influences. These biases can lead to distorted perceptions or illogical conclusions and impair decision-making and social judgements. (See also *Just World Hypothesis* and *Superiority Bias*).

Corticoids/ Corticosteroids: A collection of cholesterol-based molecules that resemble ragged chicken wire and regulate diverse functions, including sugar and mineral metabolism. Cortisol helps us handle stress but fosters weight gain at high doses and impacts stress-related disorders such as PTSD.

Dark Empath: A person afflicted with the dark triad, but who has a better understanding of the emotions of others and who can use this enhanced ability to manipulate and exert control over others. Because the dark triad is a spectral disorder, dark empaths are often subsumed under this label.

Dark Triad: A personality disorder that includes narcissism, psychopathy, and Machiavellianism. When combined, the afflicted person tends toward manipulative, antisocial behavior. Individuals with the dark triad may express each of these traits to a different extent. (See also Dark Empath and Spectral Disorder.)

Dehumanization: The psychological or physical exclusion, in part or in whole, from one's concept of humanity and human dignity. Extreme dehumanization includes torture, rape, genocide, and enslavement. Less intense forms can consist of daily hierarchical social structures and micro-aggressions. It can take different forms, including sexual dehumanization, reduction of a person to a function or an object, or reducing a person to no more than the worst thing they did during their life.

Dopamine: A hormone with strong ties to our reward circuitry. Together with oxytocin, serotonin, and endorphins, it contributes to our "feel-good" and pleasure-seeking mechanism.

Dorsal striatum: A deep brain region consisting of the putamen and the caudate nucleus. These zones are involved in motor control and reward processing. It is called a "striatum" because it is striped with grey and white matter.

Endorphins: A class of hormones linked to a general sense of well-being. The best studied is the beta-endorphin linked to exercise, but there are about 20 different ones. They are natural, physical, and emotional painkillers. Enjoyable experiences stimulate them, including laughter, positive sexual and romantic encounters, and good food.

Entorhinal Cortex: A region of the temporal lobe of the cerebrum that interacts with the memory-forming hippocampus to store and generate our autobiographical memories, our capacity to recall words (semantic memory), and to recall details about specific personal experiences (episodic memory).

Epigenetic: Biological effects caused by the silencing or activation of parts of the DNA code. Some epigenetic effects create the differences between your spleen and your big toe. Others are involved in memory and behavior patterns. Life events, such as traumas, can trigger changes in the expression of the DNA code that can span generations.

Eusocial organisms: Social organisms whose individuals combine forces to raise the young of their community, engage in a division of labor, and lose the ability to perform other tasks, making each individual function as an organ of its colony. Examples include termites, bees, and naked mole-rats. Human societies straddle the border between prosociality and the more restricted eusociality. See also: *Sociality and Prosociality*.

Fundamental Attribution Error/Bias: our innate tendency to see ourselves and our in-group as superior to those we perceive as outsiders. We ascribe character flaws to the bad behavior of others and excuse our own shortcomings as the product of circumstances.

Hippocampus: The hippocampus is involved in moving memory from short-term storage to long-term integration and spatial navigation. It is part of the limbic system and is involved in integrating emotional memory, associating events with memories.

Hormone: A molecule that carries messages that tell your body what to do. Once limited to just the stuff produced by glands, the term has developed a broader use in recent years as we recognize more production sites and learn more about the similarities between the classic hormones and other signaling molecules.

HPA axis (Hypothalamic-pituitary-adrenal axis): a communication system within the brain that interacts with the neuroendocrine (hormonal) system and which is crucial for managing stress. Damage to this system increases the risk for health problems across the lifetime.

Hubris Syndrome (Hubris Personality Disorder): A condition that arises in response to power or prestige. It is marked by excessive pride, overconfidence, and contempt for advice or criticism. Like psychopathy and sociopathy, hubris syndrome is a commonly used descriptive term, but not included in the diagnostic manuals (DSM-5 or ICD-11). (See also Narcissistic Personality Disorder).

Insula: A brain region that supports our identification of feelings and recognition of body signals. It contributes to our self-awareness. Its maturation in adolescence is one reason teens think they're the first generation to figure out life's meaning and recognize social injustices. It is also involved in the pain pathways, cravings, and enjoyment.

Just World Hypothesis: A belief that people get what they deserve and deserve what they get. This assumption can lead us to blame the victims of bullying and genocide or rationalize inequity, even when the surrounding events are unfair or random (See also Victim Blaming and Cognitive Bias).

Limbic System: A brain region that controls many of our emotional responses, especially those associated with survival. It also generates many of our basic emotions and drives.

Machiavellianism: A personality trait rendering the person strategically manipulative and self-serving. People with this trait tend to be emotionally detached and ready to use others like chess pieces to meet their own goals. "Macs" tend towards cynicism, and those high on this spectrum often lack commitment. (See also Dark Triad.)

Mirror Neuron: a brain cell (neuron) that fires both when one acts or has an experience and also when one observes the same action or condition in another.

Narcissism: An abnormal preoccupation with one's own desires and needs, often at the expense of others. People with Narcissistic Personality Disorder suffer from an unrealistic sense of grandiosity and entitlement. Because this is a spectral disorder, not everyone with the condition expresses it the same way. (See also, the *dark triad* and *narcissistic personality disorder*).

Narcissistic Personality Disorder (NPD): A recognized spectrum disorder that includes an inflated sense of self-importance, a need for admiration, and a lack of empathy for others. People who have this trait often react poorly to criticism and readily exploit others whom they feel are below them. As with borderline and antisocial disorders, some narcissistic traits can be adaptive if expressed in moderation. Along with borderline and antisocial personality disorders, ASPD has largely replaced the older term "psychopathic" disorder (See also Spectrum Disorders, Psychopathy, Narcissism, Hubris Syndrome, Antisocial Personality Disorder, and Borderline Personality Disorder).

Objectification: Treating another person as nothing more than a tool or means to attaining a goal.

Oxytocin: A brain hormone released by the pituitary gland. Although often portrayed as the "love" hormone, its impact is more selective and nuanced. It can promote warm, fuzzy feelings towards one person or group while scaffolding xenophobic responses to others.

Nucleus: A deep cluster of neurons in the brain stem or cerebrum that shares a similar function and acts as a functional unit. It usually looks like a lump of grey matter surrounded by white matter.

Nucleus Accumbens: A small but powerful region of the brain that plays a pivotal role in processing emotions and motivations. Sometimes called the "pleasure center," this region is chock full of dopamine receptors that recognize and distinguish among rewarding stimuli. These influence decision-making and motivation. It is seated by the base of the frontal lobe and is considered part of the ventral striatum. It connects to the brain stem, making some deprivations of pleasure feel like a removal of something necessary for life, like oxygen.

Pleistocene: The time (~2.6 MYBP-11,000) when our ancestors became recognizably human. *Homo habilis*, a member of our genus, dates to just over 2 million years old and diverged into other species of humans around 500 thousand years ago. Our species, *Homo sapiens*, arose about 300,000 years ago.

Prefrontal Cortex: This brain region contributes to many things, including the so-called "executive functions," activities that regulate our actions and emotions by linking to and controlling the output of other brain areas. It engages during reasoning, comprehension, creativity, and impulse control.

Prosocial organisms: Organisms that frequently engage in behaviors that benefit other members of their group, such as cooperation, sharing, or self-sacrificing actions. These actions benefit their society as a whole, resulting in enhancements to the survival and reproductive fitness of individual members. Examples include dolphins, chimpanzees, elephants, and humans. See also: *Eusocial Organisms* and *Social Organisms*.

Psychopathy: A personality trait marked by a lack of empathy and remorse, ease with cheating and lying, and impulsive, self-centered behavior. It can stem from a genetic predisposition or be triggered by environmental factors. Contrary to the popular stereotype of crazed killers, psychopathy is a spectrum disorder. Many people with psychopathy function in society, and as a mild trait, may be advantageous under some conditions. Like sociopathy and hubris syndrome, "psychopathy" is an informal descriptive term not included in the diagnostic manuals (DSM-5 or ICD-11). (See also Spectrum Disorder and Dark Triad.)

Putamen: A region of the brain that is closely aligned with the caudate nucleus. The putamen regulates learning, thinking (cognition), emotional responses, and rewards. It is involved in some major depressive disorders and sleep apnea. Together with the caudate nucleus, it forms the dorsal striatum, which is involved in motor function and reward processing.

Receptor: A cellular docking port for a hormone. The interactions of receptors and hormones aid in regulating bodily functions.

Right Temporal-parietal junction. (rTPJ): A region on the side of the brain. Along with the frontal lobes, it is critical to social behavior and understanding other people's thoughts (theory of mind).

Social Organism: Organisms who live in a cooperative group that provides benefits to its members. Examples include herds of deer, shoals of fish, and human societies. See also: Eusocial Organisms and Prosocial Organisms.

Sociopathy: Like "psychopathy," this informal term describes people who violate social norms, ignore the rights of others, and show little remorse. Its distinction from "psychopathy" is unclear; some use the two terms interchangeably, but others consider sociopaths to be a less intense form of the same conditions. This book follows a common convention and uses the terms interchangeably (See also Psychopathy, Borderline Personality Disorder, Narcissistic Personality Disorder, and Antisocial Personality Disorder.)

Spectrum Disorder: A condition or trait that ranges in severity. The best-known spectrum disorders are autism and attention-deficit hyperactivity disorder (ADHD). However, traits such as psychopathy, narcissism, and Machiavellianism also have ranges in intensity and thus have their own spectrums. Although seldom discussed, pro-social characteristics, such as generosity and empathy, also range across spectrums.

Spite: A behavior that harms another and provides no apparent benefit to the doer. Soft spite happens when the act garners no advantage. When someone engages in hard spite, it inflicts more misery on both parties. Examples of spiteful behavior include bullying, dehumanization, warfare, and genocide. Spiteful behavior tends to provide relative benefits to the spiteful party by harming the other rather than benefiting the doer.

Striatum: The primary seat of the brain's reward pathways. It is particularly active during social encounters.

Superiority Bias (= Above-Average Effect): A cognitive bias in which people overestimate their own ability, skills, or moral standing relative to others. Unlike narcissism, the superiority bias is a normal cognitive tendency, but it can still lead to overconfidence, poor judgment, and underestimating the abilities of others. Our superiority biases help explain why so many adults remember being bullied but seldom remember bullying others. (See also Cognitive Bias and Narcissism).

Testosterone (T): A much-maligned hormone that needs a better public relations agent. It is critical to sociality and maturation in males. Increased T does not necessarily improve status, but too much can spur the bullying of underlings.

Theory of Mind (ToM): The ability to form an image of another's world experience, including their intentions, beliefs, emotions, and thoughts. Once believed to be unique to humans, it is now understood to be widespread throughout vertebrates and possibly throughout the animal kingdom.

TPJ (temporoparietal junction): A region on the side of the brain where the temporal and parietal lobes of the cerebrum meet. The area is important in our "theory of mind," our understanding of another's thoughts and intentions.

Umwelt: An animal or species' experience and interpretation of its environment.

Ventral Tegmentum (a.k.a. VTA, ventral tegmental area or VTA of Tsai): A set of neurons near the lower midline of the brain that is implicated in the brain's reward system. It is sensitive to dopamine (a compound associated with reward) and helps coordinate other reward centers. It is critical in the processing of fear responses and associating them with context.

Victim Blaming: Holding victims responsible for their misfortune or degradation, such as blaming a rape victim for the rape, the victims of genocide for the persecution, or the bullied for the bullying. (See also Just World Hypothesis).

WEIRD: An acronym for **W**estern, **E**ducated, **I**ndustrialized, **R**ich, and **D**emocratic. These terms represent about 12% of the world population, yet about 96% of psychological data is collected on this fraction. This disconnect between who is studied and how most people live creates a bias in how we perceive what is "normal" in human behavior.

Works Cited

[1] Vivekananda, "The Complete Works of Swami Vivekananda/Volume 5/Epistles - First Series/IV Alasinga," 20 August 1883. [Online]. Available: https://en.wikisource.org/wiki/The_Complete_Works_of_Swami_Vivekananda/Volume_5/Epistles_-_First_Series/IV_Alasinga.

[2] R. L. Stevenson, The Strange Case of Dr. Jekyll and Mr. Hyde, CreateSpace, 1886/ 2018, p. 46.

[3] FDR Library, "October, 1940 – FDR: Day by Day," 13 October 1940. [Online]. Available: http://www.fdrlibrary.marist.edu/daybyday/event/october-1940-10/. [Accessed 25 August 2025].

[4] F. Douglass, "I have come to tell you something about slavery (1841)," in *Douglass, Speeches and Writings*, D. W. Blight, Ed., New York, New York: Library of America, 2023, p. 969.

[5] N. Machiavelli, The Prince, Oxford, Oxford: Oxford University Press, 1909/1532.

[6] R. Garg, "Ragging: A public health problem in India," *Indian Journal of Medical Sciences,* vol. 63, pp. 263-271, 2009.

[7] T. Hardy, Jude the Obscure, Chicago, IL: Mint Editions, 1895/2020, p. 384.

[8] L. R. Allen, "Solution didn't fit problem," *Star Beacon,* 11 April 2007.

[9] D. Balona and R. Mahoney, "Long-ago charge to dost man his home," *Orlando Sentinel,* 20 March 2007.

[10] C. Levitan, "You may be a sex offender and not even know it," *New York Post,* 20 May 2015.

[11] *JAMIE S. Nabozny vs. Mary Podlesny, William Davis, Thomas Blauert, et al.,* 1996.

[12] R. Oblath, J. G. Green, J. Guzmán, E. D. Felix, M. J. Furlong, M. Holt and J. Sharkey, "Retrospective perceptions of power imbalance in childhood bullying among college students," *Journal of American College Health,* vol. 68, no. 8, pp. 891-899, 2020.

[13] N. Spadafora, A. A. Volk and A. V. Dane, "Using Qualitative Methods to Measure and Understand Key Features of Adolescent Bullying: A Call to Action," *International Journal of Bullying Prevention,* vol. 4, pp. 230-241, 2022.

[14] M. Kocatürk and I. Çiçek, "Relationship between positive childhood experiences and psychological resilience in university students: The mediating role of self-esteem," *Journal of Psychologists and Counsellors in Schools,* vol. 33, no. 1, pp. 1-12, 2021.

[15] L. A. Seneca, Moral Epistles, vol. 1, Cambridge, MA: Harvard University Press, 1917.

[16] D. J. Ayre and R. K. Grosberg, "Aggression, habituation, and clonal coexistence in the sea anemone Anthopleura elegantissima," *The American Naturalist,* vol. 146, no. 3, pp. 427-453, 1995.

[17] S. M. Lane, A. J. Wilson and M. Briffa, "Analysis of direct and indirect genetic effects in fighting sea anemones," *Behavioral Ecology,* vol. 31, no. 2, pp. 540-547, 2020.

[18] R. Chandrasekaran, "Provocation Helped Set India Train Fire: Official Faults Hindu Actions, Muslim Reactions for Incident That Led to Carnage," *Washington Post,* p. A10, 6 3 2002.

[19] J. Fluck, "Why do students bully? An analysis of motives behind violence in schools," *Youth and Society,* vol. 49, no. 5, pp. 537-587, 2014.

[20] Y. Mekawi and N. R. Todd, "Focusing the lens to see more clearly: Overcoming definitional challenges and identifying new directions in racial microaggressions research," *Perspectives on Psychological Science,* vol. 16, no. 5, pp. 879-1098, 2021.

[21] A. Smeulers, Perpetrators of Mass Atrocities, First ed., New York: Routledge, 2023, p. 600.

[22] S. Santre, "Cyberbullying in adolescents: A literature review," *International Journal of Adolescent Medicine and Health,* vol. 35, no. 2, pp. 1-7, 2022.

[23] J. Lee, H. Choo, Y. Zhang, H. S. Cheung, Q. Zhang and R. P. Ang, "Cyberbullying victimization and mental health symptoms among children and adolescents: A meta-analysis of longitudinal studies," *Trauma Violence Abuse,* vol. 19, p. 15248380241313051, 2025.

[24] S. A. Sumner, B. Ferguson, B. Bason, J. Dink, E. Yard, M. Hertz, B. Hilkert, K. Holland, M. Mercado-Crespo, S. Tang and C. M. Jones, "Association of Online Risk Factors With Subsequent Youth Suicide-Related Behaviors in the US," *Journal of the American Medical Association, Network Open,* vol. 4, no. 9, p. e2125860, 1 September 2021.

[25] A. Schonfeld, D. NcNiel, T. Toyoshima and R. Binder, "Cyberbullying and adolescent suicide," *Journal of the American Academy of Psychiatry and the Law,* pp. 112-119, 2023.

[26] J. K. Chen and H. C. O. Chan, "International and interdisciplinary perspectives on school violence and bullying, cyberbullying, and cyber deviance: An introductory essay," *Behavioral Sciences and the Law,* vol. 42, no. 2, pp. 169-172, 2025.

[27] A. D. Diaz, D. A. Peeples and P. E. Weigle, "Depression and social media use in children and adolescents," *Pediatric Clinics of North America,* vol. 72, no. 2, pp. 175-187, 2025.

[28] W. Yang, L. Ning, Q. Miao, F. Xu, K. Li, X. Chen and H. Lu, "The mediating roles of anxiety, loneliness, stress, and depression in the relationship between cyberbullying and non-suicidal self-injury: propensity score matching and causal mediation analysis," *BMC Psychiatry,* vol. 25, no. 1, p. 539, 2025.

[29] J. R. W. Yarbrough, K. Sell, A. Weiss and L. R. Salazar, "Cyberbullying and the faculty victim experience: perceptions and outcomes," *International Journal of Bullying Prevention,* pp. 1-15, 2023.

[30] S. von Humboldt, G. Low and I. Leal, "From words to wounds: Cyberbullying and its influence on mental health across the lifespan," *Behavioral Sciences (Basel),* vol. 15, no. 5, p. 619, 2025.

[31] X. Liao, L. Xine and J. Ni, "Cross-sectional study of cybervictimisation and non-suicidal self-injury among college students in China: A chain mediation effect of emotion dysregulation and social exclusion," *British Medical Journal Open,* vol. 15, no. 1, p. e087346, 2025.

[32] S. Platt, Respectfully Quoted: A Dictionary of Quotations, New York: Barnes and Noble Books, 1993, p. 520.

[33] K. Peets, T. Turunen and C. Salmivalli, "Rumination mediates the longitudinal associations between elementary school victimization and adolescents' internalizing problems," *Journal of Interpersonal Violence,* vol. 37, no. 17-18, p. NP16885–NP16906, 2022.

[34] A. Kennedy and A. M. Brausch, "Emotion dysregulation, bullying, and suicide behaviors in adolescents," *Journal of Affective Disorders Reports,* vol. 15, p. 100715, 2024.

[35] M. B. Nielsen and S. V. Einarsen, "What we know, what we do not know, and what we should and could have known about workplace bullying: An overview of the literature and agenda for future research," *Aggression and Violent Behavior,* vol. 42, pp. 71-83, 2018.

[36] M. Connaughton, O. Mitchell, E. Cullen, M. O'Connor, T. Banaschewski, G. J. Barker, A. L. W. Bokde, R. Brühl, S. Desrivières, H. Flor, H. Garavan, P. Gowland, A. Grigis, A. Heinz, H. Lemaitre, J.-L. Martinot and M.-L. P. Martinot, "Bullying and early brain development: A longitudinal structural magnetic resonance imaging study from adolescence to early adulthood. (preprint)," *bioRxiv,* 2024.

[37] Y. S. Kim and B. Leventhal, "Bullying and Suicide. A Review," *International Journal of Adolescent Medicine and Health,* vol. 20, no. 2, pp. 133-154, 2008.

[38] J. Sun, Y. Ban and J. Liu, "Relationship between bullying victimization and suicide ideation among Chinese adolescents: A moderated chain mediation model," *Children and Youth Services Review,* vol. 156, p. 107304, 2024.

[39] A. B. Klomek, A. Sourander and H. Elonheimo, "Bullying by peers in childhood and effects on psychopathology, suicidality, and criminality in adulthood," *Lancet Psychiatry,* vol. 2, no. 10, pp. 930-941, 2015.

[40] N. Balluerka, J. Aliri, O. Goñi-Balentziaga and A. Gorostiaga, "Association between bullying victimization, anxiety and depression in childhood and adolescence: The mediating effect of self-esteem," *Revista de Psicodidáctica (English Edition),* vol. 28, no. 1, pp. 26-34, 2023.

[41] R. Palmu and T. Partonen, "Childhood bullying as a predictor of suicidality in young adults in a nationwide population-based health examination study," *Archives of Suicide Research,* pp. 1-10, 2024.

[42] C. Jiang and J. Shi, "The long arm of childhood bullying victimization: Associations with social isolation, gender differences, and depression in later life," *Journal of Interpersonal Violence,* vol. 39, no. 13-14, pp. 3184-3206, 2024.

[43] H. Yin, Z. Han and Y. Li, "Traditional bullying, cyberbullying, and quality of life among adolescents in 35 countries: Do cultural values matter?," *Social Science and Medicine,* vol. 340, p. 116499, 2024.

[44] T. A. Ariani, A. R. Putri, F. A. Firdausi and N. Aini, "Global prevalence and psychological impact of bullying among children and adolescents: a meta-analysis," *Journal of Affective Disorders,* vol. 385, p. 119446, 2025.

[45] Y. S. Kim, Y. J. Koh and B. L. Leventhal, "Prevalence of school bullying in Korean middle school students," *Journal of the American Medical Association, Pediatrics,* vol. 158, no. 8, pp. 737-741, 2004.

[46] M. Rana, M. Gupta, P. Malhi, S. Grover and M. Kaur, "Prevalence and correlates of bullying perpetration and victimization among school-going adolescents in Chandigarh, North India," *Indian Journal of Psychiatry,* vol. 62, no. 5, pp. 531-539, 2020.

[47] S. G. Ramya and M. L. Kulkarni, "Bullying among school children: Prevalence and association with common symptoms in childhood," *The Indian Journal of Pediatrics,* vol. 78, no. 3, p. 307–310, 2011.

[48] K. L. Modecki, J. Minchin, A. G. Harbaugh, N. G. Guerra and K. C. Runions, "Bullying prevalence across contexts: a meta-analysis measuring cyber and traditional bullying," *Journal of Adolescent Health,* vol. 55, no. 5, pp. 602-611, 2014.

[49] A. Sourander, D. Gyllenberg, A. Burnstein Klomek, L. Sillanmäki, A. M. Ilola and K. Kumpulainen, "Association of bullying behavior at 8 years of age and use of specialized services for psychiatric disorders by 29 years of age," *Journal of the American Medical Association Psychiatry,* vol. 73, no. 2, pp. 195-165, 2016.

[50] G. Erzin and S. Gülöksüz, "The exposome paradigm to understand the environmental origins of mental disorders," *Alpha Psychiatry,* vol. 22, no. 4, pp. 171-176, 2021.

[51] I. S. Durrant, "Is banter bullying or a necessary part of the police officer toolkit?," *The Police Journal: Theory, Practice and Principles,* p. 0, 2024.

[52] D. Alighieri, "The Gate of Hell. The Inefficient or Indifferent. Pope Celestine V. The Shores of Acheron. Charon. The Earthquake and the Swoon," in *Divine Comedy, Longfellow's Translation, Hell*, vol. 1, Project Gutenberg, 2009.

[53] A. Adams, Bullying at work: How to confront and overcome it, London: Virago, 1992, p. 256.

[54] J. M. León-Pérez, J. Escartín and G. Giorgi, "The presence of workplace bullying and harassment worldwide," *Handbooks of Workplace Bullying, Emotional Abuse and Harassment 1: Concepts, Approaches and Methods,* pp. 55-86, 2021.

[55] S. Reissner and E. Armitage-Chan, "Manifestations of professional identity work: an integrative review of research in professional identity formation," *Studies in Higher Education,* vol. 49, no. 12, pp. 2707-2722, 2024.

[56] Y. Wang and C.-H. Wu, "How work experiences drive personality change: the impact of work, organizational, societal, and international environment (pp 33-62)," in *Work and Personality Change*, Bristol, Bristol University Press, 2021, pp. 33-62.

[57] S. Ahmer, A. W. Yousafzai, N. Bhutto, S. Alam, A. Sarangzai and A. Iqbal, "Bullying of medical students in Pakistan: A cross-sectional questionnaire," *PLoS ONE,* vol. 3, p. e3889, 2008.

[58] S. Ahmer, A.-W. Yousafzai, M. Siddiqi, R. Faruqui, R. Khan and S. Zuberi, "Bullying of trainee psychiatrists in Pakistan: A cross-sectional questionnaire survey," *Academic Psychiatry,* vol. 33, no. 4, pp. 335-339, 2009.

[59] J. H. Coverdale, R. Balon and L. W. Roberts, "Mistreatment of trainees: Verbal abuse and other bullying behaviors," *Academic Psychiatry,* vol. 33, no. 4, 2014.

[60] M. Birks, R. P. Cant, L. M. Budden, M. Russell-Westhead, Y. S. Ü. Özçetin and S. Tee, "Uncovering degrees of workplace bullying: A comparison of baccalaureate nursing students' experiences during clinical placement in Australia and the UK," *Nurse Education in Practice,* vol. 25, pp. 14-21, 2017.

[61] C. Karatza, C. Zyga, S. Tziaferi and P. Prezerakos, "Workplace bullying and general health status among the nursing staff of Greek public hospitals," *Annals of General Psychiatry,* vol. 15, no. 7, 2016.

[62] L. Colenbrander, L. Causer and B. Haire, "'If you can't make it, you're not tough enough to do medicine': a qualitative study of Sydney-based medical students' experiences of bullying and harassment in clinical settings," *BMC Medical Education,* vol. 20, no. 86, 2020.

[63] K. L. Bairy, P. Thirumalaikolundusubramanian, G. Sivagnanam, S. Saraswathi, A. Sachidananda and A. Shalini, "Bullying among trainee doctors in Southern India: a questionnaire study," *Journal of Postgraduate Medicine,* vol. 53, no. 2, pp. 87-91, 2007.

[64] M. Birks, H. Harrison, L. Zhao, H. Wright, Y. C. Tie and N. Rathnayaka, "Nursing students' experience of bullying and/or harassment during clinical placement," *Nurse Education Today,* vol. 136, p. 106151, 2024.

[65] Y. Wang, F. Liu, W. Tong and H. Zhang, "Bullying of nursing students: A bibliometric analysis based on Web of Science database," *Journal of Professional Nursing,* vol. 58, pp. 114-123, 2025.

[66] S. R. Daugherty, D. C. J. Baldwin and B. D. Rowley, "Learning, satisfaction, and mistreatment during medical internship: A national survey of working conditions," *Journal of the American Medical Association,* vol. 279, no. 15, pp. 1194-1199, 1998.

[67] C. Jones, "Still a rite of passage? A perspective on current therapeutic attitudes and interventions in relation to cyberbullying," *Counselling and Psychotherapy Research,* vol. 24, no. 1, pp. 9-15, 2024.

[68] L. K. Barger, B. E. Cade, N. T. Ayas, J. W. R. B. Cronin, F. E. Speizer and C. A. Czeisler, "Extended work shifts and the risk of motor vehicle crashes among Interns," *New England Journal of Medicine,* vol. 352, pp. 125-134, 2005.

[69] J. Grotto-de-Souza, H. H. Pohl and D. Aguiar-Ribeiro, "Mobbing as a source of psychological harm in workers (Assédio moral como fonte de adoecimento do trabalhador)," *Revista Brasileira de Medicina do Trabalho,* vol. 20, no. 4, pp. 670-675, 2022.

[70] A. Matsson and T. Jordan, "Workplace bullying investigations: A complex endeavor for a complex problem," *Organizational Dynamics,* vol. 51, no. 2, p. 100840, 2022.

[71] K. Kamto, S. M. Nasri, D. Erwandi, F. Lestari and A. Kadir, "Workplace bullying among offshore workers of an oil and gas company," *Unnes Journal of Public Health,* vol. 11, no. 1, pp. 1-13, 2022.

[72] M. Hodgins, L. Pursell, Y. Itzkovich, S. MacCurtain and C. Rayner, "A scoping review of bullying and harassment in nonprofit and voluntary organizations," *Nonprofit Management and Leadership,* vol. 35, no. 4, pp. 813-836, 2025.

[73] J. Roper and J. Menten, "Bullying and harassment: Tips, traps and trends in the tourism industry," in *British Columbia Tourism Industry Conference*, 2017.

[74] S. T. Teo, T. Bentley and D. Nguyen, "Psychosocial work environment, work engagement, and employee commitment: A moderated, mediation model," *International Journal of Hospitality Management,* vol. 88, p. 102415, 2020.

[75] K. Xiang, J. Liu, G. Qiao, F. Gao and H. Zhang, "Does bullying reduce occupational commitment in hospitality employees? Mixed empirical evidence from resource conservation theory and embodied cognition perspectives," *Hospitality Management,* vol. 108, p. 103365, 2023.

[76] S. Singh, A. K. Subramani, R. David and N. A. Jan, "Workplace ostracism influencing turnover intentions: Moderating roles of perceptions of organizational virtuousness and authentic leadership," *Acta Psychologica,* vol. 243, p. 104136, 2024.

[77] S. E. Baker-Thompson, Workplace Mobbing, Job Satisfaction, and Intent to Leave the Information Technology (Dissertation), Minneapolis: Walden University College of Management and Human Potential, 2022, p. 166.

[78] R. McKay, D. H. Fratzl and R. Thomas, "Workplace bullying in academia: A Canadian study," *Employee Responsibilities and Rights Journal,* vol. 20, p. 77–100, 2008.

[79] T. Migliaccio, A. Rivas, B. Rivas and R. Stark, "Bullying within academia: A cultural and structural analysis," *Learning, Culture and Social Interaction,* vol. 44, p. 100783, 2024.

[80] K. Z. Young, "Workplace bullying in higher education: The misunderstood academicus," *Practicing Anthropology,* vol. 39, no. 2, pp. 14-17, 2017.

[81] A. Thomas, "Workplace bullying rife in the second, union claims," *The Times,* no. 4894, p. F1, 11 June 2010.

[82] J. Paris, "Against hope labour in the public humanities," *Public Humanities,* vol. 1, p. e40, 2025.

[83] K. A. Zemanek, Identification of Bullying Among Tenured and Non-Tenured Faculty in Colleges of Agriculture and Life Sciences at Land-Grant Universities (doctoral dissertation), College Station, Texas: Texas A & M, 2016, p. 118.

[84] C. Rayner and H. Hole, "A summary review of literature relating to workplace bullying," *Journal of Community & Applied Social Psychology,* vol. 7, no. 3, 1998.

[85] T. E. Robertson, D. Sznycer, A. W. Delton, J. Tooby and L. Cosmides, "The true trigger of shame: social devaluation is sufficient, wrongdoing is unnecessary," *Evolution and Human Behavior,* vol. 39, no. 5, pp. 566-573, 2018.

[86] K. Lambert, K. Francis and K. Tori, "Variability of terminology used to describe unwanted workplace behaviors in nursing: A scoping review," *Western Journal of Nursing Research,* vol. 47, no. 5, 2025.

[87] British Medical Association, "Bullying and Harassment of Doctors in the Workplace," British Medical Association, 2006.

[88] P. Lutgen-Sandvik, S. J. Tracy and J. K. Alberts, "Burned by bullying in the American workplace: Prevalence, perception, degree and impact," *Journal of Management Studies,* vol. 44, no. 6, pp. 837-862, 2007.

[89] M. Picard and B. S. McEwen, "Psychological stress and mitochondria: A systematic review," *Psychosomatic Medicine,* vol. 80, no. 2, pp. 141-153, 2018.

[90] D. Bürgin, V. Clemens, N. Varghese, A. Eckert, M. Huber, E. Bruttin, C. Boonmann, E. Unternährer, A. O'Donovan and M. Schmid, "Adverse and traumatic exposures, posttraumatic stress disorder, telomere length, and hair cortisol - Exploring

associations in a high-risk sample of young adult residential care leavers," *Brain, Behavior and Immunity Health,* vol. 26, p. 100524, 2022.

[91] K. Domschke, "A methylome-based signature of stress: Association with accelerated epigenetic aging and alcohol use," *Biological Psychiatry,* vol. 83, no. 4, pp. 296-297, 2023b.

[92] E. A. Ruiz-Narváez, Y. Cozier, G. Zirpoli, L. Rosenberg and J. R. Plamer, "Perceived experiences of racism in relation to genome-wide DNA methylation and epigenetic aging in the Black women's health study," *Journal of Racial and Ethnic Health Disparities,* vol. 12, no. 2, pp. 754-763, 2025.

[93] M. Kivimäki, A. Bartolomucci and I. Kawachi, "The multiple roles of life stress in metabolic disorders," *Nature Reviews,* vol. 19, pp. 10-27, 2023.

[94] L. C. Perret, M.-C. Geoffroy, E. Barr, F. Parnet, N. Provencal, M. Boivin, K. J. O'Donnell, M. Suderman, C. Power, G. Turecki and I. Ouellet-Morin, "Associations between epigenetic aging and childhood peer victimization, depression, and suicidal ideation in adolescence and adulthood: A study of two population-based samples," *Frontiers in Cell and Developmental Biology,* vol. 10, p. 1051556, 2023.

[95] H. Leymann and A. Gustafsson, "Mobbing at work and the development of post-traumatic stress disorders," *European Journal of Work and Organizational Psychology,* vol. 5, no. 2, pp. 251-275, 1996.

[96] N. Tehrani, "Bullying: A source of chronic post-traumatic stress?," *British Journal of Guidance & Counselling,* vol. 32, no. 3, pp. 357-366, 2004.

[97] N. J. Donovan and D. Blazer, "Social isolation and loneliness in older Adults: Review and commentary of a National Academies [sic]," *American Journal of Geriatric Psychiatry,* vol. 32, no. 11, pp. 1233-1244, 2020.

[98] M.-M. Lord, M.-J. Drolet, C. Viscogliosi and C. Pinard, "Elder organizational abuse in long-term care homes: An ecological perspective," *The Journal of Aging and Social Change,* vol. 13, no. 2, pp. 61-85, 2023.

[99] N. Özer and D. Tanriverdi, "Determining depression, abuse, and neglect in elderly individuals," *Psychogeriatrics,* vol. 23, no. 4, pp. 690-700, 2023.

[100] A. Duffy, M. Connolly and F. Browne, "Older people's experiences of elder abuse in residential care settings: A scoping review," *Journal of Advanced Nursing,* vol. 80, no. 6, pp. 2214-2227, 2024.

[101] S. Folorunsho and M. Okyere, "The impact of neglect, physical, and financial abuse on mental health among older adults: a systematic review," *Aging and Mental Health,* vol. 29, no. 4, pp. 567-577, 2025.

[102] R. Acierno, M. A. Hernandez, A. B. Amstadter, H. S. Resnick, K. Steve, W. Muzzy and D. G. Kilpatrick, "Prevalence and correlates of emotional, physical, sexual, and financial abuse and potential neglect in the United States: The National Elder Mistreatment Study," *American Journal of Public Health,* vol. 100, no. 2, pp. 292-297, 2010.

[103] Lifespan of Greater Rochester, Inc., Weill Cornell Medical Center, New York City Department for the Aging, "Under the Radar: New York State Elder Abuse Prevalence Study," Weill Cornell Medical Center, Cornell University, New York City, 2011.

[104] S. von Humboldt, J. A. Ribeiro-Gonçalves and I. Leal, "Bullying in old age: A qualitative study on older adults' perceptions about being bullied," *Journal of Interpersonal Violence,* vol. 37, no. 5-6, pp. 2896-2919, 2022.

[105] W. R. N. Torelli, T. A. de Bessa and B. Graeff, "Preconceito contra pessoa idosa LGBTQIA+ em Instituições de Longa Permanência para Idosos (ILPI): uma revisão de escopo. (Prejudice against the older LGBTQIA+ population in Long-Term Care Facilities (LTC): a scoping review)," *Ciência & Saúde Coletiva,* vol. 28, no. 11, pp. 3123-3135, 2023.

[106] E. P. W. Packard, Modern Persecution, or Insane Asylums Unveiled, as Demonstrated by the Report of the Investigating Committee of the Legislature of Illinois, vol. 1, Hartford: Case, Lockwood & Brainard, Printers and Binders, 1874, p. 402.

[107] I. M. Garca-Munuera, I. Méndez, C. Ruiz-Esteban and J. P. Martinez-Ramon, "Comparative study of coexistence problems among institutionalized elderly people versus senior classroom," *Revista de Psicología de la Salud (Journal of Health Psychology),* vol. 12, no. 1, pp. 197-208, 2024.

[108] C. S. Tranquilli, "De Vita XII Caesarum (The 12 Caesars), (121a.d.)," 1 3 2018. [Online]. Available: http://penelope.uchicago.edu/Thayer/L/Roman/Texts/Suetonius/12Caesars/Claudius*.html#ref30. [Accessed 13 7 2018].

[109] P. O'Connell, D. Pepler and W. Craig, "Peer involvement in bullying: Insights and challenges for intervention," *Journal of Adolescence,* vol. 22, no. 4, pp. 437-452, 1999.

[110] J. D. Unnever and D. G. Cornell, "The culture of bullying in middle school," *Journal of School Violence,* vol. 2, no. 2, pp. 5-27, 2003.

[111] R. Kurzban, "Evolutionary origins of stigmatization: The functions of social exclusion," *Psychological Bulletin,* vol. 127, no. 2, pp. 187-208, 2001.

[112] J. L. Pouwels, T. H. J. van Noorden and S. C. S. Caravita, "Defending victims of bullying in the classroom: The role of moral responsibility and social costs," *Journal of Experimental Social Psychology,* vol. 84, p. 103831, 2019.

[113] R. António, R. Guerra, L. Cameron and C. Moleiro, "Imagined and extended contact experiences and adolescent bystanders' behavioral intentions in homophobic bullying episodes," *Aggressive Behavior,* vol. 49, no. 2, pp. 110-126, 2023.

[114] S. Alberts, "Magnitude and sources of variation in male reproductive performance. pp. 412–431," in *The Evolution of Primate Societies*, Chicago, Illinois: The University of Chicago Press, 2012, p. 744.

[115] I. S. Bernstein, "Dominance, aggression and reproduction in primate societies," *Journal of Theoretical Biology,* vol. 60, no. 2, pp. 459-472, 1976.

[116] A. A. Sandel, R. B. Reddy and J. C. Mitani, "Adolescent male chimpanzees do not form a dominance hierarchy with their peers," *Primates,* vol. 58, no. 1, pp. 39-49, 2017.

[117] W. L. Vickery, J. S. Brown and G. J. FitzGerald, "Spite: Altruism's evil twin," *Oikos,* vol. 102, no. 2, pp. 413-416, 2003.

[118] J. L. Tracy, E. Mercadante, Z. Witkower and J. T. Cheng, "Chapter Two - The evolution of pride and social hierarchy," *Advances in Experimental Social Psychology,* vol. 62, pp. 52-114, 2020.

[119] J. E. Samson, M. A. Delgado, D. F. Louis and T. Ojanen, "Bullying and social goal-setting in youth: A meta-analysis," *Social Development,* vol. 31, no. 4, pp. 945-961, 2022.

[120] J. L. Bolton, J. Molet, L. Regev, Y. Chen, N. Rismanchi, E. Haddad, D. Z. Yang, A. Obenaus and T. Z. Baram, "Anhedonia following early-life adversity involves aberrant interactions of reward and anxiety circuits and is reversed by partial silencing of amygdala corticotropin-releasing hormone gene," *Biological Psychiatry,* vol. 83, no. 2, pp. 137-147, 2017.

[121] B. T. Stone, P. Antonoudiou, E. Teboul, G. Scarpa, G. Weiss and J. L. Maguire, "Early life stress impairs VTA coordination of BLA network and behavioral states," *Journal of Neuroscience,* vol. 45, no. 11, p. e0088242025, 2025.

[122] S. Sacu, M. Dubois, F. H. Hezemans, P.-M. Aggensteiner, M. Manninger, D. Brandeis, T. Banaschewski, T. U. Hauser and N. E. Holtz, "Early-life adversities are associated with lower expected value signaling in the adult brain," *Biological Psychiatry,* vol. 96, no. 12, pp. 948-958, 2024.

[123] M. S. Menken, P. J. Rodriguez Rivera, A. Isaiah, T. Ernst, C. C. Cloak and L. Chang, "Longitudinal alterations in brain morphometry mediated the effects of bullying victimization on cognitive development in preadolescents," *Developmental Cognitive Neuroscience,* vol. 61, p. 101247, 2023.

[124] X. Ma, L. L. Stewin and D. L. Mah, "Bullying in school: Nature, effects and remedies," *Research Papers in Education,* vol. 16, no. 3, pp. 247-270, 2001.

[125] R. A. Fein, B. Vossekuil, W. S. Pollack, R. Borum, W. Modzeleski and M. Reddy, "Threat Assessment in Schools: Guide," United States Secret Service and United States Department of Education, Washington DC, 2002.

[126] J. A. Fox, M. J. DeLateur and J. A. Fox, "Mass shootings in America: Moving beyond Newtown," *Crime and Justice,* vol. 18, no. 1, pp. 125-145, 2013.

[127] S. Thompson and K. Kyle, "Understanding mass school shootings: Links between personhood and power in the competitive school environment," *Journal of Primary Prevention,* vol. 26, no. 5, pp. 419-438, 2005.

[128] J. S. Hong, A. Valido and L. E. Robinson, "Understanding the Antecedents of School Shootings: The Case of the Robb Elementary School (pp 1-26)," in *Impact of Gun Violence in School Systems*, First ed., IGI Global, 2024, p. 292.

[129] I. Stewart, J. Wertz and H. Jen, "School-associated violent deaths in the United States: a comprehensive review of the literature," *Journal of School Violence,* vol. 21, no. 4, pp. 355-380, 2022.

[130] R. Oliver, J. H. Hoover and R. Hazler, "The perceived roles of bullying in small-town Midwestern schools," *Journal of Counseling and Development,* vol. 72, no. 4, pp. 416-420, 1994.

[131] P. Randall, Adult Bullying: Perpetrators and Victims, Routledge, 1997, p. 192.

[132] S. Einarsen, H. Hoel and C. Cooer, Bullying and Emotional Abuse in the Workplace: International Perspectives in Research and Practice, Milton Park, Didcot: Taylor & Francis, 2003, p. 420.

[133] R. Gupta, A. Bakhshi and S. Einarsen, "Investigating workplace bullying in India: Psychometric properties, validity, and cutoff scores of negative acts questionnaire–revised," *Sage Open,* vol. 7, no. 2, 2017.

[134] H. Hoel and S. Einarsen, "Workplace bullying," in *Wiley Encyclopedia of Management*, vol. Human Resource Management, D. Guest and D. Needle, Eds., John Wiley & Sons, Ltd, 2015.

[135] S. L. Johnson, "An ecological model of workplace bullying: A guide for intervention and research," *Nursing Forum,* vol. 46, no. 2, pp. 55-63, 2011.

[136] G. Rajesh, "Ragging: A public health problem in India," *Indian Journal of Medical Sciences,* vol. 63, no. 6, pp. 263-271, 2009.

[137] M. Ciby and R. P. Raya, "Workplace Bullying: A Review of the Defining Features, Measurement Methods and Prevalence across Continents," *IIM Kozhikode Society & Management Review,* vol. 4, no. 1, pp. 38-47, 2015.

[138] H. Leymann, "Mobbing and psychological terror at workplaces," *Violence and Victims,* vol. 5, no. 2, pp. 119-126, 1990.

[139] H. Leymann, "The content and development of mobbing at work," *European Journal of Work and Organizational Psychology,* vol. 5, no. 2, pp. 165-184, 2008.

[140] M. Priesemuth and M. Schminke, "Toxic work climates: An integrative review and development of a new construct and theoretical framework," *Journal of Applied Psychology,* vol. 109, no. 9, pp. 1344-1376, 2024.

[141] A. Berlingieri and P. D'Cruz, "Depersonalized bullying: An Emergent concern in the contemporary workplace," in *Concepts, Approaches and Methods: Handbooks of Workplace Bullying, Emotional Abuse and Harassment 1*, Singapore, Springer Nature, 2021, pp. 195-231.

[142] K. D. Pyke, "Institutional betrayal: Inequality, discrimination, bullying and retaliation in academia," *Sociological Perspectives,* vol. 61, no. 1, pp. 5-13, 2018.

[143] H. A. Washington, Medical Apartheid: The Dark History of Medical Experimentation on Black Americans from Colonial Times to the Present, New York: Doubleday, 2007, p. 512.

[144] V. Mallardi, "Le origini del consenso informato [The origin of informed consent]," *Acta Otorhinolaryngologica Italica,* vol. 25, no. 5, pp. 312-327, 2005.

[145] J. H. Jones, Bad Blood: The Tuskegee Syphilis Experiment, New and Expanded Edition, revised ed., New York, New York: Free Press (Simon & Schuster), 1993, p. 297.

[146] A. M. Hornblum, Acres of Skin, New York: Routledge, 1998, p. 320.

[147] A. M. Hornblum, Sentenced to Science, Paperback ed., H. Washington, Ed., University Park, PA: Pennsylvania State University Press, 2013, p. 232.

[148] U. Tröhler, S. Reiter-Theil and E. Herych, Ethics Codes in Medicine: Foundations and Achievements of Codification Since 1947, Routledge, 1998, p. 372.

[149] G. Corbie-Smith, "The continuing legacy of the Tuskegee syphilis study: Considerations for clinical investigation," *The American Journal of the Medical Sciences,* vol. 317, no. 1, pp. 5-8, 1999.

[150] S. M. Reverby, "The Tuskegee syphilis study," in *Oxford Research Encyclopedia of American History*, Oxford, Oxford University Press, 2022.

[151] A. M. Brandt, "Racism and research: The case of the Tuskegee syphilis study," *The Hastings Center Report,* vol. 8, no. 6, pp. 21-29, 1978.

[152] C. Lin, P. Tu and T. C. Terry, "Moving the needle on racial disparity: COVID-19 vaccine trust and hesitancy," *Vaccine,* vol. 40, no. 1, pp. 5-8, 2022.

[153] S. Shastry, J. Carpenter, A. Krotulski, J. Brent, P. Wax, K. Aldy, S. Campleman, R. Culbreth, A. Falise, A. Hughes, R. G. Hendrickson, A. Amuducci, B. Judge, C. Meaden, D. P. Calello, J. Buchanan, J. Shulman, M. Levine and E. Schwarz, "Disparities in treatment and referral after an opioid overdose among emergency department patients," *Journal of the American Medical Association,* vol. 8, no. 7, p. e2518569, 2025.

[154] H. R. Searight and L. R. Johnson, "Psychiatry and the African American community: A history of diagnosis and treatment for social control," in *The Psychology of Inequity: Motivation and Beliefs*, Santa Barbara, California: Praeger, 2022, pp. 213-236.

[155] R. Skloot, The Immortal Life of Henrietta Lacks, New York: Crown, 2010, p. 384.

[156] R. Weiss, "Volunteers at Risk in Medical Studies," *Washington Post,* 1 August 1998.

[157] E. A. Howell, T. Janevic, P. L. Herbert, N. N. Egorova, A. Balbierz and J. Zeitlin, "Differences in morbidity and mortality rates in black, white, and Hispanic very preterm infants among New York City hospitals," *Journal of the American Medical Association Pediatrics,* vol. 172, no. 3, pp. 269-277, 2018.

[158] S. S. Bajaj and F. C. Stanford, "Beyond Tuskegee--Vaccine distrust and everyday racism," *New England Journal of Medicine,* vol. 384, p. e12, 2021.

[159] L. R. Snowden and J. M. Snowden, "Coronavirus trauma and African Americans' mental health: Seizing opportunities for transformational change," *International Journal of Environmental Research and Public Health,* vol. 18, no. 7, 2021.

[160] S. L. Carson, A. Casillas, Y. Castellon-Lopez, L. N. Mansfield, D. Morris, J. Barron, E. Ntekume, R. Landovitz, S. D. Vassar, K. C. Norris, S. M. Dubinett, N. A. Garrison and A. F. Brown, "COVID-19 vaccine decision-making factors in racial and ethnic minority communities in Los Angeles, California," *Journal of the American Medical Association Network Open,* vol. 4, no. 9, pp. e2127582-e2127582, 2021.

[161] D. J. Lundberg, E. Wrigley-Field, A. Cho, R. Raquib, E. O. Nsoesie, E. Paglino, R. Chen, M. V. Kiang, A. R. Riley, Y.-H. Chen, M. L. Charpignon, K. Hempstead, S. H. Preston, I. T. Elo, M. M. Glymour and A. C. Stokes, "Geographic and temporal patterns in Covid-19 mortality by race and ethnicity in the United States from March 2020 to February 2022," *Journal of the American Medical Association,* vol. 6, no. 5, p. e2311098, 2023.

[162] R. Bromley-Dulfano, M. Barnett and L., "Racial inequities and access to COVID-19 treatment," *Journal of the American Medical Association Network,* vol. 8, no. 7, p. e2518459, 2025.

[163] L. I. Appleman, "The captive lab rat: Human medical experimentation in the carceral state," *Boston College Law Review,* vol. 61, no. 1, pp. 1-69, 2020.

[164] A. M. Hornblum, "They were cheap and available: prisoners as research subjects in twentieth century America," *British Medical Journal,* vol. 315, pp. 1437-1441, 1997.

[165] Advisory Committee on Human Radiation Experiments, Frpac, United States. Advisory Committee on Human Radiation Experiments, The Human Radiation Experiments, R. Faden, Ed., New York: Oxford University Press, USA, 1996, p. 620.

[166] R. E. Shore, R. E. Albert, M. Reed, N. Harley and B. S. Pasternack, "Skin cancer incidence among children irradiated for ringworm of the scalp," *Radiation Research,* vol. 100, no. 1, pp. 192-204, 1984.

[167] M. C. Tayade and S. D. Dalvi, "Fundamental ethical issues in unnecessary surgical procedures," *Journal of Clinical Diagnostic Research,* vol. 10, no. 4, pp. JE01-4, 2016.

[168] B. M. Tappin and R. T. McKay, "The illusion of moral superiority," *Social Psychology and Personality Science,* vol. 8, no. 6, pp. 1-9, 2016.

[169] A. M. Lindkwist, L. Koppel and G. Tinghög, "Bounded research ethicality: researchers rate themselves and their field as better than others at following good research practice," *Scientific Reports,* vol. 14, no. 1, p. 3050, 2024.

[170] S. Milgram, "Behavioral study of obedience," *Journal of Abnormal and Social Psychology,* vol. 67, no. 4, pp. 371-378, 1963.

[171] J. D. Birkmeyer, B. N. Reames, P. McCulloch, A. J. Carr, W. B. Campbell and J. E. Wennberg, "Understanding of regional variation in the use of surgery," *Lancet,* vol. 382, no. 9898, pp. 1121-1129, 2013.

[172] H. Lyu, T. Xu, D. Brotman, B. Mayer-Blackwell, M. Cooper, M. Daniel, E. C. Wick, V. Saini, S. Brownlee and M. A. Makary, "Overtreatment in the United States," *PLOS One,* vol. 12, no. 9, p. e0181970, 2017.

[173] D. Korenstein, S. Husain, R. L. Gennarelli, C. White, J. N. Masciale and B. R. Roman, "Impact of clinical specialty on attitudes regarding overuse of inpatient laboratory testing," *Journal of Hospital Medicine,* vol. 13, no. 12, pp. 844-847, 2018.

[174] United States General Accounting Office, "Medical malpractice insurance: Multiple factors have contributed to increased premium rates: Report to congressional requesters," GAO, Washington, DC, 2003.

[175] A. F. Maccioni, "Defensive response to malpractice pressure in health," *International Journal of Law and Conflict Resolution,* vol. 9, no. 1, pp. 6-8, 2021.

[176] L. L. Leape, "Unnecessary surgery," *Health Services Research,* vol. 13, pp. 363-383, 1992.

[177] B. K. Weiner and B. H. Levi, "The profit motive and spine surgery," *Spine,* vol. 29, no. 22, pp. 2588-2591, 2004.

[178] S. Ayabakan, I. Bardhan, Z. Zheng and K. Kirksey, "The impact of health information sharing on duplicate testing," *Management Information Systems Quarterly,* vol. 41, no. 4, pp. 1083-1104, 2017.

[179] S. M. Erickson, B. Outland, S. Joy, B. Rockwern, J. Serchen, R. D. Mire and J. M. Goldman, "Envisioning a better U.S. health care system for all: Health care delivery and payment system reforms," *Annals of Internal Medicine,* vol. 172, no. 2, pp. S33 - S49, 2020.

[180] É. Zola, Au Bonheur des Dames, Paris: Charpentier (reprint Hachette), 1883/1980, p. 368.

[181] T. P. Dompier, Z. Y. Kerr, S. W. Marshall, B. Hainline, E. M. Snook, R. Hayden and J. E. Simon, "Incidence of concussion during practice and games in youth, high school, and collegiate American football players," *Journal of the American Medical Association Pediatrics,* vol. 169, no. 7, p. 659–665, 2015.

[182] M. Marar, N. M. McIlvain, S. K. Fields and R. D. Comstock, "Epidemiology of concussions among United States high school athletes in 20 sports," *American Journal of Sports Medicine,* vol. 40, no. 4, pp. 747-755, 2012.

[183] S. H. Miles and S. Prasad, "Medical ethics and school football," *The American Journal of Bioethics,* vol. 16, no. 1, pp. 6-10, 2016.

[184] M. A. McCrera, A. Shah, S. Duma, S. Rowson, J. M. T. W. Harezlak, S. P. Broglio, C. C. Giza, J. C. K. L. Goldman, M. N. Houston, G. McGinty, J. .. Jackson, K. Guskiewicz, J. P. MIhalik, M. A. Brooks, P. Pasquina and B. Stemper, "Opportunities for prevention of concussion and repetitive head impact exposure in college football players: A concussion assessment, research, and education (CARE) consortium study," *Journal of The American Medical Association Neurology,* vol. 78, no. 3, p. 346–350, 2021.

[185] W. T. Tsushima, N. Shirakawa and O. Geling, "Neurocognitive functioning and symptom reporting of high school athletes following a single concussion," *Applied Neuropsychology,* vol. 2, no. 1, p. 13–16, 2013.

[186] P. R. Shankar, S. K. Fields, C. L. Collins, R. W. Dick and R. D. Comstock, "Epidemiology of high school and collegiate football injuries in the United States, 2005–2006," *American Journal of Sports Medicine,* vol. 35, no. 8, p. 1295–1303, 2007.

[187] J. Nocera and B. Strauss, Indentured: The Inside Story of the Rebellion Against the NCAA, Portfolio, 2016, p. 384.

[188] R. Clark and A. G. Stanfill, "A systematic review of barriers and facilitators for concussion reporting behavior among student athletes," *Journal of Trauma Nursing,* vol. 26, no. 6, pp. e3-e4, 2019.

[189] M. E. I. Rutledge, "Understanding holistic development in Black American male student-athletes through education models," *Negro Educational Review,* vol. 71, no. 1-4, pp. 131-163, 2020.

[190] C. J. Nowinski and R. C. Cantu, "Who will protect the brains of college football players?," *Journal of the American Medical Association Neurology,* vol. 78, no. 3, pp. 273-274, 2021.

[191] J. L. Scherer and M. L. Anson, Community Colleges and the Access Effect: Why Open Admissions Suppress Achievement, New York: Palgrave MacMillan, 2014, p. 269.

[192] L. Waller and T. Tietjen-Smith, "A national study of community college retention rates segmented by institutional degree of urbanization," *Academic Leadership,* vol. 7, 2009.

[193] A. Caumont, "6 key findings about going to college," 11 2 2014. [Online]. Available: http://www.pewresearch.org/fact-tank/2014/02/11/6-key-findings-about-going-to-college/. [Accessed 11 6 2018].

[194] W. M. Chaput, Democratic Apathy: Exploring the Roots of Millennial Democratic Fatigue (Master's Thesis), Cambridge, MA: Harvard Extension School, 2020, p. 79.

[195] T. D. Snyder, C. de Bray and S. A. Dillow, "Digest of Education Statistics 2016: NCES 2017-094," United States Department of Education, Washington DC, 2016.

[196] M. Korn, "A Smaller Share of High-School Grads Are Heading to College—Especially Poorer Ones," *Wall Street Journal,* 25 11 2015.

[197] United States Bureau of Labor Statistics, "College Enrollment and Work Activity of Recent High School and College Graduates Summary," United States Bureau of Labor Statistics, Washington DC, 2021.

[198] M. T. Nietzel, "News from U.S. Census Bureau: Number of Americans with a bachelor's degree continues to grow," *Forbes,* 22 Feb 2021.

[199] K. McElrath and M. Martin, "Bachelor's degree attainment in the United States: 2005-2019," United States Census Bureau, Washington D.C., 2021.

[200] G. Chen, S. M. Gully and D. Eden, "General self-efficacy and self-esteem: toward theoretical and empirical distinction between correlated self-evaluations," *Journal of Organizational Behavior,* vol. 25, no. 3, pp. 375-395, 2004.

[201] D. Navarro-Mateu, L. Alonso-Larza, M. Gómez-Domínguez, V. Prado-Gascó and S. Valero-Moreno, "I'm not good for anything and that's why I'm stressed: analysis of the effect of self-efficacy and emotional intelligence on student stress using SEM and QCA," *Frontiers in Psychology,* vol. 11, no. 1, p. 295, 2020.

[202] L. van Zyl, C. Olckers and L. van der Vaart, Eds., Multidisciplinary Perspectives on Grit: Contemporary Theories, Assessments, Applications and Critiques, Cham: Springer, 2021.

[203] Federal Reserve Bank of New York Research and Statistics Group, "Quarterly Report on Household Debt and Credit," Federal Reserve Bank of New York Center for Microeconomic Data, New York, 2017.

[204] Z. Friedman, "Student loan debt statistics in 2021: A record $1.7 trillion," *Forbes,* 20 Feb 2021.

[205] Federal Reserve Bank of New York, "Center for Microeconomic Data," 2021. [Online]. Available: https://www.newyorkfed.org/microeconomics/topics/student-debt. [Accessed 20 May 2021].

[206] Federal Reserve Bank of New York, "Center for Microeconomic Data," 2021. [Online]. Available: https://www.newyorkfed.org/microeconomics/topics/student-debt. [Accessed 20 May 2021].

[207] "Holy Bible, King James Version," 1 August 1989. [Online]. Available: https://www.gutenberg.org/ebooks/10. [Accessed 30 July 2025].

[208] L. T. Harris and S. T. Fiske, "Dehumanized perception: A psychological means to facilitate atrocities, torture, and genocide?," *Zeitschrift für Psychologie,* vol. 219, no. 3, pp. 175-181, 2015.

[209] E. W. deLara, Bullying Scars: The Impact on Adult Life and Relationships, New York, NY: Oxford University Press, 2016, p. 296.

[210] D. Wolke, W. E. Copeland, A. Angold and E. J. Costello, "Impact of bullying in childhood on adult health, wealth, crime, and social outcomes," *Psychological Science,* vol. 24, no. 10, pp. 1958-1970, 2013.

[211] A. Jadambaa, H. J. Thomas, J. G. Scott, N. Graves, D. Brain and R. Pacella, "The contribution of bullying victimisation to the burden of anxiety and depressive disorders in Australia," *Epidemiology and Psychiatric Sciences,* vol. 29, p. e54, 2019.

[212] R. M. B. Takizawa and L. Arseneault, "Adult health outcomes of childhood bullying victimization: Evidence from a five-decade longitudinal British birth cohort," *American Journal of Psychiatry,* vol. 171, no. 7, pp. 777-784, 2014.

[213] D. Wolke and S. T. Lereya, "Long-term effects of bullying," *Archives of Diseases in Childhood,* vol. 100, no. 9, pp. 879-885, 2015.

[214] J. A. Sweeting, D. R. Garfin, E. A. Holman and R. C. Silver, "Associations between exposure to childhood bullying and abuse and adulthood outcomes in a representative national U.S. sample," *Child Abuse & Neglect,* vol. 101, p. 104048, 2020.

[215] A. Danese and M. Tan, "Childhood maltreatment and obesity: Systematic review and meta-analysis," *Molecular Psychiatry,* vol. 19, p. 544–554, 2014.

[216] K. A. Schlauch, R. W. Read, M. Koning, I. Neveux and J. J. Grzymski, "Using phenome-wide association studies and the SF-12 quality of life metric to identify profound consequences of adverse childhood experiences on adult mental and physical health in a Northern Nevadan population," *Frontiers in Psychiatry,* vol. 13, 2022.

[217] M. al'Absi, A. Lemieux, R. Westra and S. Allen, "Early life adversity influences stress response association with smoking relapse," *Psychopharmacology,* vol. 234, no. 22, p. 3375, 2017.

[218] E. Mercandagi, A. B. Yazici and E. Yazici, "How do traumatic experiences affect relapse (Alkol ve madde kullanım bozukluklarında travmatik yaşantılar nüksleri)," *Turkish Journal of Clinical Psychiatry,* vol. 26, pp. 132-142, 2023.

[219] J. P. Davis, E. R. Pedersen, J. Tucker, M. Dunbar, A. Rodriguez, R. Seelam and E. J. D'Amico, "Childhood adversity and developmental trajectories of alcohol and cannabis co-use," *Childhood Abuse and Neglect,* vol. 141, p. 106238, 2023.

[220] B. Etain, M. Lajnef, C. Brichant-Petitjean, P. A. Geoffroy, C. Henry, S. Gard, J. P. Kahn, M. Leboyer, A. H. Young and F. Bellivier, "Childhood trauma and mixed episodes are associated with poor response to lithium in bipolar disorders," *Acta Psychiatrica Scandinavica,* vol. 135, no. 4, pp. 319-327, 2016.

[221] A. Danese and S. J. Lewis, "Psychoneuroimmunology of Early-Life Stress: The Hidden Wounds of Childhood Trauma," *Neuropsychopharmacology,* vol. 42, p. 99–114, 2017.

[222] A. Rokach and S. Clayton, "ACE and its implication over the lifespan," in *Adverse Childhood Experiences and their Life-Long Impact*, 1st ed., Academic Press, 2023, pp. 277-335.

[223] A. M. Dettmer and D. E. Chusyd, "Early life adversities and lifelong health outcomes: A review of the literature on large, social, long-lived nonhuman mammals," *Neuroscience and Biobehavioral Reviews,* vol. 152, p. 105297, 2023.

[224] E. Gorman, C. Harmon, S. Mendolia, A. Staneva and I. Walker, "Adolescent School Bullying Victimisation and Later Life Outcomes (Working Paper No 20-05)," Centre for Education Policy and Equalising Opportunities (CEPEO), UCL Institute of Education, 2020., London, 2020.

[225] F. Vergunst, M. Commisso, M.-C. Geoffroy, C. Temcheff, M. Poirier, J. Park, F. Vitaro, R. Tremblay, S. Côté and M. Orri, "Association of childhood externalizing, internalizing, and comorbid symptoms with long-term economic and social outcomes," *Journal of the American Medical Association,* vol. 6, no. 1, p. e2249568, 2023.

[226] N. Duque-Wilckens, M. Q. Steinman, M. Busnelli, B. Chini, S. Yokoyama, P. M., S. Laredo, R. Hao, A. M. Perkeybile, M. V. A., P. B. Tan, K. L. Bales and B. C. Trainor, "Oxytocin receptors in the anteromedial bed nucleus of the stria terminalis promote stress-induced social avoidance in female California mice," *Biological Psychiatry,* vol. 83, no. 3, pp. 203-213, 2018.

[227] S. L. Awalt, L. Boghean, D. Klinkebiel and R. Strasser, "A dog's life: Early life histories influence methylation of glucocorticoid (NR3C1) and oxytocin (OXTR) receptor genes, cortisol levels, and attachment styles," *Developmental Psychobiology,* vol. 66, no. 3, p. e22482, 2024.

[228] P. Gilbert, "Evolutionary Functional Analysis: the study of social mentalities, social rank and caring compassion," in *Making an Impact on Mental Health: The Applications of Psychological Research*, First ed., Routledge, 2023, p. 198.

[229] S. Mavlonova, D. Muxammadiyev and E. Jabborova, "The role of hormones of happiness and joy in human life," *Journal of Natural Science,* vol. 1, no. 10, 2023.

[230] M. Hutchinson, M. H. Vickers, L. Wilkes and D. Jackson, "'The worse you behave, the more you seem, to be rewarded': Bullying in nursing as organizational corruption," *Employee Responsibilities and Rights Journal,* vol. 21, no. 3, pp. 213-229, 2009.

[231] J. B. Koh, J. S. Wong and J.-B. Koh, "Survival of the fittest and the sexiest: Evolutionary origins of adolescent bullying," *Journal of Interpersonal Violence,* vol. 32, no. 17, pp. 2668-2690, 2015.

[232] M. Lin, D. Wolke, S. Schneider and J. Margraf, "Bullying history and mental health in university students: the mediator roles of social support, personal resilience, and self-efficacy," *Frontiers in Psychiatry,* vol. 10, no. 1, p. 960, 2020.

[233] A. A. Volk, A. V. Dane and Z. A. Marini, "What is bullying? A theoretical redefinition," *Developmental Review,* vol. 34, no. 4, p. 327–343, 2014.

[234] J. L. Pouwels, T. A. M. Lansu and A. H. N. Cillessen, "Participant roles of bullying in adolescence: Status characteristics, social behavior, and assignment criteria," *Aggressive Behavior,* vol. 42, no. 3, pp. 239-253, 2015.

[235] A. A. Volk, A. V. Dane, Z. A. Marini and T. Vaillancourt, "Adolescent bullying, dating, and mating: Testing an evolutionary hypothesis," *Evolutionary Psychology,* vol. 13, no. 4, pp. 1-11, 2015.

[236] J. Pronk, N. C. Lee, D. Sandhu, K. Kaur, S. Kaur, T. Olthof and F. A. Goossens, "Associations between Dutch and Indian adolescents' bullying role behavior and peer-group status: Cross-culturally testing an evolutionary hypothesis," *International Journal of Behavioral Development,* vol. 41, no. 6, pp. 735-742, 2016.

[237] T. Pozzoli and G. Gini, "Longitudinal relations between students' social status and their roles in bullying: the mediating role of self-perceived social status," *Journal of School Violence,* vol. 20, no. 1, pp. 76-88, 2021.

[238] M. Wiertsema, C. Vrijen, R. van der Ploeg, M. Sentse and T. Kretschmer, "Bullying perpetration and social status in the peer group: A meta-analysis," *Journal of Adolescence,* vol. 95, no. 1, pp. 34-55, 2023.

[239] S. Pabian, F. Dehue, T. Völlink and H. Vandebosch, "Exploring the perceived negative and positive long-term impact of adolescent bullying victimization: A cross-national investigation," *Aggressive Behavior,* vol. 48, no. 2, pp. 205-218, 2021.

[240] J. Lidberg, S. Berne and A. Frisén, "Challenges in emerging adulthood related to the impact of childhood bullying victimization," *Emerging Adulthood,* 4 Feb 2022.

[241] F. V. Gomes, X. Zhu and A. A. Grace, "The pathophysiological impact of stress on the dopamine system is dependent on the state of the critical period of vulnerability," *Molecular Psychiatry,* vol. 25, pp. 3278-3291, 2020.

[242] T. Vaillancourt and A. H. Farrell, "Mean kids become mean adults: Trajectories of indirect aggression from age 10 to 22," *Aggressive Behaviour,* pp. 1-11, 2021.

[243] D. G. Blanchflower and A. Bryson, "The adult consequences of being bullied in childhood," *Social Science and Medicine,* vol. 345, p. 116690, 2024.

[244] A. V. Georgiev, A. C. Klimczuk, D. M. Traficonte and D. Maestripieri, "When violence pays: A cost-benefit analysis of aggressive behavior in animals and humans," *Evolutionary Psychology,* vol. 11, no. 3, pp. 678-699, 2013.

[245] R. W. Wrangham, "Chimpanzee violence is a serious topic: A response to Sussman and Marshak's critique of demonic males: Apes and the origins of human violence," *Global Nonkilling Working Papers,* vol. 1, pp. 29-47, 2010.

[246] R. W. Wrangham, "Evolution of coalitionary killing," *American Journal of Physical Anthropology,* vol. 110, no. 29, 1999.

[247] D. L. Uliana, F. V. Gomes and A. A. Grace, "Stress impacts corticoamygdalar connectivity in an age-dependent manner," *Neuropsychopharmacology,* vol. 46, pp. 731-740, 2021.

[248] R. G. Nelson, "Patriarchy and power in anthropological world building and everyday life," *Current Anthropology,* vol. 62, no. S23, pp. S000-S000, 2021.

[249] L. M. Renner and S. D. Whitney, "Risk factors for unidirectional and bidirectional intimate partner violence among young adults," *Child Abuse & Neglect,* vol. 36, no. 1, pp. 40-52, 2012.

[250] E. Mayeza and D. Bhana, "Boys and bullying in primary school: Young masculinities and the negotiation of power," *South African Journal of Education,* vol. 41, no. 1, 2021.

[251] C. Salmivalli, A. Huttunen and K. M. J. Lagerspetz, "Peer networks and bullying in schools," *Scandinavian Journal of Psychology,* vol. 38, no. 4, 2008.

[252] A. Kaukiainen, A. Salmivalli, K. Lagerspetz, M. Tamminen, M. M. Vauras, H. Mäki and E. Poskiparta, "Learning difficulties, social intelligence, and self–concept: Connections to bully–victim problems," *Scandinavian Journal of Psychology,* vol. 43, no. 3, 2008.

[253] C. Salmivalli, "Bullying and the peer group: A review," *Aggression and Violent Behavior,* vol. 15, no. 2, pp. 112-120, 2010.

[254] M. B. Mulder and C. T. Ross, "Unpacking mating success and testing Bateman's principles in a human population," *Proceedings of the Royal Society B,* vol. 286, p. 20191516, 2019.

[255] T. Janicke, I. K. Häderer and N. Anthes, "Darwinian sex roles confirmed across the animal kingdom," *Science Advances,* vol. 2, no. 2, 2016.

[256] A. A. Volk and A. V. A.-J. E. Dane, "Is adolescent bullying an evolutionary adaptation? A 10-year review," *Evolutionary Psychology Review,* vol. 34, pp. 2351-2378, 2022.

[257] K. L. Chester, N. H. Spencer, L. Whiting and F. M. Brooks, "Association between experiencing relational bullying and adolescent health-related quality of life," *The Journal of School Health,* vol. 87, no. 11, pp. 865-872, 2017.

[258] M. Levine, "How not to play the game," *Bloomberg Business Week,* 30 December 2022.

[259] R. G. Tedeschi and L. G. Calhoun, "The posttraumatic growth inventory: Measuring the positive legacy of trauma," *Journal of Traumatic Stress,* vol. 9, no. 3, p. 455–471, 1996.

[260] P. D. Colzizzo, Director, *Britany Runs a Marathon.* [Film]. USA: Amazon Studios, 2019.

[261] K. Onoda, Y. Okamaoto, K. Nakashima, H. Nittono, S. Yoshimura, S. Yamawaki, S. Yamaguchi and M. Ura, "Does low self-esteem enhance social pain? The relationship between trait self-esteem and anterior cingulate cortex activation induced by ostracism," *Social Cognitive and Affective Neuroscience,* vol. 5, no. 4, pp. 385-392, 2010.

[262] P. C. Henry, "Life stress, explanatory style, hopelessness, and occupational stress," *International Journal of Stress Management,* vol. 12, no. 3, pp. 241-256, 2005.

[263] S. Kellens, F. Dieusaert, J. De Wilde, J. Spilt and D. Baeyens, "The impact of an interaction-based classroom program on executive function development in low-SES preschoolers: first support for effectiveness," *Frontiers in Education,* vol. 8, p. 1149977, 2023.

[264] P. Due, J. Merl, Y. Harel-Fisch, M. T. Damsgaard, B. E. H. J. Holstein, C. Currie, S. N. Gabhainn, M. G. de Matos and J. Lynch, "Socioeconomic inequality in exposure to bullying during adolescence: A comparative, cross-sectional, multilevel study in 35 countries," *American Journal of Public Health,* vol. 99, no. 5, 2009.

[265] G. M. Batsch and H. M. Knoff, "Bullies and their victims: Understanding a pervasive problem in the schools," *School Psychology Review,* vol. 23, no. 2, pp. 165-174, 1994.

[266] D. Sznycer, D. Xygalatas, E. Agey and J. Tooby, "Cross-cultural invariances in the architecture of shame," *Proceedings of the National Academy of Sciences,* vol. 115, no. 39, pp. 9702-9707, 10 September 2018.

[267] Y.-J. Wang, X. Li, C. H. Ng, D.-W. Xu, S. Hu and T.-F. Yuan, "Risk factors for non-suicidal self-injury (NSSI) in adolescents: A meta-analysis," *EClinicalMedicine,* vol. 46, p. 101350, 2022.

[268] D. M. Stone, T. R. Simon, K. A. Fowler, S. R. Kegler, K. Yuan, K. M. Holland, A. Z. Ivey-Stephenson and A. E. Crosby, "Vital signs: Trends in state suicide rates--United States, 1999-2016 and circumstances contributing to suicide--27 states, 2015," *Morbidity and Morality Weekly Report,* vol. 67, no. 22, pp. 617-624, 2018.

[269] M. C. O'Rourke, R. T. Jamil and W. Siddiqui, "Suicide screening and Prevention," *StatPearls [Internet],* p. NBK531453, 2023.

[270] R. Bhatia, "The impact of bullying in childhood and adolescence," *Current Opinion in Psychiatry,* vol. 36, no. 6, pp. 461-465, 2023.

[271] R. H. Mulder, M. López-Vicente, A. P. Cortes Hidalgo, L. R. Steenkamp, B. Güroğlu, H. Tiemeier and R. L. Muetzel, "Connecting the dots: Social networks in the classroom and white matter connections in the brain," *Journal of Child Psychology and Psychiatry, and Allied Disciplines,* vol. 63, no. 12, pp. 1622-1630, 2022.

[272] L. Lim, L. Talozzi and H. Howells, "Atypical brain structural connectivity and social cognition in childhood maltreatment and peer victimization," *BMC Psychiatry,* vol. 24, no. 1, p. 287, 2024.

[273] T. Ke, S. de Simoni, E. Barker and P. Smith, "The association between peer-victimisation and structural and functional brain outcomes: A systematic review," *Journal of Child Psychology and Psychiatry,* vol. 2, no. 2, p. e12081, 2022.

[274] I. S. Palamarchuk and T. Vaillancourt, "Integrative brain dynamics in childhood bullying victimization: Cognitive and emotional convergence associated with stress psychopathology," *Frontiers in Integrative Neuroscience,* vol. 16, p. 782154, 2022.

[275] R. Corr, S. Glier, J. Bizzell, A. Relletier-Baldelli, A. Campbell, C. Killian-Farrell and A. Belger, "Stress-related hippocampus activation mediates the association between polyvictimization and trait anxiety in adolescents," *Social Cognitive and Affective Neuroscience,* vol. 17, no. 8, pp. 767-776, 2022.

[276] X. Wen, Y. Shu, D. Qu, Y. Wang, Z. Cui, X. Zhang and R. Chen, "Associations of bullying perpetration and peer victimization subtypes with preadolescent's suicidality, non-suicidal self-injury, neurocognition, and brain development," *BMC Medicine,* vol. 21, p. 141, 2023.

[277] A. J. Pegna, D. Framorando, Z. Yu, Z. Buhmann, N. Nelson and B. J. W. Dixson, "Hierarchical status is rapidly assessed from behaviourally dominant faces," *Cognitive, Affective, and Behavioral Neuroscience,* vol. 23, pp. 1267-1280, 2023.

[278] K. Sofronoff, E. Dark and V. Stone, "Social vulnerability and bullying in children with Asperger syndrome," *Autism,* vol. 15, no. 3, pp. 355-372, 2011.

[279] H. E. Morton, "Assessment of bullying in autism spectrum disorder: systematic review of methodologies and participant characteristics," *Review Journal of Autism and Developmental Disorders,* pp. 1-16, 2021.

[280] E. Deniz and U. Toseeb, "Sibling bullying and mental health in British and Turkish autistic children and adolescents: The role of social and emotional functioning," *Research in Autism Spectrum Disorders,* vol. 114, p. 102392, 2024.

[281] L. Beckman and L. Hellström, "Cyber bullying among children with neurodevelopmental disorders: A systematic review," *Scandinavian Journal of Psychology,* vol. 61, no. 1, pp. 44-67, 2019.

[282] M. C. Cappadocia, J. A. Weiss and D. Pepler, "Bullying experiences among children and youth with autism spectrum disorders," *Journal of Autism and Developmental Disorders,* vol. 42, no. 2, pp. 266-277, 2012.

[283] L. Little, "Middle-class mothers' perceptions of peer and sibling victimization among children with Asperger's syndrome and nonverbal learning disorders," *Issues in Comprehensive Pediatric Nursing,* vol. 25, no. 1, pp. 43-57, 2002.

[284] M. Ojeda, P. Elipe and R. Del Rey, "LGBTQ+ bullying and cyberbullying: Beyond sexual orientation and gender identity," *Victims and Offenders,* vol. 19, no. 3, pp. 491-512, 2024.

[285] S. R. Kottai, "'Doctors advised to take medicines to remove these thoughts': the violence and violations in sexual mental health care in community mental health programmes in Kerala," *Community Development Journal,* vol. 59, no. 2, pp. 230-251, 2024.

[286] M. Muravyeva and A. Kondakov, "Queering family violence: Introduction to queer family violence studies," *Journal of Family Violence,* pp. https://doi.org/10.1007/s10896-024-00692-1, 2024.

[287] A. Flood, D. J. Hare and P. Wallis, "An investigation into social information processing in young people with Asperger syndrome," *Autism,* vol. 15, no. 5, pp. 601-624, 2011.

[288] B. Woodruff, M. Ternkit, J. Adams and K. Yost, "Association between affective symptoms and school bullying experiences in adults with autism spectrum disorders (P4.156)," *Neurology,* vol. 88, no. 15 Supplement, p. (p4.156), 2017.

[289] J. H. Hoover, R. Oliver and R. J. Hazler, "Bullying: Perceptions of adolescent victims in the Midwestern USA," *School Psychology International,* vol. 13, no. 1, pp. 5-16, 1992.

[290] R. A. Sansone and L. A. Sansone, "Workplace bullying: A tale of adverse consequences," *Innovations in Clinical Neuroscience,* vol. 12, no. 1-2, p. 32–37, 2015.

[291] V. W. Kshirsagar, R. Agarwal and S. B. Bavdekar, "Bullying in schools: Prevalence and short-term impact," *Indian Pediatrics,* vol. 44, no. 1, pp. 25-28, 2007.

[292] F. AlBuhairan, O. A. Abbas, D. E. Sayed, M. Badri, S. Alshahri and N. deVries, "The relationship of bullying and physical violence to mental health and academic performance: A cross-sectional study among adolescents in Kingdom of Saudi Arabia [sic]," *International Journal of Pediatrics and Adolescent Medicine,* vol. 4, no. 2, pp. 51-65, 2017.

[293] D. Lewis, "Bullying at work: The impact of shame among university and college lecturers," *British Journal of Guidance and Counselling,* vol. 32, no. 2, pp. 281-299, 2004.

[294] M. A. Vartia, "Consequences of workplace bullying with respect to the well-being of its targets and the observers of bullying," *Scandinavian Journal of Work, Environment, and Health,* vol. 27, no. 1, pp. 63-69, 2001.

[295] F. D. Karmilasari, "The susceptibility to mental health problems in the future as a serious effect of bullying on adolescent: a systematic review," *International Journal of Science and Society,* vol. 2, no. 3, pp. 295-311, 2020.

[296] W. Ruch, J. Hofmann, T. Platt and R. Proyer, "The state-of-the-art in gelotophobia research: A review and some theoretical extensions," *Humor,* vol. 27, no. 1, pp. 23-45, 2014.

[297] J. Xu, W. Troop-Gordon and K. D. Rudolph, "Within-person reciprocal associations between peer victimization and need for approval," *Developmental Psychology,* vol. 58, no. 10, pp. 1999-2001, 2022.

[298] I. Ettekal, H. Li, A. Chaudhary, W. Luo and R. J. Brooker, "Chronic, increasing, and decreasing peer victimization trajectories and the development of externalizing and internalizing problems in middle childhood," *Development and Psychopathology,* pp. 1-19, 2022.

[299] T. Vaillancourt, E. Duka, S. Becker, L. A. Schmidt, J. Nicol, C. Muir and H. MacMillan, "Peer victimization, depressive symptoms, and high salivary cortisol predict poorer memory in children," *Brain and Cognition,* vol. 77, no. 2, pp. 191-199, 2011.

[300] T. Vaillancourt, S. Hymel and P. McDougall, "The biological underpinnings of peer victimization: Understanding why and how the effects of bullying can last a lifetime," *Theory Into Practice,* vol. 52, no. 4, pp. 241-248, 2013.

[301] M. Camodeca and E. Nava, "The long-term effects of bullying, victimization, and bystander behavior on emotion regulation and its physiological correlates," *Journal of Interpersonal Violence,* vol. 37, no. 3-4, pp. NP2056-NP2075, 2020.

[302] D. Rakesh, S. Whittle, M. A. Sheridan and K. A. McLaughlin, "Childhood socioeconomic status and the pace of structural neurodevelopment: accelerated, delayed, or simply different?," *Trends in Cognitive Science,* vol. 27, no. 9, pp. 833-351, 2023.

[303] K. Ranta, R. Kaltiala-Heino, M. Pelkonen and M. Marttunen, "Associations between peer victimization, self-reported depression and social phobia among adolescents: the role of comorbidity," *Journal of Adolescence,* vol. 32, no. 1, pp. 77-93, 2009.

[304] W. Troop-Gordon, K. D. Rudolph, N. Sugimura and T. D. Little, "Peer victimization in middle childhood impedes adaptive responses to stress: A pathway to depressive symptoms," *Journal of Clinical Child and Adolescent Psychology,* vol. 44, no. 3, pp. 432-435, 2014.

[305] K. D. Rudolph, W. Troop-Gordon and M. Flynn, "Relational victimization predicts children's social-cognitive and self-regulatory responses in a challenging peer context," *Developmental Psychology,* vol. 45, no. 5, pp. 1444-1454, 2009.

[306] M. Tholander, A. Lindberg and D. Svensson, "'A freak that no one can love': difficult knowledge in testimonials on school bullying," *Research Papers in Education,* vol. 35, no. 3, pp. 359-377, 2018.

[307] K. Barchia and K. Bussey, "The psychological impact of peer victimization: Exploring social-cognitive mediators of depression," *Journal of Adolescence,* vol. 33, no. 5, pp. 615-623, 2010.

[308] K. Barchia and K. Bussey, "Individual and collective social cognitive influences on peer aggression: Exploring the contribution of aggression efficacy, moral disengagement, and collective efficacy," *Aggressive Behavior,* vol. 37, no. 2, pp. 107-120, 2011.

[309] R. H. Mulder, E. Walton, A. Neumann, L. C. Houtepen, J. F. Felix and M. J. Bakermans-Kranenburg, "Epigenomics of being bullied: changes in DNA methylation following bullying exposure," *Epigenetics,* vol. 15, no. 6-7, pp. 750-764, 2020.

[310] L. K. Scheffer, C. S. Xu, M. Januszewski, Z. T. S.-Y. Lu, K. J. Hayworth, G. B. Huang, K. Shinomiya, J. Maitlin-Shepard, S. Berg, J. Clements, P. M. Hubbard, W. Katz, L. Umayam and S. M. Plaza, "A connectome and analysis of the adult Drosophila central brain," *Computational and Systems Biology, Neuroscience,* vol. 9, p. e57443, 2020.

[311] J. L. Silberg, W. L. J. Copeland, A. A. Moore, R. Roberson-Nay and T. P. York, "Psychiatric outcomes of bullying victimization: A study of discordant monozygotic twins," *Psychological Medicine,* vol. 46, no. 9, pp. 1875-1883, 2016.

[312] P. P. Zarate-Garza, B. K. Biggs, P. Croarkin, B. Morath, J. Leffler, A. Cuellar-Barboza and S. J. Tye, "How well do we understand the long-term health implications of childhood bullying?," *Harvard Review of Psychiatry,* vol. 25, no. 2, pp. 86-95, 2017.

[313] W. Lee, T. M. Milewski, M. F. Dwortz, R. L. Young, A. D. Gaudet, L. K. Fonken, F. A. Champagne and J. P. Curley, "Distinct immune and transcriptomic profiles in dominant versus subordinate males in mouse social hierarchies," *Brain, Behavior, and Immunity,* vol. 103, pp. 130-144, 2022.

[314] J. W. Luk, J. Wang and B. G. Simons-Morton, "Bullying victimization and substance use among U.S. adolescents: Mediation by depression," *Prevention Science,* vol. 11, no. 4, p. 355–359, 2010.

[315] S. Tharp-Taylor, A. Haviland and E. J. D'Amico, "Victimization from mental and physical bullying and substance use in early adolescence," *Addictive Behaviors,* vol. 34, no. 6-7, pp. 561-567, 2009.

[316] A. E. Lyanda, B. Krishnan and T. J. Adeusi, "Epidemiology of suicidal behaviors among junior and senior high school adolescents: Exploring the interactions between bullying victimization, substance use, and physical inactivity," *Psychiatry Research,* vol. 318, p. 114925, 2022.

[317] S. M. Savell, S. R. Womack, M. N. Wilson, K. Lemery-Chalfant and D. S. Shaw, "Indirect associations between middle-childhood externalizing behaviors and adolescent substance use through late-childhood exposure to violence," *Journal of Youth and Adolescence,* vol. 51, pp. 628-642, 2022.

[318] G. A. Van Kleef, C. Oveis, A. C. Homan, I. van der Löwe and D. Keltner, "Power gets you high: The powerful are more inspired by themselves than by others," *Psychological and Personality Science,* vol. 6, no. 4, pp. 472-480, 2015.

[319] G. A. van Kleef and J. Lange, "How hierarchy shapes our emotional lives: effects of power and status on emotional experience, expression, and responsiveness," *Psychology,* vol. 33, pp. 148-153, 2020.

[320] S. Barnes, Millennials' Perceived Connection of Their Dysfunctional Households and Their Educational Attainment (Doctoral Dissertation), Capella University, 2020.

[321] V. J. Felitti, R. F. Anda, D. Nordenberg, D. F. Williamson, A. M. Spitz, V. Edwards, M. P. Koss and J. S. Marks, "Relationship of childhood abuse and household dysfunction to many of the leading causes of death in adults," *American Journal of Preventive Medicine,* vol. 14, no. 4, pp. 245-258, 1998.

[322] G. Kapur, A. F. Stenson, L. M. Chiodo, V. Delaney-Black, J. H. Hannigan, J. Janisse and H. H. Rantner, "Childhood violence exposure predicts high blood pressure in Black American young adults," *Journal of Pediatrics,* vol. 248, pp. 21-29.e1, 2022.

[323] N. Phillips, B. T. Brown, M. P. Jones, N. Magson, A. Beynon and M. S. Swain, "The association between bullying victimization and back pain in young people: a systematic literature review and meta-analysis," *Pain,* vol. 166, no. 3, pp. 502-510, 2025.

[324] D. J. Nakazawa, Childhood Disrupted: How Your Biography Becomes Your Biology, and How You Can Heal, New York, NY: Atria Books, 2016, p. 304.

[325] J. S. Womersley, J. N. S. Toikumo, S. Malan-Müller, L. L. v. d. Heuvel, N. W. McGregor, S. Seedat and S. J. Hemmings, "Childhood trauma, the stress response and metabolic syndrome: A focus on DNA methylation," *European Journal of Neuroscience,* vol. 55, no. 9-10, pp. 2253-2296, 2022.

[326] W. E. Copeland, D. Wolke, S. T. Lereya, L. Shanahan, C. Worthman and E. J. Costello, "Childhood bullying involvement predicts low-grade systemic inflammation into adulthood," *Proceedings of the National Academy of Science of the United States,* vol. 111, no. 21, pp. 7570-7575, 2014.

[327] K. Uchida, R. Ng, S. A. Vydro, J. E. Smith and D. T. Blumstein, "The benefits of being dominant: Health correlates of male social rank and age in a marmot," *Current Zoology,* p. zoab034, 2021.

[328] N. D. Simons, V. Michopoulos, M. Wilson, L. B. Barreiro and J. Tung, "Agonism and grooming behaviour explain social status effects on physiology and gene regulation in rhesus macaques," *Philosophical Transactions of the Royal Society B,* vol. 377, p. 20210132, 2022.

[329] R. Beijers, I. T. Thije, E. Bolhuis, K. J. O'Donnell, M. S. Tollenaar, I. Shalev, W. J. Hastings, J. L. MacIsaac, D. T. S. Lin, M. Meaney, M. S. Kobor, J. Belsky and C. de Weerth, "Cumulative risk exposure and child cellular aging in a Dutch low-risk community sample," *Psychophisiology,* vol. 60, no. 4, p. e14205, 2023.

[330] H. Minnis, A.-L. van Harmelen, R. Gajwani, J. Rizeq, E. Combet, R. M. Reynolds, C. Gillberg, M. Henderson, F. K. Ho, V. Mondelli, J. Pell and J. Smith, "The bio-exposome: intracellular processes, stress physiology and the environment," *Nature Mental Health,* vol. 2, pp. 132-140, 2024.

[331] S. Matovic, C. Rummel, M. Feldotto and J.-P. Gouin, "Stress, aging, and inflammation," in *Handbook of Stress Series: Stress: Immunology and Inflammation*, vol. 5, Academic Press, 2024, pp. 99-118.

[332] I. Shalev, T. E. Moffitt, K. Sugden, B. Williams, R. M. Houts, A. Danese, J. Mill, L. Arseneault and A. Caspi, "Exposure to violence during childhood is associated with telomere erosion from 5 to 10 years of age: A longitudinal study," *Molecular Psychiatry,* vol. 18, p. 576, 2013.

[333] E. M. Manczak and I. H. Gotlib, "Relational victimization and telomer length in adolescent girls," *Journal of Research on Adolescence,* vol. 30, no. 2, pp. 39-45, 2020.

[334] K. E. Rentscher, J. E. Carroll and C. Mitchell, "Psychosocial stressors and telomer length: A current review of the science," *Annual Review of Public Health,* vol. 41, pp. 223-245, 2020.

[335] M. M. M. Wilton, W. M. Craig and D. J. Pepler, "Emotional regulation and display in classroom victims of bullying: Characteristic expressions of affect, coping styles and relevant contextual factors," *Social Development,* vol. 9, no. 2, pp. 226-245, 2000.

[336] M. Delprato, K. Akyeampong and M. Dunne, "The impact of bullying on students' learning in Latin America: A matching approach for 15 countries," *Educational Development,* vol. 52, pp. 37-57, 2017.

[337] S. W. Twemlow, P. Fonagy, F. Sacco and J. R. J. Brethour, "Teachers who bully students: A hidden trauma," *International Journal of Social Psychiatry,* vol. 52, no. 3, pp. 187-198, 2006.

[338] K. S. Gusfre, J. Støen and H. Fandrem, "Bullying by teachers towards students--A scoping review," *International Journal of Bullying Prevention,* vol. 5, pp. 331-347, 2022.

[339] *Jamie S. Nabozny, Plaintiff-Appellant, v. Mary Podlesny, William Davis, Thomas Blauert, et al., Defendants-Appellees,* 1996.

[340] G. O'Farrell, C. M. Grimard, J. L. Power and R. T. Lee, "Targets of Workplace Bullying and Mistreatment: Helpless Victims or Active Provocateurs?," in *Pathways of Job-related Negative Behaviours*, first ed., Singapore, Springer, 2021, pp. 297-336.

[341] P. H. Bhuptani and T. L. Messman, "Role of blame and rape-related shame in distress among rape victims," *Psychological Trauma: Theory, Research, Practice, and Policy,* vol. 15, no. 4, pp. 557-566, 2023.

[342] S. M. A. Kazmi, R. Iftikhar and M. U. Fayyaz, "'It is all her fault': psychosocial correlates of the negative attitudes towards rape victims among the general population of Pakistan," *Egyptian Journal of Forensic Sciences,* vol. 13, no. 2, 2023.

[343] C. R. Gravelin, M. Biernat and E. Kerl, "Assessing the impact of media on blaming the victim of acquaintance rape," *Psychology of Women Quarterly,* vol. 48, no. 2, pp. 209-231, 2024.

[344] H. Ardent, Banality of Evil, New York: Viking Press, 1963, p. 312.

[345] H. Kuch, "The ritual of humiliation: Exploring symbolic vulnerability," in *Humiliation, Degradation, Dehumanization: Human Dignity Violated*, New York, Springer, 2011, pp. 37-56.

[346] G. Sereny, Into That Darkness: An Examination of Conscience, New York: Vintage Books, 1977/ 1983, p. 379.

[347] C. Rayner, "Workplace bullying: Do something!," *Journal of Occupational Health and Safety, Australia and New Zealand,* vol. 14, no. 6, pp. 581-586, 1998.

[348] C. Salmivalli, J. Karhunen and K. M. J. Lagerspetz, "How do the victims respond to bullying?," *Aggressive Behavior,* vol. 22, pp. 99-109, 1996.

[349] T. Idsoe, T. Vaillancourt, A. Dyregrov, K. A. Hagen, T. Ogden and A. Naerde, "Bullying victimization and trauma," *Frontiers In Psychiatry,* vol. 13, p. 1602, 2021.

[350] Riley, "Bullies, blades, barricades: Practical considerations for working with adolescents expressing concerns regarding gender and identity," in *Today's Transdender Youth: Health, Well-Being and Opportunities For Resilience (170 pp)*, Abingdon, Oxon: Routledge, 2019, pp. 89-97.

[351] J. B. Rotter, "Generalizing expectancies for internal versus external control of reinforcement," *Psychological Monographs,* vol. 80, no. 1, pp. 16-26, 1966.

[352] J. B. Rotter, "Some problems and misconceptions related to the construct of internal versus external control of reinforcement," *Journal of Consulting and Clinical Psychology,* vol. 43, no. 1, pp. 56-57, 1975.

[353] M. Christensen and A. Evans-Murray, "Gaslighting in nursing academia: A new or established covert form of bullying?," *Nursing Forum (Early View),* pp. 1-8, 2021.

[354] S. M. Yildiz, "The relationship between bullying and burnout: An empirical investigation of Turkish professional football players," *Sport, Business and Management,* vol. 5, no. 1, 2015.

[355] M. Boyaci and M. B. Özhan, "The role of hope and family relations of school burnout among secondary school students: A structural equation modeling," *Education and Science,* vol. 43, no. 195, pp. 137-150, 2018.

[356] J. M. Nijboer, Symptoms of Distress and Imbalance in Children: A Study About the Effects of Individual and Environmental Factors on Symptoms of Distress and Imbalance in Children, Groningen: University of Groningen, 2009, p. 117.

[357] A. Hosseini, S. H. Mousavi, F. Hajibabaee and S. Haghani, "The relationship between workplace bullying and professional self-concept in Iranian nurses," *Nursing Open,* vol. 8, no. 1, pp. 232-240, 2020.

[358] L. Fernández-Gutiérrez and M.-P. Mosteiro-Díaz, "Bullying in nursing students: An integrative literature review," *International Journal of Mental Health Nursing,* vol. 30, no. 4, p. 12854, 2021.

[359] M. Duffy, "The psychosocial impact of workplace bullying and mobbing on targets," in *Workplace Bullying and Mobbing in the United States (641 pp)*, vol. 1, Santa Barbara, CA: ABC CLIO, 2018, pp. 131-150.

[360] E. Oberle and K. A. Schonert-Reichl, "Stress contagion in the classroom? The link between classroom teacher burnout and morning cortisol in elementary school students," *Social Science & Medicine,* vol. 159, pp. 30-37, 2016.

[361] P. Choudhary, D. Pandita and M. Kunte, "It doesn't stop here: workplace bullying and family incivility mediating role of psychological distress," *Current Psychology,* vol. 42, p. 30852–30862, 2023.

[362] S. T. S. Hsiao, S. C. Ma, S. L. Guo, C. C. Kao, J. C. Tsai, M. H. Chung and H. C. Huang, "The role of workplace bullying in the relationship between occupational burnout and turnover intentions of clinical nurses," *Applied Nursing Research,* vol. 68, p. 151483, 2022.

[363] L. N. Riffle, K. M. Kelly, C. E. Demaray, C. E. Malecki, A. M. Santuzzi, D. J. Rodriguez-Harris and J. D. Emmons, "Associations among bullying role behaviors and academic performance over the course of an academic year for boys and girls," *Psychology,* vol. 86, pp. 49-63, 2021.

[364] A. W. Crescioni and R. F. Baumeister, "Alone and aggressive: Social exclusion impairs self-control and empathy and increases hostile cognition and aggression," in *Bullying, Rejection and Peer Victimization: A Social Cognitive Neuroscience Perspective. (396 pp)*, New York, New York: Springer, 2009, pp. 251-277.

[365] A. J. Thomas and B. W. Sarnecka, "Infants choose those who defer in conflicts," *Current Biology,* vol. 29, no. 3, pp. 2183-2189.e5, 2019.

[366] M. Amakusa, X. Meng and Y. Kanakogi, "Children's social evaluation toward prestige-based and dominance-based powerholders," *BMC Research Notes,* vol. 15, p. 180, 2022.

[367] A. L. Ruba, R. McMurrty, S. E. Gaither and M. P. Wilbourn, "How White American Children Develop Racial Biases in Emotion Reasoning," *Affective Science,* vol. 3, pp. 21-23, 2022.

[368] S. W. Stoughton, J. J. Noble and G. P. Alpert, Evaluating Police Uses of Force, New York: New York University Press, 2021, p. 256.

[369] C. Qu, R. Ligneul, J.-B. van der Henst and J.-C. Dreher, "An integrative interdisciplinary perspective on social dominance hierarchies (review)," *Trends in Cognitive Sciences,* vol. 21, no. 11, pp. 893-908, 2017.

[370] L. Smith-Osborne, A. Duong, A. Resendez, R. Palme and J. P. Fadok, "Female dominance hierarchies influence responses to psychosocial stressors," *Current Biology,* vol. 33, no. 8, pp. 1535-1549.e5, 2023.

[371] B. Chapais, "Alliances as a means of competition in primates: Evolutionary, developmental, and cognitive aspects," *American Journal of Physical Anthropology,* vol. 38, no. S21, pp. 115-136, 1995.

[372] F. B. M. de Waal, Chimpanzee Politics, 25th anniversary (revised) ed., Baltimore: Johns Hopkins University Press, 2007, p. 256.

[373] F. Amici, J. Call and F. Aureli, "Aversion to violation of expectations of food distribution: the role of social tolerance and relative dominance in seven primate species," *Behaviour,* vol. 149, no. 3-4, pp. 345-368, 2012.

[374] F. B. M. de Waal, "The integration of dominance and social bonding in primates," *The Quarterly Review of Biology,* vol. 61, no. 4, 1986.

[375] T. Iñiguez-Berrozpe, S. Orejudo-Hernández, L. Ruiz-Eugenio and C. Elboj-Saso, "School networks of positive relationships, attitudes against violence, and prevention of relational bullying in victim, bystander, and aggressor agents," *Journal of School Violence,* vol. 20, no. 2, pp. 212-227, 2021.

[376] A. A. Volk, D. A. Provenzano, A. H. Farrell, A. V. Dane and E. P. Shulman, "Personality and bullying: Pathways to adolescent social dominance," *Current Psychology,* 2019.

[377] M. Carlson, "I'd rather go along and be considered a man: Masculinity and bystander intervention," *The Journal of Men's Studies,* vol. 16, no. 1, pp. 3-17, 2008.

[378] J. M. Darley and B. Latané, "Bystander intervention in emergencies: Diffusion of responsibility," *Journal of Personality and Social Psychology,* vol. 8, no. 4, pt 1, p. 377–383, 1968.

[379] M. Paull, M. Omari and P. Standen, "When is a bystander not a bystander? A typology of the roles of bystanders in workplace bullying," *Asia Pacific Journal of Human Resources,* vol. 50, no. 3, pp. 351-366, 2012.

[380] D. C. Tredway, H. Yang, J. Yang, A. B. Duke and J. R. Bentley, "Strategic silencing: A political perspective on the workplace bully-bystander relationship," in *Special Topics and Particular Occupations, Professions and Sectors. Handbooks of Workplace Bullying, Emotional Abuse and Harassment*, vol. 4, Singapore, Springer, 2021, pp. 55-74.

[381] T. Jungert, P. Karataş, N. O. Lotti and S. Perrin, "Direct bullying and cyberbullying: Experimental study of bystanders' motivation to defend victims and the role of anxiety and identification with the bully," *Frontiers in Psychology,* vol. 21, 2021.

[382] K. Niven, K. Ng and H. Hoel, "The bystanders of workplace bullying," in *Bullying and Harassment in the Workplace: Theory, Research and Practice (704 pp)*, CRC Press, 2020, pp. 385-408.

[383] E. Menesini, A. Nocentini, B. E. Palladino, A. Frisén, S. Berne, R. C. J. Ortega-Ruiz, H. Scheithauer, A. Schultze-Krumbholz, P. Luik, K. Naruskov, C. Blaya, J. Berthaud and P. K. Smith, "Cyberbullying definition among adolescents: a comparison across six European countries," *Cyberpsychology, Behavior, and Social Networking,* vol. 15, no. 9, pp. 455-463, 2012.

[384] F. Sticca and S. Perren, "Is cyberbullying worse than traditional bullying? Examining the differential roles of medium, publicity, and anonymity for the perceived severity of bullying," *Journal of Youth and Adolescence,* vol. 42, no. 5, pp. 739-750, 2013.

[385] S. Cho and J. M. Lee, "Explaining physical, verbal, and social bullying among bullies, victims of bullying, and bully-victims: Assessing the integrated approach between social control and lifestyles-routine activities theories," *Children and Youth Services Review,* vol. 91, pp. 372-382, 2018.

[386] A. Ray and N. Henry, "Sextortion: A scoping review," *Trauma, Violence and Abuse,* vol. 26, no. 1, pp. 138-155, 2025.

[387] R. L. O'Malley, "Short-term and Long-term impacts of financial sextortion on victim's mental well-being," *Journal of Interpersonal Violence,* vol. 38, no. 13-14, pp. 8563-8592, 2023.

[388] T. E. Waasdorp and C. P. Bradshaw, "The overlap between cyberbullying and traditional bullying," *Journal of Adolescent Health,* vol. 56, no. 5, pp. 483-488, 2015.

[389] D. Schwartz, K. A. Dodge and J. D. Coie, "The emergence of chronic peer victimization in boys' play groups," *Child Development,* vol. 64, no. 6, pp. 1755-1772, 1993.

[390] S. Bergold, D. Kasper, H. Wendt and R. Steinmayer, "Being bullied at school: the case of high-achieving boys," *Social Psychology of Education,* vol. 23, p. 315–338, 2020.

[391] A. A. Volk, J. A. Camilleri, A. V. Dane and Z. A. Marini, "Is adolescent bullying an evolutionary adaptation?," *Aggressive Behavior,* vol. 38, no. 3, pp. 222-238, 2012.

[392] C. Jiang and S. Jiang, "Effects of adverse childhood experiences on late-life mental health: Potential mechanisms based on a nationally representative survey in China," *Archives of Gerontology and Geriatrics,* vol. 100, p. e104648, 2022.

[393] S. T. Lereya, W. E. Copeland, E. J. Costello and D. Wolke, "Adult mental health consequences of peer bullying and maltreatment in childhood: Two cohorts in two countries," *The Lancet Psychiatry,* vol. 2, no. 6, p. 524–531, 2015.

[394] H. Mohamadpour, F. F. T. Navi, S. Heysieattalab, M. Irak, A.-H. Vahabie and B. Nikzad, "How is social dominance related to our short-term memory? An EEG/ERP investigation of encoding and retrieval during a working memory task," *Heliyon,* vol. 10, no. 17, p. e37389, 2024.

[395] E. Menesini, E. Melan and B. Pignatti, "Interactional styles of bullies and victims observed in a competitive and a cooperative setting," *Journal of Genetic Psychology,* vol. 161, no. 3, pp. 261-281, 2000.

[396] D. Lantos, Y. H. Hui Lau, W. Louis and P. Molenberghs, "The neural mechanisms of threat and reconciliation efforts between Muslims and non-Muslims," *Social Neuroscience,* vol. 15, no. 4, pp. 420-434, 2020.

[397] P. Efstathopoulos, F. Andersson, P. A. Melas, L. L. Yang, J. C. Villaescusa, J. Ruegg, T. J. Ekstrom, Y. Forsell, M. R. Galanti and C. Lavebratt, "NR3C1 hypermethylation in depressed and bullied adolescents," *Translational Psychiatry,* vol. 8, p. 121, 2018.

[398] M. Comtois-Cabana, J. M. Buil, N. Provençal and I. Ouellet-Morin, "The impact of school social experiences on socioemotional and behavioral problems: The hypothesized role of DNA methylation," in *Biosocial Interplay During Elementary School*, Cham, Springer Nature, 2022, pp. 97-110.

[399] J. Heumann, S. W. Cole, M. Eisner, D. Ribeaud and E. S. M. J. Grünblatt, "Peer victimization in adolescence alters gene expression and cytokine profiles during transition to adulthood," *Brain, Behavior, and Immunity - Health,* vol. 48, p. 101025, 2025.

[400] Pie, I. et-Morin, C. C. Y. Wong, A. Danese, C. M. Pariante, A. S. Papadopoulos, J. Mill and L. Arseneault, "Increased serotonin transporter gene (SERT) DNA methylation is associated with bullying victimization and blunted cortisol response to stress in childhood: a longitudinal study of discordant monozygotic twins," *Psychological Medicine,* vol. 43, no. 9, pp. 1813-1823, 2013.

[401] G. Campanile, G. Fanelli, C. Fabbri, A. Serretti and J. Mendlewicz, "Epigenetics and aetiology of mental illness," in *Oxford Textbook of Social Psychiatry*, First ed., D. Bhugra, D. Moussaoui and T. J. Craig, Eds., Oxford, Oxford University Press, 2022, pp. 41-48.

[402] C. Beyer and D. J. Stein, "Trauma and resilience," in *Oxford Textbook of Social Psychiatry*, D. Bhugra, D. Moussaoui and T. J. Craig, Eds., Oxford, Oxford University Press, 2022, pp. 153-161.

[403] E. Wilson, H. Crudgington, C. Morgan, C. Hirsch, M. Prina and C. Gayer-Anderson, "The longitudinal course of childhood bullying victimization and associations with self-injurious thoughts and behaviors in children and young people: A systematic review of the literature," *Journal of Adolescence,* vol. 95, no. 1, pp. 5-33, 9 October 2022.

[404] X. Xu, H. Griffiths, M. Eisner, U. Hepp, D. Ribeaud and A. L. Murray, "Developmental associations between bullying victimization and suicidal ideation and direct self-injurious behavior in adolescence and emerging adulthood," *The Journal of Child Psychology and Psychiatry,* vol. 63, no. 7, pp. 820-828, 2022.

[405] M. Nemko, "From stress to genes, baboons to hormones: An interview with Robert Sapolsky," *Psychology Today,* 4 February 2017.

[406] T. Ge, Y. Lui, Q. Han, X. Cheng and Q. Jiang, "Childhood intra- and extra-familial maltreatment and later-life trajectories of depressive symptoms: evidence from China," *BMC Geriatrics,* vol. 24, p. 598, 2024.

[407] C. Musanabaganwa, A. H. Wani, J. Donglasan, S. Fatumo, S. Jansen, J. Mutabaruka, E. Rutembesa, A. Uwineza, E. J. Hermans, B. Roozendaal, D. E. Wildman, L. Mutesa and M. Uddin, "Leukocyte methylomic imprints of exposure to the genocide against the Tutsi in Rwanda: a pilot epigenome-wide analysis," *Epigenomics,* vol. 14, no. 1, pp. 11-25, 2022.

[408] H. G. Burger, "Androgen production in women," *Fertility and Sterility,* vol. 77, no. suppl. 4, pp. 3-5, 2002.

[409] G. R. Dohle, "Androgens and male fertility," *World Journal of Urology,* vol. 21, p. 341–345, 2003.

[410] S. H. Mellon, L. D. Griffin and N. A. Compagnone, "Biosynthesis and action of neurosteroids," *Brain Research News,* vol. 37, no. 1-3, pp. 3-12, 2001.

[411] *Commonwealth vs. David Garabedian,* 1987.

[412] J. K. Brewer, "Involuntary neurotoxicity, criminal responsibility, and the strange case of David Garabedian," *International Journal of Arts, Humanities and Social Science,* vol. 2, no. 8, pp. 2693-2555, 2021.

[413] J. Renner, True Crime Addict: How I Lost Myself in the Mysterious Disappearance of Maura Murray, New York: Thomas Dunne Books, 2016, p. 304.

[414] H. L. Mencken, "The Divine Afflatus," in *Prejudices: Second Series*, New York, Knopf, 1920.

[415] J. L. Loveland, A. Zemella, V. M. Jovanović, G. Möller, C. P. Sager, B. Bastos, K. A. Dyar, L. Fusaini, M. Gahr, L. M. Giraldo-Deck, W. Goymann, D. B. Lank, J. Tokarz, K. Nowick and C. Küpper, "A single gene orchestrates androgen variation underlying male mating morphs in ruffs," *Science,* vol. 387, no. 6732, pp. 406-412, 2025.

[416] J. E. Smith, C. Fitchel, R. K. Holmes, P. M. Kappeler, M. van Vugt and A. V. Jaeggi, "Sex bias in intergroup conflict and collective movements among social mammals: male warriors and female guides," *Philosophical Transactions of the Royal Society B,* 2022.

[417] D. Crews, "The (bi) sexual brain: Science & society series on sex and science," *EMBO Reports,* vol. 13, no. 9, pp. 779-784, 2012.

[418] C. A. Raper and T. J. Fowler, "Why study Schizophyllum?," *Fungal Genetics Reports,* vol. 51, p. article 12, 2004.

[419] T. J. Fowler, M. F. Mitton, L. J. Vaillancourt and C. A. Raper, "Changes in mate recognition through alterations of pheromones and receptors in multisexual mushroom fungus, Schizophyllum commune," *Genetics,* vol. 158, no. 1, pp. 1491-1503, 2001.

[420] E. Kothe, "Mating types and pheromone recognition in Homobasidiomycete Schizophyllum commune," *Fungal Genetics and Mycology,* vol. 27, no. 2-3, pp. 146-152, 1999.

[421] C. W. Petersen and E. A. Fisher, "Mating system of the hermaphroditic coral-reef fish, Serranus baldwini," *Behavioral Ecology and Sociobiology,* vol. 19, pp. 171-178, 1986.

[422] T. Kuwamura, K. Sawada, T. Sunobe, Y. Sakai and T. Kadota, Hermaphroditism and Mating Systems in Fish, Singapore: Springer, 2023, p. 259.

[423] H. N. Lawson, L. R. Wexler, H. K. Wnuk and D. S. Portman, "Dynamic, non-binary specification of sexual state in the C. elegans nervous system," *Current Biology,* vol. 30, no. 18, pp. 3617-3623, 2020.

[424] R. Wrangham and D. Peterson, Demonic Males: Apes and the Origins of Human Violence, New York, NY: Houghton Mifflin, 1996, p. 350.

[425] S. A. Juntti, J. Tollkuhn, M. V. Wu, E. Fraser, T. Soderborg, S. Tan, S.-I. Honda, N. Harada and N. M. Shah, "The androgen receptor governs the execution, but not programming, of male sexual and territorial behaviors," *Neuron,* vol. 66, no. 2, p. 260–272, 2010.

[426] D. J. Albert, M. L. Walsh and R. H. Jonik, "Aggression in humans: What is its biological foundation?," *Neuroscience & Biobehavioral Reviews,* vol. 17, no. 4, pp. 405-425, 1993.

[427] C. J. Reynolds, A. Makhanova, L. Nikonova, L. A. Eckel and P. Conway, "Testosterone and cortisol do not predict rejecting harm or maximizing outcomes in sacrificial moral dilemmas: A preregistered analysis," *Hormones and Behavior,* vol. 136, p. 105063, 2021.

[428] R. M. de Almeida, P. F. Ferrari, S. Parmigiani and K. A. Miczek, "Escalated aggressive behavior: dopamine, serotonin and GABA," *European Journal of Pharmacology,* vol. 526, no. 1-3, p. 51–64, 2005.

[429] A. C. Swann, "Neuroreceptor mechanisms of aggression and its treatment," *Journal of Clinical Psychiatry,* vol. 64, no. suppl. 4, pp. 26-36, 2003.

[430] L. Yuxin, L. Chen, L. Xiaoxia, L. Yue, L. Junjie, L. Youzhu, Z. Huiliang and L. Qicai, "Research progress on the relationship between obesity-inflammation-aromatase axis and male infertility," *Oxidative Medicine and Cellular Longevity,* p. 6612796, 2021.

[431] M. Verderame, V. Migliaccio and R. Scudiero, "Role of estrogen receptors, P450 aromatase, PCNA and p53 in high-fat-induced impairment of spermatogenesis in rats," *Comptes Rendus Biologies,* vol. 341, no. 7-8, pp. 371-379, 2018.

[432] A. M. Kelly, J. A. G. Abreu and R. R. Thompson, "Beyond sex and aggression: testosterone rapidly matches behavioural responses to social context and tries to predict the future," *The Proceedings of the Royal Academy of Sciences B,* vol. 289, p. 20220453, 2022.

[433] V. B. Shahinian, Y. F. Kuo, J. L. Freeman and J. S. Goodwin, "Risk of the 'androgen deprivation syndrome' in men receiving androgen deprivation for prostate cancer," *Archives of Internal Medicine,* vol. 166, no. 4, pp. 465-471, 2006.

[434] A. Mazur and A. Booth, "Testosterone and dominance in men," *Behavioral and Brain Sciences,* vol. 21, no. 3, pp. 353-363, 1998.

[435] M. Zitzmann, "Testosterone and the Brain," *Aging Male,* vol. 9, no. 4, pp. 195-199, 2006.

[436] C. V. Chen, J. L. Brummet, J. S. Lonstein, C. L. Jordan and S. M. Breedlov, "New knockout model confirms a role for androgen receptors in regulating anxiety-like behaviors and HPA response in mice," *Hormones and Behavior,* vol. 65, no. 3, pp. 211-218, 2014.

[437] W. H. Tong, S. Abdulai and A. Vyas, "Testosterone reduces fear and causes drastic hypomethylation of arginine vasopressin promoter in medial extended amygdala of male mice," *Frontiers in Behavioral Neuroscience,* vol. 13, p. 00033, 2019.

[438] C. V. Chen, C. L. Jordan and S. M. Breedlov, "Testosterone works through androgen receptors to modulate neuronal response to anxiogenic stimuli," *Neuroscience Letters,* vol. 753, p. 135852, 2021.

[439] P. E. Gleason, S. D. Michael and J. J. Christian, "Effects of gonadal steroids on agonistic behavior of female Peromyscus leucopus," *Hormones and Behavior,* vol. 12, no. 1, pp. 30-39, 1979.

[440] J. Ou, Y. Wu, Y. Hu, X. Gao, H. Li and P. N. Tobler, "Testosterone reduces generosity through cortical and subcortical mechanisms," *Proceedings of the National Academy of Sciences of the United States of America,* vol. 118, no. 12, p. e2021745118, 2021.

[441] E. Hogervorst, S. Bandelow, M. Combrinck and A. D. Smith, "Low free testosterone is an independent risk factor for Alzheimer's disease," vol. 39, no. 11-12, pp. 1633-1639, 2004.

[442] S. D. Moffat, A. B. Zonderman, E. J. Metter, C. Kawas, M. R. Blackman, S. M. Harman and S. M. Resnick, "Free testosterone and risk for Alzheimer disease in older men," *Neurology,* vol. 62, no. 2, p. 188–193, 2004.

[443] S. D. Moffat and E. Hampson, "A curvilinear relationship between testosterone and spatial cognition in humans: Possible influence of hand preference," *Psychoneuroendocrinology,* vol. 21, no. 3, p. 323–337, 1996.

[444] S. A. Grainger, J. K. Mead, E. J. Vanman and J. D. Henry, "The relationship between testosterone and social cognition in younger and older adults," *Biological Psychology,* vol. 161, p. 108072, 2021.

[445] M. J. Kaufman, G. Kanayama, J. I. Hudson and H. G. Pope Jr., "Supraphysiologic-dose anabolic-androgenic steroid use: A risk factor for dementia?," *Neuroscience and Behavioral Reviews,* vol. 100, pp. 108-207, 2019.

[446] A. Bjørnebekk, T. Kaufmann, L. E. Hauger, S. Klonteig, I. R. Hullstein and L. T. Westlye, "Long-term anabolic-androgenic steroid use is associated with deviant brain aging," *Biological Psychiatry: Cognitive Neuroscience and Neuroimaging,* vol. 6, no. 5, pp. 579-589, 2021.

[447] W. Corsini, A. Esteves, R. Jr., W. C., T. d. A. Hermes, B. Damião and M. R. Rodrigues, "Association between neuronal degeneration and supraphysiological doses of two types of anabolic steroids in rat brain," *Steroids,* vol. 188, p. 109121, 2022.

[448] S. J. Berg and K. E. Wynne-Edwards, "Changes in testosterone, cortisol, and estradiol levels in men becoming fathers," *Mayo Clinic Proceedings,* vol. 76, no. 6, pp. 582-592, 2001.

[449] U. Tural and D. V. Iosifescu, "Neuropeptide Y in PTSD, MDD, and chronic stress: A systematic review and meta-analysis," *Journal of Neuroscience Research,* vol. 98, no. 5, pp. 950-963, 2020.

[450] J. A. French, A. C. Mustoe, J. Cavanaugh and A. K. Birnie, "The influence of androgenic steroid hormones on female aggression in 'atypical' mammals," *Philosophical Transactions of the Royal Society B,* vol. 386, no. 1631, 2013.

[451] R. W. Wrangham, "Two types of aggression in human evolution," *Proceedings of the National Academy of Science,* vol. 115, no. 2, pp. 245-253, 2018.

[452] V. Behringer, T. Deschner, C. Deimel, J. M. G. Stevens and G. Hohmann, "Age-related changes in urinary testosterone levels suggest differences in puberty onset and divergent life history strategies in bonobos and chimpanzees," *Hormones and Behavior,* vol. 66, no. 3, pp. 525-533, 2014.

[453] S. A. Fox, M. N. Muller, N. T. González, D. K. Enigk, Z. P. Machanda, E. Otali, R. Wrangham and M. E. Thompson, "Weak, but not strong, ties support coalition formation among wild female chimpanzees," *Philosophical Transactions of the Royal Society B,* vol. 375, p. 20210427, 2022.

[454] L. Fusani, M. Gahr and J. B. Hutchison, "One display of the ring dove courtship behavior," *General and Comparative Endocrinology,* vol. 122, no. 1, pp. 23-30, 2001.

[455] M. Surbeck, T. Deschner, G. Schubert, A. Weltring and G. Hoffman, " Mate competition, testosterone and intersexual relationships in bonobos, Pan paniscus," *Animal Behaviour,* vol. 83, no. 3, pp. 659-669, 2012.

[456] K. Björkqvist, "Social defeat as a stressor in humans," *Physiology and Behavior,* vol. 73, no. 3, pp. 435-442, 2001.

[457] M. van Geel and P. Vedder, "Does cyberbullying predict internalizing problems and conduct problems when controlled for traditional bullying?," *Scandinavian Journal of Psychology,* vol. 61, no. 2, pp. 307-311, 2019.

[458] N. Hayashi, S. Ando, S. Jinde, S. Fujikawa, N. Okada, R. Toriyama, M. Masaoka, H. Sugiyama, T. Shirakawa, T. Yagi, M. Morita, R. Morishima, T. Kiyono, S. Yamasaki, A. Nishida and K. Kasai, "Social withdrawal and testosterone levels in early adolescent boys," *Psychoneuroendocrinology,* vol. 116, p. 104596, 2020.

[459] K. M. Beaver, M. G. Vaughn, M. Delisi and J. P. Wright, "Anabolic-androgenic steroid use and involvement in violent behavior in a nationally representative sample of young adult males in the United States," *American Journal of Public Health,* vol. 98, no. 12, p. 2185–2187, 2008.

[460] R. M. Sapolsky, "Social status and health in humans and other animals," *Annual Review of Anthropology,* vol. 33, no. 1, p. 393–418, 2004.

[461] E. M. George and K. A. Rosvall, "Bidirectional relationships between testosterone and aggression: A critical analysis of four predictions," *Integrative and Comparative Biology,* vol. 62, no. 3, pp. 474-486, 2022.

[462] P. J. Perry, E. C. Kutscher, L. B. C., W. R. Yates, T. L. Holman and L. Demers, "Measures of aggression and mood changes in male weightlifters with and without androgenic anabolic steroid use," *Journal of Forensic Science,* vol. 48, no. 3, p. 646–651, 2003.

[463] D. Bass, A. Aleman and R. S. Kahn, "Lateralization of amygdala activation: a systematic review of functional neuroimaging studies," *Brain Research Reviews,* vol. 45, no. 2, pp. 96-203, 2004.

[464] M. J. Kaufman, A. C. Janes, J. I. Hudson, B. P. Brennan, G. Kanayama, A. R. Kerrigan, J. E. Jensen and H. G. Pope Jr., "Brain and cognition abnormalities in long-term anabolic-androgenic steroid users," *Drug and Alcohol Dependence,* vol. 152, pp. 47-56, 2015.

[465] H. G. J. Pope, G. Kanayama, J. I. Hudson and M. J. Kaufman, "Review Article: Anabolic-androgenic steroids, violence, and crime: Two cases and literature review," *The American Journal of Addictions,* vol. 30, no. 5, pp. 423-432, 2021.

[466] P. F. Ferrari, A. M. M. Van Erp, W. Tornatzky and K. A. Miczek, "Accumbal dopamine and serotonin in anticipation of the next aggressive episode in rats: microdialysis and anticipation of aggression," *European Journal of Neuroscience,* vol. 17, no. 2, p. 371–378, 2003.

[467] B. Dai, F. Sun, X. Tong, Y. Ding, A. Kuang, T. Osakada, Y. Li and D. Lin, "Responses and functions of dopamine in nucleus accumbens core during social behaviors," *Cell Reports,* vol. 40, no. 8, p. 111246, 2022.

[468] K. K. Anstrom, K. A. Miczek and E. A. Budygin, "Increased phasic dopamine signaling in the mesolimbic pathway during social defeat in rats," *Neuroscience,* vol. 161, no. 1, pp. 3-12, 2009.

[469] S. Grillner, "Expanding the brain's terrain for reward: A previously unknown region in the brainstem controls dopamine activity," *Science,* vol. 387, no. 6732, pp. 362-363, 2025.

[470] K. Zichó, B. Z. Balog, R. Z. Sebestény, J. Brunner, V. Takács, A. M. Barth, C. Seng, A. Orosz, M. Aliczki, H. Sebők, E. Mikics, C. Földy, J. Szabadics and G. Nyiri, "Identification of the subventricular tegmental nucleus as brainstem reward center," *Science,* vol. 382, no. 6732, p. eadr2191, 2025.

[471] T. Vaillancourt, D. deCatanzaro, E. Duku and C. Muir, "Androgen dynamics in the context of children's peer relations: An examination of the links between testosterone and peer victimization," *Aggressive Behavior,* vol. 35, no. 1, pp. 103-113, 2009.

[472] D. S. Chester, C. N. DeWall, K. J. Derefinko, S. Estus, D. R. Lynam, J. R. Peters and Y. Jiang, "Looking for reward in all the wrong places: dopamine receptor gene polymorphisms indirectly affect aggression through sensation-seeking," *Social Neuroscience,* vol. 11, no. 5, pp. 487-494, 2016.

[473] S. A. Golden, M. Jin, C. Heins, M. Venniro, M. Michaelides and Y. Shaham, "Nucleus accumbens Drd1-expressing neurons control aggression self-administration and aggression seeking in mice," *Journal of Neuroscience,* vol. 39, no. 13, pp. 482-2496, 2019.

[474] W. J. Rejeski, P. Brubaker, R. A. Herb, J. R. Kaplan and D. Koritnik, "Anabolic steroids and aggressive behaviour in cynomolgus monkeys," *Journal of Behavioral Medicine,* vol. 11, no. 1, pp. 5-105, 1988.

[475] M. K. Oosthuizen, H. Viljoen and N. C. Bennett, "Effects of season and social status on gonadal function in social natal mole-rats," *Journal of Mammology,* vol. 91, no. 2, pp. 429-436, 2010.

[476] N. C. Bennett, "Reproductive suppression in social Cryptomys damarensis colonies—a lifetime of socially induced sterility in males and females (Rodentia: Bathyergidae)," *Journal of Zoology,* vol. 234, no. 1, p. 25–39, 1994.

[477] D. H. Abbott, "Behaviourally mediated suppression of reproduction in female primates," *Journal of Zoology,* vol. 213, p. 455–470, 1987.

[478] D. H. Abbott, J. K. Hodges and L. M. George, "Social status controls LH secretion and ovulation in marmoset monkeys (Callithrix jacchus)," *Journal of Endocrinology,* vol. 117, no. 3, p. 329–339, 1988.

[479] P. Stokley and J. Bro-Jørgensen, "Female competition and its evolutionary consequences in mammals," *Biological Reviews,* vol. 86, no. 2, pp. 341-366, 2011.

[480] A. Favato, H. Løvlie and O. Leimar, "Effects of social experience, aggressiveness and comb size on contest success in male domestic fowl," *Royal Society Open Science,* vol. 8, p. 201213, 2021.

[481] L. H. Apostólico and J. E. A. R. Marian, "From sneaky to bully: reappraisal of male squid dimorphism indicates ontogenetic mating tactics and striking ejaculate transition," *Biological Journal of the Linnean Society,* vol. 123, no. 3, p. 603–614, 2018.

[482] K. Witte, K. Baumgärtner, C. Röhrig and S. Nöbel, "Test of the deception hypothesis in Atlantic mollies Poecilia mexicana--Does the audience copy a pretended mate choice of others?," *Biology,* vol. 7, no. 3, p. 40, 2018.

[483] C. Garcia, S. Bouret, F. Druelle and S. Prat, "Balancing costs and benefits in primates: ecological and palaeoanthropological views," *Philosophical Transactions of the Royal Society B,* vol. 376, no. 1819, 2021.

[484] A. Dixon and J. Herbert, "Testosterone, aggressive behavior and dominance rank in captive adult male talapoin monkeys (Miopithecus talapoin)," *Physiology and Behavior,* vol. 18, no. 3, pp. 539-543, 1977.

[485] L. Bartoš, D. Schams, G. A. Bubenik, R. Kotrba and M. Tománek, "Relationship between rank and plasma testosterone and cortisol in red deer males (Cervus elaphus)," *Physiology & Behavior,* vol. 101, no. 5, pp. 628-634, 2010.

[486] P. T. Andronis, T. V. J. Layng and I. Goldiamond, "Contingency adduction of 'symbolic aggression' by pigeons," *The Analysis of Verbal Behavior,* vol. 14, pp. 5-17, 1997.

[487] P. T. Andronis, "Symbolic Aggression" by Pigeons: Contingency Coadduction. (Doctoral Dissertation), Chicago: University of Chicago, 1983.

[488] K. Kendrick and R. Drewert, "Testosterone reduces refractory period of stria terminali neurons in the rat brain," *Science,* vol. 204, no. 4395, pp. 877-879, 1979.

[489] N. E. Mbiydzenyuy and L.-A. Qulu, "Stress, hypothalamic-pituitary-adrenal axis, hypothalamic-pituitary-gonadal axis, and aggression," *Metabolic Brain Disease,* 2024.

[490] T. F. Denson, R. Ronay, W. von Hippel and M. M. Schira, "Endogenous testosterone and cortisol modulate neural responses during induced anger control," *Social Neuroscience,* vol. 8, no. 2, pp. 165-177, 2013.

[491] E. C. Jeffress and K. L. Huhman, "Copulatory and agonistic behavior in Syrian hamsters following a social defeat," *Aggressive Behavior,* vol. 39, no. 3, pp. 239-245, 2013.

[492] R. C. Kessler, "Epidemiology of women and depression," *Journal of Affective Disorders,* vol. 74, no. 1, pp. 5-13, 2003.

[493] J. J. Anker and M. E. Carroll, "Females are more vulnerable to drug abuse than males: Evidence from preclinical studies and the role of ovarian hormones," in *Current Topics in Behavioral Neuoscience: Biological Basis of Sex Differences in Psychopharmacology,* vol. 8, J. Neill and J. Kulkarni, Eds., Berlin, Springer, 2010, pp. 73-96.

[494] K. A. Rosvall, D. G. Reichard, S. M. Ferguson, D. J. Wittaker and E. D. Ketterson, "Robust behavioral effects of song playback in the absence of testosterone or corticosterone release," *Hormones and Behavior,* vol. 62, no. 4, pp. 418-425, 2012.

[495] K. A. Rosvall, "Maintenance of variation in sexually selected traits in females: A case study using intrasexual aggression in tree swallows Tachycineta bicolor," *Journal of Avian Biology,* vol. 42, no. 5, pp. 454-462, 2011.

[496] K. A. Rosvall, "Keynote Lecture: Female competition and the challenge hypothesis: Insights from a feisty female bird (abstract)," in *2018 Animal Behavior Conference*, Bloomington IN, 2018.

[497] A. B. Bentz, "Effects of Social Aggression on Tree Swallow (Tachycineta bicolor) Offspring Behavior and Physiology," Appalachian State University, Boone, NC, USA, 2012.

[498] S. Rosenbaum and L. T. Gettler, "With a little help from her friends (and family) part II: Non-maternal caregiving behavior and physiology in mammals," *Physiology and Behavior,* vol. 193, no. A, pp. 12-24, 2018.

[499] K. A. Rosvall, C. M. Bergeon Burns, J. Barske, J. L. Goodson, B. A. Schlinger, D. R. Sengelaub and E. D. Ketterson, "Neural sensitivity to sex steroids predicts individual differences in aggression: Implications for behavioural evolution," *Philosophical Transactions of the Royal Society B,* vol. 279, no. 1742, pp. 3547-3555, 2012.

[500] R. Henriksen, T. G. Groothuis and S. Rettenbacker, "Elevated plasma corticosterone decreases yolk testosterone and progesterone in chickens: Linking maternal stress and hormone-mediated maternal effects," *PLOS ONE,* vol. 6, no. 8, p. e23824, 2011.

[501] W. Goymann, A. Wittenzellner, I. Schwabl and M. Makomba, "Progesterone modulates aggression in sex-role reversed female African black coucals," *Proceedings of the Royal Society B,* vol. 275, no. 1638, pp. 1053-1060, 5 February 2008.

[502] K. V. Casto and D. A. Edwards, "Before, during, and after: How phases of competition differentially affect testosterone, cortisol, and estradiol levels in women athletes," *Adaptive Human Behavior and Physiology,* vol. 2, no. 1, pp. 11-25, 2016.

[503] G. R. Cunha, G. Risbridger, H. Wang, N. J. Place, M. Grumbach, T. J. Cunha, M. Weldele, A. J. Conley, D. Barcellos, S. Agarwal, A. Bhargava, C. Drea, G. L. Hammond, P. Siiteri, E. M. Coscia, M. J. McPhaul, L. S. Baskin and S. E. Glickman, "Development of the external genitalia: Perspectives from the spotted hyena (Crocuta crocuta)," *Differentiation,* vol. 87, no. 1-2, pp. 4-22, 2014.

[504] T. M. Yalcinkaya, P. K. Siiteri, J. L. Vigne, P. Licht, S. Pavgi, L. G. Frank and S. E. Glickman, "A mechanism for virilization of female spotted hyenas in utero," *Science,* vol. 260, no. 5116, pp. 1929-1931, 1993.

[505] J. W. Turner, A. L. Robitaille, P. S. Bills and K. E. Holekamp, "Early-life relationships matter: Social position during early life predicts fitness among female spotted hyenas," *Animal Social Networks,* vol. 90, no. 1, pp. 183-196, 2021.

[506] A. S. Flies, L. S. Mansfield, E. J. Flies, C. K. Grant and K. E. Holekamp, "Socioecological predictors of immune defences in wild spotted hyenas," *Functional Ecology,* vol. 30, no. 9, pp. 1549-1557, 2016.

[507] K. E. Holekamp and M. A. Sawdy, "The evolution of matrilineal social systems in fissiped carnivores," *Philosophical Transactions of the Royal Society B,* vol. 374, p. 20180065, 2019.

[508] A. K. Brown, M. O. H. K. E. Pioon and E. D. Strauss, "Infanticide by females is a leading source of juvenile mortality in a large social carnivore," *American Naturalist,* vol. 198, no. 5, pp. 642-652, 2020.

[509] K. McComb, A. Pusey, C. Packer and J. Grinnell, "Female lions can identify potentially infanticidal males from their roars," *Proceedings of the Royal Society: Biological Sciences,* vol. 22, no. 252, pp. 59-64, 1993.

[510] G. López, M. López-Parra, L. Fernández, G. Ruiz, J. M. Gil-Sánchez and M. A. Simón, "First evidence of non-parental male infanticide in the endangered Iberian Lynx (Lynx pardinus)," *Wildlife Biology in Practice,* vol. 6, no. 3, pp. 67-74, 2010.

[511] D. Pontier and E. Natoli, "Infanticide in rural male cats (Felis catus L.) as a reproductive mating tactic," *Aggressive Behavior,* vol. 25, no. 6, pp. 445-449, 1999.

[512] K. K. Walker, S. Foerster, C. M. Murray, D. Mjungu and A. E. Pusey, "Evaluating adaptive hypotheses for female-led infanticide in wild chimpanzees," *Animal Behaviour,* vol. 180, pp. 23-36, 2021.

[513] N. K. Kulick, S. Cheves, C. Chaves-Cordero, R. Lopez, S. R. Morales, L. M. Fedigan and K. M. Jack, "Female-committed infanticide followed by juvenile-enacted cannibalism in wild white-faced capuchins," *Primates,* vol. 62, no. 6, pp. 1037-1043, 2021.

[514] T. H. Clutton-Brock, P. N. Brotherton, R. Smith, G. M. McIlrath, R. Kansky, D. Gaynor, M. J. O'Riain and J. D. Skinner, "Infanticide and expulsion of females in a cooperative mammal," *Proceedings of the Royal Society B,* vol. 265, pp. 2291-2295, 1998.

[515] T. Schjelderup-Ebbe, "Honsenes stemme. Bidrag til honsenes psykologi," *Naturen,* pp. 262-276, 1913.

[516] T. Schjelderup-Ebbe, "Beiträge zur Sozialpsychologie des Haushuhns [Observation on the social psychology of domestic fowls]," *Zeitschrift für Psychologie und Physiologie der Sinnesorgane. Abt. 1. Zeitschrift für Psychologie,* vol. 88, pp. 225-252, 1922.

[517] J. E. Smith, C. A. Ortiz, M. T. Buhbe and M. van Vugt, "Obstacles and opportunities for female leadership in mammalian societies: A comparative perspective," *The Leadership Quarterly,* vol. 31, no. 2, p. 101267, 2020.

[518] S. Kahlenberg, M. E. Thompson and R. Wrangham, "Female competition over core areas in Pan troglodytes schweinfurthii, Kibale National Park, Uganda," *International Journal of Primatology,* vol. 29, no. 4, pp. 931-947, 2008.

[519] W. Goymann, M. L. East and H. Hofer, "Androgens and the role of female 'hyperaggressiveness' in spotted hyenas (Crocuta crocuta)," *Hormones and Behavior,* vol. 39, no. 1, pp. 83-92, 2001.

[520] C. Vullioud, E. Davidian, B. Wachter, F. Rouset, A. Courtiol and O. P. Höner, "Social support drives female dominance in the spotted hyaena," *Nature Ecology and Evolution,* vol. 3, pp. 71-76, 2019.

[521] J. B. Silk, "Hyena politics: The dynamics of dynasties," *Proceedings of the National Academy of Sciences of the United States of America (commentary),* vol. 116, no. 18, pp. 8644-8645, 2019.

[522] E. D. Strauss and K. E. Holekamp, "Social alliances improve rank and fitness in convention-based societies," *Proceedings of the National Academy of Sciences of the United States of America,* vol. 116, no. 18, pp. 8919-8924, 2019.

[523] L. Frank, S. Glickman and C. Zabel, "Ontogeny of female dominance in the spotted hyaena: Perspectives from nature and captivity. pp. 127-146," in *Symposia of the Zoological Society of London*, vol. 61, Oxford, England: Oxford University Press, 1989, p. 336.

[524] L. A. Jensen-Campbell, W. G. Graziano and S. G. West, "Dominance, prosocial orientation, and female preferences: Do nice guys really finish last?," *Journal of Personality and Social Psychology,* vol. 68, no. 3, p. 427–440, 1995.

[525] S. Jenks, M. Weldele, L. Frank and S. Glickman, "Acquisition of matrilineal rank in captive spotted hyenas: Emergence of a natural social system in peer-reared animals and their offspring," *Animal Behavior,* vol. 50, no. 4, pp. 893-904, 1995.

[526] K. C. S. Yoshida, P. E. Van Meter and K. E. Holekamp, "Variation among free-living spotted hyenas in three personality traits," *Behavior,* vol. 153, no. 13-14, p. 1665–1722, 2016.

[527] D. R. Rubinow and P. J. Schmidt, "Androgens, brain, and behaviour," *American Journal of Psychiatry,* vol. 153, no. 8, pp. 974-984, 1999.

[528] M. V. Wu, D. S. Manoli, E. J. Fraser, J. K. Coats, J. Tollkuhn, S.-I. Honda, N. Harada and N. M. Shah, "Estrogen masculinizes neural pathways and sex-specific behaviors," *Cell,* vol. 139, no. 1, pp. 61-72, 2009.

[529] A. R. Lumia and M. Y. McGinnis, "Impact of anabolic androgenic steroids on adolescent males," *Physiology & Behavior,* vol. 100, no. 3, pp. 199-204, 2010.

[530] R. L. Cunningham, A. R. Lumia and M. Y. McGinnis, "Androgenic anabolic steroid exposure during adolescence: Ramifications for brain development and behavior," *Hormones and Behavior,* vol. 64, no. 2, p. 350–356, 2013.

[531] J. R. Belthoff, A. M. Dufty and S. A. Gauthreaux, "Plumage variation, plasma steroids, and social dominance in male house finches," *The Condor,* vol. 96, pp. 614-625, 1994.

[532] S. E. Lipshutz and K. A. Rosvall, "Nesting strategy shapes territorial aggression but not testosterone: A comparative approach in female and male birds," *Hormones and Behavior,* vol. 133, p. 104995, 2021.

[533] W. Saltzman, "Reproductive skew, cooperative breeding, and eusociality in vertebrates: hormones," in *Encyclopedia of Animal Behavior*, Academic Press, 2019, pp. 448-453.

[534] G. C. Williams, "Pleiotropy, natural selection and the evolution of senescence," *Evolution,* vol. 11, no. 4, pp. 398-411, 1957.

[535] S. B. Hrdy, "Grandmothers among others," in *Mothers and others: The evolutionary origins of mutual understanding*, First ed., Belknap Press of Harvard University Press, 2009, pp. 233-272.

[536] K. Hawkes, J. F. O'Connell, B. N. G. Jones, H. Alvarez and E. L. Charnov, "Grandmothering, menopause, and the evolution of human life histories," *Proceedings of the National Academy of Sciences of the United States of America,* vol. 95, no. 3, pp. 1336-1339, 1998.

[537] A. F. Kachel, L. S. Premo and J.-J. Hublin, "Grandmothering and natural selection," *Proceedings of the Royal Society B,* vol. 278, no. 1704, pp. 384-391, 2010.

[538] L. T. Gettler, T. W. McDade, A. B. Feranil and C. W. Kuzawa, "Longitudinal evidence that fatherhood decreases testosterone in human males," *Proceedings of the National Academy of Science,* vol. 108, no. 39, pp. 16194-16199, 2011.

[539] K. Chin, W. J. Chopik, B. M. Wardecker, O. P. LaBelle, A. C. Moors and R. S. Edelstein, "Longitudinal associations between prenatal testosterone and postpartum outcomes in a sample of first-time expectant lesbian couples," *Hormones and Behavior,* vol. 125, p. 104810, 2020.

[540] A. E. Pikus, S. Guindre-Parker and D. R. Rubenstein, "Testosterone, social status and parental care in a cooperatively breeding bird," *Hormones and Behavior,* vol. 97, pp. 85-93, 2017.

[541] K. E. Wynne-Edwards and M. E. Timonin, "Paternal care in rodents: Weakening support for hormonal regulation of the transition to behavioral fatherhood in rodent animal models of biparental care," *Hormones and Behavior,* vol. 52, no. 1, pp. 114-121, 2007.

[542] R. DeAngelis, L. Dodd, A. Snyder and J. S. Rhodes, "Dynamic regulation of brain aromatase and isotocin receptor gene expression depends on parenting status," *Hormones and Behavior,* vol. 103, pp. 62-70, 2018.

[543] Y. Inoue, T. Takahashi, R. P. Burriss, S. Arai and T. Hasegawa, "Testosterone promotes either dominance or submissiveness in the Ultimatum Game depending on players' social rank," *Science Reports,* vol. 7, p. 5335, 2017.

[544] J. R. R. Tolkien, The Lord of the Rings, Houghton Mifflin Harcourt, 2005, p. 1178.

[545] J. S. Gavaler, "Alcoholic beverages as a source of estrogens," *Alcohol Health and Research World,* vol. 22, no. 3, pp. 220-228, 1998.

[546] S. Dockray, Doctoral Thesis: Cortisol, Daily Hassles and Overweight Status in Adolescence, State College, PA: Pennsylvania State University, 2007, p. 253.

[547] M. R. Amin, D. D. Pednekar, H. F. Azgomi, H. van Wietmarschen, K. Aschbacher and R. T. Faghih, "Sparse system identification of leptin dynamics in women with obesity," *Frontiers in Endocrinology,* vol. 13, p. 769951, 2022.

[548] R. Kumar, M. R. Rizvi and S. Saraswat, "Obesity and stress: A contingent paralysis," *International Journal of Preventive Medicine,* vol. 13, no. 1, p. 95, 2022.

[549] S. Cerolini, M. Vacca, A. Zegretti, A. Zagaria and C. Lombardo, "Body shaming and internalized weight bias as potential precursors of eating disorders in adolescents," *Frontiers in Psychology,* vol. 6, no. 15, p. 1356647, 2024.

[550] L. Dolezal and T. Spratt, "Fat shaming under neoliberalism and COVID-19: Examining the UK's Tackling Obesity campaign," *Sociology of Health and Illness,* vol. 45, no. 2, pp. 3-18, 2023.

[551] A. E. Moyer, J. Rodin, C. M. Grilo, N. Cummings, L. M. Larson and M. Rebuffé-Scrive, "Stress-induced cortisol response and fat distribution in women," *Obesity Research,* vol. 2, no. 3, pp. 255-262, 1994.

[552] M. D. Jennions and M. Petrie, "Why do females mate multiply? A review of the genetic benefits," *Biological Reviews of the Cambridge Philosophical Society,* vol. 75, no. 1, pp. 21-64, 2000.

[553] R. L. Trivers, "Parental investment and sexual selection," in *Sexual Selection and the Descent of Man*, B. Campbell, Ed., Chicago, Aldine, 1972, pp. 1871-1971.

[554] M. Brown, S. Donahoe and K. Boykin, "Physical strength as a cue to men's capability as protective parents," *Evolutionary Psychological Science,* vol. 8, no. 1, pp. 81-88, 2022.

[555] D. M. Buss and T. K. Shackelford, "From vigilance to violence: mate retention tactics in married couples," *Journal of Personality and Social Psychology,* vol. 73, no. 2, pp. 346-361, 1997.

[556] B. J. W. Dixson, K. R. Blake, T. F. Denson, A. Gooda-Vossos, S. M. O'Dean, D. Sulikowski, M. J. Rantala and R. C. Brooks, "The role of mating context and fecundability in women's preferences for men's facial masculinity and beardedness," *Psychoneuroendocrinology,* vol. 93, pp. 90-102, 2018.

[557] S. W. Gangestad, C. E. Garver-Apgar, J. A. Simpson and A. J. Cousins, "Changes in women's mate preferences across the ovulatory cycle," *Journal of Personality and Social Psychology,* vol. 92, no. 1, pp. 151-163, 2007.

[558] M. Peters, L. W. Simmons and G. Rhodes, "Preferences across the menstrual cycle for masculinity and symmetry in photographs of male faces and bodies," *PLoS One,* vol. 4, no. 1, p. e4138, 2009.

[559] A. C. Little and B. C. Jones, "Variation in facial masculinity and symmetry preferences across the menstrual cycle is moderated by relationship context," *Psychoneuroendocrinology,* vol. 27, no. 7, pp. 999-1008, 2012.

[560] E. G. Pillsworth and M. G. Haselton, "Male sexual attractiveness predicts differential ovulatory shifts in female extra-pair attraction and male mate retention," *Evolution and Human Behavior,* vol. 27, no. 4, pp. 247-258, 2006.

[561] C. D. Goetz, J. A. Easton and D. M. Buss, "Women's perceptions of sexual exploitability cues and their link to sexual attractiveness," *Archives of Sexual Behavior,* vol. 43, no. 5, pp. 999-1008, 2014.

[562] J. O. Strom, E. Ingberg, E. Druvefors, A. Theodorsson and E. Theodorsson, "The female menstrual cycle does not influence testosterone concentrations in male partners," *Journal of Negative Results in BioMedicine,* vol. 11, no. 1, 2012.

[563] J. O. Ström, E. Ingberg, J. K. Slezak, A. Theodorsson and E. Theodorsson, "Male testosterone does not adapt to the partner's menstrual cycle," *Journal of Sexual Medicine,* vol. 15, no. 8, pp. 1103-1110, 2018.

[564] W. Tarumi and K. Shinohara, "Women's body odour during the ovulatory phase modulates testosterone and cortisol levels in men," *PloS One,* vol. 15, no. 3, p. e0230838, 2020.

[565] S. B. Hrdy, Father Time: A Natural History of Men and Babies, Princeton: Princeton University Press, 2024, p. 432.

[566] A. E. Storey, C. J. Walsh, R. L. Quinton and K. E. Wynne-Edwards, "Hormonal correlates of paternal responsiveness in new and expectant fathers," *Evolution and Human Behavior,* vol. 21, no. 2, pp. 79-95, 2000.

[567] K. E. Wynne-Edwards and C. J. Reburn, "Behavioral endocrinology of mammalian fatherhood," *Trends in Ecology and Evolution,* vol. 15, no. 11, pp. 464-468, 2000.

[568] S. King, The Shining, New York: Doubleday, 1977, p. 447.

[569] L. Hagg, C. J. LaMere and M. Zuk, "Postcopulatory song as a mate-guarding tactic in the Pacific field cricket, Teleogryllus oceanicus," *Animal Behaviour,* vol. 207, pp. 201-207, 2024.

[570] L. Bohannan, "Shakespeare in the bush," *Natural History,* pp. 28-33, August-September 1966.

[571] J. C. Wingfield, R. E. Hegner, A. M. J. Dufty and G. F. Ball, "The challenge hypothesis: Theoretical implications for patterns of testosterone secretion, mating systems, and breeding strategies," *American Naturalist,* vol. 136, no. 6, pp. 829-846, 1990.

[572] M. Belinchón, P. Polo, C. Rodriguez-Sickert, O. Figueroa, N. Valenzuela and P. Pavez, "The effect of intergroup competition outcome on ingroup cooperation: Insights from the male warrior hypothesis," *Frontiers in Psychology,* vol. 15, p. 1303372, 2024.

[573] H. J. Krugers, H. Karst and M. Joels, "Interactions between noradrenaline and corticosteroids in the brain: from electrical activity to cognitive performance," *Frontiers in Cell Neuroscience,* vol. 6, p. 15, 2012.

[574] J. Sanguino-Gómez and H. J. Krugers, "Early-life stress impairs acquisition and retrieval of fear memories: sex-effects, corticosterone modulation, and partial prevention by targeting glucocorticoid receptors at adolescent age," *Neurobiology of Stress,* vol. 31, p. 100636, 2024.

[575] I. C. Weaver, "Epigenetic effects of glucocorticoids," *Seminars in Fetal and Neonatal Medicine,* vol. 14, no. 3, pp. 143-150, 2009.

[576] S. Oresta, C. H. Vinkers, E. F. C. van Rossum, B. W. J. H. Penninx and L. Nawijn, "How childhood trauma and recent adverse events are related to hair cortisol levels in a large adult cohort," *Psychoneuroendocrinology,* vol. 126, p. 105150, 2021.

[577] M. J. Shanahan, S. W. Cole, S. Ravi and K. M. Harris, "Socioeconomic inequalities in molecular risk for chronic diseases observed in young adulthood," *Proceedings of the National Academy of Science,* vol. 119, no. 43, p. e2103088119, 2022.

[578] A. J. Tomiyama, "Stress and obesity," *Annual Review of Psychology,* vol. 70, pp. 703-718, 2019.

[579] M. C. Evans, E. Campbell and G. M. Anderson, "Physiological regulation of leptin as an integrative signal of reproductive readiness," *Current Opinion in Pharmacology,* vol. 67, p. 102321, 2022.

[580] J. Aaseth, G. E. Roer, L. Lien and G. Bjørklund, "Is there a relationship between PTSD and complicated obesity? A review of the literature," *Biomedicine and Pharmacotherapy,* vol. 117, p. 108834, 2019.

[581] N. Watanabe and M. Yamamoto, "Neural mechanisms of social dominance," *Frontiers in Neuroscience,* vol. 9, 2015b.

[582] T. Backström and S. Winberg, "Serotonin coordinates responses to social stress—what we can learn from fish," *Frontiers in Neuroscience,* vol. 11, 2017.

[583] S. G. Reid, N. J. Bernier and S. F. Perry, "Review: The adrenergic stress response in fish: Control of catecholamine storage and release," *Comparative Biochemistry and Physiology Part C: Pharmacology, Toxicology and Endocrinology,* vol. 120, no. 1, pp. 1-27, 1998.

[584] S. Nørgaard, W. Joyce, M. F. Jensen, S. S. N. Enok and T. Wang, "Histamine exerts both direct H2-mediated and indirect catecholaminergic effects on heart rate in pythons," *Journal of Comparative Physiology B,* vol. 191, pp. 347-355, 2021.

[585] M. Sundvik, H. Puttonen, S. Semenova and P. Panula, "The bullies are the leaders of the next generation: Inherited aminergic neurotransmitter system changes in socially dominant zebrafish, Danio rerio," *Behavioural Brain Research,* vol. 409, no. 9, p. 113309, 2021.

[586] I. Babarro, J. Ibarluzea, E. Theodorsson, E. Fano, A. Lebeña, M. Guxens, J. Sunyer and A. Andiarena, "Hair cortisol as a biomarker of chronic stress in preadolescents: influence of school context and bullying," *Child Neuropsychology,* pp. 1-18, 2022.

[587] T. P. Warrington and J. M. Bostwick, "Psychiatric adverse effects of corticosteroids," *Mayo Clinic Proceedings,* vol. 81, no. 10, pp. 1361 - 1367, 2006.

[588] H. A. Kenna, A. W. Poon, C. P. de los Angeles and L. M. Koran, "Psychiatric complications of treatment with corticosteroids: Review with case report," *Psychiatry and Clinical Neurosciences,* vol. 65, no. 6, pp. 549-560, 2011.

[589] J. W. Mason, E. L. Giller, T. R. Kosten, R. B. Ostroff and L. Podd, "Urinary free-cortisol levels in posttraumatic stress disorder patients," *The Journal of Nervous and Mental Disease,* vol. 174, no. 3, pp. 145-149, 1986.

[590] R. Yehuda, B. Kahana, K. Binder-Brynes, S. M. Southwick, J. W. Mason and E. L. Giller, "Low urinary cortisol excretion in Holocaust survivors with posttraumatic stress disorder," *The American Journal of Psychiatry,* vol. 152, no. 7, pp. 982-986, 1995.

[591] L. Lindgren, J. Bergdahl and L. Nyberg, "Longitudinal evidence for smaller hippocampus volume as a vulnerability factor for perceived stress," *Cerebral Cortex,* vol. 26, no. 8, pp. 3527-3533, 2016.

[592] K. Wingenfeld and O. T. Wolf, "Stress, memory, and the hippocampus," *Frontiers of Neurology and Neuroscience,* vol. 34, pp. 109-120, 2014.

[593] Y. Watanabe, E. Gould and B. S. McEwen, "Stress induces atrophy of apical dendrites of hippocampal CA3 pyramidal neurons," *Brain Research,* vol. 588, no. 2, p. 341–345, 1992.

[594] S. S. Haas, A. Myoraku, K. Watson, T. Robakis, S. Frangou, F. Abbasi and N. Rasgon, "Lower functional hippocampal connectivity in healthy adults is jointly associated with higher levels of leptin and insulin resistance," *European Psychiatry,* vol. 65, no. 1, p. E29, 2022.

[595] C. Anacker and R. Hen, "Adult hippocampal neurogenesis and cognitive flexibility - linking memory and mood," *Nature Reviews Neuroscience,* vol. 18, no. 6, pp. 335-346, 2017.

[596] S. Tronel, L. Belnoue, N. Grosjen, J.-M. Revest, P.-V. Piazza, M. Koehl and D. N. Abrous, "Adult-born neurons are necessary for extended contextual discrimination," *Hippocampus,* vol. 22, no. 2, pp. 292-298, 2012.

[597] J. S. Snyder, A. Soumier, M. Brewer, J. Pickel and H. A. Cameron, "Adult hippocampal neurogenesis buffers stress responses and depressive behaviour," *Nature,* vol. 476, pp. 458-461, 2011.

[598] J.-M. Revest, D. Dupret, M. Koehl, C. Funk-Reiter, N. Grosjean, P.-V. Piazza and D. N. Abrous, "Adult hippocampal neurogenesis is involved in anxiety-related behaviors," *Molecular Psychiatry,* vol. 14, pp. 959-967, 2023.

[599] A. Surget and C. Belzung, "Adult hippocampal neurogenesis shapes adaptation and improves stress response: a mechanistic and integrative perspective," *Nature Molecular Psychiatry,* vol. 27, pp. 403-421, 2022.

[600] D. Montero, G. Lalumera, M. S. Izquierdo, M. J. Caballero, M. Saroglia and L. Tort, "Establishment of dominance relationships in gilthead sea bream Sparus aurata juveniles during feeding: Effects on feeding behaviour, feed utilization and fish health," *Journal of Fish Biology,* vol. 74, no. 4, p. 790–805, 2009.

[601] I. Oullet-Morin, C. Cantave, S. Paquin, M.-C. Geoffroy, M. Brendgen, F. Vitaro, R. Tremblay, M. Boivin, S. Lupien and S. Côté, "Associations between developmental trajectories of peer victimization, hair cortisol, and depressive symptoms: a longitudinal study," *Journal of Child Psychology and Psychiatry,* vol. 62, no. 1, pp. 19-27, 2021.

[602] Ø. Øverli, C. A. Harris and S. Winberg, "Short-term effects of fights for social dominance and the establishment of dominant-subordinate relationships on brain monoamines and cortisol in rainbow trout," *Brain, Behavior and Evolution,* vol. 54, no. 5, pp. 263-275, 1999.

[603] P. H. Mehta and R. A. Josephs, "Testosterone and cortisol jointly regulate dominance: Evidence for a dual-hormone hypothesis," *Hormones and Behavior,* vol. 58, no. 5, pp. 898-906, 2010.

[604] M. M. Dovom, H. Fatolahi, S. Nikbin, S. F. Noktehdan, P. Mohabbati, P. Bagheri and M. A. Azarbayjani, "Effects of official chess competition on salivary cortisol and mood swings in adolescent girls: A win-loss approach," *Applied Psychophysiology and Biofeedback,* vol. 49, p. 301–311, 2024.

[605] C. Eisenegger, R. Kumsta, M. Naef, J. Gromoll and M. Heinrichs, "Testosterone and androgen receptor gene polymorphism are associated with confidence and competitiveness in men," *Hormones and Behavior,* vol. 92, pp. 93-102, 2017.

[606] G. A. Roberts, R. Arnold, J. E. Turner, M. Colclough and J. Bilson, "A longitudinal examination of military veterans' invictus games stress experiences," *Frontiers in Psychology,* vol. 10, 2019.

[607] I.-T. Kolassa and T. Elbert, "Structural and functional neuroplasticity in relation to traumatic stress," *Current Directions in Psychological Science,* vol. 16, no. 6, pp. 321-325, 2007.

[608] F. Duclot and M. Kabbaj, "Individual differences in novelty seeking predict subsequent vulnerability to social defeat through a differential epigenetic regulation of brain-derived neurotrophic factor expression," *Journal of Neuroscience,* vol. 33, no. 27, pp. 11048-11060, 2013.

[609] . C. Weaver, N. Cervoni, F. A. Champagne, A. C. D'Alessio, S. Sharma, J. R. Seckl, S. Dymov, M. Szyf and M. J. Meaney, "Epigenetic programming by maternal behavior," *Nature Neuroscience,* vol. 7, no. 8, pp. 847-854, 2004.

[610] J. Keller, R. Gomez, G. Williams, A. Lembke, L. Lazzeroni, G. M. J. Murphy and A. F. Schatzberg, "HPA axis in major depression: cortisol, clinical symptomatology and genetic variation predict cognition," *Molecular Psychiatry,* vol. 22, p. 527–536, 2017.

[611] M. N. Muller and R. W. Wrangham, "Dominance, aggression and testosterone in wild chimpanzees: A test of the 'challenge hypothesis'," *Animal Behaviour,* vol. 67, no. 1, pp. 113-123, 2004a.

[612] M. N. Muller and R. W. Wrangham, "Dominance, cortisol and stress in wild chimpanzees (Pan troglodytes schweinfurthii)," *Behavioural Ecology and Sociobiology,* vol. 55, no. 4, pp. 332-340, 2004b.

[613] P. H. Mehta, A. C. Jones and R. A. Josephs, "The social endocrinology of dominance: Basal testosterone predicts cortisol changes and behavior following victory and defeat," *Journal of Personality and Social Psychology,* vol. 94, no. 6, pp. 1078-1093, 2008.

[614] M. L. Scherholz, N. Schlesinger and I. P. Androulakis, "Chronopharmacology of glucocorticoids," *Advanced Drug Delivery Reviews,* Vols. 151-152, pp. 245-261, 2019.

[615] G. G. Piroli, C. A. Grillo, L. R. Reznikov, S. Adams, B. S. McEwen, M. J. Charron and L. P. Reagan, "Corticosterone impairs insulin-stimulated translocation of GLUT4 in the rat hippocampus," *Neuroendocrinology,* vol. 85, no. 2, pp. 71-80, 2007.

[616] G. I. Sandle, M. J. Keir and C. O. Record, "The effect of hydrocortisone on the transport of water, sodium, and glucose in the jejunum. Perfusion studies in normal subjects and patients with coeliac disease," *Scandinavian Journal of Gastroenterology,* vol. 16, no. 5, p. 667–671, 1981.

[617] T. M. Pollard, "Physiological consequences of everyday psychosocial stress," *Collegium Antropologicum,* vol. 21, no. 1, pp. 17-28, 1997.

[618] J. Phillips and W. M. P. Klein, "Socioeconomic status and coronary heart disease risk: The role of social cognitive factors," *Social Personality Psychology Compass,* vol. 4, no. 9, pp. 704-727, 2010.

[619] M. Kivimäki, M. Virtanen, M. Vartia, M. Elovainio, J. Vahtera and L. Keltikangas-Järvinen, "Workplace bullying and the risk of cardiovascular disease and depression," *Occupational and Environmental Medicine,* vol. 60, no. 10, p. 779–783, 2003.

[620] L. Rossiter and A. Sochos, "Workplace bullying and burnout: The moderating effects of social support," *Journal of Aggression, Maltreatment and Trauma,* vol. 27, no. 4, pp. 386-408, 2018.

[621] A. Steptoe, M. Cropley, J. Griffith and C. Kirshbaum, "Job strain and anger expression predict early morning elevations in salivary cortisol," *Psychosomatic Medicine,* vol. 62, no. 2, pp. 286-292, 2000.

[622] C. F. Gillespie and C. B. Nemeroff, "Corticotropin-releasing factor and the psychobiology of early-life stress," *Current Directions in Psychological Science,* vol. 16, no. 2, pp. 85- 89, 2007.

[623] A. Erli and W. Hirst, "Flashbulb memories: An interdisciplinary research programme," *Narrative Inquiry,* vol. 33, no. 2, pp. 398-420, 2023.

[624] M. L. Stanek, K. M. Boaz, C. N. Cordes, T. D. Niese, K. E. Long, M. S. Risner, J. G. Blasco, K. N. Suzelis, K. M. Siereveld, B. R. Rorabaugh and P. R. Zoladz, "Social evaluative stress enhances central detail memory, reduces false memory, and results in intrusive memories that last for days," *Memory,* vol. 209, p. 107906, 2024.

[625] B. Roozendaal and G. Mirone, "Opposite effects of noradrenergic and glucocorticoid activation on accuracy of an episodic-like memory," *Psychoneuroendocrinology,* vol. 114, p. 104588, 2020.

[626] B. S. McEwen, C. Nasca and J. D. Gray, "Stress effects on neuronal structure: Hippocampus, amygdala, and prefrontal cortex," *Neuropsychopharmacology,* vol. 41, no. 1, p. 3–23, 2016.

[627] A. Lebedeva, A. Sundström, L. Lindgren, A. Stomby, D. W. E. Aarsland, B. Winblad, T. Olsson and L. Nyberg, "Longitudinal relationships among depressive symptoms, cortisol, and brain atrophy in the neocortex and the hippocampus," *Acta Psychiatrica Scandinavia,* vol. 137, no. 6, 2018.

[628] F. Ohl, T. Michaelis, G. K. Vollmann-Honsdorf, C. Kirschbaum and E. Fuchs, "Effect of chronic psychosocial stress and long-term cortisol treatment on hippocampus-mediated memory and hippocampal volume: A pilot-study in tree shrews," *Psychoneuroendocrinology,* vol. 25, no. 4, pp. 357-363, 2000.

[629] M. T. McAuley, R. A. Kenny, T. B. L. Kirkwood, D. J. Wilkinson, J. J. L. Jones and V. M. Miller, "A mathematical model of aging-related and cortisol induced hippocampal dysfunction," *BMC Neuroscience,* vol. 10, no. 1, pp. 26-39, 2009.

[630] E. J. Hermansa, F. P. Battaglia, P. Atsaka, L. D. de Voogda, G. Fernándeza and B. Roozendaal, "How the amygdala affects emotional memory by altering brain network properties," *Neurobiology of Learning and Memory: Special Issue: Stress and the regulation of memory: From basic mechanisms to clinical implications,* vol. 112, pp. 2-16, 2014.

[631] A. Cordonnier and O. Luminet, "Consistency and social identification: a test-retest study of flashbulb memories collected on the day of the 2016 Brussels bombings," *Memory,* vol. 29, no. 3, pp. 305-318, 2021.

[632] D. M. Diamond, A. M. Campbell, C. R. Park, J. Halonen and P. R. Zoladz, "The temporal dynamics model of emotional memory processing: A synthesis on the neurobiological basis of stress-induced amnesia, flashbulb and traumatic memories, and the Yerkes-Dodson law (review)," *Neural Plasticity,* vol. 2007, p. 60803, 2007.

[633] J. A. Quent, A. M. McCullough, M. Sazma, O. T. Wolf and A. P. Yonelinas, "Reward anticipation modulates the effect of stress-related increases in cortisol on episodic memory," *Neurobiology of Learning and Memory,* vol. 147, no. 1, pp. 65-73, 2018.

[634] Y. Watanabe, E. Gould and B. S. McEwen, "Stress induces atrophy of apical dendrites of hippocampal CA3 pyramidal neurons," *Brain Research,* vol. 588, no. 2, p. 341–345, 1992b.

[635] Z. Ben-Zion, N. Korem, N. Fine, S. Katz, M. Siddhanta, M. C. Funaro, O. Duek, T. Spiller, S. K. Danböck, I. Levy and I. Harpaz-Rotem, "Structural neuroimaging of hippocampus and amygdala subregions in posttraumatic stress disorder: A scoping review," *Biological Psychiatry Global Open Science,* vol. 4, no. 1, pp. 120-134, 2023.

[636] M. Geng, Q. Shao, J. Fu, J. Gu, L. Feng, L. Zhao, C. Liu, J. Mu, X. Zhang, M. Zhao, X. Guo, C. Song, Y. Li, H. Wang and C. Wang, "Down-regulation of MKP-1 in hippocampus protects against stress-induced depression-like behaviors and neuroinflammation," *Translational Psychiatry,* vol. 1, no. 14, p. 130, 2024.

[637] A. Keresztes, L. Raffington, A. Bender, K. Bögl, C. Heim and Y. L. Shing, "Hair cortisol concentrations are associated with hippocampal subregional volumes in children," *Scientific Reports,* vol. 10, p. 4865, 2020.

[638] C. Warhurst, D. van den Broek, R. Hall and D. Nickson, "Lookism: The new frontier of employment discrimination?," *Journal of Industrial Relations,* vol. 51, no. 1, pp. 131-136, 2009.

[639] A. Mason, What's Wrong with Lookism?: Personal Appearance, Discrimination, and Disadvantage (New Topics in Applied Philosophy), Oxford: Oxford University Press, 2023, p. 252.

[640] L. R. Bader, J. D. Carboni, C. Burleson and M. A. Cooper, "5-HT1A receptor activation reduces fear-related behavior following social defeat in Syrian hamsters," *Pharmacology and Biochemistry and Behavior,* vol. 122, pp. 182-190, 2014.

[641] Z. A. Grieb, A. P. Ross, K. E. McCan, S. Lee, M. Welch, M. G. Gomez, A. Norvella, V. Michopoulos, K. L. Huhman and H. E. Albers, "Sex-dependent effects of social status on the regulation of arginine-vasopressin (AVP) V1a, oxytocin (OT), and serotonin (5-HT) 1A receptor binding and aggression in Syrian hamsters (Mesocricetus auratus)," *Hormones and Behavior,* vol. 127, p. 104878, 2021.

[642] X. Ding, M. Yang, N. Wu and R. Song, "Blockade of dopamine D3 receptor in ventral tegmental area attenuating contextual fear memory," *Biomedicine and Pharmacotherapy,* vol. 158, p. 114179, 2023.

[643] C. Huang, Q. Lan, H. Yan, B. Huang, S. Li, C. Zhang and P. Zhou, "Role of ventral tegmental area dopamine neurons in fear memory retention in post-traumatic stress disorder model rats," *Journal of Psychiatric Research,* vol. 178, pp. 59-65, 2024.

[644] K. Tong, S. Wang, Y.-J. Zhu, Z.-P. Chen and S.-Q. Jing, "Projections from the ventral tegmental area to zona incerta regulate fear generalization in a mouse model of PTSD," *Brain Research Bulletin,* vol. 217, p. 111079, 2024.

[645] R. D. Fernald, "Communication about social status," *Current Opinion in Neurobiology,* vol. 28, pp. 1-4, 2014.

[646] H. Covington III, E. L. Newman, M. Z. Leonard and K. A. Miczek, "Translational models of adaptive and excessive fighting: an emerging role for neural circuits in pathological aggression [version 1; peer review: 3 approved]," *F1000 Research,* vol. 8, 2019.

[647] E. Chen, S. Y. Cohen and G. E. Miller, "How low socioeconomic status affects 2-year hormonal trajectories in children," *Psychological Science,* vol. 21, no. 1, pp. 31-37, 2010.

[648] E. Chen and L. Q. Paterson, "Neighborhood, family, and subjective socioeconomic status: How do they relate to adolescent health?," *Health Psychology,* vol. 25, no. 6, pp. 704-714, 2006.

[649] J. O. Lee, K. G. Hill, L. A. Hartigan, J. M. Bodenc, G. K., R. Kosterman, J. A. Bailey and R. F. Catalano, "Unemployment and substance use problems among young adults: Does childhood low socioeconomic status exacerbate the effect?," *Social Science & Medicine,* vol. 143, pp. 36-44, 2015.

[650] Y. Ni, N. Baumann and D. Woke, "Bullying victimisation in childhood and mental health in early adulthood: comparison of prospective and retrospective reports," *Current Psychology,* vol. 1, 2024.

[651] T. Deak, A. Kudinova, D. F. Lovelock, B. E. Gibb and M. B. Hennessy, "A multispecies approach for understanding neuroimmune mechanisms of stress," *Dialogues in Clinical Neuroscience,* vol. 19, no. 1, pp. 37-54, 2017.

[652] L. Cao-Lei, S. R. de Rooij, S. King, S. G. Matthews, G. A. S. Metz, T. J. Roseboom and M. Szyf, "Prenatal stress and epigenetics," *Neuroscience and Biobehavioral Reviews,* vol. 117, pp. 198-210, 2020.

[653] E. V. Todd, O. Ortega-Recalde, H. Liu, M. S. Lamm, K. M. Rutherford, H. Cross, M. Black, O. Kardailsky, J. A. M. Graves, T. A. Hore, J. R. Godwin and N. J. Gemmell, "Stress, novel sex genes, and epigenetic reprogramming orchestrate socially controlled sex change," *Science Advances,* vol. 5, no. 7, p. eaaw7006, 2019.

[654] W. T. A. F. Silva, S.-E. P., S. Torijo-Boix, R. A. C. Devaux, M. Durieux, M. J. Gómez-Torres and S. Immler, "The effects of male social environment on sperm phenotype and genome integrity," *Journal of Evolutionary Biology,* vol. 32, no. 6, pp. 535-544, 2019.

[655] C. Kent and P. Agrawal, "Regulation of social stress and neural degeneration by activity-regulated genes and epigenetic mechanisms in dopaminergic neurons," *Molecular Neurobiology,* vol. 57, no. 11, pp. 4500-4510, 2020.

[656] C. Hotzy, E. Fowler, B. Kiehl, R. Francis, J. Mason, S. Moxon, W. Rostant and T. I. S. Chapman, "Evolutionary history of sexual selection affects microRNA profiles in Drosophila sperm," *Evolution,* vol. 32, no. 6, pp. 310-319, 2022.

[657] H. Y. Shpigler, M. C. Saul, E. E. Murdoch, A. C. Cash-Ahmed, C. H. Seward, L. Sloofman, S. Chandrasekaran, S. Sinha, L. J. Stubbs and G. E. Robinson, "Behavioral, transcriptomic and epigenetic responses to social challenge in honey bees," *Genes, Brain, and Behavior,* vol. 16, no. 6, pp. 579-591, 2017.

[658] J. M. M. Kuhn, K. Meusemann and J. Korb, "Disentangling the aging gene expression network of termite queens," *BMC Genetics,* vol. 22, no. 1, p. 339, 2021.

[659] G. E. Miller, E. Chen, A. K. Fok, H. Walker, A. Lim, E. F. Nicholls, S. Cole and M. S. Kobor, "Low early-life social class leaves a biological residue manifested by decreased glucocorticoid and increased proinflammatory signaling," *Proceedings of the National Academy of Science USA,* vol. 106, no. 34, pp. 14716-14721, 2009.

[660] J. J. Chiang, P. H. Lam, E. Chen and G. E. Miller, "Psychological stress during childhood and adolescence and its association with inflammation across the lifespan: A critical review and meta-analysis," *Psychological Bulletin,* vol. 148, no. 1-2, pp. 27-66, 2022.

[661] F. Coulombe-Rozon, Effet protecteur du milieu enrichi et de l'exercice physique sur l'intégrité intestinale lors du stress chronique (Mémoire de Maître ès sciences), Québec: Bibliothèque de l'Université Laval, 2023, p. 84.

[662] A. S. M. Leal, J. A. Silvers, J. E. Carrol, S. W. Cole, K. M. Ross, S. L. Ramey, M. U. Shalowitz and C. D. Schetter, "Maternal early life stress is associated with pro-inflammatory processes during pregnancy," *Brain, Behavior, and Immunity,* vol. 109, pp. 285-291, 2023.

[663] E. Chen, G. E. Miller, M. S. Kobor and S. W. Cole, "Maternal warmth buffers the effects of low early-life socioeconomic status on pro-inflammatory signaling in adulthood," *Molecular Psychiatry,* vol. 16, no. 7, pp. 729-737, 2010.

[664] L. M. Diamond and J. Alley, "Rethinking minority stress: A social safety perspective on the health effects of stigma in sexually-diverse and gender-diverse populations," *Neuroscience and Biobehavioral Reviews,* vol. 138, p. 104720, 2022.

[665] F. Douglass, Narrative of the Life of Frederick Douglass, an American Slave, Boston: Anti-Slavery Office, 1845.

[666] United States Congress, "PART 312—CHILDREN'S ONLINE PRIVACY PROTECTION RULE," 17 January 2013. [Online]. Available: https://www.ecfr.gov/current/title-16/chapter-I/subchapter-C/part-312. [Accessed 16 May 2023].

[667] J. Bartholomaeus and P. Strelan, "The adaptive, approach-oriented correlates of belief in a just world for the self: A review of the research," *Personality and Individual Differences,* vol. 151, p. 109485, 2019.

[668] C. S. Johnston, F. Krings, C. Maggiori, L. L. Meier and M. Fiori, "Believing in a personal just world helps maintain well-being at work by coloring organizational justice perceptions," *European Journal of Work and Organizational Psychology,* vol. 25, no. 6, pp. 945-959, 2016.

[669] R. M. Sutton and K. M. Douglas, "Justice for all, or just for me? More evidence of the importance of the self-other distinction in just-world beliefs," *Personality and Individual Differences,* vol. 39, no. 3, pp. 637-645, 2005.

[670] J. R. Spence, D. L. Ferris, D. J. Brown and D. Heller, "Understanding daily citizenship behaviors: A social comparison perspective," *Journal of Organizational Behavior,* vol. 32, no. 4, pp. 547-571, 2011.

[671] I. Correia, S. Salvado and H. V. Alves, "Belief in a just world and self-efficacy to promote justice in the world predicting helping attitudes, but only among volunteers," *The Spanish Journal of Psychology,* vol. 19, p. E28, 2016.

[672] C. L. Menning, E. D. Estoya and L. K. Nunn, "Blaming the victim, the bystander, the perpetrator, or the institution: Student allocation of responsibility for sexual assault when programming hypothetical varies," *Sociological Focus,* vol. 57, no. 2, p. 94–119, 2024.

[673] S. Goodman and P. Carr, "The just world hypothesis as an argumentative resource in debates about unemployment benefits," *Journal of Community and Applied Social Psychology,* vol. 27, no. 4, pp. 312-323, 2017.

[674] C. Dalbert and H. Sallay, The Justice Motive in Adolescence and Young Adulthood, First ed., C. Dalbert and H. Sallay, Eds., New York: Routledge, 2013, p. 288.

[675] S. Nataraj-Hansen and K. Richards, "Why do fraud victims get blamed? Lerner's Belief in a Just World and its application to victims of online fraud," *Journal of Financial Crimes,* vol. 1, no. 1, 2022.

[676] E. E. Ayala, B. Kotary and M. Hetz, "Blame attributions of victims and perpetrators: Effects of victim gender, perpetrator gender, and relationship," *Journal of Interpersonal Violence,* vol. 33, no. 1, pp. 94-116, 2015.

[677] C. M. Reich, G. A. Pegel and A. B. Johnson, "Are survivors of sexual assault blamed more than victims of other crimes?," *Journal of Interpersonal Violence,* vol. 37, no. 19-20, p. NP18394–NP18416, 2022.

[678] J. Belsky and K. M. Beaver, "Cumulative-genetic plasticity, parenting and adolescent self-regulation," *The Journal of Child Psychology and Psychiatry,* vol. 52, no. 5, pp. 619-626, 2011.

[679] C. S. Widom and Brzustowicz, "MAOA and the 'cycle of violence': Childhood abuse and neglect, MAOA genotype and the risk for violent and antisocial behavior," *Biological Psychiatry,* vol. 60, no. 7, pp. 684-689, 2006.

[680] D. Laricchiuta, A. Panuccio, E. Picerni, D. Biondo, B. Genovesi and L. Petrosini, "The body keeps the score: The neurobiological profile of traumatized adolescents," *Neuroscience & Biobehavioral Reviews,* vol. 145, p. 105033, 2023.

[681] S. Zuckerman, The Social Life of Monkeys and Apes, London: Routledge & Kegan Paul, 1999/ 2015, p. 511.

[682] E. Martin and M. Taborsky, "Alternative male mating tactics in a cichlid, Pelvicachromis pulcher: A comparison of reproductive effort and success," *Behavioral Ecology and Sociobiology,* vol. 41, no. 5, pp. 311-319, 1997.

[683] K. Murlanova, M. Kirby, L. Libergod, M. Pletnikov and A. Pinhasov, "Multidimensional nature of dominant behavior: Insights from behavioral neuroscience," *Neuroscience and Biobehavioral Reviews,* vol. 132, pp. 603-620, 2022.

[684] R. Körner, B. A. Murphy, E. Zverling, A. Sha'ked and A. Schütz, "Dominance and prestige in romantic relationships: Actor and partner links to relationship quality," *Journal of Social and Personal Relationships,* vol. 41, no. 1, pp. 302-332, 2023.

[685] R. A. Hinde, "Dominance and role—two concepts with dual meanings," *Journal of Social and Biological Structures,* vol. 1, no. 1, pp. 27-38, 1978.

[686] M. J. Williams, G. Lopiano and D. Heller, "When the boss steps up: Workplace power, task responsibility, and engagement with unpleasant tasks," *Organizational Behavior and Human Decision Processes,* vol. 170, p. 104140, 2022.

[687] A. McKay, Director, *The Big Short.* [Film]. USA: Regency Enterprises; Plan B Entertainment, 2015.

[688] A. A. Filipova, "Leadership Implications Related to Bullying and Intent to Leave Among Licensed Practical Nurses," *Journal of Nursing Administration,* vol. 52, no. 5, pp. 266-272, 2022.

[689] C. Dickens, A Tale of Two Cities, CreateSpace Independent Publishing, 1859/ 2014, p. 430.

[690] R. Marvan, J. M. Stevens, A. D. Roeder, I. Mazura, M. W. Bruford and J. R. de Ruiter, "Male dominance rank, mating and reproductive success in captive bonobos (Pan paniscus)," *Folia Primatologica,* vol. 77, no. 5, pp. 364-376, 2006.

[691] M. Surbeck, R. Mundry and G. Hohmann, "Mothers matter! Maternal support, dominance status and mating success in male bonobos (Pan paniscus)," *Proceedings of the Royal Society B: Biological Sciences,* pp. 590-598, 2011.

[692] A. C. Markham, E. V. Lonsdorf, A. E. Pusey and C. M. Murray, "Maternal rank influences the outcome of aggressive interactions between immature chimpanzees," *Animal Behaviour,* vol. 100, pp. 747-755, 2015.

[693] J. Henrich and M. Muthukrishna, "The origins and psychology of human cooperation," *Annual Review of Psychology,* vol. 72, pp. 207-240, 2021.

[694] J. E. Smith, A. V. Jaeggi and R. K. S. J. B. Holmes, "Sex differences in cooperative coalitions: a mammalian perspective," *Philosophical Transactions of the Royal Society B,* vol. 378, no. 1868, 2022.

[695] T. Waters and D. Waters, Weber's Rationalism and Modern Society: New Translations on Politics, Bureaucracy, and Social Stratification, New York, NY: Palgrave Macmillan, 2015.

[696] M. Weber, "The distribution of power within the community: Classes, Stände, parties," in *Weber's Rationalism and Modern Society: New Translations on Politics, Bureaucracy, and Social Stratification*, New York, NY: Palgrave Macmillan, 1921/1915, pp. 41- 58.

[697] R. M. Sapolsky, "The influence of social hierarchy on primate health," *Science,* vol. 29, no. 1, pp. 648-652, 2005.

[698] P. J. Henry, "Low-status compensation: A theory for understanding the role of status in cultures of honor," *Journal of Personality and Social Psychology,* vol. 97, no. 3, pp. 451-466, 2009.

[699] W. T. Boyce, "Social stratification, health, and violence in the very young," *Annals of the New York Academy of Science,* vol. 1036, p. 47–68, 2004.

[700] K. Offer, D. Mischkowski, Z. Rahwan and C. Engel, "Deliberately ignoring unfairness: Responses to uncertain inequality in the ultimatum game," Max Planck Institute for Research on Collective Goods, Bonn, 2024.

[701] R. N. Sims, A. R. Burkholder and Killen, "Science resource inequalities viewed as less wrong when girls are disadvantaged," *Social Development,* vol. 32, no. 1, pp. 387-407, 2022.

[702] L. Hubber, "Ignoring the role of town planning in health perpetuates inequalities. (Opinion)," *The British Medical Journal,* vol. 382, p. 1578, 2023.

[703] N. Landry, R. Gifford, T. L. Milfont, A. Weeks and S. Arnocky, "Learned helplessness moderates the relationship between environmental concern and behavior," *Journal of Environmental Psychology,* vol. 55, no. 1, pp. 18-22, 2018.

[704] S. R. Lehner, C. Rutte and M. Taborsky, "Rats benefit from winner and loser effects," *Ethology,* vol. 117, no. 11, pp. 949-960, 2011.

[705] B. C. T. Pedroso, L. K. Deringer and A. D. V. H. F. Pereira, "Bases etológicas do infanticídio: uma abordagem sobre o estado da arte em mamíferos. (Ethological bases of infanticide: a state-of-the-art approach to mammals)," *Uningá Review,* vol. 39, 2024.

[706] T. Harano and N. Kutsukake, "The evolution of male infanticide in relation to sexual selection in mammalian carnivores," *Evolutionary Ecology,* vol. 32, pp. 1-8, 2017.

[707] V. Penteriani, I. Kojola, S. Heikkinen, S. Find'o, M. Skuban, A. Fedorca, P. García-Sánchez, M. Fedorca, A. Zarzo-Arias, J. Balbontín and M. del Mar Delgado, "Livin' on the edge: reducing infanticide risk by maintaining proximity to potentially less infanticidal males," *Animal Behaviour,* vol. 210, pp. 63-71, 2024.

[708] K. M. Bohn, C. F. Moss and Wilkinson, "Pup guarding by greater spear-nosed bats," *Behavioral Ecology and Sociobiology,* vol. 63, pp. 1693-1703, 2009.

[709] E. Bellemain, A. Zedrosser, S. Manel, L. P. Waits, P. Taberlet and J. E. Swenson, "The dilemma of female mate selection in the brown bear, a species with sexually selected infanticide," *Proceedings of the Royal Society B,* vol. 273, no. 1584, pp. 283-291, 2006.

[710] R. Dejeante, A. J. Loveridge, D. W. Macdonald, D. Madhlamoto, M. Valeix and S. Chamaillé-Jammes, "Counter-strategies to infanticide: The importance of cubs in determining lion habitat selection and social interactions," *Journal of Animal Ecology,* vol. 93, no. 2, pp. 159-170, 2024.

[711] C. B. Stanford, "The social behavior of chimpanzees and bonobos: Empirical evidence and shifting assumptions," *Current Anthropology,* vol. 39, no. 4, pp. 399-420, 1998.

[712] J. E. Lyon, S. A. Pandit, C. P. van Schaik and G. R. Pradhan, "Mating strategies in primates: A game theoretical approach to infanticide," *Journal of Theoretical Biology,* vol. 274, no. 1, pp. 103-108, 2011.

[713] A. M. Robbins, M. Gray, A. Basabose, P. Uwingeli, I. Mburanumwe, E. Kagoda and M. Robbins, "Impact of Male Infanticide on the Social Structure of Mountain Gorillas," *PLoS One,* vol. 8, no. 11, p. e78256, 2013.

[714] Z. Xiang, Y. Yu, H. Yao, Q. Hu, W. Yang and M. Li, "Female countertactics to male feticide and infanticide in a multilevel primate society," *Behavioral Ecology,* vol. 33, no. 4, 2022.

[715] C. Packer and A. E. Pusey, "Adaptations of female lions to infanticide by incoming males," *The American Naturalist,* vol. 121, no. 5, pp. 716-728, 1983.

[716] T. S. Kavana, J. J. Erinjery and M. Singh, "Male Takeover and Infanticide in Nilgiri Langurs Semnopithecus johnii in the Western Ghats, India," *Folia Primatologica,* vol. 85, pp. 164-177, 2014.

[717] S. B. Hrdy, "Male-male competition and infanticide among the langurs (Presbytis entellus) of Abu, Rajasthan," *Folia Primatologica,* vol. 22, no. 1, pp. 19-58, 1974.

[718] S. B. Hrdy, C. Janson and C. van Schaik, "Infanticide: Let's not throw out the baby with the bath water," *Evolutionary Anthropology: Issues, News, and Reviews,* vol. 3, no. 5, pp. 151-154, 1994.

[719] C. P. van Schaik, "Infanticide by male primates: the sexual selection hypothesis revisited," in *Infanticide by Males and its Implications*, Cambridge, Cambridge University Press, 2000, pp. 27-60.

[720] K. A. Widayati, O. Noerwana, R. M. Fauzi and Y. Tsuji, "Case Report of Infanticide by a Wild Long-tailed Macaque (Macaca fascicularis)," *Philippine Journal of Science,* vol. 151, no. 2, pp. 751-753, 2022.

[721] J. N. Rakotondrazandry, T. M. Sefczek, C. L. Frasier, V. L. Villanova, S. Rasoloharijaona, H. Raveloson and J. E. E. Louis, "Possible Infanticidal Event of an Aye-Aye (Daubentonia madagascariensis) in Torotorofotsy, Madagascar," *Folia Primatologica,* vol. 92, pp. 183-190, 2021.

[722] M. Broom, C. Borries and A. Koenig, "Infanticide and infant defence by males—modelling the conditions in primate multi-male groups," *Journal of Theoretical Biology,* vol. 231, no. 2, pp. 261-270, 2004.

[723] C. P. van Schaik, "Vulnerability to infanticide by males: patterns among mammals," in *Infanticide by Males and its Implications*, Cambridge, Cambridge University Press, 2000, pp. 61-72.

[724] D. P. Watts, "Infanticide in mountain gorillas: New cases and a reconsideration of the evidence," *Ethology,* vol. 81, no. 1, pp. 1-18, 1989.

[725] J. Yamagiwa, J. Kahekwa and A. K. Basabose, "Infanticide and social flexibility in the genus Gorilla," *Primates,* vol. 50, no. 4, pp. 293-303, 2009.

[726] S. Rosenbaum, V. Vecellio and T. Stoinski, "Observations of severe and lethal coalitionary attacks in wild mountain gorillas," *Scientific Reports,* vol. 16, no. 6, p. 37018, 2016.

[727] M. O. Mirville, A. R. Ridley, J. P. M. Samedi, V. Vecellio, F. Ndagijimana, T. S. Stoinski and C. C. Grueter, "Factors influencing individual participation during intergroup interactions in mountain gorillas," *Animal Behaviour,* vol. 144, pp. 75-86, 2018.

[728] A. E. Lowe, C. Hobaiter, C. Asiimwe, K. Zuberbühler and N. E. Newton-Fisher, "Intra-community infanticide in wild, eastern chimpanzees: a 24-year review," *Primates,* vol. 61, pp. 69-82, 2020.

[729] M. M. Robbins, "A demographic analysis of male life history and social structure of mountain gorillas," *Behaviour,* vol. 132, no. 1, p. 21 – 47, 1995.

[730] M. L. Manguette, T. Breuer, J. Robeyst, V. H. Kandza and M. M. Robbins, "Infant survival in western lowland gorillas after voluntary dispersal by pregnant females," *Primates,* vol. 61, 2020.

[731] J. C. Beehner and A. Lu, "Reproductive suppression in female primates: A review," *Evolutionary Anthropology,* vol. 22, no. 5, pp. 226-238, 2013.

[732] L. Digby, "Infanticide by female mammals: implications for the evolution of social systems," in *Infanticide by Males and Its Implications*, C. van Schaik and C. Janson, Eds., Cambridge, Cambridge University Press, 2000, pp. 423-446.

[733] J. M. Shopland and J. Altmann, "Fatal intragroup kidnapping in yellow baboons," *American Journal of Primatology,* vol. 13, no. 1, pp. 61-65, 1987.

[734] H. Vervaecke, J. Stevens and L. Van Elsacker, "Interfering with others: Female-female reproductive competition in Pan Paniscus," in *Sexual Selection and Reproductive Competition in Primates: New Perspectives and Directions*, C. B. Jones, Ed., Norman, OK: American Society of Primatologists, 2003, pp. 231-256.

[735] K. Ma, M. Zhu, A. Zhang, M. Zuo, Y. Huang, Y. Wan, F. Tao and Y. Sun, "Intergenerational continuation of parent-child separation and 1-year telomere length attrition among mother-offspring dyads in rural China: The moderating effects of resilience," *Journal of Affective Disorders,* vol. 368, pp. 599-606, 2025.

[736] G. Hohmann and B. Furth, "Intra- and inter-sexual aggression by bonobos in the context of mating," *Behaviour,* vol. 140, no. 11-12, pp. 1389-1413, 2003.

[737] H. Vervaecke and L. Van Elsacker, "Sexual competition in a group of captive bonobos (Pan paniscus)," *Primates,* vol. 41, no. 1, p. 109–115, 2000.

[738] C. P. Van Schaik and R. I. M. Dunbar, "The evolution of monogamy in large primates: A new hypothesis and some crucial tests," *Behaviour,* vol. 118, no. 1, p. 30 – 61, 1990.

[739] C. P. van Schaik and P. M. Kappeler, "Infanticide and the evolution of male-female associations in primates," *Proceedings of the Royal Society of London, B,* vol. 264, no. 1388, pp. 1687-1694, 1997.

[740] G. Hausfater and S. B. Hrdy, Infanticide, New York: Routledge, 2017, p. 638.

[741] H. B. Hrdy, "Infanticide among animals: A review, classification, and examination of the implications for the reproductive strategies of females," *Evolution and Human Behavior,* vol. 1, no. 1, pp. 13-40, 1979.

[742] J. A. Haines, D. W. Coltman, B. Dantzer, J. C. Gorrell, M. M. Humphries, J. E. Lane, A. G. McAdam and S. Boutin, "Sexually selected infanticide by male red squirrels in advance of a mast year," *Ecology,* vol. 99, pp. 1242-1244, 2018.

[743] C. Hashimoto, Y. Tashiro, E. Hibino, M. Mulavwa, K. Yangozene, T. Furuichi, G. Idani and O. Takenaka, "Longitudinal structure of a unit-group of bonobos: Male philopatry and possible fusion of unit-groups (pp 108-119)," in *The Bonobos: Behavior, Ecology and Conservation*, 2008, p. 327.

[744] T. Nishida, "The social structure of chimpanzees of the Mahali Mountains (pp 73-121)," in *The Great Apes*, Menlo Park, CA: Benjamin Cummings, 1996, p. 352.

[745] V. Reynolds, The Chimpanzees of the Budongo Forest, Oxford, England: Oxford University Press, 2005, p. 312.

[746] K. E. Langergraber, C. Rowney, G. Schubert, C. Crockford, C. Hobaiter, R. Wittig, R. W. Wrangham, K. Zuberbühler and L. Vigilant, "How old are chimpanzee communities? Time to the most recent common ancestor of the Y-chromosome in highly patrilocal societies," *Journal of Human Evolution,* vol. 69, no. 1, pp. 1-7, 2014.

[747] K. B. Potts, D. P. Watts and R. W. Wrangham, "Comparative feeding ecology of two chimpanzee communities in Kibale National Park, Uganda," *International Journal of Primatology,* vol. 32, no. 3, p. 669–690, 2011.

[748] J. Van Lawick-Goodall, "The behaviour of free-living chimpanzees in the Gombe Stream Reserve," *Animal Behaviour Monographs (Tutgers University),* vol. 1, no. 3, pp. 161-311, IN1-IN12, 1968.

[749] R. W. Wrangham, "The evolution of sexuality in chimpanzees and bonobos," *Human Nature,* vol. 4, no. 1, pp. 47-79, 1993.

[750] S. M. Kahlenberg, M. E. Thompson, M. N. Muller and R. W. Wrangham, "Immigration costs for female chimpanzees and male protection as an immigrant counterstrategy to intrasexual aggression," *Animal Behaviour,* vol. 76, no. 5, pp. 1497-1509, 2008.

[751] K. E. Langergraber, J. C. Mitani and L. Vigilant, "The limited impact of kinship on cooperation in wild chimpanzees," *Proceedings of the National Academy of Sciences,* vol. 104, no. 19, pp. 7786-7790, 2007.

[752] S. S. Kaburu, S. Inoue and N. E. Newton-Fisher, "Death of the alpha: Within-community lethal violence among chimpanzees of the Mahale Mountains National Park," *American Journal of Primatology,* vol. 75, no. 8, pp. 789-797, 2013.

[753] J. D. Pruetz, K. B. Ontl, E. Cleaveland, S. Lindshield, J. Marshack and E. Wessling, "Intragroup lethal aggression in West African chimpanzees (Pan troglodytes verus): Inferred killing of a former alpha male at Fongoli, Senegal," *International Journal of Primatology,* vol. 38, no. 1, pp. 31-57, 2017.

[754] F. B. M. de Waal, Chimpanzee Politics: Power and Sex Among Apes, Revised ed., Baltimore: Johns Hopkins University Press, 1983/2007, p. 256.

[755] Z. P. Machanda, I. C. Gilby and R. W. Wrangham, "Mutual grooming among adult male chimpanzees: The immediate investment hypothesis," *Animal Behaviour,* vol. 87, pp. 165-174, 2013.

[756] K. Duffy, R. Wrangham and J. Silk, "Male chimpanzees exchange political support for mating opportunities," *Current Biology,* vol. 17, no. 15, p. R586–R587, 2007.

[757] J. Mitani, D. A. Merriwether and C. Zhang, "Male affiliation, cooperation and kinship in wild chimpanzees," *Animal Behavior,* vol. 59, no. 4, pp. 885-893, 2000.

[758] T. Furuichi, "Agonistic interactions and matrifocal dominance rank in wild bonobos (Pan paniscus) at Wamba," *International Journal of Primatology,* vol. 18, no. 6, pp. 855-875, 1997.

[759] T. Gruber and T. Clay, "A Comparison Between Bonobos and Chimpanzees: A Review and Update," *Evolutionary Anthropology,* vol. 25, no. 5, 2016.

[760] A. R. Parish, "Female relationships in bonobos (Pan paniscus): Evidence for bonding, cooperation, and female dominance in a male-philopatric species," *Human Nature,* vol. 7, no. 1, pp. 61-96, 1996.

[761] T. Furuichi, "Female contributions to the peaceful nature of bonobo society," *Evolutionary Anthropology,* vol. 20, no. 4, pp. 131-142, 2011.

[762] J. M. G. Stevens, H. Vervaecke, H. de Vries and L. Elsacker, "Sex differences in the steepness of dominance hierarchies in captive bonobo groups," *International Journal of Primatology,* vol. 28, no. 6, pp. 1417-1430, 2007.

[763] M. Surbeck and G. Hohmann, "Intersexual dominance relationships and the influence of leverage on the outcome of conflicts in wild bonobos (Pan paniscus)," *Behavioral Ecology and Sociobiology,* vol. 67, no. 11, p. 1767–1780, 2013.

[764] H. Vervaecke, G. J. De Vries and L. Van Elsacker, "Dominance and its behavioral measures in a captive group of bonobos (Pan paniscus)," *International Journal of Primatology,* vol. 21, no. 1, pp. 47-68, 2000a.

[765] L. Saxon, The Naked Bonobo, CreateSpace, 2016, p. 198.

[766] M. Surbeck, C. Girard-Buttoz, C. Boesch, C. Crockford, B. Fruth, G. Hohmann, K. Z. K. E. Langergraber, R. M. Wittig and R. Mundry, "Sex-specific association patterns in bonobos and chimpanzees reflect species differences in cooperation," *Royal Society Open Science,* vol. 4, no. 5, p. 2017, 3 5 2017.

[767] C. B. Stanford, "The brutal ape vs. the sexy ape," *American Scientist,* vol. 88, no. 2, pp. 110-112, 2000.

[768] M. Surbeck, K. E. Langergraber, B. Fruth, L. Vigilant and G. Hohmann, "Male reproductive skew is higher in bonobos than chimpanzees," *Current Biology,* vol. 27, no. 13, pp. R640-R641, 2017.

[769] H. Ihobe, "Male-male relationships among wild bonobos (Pan paniscus) at Wamba, Republic of Zaire," *Primates,* vol. 33, pp. 163-179, 1992.

[770] R. I. M. Dunbar, "Neocortex size as a constraint on group size in primates," *Journal of Human Evolution,* vol. 22, no. 6, pp. 469-493, 1992.

[771] P. D. Killworth, E. C. Johnsen, H. R. Bernard, G. A. Shelley and C. McCarty, "Estimating the size of personal networks," *Social Networks,* vol. 12, no. 4, pp. 289-312, 1990.

[772] J. A. Krems, S. Claessens, M. R. Fales, M. Campenni, G. Haselton and A. Aktipis, "An agent-based model of the female rivalry hypothesis for concealed ovulation in humans," *Nature: Human Behaviour,* vol. 5, p. 726–735, 2021.

[773] E. Demuru, M. Caselli, J.-P. Guéry, C. Michelet, F. Alexieff and I. Norscia, "Female bonobos show social swelling by synchronizing their maximum swelling and increasing bonding," *Scientific Reports,* vol. 12, p. 17676, 2022.

[774] T. Furuichi, "Female contributions to the peaceful nature of bonobo society," *Evolutionary Anthropology,* vol. 20, no. 4, pp. 131-142, 2011.

[775] G. P. Murdock and S. F. Wilson, "Settlement patterns and community organization: Cross-cultural codes 3," *Ethnology,* vol. 11, no. 3, pp. 254-295, 1972.

[776] A. S. Griffin and S. A. West, "Kin selection: fact and fiction," *Trends in Ecology and Evolution,* vol. 17, no. 1, pp. 15-21, 2002.

[777] M. G. Murray and R. Gerrard, "Conflict in the neighbourhood: Models where close relatives are in direct competition," *Journal of Theoretical Biology,* vol. 111, no. 2, pp. 237-246, 1984.

[778] S. A. West, M. G. Murray, C. A. Machado, A. S. Griffin and E. A. Herre, "Testing Hamilton's rule with competition between relatives," *Nature,* vol. 409, pp. 510-513, 2001.

[779] Aristophanes, Lysistrata, Nanaimo, BC: Richer Resources Publications, 411 BCE/ 2008, p. 80.

[780] M. Surbeck, T. Deschner, A. Weltring and G. Hohmann, "Social correlates of variation in urinary cortisol in wild male bonobos (Pan paniscus)," *Hormones and Behavior,* vol. 62, no. 1, pp. 27-35, 2012.

[781] "1 Kings 3:16-28," in *The King James Bible*.

[782] J. Grimm and W. Grimm, Grimm's Complete Fairy Tales, San Diego, CA: Canterbury Classics, 1812/2011, p. 652.

[783] I. Bădescu, D. P. Watts, A. Katenberg and D. W. Sellen, "Alloparenting is associated with reduced maternal lactation effort and faster weaning in wild chimpanzees," *Royal Society Open Science,* vol. 3, p. 160577, 2016.

[784] G. E. Kennedy, "From the ape's dilemma to the weanling's dilemma: Early weaning and its evolutionary context," *Journal of Human Evolution,* vol. 48, no. 2, p. 123–145, 2005.

[785] M. A. van Noordwijk, C. W. Kuzawa and C. P. van Schaik, "The evolution of the patterning of human lactation: A comparative perspective," *Evolutionary Anthropology,* vol. 22, no. 5, pp. 202-212, 2013.

[786] M. M. Robbins and A. M. Robbins, "Variability of weaning age in mountain gorillas (Gorilla beringei beringei)," *American Journal of Biological Anthropology,* vol. 174, no. 4, pp. 776-784, 2021.

[787] W. M. Kenkel, A. M. Perkeybile and C. S. Carter, "The neurobiological causes and effects of alloparenting," *Special Issue: Oxytocin in Development and Plasticity. Developmental Neurobiology,* vol. 77, no. 2, pp. 214-232, 2017.

[788] B. Bogin, J. Bragg and C. Kuzawa, "Humans are not cooperative breeders but practice biocultural reproduction," *Annals of Human Biology,* vol. 41, no. 4, pp. 368-380, 2014.

[789] J. E. McCarroll, J. E. Fisher, S. J. Cozza and R. J. Whalen, "Child maltreatment fatality review: Purposes, processes, outcomes, and challenges," *Trauma, Violence, and Abuse,* vol. 22, no. 5, pp. 1032-1041, 2020.

[790] M. Daly and M. Wilson, "Evolutionary social psychology and family homicide," *Science,* vol. 242, no. 4878, pp. 91-105, 1988.

[791] E. Friedman and S. B. Billick, "Unintentional child neglect: Literature review and observational study," *Psychiatric Quarterly,* vol. 86, pp. 253-259, 2014.

[792] E. Putnam-Horstein, M. A. Cleves, R. Licht and Needell, "Risk of fatal injury in young children abuse allegations: Evidence from a prospective, population-based study," *American Journal of Public Health,* vol. 22, no. 3, pp. 1032-1041, 2013.

[793] P. G. Schnitzer and B. G. Ewigman, "Child deaths resulting from inflicted injuries: Household risk factors and perpetrator characteristics," *Pediatrics,* vol. 116, no. 5, p. e687–e693, 2005.

[794] U. Schjødt, H. Stødkilde-Jørgensen, A. W. Geertz and A. Roepstorff, "Rewarding prayers," *Neuroscience Letters,* vol. 443, no. 3, pp. 165-168, 2008b.

[795] M. Presser, H. J. Quiroz, E. A. Perez, J. E. Sola, N. Namias and C. M. Thorson, "Comparing fatal child abuse involving biological and surrogate parents," *Trauma and Acute Care Surgery,* vol. 92, no. 2, pp. 362-365, 2022.

[796] A. Kauppi, K. Kumpulainen, K. Karkola, T. Vanamo and J. Merikanto, "Maternal and paternal filicides: a retrospective review of filicides in Finland," *Journal of the American Academy of Psychiatry and Law,* vol. 38, no. 2, pp. 229-238, 2016.

[797] A. F. Russell and V. Lummaa, "Maternal effects in cooperative breeders: From hymenopterans to humans," *Proceedings of the Royal Society, B,* vol. 364, pp. 1143-1167, 2009.

[798] A. Widdig, "Paternal kin discrimination: the evidence and likely mechanisms," *Biological Reviews of the Cambridge Philosophical Society,* vol. 82, no. 2, pp. 19-34, 2007.

[799] P. Bressan and P. Kramer, "Human kin detection," *Wiley Interdisciplinary Review. Cognitive Science,* vol. 6, no. 3, pp. 299-311, 2015.

[800] M. G. Murray and R. Gerrard, "Putting the challenge into resource exploitation: A model of contest competition," *Journal of Theoretical Biology,* vol. 115, no. 3, pp. 367-389, 1984b.

[801] M. Barker, Director, *The Handmaid's Tale Season 1 Episode 5: Faithful.* [Film]. Canada: Daniel Wilson Productions, Inc., The Littlefield Company, White Oak Pictures; MGM Television; Hulu Originals, 2017.

[802] M. Apostolou, "The evolution of rape: The fitness benefits and costs of a forced-sex mating strategy in an evolutionary context," *Aggression and Violent Behavior,* vol. 18, no. 5, pp. 484-490, 2013.

[803] C. Mogielnicki, "Wartime forced sex as a male mating strategy," *Theory in Biosciences,* 2023.

[804] R. Thornhill, "Rape in Panorpa scorpionflies and a general rape hypothesis," *Animal Behaviour,* vol. 28, no. 1, pp. 52-59, 1980.

[805] R. Thornhill and C. T. Palmer, A Natural History of Rape: Biological Bases of Sexual Coercion, Cambridge, MA: Bradford Book, 2001a, p. 272.

[806] R. Thornhill and C. T. Palmer, "Why Men Rape," *New York Academy of Sciences,* vol. 40, no. 1, pp. 30-36, 2013.

[807] L. Sentenská, O. Šedo and S. Pekár, "Biting and binding: an exclusive coercive mating strategy of males in a philodromid spider," *Animal Behaviour,* vol. 168, pp. 59-68, 2020.

[808] J. Archer and A. E. Vaughn, "Evolutionary theories of rape," *Psychology, Evolution and Gender,* vol. 3, pp. 95-101, 2001.

[809] I. Radačić, "Rape: Does international human rights law adequately protect the dignity of women? (pp 119-132)," in *Humiliation, Degradation, Dehumanization: Human Dignity Violated*, Dordrecht, Springer, 2011, p. 280.

[810] B. A. Speziale, "Crime, violence, and the saga of Lorena Bobbitt," *AFFILIA,* vol. 9, no. 3, pp. 308-313, 1994.

[811] S. E. Ullman and R. A. Knight, "The efficacy of women's resistance strategies in rape situations," *Psychology of Women Quarterly,* vol. 17, no. 1, 2016.

[812] S. E. Ullman, "Rape resistance: A critical piece of all women's empowerment and holistic rape prevention," *Journal of Aggression, Maltreatment and Trauma,* vol. 31, no. 4, pp. 519-539, 2022.

[813] C. Anderson and G. J. Kilduff, "The pursuit of status in social groups," *Current Directions in Psychological Science,* vol. 18, no. 5, p. 295–298, 2009.

[814] B. Chapais, "Competence and the evolutionary origins of status and power in humans," *Human Nature,* vol. 26, pp. 161-183, 2015.

[815] A. Wiezel, M. Barlev, C. R. Martos and D. T. Kenrick, "Stereotypes versus preferences: Revisiting the role of alpha males in leadership," *Evolution and Human Behavior,* vol. 45, no. 3, pp. 292-308, 2024.

[816] Y. Gabriel, "The allure of strongman leaders (pp 310-323)," in *The Routledge Critical Companion to Leadership Studies*, First ed., D. Kinghts, H. Liu, O. Smolović-Jones and S. Wilson, Eds., New York, Routledge, 2024, p. 494.

[817] M. van Vugt, "Evolutionary origins of leadership and followership," *Personality and Social Psychology Review,* vol. 10, no. 4, p. 354–371, 2016.

[818] C. von Rueden, M. Gurven and H. Kaplan, "Why do men seek status? Fitness payoffs to dominance and prestige," *Proceedings of the Royal Society B,* vol. 278, no. 2715, pp. 2223-2232, 2010.

[819] W. Shakespeare, "Othello," in *The Complete Works of Shakespeare*, Garden City, NJ: Garden City Publishing, 1936.

[820] M. J. Bloomfield and C. S. G. Marvão, "Relative performance evaluation, sabotage and collusion," *Journal of Accounting and Economics,* vol. 76, no. 2-3, p. 101608, 2023.

[821] H. S. Edosomwan, T. M. Oguegbe and C. O. Joe-Akunne, "Relationship between Perceived Employability and Sabotage Behaviour: The Moderating Role of Perceived Organizational Support and Procedural Justice," *Contemporary Management Research,* vol. 19, no. 1, pp. 27-54, 2023.

[822] D. Lagziel, E. Lehrer and O. Raveh, "Resource windfalls and political sabotage: Evidence from 5.2 million political ads," *Social Science Research Network,* 2023.

[823] Y. Zhu, L. Long, Y. Xu and Y. Zhang, "Knowledge seeking or sabotage? The effect of coworker relative overqualification on employee reaction from social comparison theory," *Journal of Knowledge Management,* vol. 28, no. 3, 2024.

[824] A. Stenfors and L. Muchimba, "The anatomy of three scandals: Conspiracies, beauty contests, and sabotage in OTC markets," *Journal of Economic Issues,* vol. 57, no. 2, pp. 538-545, 2023.

[825] E. G. Dutcher, R. Oexl, D. Ryvkin and T. C. Salmon, "Do competitive bonuses ruin cooperation in heterogeneous teams?," *Journal of Economics and Management Strategy,* vol. 34, no. 1, pp. 67-101, 2025.

[826] G. A. Parker, "Assessment strategy and the evolution of fighting behaviour," *Journal of Theoretical Biology,* vol. 47, no. 1, pp. 223-242, 1974.

[827] P. L. Hurd, "Resource holding potential, subjective resource value, and game theoretical models of aggressiveness signalling," *Journal of Theoretical Biology,* vol. 241, no. 3, pp. 639-648, 2006.

[828] E. A. Tibbetts, J. Pardo-Sanchez and C. Weise, "The establishment and maintenance of dominance hierarchies," *Philosophical Transactions of the Royal Society B,* vol. 337, no. 1845, 2022.

[829] P. K. Durkee, A. W. Lukaszewski and D. M. Buss, "Psychological foundations of human status allocation," *Proceedings of the National Academy of Sciences,* vol. 117, no. 35, pp. 21235-21241, 2020.

[830] G. E. Weisfeld and L. M. Dillon, "Applying the dominance hierarchy model to pride and shame, and related behaviors," *Journal of Evolutionary Psychology,* vol. 10, no. 1, pp. 15-41, 2012.

[831] A. Avarguès-Weber, "Face recognition: Lessons from a wasp," *Current Biology,* vol. 22, no. 3, pp. R91-R93, 2012.

[832] E. A. Tibbetts, A. Mettler and S. Levy, "Mutual assessment via visual status signals in Polistes dominulus wasps," *Biology Letters,* vol. 6, no. 1, pp. 10-13, 2010.

[833] E. A. Tibbetts, E. Wong and S. Bonello, "Wasps use social eavesdropping to learn about individual rivals," *Current Biology,* vol. 30, no. 15, pp. 3007-3010, 2020.

[834] M. J. Sheehan and J. Goldberg, "Animal behavior: Socially distanced wasps learn about rivals," *Current Biology,* vol. 30, no. 15, pp. R878-R880, 2020.

[835] J. Ge, X. Zhou, Z. Ge, D. Zhu, X. Nie and X. Wang, "Phased contests allow rapid hierarchy formation in paired bumble bee workers," *Animal Behaviour,* vol. 179, pp. 125-138, 2021.

[836] E. A. Tibbetts, K. Crocker and Z. Y. Huang, "Socially selected ornaments influence hormone titers of signalers and receivers," *Proceedings of the National Academy of the Sciences,* vol. 113, no. 30, pp. 8478-8483, 2016.

[837] M. Ly, M. R. Haynes, J. W. Barter, D. R. Weinberger and C. F. Zink, "Subjective socioeconomic status predicts human ventral striatal responses to social status information," *Current Biology,* vol. 21, no. 9, pp. 794-797, 2011.

[838] R. Adolphs, "What does the amygdala contribute to social cognition?," *Annals of the New York Academy of Sciences,* vol. 1191, no. 1, pp. 42-61, 2010.

[839] M. J. Sheehan and E. A. Tibbetts, "Robust long-term social memories in a paper wasp," *Current Biology,* vol. 18, no. 18, pp. R851-R852, 2008.

[840] P. H. Crowley, L. Provencher, S. Sloane, L. A. Dugatkin, B. Spohn, L. Rogers and M. Alfieri, "Evolving cooperation: the role of individual recognition," *Biosystems,* vol. 37, no. 1-2, pp. 49-66, 1996.

[841] G. Arnott and R. W. Elwood, "Assessment of fighting ability in animal contests," *Animal Behavior,* vol. 77, no. 5, pp. 991-1004, 2009.

[842] D. Finkel, The Good Soldiers, New York: Perfection Learning, 2010, p. 313.

[843] M. B. Khaskheli, H. A. R. Saleem, S. Bibi and J. G. Mapa, "Comparative analysis of honor killing phenomena in China and Pakistan," *Journal of Law and Criminal Justice,* vol. 6, no. 2, pp. 20-31, 2018.

[844] P. Grosjean, "A history of violence: The culture of honor and homicide in the US South," *Journal of the European Economic Association,* vol. 12, no. 5, pp. 1285-1316, 2014.

[845] J. Hailé, "Honor Killing Its Causes & Consequences: Suggested Strategies for the European Parliament," European Parliament, Brussels, 2007.

[846] S. Elakkary, B. Franke, D. Shokri, S. Hartwig, M. Tsokos and K. Püschel, "Honor crimes: Review and proposed definition," *Forensic Science Medicine and Pathology,* vol. 10, no. 1, pp. 76-82, 2014.

[847] J. Beller, C. Kröger and D. Hosser, "Disentangling honor-based violence and religion: The differential influence of individual and social religious practices and fundamentalism on support for honor killings in a cross-national sample of Muslims," *Journal of Interpersonal Violence,* vol. 36, no. 19-20, pp. 19-20, 2021.

[848] Violence Policy Center, "When Men Murder Women: An Analysis of 2009 Homicide Data: Females Murdered by Males in Single Victim/Single Offender Incidents," Violence Policy Center, Washington DC, 2011.

[849] Violence Policy Center, "When Men Murder Women: An Analysis of 2020 Homicide Data," Violence Policy Center, Washington, DC, 2023.

[850] T. K. Henry, B. E. Hayes, J. D. Freilich and S. Chermak, "Comparison of honor killings to anti-LGBTQ homicides," *Journal of Aggression, Conflict and Peace Research,* vol. 10, no. 4, pp. 272-282, 2018.

[851] United States Department of Justice, "2021 Hate Crimes Statistics," 11 February 2023. [Online]. Available: https://www.justice.gov/hatecrimes/hate-crime-statistics.

[852] A. R. Flores, L. Langton, I. H. Meyer and A. P. Romero, "Victimization rates and traits of sexual and gender minorities in the United States: Results from the National Crime Victimization Survey, 2017," *Science Advances,* vol. 6, no. 40, 2020.

[853] H. Park and I. Mykhyalyshyna, "L.G.B.T. people are more likely to be targets of hate crimes than any other minority group," *New York Times,* 16 June 2016.

[854] P. Sartorio, "Seek and destroy, chasing a problem that does not exist? –A brief look at Chechnya unofficial hunt for lesbian, gay, bisexual, and transgender (LGBT) Chechens and human rights indifference," *SSRN,* 2018.

[855] M. McClain and O. Waite-Wright, "The LGBT Community in Turkey: Discrimination, Violence, and the Struggle for Equality," *Creighton International and Comparative Law Journal,* vol. 7, no. 1, p. 152, 2016.

[856] R. Umrao, "Gender Identity of LGBT and Their Position in India," *Supremo Amicus,* p. 204, 2019.

[857] J. N. da Costa, Direito Penal E Cultura: Sobre A Responsabilidade Criminal Nos Homicidos Por Motivo de Honora. (Criminal law and culture: Criminal liability in honor killings), Lisboa, 2022, p. 262.

[858] M. Rogers, "Transphobic 'honour'-based abuse: A conceptual tool," *Sociology,* vol. 51, no. 2, pp. 225-240, 2017.

[859] M. Patten, H. Carmichael, A. Moore and C. Velopulos, "Circumstances of suicide among lesbian, gay, bisexual and transgender individuals," *Journal of Surgical Research,* vol. 270, pp. 522-529, 2022.

[860] United Nations Human Rights Council 20th Session, "Report of the Special Rapporteur on violence (Agenda Item 3)," United Nations., New York, 2012.

[861] United Nations Population Fund, "The state of the world's population 2000: Lives together, worlds apart: Men and women in a time of change," United Nations, New York, 2020.

[862] M. Eisner and L. Ghuneim, "Honor killing attitudes amongst adolescents in Amman, Jordan," *Aggressive Behavior,* vol. 39, no. 5, pp. 405-417, 2013.

[863] A. Kulczycki and S. Windle, "Honor killings in the Middle East and North Africa: A systematic review," *Violence Against Women,* vol. 17, no. 11, pp. 1442-1464, 2011.

[864] F. Faquir, "Intrafamily femicide in defence of honour: The case of Jordan," *Third World Quarterly,* vol. 22, no. 1, pp. 65-82, 2001.

[865] N. Alam, S. Naveed and M. Naeem, "Frequency of female honour killing in district Peshawar, a clinical audit of autopsy conducted in forensic medicine department (2009- 2012) Khyber Medical College," *Journal of Medical Science,* vol. 25, no. 1, pp. 145-148, 2017.

[866] H. Fatima, T. F. Qadir, S. A. Hussain and R. G. Menezes, "Pakistan steps up to remove 'honour' from honour killing," *The Lancet: Global Health,* vol. 5, no. 2, p. e145, 2017.

[867] World Bank, "Pakistan Data," [Online]. Available: https://data.worldbank.org/country/pakistan?view=chart. [Accessed 2 7 2018].

[868] J. A. Fox and M. W. Zawitz, "Bureau of Justice Statistics: Homicide Trends in The United States," U.S. Department of Justice, Office of Justice Programs, Bureau of Justice Statistics, Washington, D.C., 2020.

[869] L. Hofhansel, C. Weidler, B. Clemens, U. Habel and M. Votinov, "Personal insult disrupts regulatory brain networks in violent offenders," *Cerebral Cortex,* vol. 33, no. 8, pp. 4654-4664, 2023.

[870] V. Leutgeb, A. Wabnegger, M. Leitner, T. Zussner, W. Scharmüller, D. Klug and A. Schienle, "Altered cerebellar-amygdala connectivity in violent offenders: A resting-state fMRI study," *Science Direct,* vol. 610, pp. 160-164, 2016.

[871] United States Department of Justice, "Bureau of Justice Statistics, Selected Findings: Violent Crime (NCJ-147486)," Bureau of Justice Statistics, Office of Justice Programs, Washington DC, 1994.

[872] A. L. Whaley, "The culture of gun violence: Culture of honor and code of the street," *Du Bois Review: Social Science Research on Race,* vol. 19, no. 1, pp. 193-203, 2021.

[873] H. R. McCartin, H. E. Benemann, M. Norton-Baker, T. D. Russell, D. K. Cash and A. R. King, "Boys round here: The relationship between masculine honor ideology, aggressive behavior, race, and regional affiliation," *Journal of Interpersonal Violence,* vol. 38, no. 5-6, 2022.

[874] United States Federal Bureau of Investigation, "FBI — Table 5: 2014 Crime in the United States," 17 8 2015. [Online]. Available: https://ucr.fbi.gov/crime-in-the-u.s/2014/crime-in-the-u.s.-2014/tables/table-5. [Accessed 2 7 2018].

[875] Centers for Disease Control, "Homicide mortality by state," 13 February 2023. [Online]. Available: https://www.cdc.gov/nchs/pressroom/sosmap/homicide_mortality/homicide.htm.

[876] C. Ammons-Blanfort, S. J. D'Alessio and L. Stozenberg, "Self-help and Black firearm crime," *Crime and Delinquency,* vol. 69, no. 1, pp. 178-207, 2023.

[877] T. J. Miles and A. B. Cox, "Does immigration enforcement reduce crime? Evidence from secure communities," *Journal of Law and Economics,* vol. 55, no. 4, pp. 937-973, 2014.

[878] C. Amuedo-Dorantes and M. Deza, "Police trust and domestic violence among immigrants: evidence from VAWA self-petitions," *IZA Discussion Paper No. 12868,* pp. 1-59, 2020.

[879] B. G. O'Brien, L. Collingwood and S. O. El-Khatib, "The politics of refuge: Sanctuary cities, crime, and undocumented immigration," *Urban Affairs Review,* vol. 55, no. 1, 2017.

[880] M. Ascherio, "Do sanctuary policies increase crime? Contrary evidence from a county-level investigation in the United States," *Social Science Research,* vol. 106, p. 102743, 2022.

[881] T. K. Wong, "The effects of sanctuary policies on crime and the economy," Center for American Progress, Los Angeles, 2017.

[882] M. R. Lee, M. O. Maume and G. C. Ousey, "Social isolation and lethal violence across the metro/nonmetro divide: The effects of socioeconomic disadvantage and poverty concentration on homicide," *Rural Sociology,* vol. 68, no. 1, pp. 107-131, 2009.

[883] W. B. Bankston and A. H. David, "Rural Social Areas and Patterns of Homicide: An Analysis of Lethal Violence in Louisiana," *Rural Sociology,* vol. 45, no. 2, pp. 223-237, 1980.

[884] J. D. Thomas, "Randall Collins' micro-sociology, the Southern culture of honor, and the codes of violence pathway: Toward a general theory of violence," *Social Sciences and Humanities Open,* vol. 6, no. 1, p. 100352, 2022.

[885] H. Katzman, O. Brand, S. Rapp, C. Frias-Gil, M. Rusgrove, F. Shyne and C. J. O'Dea, "Masculine honor beliefs and perceptions of intimate partner violence by men lower in relational power," *Personality and Individual Differences,* vol. 221, p. 112529, 2024.

[886] B. Wyatt-Brown, The Shaping of Southern Culture: Honor, Grace, and War, 1760s-1880s, Chapel Hill, NC: University of North Carolina Press, 2001, p. 412.

[887] D. Cohen, R. Nisbett, B. F. Bowdel and N. Schwartz, "Insult, aggression, and the southern culture of honor: An 'experimental ethnography'," *Journal of Personality and Social Psychology,* vol. 70, no. 5, pp. 945-960, 1996.

[888] J. D. Hacker, "A Census-Based Count of the Civil War Dead," *Civil War History,* vol. 57, no. 4, pp. 307-348, 2011.

[889] D. J. Sheffer, 'No Sacrifice is Too Great, Save That of Honor:' Honor, Death, and Psychological Combat Trauma in The American Civil War. (Doctoral Dissertation, University of Kansas), Lawrence KS: University of Kansas, 2009, p. 307.

[890] J. Glossner, Cavaliers and Crackers: Landless Whites in the mind of the elite antebellum South (Master's Thesis), Jackson, MS: University of Mississippi, 2016.

[891] J. Glossner, The Crisis of White Supremacy in the Antebellum South: Poor Whites, Slavery, and the Coming of the Civil War (Doctoral Dissertation, University of Mississippi), Jackson: Mississippi, 2023, p. 292.

[892] B. Wyatt-Brown, Southern Honor: Ethics and Behavior in the Old South, 25th anniversary ed., New York: Oxford University Press, 1982/2007, p. 640.

[893] B. Wyatt-Brown, A warring nation: Honor, race, and humiliation in America and abroad., Charlottesville, VA: University of Virginia Press, 2014, p. 256.

[894] B. S. Steinberg, Shame and Humiliation: Presidential Decision Making on Vietnam, Montreal: McGill-Queen's University Press, 1996, p. 368.

[895] H. Ijzerman, W. W. van Dijk and M. Gallucci, "A bumpy train ride: A field experiment on insult, honor, and emotional reactions," *Emotion,* vol. 7, no. 4, pp. 869-875, 2017.

[896] M. G. T. C. de Sousa, B. de Souza, W. Bilsky and A. Roazzi, "The culture of honor as the best explanation for the high rates of criminal homicide in Pernambuco: A comparative study with 160 convicts and non-convicts," *Anuario de Psicología Jurídica,* vol. 26, no. 1, pp. 114-121, 2016.

[897] E. Pedersen, D. E. Forster and M. E. McCullough, "Life history, code of honor, and emotional responses to inequality in an economic game," *Emotion,* vol. 14, no. 5, pp. 920-929, 2014.

[898] S. Foster, J. E. Bock, M. Carvallo, C. L. Pollet and W. Stern, "Honor-endorsing women and relational aggression: Evidence for the presence of feminine aggression norms in southern U.S. women," *Personality and Individual Differences,* vol. 194, p. 111668, 2022.

[899] T. F. Denson, W. C. Pedersen, J. Ronquillo and A. S. Nandy, "The angry brain: Neural correlates of anger, angry rumination, and aggressive personality," *Journal of Cognitive Neuroscience,* vol. 21, no. 4, pp. 734-744, 2009.

[900] C. Dickens, The Life and Adventures of Nicholas Nickleby, The Literature Page, 1839/ n.d., p. 952.

[901] R. F. Inglis, P. G. Roberts, A. Gardner and A. Buckling, "Spite and the scale of competition in Pseudomonas aeruginosa," *American Naturalist,* vol. 178, no. 2, pp. 276-85, 2011.

[902] A. Garner and S. A. West, "Spite and the scale of competition," *Journal of Evolutionary Biology,* vol. 17, no. 6, pp. 1195-1203, 2004.

[903] J. L. Baker and P. Barclay, "Local competition increases people's willingness to harm others," *Evolution of Human Behavior,* vol. 37, no. 4, pp. 315-322, 2016.

[904] R. Smead and P. Forber, "Signals and spite in fluctuating populations," *Open Philosophy,* vol. 2, no. 1, pp. 137-146, 2019.

[905] P. G. Madgwick, "Spite and the geometry of negative relatedness," *American Naturalist,* vol. 196, no. 5, pp. E119-E126, 2020.

[906] D. B. Krupp and W. Maciejewski, "The evolution of extraordinary self-sacrifice," *Science Report,* vol. 7, no. 12, p. 90, 2022.

[907] S. A. West and A. Gardner, "Altruism, spite, and greenbeards," *Science,* vol. 327, no. 5971, pp. 1341-1344, 2010.

[908] L. Lehmann, K. Bargum and M. Reuter, "An evolutionary analysis of the relationship between spite and altruism," *Journal of Evolutionary Biology,* vol. 19, no. 5, pp. 1057-1016, 2006a.

[909] Z. Fulker, P. Forber, R. Smead and C. Riedl, "Spite is contagious in dynamic networks," *Nature Communications,* vol. 12, p. 260, 2021.

[910] L. Ibanez and H. Saadaoui, "An experimental investigation on the dark side of emotions and its aftereffects," *PLoS One,* vol. 17, no. 10, p. e0274284, 2022.

[911] L. J. Morrell, P. R. Y. Blackwell and N. B. Metcalfe, "Fighting in fiddler crabs Uca mjoebergi: What determines duration?," *Animal Behaviour,* vol. 70, no. 3, pp. 653-663, 2005.

[912] J. M. McNamara and O. Leimar, Oxford Series in Ecology and Evolution: Game Theory in Biology: Concepts and Frontiers, Oxford: Oxford University Press, 2020.

[913] P. Slovic, C. K. Mertz, D. M. Markowitz, A. Quist and D. Västfjäll, "Virtuous violence from the war room to death row," *Proceedings of the National Academy of Science of the United States of America,* vol. 117, no. 34, pp. 30474-30482, 2020.

[914] L. Glowacki and R. McDermott, "Key individuals catalyse intergroup violence," *Philosophical Transactions of the Royal Society B,* vol. 277, no. 1851, 2022.

[915] J. Fooken, "Trusting when risk and ambiguity create opportunities for exploitation," *Journal of Behavioral and Experimental Economics,* vol. 103, p. 101994, 2023.

[916] M. G. Kocher, P. Martinsson, D. Matzat and C. Wollbrant, "The role of beliefs, trust, and risk in contributions to a public good," *Journal of Economic Psychology,* vol. 51, pp. 236-344, 2015.

[917] E. Fehr and S. Gächter, "Cooperation and punishment in public goods experiments (pp 134-143)," in *The Art of Experimental Economics: Twenty Top Papers Reviewed*, 1st ed., G. Charness and M. Pingle, Eds., New York, Routledge, 2021, p. 234.

[918] S. P. Hammarlund, B. D. Connelly, K. J. Dickinson and B. Kerr, "The evolution of cooperation by the Hankshaw effect," *Evolution,* vol. 70, no. 6, pp. 1367-1385, 2016.

[919] S. Fischer, S. Ataalla, X. Gabrielidis, B. Rohdmann, D. Yazar and A. Jungwirth, "Comparative conflict resolution: cooperative cichlids outperform less social species," *Animal Behaviour,* vol. 208, pp. 91-109, 2024.

[920] 水木理恵 and ミズキリエ, Childhood maltreatment and oxytocin as the physiological bases for social functioning and stress coping (abstract), Kobe: Konan University, 2023.

[921] J. L. Bronstein and H. Sridhar, "Connecting and integrating cooperation within and between species," *Philosophical Transactions of the Royal Society B,* vol. 379, no. 1909, 2024.

[922] L. Beaudrot, M. S. Palmer, T. M. Anderson and C. Packer, "Mixed-species groups of Serengeti grazers: a test of the stress gradient hypothesis," *Ecology,* vol. 101, no. 11, p. e03163, 2020.

[923] K. Stears, M. H. Schmitt, C. C. Wilmers and A. M. Shrader, "Mixed-species herding levels the landscape of fear," *Proceedings of the Royal Society B,* vol. 287, no. 1922, p. 20192555, 2020.

[924] J. Dijkstra, B. Simpson and D. M. Bakker, "Everybody herds, sometimes: cumulative advantage as a product of rational learning," *Journal of Mathematical Sociology,* vol. 48, no. 2, p. 207–271, 2024.

[925] D. R. Carrier, "The advantage of standing up to fight and the evolution of habitual bipedalism in hominins," *PLoS One,* vol. 5, p. e19630, 2011.

[926] R. B. Reddy, L. Samuni, V. Städele, L. Vigilant and M. Surbeck, "Maternal conflict intervention is more frequent in chimpanzee compared to bonobo development," *Animal Behaviour,* vol. 208, pp. 127-136, 2024.

[927] L. R. Moscovice, P. H. Douglas, L. Martinez-Iñigo, M. Surbeck, L. Vigilant and G. Hohmann, "Stable and fluctuating social preferences and implications for cooperation among female bonobos at LuiKotale, Salonga National Park, DRC," *American Journal of Physical Anthropology,* vol. 163, no. 1, pp. 158-172, 2017.

[928] L. A. Dugatkin, "The silver fox domestication experiment," *Evolution: Education and Outreach,* vol. 11, no. 16, 2018.

[929] D. K. Belyaev, A. O. Ruvinsky and L. N. Trut, "Inherited activation-inactivation of the star gene in foxes: Its bearing on the problem of domestication," *Journal of Heredity,* vol. 72, no. 4, pp. 267-274, 1981.

[930] E. E. Hecht, A. V. Kukekova, D. A. Gutman, G. M. Acland, T. M. Preuss and L. N. Trut, "Neuromorphological changes following selection for tameness and aggression in the Russian farm-fox experiment," *The Journal of Neuroscience,* vol. 41, no. 28, pp. 6144-6156, 2021.

[931] A. V. Kukekova, J. L. Johnson, X. Xiang, S. Feng, S. Liu, H. Rando, A. V. Kharlamova, Y. Herbeck, N. A. Serdyukova, Z. Xiong, V. Beklemischeva, K.-P. Koepfli, R. G. V. A. V. Gulevich, J. P. Hekman, P. L. Perelman and A. Graphodatsky, "Red fox genome assembly identifies genomic regions associated with tame and aggressive behaviours," *Nature Ecology and Evolution,* vol. 2, p. 1479–1491, 2018.

[932] C. Townsend, J. V. Ferraro, H. Habecker and M. V. Flinn, "Human cooperation and evolutionary transitions in individuality," *Philosophical Transactions of the Royal Society B,* vol. 378, no. 2872, p. 20210414, 2023.

[933] B. Hare, V. Wobber and R. Wrangham, "The self-domestication hypothesis: evolution of bonobo psychology is due to selection against aggression," *Animal Behaviour,* vol. 83, no. 3, pp. 573-585, 2012.

[934] D. Shilton, M. Breski, D. Dor and E. Jablonka, "Human social evolution: Self-domestication or self-control?," *Frontiers in Psychology,* vol. 11, p. 134, 2020.

[935] V. Y. A. E. V. Ovchinnikov, G. V. Vasilyev, S. G. Shihevich, D. V. Shepeleva and Y. E. Herbeck, "Hippocampal glucocorticoid receptor and microRNA gene expression and serum cortisol concentration in foxes selected for behavior toward humans," *Vavilov Journal of Genetics and Breeding,* vol. 22, no. 2, 2018.

[936] R. W. Wrangham, M. L. Wilson and M. N. Muller, "Comparative rates of aggression in chimpanzees and humans," *Primates,* vol. 47, pp. 14-25, 2006.

[937] R. W. Wrangham and L. Glowacki, "Intergroup aggression in chimpanzees and war in nomadic hunter-gatherers: Evaluating the chimpanzee model," *Human Nature,* vol. 23, p. 5029, 2012.

[938] N. Smith and S. J. Thomas, "Doomsday prepping during the COVID-19 Pandemic," *Frontiers in Psychology,* vol. 12, p. 659925, 2021.

[939] R. M. Post, G. S. Leverich, S. L. McElroy, R. Kupka, T. Suppes, L. L. Altshuler, W. A. Nolen, M. A. Frye, P. E. Keck, H. Grunze and M. Rowe, "Are personality disorders in bipolar patients more frequent in the US than Europe?," *European Neuropsychopharmacology,* vol. 58, pp. 47-54, 2022.

[940] P. T. de Chardin, "The evolution of chastity," in *Toward the Future*, New York, Harcourt Inc., 1936, p. 223.

[941] F. B. M. de Waal, "Apes from Venus: Bonobos and human social evolution," in *Tree of Origin: What Primate Behaviour Can Tell Us About Human Social Evolution. pp 39-68,* Cambridge, MA: Harvard University Press, 2002, p. 311.

[942] S. Huang, L. Guoliang, Y. Pan, M. Song, J. Zhao, X. Wan, C. J. Krebs, Z. Wang, W. Han and Z. Zhang, "Density-induced social stress alters oxytocin and vasopressin activities in the brain of a small rodent species," *Integrative Zoology,* vol. 16, no. 2, pp. 149-159, 2021.

[943] I. Sekol, D. P. Farrington and I. Zych, "Effects of prison crowding on prison misconduct and bullying," in *The Routledge Handbook of the Philosophy and Science of Punishment*, First ed., F. Focquaert, E. Shaw and B. N. Waller, Eds., Taylor and Francis Group, 2020.

[944] e. S. van Haeringen, C. Gerritsen and K. V. Hindriks, "Emotion contagion in agent-based simulations of crowds: a systematic review," *Autonomous Agents and Multi-Agent Systems,* vol. 37, no. 6, 2023.

[945] N. Tokuyama, "Aggressive behaviors and social dominance in bonobos. (pp 247-266)," in *Bonobos and People at Wamba: 50 Years of Research*, First ed., Singapore, Springer, 2023, p. 610.

[946] D. P. Watts, M. Muller, S. J. Amsler, G. Mbabazi and J. C. Mitani, "Lethal intergroup aggression by chimpanzees in Kibale National Park, Uganda," *American Journal of Primatology,* vol. 68, no. 2, pp. 161-180, 2006.

[947] D. L. Cheney, "Interactions and interrelationships between groups. (pp 477-490)," in *Primate Societies*, B. B. Smuts, D. L. Cheney, R. M. Seyfarth, R. W. Wrangham and T. T. Stuhsaker, Eds., Chicago, University of Chicago Press, 1987, p. 585.

[948] B. J. Ashton, P. Kennedy and A. N. Radford, "Interactions with conspecific outsiders as drivers of cognitive evolution," *Nature Communications*, vol. 11, p. 4937, 2020.

[949] S. R. T. Lemoine, L. Samuni, C. Crockford and R. M. Wittig, "Parochial cooperation in wild chimpanzees: a model to explain the evolution of parochial altruism," *Philosophical Transactions of the Royal Society B*, vol. 377, no. 1851, 23 May 2022.

[950] S. Bowles, "Group competition, reproductive leveling, and the evolution of human altruism," *Science*, vol. 314, pp. 1569-1572, 2006.

[951] S. Bowles, "Did warfare among ancestral hunter-gatherers affect the evolution of human social behaviors?," *Science*, vol. 324, pp. 1293-1298, 2009.

[952] A. N. Radford, B. Majolo and F. Aureli, "Within-group behavioural consequences of between-group conflict: a prospective review," *Proceedings of the Royal Society B.*, vol. 283, p. 20161567, 2016.

[953] A. N. Radford, "Preparing for battle? Potential intergroup conflict promotes current intragroup affiliation," *Biological Letters*, vol. 7, pp. 26-29, 2011.

[954] A. Powell, S. Shennan and M. G. Thomas, "Late Pleistocene demography and the appearance of modern human behavior," *Science*, vol. 324, no. 5932, pp. 1289-1301, 2009.

[955] R. Hames, "Pacifying Hunter-Gatherers," *Human Nature*, vol. 30, pp. 155-175, 2019.

[956] S. A. Pandit, G. R. Pradhan, H. Balashov and C. P. van Schaik, "The conditions favoring between-community raiding in chimpanzees, bonobos, and human foragers," *Human Nature*, vol. 27, no. 2, pp. 141-159, 2016.

[957] N. Tokuyama, T. Sakamaki and T. Furuichi, "Inter-group aggressive interaction patterns indicate male mate defense and female cooperation across bonobo groups at Wamba, Democratic Republic of the Congo," *American Journal of Physical Anthropology*, vol. 170, no. 4, pp. 535-550, 2019.

[958] T. Furuichi, "Factors underlying party size differences between chimpanzees and bonobos: A review and hypothesis for future studies," *Primates*, vol. 50, no. 3, pp. 197-209, 2009.

[959] G. Hohmann and B. Furth, "Dynamics in social organization of bonobos (Pan paniscus), pp 138-155," in *Behavioural Diversity in Chimpanzees and Bonobos*, New York, Cambridge University, 2002, p. 300.

[960] C. Boesch, C. Crockford, I. Herbinger, R. Wittig, Y. Moebius and E. Normand, "Intergroup conflicts among chimpanzees in Taï National Park: Lethal violence and the female perspective," *American Journal of Primatology*, vol. 70, no. 1, pp. 519-532, 2008.

[961] C. Seguin and D. Rigby, "National Crimes: A new national data set of lynchings in the United States, 1883-1941," *Socius: Sociological Research for a Dynamic World*, 2019.

[962] D. F. Jung and D. K. Cohen, Lynching and local justice: Legitimacy and accountability in weak states, Cambridge: Cambridge University Press, 2020.

[963] J. Tankebe, "Self-help, policing, and procedural justice: Ghanaian vigilantism and the rule of law," *Law and Society Review*, vol. 43, no. 2, pp. 245-270, 2024.

[964] M. Asif, D. Weenink and P. Mascini, "Engineering vengeful effervescence: Lynching rituals and religious-political power in Pakistan," *The British Journal of Criminology*, p. azac106, 2023.

[965] V. Datsenko, "In nine months, Russia spent $82 billion on the war, a quarter of its annual budget.(За дев'ять місяців Росія витратила на війну $82 млрд – це чверть її річного бюджету.)," *Forbes Ukraine*, 24 November 2022.

[966] J. G. Royde-Smith, "World War II: Human and Material Cost," [Online]. Available: https://www.britannica.com/event/World-War-II/Human-and-material-cost.

[967] S. Daggett, "Costs of Major US Wars: Congressional Research Service Report for Congress (RS22926)," Congressional Research Service, The Library of Congress, Washington, D.C., 2008.

[968] Norwich University Online, "The Cost of US Wars Then and Now," 20 October 2020. [Online]. Available: https://online.norwich.edu/academic-programs/resources/cost-us-wars-then-and-now. [Accessed 16 February 2023].

[969] B. Urlanis, War and Population, Moscow: Progress Publishers, 1971, p. 318.

[970] T. Piotrowski, Poland's Holocaust: Ethnic Strife, Collaboration with Occupying Forces and Genocide in the Second Republic, 1918-1947, Jefferson, NC: McFarland Publishing, 1998, p. 437.

[971] M. Gadzo, "Record high deaths in the Russia-Ukraine war: What you should know," 16 October 2024. [Online]. Available: https://www.aljazeera.com/news/2024/10/16/russia-ukraine-wartime-deaths. [Accessed 12 November 2024].

[972] H. Cooper, T. Gibbons-Neff, E. Schmitt and J. E. Barnes, "Troop Deaths and Injuries in Ukraine War Near 500,000, U.S. Officials Say," *The New York Times,* 18 August 2023.

[973] M. E. Wilson, C. Boesch, B. F. T. Fruth, I. C. Gilby, C. Hashimoto, C. L. Hobaiter, G. Hohmann, N. Itoh, K. Koops, J. N. Lloyd, M. Tetsuro, J. C. Mitani, D. C. Mjungu, D. Morgan, M. N. Muller, R. Mundry and M. P. J. Nakamura, "Lethal aggression in Pan is better explained by adaptive strategies than human impacts," *Nature,* vol. 513, p. 414–417, 2014.

[974] M. Schar School of Policy and Government, George Mason University, "Global BioLabs Report," King's College, London, London, 2023.

[975] M. Y. L. Wong, P. L. Munday, P. M. Buston and G. P. Jones, "Fasting or feasting in a fish social hierarchy," *Current Biology,* vol. 18, no. 9, pp. R372-R373, 2008.

[976] M. Kohda, J.-Y. Shibata, S. Awata, D. Gomagano, T. Takeyama, M. Hori and D. Heg, "Niche differentiation depends on body size in a cichlid fish: a model system of a community structured according to size regularities," *Journal of Animal Ecology,* vol. 77, no. 5, p. 859–868, 2008.

[977] I. García-Ruiz and M. Taborsky, "Group augmentation on trial: Helpers in small groups enhance antipredator defence of eggs," *Biology Letters,* vol. 18, p. 20220170, 2022.

[978] F. Termignoni-Garcia, M. I. M. Louder, C. N. Balakrishnan, L. O'Connell and S. V. Edwards, "Prospects for sociogenomics in avian cooperative breeding and parental care," *Current Zoology,* vol. 66, no. 3, pp. 293-306, 2020.

[979] K. McAuliffe, R. Wrangham, L. Glowacki and A. F. Russell, "When cooperation begets cooperation: The role of key individuals in galvanizing support," *Philosophical Transactions of the Royal Society B,* vol. 307, p. 20150012, 2015.

[980] K. A. Stiver, P. Dierkes, M. Taborsky, H. L. Gibbs and S. Balshine, "Relatedness and helping in fish: Examining the theoretical predictions," *Proceedings of the Royal Society B,* vol. 272, pp. 1593-1599, 2005.

[981] M. R. J. Heath and R. Hen, "Serotonin receptors: Genetic insights into serotonin function," *ScienceDirect,* vol. 5, no. 9, pp. 997-999, 1995.

[982] M. Klasen, D. Wolf, P. D. Eisner, U. Habel, J. Repple, I. Vernaleken, T. Schlüter, T. Eggermann, K. Zerres, F. D. Zepf and K. Mathiak, "Neural networks underlying trait aggression depend on MAOA gene alleles," *Brain Structure and Function,* vol. 223, no. 2, p. 873–881, 2018.

[983] United Nations, "Data Portal: Population Division," [Online]. Available: https://population.un.org/dataportal/home. [Accessed 23 February 2023].

[984] V. Zarulli, I. Kashnitsky and J. W. Vaupel, "Death rates at specific life stages mold the sex gap in life expectancy," *Proceedings of the National Academy of Sciences,* vol. 118, no. 20, p. e2010588118, 2021.

[985] M. E. Thompson, M. Muller, S. M. Kahlenberg and R. W. Wrangham, "Dynamics of social and energetic stress in wild female chimpanzees," *Hormones and Behavior,* vol. 58, no. 3, pp. 440-449, 2010.

[986] P. A. Nepomnaschy, K. B. Welch, D. S. McConnell, B. S. Low, B. I. Strassmann and B. G. England, "Cortisol levels and very early pregnancy loss in humans," *Proceedings of the National Academy of Science,* vol. 103, no. 10, pp. 3938-3942, 2006.

[987] N. P. Daskalakis, A. Iatrou, C. Chatzinakos, A. Jajoo, C. Snijders, D. Wylie, C. P. Dipietro, I. Tsatsani, C.-Y. Chen, C. D. Pernia, M. Soliva-Estruch, D. Arasappan and R. A. Bharadwaj, "Systems biology dissection of PTSD and MDD across brain regions, cell types, and blood," *Science,* vol. 384, no. 6698, 2024.

[988] F. R. Wendt, G. A. Pathak, J. D. Deak, F. De Angelis, D. Koller, B. Cabera-Mendoza, D. S. Lebovitch, D. F. Levey, M. B. Stein, H. R. Kranzler, K. C. Koenen, J. Gelernter, L. M. Huckins and R. Polimanti, "Using phenotype risk scores to enhance gene discovery for generalized anxiety disorder and posttraumatic stress disorder," *Molecular Psychiatry,* vol. 27, p. 2206–2215, 2022.

[989] J. E. Carroll, N. E. Mahrer, M. Shalowitz, S. Ramey and C. D. Schetter, "Prenatal maternal stress prospectively relates to shorter child buccal cell telomere length," *Psychoneuroendocrinology,* vol. 121, p. 104841, 2020.

[990] S. Joseph, B. Alava, A. Antolic, E. M. Richards, C. E. Wood and M. Keeler-Wood, "Fetal ovine skeletal and cardiac muscle transcriptomics are differentially altered by increased maternal cortisol during gestation," *Physiological Genomics,* vol. 52, no. 4, pp. 178-190, 2020.

[991] S. King, S. Kildea, M.-P. Austin, A. Brunet, V. E. Cobham, P. A. Dawson, M. Harris, E. M. Hurrion, D. P. Laplante, B. M. McDermott, H. D. McIntyre, M. W. O'Hara, N. Schmitz, H. Stapleton, S. K. Tracy, C. Vaillancourt and K. N. Dancause, "QF2011: a protocol to study the effects of the Queensland flood on pregnant women, their pregnancies, and their children's early development," *BCM Pregnancy and Childbirth,* vol. 15, p. 109, 2015.

[992] S. M. Engel, G. S. Berkowitz, M. S. Wolff and R. Yehuda, "Psychological trauma associated with the World Trade Center attacks and its effect on pregnancy outcome," *Paediatric and Perinatal Epidemiology,* vol. 19, no. 5, pp. 334-341, 2005.

[993] I. Kassotaki, G. Valsamakis, G. Mastorakos and D. K. Grammatopoulos, "Placental CRH as a signal of pregnancy adversity and impact on fetal neurodevelopment," *Frontiers in Endocrinology,* vol. 12, 2021.

[994] J. R. Seckl, "Prenatal glucocorticoids and long-term programming," *European Journal of Endocrinology,* vol. 151, no. Suppl 3, pp. U49-U62, 2004.

[995] V. M. Heine and D. H. Rowitch, "Hedgehog signaling has a protective effect in glucocorticoid-induced mouse neonatal brain injury through an 11betaHSD2-dependent mechanism," *Journal of Clinical Investigation,* vol. 119, no. 2, pp. 267-277, 2009.

[996] F. A. Bidgoli, M. Salami and S. A. Talaei, "Environmental enrichment restores impaired spatial memory and synaptic plasticity in prenatally stress exposed rats: The role of GABAergic neurotransmission," *International Journal for Developmental Neuroscience,* vol. 80, no. 7, pp. 573-585, 2020.

[997] C. E. Hostinar, "Social support can buffer against stress and shape brain activity," *AJOB Neuroscience,* vol. 6, no. 3, pp. 34-42, 2015.

[998] M. Knopfler, Composer, *The Bug.* [Sound Recording]. Vertigo. 1991.

[999] H. M. Endedijk, M. Meyer, H. Bekkering, A. H. N. Cillessen and S. Hunnus, "Neural mirroring and social interaction: Motor system involvement during action observation relates to early peer cooperation," *Developmental Cognitive Science,* vol. 24, pp. 33-41, 2017.

[1000] M. Iacoboni, "Imitation, empathy, and mirror neurons," *Annual Review of Psychology,* vol. 60, pp. 653-670, 2009.

[1001] S. Keum and H.-S. Shin, "Neural basis of observational fear learning: A potential model of affective empathy," *Neuron,* vol. 104, no. 1, pp. P78-86, 2019.

[1002] R. Viaro, E. Maggiolini, E. Farina, R. Canto, A. Iriki, A. D'Ausilio and L. Fadiga, "Neurons of rat motor cortex become active during both grasping execution and grasping observation," *Current Biology,* vol. 31, no. 19, pp. 4405-4412e4, 2021.

[1003] J. F. Prather and R. Mooney, "Mirror neurons in the songbird brain: A neural interface for learned vocal communication," in *New Frontiers in Mirror Neuron Research 1,* P. F. Ferrari and G. Rizzolatti, Eds., Oxford, Oxford University Press, 2015, p. 182–197.

[1004] F. B. M. de Waal and S. D. Preston, "Mammalian empathy: behavioural manifestations and neural basis," *Nature Reviews, Neuroscience,* vol. 18, no. 8, pp. 498-509, 2017.

[1005] S. A. Park, S. Goïame, D. A. O'Connor and J.-C. Dreher, "Integration of individual and social information for decision-making in groups of different sizes," *PLOS Biology,* vol. 16, no. 6, p. e2001958, 2017.

[1006] J. Zaki, J. Schirmer and J. P. Mitchell, "Social influence modulates the neural computation of value," *Psychological Science,* vol. 22, no. 7, pp. 894-900, 2011.

[1007] H. Wu, Y. Luo and C. Feng, "Neural signatures of social conformity: A coordinate-based activation likelihood estimation meta-analysis of functional brain imaging studies," *Neuroscience & Biobehavioral Reviews,* vol. 71, pp. 101-111, 2015.

[1008] C. Padoa-Schioppa and J. A. Assad, "Neurons in the orbitofrontal cortex encode economic value," *Nature,* vol. 441, pp. 223-226, 2006.

[1009] T. E. J. Behrens, M. W. Woolrich, M. E. Walton and M. F. S. Rushworth, "Learning the value of information in an uncertain world," *Nature Neuroscience,* vol. 10, pp. 1214-1221, 2007.

[1010] E. D. Boorman, T. E. Behrens and M. F. Rushworth, "Counterfactual choice and learning in a neural network centered on human lateral frontopolar cortex," *PLoS Biology,* vol. 9, no. 6, p. 1001093, 2011.

[1011] G. Di Pellegrino, L. Fadiga, V. Gallese and G. Rizzolatti, "Understanding motor events: a neurophysiological study," *Experimental Brain Research,* vol. 91, no. 1, pp. 176-180, 1992.

[1012] G. Rizzolatti and M. Fabbri-Destro, "Mirror neurons: from discovery to autism," *Experimental Brain Research,* vol. 200, pp. 223-237, 2010.

[1013] M. Isoda, "Understanding intentional actions from observers' viewpoints: A social neuroscience perspective," *Neuroscience Research,* vol. 112, pp. 1-9, 2016.

[1014] M. Arioli, D. Perani, S. Cappa, A. M. Proverbio, A. Zani, A. Falini and N. Canessa, "Affective and cooperative social interactions modulate effective connectivity within and between the mirror and mentalizing systems," *Human Brain Mapping,* vol. 39, no. 3, pp. 1412-1427, 2018.

[1015] F. Cushman, K. Gray, A. Gaffey and W. Mendes, "Simulating murder: The aversion to harmful action," *Emotion,* vol. 12, no. 1, pp. 2-7, 2012.

[1016] D. M. Parton and J. J. McGinley, "At the heart of harm: Cardiac substrates of action-based aversion to harm," *Personality and Individual Differences,* vol. 151, p. 109497, 2009.

[1017] J. Zapata and O. Deroy, "Ordinary citizens are more severe towards verbal than nonverbal hate-motivated incidents with identical consequences," *Scientific Reports,* vol. 13, p. 7126, 2023.

[1018] J. C. Penagos-Corzo, M. van-Hasselt, D. Escobar, R. A. Vázquez-Roque and G. Flores, "Mirror neurons and empathy-related regions in psychopathy: Systematic review, meta-analysis, and a working model," *Social Neuroscience,* vol. 17, no. 5, pp. 462-479, 2022a.

[1019] F. Hagenmuller, W. Rössler, J. Endrass, A. Rossegger and H. Haker, "Empathische Resonanzfähigkeit bei Straftätern mit psychopathischen Persönlichkeitszügen (Impaired empathic resonance in offenders with psychopathic personality traits)," *Neuropsychiatrie,* vol. 26, pp. 65-71, 2012.

[1020] Y. R. Williams, "Anatomy of an untruth: The controversy over 'Picnic' and the true cause of lynching," *Black History Bulletin,* vol. 65/66, no. 104, pp. 9-16, 2003.

[1021] J. Allen, J. Lewis, L. F. Litwack and A. H., Without Sanctuary: Lynching Photography in America, Santa Fe: Twin Palms, 2000, p. 212.

[1022] M. R. Dadds, C. Moul, A. Cauchi, C. Dobson-Stone, D. J. Hawes, J. Brennan, R. Urwin and R. E. Ebstein, "Polymorphisms in the oxytocin receptor gene are associated with the development of psychopathy," *Development and Psychopathology,* vol. 26, no. 1, pp. 21-31, 2014.

[1023] J. C. Penagos-Corzo, M. C. van Hasselt, D. Escobar, R. A. Vázquez-Roque and G. Flores, "Mirror neurons and empathy-related regions in psychopathy: Systematic review, meta-analysis, and a working model," *Social Neuroscience,* vol. 17, no. 5, pp. 462-479, 2022.

[1024] R. Ligneul, R. Girard and J.-C. Dreher, "Social brains and divides: The interplay between social dominance orientation and the neural sensitivity to hierarchical ranks," *Scientific Reports,* vol. 7, no. 1, pp. 1-9, 2017.

[1025] J. H. Beitchman, C. C. Zai, K. Muir, L. Berall, B. Nowrouzi, E. Choi and J. L. Kennedy, "Childhood aggression, callous-unemotional traits and oxytocin genes," *European Child and Adolescent Psychiatry,* vol. 21, no. 3, pp. 125-132, 2012.

[1026] C. Bell, N. Tesli, T. Gurholt, J. Rokicki, G. Hjell, T. Fisher-Vieler, I. Melle, I. Agartz, O. E. Andereassen, K. Rasmussen, R. Johansen, C. Friestad and U. K. Haukvik, "Associations between amygdala nuclei volumes, psychosis, psychopathy, and violent offending," *Psychiatric Research Neuroimaging,* vol. 319, p. 111416, 2022.

[1027] N. E. Anderson and K. A. Kiehl, "Psychopathy and aggression: when paralimbic dysfunction leads to violence," *Current Topics in Behavioral Neuroscience,* vol. 17, pp. 369-393, 2014.

[1028] B. G. Edwards, J. R. Carre and K. A. Kiehl, "A review of psychopathy and Cluster B personality traits and their neural correlates in female offenders," *Biological Psychology,* vol. 148, p. 107740, 2019.

[1029] E. Ermer, L. M. Cope, P. K. Nyalakanti, V. D. Calhoun and K. A. Kiehl, "Aberrant paralimbic gray matter in criminal psychopathy," *Journal of Abnormal Psychology,* vol. 121, no. 3, pp. 649-658, 2013.

[1030] S. Weber, U. Habel, K. Amunts and F. Schneider, "Structural brain abnormalities in psychopaths-a review," *Behavioral Science and the Law,* vol. 26, no. 1, pp. 7-28, 2008.

[1031] M. Juárez, K. A. Kiehl and V. D. Calhoun, "Intrinsic limbic and paralimbic networks are associated with criminal psychopathy," *Human Brain Mapping,* vol. 34, no. 8, pp. 1921-930, 2013.

[1032] T. J. Beckwith, K. N. Dietrich, J. P. Wright, M. Altaye and K. M. Cecil, "Reduced regional volumes associated with total psychopathy scores in an adult population with childhood lead exposure," *Neurotoxicology,* vol. 67, pp. 1-26, 2018.

[1033] L. Hofhansel, C. Weidler, M. Votinov, B. Clemens, A. Raine and U. Habel, "Morphology of the criminal brain: gray matter reductions are linked to antisocial behavior in offenders," *Brain Structure and Function,* vol. 225, no. 7, pp. 2017-2028, 2020.

[1034] S. Zeki and J. P. Romaya, "Neural correlates of hate," *PloS one,* vol. 3, no. 10, p. e3556, 2008.

[1035] E. Verona, B. Murphy and K. Bresin, "Oxytocin-related single-nucleotide polymorphisms, family environment, and psychopathic traits," *Personality Disorders,* vol. 9, no. 6, pp. 584-589, 2018.

[1036] I. Fragkaki, M. Cima, M. Verhagen, D. F. Maciejewski, M. P. Boks, P. A. C. van Lier, H. M. Koot, S. J. T. Branje and W. H. J. Meeus, "Oxytocin receptor gene (OXTR) and deviant peer affiliation: A gene-environment interaction in adolescent antisocial behavior," *Journal of Youth and Adolescence,* vol. 48, no. 1, pp. 86-101, 2019.

[1037] Y. Zhang, C. Wu, H. Chang, Q. Yan, L. Wu, S. Yuan, J. Xiang, W. Hao and Y. Yu, "Genetic variants in oxytocin receptor gene (OXTR) and childhood physical abuse collaborate to modify the risk of aggression in Chinese adolescents," *Journal of Affective Disorders,* vol. 15, no. 229, pp. 105-110, 2018.

[1038] G. B. Murrow and R. Murrow, "A hypothetical neurological association between dehumanization and human rights abuses," *Journal of Law and Biosciences,* vol. 2, no. 2, pp. 336-364, 2015.

[1039] M. M. Fourie, K. G. F. Thomas, D. M. Amodio, M. R. Warton and E. M. Meintjes, "Neural correlates of experienced moral emotion: an fMRI investigation of emotion in response to prejudice feedback," *Social Neuroscience,* vol. 9, no. 2, pp. 203-208, 2014.

[1040] L. Leets, "Experiencing hate speech: Perceptions and responses to anti-Semitic and antigay speech," *Journal of Social Issues,* vol. 58, no. 2, pp. 341-361, 2002.

[1041] R. Keighley, "Hate hurts: Exploring the impact of online hate on LGBTQ+ young people," *Women and Criminal Justice,* vol. 32, no. 1-2, pp. 29-48, 2022.

[1042] P. A. Goff, J. L. Eberhardt, M. J. Williams and M. C. Jackson, "Not yet human: Implicit knowledge, historical dehumanization, and contemporary consequences," *Journal of Personality and Social Psychology,* vol. 94, no. 2, p. 292–306, 2008.

[1043] M. Kaakinen, T. Keipi, A. Oksanen and P. Räsänen, "How does social capital associate with being a victim of online hate?," *Policy and Internet 10,* vol. 10, no. 3, pp. 302-323, 2018.

[1044] M. Bilewicz and W. Soral, "Hate speech epidemic. The dynamic effects of derogatory language on intergroup relations and political radicalization," *Political Psychology,* vol. 41, no. S1, pp. 3-33, 2020.

[1045] P. Semi, K. Blee, M. DeMichele and S. Windisch, "Addicted to hate: Identity residual among former white supremacists," *American Sociological Review,* vol. 82, no. 6, pp. 1167-1187, 2017.

[1046] T. Szanto, "In hate we trust: The collectivization and habitualization of hatred," *Phenomenology and Cognitive Sciences,* vol. 19, pp. 453-480, 2018.

[1047] S. Terbeck, "The neuroscience of prejudice," in *The Social Neuroscience of Intergroup Relations: Prejudice, Can We Cure It? (pp 29-49),* Cham, Springer, 2016, p. 98.

[1048] D. M. Amodio, J. T. Kubota, E. Harmon-Jones and P. G. Devine, "Alternative mechanisms for regulating racial responses according to internal vs external cues," *Social Cognitive and Affective Neuroscience,* vol. 1, no. 1, pp. 26-36, 2006.

[1049] J. Altman, Baboon Mothers and Infants, revised ed., Cambridge, MA: Harvard University Press, 2021, p. 240.

[1050] K. Fawcett and G. Muhumuza, "Death of a wild chimpanzee community member: Possible outcome of intense sexual competition," *American Journal of Primatology,* vol. 51, no. 4, pp. 243-247, 2000.

[1051] T. Nishida, "Reports from the Field: Mahale, Tanzania: The Death of Ntologi, The Unparalleled Leader of M Group," *Pan African News,* p. 4, 1996.

[1052] D. P. Watts, "Intracommunity coalitionary killing of an adult male chimpanzee at Ngogo, Kibale National Park, Uganda," *International Journal of Primatology,* vol. 25, p. 507–521, 2004.

[1053] T. Nishida, K. Hosaka, M. Nakamura and M. Hamai, "A within-group gang attack on a young adult male chimpanzee: Ostracism of an ill-mannered member?," *Primates,* vol. 36, pp. 207-211, 1995.

[1054] K. Jensen, "Punishment and spite, the dark side of cooperation," *Philosophical Transactions of the Royal Society of London B: Biological Sciences,* vol. 365, no. 1553, pp. 2635-2650, 2010.

[1055] D. Watts, "Reciprocity and interchange in the social relationships of wild male chimpanzees," *Behaviour,* vol. 139, no. 2, pp. 343-370, 2002.

[1056] H. D. Rijksen, A Field Study on Sumatran Orang Utans (Pongo pygmaeus abelii Lesson 1827): Ecology, Behaviour and Conservation (Dissertation), Wageningen: Wageningen University and Research ProQuest Dissertations Publishing, 1988, p. 24.

[1057] R. W. Wrangham and F. J. White, "Feeding competition and patch size in the chimpanzee species Pan paniscus and Pan troglodytes," *Behaviour,* vol. 105, no. 1-2, pp. 148-164, 1988.

[1058] K. Aoki, "Evolution of alliance in primates: A population genetic model," *Journal of Ethology,* vol. 2, pp. 55-61, 1984.

[1059] K. Hosaka and T. Nishida, "Long-term changes in hunting behavior of Mahale chimpanzees," *Primate Research, Supplement International Primatological Society,* pp. 668-668, 2010.

[1060] T. Nishida, "Alpha status and agonistic alliance in wild chimpanzees (Pan troglodytes schweinfurthii)," *Primates,* vol. 24, pp. 318-336, 1983.

[1061] F. B. M. de Waal, "Sex differences in the formation of coalitions among chimpanzees," *Ethology and Sociobiology,* vol. 5, no. 4, pp. 239-255, 1984.

[1062] F. B. M. de Waal and J. A. Hoekstra, "Contexts and predictability of aggression in chimpanzees," *Animal Behaviour,* vol. 28, no. 3, pp. 929-937, 1980.

[1063] J. Scott and J. S. Lockard, "Competition coalitions and conflict interventions among captive female gorillas," *International Journal of Primatology,* vol. 28, pp. 761-781, 2007.

[1064] T. S. Kraft, D. K. Cummings, V. V. Venkataraman, S. Alami, B. Beheim, P. Hooper, E. Seabright, B. C. Trumbel, J. Stieglitz, H. Kaplan, K. L. Endicott, K. M. Endicott and M. Gurven, "Female cooperative labour networks in hunter-gatherers and horticulturalists," *Philosophical Transactions of the Royal Society of London B.,* vol. 378, no. 1868, p. 20210431, 2021.

[1065] A. J. Stier, K. E. Schertz, N. W. Rim and M. G. Berman, "Evidence and theory for lower rates of depression in larger US urban areas," *Proceedings of the National Academy of Sciences,* vol. 118, no. 31, p. e2022472118, 2021.

[1066] L. Gutiérrez-Rojas, A. Porras-Segovia, H. Dunne, N. Andrade-González and J. A. Cervilla, "Prevalence and correlates of major depressive disorder: a systematic review," *Brazilian Journal of Psychiatry,* vol. 42, no. 6, 2020.

[1067] R. H. Salk, J. S. Hyde and L. Y. Abramson, "Gender differences in depression in representative national samples: Meta-analyses of diagnoses and symptoms," *Psychological Bulletin,* vol. 143, no. 8, pp. 783-822, 2017.

[1068] F. Botha and M. Bower, "Predictors of male loneliness across life stages: an Australian study of longitudinal data," *BMC Health,* vol. 24, p. 1285, 2024.

[1069] J. Ratcliffe, P. Galdas and M. Kanaan, "Older men and loneliness: a cross-sectional study of sex differences in the English Longitudinal Study of Ageing," *BMC Health,* vol. 24, p. 354, 2024.

[1070] A. Sannen, L. V. Elsacker and M. Eens, "Effect of spatial crowding on aggressive behavior in a bonobo colony," *Zoo Biology,* vol. 23, no. 5, pp. 383-395, 2004.

[1071] K. Nieuwenhuijsen and F. B. M. de Waal, "Effects of spatial crowding on social behavior in a chimpanzee colony," *Zoo Biology,* vol. 1, no. 1, pp. 5-28, 1982.

[1072] F. Aureli and F. B. M. de Waal, "Inhibition of social behavior in chimpanzees under high-density conditions," *American Journal of Primatology,* vol. 41, no. 3, pp. 213-228, 1999.

[1073] M. Schläpfer, L. M. A. Bettencourt, S. Grauwin, M. Raschke, R. Claxton, Z. Smoreda, G. B. West and C. Ratti, "The scaling of human interactions with city size," *Journal of the Royal Society Interface,* vol. 11, no. 98, 2014.

[1074] J. Saramäki, E. A. Leicht and E. López, "Persistence of social signatures in human communication," *Proceedings of the National Academy of Science,* vol. 111, no. 3, pp. 942-947, 2014.

[1075] R. W. Wrangham, "Targeted conspiratorial killing, human self-domestication and the evolution of groupishness," *Evolutionary Human Sciences,* vol. 3, p. E26, 2021.

[1076] R. Cervo, L. Dapporto, L. Beani, J. E. Strassmann and S. Turillazzi, "On status badges and quality signals in the paper wasp Polistes dominulus: Body size, facial colour patterns and hierarchical rank," *Proceedings of the Royal Society B,* vol. 275, no. 1639, p. 1189–1196, 2008.

[1077] E. Lenning, S. Brightman and C. L. Buist, "The trifecta of violence: A socio-historical comparison of lynching and violence against transgender women," *Critical Criminology,* vol. 29, p. 29, 2020.

[1078] J. Valcore, J. Rodgers and N. L. Asquith, "'We will build a great wall': Domination, criminalization, and expatriation in Trump campaign and rally speeches," *International Journal for Crime, Justice, and Social Democracy,* vol. 10, no. 1, pp. 143-158, 2021.

[1079] B. de Mesquita and A. Smith, The Dictator's Handbook: Why Bad Behavior is Almost Always Good Politics, Reprint ed., PublicAffairs, 2012, p. 352.

[1080] G. Thomas and G. Lewis, "Götterdämmerung," in *Defying Hitler: The Germans Who Resisted Nazi Rule*, New York, Dutton Caliber, 2019, p. 560.

[1081] C. Kurzman, "Muslim-American Terrorism Since 9/11: An Accounting," University of North Carolina, Chapel Hill, Chapel Hill, 2011.

[1082] A. R. DiMaggio, "Conspiracy theories and the manufacture of dissent: QAnon, the 'Big Lie,' Covid-19 and the Rise of Rightwing Propaganda," *Critical Sociology,* vol. 48, no. 6, pp. 1025-1048, 2022.

[1083] M. Bakry, A. Syatar, I. Haq, C. Mundzir, M. Arif and M. M. Amiruddin, "Arguing Islamophobia during Covid-19 Outbreaks: A consideration using Khuṣūṣ Al-Balwā," *International Journal of Criminology and Sociology,* vol. 9, pp. 2757-2765, 2020.

[1084] K. A. Beydoun, "Exporting Islamophobia in the global 'War on Terror'," *New York University Law Review Online,* vol. 95, pp. 81-100, 2020.

[1085] A. Kundnani, The Muslims are Coming!: Islamaphobia, Extremism, and the Domestic War on Terror, New York: Verso, 2014, p. 335.

[1086] P. Chiril, E. W. Pamungkas, F. Benamara, V. Moriceau and V. Patti, "Emotionally informed hate speech detection: A multi-target perspective," *Cognitive Computation,* vol. 14, pp. 322-352, 2022.

[1087] C. Allen, Islamophobia, London: Routledge, 2011, p. 218.

[1088] M. P. Choudhury, "The sacred and the secular: Influence of religion on George W. Bush's foreign policy," *Jadavpur Journal of International Relations,* vol. 19, no. 2, pp. 159-183, 2016.

[1089] E. Miller, "Profiles of Perpetrators of Terrorism in the United States, 1970-2013: Final Report to Resilient Systems Division, DHS Science and Technology Directorate," START, College Park, 2014.

[1090] S. G. Jones, C. Doxsee and N. Harrington, "The Escalating Terrorism Problem in the United States," Center for Strategic and International Studies, Washington, D. C., 2020.

[1091] K. A. Boon, "Heroes, metanarratives, and the paradox of masculinity in contemporary Western Culture," *Journal of Men's Studies,* vol. 13, no. 3, pp. 301-312, 2005.

[1092] K. Rapoza and M. Clapp, "Torture: Integrative Chapter pp 663-677," in *International Handbook of War, Torture, and Terrorism. Peace Psychology Book Series*, 2012, p. 706.

[1093] D. P. Watts, "Agonistic interactions in wild mountain gorilla groups," *Behaviour,* vol. 134, no. 1-2, pp. 23-57, 1997.

[1094] D. Heg and M. Taborsky, "Helper response to experimentally manipulated predation risk in the cooperatively breeding cichlid Neolamprologus pulcher," *PLoS One,* vol. 5, p. e10784, 2010.

[1095] I. García-Ruiz, A. Quiñones and M. Taborsky, "The evolution of cooperative breeding by direct and indirect fitness effects," *Science Advances,* vol. 8, p. eabl7853, 2022.

[1096] Z. Li, S. Chen, S. Wei, J. Komdeur and X. Lu, "Should sons breed independently or help? Local relatedness matters," *Journal of Animal Ecology,* vol. 92, no. 11, pp. 2189-2200, 2023.

[1097] C. H. Wheat, E. O'Connor, P. A. Downing, A. Griffin and C. K. Cornwallis, "Fluctuating environments favor cooperation among non-kin in birds (preprint)," *EcoEvo Rxiv,* 2024.

[1098] A. S. Pereira, D. De Moor, C. Casanova and L. J. N. Brent, "Kinship composition in mammals," *Royal Society Open Science,* vol. 10, no. 7, p. 230486, 2023.

[1099] H. Fan, W. Guo and X. Lu, "How do helpers slow senescence in a Tibetan cooperatively breeding bird?," *Ecosphere,* vol. 13, no. 12, p. e4314, 2022.

[1100] L. Samuni, A. Preis, A. Mielke, T. Deschner, R. M. Wittig and C. Crockford, "Social bonds facilitate cooperative resource sharing in wild chimpanzees," *Proceedings of the Royal Society B,* vol. 285, no. 1888, p. 20181643, 2018.

[1101] G. S. Wilkinson, G. G. Carter, K. M. Bohn and D. M. Adams, "Non-kin cooperation in bats," *Philosophical Transactions of the Royal Society of London B: Biological Sciences,* vol. 5, p. 20150095, 2016.

[1102] Z. Li, X. Da and X. Lu, "Complementary interactions between indirect and direct fitness in a cooperatively breeding bird," *Current Zoology,* vol. 69, no. 1, pp. 76-81, 2023.

[1103] C. D. Michener, "Castes in Xylocopine Bees (pp 123–146)," in *Social Insects: An Evolutionary Approach to Castes and Reproduction*, W. Engel, Ed., Berlin, Springer Verlag, 1990.

[1104] A. Kodric-Brown, "Satellites and sneakers: Opportunistic male breeding tactics in pupfish (Cyprinodon pecosensis)," *Behavioral Ecology and Sociobiology,* vol. 19, no. 6, p. 425–432, 1986.

[1105] M. Asher, T. Lippman, J. T. Epplen, C. Kraus, F. Trillmich and N. Sachser, "Large males fominate: Ecology, social organization, and mating system of wild cavies, the ancestors of the guinea pig," *Behavioral Ecology and Sociobiology,* vol. 62, no. 9, pp. 1509-1521, 2008.

[1106] J. Ortega, J. A. Guerrero and J. E. Maldonado, "Aggression and tolerance by dominant males of Artibeus jamaicensis: Strategies to maximize fitness in harem groups," *Journal of Mammalogy,* vol. 86, no. 6, pp. 1372-1378, 2008.

[1107] S. H. Alonzo, "Learning from our study organisms about sexual selection: Lessons from the ocellated wrasse," *The American Naturalist,* vol. 201, no. 3, 2023.

[1108] M. J. Borger, L. E. Johnson, N. O. Salazar, C. L. Dreghorn, J. Komdeur and T. G. Murphy, "The influence of social-grouping on territorial defense behavior in the black-crested titmouse (Baeolophus atricristatus)," *Behavioral Ecology and Sociobiology,* vol. 74, p. 141, 2020.

[1109] N. B. Davies and A. I. Houston, "Owners and satellites: The economics of territory defence in the pied wagtail, Motacilla alba," *Journal of Animal Ecology,* pp. 157-180, 1981.

[1110] R. Bergmüller, R. A. Johnstone, A. F. Russell and R. Bshary, "Integrating cooperative breeding into theoretical concepts of cooperation," *Behavioural Processes,* vol. 76, no. 2, pp. 61-72, 2007.

[1111] C. Boehm and J. C. Flack, "The emergence of simple and complex power structures through social niche construction, pp 46-86," in *The Social Psychology of Power*, New York, Guilford Press, 2010, p. 469.

[1112] A. Giddens, "'Power' in the recent writings of Talcott Parsons," *Sociology,* vol. 2, no. 3, pp. 257-272, 1968.

[1113] C. Boehm, Hierarchy in the Forest: The Evolution of Egalitarian Behavior, Cambridge, MA: Harvard University Press, 2009, p. 304.

[1114] J. G. Flanagan, "Hierarchy in simple 'egalitarian' societies," *Annual Review of Anthropology,* vol. 18, pp. 245-266, 1989.

[1115] C. L. Apicella, F. W. Marlowe, J. H. Fowler and N. A. Christakis, "Social networks and cooperation in hunter-gatherers," *Nature,* vol. 481, no. 7382, pp. 479-501, 2012.

[1116] M. Ananth, "Psychological altruism vs. biological altruism: Narrowing the gap with the Baldwin effect," *Acta Biotheoretica,* vol. 53, no. 3, p. 217–239, 2005.

[1117] P. Hawley, "Ontogeny and social dominance: A developmental view of human power patterns," *Evolutionary Psychology,* vol. 12, no. 2, 2014.

[1118] R. B. Lee, The !Kung San: Men, Women and Work in a Foraging Society, Cambridge: Cambridge University Press, 1979, p. 526.

[1119] M. van Vugt, R. Hogan and R. B. Kaiser, "Leadership, followership, and evolution: Some lessons from the past," *American Psychologist,* vol. 63, no. 3, pp. 182-196, 2008.

[1120] P. Wiessner, "Leveling the hunter: Constraints on the status quest in foraging societies," in *Food and the Status Quest: An Interdisciplinary Perspective (Anthropology of Food & Nutrition)*, New York, Berghahn Books, 1996, p. 288.

[1121] K. Takács, "A reputation-centered theory of human cooperation and social organization," *International Journal for Sociological Debate,* vol. 16, no. 2, pp. 11-51, 2022.

[1122] C. Boehm, "Forager hierarchies, innate dispositions, and the behavioral reconstruction of prehistory. Center for Archaeological Investigations, Occasional Paper No. 27," in *Hierarchies in Action: Cui Bono?*, vol. 27, Carbondale, Center for Archaeological Investigations, SIU Press, 1999a, pp. 31-58.

[1123] F. B. M. de Waal, Good Natured: The Origins of Right and Wrong in Humans and Other Animals, Cambridge, MA: Harvard University Press, 1996, p. 368.

[1124] F. de Waal, Primates and Philosophers: How Morality Evolved, Princeton: Princeton University Press, 2016, p. 232.

[1125] F. de Waal, Our Inner Ape: The Best and Worst of Human Nature, London: Granta Books, 2005, p. 300.

[1126] C. Boehm, "The evolutionary development of morality as an effect of dominance behavior and conflict interference," *Journal of Social and Biological Structures,* vol. 5, no. 4, pp. 413-421, 1982.

[1127] C. Boehm, "Egalitarian society and reverse dominance hierarchy," *Current Anthropology,* vol. 34, no. 3, pp. 227-254, 1993.

[1128] C. Boehm, "Purposive social selection and the evolution of human altruism," *Cross-Cultural Research,* vol. 42, no. 4, pp. 319-352, 2008.

[1129] M. Luther, "Whether soldiers, too, can be saved (1526) pp 101-120," in *Luther: Selected Political Writings*, Eugene, OR: Wipf & Stock Publishing, 2003, p. 160.

[1130] K.-E. Sveiby, "Collective leadership with power symmetry: Lessons from Aboriginal prehistory," *Leadership,* vol. 7, no. 4, pp. 385-414, 2011.

[1131] I. Gram-Hanssen, "Individual and collective leadership for deliberate transformations: Insights from Indigenous leadership," *Leadership,* vol. 17, no. 5, 2021.

[1132] M. J. Sandstrom and M. Bartini, "Do perceptions of discrepancy between self and group norms contribute to peer harassment at school?," *Basic and Applied Social Psychology,* vol. 32, no. 3, pp. 217-225, 2010.

[1133] M. Sandstroma, H. Makovera and M. Bartinib, "Social context of bullying: Do misperceptions of group norms influence children's responses to witnessed episodes?," *Social Influence,* vol. 8, no. 2-3, pp. 196-215, 2013.

[1134] S. Einarsen, M. S. Aasland and A. Skogstad, "Destructive leadership behaviour: A definition and conceptual model," *The Leadership Quarterly,* vol. 18, no. 3, pp. 207-216, 2007.

[1135] M. Bermejo, "Home-range use and intergroup encounters in western gorillas (Gorilla g. gorilla) at Lossi forest, North Congo," *American Journal of Primatology,* vol. 64, no. 2, pp. 223-232, 2004.

[1136] R. W. Wrangham, "Evolution of coalitionary killing," *American Journal of Physical Anthropology,* vol. 110, no. 29, 1999b.

[1137] Thucydides, "Sixteenth Year of the War - The Melian Conference - Fate of Melos, (pp. 362- 401)," in *History of the Peloponnesian War*, vol. 5, London, J. M. Dent and Sons Ltd., 431 BCE/ 1914, p. 614.

[1138] M. Teschler-Nicola, F. Gerold, M. Bujatti-Narbeshuber, T. Prohaska, C. Latkoczy, G. Stingeder and M. Watkins, "Evidence of genocide 7000 BP-Neolithic paradigm and geo-climatic reality," *Collegium Anthropologicum,* vol. 23, no. 2, pp. 437-450, 1999.

[1139] R. R. Paine, D. Mancinelli, M. Ruggieri and A. Coppa, "Cranial trauma in Iron Age Samnite agriculturists, Alfedena, Italy: Implications for biocultural and economic stress," *American Journal of Biological Anthropology,* vol. 132, no. 1, pp. 48-58, 2006.

[1140] L. L. Cavalli-Sforza and E. Minch, "Paleolithic and Neolithic lineages in the European mitochondrial gene pool," *American Journal of Human Genetics,* vol. 61, no. 1, pp. 247-254, 1997.

[1141] Remembering Srebrenica, UK, "Sexual Violence in Bosnia," [Online]. Available: https://srebrenica.org.uk/what-happened/sexual-violence-bosnia. [Accessed 26 February 2023].

[1142] T. da Silva and L. Harkins, "Multiple perpetrator rape," in *Handbook on the Study of Multiple Perpetrator Rape: A Multidisciplinary Response to an International Problem*, M. Horvath and J. Woodham, Eds., Milton Park, Oxfordshire: Routledge, 2013, p. 300.

[1143] E. J. Wood, "Multiple perpetrator rape during war," in *Handbook on the Study of Multiple Perpetrator Rape: A Multidisciplinary Response to an International Problem (Issues in Forensic Psychology)*, M. Horvath and J. Woodhams, Eds., Milton Park, Oxfordshire: Routledge, 2013, p. 300.

[1144] C. M. Cameron, "How people moved among ancient societies: Broadening the view," *American Anthropologist,* vol. 115, no. 2, pp. 218-231, 2013a.

[1145] C. M. Cameron, "Captives and cultural change," *Current Anthropology,* vol. 52, no. 2, p. 169–209, 2011.

[1146] I. Chang, W. C. Kirby and S. Sloan, The Rape of Nanking: The Forgotten Holocaust of World War II, New York: Penguin, 2012, p. 330.

[1147] E. J. Wood, "Variation in sexual violence during war," *Politics and Society,* vol. 34, no. 3, pp. 307-341, 2006.

[1148] J. S. Cybulski, "Culture change, demographic history, and health and disease on the northwest coast (pp 75-85)," in *In the Wake of Contact: Biological Responses to Conquest*, First ed., C. S. Larsen and G. R. Milner, Eds., New York, Wiley-Liss, 1994, p. 206.

[1149] J. S. Cybulski, "Conflict on the northern northwest coast: 2,000 years plus of bioarchaeological evidence (pp 415-451)," in *The Routledge Handbook of the Bioarchaeology of Human Conflict*, C. Knüsel and M. Smith, Eds., London, Routledge, 2014, p. 752.

[1150] P. Willey, Prehistoric Warfare on the Great Plains: Skeletal Analysis of the Crow Creek Massacre Victims, New York: Routledge, 1990.

[1151] J. C. Ewers, "Women's roles in Plains Indian warfare (pp 325-331)," in *Skeletal Biology in the Great Plains: Migration, Warfare, Health and Subsistence*, Washington DC, Smithsonian Institution Press, 1994, p. 415.

[1152] T. A. Kohler and T. K. Kramer, "Raiding for women in the pre-Hispanic northern Pueblo Southwest? A pilot examination," *Current Anthropology,* vol. 47, no. 6, pp. 1035-1045, 2006.

[1153] B. Madley, "Patterns of frontier genocide 1803–1910: The Aboriginal Tasmanians, the Yuki of California, and the Herero of Namibia," *Journal of Genocide Research,* vol. 6, no. 2, p. 167–192, 2004.

[1154] J. Sarkin, Germany's Genocide of the Herero: Kaiser Wilhelm II, His General, His Settlers, His Soldiers, Cape Town: University of Cape Town Press, 2011, p. 288.

[1155] B. Madley, "Reexamining the American genocide debate: Meaning, historiography, and new methods," *American Historical Review,* vol. 120, no. 1, p. 98–139, 2015.

[1156] I. Chatterjee, Unfamiliar Relations: Family and History in South Asia, New Brunswick, NJ: Rutgers University Press, 2004, p. 272.

[1157] S. Kovner, Occupying Power: Sex Workers and Servicemen in Postwar Japan. Series: Studies of the Weatherhead East Asian Institute, Stanford, CA: Stanford University Press, 2012, p. 240.

[1158] Y. Tanaka, Japan's Comfort Women. Series: Asia's Transformations, New York, NY: Routledge, 2002, p. 232.

[1159] J. Zenn and E. Pearson, "Women, Gender and the evolving tactics of Boko Haram," *Journal of Terrorism Research,* vol. 5, no. 1, 2014.

[1160] J. Mannell, "Sexual violence in Ukraine," *The British Medical Journal,* vol. 377, p. o1016, 2022.

[1161] C. M. Horne, "Accountability for atrocity crimes in Ukraine: Gendering transitional justice," *Women's Studies International Forum,* vol. 96, p. 102666, 2023.

[1162] United Nations Security Council, "Reasonable grounds to believe conflict-related sexual violence occurred in Israel during 7 October Attacks, senior UN official tells Security Council: However, urges mandate is not 'is without rape' but 'world without war'," 11 March 2024. [Online]. Available: https://press.un.org/en/2024/sc15621.doc.htm. [Accessed 28 April 2024].

[1163] United Nations, Office of the Special Representative of the Secretary-General, "Sexual Violence in Conflict," 4 March 2024. [Online]. Available: https://www.un.org/sexualviolenceinconflict/press-release/israel-west-bank-mission/. [Accessed 28 April 2024].

[1164] Al Jazeera, "UN experts say grounds to believe rape occurred in Hamas attack on Israel: Palestinian group has repeatedly denied allegations its fighters committed acts of sexual violence during the October 7 attack," 4 March 2024. [Online]. Available: https://www.aljazeera.com/news/2024/3/4/reasonable-grounds-to-believe-hamas-committed-sexual-violence-un. [Accessed 28 April 2024].

[1165] United Nations, "Rights experts call for probe into alleged violations against Palestinian women and girls," 19 February 2024. [Online]. Available: https://news.un.org/en/story/2024/02/1146667. [Accessed 28 April 2024].

[1166] J. G. Green, M. Ramirez, G. J. Merrin and M. K. Holt, "Bias-based harassment among US adolescents (publisher's preview)," *School Mental Health,* p. s12310, 2024.

[1167] W. D. Hamilton, "The genetical evolution of social behaviour. I," *Journal of Theoretical Biology,* vol. 7, no. 1, pp. 1-16, 1964a.

[1168] W. D. Hamilton, "The genetical evolution of social behaviour. II," *Journal of Theoretical Biology,* vol. 7, no. 1, pp. 17-52, 1964b.

[1169] J. García, M. van Veelen and A. Traulsen, "Evil green beards: Tag recognition can also be used to withhold cooperation in structured populations," *Journal of Theoretical Biology,* vol. 360, pp. 181-186, 2014.

[1170] R. A. Laird, "Green-beard effect predicts the evolution of traitorousness in the two-tag Prisoner's dilemma," *Journal of Theoretical Biology,* vol. 288, pp. 84-91, 2011.

[1171] V. Nehring, S. E. F. Evison, L. A. Santorelli, P. d'Ettorre and W. O. H. Hughes, "Kin-informative recognition cues in ants," *Proceedings of the Royal Society B: Biological Sciences,* vol. 278, no. 1714, p. 1942–1948, 2011.

[1172] R. Pracana, I. Levantis, C. Martínez-Ruiz, E. Stolle, A. Priyam and Y. Wurm, "Fire ant social chromosomes: Differences in number, sequence and expression of odorant binding proteins," *Evolution Letters,* vol. 1, no. 4, pp. 199-210, 2017.

[1173] M. D. Breed, "Recognition pheromones of the honey bee," *BioScience,* vol. 48, no. 6, pp. 463-470, 1998.

[1174] S. I. Jimenez, R. Gries, H. Zhai, N. Derstine, S. McCann and G. Gries, "Evidence for a nest defense pheromone in bald-faced hornets, Dolichovespula maculata, and identification of components," *Journal of Chemical Ecology,* vol. 42, p. 414–424, 2016.

[1175] J. Shonfield, J. C. Gorrell, D. W. Coltman, S. Boutin, M. M. Humphries, D. R. Wilson and A. G. McAdam, "Using playback of territorial calls to investigate mechanisms of kin discrimination in red squirrels," *Behavioral Ecology,* vol. 28, no. 2, p. 382–390, 2017.

[1176] T. E. R. Belpaire, J. Pešek, B. Lories, K. J. Verstrepen, H. P. Steenackers, H. Ramon and B. Smeets, "Permissive aggregative group formation favors coexistence between cooperators and defectors in yeast," *Interdisciplinary Journal of Microbial Ecology,* vol. 16, pp. 2305-2312, 2022.

[1177] A. Dragoš and A. T. Kovács, "The peculiar functions of the bacterial extracellular matrix," *Trends in Microbiology,* vol. 25, no. 4, pp. 257-266, 2017.

[1178] S. Smukalla, M. Caldara, N. Poche, A. Beauvais, S. Guadagnini, C. Yan, M. D. Vinces, A. Jansen, M. C. Prevost, J.-P. Latgé, G. R. Fink, K. R. Foster and K. J. Verstrepen, "FLO1 is a variable green beard gene that drives biofilm-like cooperation in budding yeast," *Cell,* vol. 135, no. 4, pp. 726-737, 2008.

[1179] S. Sathe and R. Kümmerli, "Antagonistic interactions subdue inter-species green-beard cooperation in bacteria," *Journal of Evolutionary Biology,* vol. 33, no. 9, pp. 1245-1255, 2020.

[1180] R. Sear and R. Mace, "Who keeps children alive? A review of the effects of kin on child survival," *Evolution and Human Behavior,* vol. 29, no. 1, pp. 1-18, 2008.

[1181] M. Daly and G. Perry, "Matrilateral bias in human grandmothering," *Frontiers in Sociology,* vol. 2, 2017.

[1182] G. Perry and M. Daly, "A model explaining the matrilateral bias in alloparental investment," *Proceedings of the National Academy of Sciences,* vol. 114, no. 35, pp. 9290-9295, 2017.

[1183] E. Seabright, S. Alami, T. S. Kraft, H. Davis, A. E. Caldwell, P. Hooper, L. McAllister, S. Mulville, A. Veile, C. von Rueden, B. Trumble, J. Stieglitz, M. Gurven and H. Kaplan, "Repercussions of patrilocal residence on mothers' social support networks among Tsimane forager–farmers," *Philosophical Transactions of the Royal Society B,* vol. 278, no. 1868, p. 20210442, 2022.

[1184] J. J. Helton, B. B. Boutwell and M. DiBernardo, "The relative safety of paternal, maternal, and traditional foster care placements," *Child Abuse and Neglect,* vol. 70, pp. 1-10, 2017.

[1185] S. Ayaya, A. DeLong, L. Embleton, D. Ayukau, E. Sang, J. Hogan, A. Kamanda, L. Atwoli, D. Makori, M. A. Ott, C. Ombok and P. Braitstein, "Prevalence, incidence and chronicity of child abuse among orphaned, separated, and street-connected children and adolescents in western Kenya: What is the impact of care environment?," *Child Abuse and Neglect,* vol. 139, p. 104920, 2023.

[1186] N. Hallett, J. Garstang and J. Taylor, "Kinship care and child protection in high-income countries: A scoping review," *Trauma, Violence & Abuse,* vol. 24, no. 2, pp. 632-645, 2023.

[1187] G. Roberts, "Helping those who help others: The roles of indirect reciprocity and relatedness," *Ethology,* vol. 130, no. 4, p. e13453, 2024.

[1188] D. J. Rankin and M. Taborsky, "Assortment and the evolution of generalized reciprocity," *Evolution,* vol. 63, no. 7, pp. 1913-1922, 2009.

[1189] B. Majolo, L. Maréchal, F. Igali and J. van de Vyver, "Cooperation and group similarity in children and young adults in the UK," *Evolutionary Human Sciences,* vol. 5, p. e29, 2023.

[1190] D. P. Nguyen, B. K. Desai and M. Falgiani, "Visual diagnosis: Pearling: A case study," *International Journal of Emergency Medicine,* vol. 4, p. 74, 2011.

[1191] S. Tsunenari, T. Idaka, M. Kanda and Y. Koga, "Self-mutilation. Plastic spherules in penile skin in yakuza, Japan's racketeers," *American Journal of Forensic Medicine and Pathology,* vol. 2, no. 3, pp. 203-207, 1981.

[1192] S. Tsunenari, K. Yonemitsu, T. Kanbe and M. Kanda, "How to identify the Yakuza, Japanese racketeers--their sociology, criminology and physical characteristics," *Annals of the Academy of Medicine, Singapore,* vol. 13, no. 1, pp. 25-31, 1984.

[1193] M. A. Rothschild, E. Ehrlich, W. A. Klevno and V. Schneider, "Self-implanted subcutaneous penile balls – a new phenomenon in Western Europe," *International Journal of Legal Medicine,* vol. 110, no. 2, pp. 88-91, 1997.

[1194] L. B. Jorde and S. P. Wooding, "Genetic variation, classification and 'race'," *Nature Genetics,* vol. 36, no. 11, p. S28–S33, 2004.

[1195] I. Thanseem, K. Thangaraj, G. Chaubey, V. K. Singh, L. V. S. Bhaskar, B. M. Reddy, A. G. Reddy and L. Singh, "Genetic affinities among the lower castes and tribal groups of India: Inference from Y chromosome and mitochondrial DNA," *BioMed Central Genetics,* vol. 7, p. 42, 2006.

[1196] J. Rodriguez-Luis, L. Palencia-Madrid, K. Deshpande, M. A. Alfonso-Sanchez, J. A. Peña, M. M. de Pancorbo, R. Garcia-Bertrand and R. J. Herrera, "On the Y chromosome of Chennai, Tamil Nadu and the Indian subcontinent," *Gene,* vol. 859, p. 147175, 2023.

[1197] A. de Hoyes, S. B. Morris and A. Bernheim, Freemasonry in Context: History, Ritual, Controversy, Lanham: Lexington Books, 2005, p. 368.

[1198] A. J. Stewart and N. Raihani, "Group reciprocity and the evolution of stereotyping," *Proceedings of the Royal Society B,* vol. 290, p. 20221834, 2023.

[1199] T. Pozzoli and G. Gianluca, "Friend similarity in attitudes toward bullying and sense of responsibility to intervene," *Social Influence,* vol. 8, no. 2-3, pp. 161-176, 2013.

[1200] O. B. Ayoko, V. J. Callan and E. J. Härtel, "Workplace conflict, bullying, and counterproductive behaviors," *International Journal of Organizational Analysis,* vol. 11, no. 4, pp. 283-301, 2003.

[1201] D. Lewis, S. Giga and H. Hoel, "Discrimination and bullying p 267-282," in *Bullying and Harassment in the Workplace: Developments in Theory, Research, and Practice, Second Edition*, London, CRC Press, 2010, p. 512.

[1202] J. H. Neuman and R. A. Baron, "Social antecedents of bullying: A social interactionist perspective pp 185-202," in *Bullying and Emotional Abuse in the Workplace*, London, CRC Press, 2003, p. 420.

[1203] S. Ramsay, A. Troth and S. Branch, "Work-place bullying: A group processes framework," *Journal of Occupational and Organizational Psychology,* vol. 84, pp. 799-816, 2010.

[1204] L. J. Lambe, V. D. Cioppa, I. K. Hong and W. M. Craig, "Standing up to bullying: A social ecological review of peer defending in offline and online contexts," *Aggression and Violent Behavior,* vol. 45, pp. 51-74, 2019.

[1205] R. van der Ploeg, T. Kretschmer, C. Salmivalli and R. Veenstra, "Defending victims: What does it take to intervene in bullying and how is it rewarded by peers?," *Journal of School Psychology,* vol. 65, pp. 1-10, 2017.

[1206] G. Thomas and G. Lewis, Defying Hitler: The Germans Who Resisted Nazi Rule, New York: Dutton Caliber, 2019b, p. 560.

[1207] E. A. Johnson and K.-H. Reuband, What We Knew: Terror, Mass Murder, and Everyday Life in Nazi Germany, Cambridge, MA: Basic Books, 2006, p. 464.

[1208] J. E. Bartlett II and M. E. Bartlett, "Workplace bullying: An integrative literature review," *Advances in Developing Human Resources,* vol. 13, no. 1, pp. 69-84, 2011.

[1209] J. Escartína, J. Ullrich, D. Zapf, E. Schlüter and R. van Dick, "Individual- and group-level effects of social identification on workplace bullying," *European Journal of Work and Organizational Psychology,* vol. 22, no. 2, pp. 182-193, 2013.

[1210] Aaroncynic, "CPS Parents 'Walk the Walk' to Highlight Safety Issues of School Closings," *Chicagoist,* p. 1, 10 April 2013.

[1211] Midwest Coalition for Human Rights, "Letter of Allegation Regarding the Closing of 49 Public Elementary Schools in Chicago, Illinois, United States of America," Midwest Coalition for Human Rights, Chicago, 2013.

[1212] M. F. Gordon, M. de la Torre, J. R. Gowhy, P. T. Moore, L. Sartain and D. Knight, "School Closings in Chicago: Staff and Student Experiences and Academic Outcomes," University of Chicago Consortium on School Research, Chicago, 2018.

[1213] N. Ahmed-Ullah, "CPS releases maps of 'safe passage' routes," *Chicago Tribune,* 9 August 2013.

[1214] M. de la Torre and J. Gwynne, "When Schools Close: Effects on Displaced Students in Chicago Public Schools," University of Chicago Consortium on Chicago School Research, Chicago, 2009.

[1215] L. Lutton, B. Vevea, S. Karp, A. Cardona-Maguigad and K. McGee, "A Generation of School Closings," 3 December 2018. [Online]. Available: https://interactive.wbez.org/generation-school-closings/. [Accessed 1 March 2023].

[1216] WBEZ Chicago, "A Generation of School Shakups (Public Data)," 3 December 2018. [Online]. Available: https://docs.google.com/spreadsheets/u/1/d/e/2PACX-1vRmKox-lDNqhtUNL4WLl8x6DljIi3b0k1pYEmD7adfCwX-rGYyFw0XIjxNWkOfL6og3CHfUEHPMwv6k/pubhtml#. [Accessed 2 March 2023].

[1217] U.S. News and World Report, "Chicago Public Schools," 2 March 2023. [Online]. Available: https://www.usnews.com/education/k12/illinois/districts/city-of-chicago-sd-299-110570.

[1218] J. Jackson and J. Foster, "The Politics of Public Budgeting in Illinois," The Paul Simon Policy Institute, Carbondale, 2019.

[1219] R. B. Lee, Subsistence ecology of !KungBushmen (dissertation), Berkeley: University of California, 1966, p. 418.

[1220] R. B. Lee, "!Kung Bushmen subsistence: An Input-output analysis," in *Environment and cultural behavior*, New York, Natural History Press, 1969.

[1221] M. D. Sahlins, "Notes on the original affluent society (p 85-89)," in *Man the Hunter*, Chicago, Aldine, 1968, p. 432.

[1222] M. D. Sahlins, Stone Age Economics, London: Routledge, 1972/ 2017, p. 376.

[1223] A. E. Page, S. Viguier, M. Dyble, D. Smith, N. Chaudhary, G. D. Salali, J. Thompson, L. Vinicius, R. Mace and A. B. Migliano, "Settled Agta foragers trade health for fertility," *Proceedings of the National Academy of Sciences,* vol. 113, no. 17, pp. 4694-4699, 2016.

[1224] R. H. MacArthur and E. R. Pianka, "On optimal use of a patchy environment," *The American Naturalist,* vol. 100, no. 916, pp. 603-609, 1966.

[1225] H. Richardson and N. A. Verbeek, "Diet selection by yearling northwestern crows (Corvus caurinus) feeding on littleneck clams (Venerupis japonica)," *The Auk,* vol. 104, no. 2, pp. 263-269, 1987.

[1226] R. H. Heineman, R. Springman and J. J. Bull, "Optimal foraging by bacteriophages through host avoidance," *The American Naturalist,* vol. 171, no. 4, pp. 149-157, 2008.

[1227] G. E. Belovsky, "Insights for caribou/reindeer management using optimal foraging theory," *Rangifer,* vol. 11, no. 7, pp. 7-23, 1991.

[1228] M. E. Ritchie, "Individual variation in the ability of Columbian ground squirrels to select an optimal diet," *Evolutionary Ecology,* vol. 2, no. 3, pp. 232-252, 1988.

[1229] S. D. Fretwell, Populations in Seasonal Environment, Princeton, NJ: Princeton University Press, 1973, p. 224.

[1230] S. D. Fretwell and H. L. J. Lucas, "On territorial behavior and other factors influencing habitat distribution in birds. I. Theoretical development," *Acta Biotheoretica,* vol. 19, pp. 16-36, 1969.

[1231] P. M. Yaworsky, S. T. Hussain and F. Riede, "The effects of climate and population on human land use patterns in Europe from 22ka to 9ka ago," *Quaternary Science Reviews,* vol. 344, p. 108956, 2024.

[1232] D. W. Morris, "Adaptive responses to habitat change: Theory and tests with field experiments," *Ecology,* vol. 105, no. 7, p. e4333, 2024.

[1233] S. D. Held and M. Špinka, "Animal play and animal welfare," *Animal Behaviour,* vol. 81, no. 5, pp. 891-899, 2011.

[1234] K. D. Evans, L. Swedell and S. Chowdhury, "Respect for males amid suppression of females: Selective use of aggression and fitness correlates in the male-dominated society of hamadryas baboons," *Journal of Human Evolution,* vol. 165, p. 103151, 2022.

[1235] F. B. M. de Waal, Peacemaking Among Primates, Cambridge, MA: Harvard University Press, 1989, p. 308.

[1236] B. Christopher, "The massacre at Monkey Hill," *Priceonomics,* 2016.

[1237] R. D. Estes, The Behavior Guide to African Mammals, Oakland: University of California Press, 1991, p. 660.

[1238] M. G. Forero, J. L. Tella, K. A. Hobson, M. Bertellotti and G. Blanco, "Conspecific food competition explains variability in colony size: A test in Magallanic penguins," *Ecology,* vol. 83, no. 12, pp. 3466-3475, 2002.

[1239] I. Karsai and J. W. Wenzel, "Productivity, individual-level and colony-level flexibility, and organization of work as consequences of colony size," *Proceedings of the National Academy of Sciences USA,* vol. 98, no. 15, pp. 8665-8669, 1998.

[1240] A. Dussutour, J. L. Deneubourg and V. Fourcassie, "Amplification of individual preferences in a social context: The case of wall-following in ants," *Proceedings of the Royal Society of London B,* vol. 272, no. 1564, pp. 705-714, 2005.

[1241] R. Jeanson, J. L. Deneubourg, A. Grimal and G. Theraulaz, "Modulation of individual behaviour and collective decision-making aggregation site selection by the ant Messor barbarus," *Behavioural Ecology and Sociobiology,* vol. 55, no. 4, pp. 388-394, 2004.

[1242] S. Artinger and N. Vulkan, "Does group size matter for behavior in online trust dilemmas?," *PLoS ONE,* vol. 11, no. 11, p. e0166279, 2016.

[1243] A. M. Q'Moore, C. Kirkham and M. Smith, "Bullying behaviour in Irish schools: A nationwide study," *Irish Journal of Psychology,* vol. 18, no. 2, pp. 141-169, 1997.

[1244] C. F. Grandeau, T. Yanagida, M. M. Vermande, D. Strohmeier and C. Salmivalli, "Prevalence of bullying and victimization: Testing three explanations for the negative association," *Frontiers in Psychology,* vol. 10, 2019.

[1245] U. S. Department of Education, "Student Reports of Bullying: Results From the 2017 School Crime Supplement to the National Crime Victimization Survey," U. S. Department of Education, 2019.

[1246] L. T. Evans, "Field study of the social behavior of the black lizard, Ctenosaura pectinata," *American Museum Novitates,* vol. 1493, pp. 24-26, 1951.

[1247] K. D. Wells, "Territoriality and male mating success in the green frog (Rana clamitans)," *Ecology,* vol. 58, no. 4, p. 750–762, 1977.

[1248] P. Wirtz, "Territorial defence and territory take-over by satellite males in the waterbuck Kobus ellipsiprymnus (Bovidae)," *Behavioral Ecology and Sociobiology,* vol. 8, no. 2, pp. 161-162, 1981.

[1249] K. E. Langergraber, J. C. Mitani, D. P. Watts and L. Vigilant, "Male-female socio-spatial relationships and reproduction in wild chimpanzees," *Behavioral Ecology and Sociobiology,* vol. 67, no. 6, p. 861–873, 2013.

[1250] G. Ossendorf, A. R. Gross, T. Bromm, G. Tekelemariam, B. Glaser, J. Lesure, J. Schmidt, N. Akça, T. Bekele, A. Beldados, S. Demissew, T. H. Kahsay, B. P. Nash, T. Nauss, A. Negash, S. Nemomissa, R. Vogelsang, Z. Woldu and W. Zech, "Middle Stone Age foragers resided in high elevations of the glaciated Bale Mountains, Ethiopia," *Science,* vol. 365, no. 6453, pp. 583-587, 2019.

[1251] Martin-Luther-Universität Halle-Wittenberg, "Why humans in Africa fled to the mountains during the last ice age," *Science News,* 8 August 2019.

[1252] J. C. Mitani, D. P. Watts and S. J. Amsler, "Lethal intergroup aggression leads to territorial expansion in wild chimpanzees," *Current Biology,* vol. 20, no. 12, pp. R507 - R508, 2010.

[1253] M. M. McDonald, C. D. Navarrete and M. & Van Vugt, "Evolution and the psychology of intergroup conflict: The male warrior hypothesis," *Philosophical Transactions of the Royal Society B: Biological Sciences,* vol. 367, no. 1589, p. 670–679, 2012.

[1254] T. Livy, "A History of Rome: Book I, Chapters 9-13 (pp 33-51)," in *Livy in Fourteen Volumes*, vol. 1, Cambridge, MA: Harvard University Press, ca 27-9BCE/1976, p. 447.

[1255] Holy Bible, King James Version, "Judges 20," nd. [Online]. Available: 071001180.flex.online-banking-services.com/. [Accessed 12 September 2025].

[1256] R. A. Gabriel, The Military History of Ancient Israel, Westport, CT: Praeger, 2003, p. 334.

[1257] Athenaeus, The Deipnosophists Or Banquet Of The Learned Of Athenaeus., vol. 14, London: Henry G. Bohn, 1854.

[1258] B. D. Cross, "Blackhead Signpost Road," 26 May 2023. [Online]. Available: https://www.hmdb.org/m.asp?m=224336.

[1259] D. A. Blackmon, Slavery by Another Name: The Re-Enslavement of Black Americans from the Civil War to World War II, New York: Doubleday, 2008, p. 480.

[1260] O. Patterson, Slavery and Social Death: A Comparative Study, With a New Preface, Cambridge, MA: Harvard University Press, 2018/1982, p. 560.

[1261] A. Nikiforuk, The Energy of Slaves, Toronto: Greystone Books, 2012, p. 272.

[1262] D. J. Wyatt, Slavery in East Asia, Cambridge: Cambridge University Press, 2023, p. 75.

[1263] Waitangi Tribunal, "Rekohu: A report on Moriori and Ngati Mutunga claims in the Chatham Islands," Wellington Legislation District, Wellington, 2001.

[1264] S. J. King, "Ending hereditary slavery in Mauritania: Bidan (Whites and Black 'slaves' in 2021," 26 August 2021. [Online]. Available: https://www.arab-reform.net/publication/ending-hereditary-slavery-in-mauritania-bidan-whites-and-black-slaves-in-2021/. [Accessed 5 March 2023].

[1265] A. E. Roberts, "Black slaves and Indian owners: The continuous rediscovery of indian territory," *The Journal of the Civil War Era*, vol. 13, no. 1, pp. 87-104, 2023.

[1266] W. G. McLoughlin, "Red Indians, Black slavery and White racism: America's slaveholding Indians," *American Quarterly,* vol. 26, no. 4, pp. 367-385, 1974.

[1267] W. C. MacLeod, "Some aspects of primitive chattel slavery," *Social Forces,* vol. 4, no. 1, pp. 137-141, 1925.

[1268] A. Little, "Native American captivity and slavery in North America, 1492-1848," in *Oxford Research Encyclopedia of American History*, Oxford, Oxford University, 2022.

[1269] E. A. Smith and B. F. Codding, "Ecological variation and institutionalized inequality in hunter-gatherer societies," *Proceedings of the National Academy of Sciences,* vol. 118, no. 13, p. e2016134118, 2021.

[1270] A. Underwood, "Which countries have the highest rates of modern slavery and most victims?," 30 July 2018. [Online]. Available: https://news.trust.org/item/20180730000101-aj7ui. [Accessed 6 March 2023].

[1271] International Labor Organization, "Global Estimates of Modern Slavery: Forced Labour and Forced Marriage," International Labor Organization, Geneva, 2022.

[1272] N. Vardi, "The World's Biggest Illicit Industries," *Forbes,* 4 June 2010.

[1273] C. May, "Transnational Crime and the Developing World," Global Financial Integrity, Washington DC, 2017.

[1274] C. Mavrellis, "The Business of Transnational Crime," Global Financial Integrity, Washington DC, 2017.

[1275] S. B. Hrdy, The Woman That Never Evolved: With a New Preface and Bibliographical Updates, Revised Edition, Cambridge: Harvard University Press, 1999, p. 304.

[1276] T. Nishida, "The social structure of chimpanzees of the Mahali mountains," *Primates,* vol. 9, no. 3, pp. 167-224, 1968.

[1277] T. Nishida, M. Hiraiwa-Hasegawa, T. Hasegawa and Y. Takahata, "Group extinction and female transfer in wild chimpanzees in the Mahale National Park, Tanzania," *Zeitschrift für Tierpsychologie,* vol. 67, no. 1-4, p. 284–301, 1985.

[1278] A. Maryanski and J. H. Turner, The Social Cage: Human Nature and the Evolution of Society, Stanford, CA: Stanford University Press, 1992, p. 214.

[1279] Y. Sugiyama, "Socioecological factors of male chimpanzee migration at Bossou, Guinea," *Primates,* vol. 40, no. 1, pp. 61-68, 1999.

[1280] R. W. Wrangham, "The significance of African apes for reconstructing human social evolution," in *The Evolution of Human Behavior: Primate Models*, W. G. Kinzey, Ed., Albany, NY: State University of New York, 1986, pp. 51-71.

[1281] R. F. Fox, Kinship and Marriage, Cambridge: Cambridge University Press, 1984, p. 288.

[1282] D. H. Price, Atlas of World Cultures: A Geographical Guide to Ethnographic Literature, Caldwell, NJ: Blackburn Press, 2004, p. 156.

[1283] R. A. Foley, Hominid Evolution and Community Ecology: Prehistoric Human Adaptation in Biological Perspective, Cambridge, MA: Academic Press, 1984, p. 296.

[1284] R. A. Foley and P. Lee, "Finite social space, evolutionary pathways, and reconstructing hominid behavior," *Science,* vol. 243, no. 4893, pp. 901-906, 1989.

[1285] B. Gibb, R. Gibb and M. Gibb, Composers, *Stayin' Alive.* [Sound Recording]. Criteria. 1977.

[1286] L. E. O. Kennair, T. J. Wade, M. T. Tallaksen, T. V. Grøntvedt, A. M. Kessler, R. L. Burch and M. Bendixen, "Perceived effectiveness of flirtation tactics: The effects of sex, mating context and individual differences in US and Norwegian samples," *Evolutionary Psychology,* vol. 20, no. 1, 2022.

[1287] C. Beckwith, *Niger's Wodaabe: "People of the Taboo"--Their Elaborate Male Beauty Contests Lighten Their Lives During the Rainy Season in The Sahara (Pamphlet),* National Geographic Magazine, 1983.

[1288] J. E. Lycett and R. I. Dunbar, "Mobile phones are lekking devices among human males," *Human Nature,* vol. 11, no. 1, pp. 93-104, 2000.

[1289] J. Braithwaite, "Lekking displays in contemporary organizations: ethologically oriented, evolutionary and cross-species accounts of male dominance," *Journal of Health Organization and Management,* vol. 22, no. 5, pp. 529-559, 2008.

[1290] P. Buirski, "Some contributions of ethology to group therapy: Dominance and hierarchies," *International Journal of Group Psychotherapy,* vol. 25, no. 2, pp. 227-235, 2015.

[1291] S. Kanazawa and R. L. Frerichs, "Why single men might abhor foreign cultures," *Social Biology,* vol. 48, no. 3-4, pp. 321-328, 2001.

[1292] A. J. Marsella and E. Ring, "Human migration and immigration: An overview (pp 3-22)," in *Migration: Immigration and Emigration in International Perspective*, L. L. Adler and U. P. Gielen, Eds., Westport, CT: Praeger, 2003, p. 384.

[1293] J. A. Gardner, "Can party politics be virtuous?," *Columbia Law Review,* vol. 100, no. 3, pp. 667-701, 2000.

[1294] C. de Weerth and A. P. Kalma, "Female aggression as a response to sexual jealousy: A sex role reversal?," *Aggressive Behavior,* vol. 19, no. 4, p. 265–279, 1993.

[1295] S. Davis, "Men as success objects and women as sex objects: A study of personal advertisements," *Sex Roles,* vol. 23, pp. 43-50, 1990.

[1296] M. Twain, "My first lie and how I got out of it," in *The Man that Corrupted Hadleyburg, And Other Essays and Stories*, New York, Harper & Brothers Publishers, 1903, p. 365.

[1297] J. Milton, Paradise Lost, 1674.

[1298] H. Vervaecke, H. de Vries and L. van Elsacker, "Function and distribution of coalitions in captive bonobos (Pan paniscus)," *Primates,* vol. 41, no. 3, pp. 249-265, 2000b.

[1299] Z. Liang, "Reexamination of individual knowledge and common behavior rules: A cross-disciplinary view based on empirical evidences," *MPRA Paper,* vol. 20050, pp. 1-21, 2009.

[1300] C. H. Vanderwolf, The Evolving Brain: The Mind and the Neural Control of Behavior, New York: Springer, 2007, p. 104.

[1301] P. Roberts and B. A. Stewart, "Defining the 'generalist specialist' niche for Pleistocene Homo sapiens," *Nature Human Behaviour,* vol. 2, no. 8, pp. 542-550, 2018.

[1302] S. Thomson and E. C. Ip, "COVID-19 emergency measures and the impending authoritarian pandemic," *Journal of Law and the Biosciences,* vol. 29, no. 7, p. lsaa064, 2020.

[1303] N. Lupu and E. J. Zechmeister, "The early COVID-19 pandemic and democratic attitudes," *PLoS One,* vol. 22, no. 16, p. e0253485, 2021.

[1304] M. Leach, "Epidemics and the politics of knowledge: Contested narratives in Egypt's H1N1 response," *Medical Anthropology,* vol. 33, no. 3, pp. 240-254, 2014.

[1305] S. Qazi and M. Readey, "Fundamental Behavioural, Sociological and Evolutionary Understanding of Covid-19 as a Pandemic: An Introductory Review," *Journal of Environmental Peace,* vol. 11, 2021.

[1306] V. A. Boese-Schlosser, N. Alizada, M. Lundstedt, K. Morrison, N. Natsika, Y. Sato, H. Tai and S. I. Lindberg, "Democracy Report 2022: Autocratization Changing Nature?," V-Dem Institute at the University of Gothenburg, Gothenburg, 2022.

[1307] B. Majolo, J. Lehmann, A. de Bortoli Vizioli and G. Schino, "Fitness-related benefits of dominance in primates," *American Journal of Physical Anthropology,* vol. 147, no. 4, p. 652–660, 2012.

[1308] S. H. Lee and S. Yamamoto, "The evolution of prestige: Perspectives and hypotheses from comparative studies," *New Ideas in Psychology,* vol. 68, p. 100987, 2023.

[1309] J. T. Cheng and J. L. Tracy, "The impact of wealth on prestige and dominance rank relationships," *Psychological Inquiry,* vol. 24, no. 2, pp. 102-108, 2013.

[1310] C. R. Ember, "Residential variation among hunter-gatherers," *Behavior Science Research,* vol. 10, no. 3, pp. 119-227, 1975.

[1311] M. M. Jones, Intraguild Predation Between the Larvae of an Indigenous and Introduced Species of Ladybird Beetle (Coleoptera: Coccinellidae) when Prey are Not Limiting: A Thesis Presented to the Faculty of the Department of Biology (Master's Thesis), Chicago: Northeastern Illinois University, 2006.

[1312] C. M. Oliveira, C. S. A. Silva-Torres, J. B. Torres and N. De La Pava, "Cannibalism and intraguild predation between endemic and introduced coccidophagous lady beetles (Coleoptera: Coccinellidae)," *Austral Entomology,* vol. 61, no. 2, pp. 225-235, 2002.

[1313] L. A. Coticchio, R. Vetter and D. L. Cassill, "Predation by the introduction of brown widow spiders (Araneae: Theridiidae) may explain local extinctions of native black widows in urban habitats," *Annals of the Entomological Society of America,* vol. 20, no. 20, pp. 1-10, 2023.

[1314] J. Ji, G. Lin and L. Wang, "Effects of intraguild prey dispersal driven by intraguild predator-avoidance on species coexistence," *Applied Mathematical Modeling,* vol. 103, pp. 51-67, 2022.

[1315] K. T. Quach and S. H. Chalasani, "Intraguild predation between Pristionchus pacificus and Caenorhabditis elegans: a complex interaction with the potential for aggressive behaviour," in *Nature's Gift to Neuroscience: A Tribute to Sydney Brenner and John Sulston*, First ed., C. Wu and J. Alcedo, Eds., CRC, 2022.

[1316] P. J. Manlick and J. N. Pauli, "Human disturbance increases trophic niche overlap in terrestrial carnivore communities," *Proceedings of the National Academy of Sciences,* vol. 117, no. 43, pp. 2684-26848, 2020.

[1317] L. M. Southern, T. Deschner and S. Pika, "Lethal coalitionary attacks of chimpanzees (Pan troglodytes troglodytes) on gorillas (Gorilla gorilla gorilla) in the wild," *Science Reports,* vol. 11, no. 1, p. 14673, 2021.

[1318] K. Morris and J. Goodall, "Competition for meat between chimpanzees and baboons of the Gombe National Park," *Folia Primatologica,* vol. 28, pp. 109-121, 1977.

[1319] V. Reynolds, "How wild are Gombe chimpanzees," *Man,* vol. 10, no. 1, pp. 123-125, 1975.

[1320] H. Nishie, "Increased hunting of yellow baboons (Papio cynocephalus) by M group chimpanzees at Mahale," *Pan African News,* vol. 11, no. 2, pp. 10-12, 2004.

[1321] G. Hohmann, "Association and social interaction between strangers and residents in Bonobos (Pan paniscus)," *Primates,* vol. 42, no. 1, pp. 91-99, 2001.

[1322] J. M. Satterfield and M. E. P. Seligman, "Military aggression and risk predicted by explanatory style," *Psychological Science,* vol. 5, no. 2, pp. 77-82, 1994.

[1323] M. A. Nowak and R. M. May, "Evolutionary games and spatial chaos," *Nature,* vol. 359, p. 826–829, 1992.

[1324] S. Sinek, Leaders Eat Last: Why Some Teams Pull Together and Others Don't, New York: Portfolio/Penguin, 2014.

[1325] C. Borries, K. Launhardt, C. Epplen, J. T. Epplen and P. Winkler, "Males as infant protectors in Hanuman langurs (Presbytis entellus) living in multimale groups – defence pattern, paternity and sexual behaviour," *Behavioral Ecology and Sociobiology,* vol. 46, pp. 350-356, 1999.

[1326] J. van der Toorn, T. R. Tyler and J. T. Jost, "More than fair: Outcome dependence, system justification, and the perceived legitimacy of authority figures," *Journal of Experimental Social Psychology,* vol. 7, no. 1, pp. 127-138, 2010.

[1327] J. B. Silk, S. F. Brosnan, J. Henrich, S. P. Lambeth and S. J. Shapiro, "Chimpanzees share food for many reasons: the role of kinship, reciprocity, social bonds and harassment on food transfers," *Animal Behaviour,* vol. 85, no. 5, pp. 941-947, 2013.

[1328] B. M. F. Galdikas, "Subadult male orangutan sociality and reproductive behavior at Tanjung Puting," *American Journal of Primatology,* vol. 8, no. 2, pp. 87-99, 1985.

[1329] C. D. Knott, M. E. Thompson, R. M. Stumpf and M. H. McIntyre, "Female reproductive strategies in orangutans, evidence for female choice and counterstrategies to infanticide in a species with frequent sexual coercion," *Proceedings of the Royal Society B,* vol. 277, no. 1678, pp. 105-113, 2010.

[1330] J. A. Kunz, G. J. Duvot, A. M. Ashbury, E. P. Willems, B. Spillmann, L. P. Dunkel, M. bin Abdullah, C. Schuppli, E. R. Vogel, S. S. U. Atmoko, M. A. van Noordwijk and C. P. van Schaik, "Alternative reproductive tactics of unflanged and flanged male orangutans revisited," *American Journal of Primatology,* vol. 85, no. 9, p. e23535, 2023.

[1331] D. P. Watts, "Male chimpanzee sexual coercion and mating success at Ngogo," *American Journal of Primatology,* vol. 84, no. 2, p. e23361, 2022.

[1332] D. K. Enigk, M. E. Thompson, Z. P. Machanda, R. W. Wrangham and M. N. Muller, "Female-directed aggression by adolescent male chimpanzees primarily constitutes dominance striving, not sexual coercion," *American Journal of Biological Anthropology,* vol. 176, no. 1, pp. 66-79, 2021.

[1333] G. S. Vance, V. Zeigler-Hill, J. Vonk and T. K. Shackelford, "Inevitable or preventable? The biosocial theory of wartime rape," *Evolutionary Psychological Science,* vol. 10, pp. 135-154, 2024.

[1334] M. E. Baaz and S. M., "Why do soldiers rape? Masculinity, violence, and sexuality in the armed forces in the Congo (DRC)," *International Studies Quarterly,* vol. 53, no. 2, pp. 495-518, 2009.

[1335] S. Lindsey and C. Koos, "Legacies of wartime sexual violence: Survivors, psychological harms, and mobilization," *American Political Science Review,* pp. 1-17, 2024.

[1336] L. M. Carter, L. B. Goodmon, M. Urs and H. Rutledge-Jukes, "The effects of victim gender identity, juror gender, and judicial instructions on victim blaming, crime severity ratings, and verdicts in sexual assault trials," *Journal of Homosexuality,* vol. 70, no. 6, pp. 1187-1224, 2023.

[1337] C. Crookston, "Off the Clock: Is drag 'just a job'?," *Queer Studies in Media and Popular Culture,* vol. 3, no. 1, pp. 101-115, 2018.

[1338] W. C. Andresen, "Gay and trans panic defense: Focusing on the homicides, not the court room strategy (sic)," *Homicide Studies,* vol. 28, no. 3, pp. 270-291, 2023.

[1339] C. Boesch and H. Boesch-Achermann, The Chimpanzees of the Taï Forest: Behavioural Ecology and Evolution, First ed., Cambridge: Cambridge University Press, 2020, p. 516.

[1340] P. Tjaden and N. Thoennes, "Prevalence and consequences of male-to-female and female-to-male intimate partner violence as measured by the national violence against women survey," *Violence Against Women,* vol. 6, no. 2, pp. 142-161, 2000.

[1341] J. M. G. Stevens, H. Vervaecke, H. de Vries and L. Elsacker, "Sex differences in the steepness of dominance hierarchies in captive bonobo groups," *International Journal of Primatology,* vol. 28, no. 6, pp. 1417-1430, 2007.

[1342] F. J. White and K. D. Wood, "Female feeding priority in bonobos, Pan paniscus, and the question of female dominance," *American Journal of Primatology,* vol. 69, no. 8, p. 37–850, 2007.

[1343] J. M. G. Stevens, H. Vervaecke, H. de Vries and L. Elsacker, "The influence of steepness of dominance hierarchies on reciprocity and interchange in bonobos (Pan paniscus)," *Behaviour,* vol. 142, no. 7, p. 941–960, 2005.

[1344] F. Dahlberg, Woman The Gatherer, New Haven, Connecticut: Yale University Press, 1981, p. 254.

[1345] V. Wobber, B. Hare, J. Maboto, S. Lipson, R. Wrangham and P. T. Ellison, "Differential changes in steroid hormones before competition in bonobos and chimpanzees," *Proceedings of the National Academy of Science USA,* vol. 107, no. 28, pp. 12457-12462, 2010.

[1346] P. A. Corning, "Synergy goes to war: A bioeconomic theory of collective violence," *Journal of Bioeconomics,* vol. 9, no. 2, pp. 109-144, 2007.

[1347] A. J. King, D. D. P. Johnson and M. van Vugt, "The origins and evolution of leadership," *Current Biology,* pp. R911-R916, 2009.

[1348] C. K. W. De Dreu, J. Gross and A. Romano, "Group formation and the evolution of human social organization," *Perspectives on Psychological Science,* vol. 19, no. 2, pp. 320-334, 2024.

[1349] H. Johannes and S. Schnettler, "Zur Evolution der sozialen Ungleichheit (pp1-36)," in *Handbuch Sozialstrukturanalyse*, P. Böhnke and D. Konietzka, Eds., Wiesbaden, Springer, 2024.

[1350] C. Longhi, L. A. Martinez-Vaquero, F. W. Vázquez and V. Trianni, "Strong leaders don't cheat: An evolutionary appraisal of population heterogeneity and leadership," *ALIFE 2024: Proceedings of the 2024 Artificial Life Conference, MIT,* 2024.

[1351] S. Rosefielde, "Excess mortality in the Soviet Union: A reconsideration of the demographic consequences of forced industrialization," *Soviet Studies,* vol. 35, no. 3, p. 385–409, 1983.

[1352] K. Tsuno and N. Kawakami, "Multifactor leadership styles and new exposure to workplace bullying: a six-month prospective study," *Industrial Health,* vol. 53, no. 2, pp. 139-151, 2015.

[1353] S. Milgram, Obedience to Authority, New York: Harper & Row, 1974, p. 256.

[1354] A. G. Miller, The Obedience Experiments: A Case Study of Controversy in Social Science, New York: Praeger Publishers, 1986, p. 305.

[1355] J. M. Burger, "Replicating Milgram: Would people still obey today?," *American Psychologist,* vol. 64, no. 1, p. 1–11, 2009.

[1356] L. Bègue and K. Vezirian, "The blind obedience of others: a better than average effect in a Milgram-like experiment," *Ethics and Behavior,* pp. 1-11, 2023.

[1357] C. Keysers, E. Knapska, M. A. Moita and V. Gazzola, "Emotional contagion and prosocial behavior in rodents," *Trends in Cognitive Sciences,* vol. 26, no. 8, pp. 688-706, 2022.

[1358] L. Bonini, C. Rotunno, E. Arcuri and V. Gallese, "Mirror neurons 30 years later: implications and applications," *Trends in Cognitive Sciences,* vol. 26, no. 9, pp. 767-781, 2022.

[1359] E. A. Caspar, K. Loumpa, C. Keysers and V. Gazzola, "Obeying orders reduces vicarious brain activation towards victims' pain," *NeuroImage,* vol. 222, p. 117251, 2020.

[1360] S. A. Haslam, S. D. Reicher, K. Millard and R. McDonald, "'Happy to have been of service': The Yale archive as a window into the engaged followership of participants in Milgram's 'obedience' experiments," *British Journal of Social Psychology,* vol. 54, no. 1, pp. 55-83, 2015.

[1361] S. Dong, K. Tan and J. C. Nieh, "Visual contagion in prey defence signals can enhance honest defence," *The Journal of Animal Ecology,* vol. 90, no. 3, pp. 594-601, 2021.

[1362] M. Detoni, X. Feás, R. J. Jeanne, K. J. Loope, S. O'Donnell, D. Santoro, S. Sumner and J. M. Jandt, "Evolutionary and ecological pressures shaping social wasps collective defenses," *Annals of the Entomological Society of America,* vol. 114, no. 5, pp. 581-595, 2021.

[1363] K. Kareklas and R. F. Oliveira, "Emotional contagion and prosocial behaviour in fish: An evolutionary and mechanistic approach," *Neuroscience and Biobehavioral Reviews,* vol. 163, p. 105780, 2024.

[1364] S. Kikuchi, N. Kondo and H. Koda, "Temporal regularities of vocal exchange in Java sparrows," *PLOS One,* vol. 20, no. 7, p. 0327024, 2025.

[1365] G. Kaplan, "The evolution of social play in songbirds, parrots and cockatoos - emotional or highly complex cognitive behaviour or both?," *Neuroscience and Biobehavioral Reviews,* vol. 161, p. 105621, 2024.

[1366] J. M. Plotnik and F. B. M. de Waal, "Asian elephants (Elephas maximus) reassure others in distress," *PeerJ,* vol. 2, p. e278, 2014.

[1367] G. Cordoni, M. Hecker, V. Crippa, B. G. Aldama, S. B. Satos and I. Norscia, "Rapid mimicry of trunk and head movements during play in African Savanna elephants (Loxodonta africana)," *Scientific Reports,* vol. 15, p. 16263, 2025.

[1368] A. Brosnan and E. Knapska, "Cheerful tails: Delving into positive emotional contagion," *Neuroscience and Biobehavioral Reviews,* vol. 161, p. 105674, 2024.

[1369] E. A. Madsen, T. Persson, S. Sayehli and S. S. G. Lenninger, "Chimpanzees show a developmental increase in susceptibility to contagious yawning: a test of the effect of ontogeny and emotional closeness on yawn contagion," *PLoS One,* vol. 8, no. 10, p. e76266, 2013.

[1370] M. W. Campbell and F. B. M. de Waal, "Chimpanzees empathize with group mates and humans, but not with baboons or unfamiliar chimpanzees," *Proceedings of the Royal Society B,* vol. 381, no. 1782, p. 20140013, 2014.

[1371] I. Norscia, M. Caselli, G. De Meo, G. Cordoni, J.-P. Guéry and E. Demuru, "Yawn contagion in bonobos: Another group, another story," *American Journal of Primatology,* vol. 84, no. 3, p. e23366, 2022.

[1372] C. F. I. Watson and C. A. Caldwell, "Neighbor effects in marmosets: social contagion of agonism and affiliation in captive Callithrix jacchus," *American Journal of Primatology,* vol. 72, no. 6, pp. 549-558, 2010.

[1373] E. N. Videan, J. Fritz, M. Schwandt and S. Howell, "Neighbor effect: evidence of affiliative and agonistic social contagion in captive chimpanzees (Pan troglodytes)," *American Journal of Primatology,* vol. 66, no. 2, pp. 131-144, 2005.

[1374] Z. Clay, "The evolution of empathy and its expression in bonobos," in *Bonobos and People at Wamba: 50 Years of Research*, T. Furuichi, G. Idani, D. Kimura, H. Ihobe and C. Hashimoto, Eds., Singapore, Springer, 2023, pp. 497-519.

[1375] W. P. Lemes and F. Amici, "Contagious yawning and scratching in captive lemurs," *Scientific Reports,* vol. 14, p. 26672, 2024.

[1376] J. S. Brooker, C. E. Webb, F. B. M. de Waal and Z. Clay, "The expression of empathy in human's closest relatives, bonobos and chimpanzees: current and future directions," *Biological Reviews,* vol. 99, no. 4, pp. 1556-1575, 2024.

[1377] L. J. Hardy, "Connection, Contagion, and Covid-19," *Medical Anthropology,* vol. 39, no. 8, pp. 655-659, 2020.

[1378] C. Landis, "Studies of emotional reactions. II. General behavior and facial expression," *Journal of Comparative Psychology,* vol. 4, no. 5, pp. 447-510, 1924.

[1379] P. V. Butler, "Destructive obedience in 1924: Landis' 'Studies of emotional reactions' as a prototype of the Milgram paradigm," *Irish Journal of Psychology,* vol. 19, no. 2-3, pp. 236-247, 1998.

[1380] D. Dolinski and D. Grzyb, "Obedience to authority as a function of the physical proximity of the student, teacher, and experimenter," *Journal of Social Psychology,* p. 1–13, 2024.

[1381] C. K. Hofling, E. Brotzman, S. Dalrymple, N. Graves and C. M. Pierce, "An experimental study in nurse-physician relationships," *Journal of Nervous and Mental Disease,* vol. 143, pp. 171-180, 1966.

[1382] W. Kilham and L. Mann, "Level of destructive obedience as a function of transmitter and executant roles in the Milgram obedience paradigm," *Journal of Personality and Social Psychology,* vol. 2, no. 5, pp. 696-702, 1974.

[1383] K. S. Larsen, D. Coleman, J. Forbes and R. Johnson, "Is the subject's personality or the experimental situation a better predictor of a subject's willingness to administer shock to a victim?," *Journal of Personality and Social Psychology,* vol. 22, no. 3, pp. 287-295, 1972.

[1384] S. G. Rank and C. K. Jacobson, "Hospital nurses' compliance with medication overdose orders: A failure to replicate," *Journal of Health and Social Behavior,* vol. 18, no. 2, pp. 147-156, 1977.

[1385] C. Gelpi, The Power of Legitimacy: Assessing the Role of Norms in Crisis Bargaining, Princeton: Princeton University Press, 2003, p. 224.

[1386] L. J. Skitka, C. W. Bauman and B. L. Lytle, "Limits on legitimacy: Moral and religious convictions as constraints on deference to authority," *Journal of Personality and Social Psychology,* vol. 97, no. 4, p. 567–578, 2009.

[1387] T. R. Tyler, "Psychological perspectives on legitimacy and legitimation," *Annual Review of Psychology,* vol. 57, no. 1, p. 376–400, 2006.

[1388] J. L. Beauvois, D. Courbet and D. Oberlé, "Le pouvoir prescriptif de l'animateur de télévision. Une transposition du paradigme de l'obéissance de Milgram dans le contexte d'un jeu télévisé," *European Review of Applied Psychology,* vol. 62, no. 3, pp. 111-119, 2012.

[1389] W. H. J. Meeus and Q. A. W. Raaijmakers, "Obedience in modern society: The Utrecht studies," *Journal of Social Issues,* vol. 51, no. 3, p. 155–175, 1995.

[1390] P. Bocchiaro and P. G. Zimbardo, "Defying unjust authority: An exploratory study," *Current Psychology: A Journal for Diverse Perspectives on Diverse Psychological Issues,* vol. 29, no. 2, pp. 155-170, 2010.

[1391] L. Bègue, J.-L. Beauvois, D. Courbet, D. Oberlé, J. Lepage and A. A. Duke, "Personality Predicts Obedience in a Milgram Paradigm," *Journal of Personality,* vol. 83, no. 3, pp. 299-306, 2014.

[1392] T. Grzyb and D. Dolinski, "You and I are alike, so I will hold back – The effect of directed empathy on the behavior of participants of Stanley Milgram's obedience paradigm," *Acta Psychologica,* vol. 234, p. 103859, 2023.

[1393] E. A. Caspar, D. Gishoma and A. M. Magalhaes de Saldanha da Gama, "On the cognitive mechanisms supporting prosocial disobedience in a post-genocidal context," *Science Reports,* vol. 12, p. 21875, 2022.

[1394] M. Gilbert, The Second World War: A Complete History, New York, NY: Holt, 2004, p. 846.

[1395] W. D. Brant, "The effects of race and social distance on obedience," *The Journal of Social Psychology,* vol. 112, no. 2, pp. 229-235, 1980.

[1396] A. Machcewicz, "Obedience to authority: Attitudes of prison officers in Stalinist Poland, 1944-1955," *Eastern European Politics and Societies,* vol. 37, no. 4, pp. 1180-1203, 2022.

[1397] M. E. Wheeler and S. T. Fiske, "Controlling racial prejudice: social-cognitive goals affect amygdala and stereotype activation," *Psychological Science,* vol. 16, no. 1, pp. 56-63, 2005.

[1398] E. C. Willadsen-Jensen and T. Ito, "A foot in both worlds: Asian Americans' perceptions of Asian, White, and racially ambiguous faces," *Group Processes & Intergroup Relations,* vol. 11, no. 2, pp. 182-200, 2008.

[1399] D. M. Amodio and M. Cikara, "The social neuroscience of prejudice," *Annual Review of Psychology,* vol. 72, pp. 439-469, 2021.

[1400] J. Correll, G. R. Urland and T. A. Ito, "Event-related potentials and the decision to shoot: the role of threat perception and cognitive control," *Journal of Experimental Social Psychology,* vol. 41, no. 1, pp. 120-128, 2006.

[1401] J. P. M. Balagtas, S. Tolomeo, B. L. Ragunath, P. Rigo, M. H. Bornstein and G. Esposito, "Neuroanatomical correlates of system-justifying ideologies: a pre-registered voxel-based morphometry study on right-wing authoritarianism and social dominance orientation. (Registered Report.)," *The Royal Society Open Science,* 2023.

[1402] R. A. Lavine, "Personality traits across cultures and research on obedience," *American Psychologist,* vol. 54, no. 7, p. 620–625, 2009.

[1403] W. Baryla, B. Wojciszke and A. Cichoka, "Legitimization and delegitimization of social hierarchy," *Social Psychology and Personality Science,* vol. 6, no. 6, pp. 669-676, 2015.

[1404] J. T. Jost and M. R. Banaji, "The role of stereotyping in system-justification and the production of false consciousness," *British Journal of Social Psychology,* vol. 33, no. 1, pp. 1-27, 1994.

[1405] A. C. Elms, "Obedience in retrospect," *Journal of Social Issues,* vol. 51, no. 3, pp. 21-31, 1995.

[1406] K. S. Masters, "Milgram, stress research, and the institutional review," *American Psychologist,* vol. 67, no. 7, pp. 621-622, 2009.

[1407] E. A. Caspar, J. F. Christensen, A. Cleeremans and P. Haggard, "Coercion changes the sense of agency in the human brain," *Current Biology,* vol. 26, no. 5, pp. P585-592, 2016.

[1408] J. M. Doris, L. Niemi and E. Machery, "True Believers: The Incredulity Hypothesis and the Enduring Legacy of the Obedience Experiments," *Philosophia Scientiæ,* vol. 28, no. 2, pp. 53-89, 2024.

[1409] J. M. Darley and C. D. Batson, "'From Jerusalem to Jericho': A study of situational and dispositional variables in helping behavior," *Journal of Personality and Social Psychology,* vol. 27, no. 1, pp. 100-108, 1973.

[1410] E. A. Caspar, F. Beyer, A. Cleeremans and P. Haggard, "The obedient mind and the volitional brain: A neural basis for preserved sense of agency and sense of responsibility under coercion," *PLoS One,* vol. 16, no. 10, p. e0258884., 2021.

[1411] K. A. Schwarz, S. Tonn, J. Büttner, W. Kunde and R. Pfister, "Sense of agency in social hierarchies," *Journal of Experimental Psychology: General,* vol. 152, no. 10, p. 2957–2976, 2023.

[1412] B. Latané and J. Rodin, "A lady in distress: Inhibiting effects of friends and strangers on bystander intervention," *Journal of Experimental Social Psychology,* vol. 5, no. 2, pp. 189-202, 1969.

[1413] E. A. Caspar, K. Loumpa, I. Arnaldo, L. Di Angelis, V. Gazzola and C. Keysers, "Commanding or Being a Simple Intermediary: How Does It Affect Moral Behavior and Related Brain Mechanisms?," *eNeuro,* vol. 9, no. 5, 2022.

[1414] *BROWN ET AL. v. BOARD OF EDUCATION OF TOPEKA ET AL. APPEAL FROM THE UNITED STATES DISTRICT COURT FOR THE DISTRICT OF KANSAS. No. 1,* 1954.

[1415] D. Bolton, "A revitalized biopsychosocial model: core theory, research paradigms, and clinical implications," *Physiological Medicine,* vol. 53, no. 16, pp. 7504-7511, 2023.

[1416] M. Giachero, A. B. Sacson, M. B. Vitullo, P. Bekinschtein and N. Weisstaub, "Targeting fear memories: Examining pharmacological disruption in a generalized fear framework," *Neurobiology of Learning and Memory,* vol. 213, p. 107960, 2024.

[1417] F. M. Bercum, M. J. Navarro Gomez and M. P. Saddoris, "Prefrontal cortex neurons in adult rats exposed to early life stress fail to appropriately signal the consequences of motivated actions," *Physiology and Behavior,* vol. 263, no. 1, p. 114107, 2023.

[1418] Y. Li, X. Lu, W. Zheng and J. Luo, "The role of the mPFC in the social influence of majority and expert opinions," *Neuropsychologia,* vol. 159, p. 107951, 2021.

[1419] A. N. Fassett-Carman, A. D. Moser, L. Ruzic, C. Neilson, J. Jones, S. Barnes-Horowitz, C. D. Schneck and R. H. Kaiser, "Amygdala and nucleus accumbens activation during reward anticipation moderates the association between life stressor frequency and depressive symptoms," *Journal of Affective Disorders,* vol. 330, no. 1, pp. 309-318, 2003.

[1420] Z. Tian, J. Song, X. Zhao, Y. Zhou, X. Chen, L. Q., F. Wang, L. Ma and X. Liu, "The interhemispheric amygdala-accumbens circuit encodes negative valence in mice," *Science,* vol. 386, no. 6722, p. eadp7520, 2024.

[1421] M. E. P. Seligman, Helplessness: On Depression, Development, and Death, San Francisco: W.H. Freeman, 1975, p. 250.

[1422] M. E. P. Seligman and S. F. Maier, "Failure to escape traumatic shock," *Journal of Experimental Psychology,* vol. 74, no. 1, p. 1–9, 1967.

[1423] B. Brown, L. T. Nguyen, I. Morales, E. M. Cardinale, W.-L. Tseng, C. C. McKay, K. Kircanski, M. A. Brotman, D. S. Pine, E. Leibenluft and J. O. Linke, "Associations between neighborhood resources and youths' response to reward omission in a task modeling negatively biased environments," *Journal of the American Academy of Child and Adolescent Psychiatry,* vol. 64, no. 4, pp. 463 - 474, 2025.

[1424] M. D. Alicke, "Global self-evaluation as determined by the desirability and controllability of trait adjectives," *Journal of Personality and Social Psychology,* vol. 49, no. 6, pp. 1621-1630, 1985.

[1425] P. Harrison, A. J. Lawrence, S. Wang, S. Liu, G. Xie, X. Yang and R. Zahn, "The psychopathology of worthlessness in depression," *Frontiers in Psychiatry,* vol. 19, p. 818542, 2022.

[1426] D. G. Dutton and S. Painter, "The battered woman syndrome: Effects of severity and intermittency of abuse," *American Journal of Orthopsychiatry,* vol. 63, no. 4, p. 614–622, 1993.

[1427] B. Pugh, L. Li and I. V. Sun, "Perceptions of why women stay in physically abusive relationships: A comparative study of Chinese and U.S. college students," *Journal of Interpersonal Violence,* vol. 36, no. 7-8, pp. 3778-3813, 2021.

[1428] T. W. Miller and A. Beane, "Loss of the safety signal in childhood and adolescent trauma. pp 377-388," in *Handbook of Stressful Transitions Across the Lifespan*, New York, Springer, 2010, p. 691.

[1429] A. Crandall, G. L. Castaneda, M. J. Barlow and B. M. Magnusson, "Do positive childhood and adult experiences counter the effects of adverse childhood experiences on learned helplessness?," *Frontiers in Child and Adolescent Psychiatry,* vol. 2, p. 1249529, 2024.

[1430] A. Rain, D. B. Roger and P. H. Venables, "Locus of control and socialization," *Journal of Research in Personality,* vol. 16, no. 2, pp. 147-156, 1982.

[1431] J. Rajchert, W. Molińska and W. Vonrath, "Social exclusion effect on self-efficacy. (Online ahead of print)," *Psychological Reports,* 2024.

[1432] D. Olweus, "Bully/victim problems among schoolchildren: Basic facts and effects of a school-based intervention program (pp. 411-448)," in *The Development and Treatment of Childhood Aggression*, D. J. Pepler and K. H. Rubin, Eds., Hillsdale, NJ: Psychology Press, 1991, p. 488.

[1433] D. Olweus, "Bullying among schoolchildren: Intervention and prevention. pp. 100-125," in *Aggression and Violence Throughout the Life Span*, Newbury Park, CA: Sage, 1992, p. 360.

[1434] D. Olweus, "Bully/victim problems in school: Facts and intervention," *European Journal of Psychology of Education,* vol. 12, pp. 495-510, 1997.

[1435] H. Arendt, Eichmann in Jerusalem: A Report on the Banality of Evil, New York: Viking Press, 1963, p. 273.

[1436] W. H. J. Meeus and Q. A. W. Raaijmakers, "Administrative obedience: Carrying out orders to use psychological-administrative violence," *European Journal of Social Psychology,* vol. 16, no. 4, p. 311–324, 1986.

[1437] W. H. J. Meeus and Q. A. W. Raaijmakers, "Autoritätsgehorsam in Experimenten des Milgram-Typs: Eine Forschungsübersicht [Obedience to authority in Milgram-style studies: A research review]," *Zeitschrift für Sozialpsychologie,* vol. 20, no. 2, pp. 155-175, 1989.

[1438] J. L. Embree and A. H. White, "Concept analysis: Nurse-to-nurse lateral violence," *Nursing Forum,* vol. 45, no. 3, pp. 241-245, 2010.

[1439] X. Peng, Y. Gan, Z. M. M. Qingsong, L. Xiong, F. Zhang, H. Xiong, H. Chang, Y. Chen, C. Guan, J. Wang and Y. Liu, "Nurse-to-nurse horizontal violence in Chinese hospitals and the protective role of head nurse's caring and nurses' group behaviour on it: A cross-sectional study," *Journal of Nursing Management,* vol. 30, no. 6, pp. 1590-1599, 2021.

[1440] D. Pina, M. Vidal-Alves, E. Puente-López, A. Luna-Maldonado, A. Ruiz-Cabello, T. Magalhães, B. Llor-Esteban, J. A. Ruiz-Hernández and B. Martínez-Jarreta, "Profiles of lateral violence in nursing personnel of the Spanish public health system," *PLOS One,* vol. 17, no. 5, p. e0268636, 2022.

[1441] S. S. Tikva, G. Gabay, L. Asraf, A. N. Kluger and Y. Lerman, "Experiencing and witnessing disruptive behaviors toward nurses in COVID-19 teams, patient safety, and errors in care," *Journal of Nursing Scholarship,* vol. 55, no. 1, pp. 253-261, 2023.

[1442] Y. Boddez, P. Van Dessel and J. De Houwer, "Learned helplessness and its relevance for psychological suffering: a new perspective illustrated with attachment problems, burn-out, and fatigue complaints," *Cognition and Emotion,* vol. 36, no. 6, pp. 1027-1036, 2022.

[1443] K. R. Scherer, "Learned helplessness revisited: biased evaluation of goals and action potential are major risk factors for emotional disturbance," *Cognition and Emotion,* vol. 36, no. 6, pp. 1021-1026, 2022.

[1444] C. Buzzai, L. Sorrenti, F. Tripicano, S. Orecchio and P. Filippello, "School alienation and academic achievement: The role of learned helplessness and mastery orientation," *School Psychology,* vol. 36, no. 1, pp. 17-23, 2021.

[1445] W. Schaufeli, "The burnout enigma solved?," *Scandinavian Journal of Work, Environment and Health,* vol. 47, no. 3, pp. 169-170, 2021.

[1446] L. Mutesa, "Predictive evidence of the relevance of epigenetics to PTSD," *Nature Reviews Genetics,* vol. 24, p. 807, 2023.

[1447] A. J. Isaac, S. J. Bufferd and Y. Mekawi, "Racism and hypothalamic-pituitary-adrenal axis functioning in childhood as risks for health disparities across the lifespan," *Psychoneuroendocrinology,* vol. 176, p. 107416, 2025.

[1448] H. J. Freudenberger, "Staff burn-out," *Journal of Social Issues,* vol. 30, no. 1, pp. 159-165, 1974.

[1449] W. B. Schaufeli, Professional Burnout: Recent Developments in Theory and Research, vol. 33, W. B. Schaufeli, C. Maslach and T. Marek, Eds., New York: Routledge, 2018, p. 308.

[1450] P. Moreno-Jiménez and M. C. Hidalgo, "Prediction of burnout in volunteers," *Journal of Applied Social Psychology,* vol. 40, no. 7, pp. 1798-1818, 2010.

[1451] D. C. Duru and M. S. Hammoud, "Identifying effective retention strategies for front-line nurses," *Nursing Management (Harrow),* vol. 29, no. 1, pp. 17-24, 2022.

[1452] D. Larson, "Why Scandinavian prisons are superior," *The Atlantic,* vol. 24, no. 1, pp. 3-9, 24 September 2013.

[1453] P. Zimbardo, C. Haney, W. C. Banks, D. Jaffe and C. Prescott, *The Stanford Prison Experiment: A Simulation Study of the Psychology of Imprisonment,* Stanford, CA: Stanford University, 1971.

[1454] S. D. Reicher, J. J. van Bavel and H. S. Alexander, "Debate around leadership in the Stanford Prison Experiment: Reply to Zimbardo and Haney (2020) and Chan et al. (2020)," *American Psychologist,* vol. 75, no. 3, pp. 406-407, 2020.

[1455] H. Anderson, The Study of Language in Laboratory Based Interactions- The Stanford Prison Experiment, York: University of York, 2022.

[1456] T. Le Texier, "Debunking the Stanford Prison Experiment," *American Psychologist,* vol. 74, no. 7, pp. 823-839, 2019.

[1457] S. A. Haslam, S. D. Reicher and J. J. Van Bavel, "Rethinking the nature of cruelty: The role of identity leadership in the Stanford Prison Experiment," *American Psychologist,* vol. 74, no. 5, pp. 809-822, 2019.

[1458] P. G. Zimbardo, "The psychological power and pathology of imprisonment. A statement prepared for the U. S. House of Representatives Committee on the Judiciary," U. S. House of Representatives Committee on the Judiciary, Subcommittee No. 3. Hearing on Prison Reform, San Francisco, 1971.

[1459] P. G. Zimbardo, "Pathology of imprisonment," *Society,* vol. 9, no. 6, pp. 4-8, 1972a.

[1460] H. Perlstat, "The Stanford Prison Experiment: The power of the situation (pp 183-204)," in *Assessing Social Science Research Ethics and Integrity. Clinical Sociology: Research and Practice*, First ed., Cham, Springer, 2023, p. 266.

[1461] Chicago Tribune Editorial Board, "Rekia Boyd shooting was 'beyond reckless,' so cop got a pass," *Chicago Tribune,* 22 April 2015.

[1462] E. Hill, Tiefenthäler, C. Triebert, D. Jordan, H. Willis and R. Stein, "How George Floyd was killed in police custody," *The New York Times,* 24 January 2022 (updated).

[1463] S. E. Wolfe, K. Mclean, G. P. Alpert and J. Rojek, "Us versus them? The problem of cognitive distortions in policing," *Police Quarterly,* vol. 27, no. 4, pp. 532-560, 2024.

[1464] L. A. Fridell and C. J. Marier, "The impact of suspect race and precipitating incident on community members' assessment of deadly force reasonableness," *Homicide Studies,* vol. 27, no. 1, pp. 242-166, 2023.

[1465] K. Stafford and H. Fingerhut, "AP-NORC poll: Sweeping change in US views of police violence," Associated Press, New York, 2020.

[1466] R. VerBruggen, "Fatal Police Shootings and Race: A Review of the Evidence and Suggestions for Future Research," Manhattan Institute for Policy Research, New York, 2022.

[1467] D. Fatheree, "Trashing their rights: Alabama town uses 'debtor' prison' for people who fall behind on garbage bills," 31 March 2023. [Online]. Available: https://www.splcenter.org/news/2023/03/31/alabama-town-debtors-prison-garbage-bills. [Accessed 2 April 2023].

[1468] M. Hoekstra and C. W. Sloan, "National Bureau of Economic Research Working Paper 26774: Does Race Matter for Police Use of Force? Evidence From 911 Calls," National Bureau of Economic Research, Washington, DC, 2020.

[1469] A. Braga, J. R. Coldren Jr., W. Sousa, D. Rodriguez and O. Alper, "The Benefits of Body-Worn Cameras: New Findings from a Randomized Controlled Trial at the Las Vegas Metropolitan Police Department," Office of Justice Programs' National Criminal Justice Reference, Washington, DC, 2017.

[1470] R. H. Mulder, R. Kraaij, I. K. Schuurmans, C. Frances-Cuesta, Y. Sanz, C. Medina-Gomez, L. Duijts, F. Rivadereira, H. Tiemeier, V. W. V. Jaddoe, J. F. Felix and C. A. M. Cecil, "Early-life stress and the gut microbiome: A comprehensive population-based investigation," *Brain, Behavior, and Immunity,* vol. 118, pp. 117-127, 2024.

[1471] R. García-Cabrerizo and J. F. Cryan, "A gut (microbiome) feeling about addiction: Interactions with stress and social systems," *Neurobiology of Stress,* vol. 30, p. 100629, 2024.

[1472] J. Quin, R. Li, J. Raes, M. Arumugam, K. S. Burgdorf, C. Manichanh, T. Nielsen, N. Pons, F. Levenez, T. Yamada, D. R. Mende, J. Li, J. Xu, S. Li, D. Li, J. Cao, B. Wang, H. Liang, H. Zheng, Y. Xie, J. Tap, P. Lepage and e. al., "A human gut microbial gene catalogue established by metagenomic sequencing," *Nature,* vol. 464, no. 7285, pp. 59-65, 2010.

[1473] P. Luczynski, K.-A. McVey Neufeld, C. S. Oriach, G. Clarke, T. G. Dinan and J. F. Cryan, "Growing up in a bubble: Using germ-free animals to assess the influence of the gut microbiota on brain and," *International Journal of Neuropsychopharmacology,* vol. 19, no. 8, p. pyw020, 2016.

[1474] D. R. Simkin, "Microbiome and mental health, specifically as it relates to adolescents," *Current Psychiatric Reports,* vol. 21, pp. 1-12, 2019.

[1475] R. EL Rawas, I. M. Amarala and A. Hofer, "Social interaction reward: A resilience approach to overcome vulnerability to drugs of abuse," *European Journal of Neuropsychopharmacology,* vol. 37, pp. 12-28, 2020.

[1476] R. García-Cabrerizo, C. Carbia, K. J. O'Riodan, H. Schellekens and J. F. Cryan, "Microbiota-gut-brain axis as a regulator of reward processes," *Journal of Neurochemistry,* vol. 157, no. 5, pp. 1495-1524, 2020.

[1477] S. Chbeir and V. Carrión, "Resilience by design: How nature, nurture, environment, and microbiome mitigate stress and allostatic load," *World Journal of Psychiatry,* vol. 13, no. 5, pp. 144-159, 2023.

[1478] Y. Cao, Y. Cheng, W. Pan, J. Diao, L. Sun and M. & Meng, "Gut microbiota variations in depression and anxiety: a systematic review," *BMC Psychiatry,* vol. 21, no. 1, p. 443, 2025.

[1479] J. P. Oyarzun, T. M. Kuntz, Y. Stussi, O. T. Karaman, S. Vranos, B. L. Callaghan, C. Huttenhower, J. E. LeDoux and E. A. Phelps, "Human threat learning is associated with gut microbiota composition," *PNAS Nexus,* vol. 1, no. 5, p. pgac271, 2022.

[1480] E. A. Mayer, K. Tillisch and A. Gupta, "Gut/brain axis and the microbiota," *Journal of Clinical Investigation,* vol. 125, no. 3, pp. 926-938, 2015.

[1481] E. E. Fröhlich, A. Farzi, R. Mayerhofer, F. Reichmann, A. Jačan, B. Wagner, E. Zinser, N. Bordag, C. Magnes, E. Fröhlich, K. Kashofer, G. Gorkiewicz and P. Holzer, "Cognitive impairment by antibiotic-induced gut dysbiosis: Analysis of gut microbiota-brain communication," *Brain, Behavior, and Immunity,* vol. 56, pp. 140-155, 2016.

[1482] E. A. Mayer, D. Padua and K. Tillisch, "Altered brain-gut axis in autism: Comorbidity or causative mechanisms?," *BioEssays,* vol. 36, no. 10, pp. 933-939, 2014.

[1483] M. B. Cooke, S. Catachlove and K. L. Tooley, "Examining the influence of the human gut microbiota on cognition and stress: A systematic review of the literature," *Nutrients,* vol. 14, no. 21, p. 4623, 2022.

[1484] K. Rathore, N. Shukla, S. Naik, K. Sambhav, K. Dange, D. Bhuyan and Q. M. Haq, "The bidirectional relationship between the gut microbiome and mental health: A comprehensive review," *Cureus,* vol. 17, no. 3, p. e80810, 2025.

[1485] A. J. Montiel-Castro, R. M. González-Cervantes, G. Bravo-Ruiseco and Pacheo-López, "The microbiota-gut-brain axis: neurobehavioral correlates, health and sociality," *Frontiers in Integrative Neuroscience,* vol. 7, p. 70, 2013.

[1486] M. Abavisani, N. Faraji, N. Ebadpour, P. Kesharwani and A. Sahebkar, "Beyond digestion: Exploring how the gut microbiota modulates human social behaviors," *Neuroscience,* vol. 565, pp. 52-62, 2025.

[1487] C. M. Lynch, C. S. M. Cowan, T. F. S. Bastiaanssen, G. M. Moloney, N. Theune, M. van de Wouw, E. F. Zanuy, A. P. Ventura-Silva, M. G. Codagnone, F. Villalobos-Manríquez, M. Segalla, F. Koc, C. Stanton, P. Ross, T. G. Dinan and G. Clarke, "Critical windows of early-life microbiota disruption on behaviour, neuroimmune function, and neurodevelopment," *Brain, Behavior, and Immunity,* vol. 108, pp. 309-327, 2023.

[1488] W. E. Gato, C. Posick, A. Williams and C. Mays, "Examining the link between the human microbiome and antisocial behavior: Why criminologists should care about biochemistry, too," *Deviant Behavior,* vol. 39, no. 9, pp. 1191-1201, 2018.

[1489] R. Debray, N. Snyder-Mackler, J. N. Kohn, M. E. Wilson, L. B. Barreiro and J. Tung, "Social affiliation predicts mitochondrial DNA copy number in female rhesus macaques," *Biology Letters,* vol. 15, no. 1, p. 20180643, 2019.

[1490] B. N. Dulka, A. K. Bourdon, C. T. Clinard, M. B. K. Muvvala, S. R. Campagna and M. A. Cooper, "Metabolomics reveals distinct neurochemical profiles associated with stress resilience," *Neurobiology of Stress,* vol. 7, pp. 103-112, 2017.

[1491] F. Hollis, B. S. Pope, E. Gorman-Sandler and S. K. Wood, "Neuroinflammation and mitochondrial dysfunction link social stress to depression," *Current Topics in Behavioral Neurosciences,* vol. 54, pp. 59-93, 2022.

[1492] R. K. Naviaux, "Metabolic features of the cell danger response," *Mitochondria,* vol. 16, pp. 7-17, 2014.

[1493] H. E. Lapp, S. Ahmed, C. L. Moore and R. G. Hunger, "Toxic stress history and hypothalamic-pituitary-adrenal axis function in a social stress task: Genetic and epigenetic factors," *Neurotoxicology and Teratology,* vol. 71, pp. 41-49, 2019.

[1494] D. H. Ülgen, S. R. Ruigrok and C. Sandi, "Powering the social brain: Mitochondria in social behaviour," *Current Opinion in Neurobiology,* vol. 79, p. 102675, 2023.

[1495] A. L. M. Reis, J. M. Hammond, I. Stevanovski, J. C. Arnold, I. S. McGregor, I. W. Deveson and A. Gururajan, "Sex-specific transcriptomic and epitranscriptomic signatures of PTSD-like fear acquisition," *iScience,* vol. 25, no. 9, p. 104861, 2022.

[1496] S. Alhassen, S. Chen, L. Alhassen, A. Phan, M. Khoudari, A. De Silva, H. Barhoosh, Z. Wang, C. Parrocha, E. Shapiro, C. Henrich, Z. Wang, L. Mutesa, P. Baldi, G. W. Abbott and A. Alachkar, "Intergenerational trauma transmission is associated with brain metabotranscriptome remodeling and mitochondrial dysfunction," *Communications Biology,* vol. 4, p. 783, 2021.

[1497] A. Larosa and T. P. Wong, "The hippocampus in stress susceptibility and resilience: Reviewing molecular and functional markers," *Psychopharmacology and Biological Psychiatry,* vol. 119, p. 110601, 2022.

[1498] M. E. Smith, "Bilateral hippocampal volume reduction in adults with post-traumatic stress disorder: a meta-analysis of structural MRI studies," *Hippocampus,* vol. 15, no. 6, pp. 798-807, 2005.

[1499] P. Videbech and B. Ravnkilde, "Hippocampal volume and depression: a meta-analysis of MRI studies," *American Journal of Psychiatry,* vol. 161, no. 11, pp. 1957-1966, 2004.

[1500] J. D. Bremner, M. Narayan, E. R. Anderson, L. H. Staib, H. L. Miller and D. S. Charney, "Hippocampal volume reduction in major depression," *American Journal of Psychiatry,* vol. 157, no. 1, pp. 115-118, 2000.

[1501] Y. Shang, Y. Yang, G. Zheng, Z. Zhao, Y. Wang, L. Yang, L. Han, Z. Yao and B. Hu, "Aberrant functional network topology and effective connectivity in burnout syndrome," *Clinical Neurophysiology,* vol. 138, pp. 163-172, 2022.

[1502] Y. C. Ho and S. Wang, "Adult neurogenesis is reduced in the dorsal hippocampus of rats displaying learned helplessness behavior," *Neuroscience,* vol. 171, no. 1, pp. 153-161, 2010.

[1503] T. Hajszan, A. Dow, J. L. Warner-Schmidt, K. Szigeti-Buck, N. L. Sallam, A. Parducz and C. D. S. R. Leranth, "Remodeling of hippocampal spine synapses in the rat learned helplessness model of depression," *Biological Psychiatry,* vol. 65, no. 5, p. 392–400, 2009.

[1504] D. Liu, Z. Wang, Z. Gao, K. Xie, Q. J. H. Zhang and Q. Pang, "Effects of curcumin on learning and memory deficits, BDNF, and ERK protein expression in rats exposed to chronic unpredictable stress," *Behavioural Brain Research,* vol. 271, no. 1, pp. 116-121, 2014.

[1505] P. S. Eriksson and L. Wallin, "Functional consequences of stress-related suppression of adult hippocampal neurogenesis - a novel hypothesis on the neurobiology of burnout," *Acta Neurologica Scandinavica,* vol. 110, no. 5, pp. 275-280, 2004.

[1506] L. M. Shin, S. L. Rauch and R. K. Pitman, "Amygdala, medial prefrontal cortex and hippocampal function in PTSD," *Annals of the New York Academy of Sciences,* vol. 1071, no. 1, pp. 67-79, 2006.

[1507] S. Liu, Y. Wang, W. He, Y. Chen and Q. Qang, "The effect of students' effort–reward imbalance on learning engagement: The mediating role of learned helplessness and the moderating role of social support," *Frontiers in Psychology,* vol. 15, p. e1329664, 2024.

[1508] K. E. McCann and K. L. Huhman, "Social defeat in Syrian hamsters: controllability, duration, and social recognition," in *Proceedings: Animal Behavior Conference. Center for the Integrative Study of Animal Behavior*, Bloomington IN, 2013.

[1509] K. E. McCann, C. N. Bicknese, A. Norvelle and K. L. Huhman, "Effects of inescapable versus escapable social stress in Syrian hamsters: The importance of stressor duration versus escapability," *Physiology and Behavior,* vol. 22, no. 129, pp. 25-29, 2014.

[1510] V. Weihgold, "Moral submissiveness: social origin as a vulnerability for well-being on a warming planet," *Frontiers in Psychology,* vol. 15, p. e1355736, 2024.

[1511] L. A. Pavcovich and O. A. Ramirez, "Effects of chronic desipramine administration on the locus coeruleus neuronal activity in the learned helplessness paradigm," *Brain Research Bulletin,* vol. 32, no. 2, p. Brain Research Bulletin, 1993.

[1512] G. E. Brown, D. A. Davenport and A. R. Howe, "Naloxone blocks learned helplessness in the slug (Limax Maximus)," *Psychological Reports,* vol. 76, no. 1, p. 14, 1995.

[1513] P. S. Wai, Religious Orientation, Locus of Control and Learned Helplessness (dissertation), Jos: Department of General and Applied Psychology, University of Jos, 2008, p. 222.

[1514] D. Finkelhor, H. Turner and R. Ormrod, "Kid's stuff: The nature and impact of peer and sibling violence on younger and older children," *Child Abuse and Neglect,* vol. 30, no. 12, pp. 1401-1421, 2006.

[1515] B. F. Chorpita and D. H. Barlow, "The development of anxiety: The role of control in the early environment," *Psychological Bulletin,* vol. 124, no. 1, pp. 3-21, 1998.

[1516] R. G. Bradley, E. B. Binder, M. P. Epstein, Y. Tang, H. P. Nair, W. Liu, C. F. Gillespi, T. Berg, M. Evces, J. Newport, Z. N. Stowe, C. M. Heim, C. B. Nemeroff, A. Schwartz, J. F. Cubelles and K. J. Ressler, "Influence of child abuse on adult depression: Moderation by the corticotropin-releasing hormone receptor gene," *Journal of the American Medical Association,* vol. 65, no. 2, p. 190–200, 2008.

[1517] M. Chapman, "The epistemic triangle: Operative and communicative components of cognitive development (pp. 209–228)," in *Criteria for Competence: Controversies in the Conceptualization and Assessment of Children's Abilities*, Mahwah, Erlbaum, 1991, p. 286.

[1518] M. Chapman, "Constructivism and the problem of reality," *Journal of Applied Development Psychology,* vol. 20, no. 1, p. 31–43, 1999.

[1519] J. Winterson, Oranges Are Not the Only Fruit, New York: Grove Press, 1985/ 1997, p. 176.

[1520] C. Levi-Strauss, The Raw and the Cooked. Introduction to a Science of Mythology. Volume I, New York: Harper and Row, 1969.

[1521] T. Ingold, The Perception of the Environment, New York: Routledge, 2021, p. 602.

[1522] R. Kelly, The Lifeways of Hunter-Gatherers: The Foraging Spectrum, Second Revised ed., Cambridge: Cambridge University Press, 2013, p. 376.

[1523] M. J. Winkelman, "An ethnological analogy and biogenic model for interpretation of religion and ritual in the past," *Journal of Archaeological Method and Theory,* vol. 29, pp. 335-389, 2021.

[1524] C. S. Alcorta and R. Sosis, "Ritual, emotion, and sacred symbols: The evolution of religion as an adaptive complex," *Human Nature,* vol. 16, no. 4, pp. 323-359, 2005.

[1525] Pascua Yaqui Tribe, "Pascua Yaqui Tribe," [Online]. Available: https://www.pascuayaqui-nsn.gov/culture/yaqui-ceremonies/. [Accessed 13 April 2023].

[1526] J. Fuster, "Frontal lobe and cognitive development," *Journal of Neurocytology,* vol. 31, no. 3-5, pp. 373-385, 2002.

[1527] J. D. Maser and G. G. Gallup Jr., "Theism as a by-product of natural selection," *The Journal of Religion,* vol. 70, no. 4, pp. 515-532, 1990.

[1528] E. Voland and W. Schiefenhövel, Eds., The Biological Evolution of Religious Mind and Behavior (The Frontiers Collection), New York: Springer, 2009, p. 305.

[1529] P. McNamara, Where God and Science Meet: How Brain and Evolutionary Studies Alter Our Understanding of Religion, New York: Greenwood Publishers, 2006, p. 928.

[1530] J. Fest, Hitler, New York: Vintage, 1973, p. 844.

[1531] R. Ponce de Leon and A. C. Kay, "Political ideology and compensatory control mechanisms," *Current Opinion in Behavioral Sciences,* vol. 34, pp. 112-117, 2020.

[1532] I. Pyysiäinen and M. Hauser, "The origins of religion: evolved adaptation or by-product? (opinion)," *Trends in Cognitive Sciences,* vol. 14, no. 2, pp. 104-109, 2010.

[1533] A. Norenzayan, A. F. Shariff, W. M. Gervais, A. K. Willard, R. A. McNamara, E. Slingerland and J. Henrich, "The cultural evolution of prosocial religions," *The Behavioral and Brain Sciences,* vol. 39, p. e1, 2016.

[1534] A. C. Gaw, "Religious belief at the level of the brain: Neural correlates and the influence of culture," *The Journal of Nervous and Mental Disease,* vol. 207, no. 7, pp. 604-610, 2019.

[1535] E. Appelros, "Ritual dance from a philosophical perspective," in *Handbook of Rituals in Contemporary Studies of Religion: Exploring Ritual Creativity in the Footsteps of Anne-Christine Hornborg*, vol. 22, I. Brissman, P. Linjamaa and T. T. Makeeff, Eds., Brill Academic Publishing, 2024, pp. 179-192.

[1536] A. Rosenfeldt, "The 'spirit' of new atheism and religious activism in the post-9/11 God debate," *Human Studies,* 29 August 2024.

[1537] S. Muralidharan, C. La Ferle and O. Roth-Cohen, "How religiosity and spirituality influences the ecologically conscious consumer psychology of Christians, the non-religious, and atheists in the United States," *Archives for the Psychology of Religion,* vol. 46, no. 1, pp. 71-87, 2024.

[1538] J. Haidt, "Misunderstanding Moral Psychology," in *The Morality Wars: The Ongoing Debate Over The Origin of Human Goodness*, 2021, p. 77.

[1539] M. J. Winkelman, "Mimesis and the origins of religion," *Religion, Brain and Behavior,* vol. 14, no. 1, pp. 90-98, 2023.

[1540] U. Schjødt, H. Stødkilde-Jørgensen, A. W. Geertz and A. Roepstorff, "Rewarding prayers," *Neuroscience Letters,* vol. 443, no. 3, pp. 165-168, 2008.

[1541] C. M. Adler, P. McDonough-Ryan, K. W. Sax, S. K. Holland, S. Arndt and S. M. Strakowski, "fMRI of neuronal activation with symptom provocation in unmedicated patients with obsessive compulsive disorder," *Journal of Psychiatric Research,* vol. 34, no. 4-5, pp. 327-324, 2000.

[1542] R. Durrett and S. A. Levin, "Can stable social groups be maintained by homophilous imitation alone?," *Journal of Economic Behavior & Organization,* vol. 57, no. 3, p. 267–286, 2005.

[1543] A. C. Kay, D. Gaucher, I. McGregor and K. Nash, "Religious belief as compensatory control," *Personality and Social Psychology Review,* vol. 14, no. 1, pp. 37-48, 2010.

[1544] J. P. Rushton, R. J. H. Russell and P. A. Wells, "Genetic similarity theory: Beyond kin selection," *Behavior Genetics,* vol. 14, no. 3, pp. 179-193, 1984.

[1545] A. J. Stewart, "Group reciprocity and the evolution of stereotyping," *Proceedings of the Royal Society B,* vol. 290, no. 1991, p. 20221834, 2023.

[1546] H. A. Viciana, C. Loverdo and A. Gomila, "Credibility, credulity, and redistribution," *Behavioral and Brain Science,* vol. 39, p. e25, 2016.

[1547] J. W. Dow, "Is religion an evolutionary adaptation?," *Journal of Artificial Societies and Social Simulation,* vol. 11, no. 22, 2008.

[1548] B. G. Purzycki and R. Sosis, "The religious system as adaptive: Cognitive flexibility, public displays, and acceptance. pp 243-256," in *The Biological Evolution of Religious Mind and Behavior*, E. Voland and W. Schiefenhövel, Eds., New York, Springer, 2009, p. 305.

[1549] R. Sosis and C. Alcorta, "Ritual, emotion, and sacred symbols: The evolution of religion as an adaptive complex," *Human Nature,* vol. 16, no. 4, pp. 323-359, 2005.

[1550] R. Sosis and B. Ruffle, "Ideology, religion and the Evolution of cooperation: Field experiments on Israeli kibbutzim," *Research in Economic Anthropology,* vol. 23, pp. 89-117, 2004.

[1551] A. Kaznatcheev, "Bifurcation of cooperation and inviscid ethnocentrism," 26 6 2012. [Online]. Available: https://egtheory.wordpress.com/2012/06/26/bifurcation-and-ethnocentrism/. [Accessed 26 6 2018].

[1552] I. H. Y. Koh and N. P. Li, "Costly Signaling Theory," in *Encyclopedia of Religious Psychology and Behavior*, Cham, Springer, 2025.

[1553] M. J. Gantley, J. P. McKeown and A. V. Jimenez, "The social brain hypothesis: Dunbar's number and the stability of religious social networks (pp 111-129)," in *The Rutledge Handbook of Evolutionary Approaches to Religion*, Milton Park, Routledge, 2022, p. 524.

[1554] J. Kantner and K. J. Vaughn, "Pilgrimage as costly signal: Religiously motivated cooperation in Chaco and Nasca," *Journal of Anthropological Archaeology,* vol. 31, no. 1, pp. 66-82, 2012.

[1555] D. L. Bermudez, M. E. Garcia, V. F. Iturriaga, R. E. Rodriguez and M. Gonzales-Backen, "Latinx youth in rural settings: Understanding the links between ethnic-racial identity, neighborhood risks, perceived discrimination, and depressive symptoms," *Journal of Adolescence,* vol. 97, no. 4, pp. 1103-1112, 2025.

[1556] J. Bulbulia, "Religious costs as adaptations that signal altruistic intention," *Evolution and Cognition,* vol. 10, no. 1, pp. 19-42, 2004.

[1557] R. Sosis, "Costly Signaling: The ABCs of signal theory and religion (pp 209-226)," in *The Routledge Handbook of Evolutionary Approaches to Religion (Routledge Handbooks in Religion)*, L. Yair and J. Lane, Eds., Milton Park, Routledge, 2023, p. 524.

[1558] P. Boyer, The Naturalness of Religious Ideas: A Cognitive Theory of Religion, Berkeley: University of California Press, 1994, p. 344.

[1559] J. Bulbulia, R. Sosis, E. Harris, R. Genet, C. Genet and C. Wyman, Eds., The Evolution of Religion: Studies, Theories, and Critiques. 1st ed, Santa Margarita, CA: Collins Foundation Press, 2008, p. 406.

[1560] F. Pazhoohi, A. Kingstone and D. Xygalatas, "Does the conservativeness of religious veiling signal trust and in-group commitment?," *Evolutionary Psychological Science,* vol. 11, pp. 76-84, 2025.

[1561] A. Kaznatcheev and T. R. Schultz, "Ethnocentrism maintains cooperation, but keeping one's children close fuels it," in *Proceedings of the 33rd Annual Conference of the Cognitive Science Society*, Merced CA, 2011.

[1562] R. B. Glassman, E. W. Packel and D. L. Brown, "Greenbeards and kindred spirits: A preliminary mathematical model of altruism toward nonkin who bear similarities to the giver," *Ethology and Sociobiology,* vol. 7, no. 2, pp. 107-115, 1986.

[1563] R. A. Hammond and R. Axelrod, "Evolution of contingent altruism when cooperation is expensive," *Theoretical Population Biology,* vol. 69, no. 3, p. 333–338, 2006.

[1564] J. C. Cohrs and F. Asbrock, "Right-wing authoritarianism, social dominance orientation and prejudice against threatening and competitive ethnic groups," *European Journal of Social Psychology,* vol. 39, no. 2, pp. 270-389, 2008.

[1565] S. Arieli, A. M. Grant and L. Sagiv, "Convincing yourself to care about others: An intervention for enhancing benevolence values," *Journal of Personality,* vol. 82, no. 1, pp. 15-24, 2014.

[1566] D. Malhotra, "(When) are religious people nicer? Religious salience and the “Sunday Effect” on pro-social behavior," *Judgement and Decision Making,* vol. 5, no. 2, pp. 138-143, 2010.

[1567] Г. В. Ожиганова, *Вестник РУДН. Серия: Психология и педагогика [RUDN Journal of Psychology and Pedagogics],* vol. 19, no. 4, pp. 744-764, 2022.

[1568] X. Ma and Z.-X. Wu, "Do sacred texts glorifying Allah facilitate Muslims' trust and trustworthiness? Evidence from field experiments in China," *Journal of Behavioral and Experimental Economics,* vol. 83, p. 101466, 2019.

[1569] L. B. Koenig, M. McGue, R. F. Krueger and B. J. T. J., "Religiousness, antisocial behavior, and altruism: Genetic and environmental mediation," *Journal of Personality,* vol. 75, no. 2, pp. 265-290, 2007.

[1570] E. F. Lomash, T. D. Brown and M. P. Galupo, "'A whole bunch of love the sinner hate the sin': LGBTQ microaggressions experienced in religious and spiritual context," *Journal of Homosexuality,* vol. 66, pp. 1495-1511, 2019.

[1571] R. S. Rajan, "The politics of Hindu 'tolerance'," *Boundary 2,* vol. 38, no. 3, pp. 67-86, 2011.

[1572] G. Omvedt, Dalit Visions: The Anti-Caste Movement and the Construction of an Indian Identity, Rev. Edition, Pa, revised ed., Himayatnagar, Hyderabad: Orient BlackSwan, 2006, p. 108.

[1573] B. Laythe, D. G. Finkel, R. G. Bringle and L. A. Kirkpatrick, "Religious fundamentalism as a predictor of prejudice: a two-component model," *Journal for the Scientific Study of Religion,* vol. 41, no. 4, pp. 623-635, 2002.

[1574] B. Laythe, D. Finkel and L. A. Kirkpatrick, "Predicting prejudice from religious fundamentalism and right-wing authoritarianism: A multiple-regression approach," *Journal for the Scientific Study of Religion,* vol. 40, no. 1, pp. 1-10, 2002.

[1575] D. Saunders, "The US got what it deserved. Falwell says," *Toronto Globe and Mail,* p. A2, 15 September 2001.

[1576] R. E. Edmonson, "Robert Edward Edmondson anti-communist and antisemitic propaganda fliers, 1935-1939," 1935-1939. [Online]. Available: https://digital.library.temple.edu/digital/collection/p16002coll31/id/251. [Accessed 9 April 2023].

[1577] M. Basedau and C. Koos, "When do religious leaders support faith-based violence? Evidence from a survey poll in South Sudan," *Political Research Quarterly,* vol. 68, no. 4, pp. 760-772, 2015.

[1578] A. Matsuo and C. M. Brown, "Culture points the moral compass: Shared basis of culture and morality," *Culture and Brain,* vol. 10, pp. 113-139, 2022.

[1579] J. Haidt, The Righteous Mind: Why Good People are Divided by Politics and Religion, Reprint edition ed., Vintage, 2013, p. 528.

[1580] J. Haidt and C. Joseph, "Intuitive ethics: How innately prepared intuitions generate culturally variable virtues," *Daedalus,* vol. 133, no. 4, pp. 55-66, 2004.

[1581] K. Gray and S. Pratt, "Morality in our mind and across cultures and politics," *Annual Review of Psychology,* vol. 76, pp. 663-691, 2025.

[1582] M. Atari and A. Zewail, "Morality across time and space," *European Review of Social Psychology,* pp. 1-36, 2025.

[1583] S. B. Northover, T. Quillien, D. Conroy-Beam and A. B. Cohen, "Religious signaling and prosociality: A review of the literature," *Evolution and Human Behavior,* vol. 45, no. 5, p. 106593, 2024.

[1584] E. Fehr and S. Gächter, "Cooperation and punishment in public goods experiments," *The American Economic Review,* vol. 90, no. 4, pp. 980-994, 2000.

[1585] E. Ostrom, J. Walker and R. Gardner, "Covenants with and without a sword: Self-governance is possible," *American Political Science Review,* vol. 86, no. 2, pp. 404-417, 1992.

[1586] G. Allport, "The religious context of prejudice," *Journal for the scientific study of religion,* vol. 5, no. 3, pp. 447-445, 1966.

[1587] J. A. Roberts and M. E. David, "Holier than thou: Investigating the relationship between religiosity and charitable giving," *Journal of Philanthropy and Marketing,* vol. 24, no. 1, p. e1619, 2018.

[1588] C. L. Park, "Intrinsic and extrinsic religious motivation: Retrospect and prospect," *The International Journal for the Psychology of Religion,* vol. 31, no. 3, pp. 213-222, 2021.

[1589] K. A. Johnson, A. B. Cohen and M. A. Okun, "Intrinsic religiosity and volunteering during emerging adulthood: A comparison of Mormons with Catholics and Non-Catholic Christians," *Journal for the Scientific Study of Religion,* vol. 52, no. 4, pp. 842-851, 2013.

[1590] A. A. Agbo, C. N. Ngwu, N. G. Onah and I. E. Iwundu, "Religiosity and well-being among Nigerians: Exploring the mediating role of fatalism," *Journal of Psychology in Africa,* vol. 34, no. 1, p. 26–35, 2024.

[1591] A. M. Rejosumarto, M. Bulut, V. S. Farrasyanti and A. Anggraeni, "The effects of religiosity and family tradition on altruistic behavior: Evidence in Indonesia," *Asian Journal of Management Entrepreneurship and Social Science,* vol. 4, no. 3, pp. 1698-1720, 2024.

[1592] A. G. Walker, J. W. Smither and J. DeBode, "The effects of religiosity on ethical judgements," *Journal of Business Ethics,* vol. 106, p. 437–452, 2012.

[1593] F. Maleki and S. M. Hosseini, "Charity donation intention via m-payment apps: donor-related, m-payment system-related, or charity brand-related factors, which one is overkill?," *International Review on Public and Nonprofit Marketing,* vol. 17, pp. 409-443, 2020.

[1594] F. Septianto, F. Tjiptono, W. Paramita and T. M. Chiew, "The interactive effects of religiosity and recognition in increasing donation," *European Journal of Marketing,* vol. 55, no. 1, pp. 1-26, 2020.

[1595] M. Aydemir and Ö. Eğilmez, "An important antecedent of ethical/unethical behavior: Religiosity," *Eurasian Journal of Business and Economics,* vol. 3, no. 6, pp. 71-84, 2010.

[1596] S. M. Hassan and Z. Rahman, "Curbing unethical consumer behaviour: the role of religiosity, consumer ethical beliefs and anticipated guilt," *International Journal of Ethics and Systems,* vol. 40, no. 2, p. 340–361, 2024.

[1597] R. M. M. Wong, C. K. Owuamalam and S. Stewart-Williams, "Right-leaning egalitarians are just as susceptible to social justice-induced product patronage! Evidence from the US and Malaysia," *Acta Psychologica,* vol. 237, p. 103935, 2023.

[1598] E. M. Sulastri, A. E. Putri, A. M. Nahruddien and S. Suhono, "Are intrinsic religiosity and extrinsic religiosity able to moderate money ethics on tax expenditures (Case study of individual taxpayers registered at KPP Karawang)," *Jurnal Pendidikan Tambusai,* vol. 8, no. 1, pp. 16566-16577, 2024.

[1599] G. W. Alloport and J. M. Ross, "Personal religious orientation and prejudice," *Journal of Personality and Social Psychology,* vol. 5, no. 4, pp. 432-443, 1967.

[1600] J. Maltby, "The internal structure of a derived, revised and amended measure of the religious orientation scale: The 'age-universal' I-E scale 12," *Social Behavior and Personality,* vol. 27, no. 4, pp. 407-412, 1999.

[1601] J. Maltby and L. Day, "Religious experience, religious orientation and schizotypy," *Mental Health, Religion and Culture,* vol. 5, no. 2, pp. 163-174, 2002.

[1602] B. Hunsberger, "Religion and prejudice: The role of religious fundamentalism, quest, and right-wing authoritarianism," *Journal of Social Issues,* vol. 51, no. 2, pp. 113-129, 1995.

[1603] B. Hunsberger and L. M. Jackson, "Religion, meaning, and prejudice," *Journal of Social Issues,* vol. 61, no. 4, pp. 807-826, 2005.

[1604] C. L. Beck and E. A. Plant, "The implications of right-wing authoritarianism for non-Muslims' aggression toward Muslims in the United States," *Analysis of Social Issues and Public Policy,* vol. 18, no. 1, pp. 353-377, 2018.

[1605] P. O. Jacquet, F. Pazhoohi, C. Findling, H. Mell, C. Chevallier and N. Baumard, "Predictive modeling of religiosity, prosociality, and moralizing in 295,000 individuals from European and non-European populations," *Human and Social Sciences Communications,* vol. 8, no. 9, 2021.

[1606] M. J. Donahue, "Intrinsic and extrinsic religiousness: Review and meta-analysis," *Journal of Personality and Social Psychology,* vol. 48, no. 2, pp. 400-419, 1985.

[1607] C. D. Batson, P. Schoenrade and W. I. Ventis, Religion and the Individual: A Social Psychological Perspective, Oxford: Oxford University Press, 1993, p. 440.

[1608] W. C. Rowatt and L. M. Franklin, "Christian orthodoxy, religious fundamentalism, and right-wing authoritarianism as predictors of implicit racial prejudice," *The International Journal for the Psychology of Religion,* vol. 14, no. 2, pp. 125-138, 2009.

[1609] G. M. Herek, "Religious orientation and prejudice: A comparison of racial and sexual attitudes," *Personality and Social Psychology Bulletin,* vol. 13, no. 1, pp. 34-44, 1987.

[1610] E. Leach and J. Gore, "Culture, religion, and homonegativity: a multi-level analysis," *Mental Health, Religion and Culture,* vol. 25, no. 1, pp. 85-98, 2022.

[1611] S. U. Salah, "Religious orientation and its relationship with intolerance trends in secondary school students after displacement," *Al-Anbar University Journal For Humanities,* vol. 2018, no. 5, pp. 316-335, 2022.

[1612] L. M. Hempel, T. Mathews and J. Bartkowski, "Trust in a 'fallen world': The case of Protestant theological conservatism," *Journal for the Scientific Study of Religion,* vol. 51, no. 3, pp. 522-541, 2012.

[1613] K. B. Kristensen, D. M. Pedersen and R. N. Williams, "Profiling religious maturity: The relationship of religious attitude components to religious orientations," *Journal for Scientific Study of Religion,* vol. 40, no. 1, pp. 75-86, 2001.

[1614] S. Joseph, D. Smith and D. Diduca, "Religious orientation and its association with personality, schizotypal traits and manic-depressive experiences," *Mental Health, Religion and Culture,* vol. 5, no. 1, p. 73 – 81, 2002.

[1615] F. M. Bernt, "Being religious and being altruistic: A study of college service volunteers," *Personality and Individual Differences,* vol. 10, no. 6, pp. 663-669, 1989.

[1616] D. E. Hansen, B. Vandenberg and M. L. Patterson, "The effects of religious orientation on spontaneous and nonspontaneous helping behaviors," *Personalities and Individual Differences,* vol. 19, no. 1, pp. 101-104, 1995.

[1617] J. Goldfried and M. Miner, "Quest religion and the problem of limited compassion," *Journal for the Scientific Study of Religion,* vol. 41, no. 4, pp. 685-695, 2002.

[1618] C. D. Batson, S. H. Eidelman, S. L. Higley and S. Russell, "'And who is my neighbor?' II: Quest religion as a source of universal compassion," *Journal for the Scientific Study of Religion,* vol. 40, no. 1, pp. 39-50, 2002.

[1619] E. P. Shafranske, Religion and the Clinical Practice of Psychology, Washington DC: American Psychological Association, 1996, p. 617.

[1620] T. R. Hettler and L. H. Cohen, "Intrinsic religiousness as a stress moderator for adult Protestant churchgoers," *Journal of Community Psychology,* vol. 26, no. 6, pp. 597-609, 1998.

[1621] T. B. Smith, M. E. McCullough and J. Poll, "Religiousness and depression: Evidence for a main effect and the moderating influence of stressful life events," *Psychological Bulletin,* vol. 129, no. 4, pp. 614-636, 2003.

[1622] Y. Kim and L. Seidlitz, "Spirituality moderates the effect of stress on emotional and physical adjustment," *Personality and Individual Differences,* vol. 32, no. 8, pp. 1377-1390, 2002.

[1623] N. Ghorbani, P. J. Watson, A. F. Ghramaleki, R. J. Morris and R. W. J. Hood, "Muslim-Christian religious orientation scares. Distinction, correlation and cross-cultural analysis in Iran and the United States," *International Journal for the Psychology of Religion,* vol. 12, no. 2, pp. 69-91, 2002.

[1624] C. T. Burris, C. D. Batson, M. Altstaedten and K. Stephens, "'What a Friend...': Loneliness as a motivator of intrinsic religion," *Journal for the Scientific Study of Religion,* vol. 33, no. 4, pp. 326-334, 1994.

[1625] C. T. Burris, "Curvilinearity and religious types: A second look at intrinsic, extrinsic, and quest relations," *The International Journal for the Psychology of Religion,* vol. 4, no. 2, pp. 245-260, 1994.

[1626] D. Buhagar, R. Piedmont and K. O'Grady, "The virtue of religious faith and its relationship to posttraumatic stress disorder in victims of torture: the unique outcomes of men and women," *Mental Health, Religion and Culture,* pp. 1-21, 2023.

[1627] D. F. Gontijo and D. M. R. Silva, "Religiosity/spirituality and mental health: Evidence of curvilinear relationships in a sample of religious people, spirituals, atheists, and agnostics," *Archive for the Psychology of Religion,* vol. 44, no. 2, pp. 69-90, 2022.

[1628] D. Speed and P. Edgell, "Eternally damned, yet socially conscious? The volunteerism of Canadian atheists," *Sociology of Religion,* p. srac035, 2003.

[1629] R. Stark, "Religion and conformity: reaffirming a sociology of religion," *Sociological Analysis,* vol. 45, no. 4, pp. 273-282, 1984.

[1630] E. J. Wielenberg, "Atheism and morality (pp 89-103)," in *The Oxford Handbook of Atheism*, Oxford, Oxford University Press, 2013/2016, p. 784.

[1631] S. Maitzen, "Ordinary morality implies atheism," *European Journal of Philosophy of Religion,* vol. 1, no. 2, pp. 107-126, 2009.

[1632] R. Stansfield, T. O'Connor, J. Duncan and S. Hall, "Comparing recidivism of sexual and nonsexual offenders: The role of humanist, spiritual, and religious involvement," *Sexual Abuse,* vol. 32, no. 6, pp. 634-656, 2020.

[1633] W. C. Rowatt and R. L. Al-Kire, "Dimensions of religiousness and their connection to racial, ethnic, and atheist prejudices," *Current Opinion in Psychology,* vol. 40, pp. 86-91, 2021.

[1634] W. C. Rowatt, J. LaBouff, M. Johnson, P. Froese and J.-A. Tsang, "Associations among religiousness, social attitudes, and prejudice in a national random sample of American adults," *Psychology of Religion and Spirituality,* vol. 1, no. 1, pp. 14-24, 2009.

[1635] J. Anderson and Y. Koc, "Exploring patterns of explicit and implicit anti-gay attitudes in Muslims and Atheists," *European Journal of Social Psychology,* vol. 45, no. 6, pp. 687-701, 2015.

[1636] S. Doebler, "Relationships between religion and intolerance towards Muslims and Immigrants in Europe: A multilevel analysis," *Review of Religious Research,* vol. 56, pp. 61-86, 2013.

[1637] S. Ensa and P. J. Jankowski, "Religiousness and rape myth acceptance: Risk and protective effects," *Journal of Interpersonal Violence,* vol. 35, no. 7-8, pp. 1671-1693, 2020.

[1638] M. C. Inechi, J. U. Onu, J. U. Achor and V. N. Ubochi, "Religiosity and religious coping among outpatients with schizophrenia: Association of severity of illness and psychosocial functioning," *Journal of Nigerian Academy of Medicine,* vol. 1, no. 1, pp. 14-20, 2022.

[1639] J. Onu and M. C. Inechi, "Religiosity and religious coping among out-patients with schizophrenia: Association with severity of illness and psychosocial functioning," *Biological Psychiatry,* vol. 91, no. 9, p. S293, 2022.

[1640] L. Zhan, L. Lin, X. Wang, J. Zhang and L. Zhang, "Internal but not external locus of control predicts acute stress response," *Psychophysiology,* vol. 60, no. 4, p. e14220, 2023.

[1641] E. Dingemans and E. Ingen, "Does religion breed trust? A cross-national study of the effects of religious involvement, religious faith, and religious context on social trust," *Journal for the Scientific Study of Religion,* vol. 54, no. 4, pp. 739-755, 2015.

[1642] J. Diamond, The World Until Yesterday: What Can We Learn from Traditional Societies?, New York, NY: Viking Press, 2012, p. 499.

[1643] M. Pina e Cunha, A. Rego and S. R. Clegg, "Obedience and evil: From Milgram and Kampuchea to normal organizations," *Journal of Business Ethics,* vol. 97, no. 2, pp. 291-309, 2010.

[1644] R. Lemkin, Axis Rule in Occupied Europe: Laws of Occupation, Analysis of Government, Proposals for Redress, New York: H. Fertig; Carnegie Endowment for International Peace, Division of International Law, 1944/ 1973, p. 712.

[1645] C. H. Anderton, "Choosing genocide: Economic perspectives on the disturbing rationality of race murder," *Defence and Peace Economics,* vol. 21, no. 5-6, pp. 459-486, 2010.

[1646] J. L. Maynard, Ideology and Mass Killing: The Radicalized Security Politics of Genocides and Deadly Atrocities, Oxford: Oxford University Press, 2022, p. 394.

[1647] R. A. Smith, "The scalp hunt in Chihuahua--1849," *New Mexico Historical Review,* vol. 40, no. 2, pp. 116-140, 1965.

[1648] R. A. Smith, "Scalp hunting: A Mexican experiment in warfare," *Great Plains Journal,* vol. 23, p. 41, 1984.

[1649] B. A. Mann, George Washington's War on Native America, Westport, CT: Praeger, 2005, p. 316.

[1650] C. Strobel, War and Colonization in the Early American Northeast, New York: Routledge, 2023, p. 154.

[1651] D. MacDonald, "Genocide of indigenous peoples in North America (pp 28-41)," in *Handbook of Genocide Studies*, D. J. Simon and L. Kahn, Eds., Northampton, MA: Edward Elgar, 2023, p. 320.

[1652] D. N. Paul, We Were Not the Savages: First Nations History? Collision Between European and Native American Civilizations, Halifax, NS: Fernwood, 2006, p. 368.

[1653] J. Tattrie, Cornwallis: A violent birth of Halifax, Lawrencetown, NS: Pottersfield Press, 2013, p. 238.

[1654] E. Larson, Dead Wake: The Last Crossing of the Lusitania, first ed., New York, NY: Crown, 2015, p. 430.

[1655] P. Gourevitch, We Wish to Inform You that Tomorrow We Will Be Killed with Our Families: Stories From Rwanda, New York: Farrar, Straus and Giroux, 1998.

[1656] United States House of Representatives 108th Congress, "Serial No. 108–96: Rwanda's Genocide: Looking Back. Hearing Before the Subcommittee on Africa of the Committee on International Relations," United States House of Representatives, Washington DC, 2004.

[1657] C. L. Kellow and H. L. Steeves, "The role of radio in the Rwandan genocide," *Journal of Communication,* vol. 48, no. 3, p. 107–128, 1998.

[1658] Human Rights Watch, "The Genocide: The Strategy of Ethnic Division," 19 7 2017. [Online]. Available: https://www.hrw.org/reports/1999/rwanda/Geno1-3-02.htm. [Accessed 26 6 2018].

[1659] A. L. Des Forges, Leave None to Tell the Story: Genocide in Rwanda, Human Rights Watch, 1999, p. 789.

[1660] L. Melvern, Conspiracy to Murder: The Rwandan Genocide, London: Verso, 2006, p. 380.

[1661] Human Rights Watch, "Law and Reality: Progress in Judicial Reform in Rwanda," 26 6 2018. [Online]. Available: https://www.hrw.org/report/2008/07/25/law-and-reality/progress-judicial-reform-rwanda. [Accessed 26 6 2018].

[1662] S. Straus, "How many perpetrators were there in the Rwandan genocide? An estimate," *Journal of Genocide Research,* vol. 6, no. 1, p. 85–98, 2004.

[1663] J. McCarthy, "Concentration camps run by holy men," *Tribune News,* p. 1, 24 5 2011.

[1664] E. C. Urbano and M. I. R. Riuz, "Trauma after a life of torture in Irish Magdalene Laundries (pp 54-64)," in *Trauma, Memory and Silence of the Irish Women in Contemporary Literature: Wounds of the Body and the Soul*, First ed., M. Armie and V. Membrive, Eds., New York, Routledge, 2023, p. 222.

[1665] L. Black and S. Ring, "Historical gendered institutional violence: A research agenda for criminologists," *Journal of Contemporary Criminal Justice,* vol. 39, no. 1, 2023.

[1666] I. O'Donnell and E. O'Sullivan, "'Coercive confinement': An idea whose time has come?," *Incarceration,* vol. 1, no. 1, pp. 1-20, 2020.

[1667] D. Dizdarević, R. Leskovar and G. Vukovič, "Primus inter pares effect in high schools," *Frontiers in Psychology,* vol. 11, p. 1382062, 2024.

[1668] H. Knof, M. Berndt and T. Shiozawa, "Prevalence of Dunning-Kruger effect in first semester medical students: A correlational study of self-assessment and actual academic performance," *BMC Medical Education,* vol. 24, p. 1210, 2024.

[1669] S. T. Allison, D. M. Messick and G. R. Goethals, "On being better but not smarter than others: the Muhammad Ali effect," *Social Cognition,* vol. 7, no. 3, pp. 275-95, 2011.

[1670] P. A. M. van Lange and C. Sedikides, "Being more honest but not necessarily more intelligent than others: generality and explanations for the Muhammad Ali effect," *European Journal of Social Psychology,* vol. 28, no. 4, p. 675–680, 1998.

[1671] E. Zell, J. E. Strickhouser, C. Sedikides and M. D. Alicke, "The better-than-average effect in comparative self-evaluation: A comprehensive review and meta-analysis," *Psychological Bulletin,* vol. 146, no. 2, pp. 118-149, 2020.

[1672] Centrum Badania Opinii Społecznej, "Komunikat z badań," *Centrum Badania Opinii Społecznej,* no. 19-37, 2008.

[1673] T. Kim and L. P. Jordan, "A systematic review on intergenerational transmission of genocide-related trauma," *Trauma Violence and Abuse,* vol. (Online ahead of print), 2025.

[1674] C. Musanabaganwa, S. Jansen, A. Wani, A. Rugamba, J. Mutabaruka, E. Rutembesa, A. Uwineza, S. Fatumo, E. J. Hermans, J. Souopgui, D. E. Wildman, M. Uddin, B. Roozendaal, R. Njeminiss and L. Mutesass, "Community engagement in epigenomic and neurocognitive research on post-traumatic stress disorder in Rwandans exposed to the 1994 genocide against the Tutsi: lessons learned," *Epigenomics,* vol. 14, no. 15, 2022a.

[1675] E. M. Weiss, W. Parson, H. Niederstätter, J. Marksteiner and A. Lampe, "Genetische Grundlagen der Posttraumatischen Belastungsstörung [Genetics of Posttraumatic Stress Disorder]," *Psychotherapie, Psychosomatik, medizinische Psychologie,* vol. 69, no. 7, pp. 266-274, 2019.

[1676] V. Vukejevic, I.-T. Kolassa, M. Fastenrath, L. Gschwind, K. Spalek, A. Milnik, A. Heck, C. Vogler, S. Wilker, P. Demougin, F. Peter, E. Atucha, A. Stetak, B. Roozendaal, T. Elbert, A. Papassotiropoulos and D. J.-F. de Quervain, "Epigenetic modification of glucocorticoid receptor gene is linked to traumatic memory and post-traumatic stress disorder risk in genocide survivors," *The Journal of Neuroscience,* vol. 34, no. 31, pp. 10274-10284, 2014.

[1677] D. Bhavya, "A coupled HPA-mirror neuron model of chronic maternal stress and autism susceptibility (preprint)," *TechRxiv,* 2025.

[1678] C. Musanabaganwa, A. H. Wani, J. Donglasan, S. Fatumo, S. Jansen, J. Mutabaruka, E. Rutembesa, A. Uwineza, E. J. Hermans, B. Roozendaal, D. E. Wildman and L. U. M. Mutesa, "Leukocyte methylomic imprints of exposure to the genocide against the Tutsi in Rwanda: a pilot epigenome-wide analysis," *Epigenomics,* vol. 14, no. 1, pp. 11-25, 2022.

[1679] U. M. C. C. Thivisol, P. Ho, B. Li, M. Trompke, L. B. Hoffmann, A. J. Hannan and T. Y. Pang, "Paternal early life stress exerts intergenerational effects on male C57Bl/6J offspring risk-taking behaviors and predator scent-induced c-Fos expression," *Neuronal Signaling,* vol. 7, p. NS20220097, 2023.

[1680] C. L. Dent, K. D. A. Rienecker, A. Ward, J. F. Wilkins, T. Humby and A. R. Isles, "Mice lacking paternal expression of imprinted Grb10 are risk-takers," *Genes, Brain and Behavior,* vol. 19, no. 7, p. e12679, 2020.

[1681] R. Yehuda, L. M. Bierer, J. Schmeidler, D. H. Aferiat, I. Breslau and S. Dolan, "Low cortisol and risk for PTSD in adult offspring of Holocaust survivors," *American Journal of Psychiatry,* vol. 157, no. 8, pp. 1252-1259, 2000.

[1682] L. M. Bierer, H. N. Bader, N. P. Daskalakis, A. Lehrner, N. Provencal, T. Wiechmann, T. Klengel, I. Makotkine, E. B. Binder and R. Yehuda, "Intergenerational effects of maternal Holocaust exposure on FKBP5 methylation," *American Journal of Psychiatry,* vol. 177, no. 8, pp. 744-753, 2020.

[1683] N. Perroud, E. Rutembesa, A. Paoloni-Giacobino, J. Mutabaruka, L. Mutesa, L. Stenz, A. Malofosse and F. Karege, "The Tutsi genocide and transgenerational transmission of maternal stress: epigenetics and biology of the HPA axis," *World Journal of Biological Psychiatry,* vol. 15, no. 4, pp. 334-345, 2014.

[1684] L. M. Bierer, H. N. Bader, N. P. Daskalakis, A. L. Lehrner, I. Makotkine, J. R. Seckl and R. Yehuda, "Elevation of 11β-hydroxysteroid dehydrogenase type 2 activity in Holocaust survivor offspring: Evidence for an intergenerational effect of maternal trauma exposure," *Psychoneuroendocrinology,* vol. 48, pp. 1-10, 2014.

[1685] R. Yehuda, N. P. Daskalakis, A. Lehrner, F. Desarnaud, H. N. Bader, I. Makotkine, J. D. Flory, L. M. Bierer and M. J. Meaney, "Influences of maternal and paternal PTSD on epigenetic regulation of the glucocorticoid receptor gene in Holocaust survivor offspring," *The American Journal of Psychiatry,* vol. 171, no. 8, pp. 872-880, 2014.

[1686] J. P. Curley, R. Mashoodh and F. A. Champagne, "Transgenerational Epigenetics," in *Handbook of Epigenetics*, Cambridge MA, Academic Press, 2023, pp. 465-478.

[1687] G. P. Pech, D. Gishoma and E. A. Caspar, "A novel electroencephalography-based paradigm to measure intergroup prosociality: An intergenerational study in the aftermath of the genocide against Tutsis in Rwanda," *Journal of Experimental Psychology: General,* vol. 153, no. 1, p. 241–254, 2024.

[1688] M. M. Hollander, "The repertoire of resistance: Non-compliance with directives in Milgram's 'obedience' experiments," *British Journal of Social Psychology,* vol. 54, no. 3, pp. 425-444, 2015.

[1689] E. Hoffman, N. R. Myerberg and J. G. Morawski, "Acting otherwise: Resistance, agency, and subjectivities in Milgram's studies of obedience," *Theoretical Psychology,* vol. 25, no. 5, pp. 670-689, 2015.

[1690] S. Milgram, "Liberating effects of group pressure," *Journal of Personality and Social Psychology,* vol. 1, no. 2, pp. 127-134, 1965.

[1691] G. A. Matthews and K. M. Tye, "Neural mechanisms of social homeostasis," *Annals of the New York Academy of Sciences,* 2019.

[1692] C. R. Browning, Ordinary Men: Reserve Police Battalion 101 and the Final Solution in Poland, Second ed., New York: Harper Perennial, 2017, p. 384.

[1693] E. Caspar, "Understanding individual motivations and desistance: Interviews with genocide perpetrators from Rwanda and Cambodia," *Journal of Perpetrator Research,* vol. 6, no. 2, p. 105–144, 2024.

[1694] C. Pscherer, A. Bluschke, A. Prochnow, E. Eggert, M. Mückschel and C. Beste, "Resting theta activity is associated with specific coding levels in event-related theta activity during conflict monitoring," *Human Brain Mapping,* vol. 41, no. 18, pp. 5114-5127, 2020.

[1695] D. M. Amodio, P. G. Devine and E. Harmon-Jones, "Individual differences in the regulation of intergroup bias: The role of conflict monitoring and neural signals for control," *Journal of Personality and Social Psychology,* vol. 94, no. 1, pp. 60-74, 2008.

[1696] E. A. Caspar and G. P. Pech, "Obedience to authority reduces cognitive conflict before an action," *Social Neuroscience,* vol. 19, no. 2, pp. 94-105, 2024.

[1697] N. K. Steffens, S. A. Haslam and S. D. Reicher, "Up close and personal: Evidence that shared social identity is a basis for the 'special' relationship that binds followers to leaders," *The Leadership Quarterly,* vol. 25, no. 2, pp. 296-313, 2014.

[1698] E. Caspar, "A novel experimental approach to study disobedience to authority," *Scientific Reports,* vol. 11, p. 22927, 2021.

[1699] L. Bègue, J.-L. Beauvois, D. Courbet, D. Oberlé, J. LePage and A. A. Duke, "Personality predicts obedience in a Milgram Paradigm," *Journal of Personality,* vol. 83, no. 3, pp. 299-306, 2014.

[1700] M. J. Gross, "Jewish rescue in Holland and France during the Second World War: Moral cognition and collective action," *Social Forces,* vol. 73, no. 2, pp. 463-496, 1994.

[1701] J. Shay, "Moral injury," *Psychoanalytic Psychology,* vol. 31, no. 2, pp. 182-191, 2014.

[1702] H. Von Fallersleben, "Aus Neukirch bei Schönau," in *Schlesische Volkslieder mit Melodien*, Nabu Press, 1842/ 2012, p. 278.

[1703] J. D. Woolley, L. Ma and G. Lopez-Mobilia, "Development of the use of conversational cues to assess reality status," *Journal of Cognition and Development,* vol. 12, no. 4, pp. 537-555, 2011.

[1704] N. Darling, P. Cumsille and M. L. Martinez, "Individual differences in adolescents' beliefs about the legitimacy of parental authority and their own obligation to obey: A longitudinal investigation," *Child Development,* vol. 79, no. 4, pp. 1103-1118, 2008.

[1705] P. Slurink, "Wat is menselijk? Wat is wenselijk? Evolutionaire ethiek en het ecologisch humanisme," *Krisis,* vol. 26, no. 1, 2006.

[1706] A. Modigliani and F. Rochat, "The role of interaction sequences and the timing of resistance in shaping obedience and defiance to authority," *Journal of Social Issues,* vol. 51, no. 3, pp. 107-123, 1995.

[1707] M. M. Hollander and J. Turowetz, "Normalizing trust: Participants' immediately post-hoc explanations of behaviour in Milgram's 'obedience' experiments," *British Journal of Social Psychology,* vol. 56, pp. 655-674, 2017.

[1708] S. A. Haslam and S. D. Reicher, "Commentary: A truth that does not always speak its name: How Hollander and Turowetz's finding confirm and extend the engaged followership analysis of harm-doing in the Milgram paradigm," *British Journal of Social Psychology,* vol. 57, no. 2, p. 292–300, 2018.

[1709] R. Kahn, "COVID masks as semiotic expressions of hate," *International Journal for the Semiotics of Law,* vol. 35, pp. 2391-2407, 2022.

[1710] K. Shukla, V. Phalke, P. K. Gupta, K. A. Bandawane, A. Dixit and H. Chadra, "Perceptions and factors associated with Covid appropriate behavior (CAB) obedience among general public in India: A mixed methods research," *Journal of Education and Health Promotion,* vol. 12, no. 1, p. 355, 2023.

[1711] M. L. Simner, "Newborn's response to the cry of another infant," *Developmental Psychology,* vol. 5, no. 1, pp. 136-150, 1971.

[1712] M. Paulus, T. Becher, N. Christner, M. Kammermeier, B. Gniewosz and C. Pletti, "When do children begin to care for others? The ontogenetic growth of empathic concern across the first two years of life," *Cognitive Development,* vol. 70, p. 101439, 2024.

[1713] B. Lante and J. M. Darley, "Group inhibition of bystander intervention in emergencies," *Journal of Personality and Social Psychology,* vol. 10, no. 3, pp. 215-221, 1968.

[1714] C. M. Stolle and Y. Huang, "The mere reminder of others: A cognitive modelling approach to the implicit bystander effect," *Proceedings of the Annual Meeting of the Cognitive Science Society,* vol. 46, 2024.

[1715] H. Kono, K. Takaishi, M. Onuma, M. Fukushima and R. Takeuchi, "Bystanders' willingness to assist using automated external defibrillators during cardiac arrest," *Heliyon,* vol. 10, no. 17, p. e37316, 2024.

[1716] P. Fischer, J. I. Krueger, T. Greitemeyer, C. Vogrincic, A. Kastenmüller, F. D., M. Heene, M. Wicher and M. Kainbacher, "The bystander-effect: A meta-analytic review on bystander intervention in dangerous and non-dangerous emergencies," *Psychological Bulletin,* vol. 137, no. 4, pp. 517-537, 2011.

[1717] D. K. Campbell-Meiklejohn, D. R. Bach, A. Roepstorff, R. J. Dolan and C. D. Frith, "How the opinion of others affects our valuation of objects," *Current Biology,* vol. 20, no. 13, pp. 1165-1170, 2010.

[1718] L. E. Sherman, A. A. Payton, L. M. Hernandez, P. M. Greenfield and M. Dapretto, "The power of the Like in adolescence: Effects of peer influence on neural and behavioral responses to social media," *Psychological Science,* vol. 27, no. 7, pp. 1027-1035, 2016.

[1719] J. W. Kable and P. W. Glimcher, "The neural correlates of subjective value during intertemporal choice," *Nature Neuroscience,* vol. 10, no. 12, pp. 1625-1633, 2007.

[1720] A. Shenhav, M. A. Straccia, J. D. Cohen and M. M. Botvinick, "Anterior cingulate engagement in a foraging context reflects choice difficulty, not foraging value," *Nature Neuroscience,* vol. 17, no. 9, pp. 1249-1254, 2014.

[1721] H. Wu, Y. Luo and C. Feng, "Neural signatures of social conformity: A coordinate-based activation likelihood estimation meta-analysis of functional brain imaging studies," *Neuroscience & Biobehavioral Reviews,* vol. 71, pp. 101-111, 2016.

[1722] C. R. Browning, "The span of collective efficacy: Extreme social disorganization theory to partner violence," *Journal of Marriage and Family,* vol. 64, no. 4, pp. 833-850, 2004.

[1723] W. Shakespeare, "Henry VI (Part 2)," in *The Complete Works of Shakespeare*, Garden City, Garden City Publishing, 1596-1599/1936.

[1724] M. Marrus, The Holocaust in History, New York: New American Library, 1987, p. 285.

[1725] J. M. Twenge, "Change over time in obedience: The jury's still out, but it might be decreasing," *American Psychologist,* vol. 64, no. 1, pp. 28-31, 2009.

[1726] H. Triandis, "Collectivism v. individualism: A reconceptualization of a basic concept in cross-cultural social psychology," in *Cross-Cultural Studies of Personality, Attitudes and Cognition*, G. K. B. C. Verma, Ed., London, Palgrave Macmillan, 1988, p. 60–95.

[1727] A. Käffer and J. Mahlich, "Culture and COVID-19-related mortality: a cross-sectional study of 50 countries," *Journal of Public Health Policy,* vol. 43, pp. 413-430, 2022.

[1728] C. Grothaus, "Coping strategies and social representations of bullies among bullying victims from individualistic and collectivist societies," *Culture and Psychology,* vol. 30, no. 1, pp. 115-136, 2023.

[1729] H. C. Triandis, C. McCusker and C. H. Hui, "Multimethod probes of individualism and collectivism," *Journal of Personality and Social Psychology,* vol. 59, no. 5, pp. 1006-1020, 1990.

[1730] S. Shapouri, "Of germs and culture; parasite stress as the origin of individualism-collectivism," *Evolutionary Psychological Science,* vol. 9, pp. 82-89, 2022.

[1731] H. Mynard, S. Joseph and J. Alexander, "Peer-victimisation and posttraumatic stress in adolescents," *Personality and Individual Differences,* vol. 20, no. 5, pp. 815-821, 2000.

[1732] P. C. F. Law, L. S. Too, P. Butterworth, K. Witt, N. Reavley and A. J. Milner, "A systematic review on the effect of work-related stressors on mental health of young workers," *International Archives of Occupational and Environmental Health,* vol. 93, p. 611–622, 2020.

[1733] L. Kollerová, P. Květon, K. Zábrodská and P. Janošová, "Teacher exhaustion: The effects of disruptive student behaviors, victimization by workplace bullying, and social support from colleagues," *Social Psychology of Education,* pp. 885-902, 2023.

[1734] J. L. Grayson and H. K. Alvarez, "School climate factors relating to teacher burnout: A mediator model," *Teaching and Teacher Education,* vol. 24, no. 5, pp. 1349-1363, 2008.

[1735] R. M. Klassen and M. M. Chiu, "Effects on teachers' self-efficacy and job satisfaction: Teacher gender, years of experience, and job stress," *Journal of Educational Psychology,* vol. 102, no. 3, p. 741–756, 2010.

[1736] N. A. Malik and K. Björkqvist, "Workplace bullying and occupational stress among university teachers: Mediating and moderating factors," *Europe's Journal of Psychology,* vol. 15, no. 2, p. 240–259, 2019.

[1737] S. D. McMahon, E. M. Anderman, R. A. Astor, D. L. Espelage, A. Martinez, L. A. Reddy and F. C. Worrell, "Violence Against Educators and School Personnel: Crisis During COVID. Technical Report," American Psychological Association, Washington DC, 2022.

[1738] United States Department of Education, "Teachers Threatened With Injury or Physically Attacked by Students. Condition of Education," National Center for Education Statistics, Washington, DC, 2023.

[1739] "Ecclesiastes 9:11," in *The New King James Version Bible*, 1769/2008.

[1740] P. Mudhovozi, M. Gumani, L. Maunganidze and T. Sodi, "Causal attribution: Actor-observer bias in academic achievement among students at an institution of higher learning," *South African Journal of Education,* vol. 24, no. 4, pp. 585-601, 2010.

[1741] S. Simon, M. Soilleux, C. Wyland and L. T. O'Brien, "Does he see what she sees? The gender gap in perceptions of institutional sexism," *The Journal of Social Psychology,* pp. 1-18, 2022.

[1742] A. Dunbar and M. A. Holbrook, "A color-blind Lens: public perceptions of systemic racism in the criminal justice system," *Crime, Law, and Social Change,* vol. 79, pp. 1-20, 2023.

[1743] R. Rothstein, The Color of Law: A Forgotten History of How Our Government Segregated America, New York: Liveright Publishing Corporation, 2018, p. 368.

[1744] A. Elias, "Racism as neglect and denial," *Ethnic and Racial Studies,* vol. 47, no. 3, p. 2181668, 2024.

[1745] H. Jun, "Heterosexism (pp 205–239)," in *Social Justice, Multicultural Counseling, and Practice*, Third ed., Cham, Springer, 2024, p. 481.

[1746] E. R. Carter and M. C. Murphy, "Group-based differences in perceptions of racism: What counts, to whom, and why?," *Social and Personality Psychology Compass,* vol. 9, no. 6, pp. 269-280, 2015.

[1747] S. G. Hauck, "Weiße Deutungshoheit statt Objektivität: Der ‚objektive Dritte' und die systematische Abwertung rassismuserfahrener Perspektiven," *Zeitschrift für Rechtssoziologie,* vol. 42, no. 2, pp. 153-175, 2022.

[1748] C. K. Marshburn, B. A. Reinkensmeyer and E. D. Knowles, "Dominance motivated delusions: Whites with high social dominance orientation perceive equal amounts of institutional racism between Blacks and Whites," *Group Processes and Intergroup Relations,* p. 136843022211039, 2022.

[1749] K. Greenland, K. West and C. van Laar, "Definitional boundaries of discrimination: Tools for deciding what constitutes discrimination (and what doesn't)," *Journal of Applied Social Psychology,* vol. 52, no. 10, pp. 945-964, 2022.

[1750] C. W. Leach and C. P. Teixeira, "Understanding sentiment toward 'Black Lives Matter'," *Social Issues and Policy Review,* vol. 16, no. 1, pp. 3-32, 2022.

[1751] A. N. Edgar and A. E. Johnson, The Struggle over Black Lives Matter and All Lives Matter (Rhetoric, Race, and Religion), Lanham: Lexington Books, 2018, p. 164.

[1752] C. Haney and P. G. Zimbardo, "Persistent dispositionalism in interactionist clothing: Fundamental attribution error in explaining prison abuse," *Personality and Social Psychology Bulletin,* vol. 35, no. 6, 2009.

[1753] C. Denke, M. Rotte, H.-J. Heinze and M. Schaefer, "Belief in a just world is associated with activity in insula and somatosensory cortices as a response to the perception of norm violations," *Social Neuroscience,* vol. 9, no. 5, pp. 514-521, 2014.

[1754] J. de Vries, M. Byrne and E. Kehoe, "Cognitive dissonance induction in everyday life: An fMRI study," *Social Neuroscience,* vol. 10, no. 3, pp. 268-281, 2015.

[1755] V. Wilkins and J. B. Wenger, "Belief in a just world and attitudes toward affirmative action," *Policy Studies Journal,* vol. 10, no. 6, pp. 790-812, 2014.

[1756] Z. Rubin and A. Peplau, "Belief in a just world and reactions to another's lot: A study of participants in the national draft lottery," *Journal of Social Issues,* vol. 29, no. 4, pp. 73-93, 1973.

[1757] S. Halabi, Y. Statman and J. F. Dovidio, "Attributions of responsibility and punishment for ingroup and outgroup members: The role of just world beliefs," *Group Processes and Intergroup Relations,* vol. 18, no. 1, pp. 104-105, 2015.

[1758] L. H. Keeley, War Before Civilization, New York: Oxford University Press, 1996, p. 272.

[1759] World Bank, "Mortality rate, adult, male (per 1,000 male adults)," 2022a. [Online]. Available: https://data.worldbank.org/indicator/SP.DYN.AMRT.MA. [Accessed 16 April 2023].

[1760] World Bank, "Mortality rate, adult, female (per 1,000 female adults) - Serbia, Bosnia and Herzegovina, United States, Netherlands," 2022b. [Online]. Available: https://data.worldbank.org/indicator/SP.DYN.AMRT.FE?end=2005&locations=RS-BA-US-NL&start=1989&view=chart. [Accessed 17 April 2023].

[1761] R. Sheshadri, "Pygmies in the Congo Basin and Conflict," Inventory of Conflict and Environment (ICE), American University School of Public Affairs, Washington, DC, 2005.

[1762] Minority Rights Group International, "Erasing the Board: Report of the international research mission into crimes under international law committed against the Bambuti Pygmies in the eastern Democratic Republic of Congo," Minority Rights Group International, 2004.

[1763] M. Verpoorten, "The death toll of the Rwandan genocide: A detailed analysis of Gikongoro Provence," *Population,* vol. 60, no. 4, pp. 331-367, 2005.

[1764] S. Bowles, J.-K. Choi and A. Hopfensitz, "The co-evolution of individual behaviors and social institutions," *Journal of Theoretical Biology,* vol. 223, no. 2, pp. 135-147, 2003.

[1765] R. Mace, "On becoming modern: Human social evolution is determined by demography," *Science,* vol. 324, no. 5932, pp. 1280-1281, 2009.

[1766] D. Sznycer, A. Sell and K. E. G. Williams, "Justice-making institutions and the ancestral logic of conflict," *Evolution and Human Behavior,* vol. 44, no. 3, pp. 184-194, 2023.

[1767] H. Mercier and P. Boyer, "Truth-making institutions: From divination, ordeals and oaths to judicial torture and rules of evidence," *Evolution and Behavior,* vol. 42, no. 3, pp. 259-267, 2021.

[1768] M. C. Ingabire, S. Caparos, E. Rutembesa, E. Habimana, E. Ndushabandi and I. Blanchette, "Intergenerational transmission of trauma and its association with attitudes toward reconciliation," *Trauma Psychology,* vol. 15, no. S2, p. S393–S400, 2022.

[1769] I. Borinca, J. Van Assche and Y. Koc, "How meta-humanization leads to conciliatory attitudes but not intergroup negotiation: The mediating roles of attribution of secondary emotions and blatant dehumanization," *Current Research in Ecological and Social Psychology,* vol. 7, p. 100198, 2024.

[1770] A. E. Quiñones, G. S. van Doorn, I. Pen, F. J. Weissing and M. Taborsky, "Negotiation and appeasement can be more effective drivers of sociality than kin selection," *Philosophical Transactions of the Royal Society,* vol. 371, no. 1687, p. 20150089, 2016.

[1771] S. Gharibyan, "Collective vs. family rememberances: Evidence from two Russian betrayals," *CERGE-EI Working Paper Series No 787,* 2024.

[1772] A. Acharya, M. Blackwell and M. Sen, "The political legacy of American Slavery," *The Journal of Politics,* vol. 78, no. 3, 2024.

[1773] M. Alsan and M. Wanamaker, "Tuskegee and the health of Black men," *The Quarterly Journal of Economics,* vol. 133, no. 1, pp. 407-455, 2018.

[1774] B. Der Matossian, Ed., Denial of Genocides in the Twenty-First Century, Lincoln, NB: University of Nebraska Press, 2023, p. 416.

[1775] N. Koposov, Memory Laws, Memory Wars: The Politics of the Past in Europe and Russia (New Studies in European History), Cambridge: Cambridge University Press, 2017, p. 338.

[1776] Future Ed, "FutureEd Analysis of Legislation Limiting the Teaching of Racism," 10 February 2024. [Online]. Available: https://www.quorum.us/spreadsheet/external/KBYQbCxNgAheAgQCtQZC/.

[1777] L. F. Richardson, "War moods," *Psychometrica,* vol. 13, no. 3, pp. 147-174, 1948.

[1778] P. F. Diehl, "What Richardson got right (and wrong) about arms races and wars," in *Lewis Fry Richardson: His Intellectual Legacy and Influence in the Social Sciences*, vol. 27, N. P. Gleditsch, Ed., Cham, Springer, 2020, pp. 35-43.

[1779] P. Catapano, "Freedom to Inflame: don't encourage him by paying attention. Just ignore him," *New York Times,* 8 4 2011.

[1780] L. Alvarez, "Pastor Who Burned Koran Says He Was Duped Into Holding Back," *New York Times,* 11 4 2011.

[1781] T. Erdbrink, "Iran Denounces Florida Pastor Over Koran Burning," *New York Times,* 30 April 2011.

[1782] NPR Staff and Wires, "Riots in Afghanistan follow Fla. Quran burning," 2 April 2011. [Online]. Available: https://www.npr.org/2011/04/02/135063155/violence-in-afghanistan-after-quran-burned-in-u-s.

[1783] T. Shah and R. Norland, "Protests Over Koran Burning Reach Kandahar," *New York Times,* 2 4 2011a.

[1784] A. Al-Rawi, "Burning the Quran," in *Islam on YouTube*, First ed., London, Palgrave Macmillan, 2017, pp. 59-70.

[1785] A. Saarinen, I. P. Jääskeläinen, V. Harjunen, L. Keltikangas-Järvinen, I. Jasinskaja-Lahti and N. Ravaja, "Neural basis of in-group bias and prejudices: A systematic meta-analysis," *Neuroscience and Behavioral Reviews,* vol. 131, pp. 1214-1227, 2021.

[1786] A. M. Chekroud, J. A. C. Everett, H. Bridge and M. Hewstone, "A review of neuroimaging studies of race-related prejudice: does amygdala response reflect threat?," *Frontiers in Human Neuroscience,* vol. 8, p. 179, 2014.

[1787] D. K. Grey, J. B. Purcell, K. N. Buford, M. A. Schuster, M. N. Elliott, S. T. Emery, S. Mrug and D. C. Knight, "Discrimination exposure, neural reactivity to stress, and psychological distress," *The American Journal of Psychiatry,* 2024.

[1788] M. Dawson, D. J. Terstege, N. Jamani, M. Tsutsui, D. Pavlov, R. Bugescu, J. R. Epp, G. M. Leinninger and D. Sargin, "Hypocretin/orexin neurons encode social discrimination and exhibit a sex-dependent necessity for social interaction," *Cell Reports,* vol. 42, no. 7, p. 112815, 2023.

[1789] C. K. W. de Dreu, L. L. Greer, G. A. van Kleef and M. J. J. Handgraff, "Oxytocin promotes human ethnocentrism," *Proceedings of the National Academy of Sciences,* vol. 108, no. 4, pp. 1262-1266, 2011.

[1790] T. L. Procyshyn, L.-A. L. Bédard, B. J. Crespi and J. A. Bartz, "CD38 genetic variation is associated with increased personal distress to an emotional stimulus," *Scientific Reports,* vol. 14, p. 2571, 2024.

[1791] Q. Duan, L. Fan, Y. Zhou, S. Luo and S. Han, "The oxytocinergic system and racial ingroup bias in empathic neural activity," *Neuropharmacology,* vol. 261, p. 110151, 2024.

[1792] N. Marsh, D. Scheele, J. S. Feinstein, H. Gerhardt, S. Strang, W. Maier and R. Hurlemann, "Oxytocin-enforced norm compliance reduces xenophobic outgroup rejection," *Proceedings of the National Academy of the Sciences of the USA,* vol. 114, no. 35, pp. 9314-9319, 2017.

[1793] L. Kesner, I. Fajnerová, P. Adámek, M. Buchtík, G. D., J. Hlinka, A. P. Kozelk, N. T., F. Španiel, J. Tintěra, A. Alexová, D. Greguš and J. Horáček, "Fusiform activity distinguishes between subjects with low and high xenophobic attitudes toward refugees," *Frontiers in Behavioral Neurology,* vol. 14, p. 98, 2020.

[1794] J. Wacker, M.-L. Chavanon, A. Leue and G. Stemmler, "Is running away right? The behavioral activation-behavioral inhibition model of anterior asymmetry," *Emotion,* vol. 8, no. 2, pp. 232-249, 2008.

[1795] A. M. Perkins, S. E. Kemp and P. J. Corr, "Fear and anxiety as separable emotions: an investigation of the revised reinforcement sensitivity theory of personality," *Emotion,* vol. 7, no. 2, pp. 252-261, 2007.

[1796] D. Agroskin, E. Jonas, J. Klackl and M. Prentice, "Inhibition underlies the effect of high need for closure on cultural closed-mindedness under moral salience," *Frontiers in Psychology,* vol. 7, p. 1583, 2016.

[1797] L. Lugli, A. C. Obertis and A. M. Borghi, "Hitting is male, giving is female: automatic imitation and complementarity during action observation," *Psychological Research,* vol. 81, no. 6, pp. 1180-1191, 2017.

[1798] C. A. Jutzi, R. Willardt, J. Klackl, J. Stollberg, S. Reisse and E. Jonas, "A threat-and-defense perspective on the psychological dynamics behind the Covid-19 pandemic," *International Review of Social Psychology,* vol. 36, no. 2, pp. 1-15, 2023.

[1799] R. M. Carter and S. A. Huettel, "A nexus model of the temporal-parietal junction," *Trends in Cognitive Science,* vol. 17, no. 7, pp. 328-336, 2013.

[1800] A. M. Proverbio, A. Orlandi and E. Bianchi, "Electrophysiological markers of prejudice related to sexual gender," *Neuroscience,* vol. 358, no. 1-12, pp. 1-12, 2017.

[1801] A. Pluta, J. Mazurek, J. Wojciechowski, T. Wolak, W. Soral and M. Bilewicz, "Exposure to hate speech deteriorates neurocognitive mechanisms of the ability to understand others' pain," *Scientific Reports,* vol. 13, no. 1, p. 4127, 2023.

[1802] M. Kutas and S. A. Hillyard, "Reading senseless sentences: Brain potentials reflect semantic incongruity," *Science,* vol. 207, no. 4427, pp. 203-205, 1980.

[1803] A. Brusa, A. Pesič and A. M. Proverbio, "Learning positive social information reduces racial bias as indexed by N400 response," *PLoS One,* vol. 16, no. 11, p. e0260540, 2021.

[1804] L. Wang, Q. Ma, Z. Song, Y. Shi, Y. Wang and L. Pfotenhaur, "N400 and the activation of prejudice against rural migrant workers in China," *Brain Research,* vol. 1375, pp. 103-110, 2011.

[1805] L. Chen, W. Li, X. Shi and M. Han, "Cognitive processing differences between stereotype activation and semantic activation," *Applied Neuropsychology: Adult,* vol. 31, no. 6, pp. 1463-1473, 2024.

[1806] L. Serafini and F. Pesciarelli, "The influence of gender stereotypical primes on the neural processing of words and faces," *Social Cognitive and Affective Neuroscience,* vol. 20, no. 1, p. nsaf031, 2025.

[1807] Y. Huang, K. M. Kendrick and R. Yu, "Social conflicts elicit an N400-like component," *Neuropsychologia,* vol. 65, pp. 211-220, 2014.

[1808] B. W. Allport, K. Clark and T. Pettigrew, The Nature of Prejudice: 25th Anniversary Edition, unabridged, Ed., New York: Basic Books, 1979, p. 575.

[1809] A. Brusa, G. Bordone and A. M. Proverbio, "Measuring implicit mental representations related to ethnic stereotypes with ERPs: An exploratory study," *Neuropsychologia,* vol. 155, p. 107808, 2021.

[1810] G. V. Bodenhausen, N. Schwarz, H. Bless and M. Wänke, "Effects of atypical exemplars on racial beliefs: Enlightened racism or generalized appraisals," *Journal of Experimental Social Psychology,* vol. 31, no. 1, pp. 48-63, 1995.

[1811] K. Kawakami, J. F. Dovidio, J. Moll, S. Hermsen and A. Russin, "Just say no (to stereotyping): Effects of training in the negation of stereotypic associations on stereotype activation," *Journal of Personality and Social Psychology,* vol. 78, no. 5, pp. 871-888, 2000.

[1812] J. W. J. Beenjes, M. van Oordt and T. H. A. van der Voort, "How television commentary affects children's judgments on soccer fouls," *Communications Research,* vol. 29, no. 1, 2002.

[1813] S. Ramasubramanian and M. B. Oliver, "Activating and suppressing hostile and benevolent racism: Evidence for comparative media stereotyping," *Media Psychology,* vol. 9, no. 3, pp. 623-646, 2007.

[1814] N. Hellbernd and D. Sammler, "Neural bases of social communicative intentions in speech," *Social Cognitive and Affective Neuroscience,* vol. 13, no. 6, pp. 604-615, 2018.

[1815] B. Naderer, D. Rieger and U. Schwertberger, "An online world of bias. The mediating role of cognitive biases on extremist attitudes," *Communications,* vol. 49, no. 1, pp. 51-73, 2023.

[1816] L. J. Gasnier, Director, *Reefer Madness (= Tell Your Children).* [Film]. United States: G&H Productions, 1936.

[1817] B. Sparks, Go Ask Alice, Reprint ed., New York: Simon & Schuster Books for Young Readers, 1971/2006, p. 224.

[1818] M. Rothchild, Jewish Space Lasers: The Rothschilds and 200 Years of Conspiracy Theories, Brooklyn: Melville House, 2023, p. 336.

[1819] D. Icke, Interviewee, *David Icke on 9/11 and lizards in Buckingham Palace theories.* [Interview]. 20 May 2016.

[1820] T. Winter, M. Kosnar and W. Wong, "Feds probing whether Nashville bomber believed in lizard people conspiracy," 30 December 2020. [Online]. Available: https://www.nbcnews.com/news/us-news/girlfriend-nashville-bomber-warned-police-he-was-building-explosives-2019-n1252536.

[1821] P. A. Orsagos and M. Rubinkam, "Ohio state police to protect schools after furor over Haitian immigrants in Springfield," 16 September 2024. [Online]. Available: https://apnews.com/article/trump-vance-haitian-immigrants-springfield-ohio-threats-d74b7ff56f9a45d9389d8ebee4af1652. [Accessed 17 September 2024].

[1822] M. Kinnarad, "Trump and Vance are still stoking fears of Haitian migrants, as Ohio community faces bomb threats," 13 September 2024. [Online]. Available: https://apnews.com/article/trump-vance-haitian-immigrants-ohio-threats-d8efef1d59cbc839534bf3c02f1f8574. [Accessed 17 September 2024].

[1823] J. Grèzes, S. Berthoz and R. E. Passingham, "Amygdala activation when one is the target of deceit: Did he lie to you or to someone else?," *Neuroimaging,* vol. 30, no. 2, pp. 601-608, 2006.

[1824] T. Harada, S. Itakura, F. Xu, K. Lee, S. Nakashita, D. N. Saito and N. Sadato, "Neural correlates of the judgment of lying: A functional magnetic resonance imaging study," *Neuroscience Research,* vol. 63, no. 1, pp. 24-34, 2009.

[1825] T. Brosch, D. Shiller, R. Mojdehbakhsh, J. S. Uleman and E. A. Phelps, "Neural mechanisms underlying the integration of situational information into attribution outcomes," *Social Cognitive and Affective Neuroscience,* vol. 8, no. 6, pp. 640-646, 2013.

[1826] A. K. Martin, J. Huang, A. Hunold and M. Meinzer, "Dissociable Roles Within the Social Brain for Self–Other Processing: A HD-tDCS Study," *Cerebral Cortex,* vol. 29, no. 8, p. 3642–3654, 2019.

[1827] A. Critchfield, "Is There Room for Grace in the American Dream?," *Sigma: Journal of Political and International Studies,* vol. 41, p. Article 6, 2024.

[1828] L. Ross, "The intuitive psychologist and his shortcomings: Distortions in the attribution process pp. 173–220," in *Advances in Experimental Social Psychology*, vol. 10, New York, Academic Press, 1977, p. 351.

[1829] S. E. Taylor and J. D. Brown, "Illusion and well-being: A social psychological perspective on mental health," *Psychological Bulletin,* vol. 103, no. 2, p. 193–210, 1988.

[1830] E. Steuter and D. Wills, "'The vermin have struck again': dehumanizing the enemy in post 9/11 media representations," *Media, War & Conflict,* vol. 3, no. 2, pp. 152-167, 2010.

[1831] CNN, "3 whites indicted in dragging death of a black man in Texas: Trial not likely until 1999," 6 July 1998. [Online]. Available: http://edition.cnn.com/US/9807/06/dragging.death.02/. [Accessed 7 July 2022].

[1832] G. Lopez, "Here are the members of Congress who voted against protecting gay people from hate crimes," 13 June 2016. [Online]. Available: https://www.vox.com/2016/6/12/11912076/orlando-florida-mass-shooting-gay-hate-crime-law.

[1833] R. W. Wrangham, "Chimpanzee violence is a serious topic: A response to Sussman and Marshak's critique of demonic males: Apes and the origins of human violence," *Global Nonkilling Working Papers,* vol. 1, pp. 29-47, 2010.

[1834] R. W. Sussman and J. L. Marshack, "Are humans inherently killers?," *Global Nonkilling Working Papers,* pp. 1-28, 2010.

[1835] L. T. Price, "Violence, impunity and their impact on press Freedom (pp 296-306)," in *The Routledge Companion to Freedom of Expression and Censorship*, First ed., New York, Routledge, 2023, p. 420.

[1836] J. M. Gómez, M. Verdú, A. González-Megías and M. Méndez, "The phylogenetic roots of human lethal violence," *Nature,* vol. 538, p. 233–237, 2016.

[1837] Hammurabi, L. W. King and C. F. Horne, The Code of Hammurabi, CreateSpace, 2015, p. 60.

[1838] M. Demker, A. Towns, G. Duus-Otterström and J. Sebring, "Fear and punishment in Sweden," *Punishment and Society,* vol. 10, no. 3, pp. 319-332, 2008.

[1839] J. Billingsley and E. A. R. Losin, "The neural systems of forgiveness: An evolutionary psychological perspective," *Frontiers in Psychology,* vol. 8, no. 737, p. 737, 2017.

[1840] G. B. Niebuhr, Rome. Various Traditions Concerning the Founding of the City: The Beginnings of Rome and its Earliest Tribes, vol. 1, Philadelphia, PA: Thomas Wardle, 1835, p. 475.

[1841] W. D. Hamilton, "Selfish and spiteful behaviour in an evolutionary model," *Nature,* vol. 228, no. 5277, p. 1218–1220, 1970.

[1842] G. A. van Kleef, A. C. Homan, C. Finkenauer, N. M. Blaker and M. W. Heerdink, "Prosocial norm violations fuel power affordance," *Journal of Experimental Social Psychology,* vol. 48, no. 4, pp. 937-942, 2012.

[1843] G. A. van Kleef, "When and how norm violators gain influence: Dominance, prestige, and the social dynamics of (counter) normative behavior," *Social and Personality Psychology,* vol. 17, no. 5, p. e12745, 2023.

[1844] G. A. van Kleef, F. Wanders, A. E. M. van Vianen, R. L. Dunham, X. Du and A. C. Homan, "Rebels with a cause? How norm violations shape dominance, prestige, and influence granting," *PLOS One,* vol. 18, no. 11, p. e0294019, 2023.

[1845] K. R. Foster, T. Wenselers and F. L. W. Ratnieks, "Spite: Hamilton's unproven theory," *Annales Zoologici Fennici,* vol. 38, no. 3-4, p. 229–238, 2001.

[1846] M. Koenigs and D. Tranel, "Irrational economic decision-making after ventromedial prefrontal damage: Evidence from the ultimatum game," *Journal of Neuroscience,* vol. 27, no. 4, pp. 951-956, 2007.

[1847] W. van den Bos and B. Güroğlu, "The role of the ventral medial prefrontal cortex in social decision making," *The Journal of Neuroscience,* vol. 29, no. 24, pp. 7631-7632, 2009.

[1848] M. Hoven, G. Brunner, N. S. de Boer, A. E. Goudriaan, D. Denys, R. J. van Holst, J. Luigjes and M. Lebreton, "Motivational signals disrupt metacognitive signals in the human ventromedial prefrontal cortex," *Communications Biology,* vol. 5, p. 244, 2022.

[1849] A. R. Damasio, D. Tranel and H. Damasio, "Somatic Markers and the Guidance of Behavior: Theory and Preliminary Testing (pp 217-229)," in *Frontal Lobe Function and Dysfunction*, First ed., H. S. Levin, H. M. Eisenberg and A. L. Benton, Eds., Oxford, Oxford University Press, 1991, p. 448.

[1850] L. Li, A. D. Isherwood and J. Zhou, "The role of positive emotions in education: A neuroscience perspective," *Mind, Brain, and Education,* vol. 14, no. 3, pp. 220-234, 2020.

[1851] M. Koenigs, "The role of prefrontal cortex in psychopathy," *Reviews in the Neurosciences,* vol. 23, no. 3, pp. 253-262, 2012.

[1852] V. Nejati, M. Majidinezhad and M. Nitsche, "The role of the dorsolateral and ventromedial prefrontal cortex in emotion regulation in females with major depressive disorder (MDD): A tDCS study," *Journal of Psychiatric Research,* vol. 148, pp. 149-158, 2022.

[1853] A. D. Boes, A. Bechara, D. Tranel, S. W. Anderson, L. Richman and P. Nopoulos, "Right ventromedial prefrontal cortex: a neuroanatomical correlate of impulse control in boys," *Social Cognitive and Affective Neuroscience,* vol. 4, no. 1, pp. 1-9, 2009.

[1854] J. C. Motzkin, J. P. Newman, K. A. Kiehl and M. Koenigs, "Reduced prefrontal connectivity in psychopathy," *Journal of Neuroscience,* vol. 31, no. 48, pp. 17348-17357, 2011.

[1855] G. P. Hodgkinson, J. Langan-Fox and E. Sadler-Smith, "Intuition: A fundamental bridging construct in the behavioural sciences," *British Journal of Psychology,* vol. 99, no. 1, pp. 1-27, 2010.

[1856] J. Gore and E. Sadler-Smith, "Unpacking intuition: A process and outcome framework," *Review of General Psychology,* vol. 15, no. 4, pp. 304-316, 2011.

[1857] J. Woodward and J. Allman, "Moral intuition: Its neural substrates and normative significance," *Journal of Physiology-Paris,* vol. 101, no. 4-6, pp. 179-202, 2007.

[1858] A. Bechara, D. Tranel and H. Damasio, "Characterization of the decision-making deficit of patients with ventromedial prefrontal cortex lesions," *Brain,* vol. 123, no. 11, p. 2189–2202, 2000.

[1859] A. R. Damasio, Descartes' Error: Emotion, Reason, and the Human Brain, New York: G.P. Putnam's, 2004, p. 312.

[1860] A. R. Damasio, "Toward a neurobiology of emotion and feeling: operational concepts and hypotheses," *Neuroscientist,* vol. 1, no. 1, p. 19–25, 1995.

[1861] M. Koenigs, L. Young, R. Adolphs, D. Tranel, F. Cushman, M. Hauser and A. Damasio, "Damage to the prefrontal cortex increases utilitarian moral judgements," *Nature,* vol. 446, pp. 908-911, 2007.

[1862] E. T. Rolls, "The orbitofrontal cortex and reward," *Cerebral Cortex,* vol. 10, no. 3, p. 284–294, 2000.

[1863] E. T. Rolls, The Brain and Emotion, Cary: Oxford University Press, 2018, p. 336.

[1864] E. T. Rolls, J. Hornak, D. Wade and J. McGrath, "Emotion-related learning in patients with social and emotional changes associated with frontal lobe damage," *Journal of Neurology and Neurosurgery and Psychiatry,* vol. 57, no. 12, p. 1518–1524, 1994.

[1865] S. F. Brosnan, H. C. Schiff and F. B. M. de Wall, "Tolerance for inequity may increase with social closeness in chimpanzees," *Proceedings of the Royal Society of London B,* vol. 272, no. 1560, p. 253–258, 2005.

[1866] D. Proctor, R. A. Williamson, F. B. M. de Waal and S. F. Brosnan, "Chimpanzees play the ultimatum game," *PNAS,* vol. 110, no. 6, pp. 2070-2075, 2013.

[1867] A. Loi, "Oltre l'umano: la giustizia nel mondo animale," *Journal of Ethics and Legal Technologies,* vol. 5, no. 1, pp. 75-101, 2023.

[1868] G. Wild, V. J. Flear and G. J. Thompson, "A kin-selection model of fairness in heterogeneous populations," *Journal of Theoretical Biology,* vol. 565, p. 111469, 2023.

[1869] S. F. Brosnan and F. B. M. de Waal, "Monkeys reject unequal pay," *Nature,* vol. 425, p. 297–299, 2003.

[1870] J. Henrich, R. Boyd, S. Bowles, C. Camerer, E. Fehr, H. Gintis and R. McElreath, "In search of Homo economicus: Behavioral experiments in 15 small-scale societies," *American Economic Review,* vol. 91, no. 2, pp. 73-78, 2001.

[1871] M. K. Schweinfurth and J. Call, "Capuchins (Sapajus apella) and their aversion to inequity," in *Comparative Cognition*, J. Anderson and H. Kuroshima, Eds., Singapore, Springer, 2021, p. 173–195.

[1872] K. Jensen, B. Hare, J. Call and M. Tomasello, "What's in it for me? Self-regard precludes altruism and spite in chimpanzees," *Proceedings of the Royal Society, B,* vol. 273, no. 1589, pp. 1013-1021, 2006.

[1873] J. Verspeek, E. J. C. van Leeuwen, D. W. Laméris, N. Staes and J. M. G. Stevens, "Adult bonobos show no prosociality in both prosocial choice task and group service paradigm," *Zoological Science,* vol. 10, p. e12849, 2022.

[1874] M. M. Pillutla and J. K. Murnighan, "Unfairness, anger, and spite: Emotional rejections of ultimatum offers," *Organizational Behavior and Human Decision Processes,* vol. 68, no. 3, pp. 208-224, 1996.

[1875] S. Prediger, B. Bollan and B. Herrmann, "Resource scarcity and antisocial behavior," *Journal of Public Economics,* vol. 119, pp. 1-9, 2014.

[1876] W. Chen, Q. Zhu and T. Wu, "Unfairness promotes the evolution of cooperation," *Applied Mathematics and Computation,* vol. 438, p. 127 578, 2023.

[1877] R. Bosman and F. (. Van Winden, "Emotional hazard in a power-to-take experiment," *The Economic Journal,* vol. 112, no. 476, p. 147–169, 2002.

[1878] E. F. Fabre, R. Rumiati, C. Cacciari, S. Borau, M. Causse and L. Lotto, "Investigating the impact of offer frame manipulations on responders playing the ultimatum game," *International Journal of Psychophysiology,* vol. 182, pp. 129-141, 2022.

[1879] W. Güth, "On ultimatum bargaining experiments–A personal review," *Journal of Economic Behavior and Organization,* vol. 27, no. 3, pp. 329-344, 1995.

[1880] W. Güth and R. Tietz, "Ultimatum bargaining behavior: A survey and comparison of experimental results," *Journal of Economic Psychology,* vol. 11, no. 3, pp. 417-449, 1990.

[1881] D. A. Kravitz and S. Gunto, "Decisions and perceptions of recipients in ultimatum bargaining games," *The Journal of Socio-Economics,* vol. 21, no. 1, pp. 65-84, 1992.

[1882] C. F. Camerer and R. H. Thaler, "Anomalies: Ultimatums, dictators, and manners," *Journal of Economic Perspectives,* vol. 9, no. 2, pp. 209-219, 1995.

[1883] J. Henrich, R. McElreath, A. Barr, J. Ensminger, C. Barrett, A. Bolyanatz, J. C. Cardenas, M. Gurven, E. Gwako, N. Henrich, C. Lesorogol, F. Marlowe, D. Tracer and J. Ziker, "Costly punishment across human societies," *Science,* vol. 312, no. 5781, pp. 1767-1770, 2006.

[1884] E. Pulcu and M. Haruno, "Value computations underlying human proposer behavior in the ultimatum game," *Journal of Experimental Psychology: General,* vol. 149, no. 2, pp. 125-137, 2020.

[1885] D. Kahneman, J. L. Knetsch and R. H. Thaler, "Fairness and the assumptions of economics," *The Journal of Business,* vol. 59, no. 4 part 2, pp. S285-300, 1986.

[1886] D. Procter, S. F. Brosnan and F. B. M. de Waal, "How fairly do chimpanzees play the ultimatum game?," *Communicative and Integrative Biology,* vol. 6, no. 3, p. e23819, 2013.

[1887] A. Valenzuela and J. Srivastava, "Role of information asymmetry and situational salience in reducing intergroup bias: the case of ultimatum games," *Personality and Social Psychology Bulletin,* vol. 38, no. 12, pp. 1671-1683, 2012.

[1888] L. Prétôt, Q. Taylor and K. McAuliffe, "Children cooperate more with in-group members than with out-group members in an iterated face-to-face Prisoner's Dilemma Game," *Journal of Experimental Child Psychology,* vol. 214, p. 105858, 2024.

[1889] S.-H. Chuah, R. Hoffmann, M. Jones and G. Williams, "Do cultures clash? Evidence from cross-national ultimatum game experiments," *Journal of Economic Behavior and Organization,* vol. 64, no. 1, pp. 35-48, 2007.

[1890] E. F. Fabre, M. Causse, F. Pesciarelli and C. Cacciari, "The responders' gender stereotypes modulate the strategic decision-making of proposers playing the ultimatum game," *Frontiers in Psychology,* vol. 7, no. 12, p. 12, 2016.

[1891] E. F. Fabre, M. Causse, F. Pesciarelli and C. Cacciari, "Sex and the money--How gender stereotypes modulate economic decision-making: An ERP study," *Neuropsychologia,* vol. 75, pp. 221-232, 2015.

[1892] G. Saad and T. Gill, "Sex differences in the ultimatum game: An evolutionary psychology perspective," *Journal of Bioeconomics,* vol. 3, pp. 171-193, 2001.

[1893] C. C. Eckel and P. J. Grossman, "Chivalry and solidarity in ultimatum games," *Economic Inquiry,* vol. 39, no. 2, pp. 171-188, 2007.

[1894] B. Kubera, J. Klement, C. Wagner, C. Rädel, J. Eggeling, S. Füllbrunn, M. C. Kaczmarek, R. Levinsky and A. Peters, "Differences in fairness and trust between lean and corpulent men," *International Journal of Obesity,* vol. 40, no. 11, pp. 1802-1808, 2016.

[1895] Q. Ma, D. Qian, L. Hu and L. Wang, "Hello handsome! Male's facial attractiveness gives rise to female's fairness bias in Ultimatum Game scenarios-An ERP study," *PLoS One,* vol. 12, no. 7, p. e0180459, 2017.

[1896] S. J. Solnick and M. E. Schweitzer, "The influence of physical attractiveness and gender on ultimatum game decisions," *Organizational Behavior and Human Decision Processes,* vol. 79, no. 3, pp. 199-215, 1999.

[1897] B. J. Ruffle and Z. Shtudiner, "Are good-looking people more employable?," *Management Science,* vol. 61, no. 8, p. 1741–2011, 2014.

[1898] O. Hübler, "The nonlinear link between height and wages in Germany, 1985–2004," *Economics and Human Biology,* vol. 7, no. 2, p. 191–199, 2009.

[1899] M. Koenigs, M. Kruepke and J. P. Newman, "Economic decision-making in psychopathy: a comparison with ventromedial prefrontal lesion patients," *Neuropsychologia,* vol. 48, no. 7, pp. 2198-2204, 2010.

[1900] B. C. Clark, C. B. Thorne, P. S. Hendricks, C. Sharp, S. K. Clark and K. L. Cropsey, "Individuals in the criminal justice system show differences in cooperative behaviour: Implications from cooperative games," *Criminal Behaviour and Mental Health,* vol. 25, no. 3, pp. 168-180, 2015.

[1901] L. Balafoutas, A. García-Gallego, N. Georgantizis, T. Jaber-Lopez and E. Mitrokostas, "Psychopathy and economic behavior among prison inmates: An experiment," *Frontiers in Psychology,* vol. 12, p. 732184, 2021.

[1902] A. Benozio, B. R. House and M. Tomasello, "Apes reciprocate food positively and negatively," *Proceedings of the Royal Society B.,* vol. 290, p. 20222541, 2024.

[1903] G. G. Carter, "Modeling the evolution and formation of animal friendship," *Proceedings of the National Academy of Sciences,* vol. 121, no. 15, p. e2403318121, 2024.

[1904] S. C. Engelhardt, R. B. Weladji, Ø. Holand, K. H. Røed and M. Nieminen, "Evidence suggesting that reindeer mothers allonurse according to the direct reciprocity and generalized reciprocity decision rules," *PLOS One,* vol. 18, no. 12, p. e0295497, 2023.

[1905] W. D. Koenig, "Reciprocal altruism in birds: A critical review," *Ethology and Sociobiology,* vol. 9, no. 2-4, pp. 73-84, 1988.

[1906] L. Horn, T. Bugnyar, M. Griesser, M. Hengl, E.-I. Izawa, T. Oortwijn, C. Rössler, C. Scheer, M. Schiestl, M. Suyama, A. H. Taylor, L. C. Vanhooland, A. M. P. von Bayern, Y. Zürcher and J. J. M. Massen, "Sex-specific effects of cooperative breeding and colonial nesting on prosociality in corvids," *eLife,* vol. 9, p. 235–244, 2020.

[1907] J. J. M. Massen, S. M. Haley and T. Bugnyar, "Azure-winged magpies' decisions to share food are contingent on the presence or absence of food for the recipient," *Scientific Reports,* vol. 10, no. 1, p. 16147, 2020.

[1908] G. A. van Kleef, C. K. W. de Dreu and A. S. R. Manstead, "The interpersonal effects of emotions in negotiations: A motivated information processing approach," *Journal of Personality and Social Psychology,* vol. 87, no. 4, p. 510–528, 2004.

[1909] N. Weinmann and M. Voeth, "Priming the future: Navigating ethical transformations in negotiation landscapes," in *Business Ethics - The Competitive Advantage of Trust and Reputation*, IntechOpen, 2024, p. e114812.

[1910] N. Ashraf, I. Bohnet and N. Piankov, "Is trust a bad investment? KSG Working Paper No. RWP03-047," Harvard University, Cambridge MA, 2003.

[1911] J. Berg, J. Dickhaut and K. A. McCabe, "Trust, reciprocity, and social history," *Games and Economic Behavior,* vol. 10, no. 1, pp. 290-307, 1995.

[1912] C. C. Eckel and R. K. Wilson, "Is trust a risky decision?," *Journal of Economic Behavior and Organization,* vol. 55, no. 4, pp. 447-465, 2004.

[1913] I. Bohnet and B. S. Frey, "The sound of silence in prisoner's dilemma and dictator games," *Journal of Economic Behavior and Organization,* vol. 38, p. 43–57, 2004.

[1914] I. Bohnet and R. Zeckhauser, "Trust, risk and betrayal," *Journal of Economic Behavior and Organization,* vol. 55, pp. 467-484, 2004.

[1915] E. L. Glaeser, D. I. Laibson, J. A. Scheinkman and C. L. Soutter, "Measuring trust," *Quarterly Journal of Economics,* vol. 115, no. 3, pp. 811-846, 2000.

[1916] H. Ittefaq, N. Akhtar, U. I. Siddiqui, T. Islam and A. Kuzior, "The betrayal puzzle: Unraveling the connection between inauthenticity, corporate wrongdoing and brand betrayal with avoidance and reparation," *Journal of Retailing and Consumer Services,* vol. 76, p. 103597, 2024.

[1917] M.-E. Christl, K.-C. T. Pham, A. Rosenthal and A. P. DePrince, "When institutions harm those who depend on them: A scoping review of institutional betrayal," *Trauma, Violence, and Abuse,* vol. 0, no. 0, 2024.

[1918] G. del Toro, Director, *The Shape of Water.* [Film]. United States: TSG Entertainment; Double Dare You Productions, 2017.

[1919] R. Gadagkar, "Can animals be spiteful?," *Trends in Ecology and Evolution,* vol. 8, no. 7, pp. 232-234, 1993.

[1920] K. R. Foster, F. L. W. Ratnieks and T. Wenseleers, "Spite in social insects," *Trends in Ecology and Evolution,* vol. 15, no. 11, pp. 469-470, 2000.

[1921] K. Foster, T. Wensellers and F. L. W. Ratnieks, "Spite: Hamilton's unproven theory," *Annales Zoologici Fennici,* vol. 38, no. 3/4, pp. 229-238, 2001.

[1922] F. S. Dobson, R. K. Chesser and B. Zinner, "The evolution of infanticide: genetic benefits of extreme nepotism and spite," *Ethology, Ecology, and Evolution,* vol. 12, no. 2, pp. 131-148, 2000.

[1923] G. J. FitzGerald, "Egg cannibalism by sticklebacks: spite or selfishness?," *Behavioral Ecology and Sociobiology,* vol. 30, pp. 201-206, 1992.

[1924] A. R. Brereton, "Return-benefit spite hypothesis: An explanation for sexual interference in stumptail macaques (Macaca arctoides)," *Primates,* vol. 35, p. 123–136, 1994.

[1925] T. Saijo and H. Nakamura, "The 'spite' dilemma in voluntary contribution mechanism experiments," *Journal of Conflict Resolution,* vol. 39, no. 3, pp. 535-560, 1995.

[1926] T. Dietz, E. Ostrom and P. C. Stern, "The struggle to govern the commons," *Science,* vol. 302, no. 5652, pp. 1907-1912, 2003.

[1927] P. Forber and R. Smead, "The evolution of fairness through spite," *Proceedings of the Royal Society B,* vol. 281, no. 1780, 2014.

[1928] S. Kurokawa, "Successful conditional altruistic strategies and successful conditional spiteful strategies are different," *Animal Behaviour,* vol. 209, pp. 143-153, 2024.

[1929] E. Fehr and S. Gächter, "Altruistic punishment in humans," *Nature,* vol. 415, p. 137–140, 2002.

[1930] Z. Barlas, W. E. Hockley and S. S. Obhi, "Effects of free choice and outcome valence on the sense of agency: Evidence from measures of intentional binding and feelings of control," *Experimental Brain Research,* vol. 236, pp. 129-139, 2018.

[1931] M. Rashidi, M. M. Schmitgen, M. Weisbrod, K. Schnell, R. C. Wolf and M. A. Whittington, "Right parietotemporal activity predicts sense of agency under uncertain delays of sensory outcomes," *Journal of Neurophysiology,* vol. 125, pp. 903-914, 2021.

[1932] M. Archetti and I. Scheuring, "Coexistence of cooperation and defection in public goods games," *Evolution,* vol. 65, no. 4, pp. 1140-1148, 2010.

[1933] M. Archetti, I. Scheuring, M. Hoffman, M. E. Frederickson, M. E. Pierce and D. W. Yu, "Economic game theory for mutualism and cooperation," *Ecology Letters,* vol. 14, no. 12, pp. 1300-1312, 2011.

[1934] C. Gilligan, "States with the highest rates of SNAP recipients," *U.S. News and World Report,* 16 March 2023.

[1935] B. Keith-Jennings and R. Chaudhry, "Most working-age SNAP participants work, but often in unstable jobs," Center on Budget and Policy Priorities, Washington DC, 2018.

[1936] L. Hall and C. Nchako, "A Closer Look at Who Benefits from SNAP: State-by-State Fact Sheets," Center on Budget and Policy Priorities, Washington DC, 2023.

[1937] K. Cronquist, "Characteristics of Supplemental Nutrition Assistance Program Households: Fiscal Year 2019," U.S. Department of Agriculture, Food and Nutrition Service, Office of Policy Support, Alexandria, VA, 2021.

[1938] E. Reuben and A. Riedl, "Public goods provision and sanctioning in privileged groups," *Journal of Conflict Resolution,* vol. 53, no. 3, pp. 72-93, 2008.

[1939] J. Bednar, "Is full compliance possible? Conditions for shirking with imperfect monitoring and continuous action spaces," *Journal of Theoretical Politics,* vol. 18, no. 3, pp. 347-375, 2006.

[1940] T. Clutton-Brock and G. Parker, "Punishment in animal societies," *Nature,* vol. 373, no. 6511, p. 209–216, 1995.

[1941] J. Henrich and R. Boyd, "Why people punish defectors," *Journal of Theoretical Biology,* vol. 208, no. 1, p. 79–89, 2001.

[1942] M. G. L. Mills, Kalahari Hyenas, Dordrecht: Springer, 1990, p. 304.

[1943] F. B. M. de Waal, "Intentional deception in primates," *Evolutionary Anthropology: Issues, News, and Reviews,* vol. 1, no. 3, p. 86–92, 1992.

[1944] I. Wobker, I. M. Maria Kopton and P. Kenning, "What about me? Empirical evidence of consumer envy and destructive envy behavior," *ACR North American Advances,* vol. 41, 2013.

[1945] T. David-Barrett, "Network effects of demographic transition," *Scientific Reports,* vol. 9, no. 1, p. 2361, 2019.

[1946] M. Vogelsang, K. Jensen, S. Kirschner, C. Tennie and M. Tomasello, "Preschoolers are sensitive to free riding in a public goods game," *Frontiers in Psychology,* vol. 5, no. 5, p. 729, 2014.

[1947] F. Yang, Y. J. Choi, A. Misch, X. Yang and Y. Dunham, "In defense of the commons: Young children negatively evaluate and sanction free riders," *Psychological Science,* vol. 29, no. 10, 2018.

[1948] H. Hermes, F. Hett, M. Mechtel, F. Schmidt, D. Schunk and V. Wagner, "Do children cooperate conditionally? Adapting the strategy method for first-graders," *Journal of Economic Behavior and Organization,* vol. 179, pp. 638-652, 2020.

[1949] Y. Ohtsubo and A. Yagi, "Relationship value promotes costly apology-making: Testing the valuable relationships hypothesis from the perpetrator's perspective," *Evolution and Human Behavior,* vol. 36, no. 3, pp. 232-239, 2015.

[1950] W. R. Louil, J. M. Duck, D. J. Terry, R. A. Schuller and R. N. Lalonde, "Why do citizens want to keep refugees out? Threats, fairness and hostile norms in the treatment of asylum seekers," *European Journal of Social Psychology,* vol. 37, no. 1, pp. 53-73, 2006.

[1951] K. Jensen, J. Call and M. Tomasello, "Chimpanzees are vengeful but not spiteful," *Proceedings of the National Academy of Sciences USA.,* vol. 104, no. 32, pp. 13046-13050, 2007.

[1952] J. B. Silk, "Chimps don't get mad, they get even," *Proceedings of the National Academy of Science USA,* vol. 104, no. 34, pp. 13537-13538, 2007.

[1953] N. Mendes, N. Steinbeis, N. Bueno-Guerra, J. Call and T. Singer, "Preschool children and chimpanzees incur costs to watch punishment of antisocial others," *Nature Human Behaviour,* vol. 2, no. 1, pp. 45-51, 2018.

[1954] R. White, "Elephants trample man to death after villagers killed one-year-old calf," *Newsweek,* 24 October 2022.

[1955] M. Pates, "Deer defense: State steps in when depredation is too much," *AG Week,* 8 March 2010.

[1956] E. B. Mallon and P. Schmid-Hempel, "Behavioural interactions, kin and disease susceptibility in the bumblebee Bombus terrestris," *Journal of Evolutionary Biology,* vol. 17, no. 4, pp. 829-833, 2004.

[1957] L. Rózsa, "Spite, xenophobia and collaboration between hosts and parasites," *Oikos,* vol. 91, no. 2, p. 396–400, 2000.

[1958] D. Giron, D. W. Dunn, I. C. Hardy and M. R. Strand, "Aggression by polyembryonic wasp soldiers correlates with kinship but not resource competition," *Nature,* vol. 430, no. 7000, pp. 676-679, 2004.

[1959] A. Gardner and S. A. West, "Ecology. Spite among siblings," *Science,* vol. 305, no. 5689, pp. 1413-1414, 2004.

[1960] S. A. West, A. S. Griffin and A. Gardner, "Social semantics: Altruism, cooperation, mutualism, strong reciprocity and group selection," *Journal of Evolutionary Biology,* vol. 20, no. 2, p. 415–432, 2007.

[1961] D. B. Krupp, "How to distinguish altruism from spite (and why we should bother)," *Journal of Evolutionary Biology,* vol. 26, no. 12, pp. 2746-2749, 2013.

[1962] S. A. West, I. Pen and A. S. Griffin, "Cooperation and competition between relatives," *Science,* vol. 296, no. 5565, pp. 72-75, 2002.

[1963] L. B. Bushin, B. C. Covington, B. E. Rued, M. J. Federle and M. R. Seyedsayamdost, "Discovery and biosynthesis of streptosactin, a sactipeptide with an alternative topology encoded by commensal bacteria in the human microbiome," *Journal of the American Chemical Society,* vol. 142, no. 38, pp. 16265-16275, 2020.

[1964] V. C. Thomas and L. E. Hancock, "Suicide and fratricide in bacterial biofilms," *The International Journal of Artificial Organs,* vol. 32, no. 9, 2018.

[1965] H. Wei and L. S. Håvarstein, "Fratricide is essential for efficient gene transfer between pneumococci in biofilms," *Applied and Environmental Microbiology,* vol. 78, no. 16, 2012.

[1966] C. Ingram, "The importance of juvenile cannibalism in the breeding biology of certain birds of prey," *The Auk,* vol. 76, no. 2, pp. 218-226, 1959.

[1967] R. J. O'Connor, "Brood reduction in birds: Selection for fratricide, infanticide and suicide?," *Animal Behaviour,* vol. 26, no. 1, pp. 79-96, 1978.

[1968] C. H. Stinton, "On the selective advantage of fratricide in raptors," *Evolution,* vol. 33, no. 4, pp. 1219-1225, 1979.

[1969] V. Gargett, "Black eagle experiment No. 2," *Bokmakierie,* vol. 22, pp. 32-35, 1970.

[1970] V. Gargett, "Sibling aggression in the black eagle in the Matopos, Rhodesia," *Ostrich,* vol. 49, no. 2, pp. 57-63, 1978.

[1971] B. Meyburg, "Sibling aggression and mortality among nestling eagles," *Ibis,* vol. 116, no. 2, pp. 224-228, 1974.

[1972] M. W. Colopy and T. C. J. Edwards, "Obligate and facultative brood reduction in eagles: An examination of factors that influence fratricide," *The Auk,* vol. 100, no. 3, pp. 630-635, 1983.

[1973] D. W. Dickins and R. A. Clark, "Games theory and siblicide in the kittiwake gull, Rissa tridactyla," *Journal of Theoretical Biology,* vol. 125, no. 3, pp. 301-305, 1987.

[1974] D. W. Dickins, "'Punishing' pecks and 'siblicidal' pecks in kittiwake (Rissa tridactyla) chicks," *Learning and Motivation,* vol. 73, p. 101695, 2021.

[1975] V. Morandini and M. Ferrer, "Sibling aggression and brood reduction: a review," *Ethology Ecology and Evolution,* vol. 27, no. 1, pp. 2-16, 2015.

[1976] L. G. Frank, S. E. Glickman and P. Licht, "Fatal sibling aggression, precocial development, and androgens in neonatal spotted hyenas," *Science,* vol. 252, no. 5006, pp. 702-704, 1991.

[1977] A. L. Antonevich and S. V. Naĭdenko, "[Early sibling aggression in mammals and its hormonal correlates] (abstract)," *Zhurnal Obshchei Biologii,* vol. 78, no. 2, pp. 74-80, 2017.

[1978] J. van Heerden, R. Verster, M. L. Penrith and I. Espie, "Disease and mortality in captive wild dogs (Lycaon pictus)," *Journal of the South African Veterinary Association,* vol. 67, no. 3, pp. 141-145, 1996.

[1979] H. J. Landy, L. Keith and D. Keith, "The vanishing twin," *Acta Geneticae Medicae et Gemellogiae (Roma),* vol. 31, no. 3-4, pp. 179-194, 1982.

[1980] D. Berry, P. Barzman, R. Martin, C. Borris, T. J. Wright, P. Ellis and R. Austin, Directors, *The Highlander.* [Film]. USA: Davis-Panzer Productions; Filmline International; Gaumont Television, 1992-1998.

[1981] J. Dunn, D. W. Dunn, M. R. Strand and I. C. W. Hardy, "Higher aggression towards closer relatives by soldier larvae in a polyembryonic wasp," *Biology Letters,* vol. 10, no. 5, p. 20140229, 2014.

[1982] J. L. Sanderson, J. Wang, E. Vitikainen, M. A. Cant and H. J. Nichols, "Banded mongooses avoid inbreeding when mating with members of the same natal group," *Molecular Ecology,* vol. 24, pp. 3738-3751, 2015.

[1983] F. J. Thompson, M. A. Cant, H. H. Marshall and R. A. Johnstone, "Explaining negative kin discrimination in a cooperative mammal society," *Proceedings of the National Academy of Sciences,* vol. 114, no. 20, pp. 5207-5212, 2017.

[1984] D. Mehlhausen-Hassoen, "The gender perspective on parent-to-child aggression: Gender and Gender-constellations," *Journal of Interpersonal Violence,* vol. 37, no. 11-12, pp. NP8890-NP8915, 2022.

[1985] K. Ackermann, M. Kirchner, A. Bernhard, A. Martinelli, C. Anomitri, R. Baker, S. Baumann, R. Dochnal, A. Fernandez-Rivas, K. Gonzalez-Madruga, B. Herpertz-Dahlmann, A. Hervas, L. Jansen, K. Kapornai, L. Kersten, G. Kohls and R. Limprecht, "Relational aggression in adolescents with conduct disorder: Sex differences and behavioral correlates," *Journal of Abnormal Child Psychology,* vol. 47, pp. 1625-1637, 2019.

[1986] A. D. de Haan, B. Soenens, ,. M. Deković and P. Prinzie, "Effects of childhood aggression on parenting during adolescence: the role of parental psychological need satisfaction," *Journal of Clinical Child and Adolescent Psychology,* vol. 42, no. 3, pp. 393-404, 2013.

[1987] L. Rozsa, "Spite, xenophobia, and collaboration between hosts and parasites," *Oikos,* vol. 91, no. 2, pp. 396-410, 2000.

[1988] C. G. Calloway, The Scratch of a Pen: 1763 and the Transformation of North America (Pivotal Moments in American History, Reprint ed., New York: Oxford University Press, 2007, p. 240.

[1989] E. A. Fenn, "Biological warfare in eighteenth-century North America: Beyond Jeffery Amherst," *The Journal of American History,* vol. 86, no. 4, pp. 1552-1580, 2000.

[1990] F. Frischknecht, "The history of biological warfare: Human experimentation, modern nightmares and lone madmen in the twentieth century," *EMBO Reports,* vol. 4, pp. S37-S52, 2003.

[1991] Barenblatt, A Plague upon Humanity: The Secret Genocide of Axis Japan's Germ Warfare Operation, New York: Harper, 2004, p. 288.

[1992] J. A. Lockwood, Six-Legged Soldiers: Using Insects as Weapons of War, Oxford: Oxford University Press, 2009, p. 377.

[1993] M. Jog and M. Watve, "Role of parasites and commensals in shaping host behaviour," *Current Science,* vol. 89, no. 7, pp. 1184-1191, 2005.

[1994] F. Dionisio, "Selfish and spiteful behaviour through parasites and pathogens," *Evolutionary Ecology Research,* vol. 9, no. 7, p. 1199–1210, 2007.

[1995] A. S. Griffin, S. A. West and A. Buckling, "Cooperation and competition in pathogenic bacteria," *Nature,* vol. 430, no. 7003, pp. 1024-1027, 2004.

[1996] S. Mridha and R. Kümmerli, "Enforced specialization fosters mutual cheating and not division of labour in the bacterium Pseudomonas aeruginosa," *Journal of Evolutionary Biology,* vol. 35, no. 5, pp. 719-730, 2022.

[1997] C. Achebe, Things Fall Apart, New York: Everyman's Library, 1958/ 1994, p. 181.

[1998] M. S. Boguski, "Comparative genomics: The mouse that roared," *Nature,* vol. 420, no. 6915, pp. 515-516, 2002.

[1999] Z. Mirzoyan, M. Sollazzo, M. Allocca, A. M. Valenza, D. Grifoni and P. Bellosta, "Drosophila melanogaster. A model organism to study cancer," *Frontiers in Genetics,* vol. 10, 2019.

[2000] W. Lui, L. Li, H. Chen, W. Shen, Y. Zhong, T. Tian and H. He, "From Saccharomyces cerevisiae to human: The important gene co-expression modules," *Biomedical Reports,* vol. 7, no. 2, pp. 153-158, 2017.

[2001] S. J. Wheelan, M. S. Boguski, L. Duret and W. G. Makalowski, "Human and nematode orthologs—Lessons from the analysis of 1800 human genes and the proteome of Caenorhabditis elegans," *Gene,* vol. 238, no. 1, pp. 163-170, 1999.

[2002] W. Makalowski and M. S. Boguski, "Evolutionary parameters of the transcribed mammalian genome: An analysis of 2,820 orthologous rodent and human sequences," *Proceedings of the National Academy of Science USA,* vol. 95, no. 16, pp. 9407-9412, 1998.

[2003] R. L. Cann, M. Stoneking and A. C. Wilson, "Mitochondrial DNA and human evolution," *Nature,* vol. 325, no. 6099, pp. 31-36, 1987.

[2004] J. Chang, "Recent common ancestors of all present-day individuals," *Advances in Applied Probability,* vol. 31, no. 4, pp. 1002-1026, 1999.

[2005] F. Mendez, T. Krahn, B. Schrack, A.-M. Krahn, K. Veeramah, A. Woerner, F. Fomine, L. M., N. Bradman and M. Thomas, "An African American paternal lineage adds an extremely ancient root to the human Y chromosome phylogenetic tree," *American Journal of Human Genetics,* vol. 92, no. 3, p. 454–459, 2013.

[2006] G. D. Poznik, B. M. Henn, M.-C. Yee, E. Sliwerska, G. M. Euskirchen, A. A. Lin, M. Snyder, L. Quintana-Murci, J. M. Kidd, P. A. Underhill and C. D. Bustamante, "Sequencing Y chromosomes resolves discrepancy in time to common ancestor of males versus females," *Science,* vol. 341, no. 6145, p. 562–565, 2013.

[2007] L. Vigilant, M. Stoneking, H. Harpending, K. Hawkes and A. C. Wilson, "African populations and the evolution of human mitochondrial DNA," *Science,* vol. 253, no. 5027, pp. 1503-1507, 1991.

[2008] V. M. Cabrera, "Following the evolution of Homo Sapiens across Africa using a uniparental genetic guide," *Medical Research Archives,* vol. 11, no. 1, pp. 1-20, 2023.

[2009] K. Skorecki, S. Selig, S. Blazer, R. Bradman, N. Bradman, P. J. Waburton, M. Ismajlowicz and M. F. Hammer, "Y chromosomes of Jewish priests," *Nature,* vol. 385, no. 6611, p. 32, 1997.

[2010] D. L. T. Rohde, S. Olson and J. T. Chang, "Modelling the recent common ancestor of all living humans," *Nature,* vol. 431, pp. 562-566, 2004.

[2011] C. L. Raak-van den Berg, H. J. De Lange and J. C. Van Lenteren, "Interguild predation behaviour of ladybirds in semi-field experiments explains invasion success of Harmonia axyridis," *PLOS One,* vol. 7, no. 7, p. e40681, 2012.

[2012] A. Rasekh and N. Osawa, "Direct and indirect effect of cannibalism and intraguild predation in the two sibling Harmonia ladybird beetles," *Ecology and Evolution,* vol. 10, no. 12, pp. 5899-5912, 2020.

[2013] W. A. Foster and J. E. Treherne, "Evidence for the dilution effect in the selfish herd from fish predation on a marine insect," *Nature,* vol. 293, pp. 466 - 467, 1981.

[2014] R. S. Ezekiel, The Racist Mind, New York: Penguin, 1996, p. 330.

[2015] R. Cintron, "Listening to what the streets say: Vengeance as ideology?," *The Annals of the American Academy of Political and Social Science,* vol. 567, no. 1, pp. 42-53, 2000.

[2016] R. Jasnow, A Late Period Hieratic Wisdom Text: (P. Brooklyn 47.218.135), Chicago: University of Chicago Press, 2007, p. 217.

[2017] D. Hinton and D. Bullen, Analects: Confucius, Berkeley, CA: Counterpoint, 2014, p. 177.

[2018] R. M. Perloff, "The press and lynchings of African Americans," *Journal of Black Studies,* vol. 30, no. 3, pp. 315-330, 2000.

[2019] A. F. Raper, The Tragedy of Lynching, Mineola, NY: Dover Publications, 1933/2003, p. 507.

[2020] A. Weiss, J. T. Feldblu, D. M. Altschul, D. A. Collins, S. Kamenya, D. Mjungu, S. Foerster, I. C. Gilby, M. L. Wilson and A. E. Pusey, "Personality traits, rank attainment, and siring success throughout the lives of male chimpanzees of Gombe National Park," *PeerJ,* vol. 11, p. e15083, 2023.

[2021] E. O. Wilson, The Social Conquest of Earth, New York: Liveright, 2012, p. 330.

[2022] S. E. Seymour, P. A. Reuter-Lorenz and M. S. Gazzaniga, "The disconnection syndrome: Basic findings reaffirmed," *Brain,* vol. 117, no. 1, pp. 105-115, 1994.

[2023] A. T. Hale, A. S. Barkley and J. P. Blount, "Corpus callostomy is a safe and effective procedure for medically resistant epilepsy," in *Advances and Technical Standards in Neurosurgery*, vol. 48, Singapore, Springer, 2023, pp. 355-369.

[2024] E. Malloy and J.-F. Delvenne, "Investigating the effects of ageing on interhemispheric communication using visual short-term memory binding (preprint)," p. 43, 2024.

[2025] J.-F. Delvenne and E. Malloy, "Functional implications of age-related atrophy of the corpus callosum," *Neuroscience and Biobehavioral Reviews,* vol. 169, p. 105982, 2025.

[2026] M. S. Gazzaniga, "The split brain revisited," *Scientific American,* vol. 279, no. 1, pp. 35-39, 1998.

[2027] O. Sacks, The Man Who Mistook His Wife For a Hat and Other Clinical Tales, New York: Summit Books, 1985, p. 233 .

[2028] S. Lux, M. Kurthen, C. Helmstaedter, W. Hartje, M. Reuber and C. E. Elger, "The localizing value of ictal consciousness and its constituent functions: A video-EEG study in patients with focal epilepsy," *Brain,* vol. 125, no. 12, pp. 2691-2698, 2002.

[2029] L. I. Benowitz, S. Finkelstein, D. N. Levine and K. Moya, "The role of the right cerebral hemisphere in evaluating configurations (pp 320-333)," in *Brain Circuits and Functions of the Mind: Essays in Honor of Roger W. Sperry* , C. Trevarthen, Ed., Cambridge University Press, 1990, p. 410.

[2030] R. McKay, R. Langdon and M. Coltheart, "'Sleights of mind': Delusions and self-deception; Macquarie Monographs in Cognitive Science (pp 165-196)," in *Delusion and Self-Deception: Affective and Motivational Influences on Belief Formation*, Abingdon-on-Thames, Taylor Francis, 2015, p. 310.

[2031] S. Schachter and J. E. Singer, "Cognitive, social, and physiological determinants of emotional state," *Psychological Review,* vol. 69, no. 5, pp. 379-399, 1962.

[2032] E. Larson, In the Garden of Beasts: Love, Terror, and an American Family in Hitler's Berlin, New York, NY: Crown, 2011, p. 464.

[2033] D. Hecht, "An inter-hemispheric imbalance in the psychopath's brain," *Personality and Individual Differences,* vol. 51, no. 1, pp. 3-10, 2011.

[2034] J. Pujol, B. J. Harrison, O. Contreras-Rodriguez and N. Cardoner, "The contribution of brain imaging to the understanding of psychopathy," *Psychological Medicine,* vol. 49, no. 1, pp. 20-31, 2019.

[2035] M. Sommer, G. Hajak, K. Döhnel, J. Schwerdtner, J. Meinhardt and J. L. Müller, "Integration of emotion and cognition in patients with psychopathy," *Progress in Brain Research,* vol. 156, pp. 457-466, 2006.

[2036] M. J. Hofmann, A. Mokros and S. Schneider, "The joy of being frightened: Fear experience in psychopathy," *Journal of Personality,* vol. 92, no. 2, pp. 321-341, 2024.

[2037] A. Voulgaris, S. V. Biedermann, S. Bründl, L. Roth, C. Wiessner and J. Fuss, "The impact of psychopathic traits on anxiety-related behaviors in a mixed reality environment," *Scientific Reports,* vol. 14, p. 11832, 2024.

[2038] E. R. Kimonis, P. J. Frick, H. Fazekas and B. R. Loney, "Psychopathy, aggression, and the processing of emotional stimuli in non-referred girls and boys," *Behavioral Sciences and the Law,* vol. 24, no. 1, pp. 21-37, 2006.

[2039] A. Y. Kim and Y. Y. Kim, "Reduced late positive potentials to distress in individuals with high psychopathic traits during pain judgment tasks," *Biological Psychiatry,* vol. 190, p. 108810, 2024.

[2040] P. Sparks, D. C. Jessop, J. Chapman and K. Holmes, "Pro-environmental actions, climate change, and defensiveness: Do self-affirmations make a difference to people's motives and beliefs about making a difference?," *British Journal of Social Psychology,* vol. 49, no. 3, pp. 553-568, 2010.

[2041] M. Ly, J. C. Motzkin, C. L. Philippi, G. R. Kirk, J. P. Newman, K. A. Kiehl and M. Koenigs, "Cortical thinning in psychopathy," *American Journal of Psychiatry,* vol. 169, no. 7, pp. 743-749, 2012.

[2042] J. Decety and P. L. Jackson, "The functional architecture of human empathy," *Behavioral and Cognitive Neuroscience Reviews,* vol. 3, no. 2, pp. 71-100, 2004.

[2043] P. J. Gianaros, S. W. G. Derbyshire, J. C. May, G. J. Siegle, M. A. Gamalo and J. R. Jennings, "Anterior cingulate activity correlates with blood pressure during stress," *Psychophysiology,* vol. 42, no. 6, p. 627–635, 2005.

[2044] A. G. Sanfey, J. K. Rilling, J. A. Aronson, L. E. Nystrom and J. D. Cohen, "The neural basis of economic decision-making in the ultimatum game," *Science,* vol. 300, no. 5626, p. 1755–1758, 2003.

[2045] T. Singer, "The neuronal basis of empathy and fairness," in *Empathy and Fairness: Novartis Foundation Symposium 278*, vol. 278, G. Bock and J. Goode, Eds., Novartis Foundation, 2006, pp. 20-40.

[2046] Q. Li, X. Lai, T. Li, K. H. Madsen, J. Xiao, K. Hu, C. Feng, D. Fu and X. Liu, "Brain responses to self- and other- unfairness under resource distribution context: Meta-analysis of fMRI studies," *Neuroimaging,* vol. 297, p. 120707, 2024.

[2047] L. Luo, H. Xu, X. Tian, Y. Zhao, R. Xiong, H. Dong, X. Li, Y. Wang, Y.-J. Lou and C. Feng, "The neurocomputational mechanism underlying decision-making on unfairness to self and others," *Neuroscience Bulletin,* 2024.

[2048] M. R. Privitera and J. Arnetz, "Effect of workplace violence on staff, institutions, and quality of patient care (pp 29-43)," in *Workplace Violence in Mental and General Health Settings*, Burlington, MA: Jones & Bartlett Learning, 2010, p. 330.

[2049] L. Bernstein and S. Sondheim, Composers, *The Jets Song.* [Sound Recording]. Amberson Holdings LLC and Stephen Sondheim. Leonard Bernstein Music Publishing Company LLC, Publisher. 1956, 1957.

[2050] C. H. W. Johns, Oldest Code of Laws in the World: The Code of Laws Promulgated by Hammurabi, King of Babylon B.C. 2285-2242, Project Gutenberg Literary Archive Foundation, 1903.

[2051] United States Government Printing Office, "18 U.S. Code § 1385 - Use of Army and Air Force as Posse Comitatus," 15 6 2015. [Online]. Available: https://www.law.cornell.edu/uscode/text/18/1385.

[2052] M. Buber, I and Thou, Mansfield Centre, CT: Martino Publishing, 1923/2010, p. 123.

[2053] B. Ganguly and R. Hansell, "The dynamics of peace," *Journal of Environmental Peace,* vol. 1, no. 1, 2002.

[2054] J. Galtung, "Cultural violence," *Journal of Peace Research,* vol. 27, no. 3, pp. 291-305, 1990.

[2055] S. E. Asch, "Studies of independence and conformity: a minority of one against a unanimous majority," *Psychological Monographs, no. 416,* vol. 70, no. 9, pp. 1-70, 1956.

[2056] R. A. Crutchfield, "Conformity and character," *American Psychologist,* vol. 10, no. 5, pp. 191-198, 1955.

[2057] D. Krech, R. A. Crutchfield and E. I. Ballachey, Individual in Society, New York: McGraw-Hill, 1962, p. 564.

[2058] T. Väyrynen and S. Laari-Salmela, "Men, mammals, or machines? Dehumanization embedded in organizational practices," *Journal of Business Ethics,* vol. 147, no. 1, p. 95–113, 2018.

[2059] N. Cartwright, "Setting up and Sustaining Peer Support Systems in a Range of Schools over 20 Years," *Pastoral Care in Education,* vol. 23, no. 2, pp. 45-50, 2005.

[2060] Ontario Ministry of Education, Effective Educational Practices for Students with Autism Spectrum Disorders: A Resource Guide, Ottawa: Queen's Printer for Ontario, 2007.

[2061] B. Saggers, M. Campbell and G. Mann, "Supporting a sense of belonging and peer relationships (pp 302-214)," in *Inclusive Education for the 21st Century*, Second ed., L. J. Grahm, Ed., London, Routledge, 2023, p. 418.

[2062] C. Wang and T. S. Goldberg, "Using children's literature to decrease moral disengagement and victimization among elementary school students," *Psychology in the Schools,* vol. 54, no. 9, pp. 918-931, 2017.

[2063] V. Sánchez-Jiménez, M. L. Rodríguez-de Arriba, J. Ortega-Rivera and N. Muñoz-Fernández, "Can Virtual Reality be Used for the Prevention of Peer Sexual Harassment in Adolescence? First Evaluation of the Virtual-PRO Program," *Psychosocial Intervention,* vol. 33, no. 1, pp. 29-42, 2024.

[2064] D. Falla, R. Ortega-Ruiz and E. M. Romera, "Minimizing responsibility in the aggressive dynamics of bullying and its impact on other strategies of moral disengagement: a longitudinal study," *Current Psychology,* vol. 42, p. 32512–32523, 2023.

[2065] T. Engels, I. J. Traast, B. Doosje, D. M. Amodio and D. Sauter, "Moral elevation mitigates dehumanization of ethnic outgroups," *Current Research in Ecological and Social Psychology,* vol. 6, p. 100187, 2024.

[2066] "Online Etymological Dictionary," 18 7 2018. [Online]. Available: https://www.etymonline.com/.

[2067] R. N. Felig, E. P. Cortney, K. M. Ligman, K. J. Lee and J. L. Goldenberg, "Objects do not suffer: An impact of mechanistic dehumanization on perceptions of women's suffering and lack of justice in domestic assault," *Journal of Interpersonal Violence,* vol. 39, no. 5-6, pp. 1245-1267, 2024.

[2068] J. C. Deska, J. Kunstman, E. P. Lloyd, S. M. Almaraz, M. J. Bernstein, J. P. Gonzales and K. Hugenberg, "Race-based biases in judgments of social pain," *Journal of Experimental Social Psychology,* vol. 88, p. 103964, 2020.

[2069] K. M. Hoffman, S. Trawalter, J. R. Azt and M. N. Oliver, "Racial bias in pain assessment and treatment recommendations, and false beliefs about biological differences between blacks and whites," *Proceedings of the National Academy of Sciences,* vol. 113, no. 16, pp. 4296-4301, 2016.

[2070] J.-P. Leyens, P. M. Paladino, R. Rodriguez-Torres, J. Vaes, S. Demoulin, A. Rodriguez-Perez and R. Gaunt, "The emotional side of prejudice: The attribution of secondary emotions to ingroups and outgroups," *Personality and Social Psychology Review,* vol. 4, no. 2, pp. 186-197, 2000.

[2071] H. G. Kelman, "Violence without moral restraint: Reflections on the dehumanization of victims and victimizers," *Journal of Social Issues,* vol. 29, no. 4, pp. 21-61, 1973.

[2072] N. S. Kteily and A. P. Landry, "Dehumanization: trends, insights, and challenges," *Trends in Cognitive Sciences,* vol. 26, no. 3, pp. 222-240, 2022.

[2073] D. Schulman-Green, "Coping mechanisms of physicians who routinely work with dying patients," *Omega,* vol. 47, no. 3, pp. 253-264, 2003.

[2074] R. Wollast, E. Puvia, P. Bernard, P. Tevichapong and O. Klein, "How sexual objectification generates dehumanization in Western and Eastern cultures: A comparison between Belgium and Thailand," *Swiss Journal of Psychology,* vol. 77, no. 2, pp. 69-82, 2018.

[2075] A. Gupta and A. Ailawadi, "Childhood and adolescent sexual abuse and incest: experiences of women survivors in India (pp 171-185)," in *Sex Without Consent: Young People in Developing Countries*, S. J. Jejeebhoy, I. Shah and S. Thapa, Eds., New York, Zed Books, 2005, p. 384.

[2076] D. J. Wilkinson, L. S. Bearup and T. Soprach, "Youth gang rape in Phnom Penh (pp 158-168)," in *Sex Without Consent: Young People in Developing Countries*, S. J. Jejeebhoy, I. Shah and S. Thapa, Eds., New York, Zed Books, 2005.

[2077] A. P. Gieseke, "'The New Weapon of Choice': Law's Current Inability to Properly Address Deepfake Pornography," *Vanderbilt Law Review,* vol. 73, no. 1479, p. 1479, 2020.

[2078] S. Maddocks, "'A deep fake porn plot intended to silence me': Exploring continuities between pornographic and 'political' deep fakes," *Porn Studies,* vol. 7, no. 4, pp. 415-423, 2020.

[2079] R. Cover, "Deepfake culture: the emergence of audio-video deception as an object of social anxiety and regulation," *Continuum Journal of Media and Cultural Studies,* vol. 36, no. 4, pp. 609-621, 2022.

[2080] M. Gámez-Guadix, E. Mateos-Pérez, S. Wachs, M. Wright, J. Martínez and D. Íncera, "Assessing image-based sexual abuse: Measurement, prevalence, and temporal stability of sextortion and nonconsensual sexting ("revenge porn") among adolescents," *Journal of Adolescence,* vol. 94, no. 5, pp. 789-799, 2022.

[2081] W. A. Walsh and D. Tener, "'If you don't send me five other pictures I am going to post the photo online': A qualitative analysis of experiences of survivors of sextortion," *Journal of Child Sexual Abuse,* vol. 31, no. 4, pp. 447-465, 2022.

[2082] M. C. Nussbaum, Sex and Social Justice, First ed., Oxford: Oxford University Press, 2000, p. 488.

[2083] L. M. Warad, "Media and sexualization: State of empirical research, 1995-2015," *Journal of Sex Research,* vol. 53, no. 4-5, pp. 560-577, 2016.

[2084] B. L. Fredrickson and T.-A. Roberts, "Objectification theory: Toward understanding women's lived experiences and mental health risks," *Psychology of Women,* vol. 21, no. 2, pp. 173-206, 1997.

[2085] R. M. Calogero and J. K. Thompson, "Potential implications of the objectification of women's bodies for women's sexual satisfaction," *Body Image,* vol. 6, no. 2, pp. 145-148, 2009.

[2086] R. M. Calogero, "A test of objectification theory: The effect of the male gaze on appearance concerns in college women," *Psychology of Women Quarterly,* vol. 28, no. 1, pp. 16-21, 2004.

[2087] C. McGettrick, K. O'Donnell, M. O'Rourke, J. M. Smith and M. Steed, Ireland and the Magdalene Laundries: A Campaign for Justice, London: I. B. Tauris, 2021, p. 304.

[2088] J. M. Smith, Ireland's Magdalen Laundries and the Nation's Architecture of Containment, First ed., Notre Dame, Indiana: University of Notre Dame Press, 2007, p. 312.

[2089] P. Hosford, "'The nun made us put our hands on electrical panels to get an electric shock': Abandoned, abused and abducted-- this is one woman's Magdalene story," *The Journal,* 9 June 2018.

[2090] D. J. Angelone, N. Cantor, T. Marcantonio and M. Joppa, "Does Sexism mediate the gender and reape myth acceptance relationship?," *Violence Against Women,* vol. 27, no. 6-7, pp. 748-765, 2020.

[2091] J. M. Hockett, D. A. Saucier, B. H. Hoffman, S. J. Smith and A. W. Craig, "Oppression through acceptance?: Predicting rape myth acceptance and attitudes toward rape victims," *Violence Against Women,* vol. 15, no. 8, pp. 877-897, 2009.

[2092] T. G. Viki, P. Chiroro and D. Abrams, "Hostile sexism, type of rape, and self-reported rape proclivity within a sample of Zimbabwean males," *Violence against women,* vol. 12, no. 8, pp. 789-800, 2006.

[2093] J. P. Sheldon and S. L. Parent, "Clergy's attitudes and attributions of blame toward female rape victims," *Violence Against Women,* vol. 8, no. 2, pp. 233-256, 2002.

[2094] R. Condry, Families Shamed: The Consequences of Crime for Relatives of Serious Offenders, Milton Mark, Oxfordshire: Routledge, 2007, p. 232.

[2095] D. Jackson and J. Mannix, "Giving voice to the burden of blame: A feminist study of mothers' experiences of mother blaming," *International Journal of Nursing Practice,* vol. 10, no. 4, pp. 150-158, 2004.

[2096] D. P. Sommerfeld, "The origins of mother blaming: Historical perspectives on childhood and motherhood," *Infant Mental Health Journal,* vol. 10, no. 1, pp. 14-24, 1989.

[2097] P. Caplan and I. Hall-McCorquodale, "Mother-blaming in major clinical journals," *American Journal of Orthopsychiatry,* vol. 55, no. 3, pp. 345-353, 1985.

[2098] K. Gueta and R. Condry, "Centering mothers in feminist criminology: A critical review of literature on mothers of victims and offenders," *Feminist Criminology,* vol. 19, no. 5, p. 15570851241241069, 2024.

[2099] V. Phares, "Where's poppa? The relative lack of attention to the role of fathers in child and adolescent psychopathology," *American Psychologist,* vol. 47, no. 5, pp. 656-664, 1992.

[2100] B. A. Farber and S. T. Azar, "Blaming the helpers: The marginalization of teachers and parents of the urban poor," *American Journal of Orthopsychiatry,* vol. 69, no. 4, pp. 515-528, 2010.

[2101] National Center for Educational Statistics, "Characteristics of Public School Teachers (Annual Report)," National Center for Educational Statistics, Washington DC, 2021.

[2102] National Center for Education Statistics, "Digest of Education Statistics," National Center for Education Statistics, Washington, D. C., 2019.

[2103] C. Goldin, "Marriage Bars: Discrimination Against Married Women Workers, 1920's to 1950's," National Bureau of Economic Research, Washington DC, 1988.

[2104] C. Goldin, "The bridge group (pp 63-84)," in *Career and Family: Women's Century-Long Journey toward Equity*, First ed., Princeton, Princeton University Press, 2021, p. 344.

[2105] M. V. Harris-Perry, Sister Citizen: Shame, Stereotypes, and Black Women in America, Reprint ed., New Haven, CT: Yale University Press, 2013, p. 400.

[2106] T. G. Townsend, T. B. Neilands, A. J. Thomas and T. R. Jackson, "I'm no Jezebel; I am young, gifted, and Black: Identity, sexuality, and Black Girls," *Psychology of Women,* vol. 34, no. 3, pp. 273-285, 2010.

[2107] D. Merskin, "Three faces of Eva: Perpetuation of the Hot-Latina stereotype in Desperate Housewives," *Howard Journal of Communications,* vol. 18, no. 2, pp. 133-151, 2007.

[2108] V. Rojas, "The gender of Latinidad: Latinas speak about Hispanic television," *The Communication Review,* vol. 7, no. 2, pp. 125-153, 2004.

[2109] B. McLaughlin, N. S. Rodriguez, J. A. Dunn and J. Martinez, "Stereotyped identification: How identifying with fictional Latina characters increases acceptance and stereotyping," *Mass Communication and Society,* vol. 21, no. 5, pp. 585-605, 2018.

[2110] S. Prasso, The Asian Mystique: Dragon Ladies, Geisha Girls, and Our Fantasies of the Exotic Orient, New York: Public Affairs, 2005, p. 464.

[2111] S. Mukkamala and K. L. Suyemoto, "Racialized sexism/sexualized racism: A multimethod study of intersectional experiences of discrimination for Asian American women," *Asian American Journal of Psychology,* vol. 9, no. 1, pp. 32-46, 2018.

[2112] C. P. Shimizu, The Hypersexuality of Race: Performing Asian/American Women on Screen and Scene, Durham: Duke University Press, 2007, p. 352.

[2113] R. Endo, "Asian/American women scholars, gendered orientalism, and racialized violence: Before, during, and after the 2021 Atlanta Massacre," *Cultural Studies, Critical Methodologies,* vol. 21, no. 4, pp. 344-350, 2021.

[2114] D. C. Hine, "Rape and the inner lives of Black women in the Middle West," *Signs: Journal of Women in Culture and Society,* vol. 14, no. 4, 1989.

[2115] D. Apel, Imagery of Lynching: Black Men, White Women, and the Mob, New Brunswick, New Jersey: Rutgers University Press, 2004, p. 304.

[2116] H. V. Carby, "'On the threshold of women's era': Lynching, empire and sexuality in Black feminist theory," *Critical Inquiry,* vol. 12, no. 1, pp. 262-277, 1985.

[2117] T. N. Page, "The lynching of negros: Its cause and its prevention," *The North American Review,* vol. 178, no. 566, pp. 33-48, 1904.

[2118] A. Anzani, L. Lindley, G. Tognasso, M. P. Galupo and A. Prunas, "'Being talked to like I was a sex toy, like being transgender was simply for the enjoyment of someone else': Fetishization and sexualization of transgender and nonbinary individuals," *Archives of Sexual Behavior,* vol. 50, pp. 897-911, 2021.

[2119] L. B. Watson, R. S. Craney, S. K. Greenwalt, M. Beaumont, C. Whitney and M. J. Flores, "'I was a game or a fetish object': Diverse bisexual women's sexual assault experiences and the effects on bisexual identity," *Journal of Bisexuality,* vol. 21, no. 2, pp. 225-261, 2021.

[2120] N. L. Johnson and M. Grove, "Why us? Toward an understanding of bisexual women's vulnerability for and negative consequences of sexual violence," *Journal of Bisexuality,* vol. 17, no. 4, pp. 435-450, 2017.

[2121] C. H. Nelson, "Sexualized violence against lesbians," *Peace Review,* vol. 17, no. 2-3, pp. 163-180, 2005.

[2122] J. Bond, "A decade after Abu Ghraib: Lessons in softening up the enemy and sex-based humiliation," *Minnesota Journal of Law and Inequality,* vol. 31, no. 1, pp. 1-36, 2013.

[2123] H. J. Johnson, "Bisexuality, mental health, and media representation," *Journal of Bisexuality,* vol. 16, no. 3, pp. 378-396, 2015.

[2124] R. Randazzo, K. Farmer and S. Lamb, "Queer women's perspectives on sexualization of women in media," *Journal of Bisexuality,* vol. 15, no. 1, pp. 99-129, 2015.

[2125] S. C. Gomillion and T. A. Giuliano, "The influence of media role models on gay, lesbian, and bisexual identity," *Journal of Homosexuality,* vol. 58, no. 3, pp. 330-354, 2011.

[2126] B. Amber, B. A. Raley and J. L. Lucas, "Stereotype or success? Prime-time television's portrayals of gay male, lesbian, and bisexual characters," *Journal of Homosexuality,* vol. 51, no. 2, pp. 19-38, 2023.

[2127] C. Serpe, C. Brown, S. Criss, K. Lamkins and L. Watson, "Bisexual women: Experiencing and coping with objectification, prejudice, and erasure," *Journal of Bisexuality,* vol. 20, no. 4, pp. 456-492, 2020.

[2128] B. Gerrard, J. Morandini and I. Dar-Nimrod, "Gay and straight men prefer masculine-presenting gay men for high-status role: Evidence from an ecologically valid experiment," *Sex Roles,* vol. 88, pp. 119-129, 2023.

[2129] A. M. Koenig, A. H. Eagly and A. A. Mitchell, "Are leader stereotypes masculine? A meta-analysis of three research paradigms," *Psychological Bulletin,* vol. 137, no. 4, pp. 616-642, 2011.

[2130] Reuters, "Photos show rape and sex abuse in Iraq jails- paper," *Reuters,* 27 May 2009.

[2131] S. F. Eisenman, The Abu Ghraib Effect, London: Reaktion Books, 2010, p. 144.

[2132] S. Loughnan, A. Pina, E. A. Vasquez and E. Puvia, "Sexual objectification increases rape victim blame and decreases perceived suffering," *Psychology of Women,* vol. 37, no. 4, pp. 455-461, 2013.

[2133] K. L. Morris, J. Goldenberg and P. Boyd, "Women as animals, women as objects: Evidence for two forms of objectification," *Personality and Social Psychology Bulletin,* vol. 44, no. 9, pp. 1302-1314, 2018.

[2134] V. Hoorens, "Self-enhancement and superiority biases in social comparison," *European Review of Social Psychology,* vol. 4, no. 1, p. 113–139, 1993.

[2135] J. Kruger and D. Dunning, "Unskilled and unaware of it: How difficulties in recognizing one's own incompetence lead to inflated self-assessments," *Journal of Personality and Social Psychology,* vol. 77, no. 6, p. 1121–1134, 1999.

[2136] Aristotle, "Animal rights and obligation," in *Animals and Slavery. Politics Book I,* T. Regan and P. Singer, Eds., Englewood Cliffs, NJ: Prentice Hall, 1954/1989.

[2137] B. Ginsburg and W. C. Allee, "Some effects of conditioning on social dominance and subordination of inbred strains of mice," *Physiological Zoology,* vol. 15, no. 4, pp. 485-506, 1942.

[2138] W. M. Kahn, "The effect of severe defeat at various age levels on aggressive behavior of mice," *Journal of Genetic Psychology,* vol. 79, pp. 117-130, 1951.

[2139] J. F. McCarthy and B. R. Kelly, "Aggressive behavior and its effect on performance over time in ice hockey athletes: An archival study," *International Journal of Sport Psychology,* vol. 9, no. 2, pp. 90-96, 1978a.

[2140] J. F. McCarthy and B. R. Kelly, "Aggression, performance variables, and anger self-report in ice hockey players," *Journal of Psychology,* vol. 99, no. 1, pp. 97-101, 1978b.

[2141] J. Z. Rubin, "Some wise and mistaken assumptions about conflict and negotiation," *Journal of Social Issues,* vol. 45, no. 2, pp. 195-209, 1989.

[2142] J. A. M. Reif and F. C. Brodbeck, "When do people initiate a negotiation? The role of discrepancy, satisfaction, and ability beliefs," *Negotiation and Conflict Management Research,* vol. 10, no. 4, pp. 252-26, 2017.

[2143] L. Oldham, Affective and Cognitive Components of Agonistic Behaviour in Pigs: Implications for Animal Welfare and Social Behaviour (Dissertation), Edinburgh: University of Edinburgh, 2024, p. 278.

[2144] N. M. T. Smith and R. Dukas, "Winner and loser effects in humans: evidence from randomized trials," *Animal Behaviour,* vol. 207, pp. 101-107, 2024.

[2145] J. X. Kuske, A. S. Godoy, A. V. Ramirez and B. C. Trainor, "Sex differences in responses to aggressive encounters among California mice," *Hormones and Behavior,* vol. 162, p. 105537, 2024.

[2146] S. L. Peters, K. van den Bos and D. R. Bobocel, "The moral superiority effect: Self versus other differences in satisfaction with being overpaid," *Social Justice Research,* vol. 17, no. 3, pp. 257-273, 2004.

[2147] R. G. Frey, "Rights, interests, desires, and beliefs," *American Philosophical Quarterly,* vol. 16, no. 3, pp. 233-239, 1979.

[2148] R. G. Frey, Interests and Rights: The Case Against Animals, Oxford: Clarendon Press, 1980, p. 188.

[2149] D. S. Cary, Jackie Coogan: The World's Boy King: A Biography of Hollywood's Legendary Child Star, Lanham, MD: Scarecrow Press, 2003.

[2150] T. Hobbes, Leviathan, London: Wordsworth Editions, 1651/2014, p. 592.

[2151] J.-J. Rousseau, The Social Contract and Discourses, London: Everyman's Library, 1761/ 1993, p. 472.

[2152] J. Locke, Two Treatises of Government (Annotated), Whitefish, MT: Kessinger Publishing, 1690/2010, p. 84.

[2153] ACTRA, "Minor's Trust," 15 7 2018. [Online]. Available: http://www.actra.ca/prs/minors-trust/. [Accessed 15 7 2018].

[2154] G. Vecesy, "Moceanu sues parents, alleging mismanagement," 22 October 1998. [Online]. Available: Moceanu sues parents, alleging mismanagement.

[2155] UPI, "LeAnn Rimes sues father, ex-manager," 11 May 2000. [Online]. Available: LeAnn Rimes sues father, ex-manager. [Accessed 23 July 2025].

[2156] *Jasmine Phillips as trustee v. Bank of America,* 2016.

[2157] "Children's Online Privacy Protection Rule," 17 January 2013. [Online]. Available: https://www.federalregister.gov/documents/2013/01/17/2012-31341/childrens-online-privacy-protection-rule. [Accessed 16 May 2023].

[2158] S. Holiday, M. S. Norman and R. L. Densley, "Sharenting and the extended self: self-representation in parents' Instagram presentations of their children," *Popular Communication: The International Journal of Media and Culture,* vol. 20, no. 1, pp. 1-15, 2022.

[2159] K. Kopecky, R. Szotkowski, I. Aznar-Díaz and J.-M. Romero-Rodríguez, "The phenomenon of sharenting and its risks in the online environment. Experiences from Czech Republic and Spain," *Children and Youth Services Review,* vol. 110, p. 104812, 2020.

[2160] C. S. Mott Children's Hospital, "Mott Poll Report: Parents on Social Media: Likes and Dislikes of Sharenting," University of Michigan Health System, Ann Arbor, 2015.

[2161] K. Verswijvel, M. Walrave, K. Hardies and W. Heirman, "Sharenting, is it a good or a bad thing? Understanding how adolescents think and feel about sharenting on social network sites," *Children and Youth Services Review,* vol. 104, p. 104401, 2019.

[2162] B. Shmueli and A. Blecher-Prigat, "Privacy for Children," *Columbia Human Rights Law Review,* vol. 42, pp. 759-795, 2011.

[2163] E. Nottingham, "'Dad! Cut that part out!' Children's rights to privacy in the age of 'generation tagged': Sharenting, digital kidnapping, and the child micro-celebrity (pp 183-191)," in *The Routledge International Handbook of Young Children's Rights*, First ed., London, Routledge, 2019, p. 604.

[2164] K. Haley, "Sharenting and the (potential) right to be forgotten," *Indiana Law Journal,* vol. 95, no. 3, pp. 1005-1020, 2020.

[2165] G. Frosio, "Right to be forgotten: Much ado about nothing," *Colorado Technology Law Journal,* vol. 15, no. 2, pp. 307-336, 2017.

[2166] B. Holfeld and R. Baitz, "The mediating and moderating effects of social support and school climate on the association between cyber victimization and internalizing symptoms," *Journal of Youth and Adolescence,* vol. 49, p. 2214–2228, 2020.

[2167] J. Lee, "Pathways from childhood bullying victimization to young adult depressive and anxiety symptoms," *Child Psychiatry and Human Development,* vol. 52, pp. 129-140, 2021.

[2168] S. Back and E. S. Suh, "The effects of capable guardians on the risk of cyberbullying victimization," *Journal of Criminology and Criminal Justice Studies,* vol. 1, no. 2, pp. 185-201, 2023.

[2169] G. S. O'Keeffe and K. Clarke-Pearson, "Clinical report--the impact of social media on children, adolescents, and families," *Pediatrics,* vol. 127, no. 4, pp. 800-804, 2011.

[2170] B. E. Keith and S. Steinberg, "Parental sharing on the internet: Child privacy in the age of social media and the pediatrician's role," *Journal of the American Medical Association,* vol. 171, no. 5, pp. 413-414, 2017.

[2171] B. Bosker, "Instamom: The enviable, highly profitable life of Amber Fillerup Clark, perfect mother and social-media influencer," *The Atlantic,* March 2017.

[2172] E. E. O'Neill, "Influencing the future: Compensating children in the age of social-media influencer marketing," *Stanford Law Review Online,* vol. 72, pp. 42-52, 2019.

[2173] S. K. McGrath, "Hollywood at home: Applying federal child labor laws to traditional and modern child performers," *Washington and Lee Journal of Civil Rights and Social Justice,* vol. 29, no. 3, pp. 291-327, 2023.

[2174] J. Alanis and S. Rubio, "California Assembly Bill 1880," 26 September 2024. [Online]. Available: https://legiscan.com/CA/text/AB1880/id/3022805. [Accessed 1 June 2025].

[2175] C. A. Salmon and J. Malcom, "Parent offspring conflict," in *The Oxford Handbook of Evolutionary Family Psychology*, New York, Oxford University Press, 2011, p. 432.

[2176] Durkheim, The Division of Labor in Society, New York, NY: Free Press, 1933/2014, p. 416.

[2177] P. T. Leeson, "The calculus of piratical consent: The myth of the myth of social contract," *Public Choice,* vol. 139, no. 3-4, pp. 443-459, 2009.

[2178] H. Goering, "Translation of Document 1816-PS, Stenographic Report of the Meeting on 'The Jewish Question' Under the Chairmanship of Field Marshall Goering In the Reichs Air Force," 1938.

[2179] C. S. Wren, "Insurers Swindled Jews, Nazi Files Show," *The New York Times,* 18 5 1998.

[2180] H. A. Wallace, New Frontiers, New York: Greenwood Press, 1934/1969, p. 320.

[2181] M. S. Venkataramani, "Norman Thomas, Arkansas sharecroppers, and the Roosevelt agricultural policies, 1933-1937," *The Mississippi Valley Historical Review,* vol. 47, no. 2, pp. 225-246, 1960.

[2182] C. Love, "Non-tenure-track faculty and shared governance," *Profession,* pp. 132-137, 2000.

[2183] A. Kezar, "Departmental cultures and non-tenure-track faculty: Willingness, capacity, and opportunity to perform at four-year institutions," *The Journal of Higher Education,* vol. 84, no. 2, pp. 153-188, 2013.

[2184] N. F. Alleman and D. Haviland, "'I expect to be engaged as an equal': collegiality expectations of full-time, non-tenure-track faculty members," *Higher Education,* vol. 74, no. 3, pp. 527-542, 2017.

[2185] S. T. Coleridge, "The Rime of the Ancient Mariner (text of 1834)," 1834. [Online]. Available: https://www.poetryfoundation.org/poems/43997/the-rime-of-the-ancient-mariner-text-of-1834. [Accessed 15 7 2018].

[2186] P. Levi, The Drowned and the Saved, New York: Simon and Schuster, 2017, p. 208.

[2187] C. F. Wilkinson, Blood Struggle: The Rise of Modern Indian Nations, New York: W. W. Norton and Company, 2006, p. 562.

[2188] J. McCain and L. Kyle, "H.R. 4067--112th Congress: Navajo-Hopi Little Colorado River Water Rights Settlement Act of 2012," United States Senate, Washington DC, 2012.

[2189] E. Becenti, "Senate Bill 2109 Seeks to Extinguish Navajo and Hopi Water Rights. Guest Commentary in Native Condition: Analysis and Opinion in Native News Network," 4 4 2012. [Online]. Available: http://www.nativenewsnetwork.com/senate-bill-2109-seeks-to-extinguish-navajo-and-hopi-water-rights.html. [Accessed 17 5 2012].

[2190] C. F. Wilkinson, "Home dance, the Hopi, and Black Mesa coal: Conquest and endurance in the American Southwest," *Brigham Young University Law Review,* pp. 449-482, 1996.

[2191] E. Ionesco, Rhinoceros, Folio ed., Schoenhofs Foreign Books, 1959, p. 246.

[2192] J. A. Getty, G. T. Rittersporn and V. N. Zemskov, "Victims of the Soviet penal system in the pre-war years: A first approach on the basis of archival evidence," *The American Historical Review,* vol. 98, no. 4, pp. 1017-1049, 1993.

[2193] M. Massie, "Locked up and trafficked out: Prison labor and the Thirteenth Amendment," *University of St. Thomas Law Journal,* vol. 19, no. 2, pp. 498-514, 2023.

[2194] D. A. Blackmon, Slavery by Another Name: The Re-Enslavement of Black Americans from the Civil War to World War II, New York: Doubleday, 2008, p. 480.

[2195] A. Lichtenstein, "Georgia history in fiction: Chain gangs, communism, and the 'Negro question': John L. Spivak's Georgia Nigger," *The Georgia Historical Quarterly,* vol. 79, no. 3, pp. 633-658, 1995.

[2196] J. L. Spivak, Georgia Nigger (Republished as Hard Times on a Southern Chain Gang), Columbia: University of South Carolina Press, 1932a/2012, p. 320.

[2197] H. B. Stowe, Uncle Tom's Cabin, or, Life among the Lowly, CreateSpace Independent Publishing Platform, 1851-1852/ 2015, p. 610.

[2198] J. L. Spivak, On the Chain Gang (pamphlet), vol. No. 32, International Pamphlets, 1932b.

[2199] H. C. Lea, A History of the Inquisition of the Middle Ages, Amazon Digital Services, 2012, p. 2709.

[2200] Gregorii IX, Decretalium D. Gregorii Papae Ix. Compilatio Liber Quintus: Titulus Vi. De Iudaeis, Sarracenis, Et Eorum Servis. Cap 13, Vatican City: Vatican, 1212.

[2201] P. Dawe, Artist, *The Bostonians Paying the Excise-man or Tarring and Feathering.* [Art]. Printed for Robert Sayer and John Bennett, map and printseller of 53 Fleet Street, London, England, UK, c1774.

[2202] Maine Historical Society, Maine: A History, vol. 1, C. L. Hatch, Ed., New York: American Historical Society, 1919/ 2016, p. 304.

[2203] R. L. Bushman, Joseph Smith: Rough Stone Rolling, New York: Vintage, 2007, p. 784.

[2204] T. D. Clark, The Southern Country Editor, Columbia: University of South Carolina Press, 1948/ 1991, p. 382.

[2205] P. Devi, I Phoolan Devi: The Autobiography of India's Bandit Queen, London: Sphere, 1997, p. 521.

[2206] A. Byantnal, "Dalit woman paraded naked in Chavan's hometown," *The Hindu,* 11 1 2012.

[2207] J. Arquilla, Insurgents, Raiders, and Bandits: How Masters of Irregular Warfare Have Shaped Our World, Chicago: Ivan R. Dee, 2011, p. 336.

[2208] S. Naim, "Kshatriya Samaj to honour Phoolan's killer," *The Tribune, Chandigarh,* p. 21, 5 01 2006.

[2209] J. Stradling, "Chapter 12: Phoolan Devi," in *Small History: Good Girls Don't Make History: Femmes Fatales, Different Drummers & Dangerous Renegades*, Sidney, NSW: Pier 9, 2011, p. 223.

[2210] K. Ohbuchi and T. Kambara, "Attacker's intent and awareness of outcome, impression management, and retaliation," *Journal of Experimental Social Psychology,* vol. 21, no. 4, pp. 321-330, 1985.

[2211] S. P. Taylor and R. Pissano, "Physical aggression as a function of frustration and physical attack," *Journal of Social Psychology,* vol. 54, no. 2, pp. 261-267, 1971.

[2212] M. G. Findley, "Agents and conflict: adaptation and the dynamics of war," *Complexity,* vol. 14, no. 1, pp. 22-35, 2009.

[2213] Voltaire, Candide, Third ed., New York: W. W. Norton and Company, 1769/ 2016, p. 264.

[2214] G. W. Leibniz, P. Remnant and J. F. Bennett, New Essays on Human Understanding 1645-1761 Book II: Of Ideas., Cambridge: Cambridge University Press, 1996, p. 467.

[2215] H. S. Hernandez, A. Hovasapian and B. Campos, "Displaying pride: Variation by social context, ethnic heritage, and gender?," *PLoS One,* vol. 18, no. 4, p. e0285152, 2023.

[2216] R. Saju, Y. Castellon-Lopez, N. Turk, T. Moin, C. M. Mangione, K. C. Norris, A. Vu, R. Maranon, J. Fu, F. Cheng and O. K. Duru, "Differences in weight loss by race and ethnicity in the PRIDE trial: A qualitative analysis of participant perspectives," *Journal of General Internal Medicine,* vol. 37, pp. 3715-3722, 2022.

[2217] Y. Ouyang, D. E. Rast III, J. D. Hackett and Z. P. Hohman, "The American South: Explorations on southern attachments and personal values," *Journal of Social Psychology,* vol. 160, no. 2, pp. 137-149, 2020.

[2218] A. Nexhip, M. Riley and K. Robinson, "Professional identity and workplace motivation: A case study of health information managers," *Health Information Management,* vol. 15, pp. 76-84, 2022.

[2219] E. Cohen, "The Nazification of German physicians, 1918-1937," *Annals of the Royal College of Physicians and Surgeons Canada,* pp. 336-340, 1998.

[2220] E. Hoffer, The True Believer, New York: Perennial Classics, 1951/2010, p. 192.

[2221] Y. Kongjareon, N. Samoh, P. Peerawaranun and T. E. Guadamuz, "Pride-based violence, intoxicated sex and poly-drug use: a vocational school-based study of heterosexual and LGBT students in Bangkok," *BMC Psychiatry,* vol. 22, p. 148, 2022.

[2222] D. Owen, "Hubris syndrome," *Clinical Medicine,* vol. 8, no. 4, p. 428, 2008.

[2223] L. D. Owen, "Hubris and nemesis in heads of government," *Journal of the Royal Society of Medicine,* vol. 99, no. 11, pp. 548-551, 2006.

[2224] D. Owen and J. Davidson, "Hubris syndrome: An acquired personality disorder? A study of US Presidents and UK Prime Ministers over the last 100 years," *Brain,* vol. 132, no. 5, pp. 1396-1406, 2009.

[2225] R. Cools, M. Sheridan, E. Jacobs and M. D'Esposito, "Impulsive personality predicts dopamine-dependent changes in frontostriatal activity during component process of working memory," *Journal of Neurosciences,* vol. 27, no. 20, pp. 5506-5514, 2007.

[2226] M. I. Froböse, J. C. Swart, Cook, J. L., D. E. M. Geurts, H. E. M. den Ouden and R. Cools, "Catecholaminergic modulation of the avoidance of cognitive control," *Journal of Experimental Psychology: General,* vol. 147, no. 12, pp. 1763-1781, 2018.

[2227] C. C. Liapis, "From ancient hatei to hubris syndrome in SARS-CoV-2 health crisis management," *Psychiatria Danubina,* vol. 34, no. 1, pp. 111-115, 2022.

[2228] C. F. Zinc, Y. Tong, Q. Chen, D. S. Bassett, J. L. Stein and A. Meyer-Lindenberg, "Know your place: Neural processing of social hierarchy in humans," *Neuron,* vol. 58, no. 2, pp. 273-283, 2008.

[2229] M. Koenigs, M. Kruepke, J. Zeier and J. P. Newman, "Utilitarian moral judgment in psychopathy," *Social Cognitive and Affective Neuroscience,* vol. 7, no. 6, pp. 708-714, 2012.

[2230] R. Carron, E. Brigaud, R. Anders and N. Blanc, "Being blind (or not) to scenarios used in sacrificial dilemmas: the influence of factual and contextual information on moral responses," *Frontiers in Psychology,* vol. 15, p. 1477825, 2024.

[2231] L. Young, M. Koenigs, M. Kruepke and J. P. Newman, "Psychopathy increases perceived moral permissibility of accidents," *Journal of Abnormal Psychology,* vol. 121, no. 3, pp. 659-667, 2012.

[2232] R. J. P. Rijnders, S. van den Hoogen, J. van Honk, D. Terburg and M. M. Kempes, "Would you? Effects of oxytocin on moral choices in forensic psychopathic patients," *Comprehensive Psychoneuroendocrinology,* vol. 18, p. 100245, 2024.

[2233] S. Murray, W. Jiménez-Leal and S. Amaya, "Within your rights: Dissociating wrongness and permissibility in moral judgement," *British Journal of Social Psychology,* vol. 63, no. 1, pp. 340-361, 2023.

[2234] M. P. Walker and E. van der Helm, "Overnight therapy? The role of sleep in emotional brain processing," *Psychological Bulletin,* vol. 135, no. 5, p. 731–748, 2009.

[2235] N. Cellini, M. Mercurio and M. Sarlo, "Sleeping over moral dilemmas modulates utilitarian decision-making," *Current Psychology,* vol. 42, pp. 8244-8254, 2021.

[2236] J. B. Calhoun, "Population density and social pathology," *Scientific American,* vol. 206, no. 2, pp. 139-148, 1962.

[2237] I. Fleming, A. Baum and L. Weiss, "Social density and perceived control as mediators of crowding stress in high-density residential neighborhoods," *Journal of Personality and Social Psychology,* vol. 52, no. 5, pp. 899-906, 1987.

[2238] R. Gifford and J. Peacock, "Crowding: More fearsome than crime-provoking? Comparison of an Asian city and a North American city," *Psychologia,* vol. 22, no. 2, pp. 79-83, 1979.

[2239] T. Curry, "Fraternal bonding in the locker room: A profeminist analysis of talk about competition and women," *Sociology of Sport Journal,* vol. 8, no. 2, pp. 119-135, 1991.

[2240] A. Bierce, The Devil's Dictionary, Salt Lake City, UT: Project Gutenberg, 1881-1906/ 2015.

[2241] A. Schweitzer, "The ethic of reverence for life," in *Animal Rights and Obligations*, Englewood Cliffs, NJ: Macmillan Publishing Co, 1923/1989, pp. 32-37.

[2242] P. Singer, "All animals are equal," *Philosophical Exchange,* vol. 5, no. 1, pp. 103-116, 1974 1974.

[2243] L. Berkowitz, "Aversively stimulated aggression: Some parallels and differences in research with animals and humans," *American Psychologist,* vol. 98, no. 11, pp. 1135-1144, 1983.

[2244] L. Berkowitz, "Frustration, appraisals, and aversively stimulated aggression," *Aggressive Behavior,* vol. 14, no. 1, pp. 3-11, 1988.

[2245] N. H. Azrin, H. B. Rubin and R. R. Hutchinson, "Biting attack by rats in response to aversive shock," *Journal of the Experimental Analysis of Behavior,* vol. 11, no. 5, pp. 633-639, 1968.

[2246] K. G. Anderson, S. S. Deschênes and M. J. Dugas, "Experimental manipulation of avoidable feelings of uncertainty: Effects on anger and anxiety," *Journal of Anxiety Disorders,* vol. 41, pp. 50-58, 2016.

[2247] R. N. Carleton, "Fear of the unknown: One fear to rule them all?," *Journal of Anxiety Disorders,* vol. 41, pp. 2-21, 2016.

[2248] R. Krieglmeyer, D. Wittstadt and F. Strack, "How attribution influences aggression: Answers to an old question by using an implicit measure of anger," *Journal of Experimental Social Psychology,* vol. 45, no. 2, pp. 379-385, 2009.

[2249] J. A. Kulik and R. Brown, "Frustration, attribution of blame, and aggression," *Journal of Experimental Social Psychology,* vol. 15, no. 2, pp. 183-194, 1979.

[2250] BBC, "Irwin fans 'in revenge attacks'," 12 10 2006. [Online]. Available: http://news.bbc.co.uk/2/hi/asia-pacific/5338118.stm. [Accessed 10 7 2018].

[2251] Voice of America, "Afghans Continue Protests Over US Quran Burning," 4 4 2011. [Online]. Available: http://www.voanews.com/english/news/asia/Afghans-Continue-Protests-Over-US-Quran-Burning-119173679.html. [Accessed 18 6 2018].

[2252] W. Griffitt, "Environmental effects on interpersonal affective behavior. Ambient effective temperature and attraction," *Journal of Personality and Social Psychology,* vol. 15, no. 3, pp. 240-244, 1970.

[2253] W. Griffitt and R. Veitch, "Hot and crowded: Influences on population density and temperature on interpersonal affective behavior," *Journal of Personality and Social Psychology,* vol. 17, no. 1, pp. 92-98, 1971.

[2254] J. M. Carlsmith and C. A. Anderson, "Ambient temperature and the occurrence of collective violence: A new analysis," *Journal of Personality and Social Psychology,* vol. 37, no. 3, pp. 337-344, 1979.

[2255] K. D. Harries and S. J. Stadler, "Heat and violence: New findings from Dallas field data, 1980-1981," *Journal of Applied Social Psychology,* vol. 18, no. 2, pp. 129-138, 1988.

[2256] C. A. Anderson, "Heat and violence," *Current Directions in Psychological Science,* vol. 10, no. 1, pp. 33-38, 2001.

[2257] J. Rotton and J. Frey, "Air pollution, weather, and violent crimes: Concomitant time-series analysis of archival data," *Journal of Personality and Social Psychology,* vol. 49, no. 5, pp. 1207-1220, 1985.

[2258] C. A. Anderson, W. E. Deuser and K. M. DeNeve, "Hot temperatures, hostile affect, hostile cognition, and arousal: Tests of a general model of affective aggression," *Personality and Social Psychology Bulletin,* vol. 21, no. 5, pp. 434-448, 1995.

[2259] A. S. Reifman, R. Larrick and S. Fein, "Temper and temperature on the diamond: The heat-aggression relationship in Major League Baseball," *Personality and Social Psychology,* vol. 17, no. 5, pp. 580-585, 1991.

[2260] R. M. Brown, The Rubyfruit Jungle, reprint ed., New York: Bantam Books, 1973/ 2015, p. 240.

[2261] C. Oveis, A. Spectre, P. K. Smith, M. Y. Liu and D. Keltner, "Laughter conveys status," *Journal of Experimental Social Psychology,* vol. 65, pp. 109-115, 2016.

[2262] K. H. K. E. K. Greenaway, S. C. Murphy and T. McIlroy, "Winners are grinners: Expressing authentic positive emotion enhances status in performance contexts," *Journal of Experimental Social Psychology,* vol. 78, pp. 168-180, 2018.

[2263] C. J. Ferguson, R. Maguire and S. Lernar, "Pick your poison: Choice of activity determines mood management following a stressful task," *Emotional Regulation and Stress,* vol. 27, no. 3, pp. 332-346, 2017.

[2264] L. M. Bylsma, M. A. Croon, A. J. J. M. Vingerhoets and J. Rottenberg, "When and for whom does crying improve mood? A daily diary study of 1004 crying episodes," *Journal of Research in Personality,* vol. 45, no. 4, pp. 385-392, 2011.

[2265] E. B. Ebbesen, B. Duncan and V. J. Konecni, "Effects of content of verbal aggression on future verbal aggression: A field experiment," *Journal of Experimental Social Psychology,* vol. 11, no. 2, pp. 192-204, 1975.

[2266] V. J. Konecni and E. B. Ebbesen, "Disinhibition versus the cathartic effect: Artifact and substance," *Journal of Personality and Social Psychology,* vol. 34, no. 3, pp. 352-365, 1976.

[2267] M. Bauer, A. Cassar, J. Chytilová and J. Henrich, "War's enduring effects on the development of egalitarian motivations and in-group biases," *Psychological Science,* vol. 25, no. 1, pp. 47-57, 2014.

[2268] D. Archer and R. Gartner, "Violent acts and violent times: a comparative approach to postwar homicide rates," *American Sociological Review,* vol. 41, no. 6, pp. 937-963, 1976.

[2269] J. Di Salvatore, "Peacekeepers against Criminal Violence—Unintended Effects of Peacekeeping Operations?," *American Journal of Political Science,* vol. 63, no. 4, pp. 840-858, 2019.

[2270] C. Ransford, R. B. Decker, G. M. Cruz, F. Sánchez and G. Slutkin, "El modelo Cure Violence: reducción de la violencia en San Pedro Sula (Honduras)/ The Cure Violence model: violence reduction in San Pedro Sula (Honduras)," *Revista CIDOB d'Afers Internacionals,* vol. 116, pp. 179-204, 2017.

[2271] G. Russell, "Athletes as targets of aggression (pp 221-252)," in *Advances in Psychology,* vol. 76, R. Baenninger, Ed., Elsevier, 1991, p. 476.

[2272] C. Stott, O. Adang, A. Livingstone and M. Schreiber, "Variability in the collective behaviour of England fans at Euro2004: 'Hooliganism', public order policing and social change," *European Journal of Social Psychology,* vol. 37, no. 1, p. 75–100, 2007.

[2273] E. F. Coccaro, B. Beresford, P. Minar, J. Kaskow and T. Geracioti, "CSF testosterone: Relationship to aggression, impulsivity, and venturesomeness in adult males with personality disorder," *Journal of Psychiatric Research,* vol. 41, no. 6, pp. 488-492, 2007.

[2274] R. L. Arms, G. W. Russell and M. L. Sandilands, "Effects on the hostility of spectators of viewing aggressive sports," *Social Psychology Quarterly,* vol. 42, no. 3, pp. 275-279, 1979.

[2275] G. W. Russell, "Psychological issues in sports aggression," *Sports Violence,* pp. 157-181, 1983.

[2276] C. A. Anderson and B. J. Bushman, "Effects of violent video games on aggressive behavior, aggressive cognition, aggressive affect, physiological arousal, and prosocial behavior: A meta-analytic review of the scientific literature," *Psychological Science,* vol. 12, p. 353–359, 2001.

[2277] L. A. Fehr, "Media violence and catharsis in college females," *Journal of Social Psychology,* vol. 109, no. 2, pp. 307-308, 1979.

[2278] R. G. Green, G. L. Stonner and G. L. Shope, "The facilitation of aggression by aggression: evidence against the catharsis hypothesis," *Journal of Personality and Social Psychology,* vol. 21, no. 4, pp. 721-726, 1975.

[2279] J. Strayhorn Jr., "Editorial: Modeling and fantasy rehearsal: From self-injruy to self-efficacy," *Journal of the American Academy of Child and Adolescent Psychiatry,* vol. 62, no. 6, pp. 656-664, 2023.

[2280] R. Kersten and T. Greitemeyer, "Why do habitual violent video game players believe in the cathartic effects of violent video games? A misinterpretation of mood improvement as a reduction in aggressive feelings," *Aggressive Behavior,* vol. 48, no. 2, pp. 219-231, 2022.

[2281] Y. Hasan, L. Bègue and B. J. Bushman, "Violent video games stress people out and make them more aggressive," *Aggressive Behavior,* vol. 39, no. 1, pp. 64-70, 2012.

[2282] T. Greitemeyer, "The contagious impact of playing violent video games on aggression: Longitudinal evidence," *Aggressive Behavior,* vol. 45, no. 6, pp. 635-642, 2019.

[2283] C. A. Anderson, "Violent video games increase aggression and violence," U.S. Senate Commerce Committee, Washington, 2000.

[2284] B. J. Bushman and C. A. Anderson, "Violent video games and hostile expectations: A test of the general aggression model," *Personality and Social Psychology Bulletin,* pp. 1679-1686, 2002.

[2285] J. L. Whitaker and B. J. Bushman, "A review of the effects of violent video games on children and adolescents," *Washington and Lee Law,* vol. 66, no. 3, pp. 1003-1051, 2009.

[2286] E. Hurel, M. Grall-Bronnec, O. Bouillard, M. Chirio-Espitalier, M. Barrangou-Poueys-Darlas and G. Challet-Bouju, "Systematic review of gaming and neuropsychological assessment of social cognition," *Neuropsychological Review,* pp. 1-30, 2023.

[2287] J. J. Lindsay and C. A. Anderson, "From antecedent conditions to violent actions: A general affective aggression model," *Personality and Social Psychology Bulletin,* vol. 26, no. 5, pp. 533-547, 2000.

[2288] A. Vijayalakshmi and R. N. Laczniak, "What happens when parents find violence acceptable? A case of violent-humorous commercials targeted at children," *Journal of Advertising,* pp. 1-17, 2024.

[2289] C. A. Anderson and C. M. Ford, "Affect of the game player: Short-term effects of highly and mildly aggressive video games," *Personality and Social Psychology Bulletin,* vol. 12, no. 4, pp. 390-402, 1986.

[2290] B. E. Sheese and W. G. Graziano, "Deciding to defect: The effects of video-game violence on cooperative behavior," *Psychological Science,* vol. 16, no. 5, pp. 354-357, 2005.

[2291] T. Greitemeyer and N. McLatchie, "Denying Humanness to Others: A Newly Discovered Mechanism by Which Violent Video Games Increase Aggressive Behavior," *Psychological Science,* vol. 22, no. 5, pp. 659-665, 2011.

[2292] T. R. A. Kral, D. E. Stodola, R. M. Birn, J. A. Mumford, E. Solis, L. Flook, E. G. Patsenko, C. G. Anderson, C. Steinkuehler and R. J. Davidson, "Neural correlates of video game empathy training in adolescents: a randomized trial," *NPJ Science of Learning,* vol. 3, p. 13, 2018.

[2293] D. Z. Kádár, Politeness, Impoliteness and Ritual: Maintaining the Moral Order in Interpersonal Interaction, Cambridge: Cambridge University Press, 2017, p. 283.

[2294] N. Haslam, S. Loughnam and P. Sun, "Beastly: What makes animal metaphors offensive?," *Journal of Language and Social Psychology,* vol. 30, no. 3, pp. 311-325, 2012.

[2295] N. Haslam, "Dehumanization: An integrative review," *Personality and Social Psychology Review,* vol. 10, no. 3, pp. 252-264, 2006.

[2296] S. K. Herrmann, "Social exclusion: Practices of misrecognition. (pp 133-149)," in *Humiliation, Degradation, Dehumanization*, First ed., P. Kaufmann, H. Kuch, C. Neuhauser and E. Webster, Eds., Springer, 2011, pp. 978-9048196609.

[2297] P. Schaber, "Absolute poverty: Human dignity, self-respect, and dependency. (pp 151-158)," in *Humiliation, Degradation, Dehumanization*, P. Kaufmann, H. Kuch, C. Neuhauser and E. Webster, Eds., Dordrecht, Springer, 2011, p. 280.

[2298] A. L. Wood, "The spectacle of lynching: Rituals of White supremacy in the Jim Crow South," *American Journal of Economics and Sociology,* vol. 77, no. 3-4, pp. 757-788, 2018.

[2299] A. S. Buckser, "Lynching as Ritual in the American South," *Berkeley Journal of Sociology,* vol. 37, pp. 11-28, 1992.

[2300] D. Matthews, "The Southern Rite of Human Sacrifice," *Black History Bulletin,* vol. 64, no. 3-4, pp. 20-47, 2002.

[2301] D. G. Matthews, "The Southern Rite of Human Sacrifice. Lynching in the American South," *The Mississippi Quarterly,* vol. 61, no. 1-2, pp. 27-70, 2008.

[2302] J. P. Wallis and J. Mechling, "Devil dogs and dog piles," *Western Folklore,* vol. 74, no. 3/4, pp. 275-308, 2015.

[2303] A. Hawke and O. Ness, "Rituals that helped heal a nation after a terror attack," *Scandinavian Psychologist,* vol. 4, p. e6, 2017.

[2304] E. G. D'Aquili, "Human ceremonial ritual and the modulation of aggression," *Journal of Religion and Science,* vol. 20, no. 1, pp. 21-30, 1985.

[2305] A. D. Tian, J. Schroeder, G. Häubl, J. L. Risen, M. I. Norton and F. Gino, "Enacting rituals to improve self-control," *Journal of Personality and Social Psychology,* vol. 114, no. 6, pp. 851-876, 2018.

[2306] E. D. Brocato, D. A. Gentile, R. N. Laczniak, J. A. Maier and M. Ji-Song, "Television Commercial Violence: Potential Effects on Children (pp 161-178)," in *Advertising and Violence*, First ed., New York, Routledge, 2014, p. 320.

[2307] N. Machiavelli, The Prince, Project Gutenberg, 1513/ 2006.

[2308] W. Shakespeare, Hamlet, Third ed., Cambridge: Cambridge University Press, ca 1600/2014, p. 296.

[2309] A. P. Brief, J. Dietz, R. R. Cohen, S. D. Pugh and J. B. Vaslow, "Just doing business: Modern racism and obedience to authority as explanations for employment discrimination," *Organizational Behavior and Human Decision Processes,* vol. 81, no. 1, p. 72–97, 2000.

[2310] K. J. Anderson and D. Beason, Ill Wind, New York: Tom Doherty Associates, 1995/ 2007, p. 576.

[2311] D. Schwartz, J. E. Landsford, K. A. Dodge, G. S. Pettit and J. E. Bates, "Peer victimization during middle childhood as a marker of attenuated risk for adult arrest," *Journal of Abnormal Child Psychology,* vol. 46, pp. 57-65, 2018.

[2312] A. Sourander, J. Ronning, A. Brunstein-Klomek, D. Gyllenberg, K. Kumpulainen, S. Niemela, H. Helenius, L. Sillanmaki, T. Ristkari, T. Tamminen, I. Moilanen, J. Piha and F. Almqvist, "Childhood bullying behavior and later psychiatric hospital and psychopharmacologic treatment findings from the Finnish 1981 birth cohort study," *Archives of General Psychiatry Volume,* vol. 66, no. 9, pp. 1005-1012, 2009.

[2313] Q. Song, T. Yuan, Y. Hu, X. Liu, J. Fei, X. Zhao, R. Gao, J. Yue and S. Mei, "The effect of peer victimization during adolescence on depression and gender differences: A systematic review and meta-analysis," *Trauma, Violence, and Abuse,* vol. 25, no. 4, pp. 2862-2876, 2024.

[2314] A.-H. Luukkonen, P. Rasanen, H. Hakko and K. Riala, "Bullying behavior is related to suicide attempts but not to self-mutilation among psychiatric inpatient adolescents," *Psychopathology,* vol. 42, no. 2, pp. 131-138, 2009.

[2315] R. Pauli and P. L. Lockwood, "The computational psychiatry of antisocial behaviour and psychopathy," *Neuroscience and Biobehavioral Reviews,* vol. 145, p. 104995, 2023.

[2316] K. Rigby, "Addressing Bullying in Schools: Theory and Practice," Australian Institute of Criminology, Canberra, 2003b.

[2317] F. Thériault-Couture, J. Agnew-Blais, V. C. Leno, A. Danese, K. Ganaesan, T. Matthews, G. Morneau-Vallancourt, K. N. Thompson, S. Shakor and L. Arseneault, "Trajectories of childhood bullying behaviors and conduct problems: Associations with cognitive functioning in a nationally representative cohort study," *Developmental Psychology,* vol. 61, no. 2, p. 351–365, 2025.

[2318] K. Rigby, "Consequences of bullying in schools," *The Canadian Journal of Psychiatry,* vol. 48, no. 9, pp. 583-590, 2003.

[2319] T. R. Nansel, "Relationship between bullying and violence among US youth," *Archives of Pediatrics and Adolescent Medicine,* vol. 157, no. 4, p. 348–353, 2003.

[2320] J. F. Sigurdson, A. M. Undheim, J. L. Wallander, S. Lydersen and A. M. Sund, "The long-term effects of being bullied or a bully in adolescence on externalizing and internalizing mental health problems in adulthood," *Child and Adolescent Psychiatry and Mental Health,* vol. 9, p. 42, 2015.

[2321] D. P. Farrington and M. M. Ttofi, "Bullying as a predictor of offending, violence and later life outcomes," *Criminal Behaviour and Mental Health,* vol. 21, no. 2, pp. 90-98, 2011.

[2322] C. Vrijen, M. Wiertsema, M. A. Ackermans and R. K. T. van der Ploeg, "Childhood and adolescent bullying perpetration and later substance use: A meta-analysis," *Pediatrics,* vol. 147, no. 3, p. e2020034751, 2021.

[2323] L. Arseneault, E. Walsh, K. Trzesniewski, R. Newcombe, A. Caspi and T. Moffitt, "Bullying victimization uniquely contributes to adjustment problems in young children: A nationally representative cohort study," *Pediatrics,* vol. 118, no. 1, p. 130–138, 2006.

[2324] D. S. J. Hawker and M. J. Boulton, "Twenty years' research on peer victimization and psychosocial maladjustment: A meta-analytic review of cross-sectional studies," *Journal of Personality and Social Psychology,* vol. 41, no. 4, pp. 677-685, 1999.

[2325] A. R. Pollastri, E. V. Cardemil and E. H. O'Donnell, "Self-esteem in pure bullies and bully/victims: A longitudinal analysis," *Journal of Interpersonal Violence,* vol. 25, no. 8, pp. 1489-1502, 2010.

[2326] C. Samivalli, A. Kaukiainen, L. Kastaniemi and K. M. J. Lagerspetz, "Self-evaluated self-esteem, peer-evaluated self-esteem, and defensive egotism as predictors of adolescents' participation in bullying situations," *Personality and Social Psychology Bulletin,* vol. 25, no. 10, pp. 1268-1278, 1999.

[2327] M. O'Moore and C. Kirkham, "Self-esteem and its relationship to bullying behaviour," *Aggressive Behavior,* vol. 27, no. 4, pp. 269-283, 2001.

[2328] C. Tritte and R. D. Duncan, "Research: The relationship between childhood bullying and young adult self-esteem and loneliness," *Humanistic Counseling,* vol. 36, no. 1, pp. 345-44, 2011.

[2329] K. Hosseini, The Kite Runner, New York, NY: Riverhead, 2003, p. 371.

[2330] J. A. Rambaran, J. K. Dijkstra and R. Veenstra, "Bullying as a group process in childhood: A longitudinal social network analysis," *Child Development,* vol. 91, no. 4, pp. 1336-1352, 2020.

[2331] G. S. Vance, V. Zeigler-Hill, J. Vonk and T. K. Shackelford, "Inevitable or preventable? The biosocial theory of wartime rape," *Evolutionary Psychological Science,* 2024b.

[2332] D. M. Blonigen, B. M. Hicks, R. F. Krueger, C. J. Patrick and W. G. Iacono, "Psychopathic personality traits: Heritability and genetic overlap with internalizing and externalizing psychopathology," *Psychological Medicine,* vol. 35, no. 5, pp. 637-648, 2005.

[2333] J. Hill, "Early identification of individuals at risk for antisocial personality disorder," *The British Journal of Psychiatry,* vol. 182, no. S44, pp. s11-s14, 2018.

[2334] E. Brontë, Wuthering Heights, Toronto, ON: Penguin Books, 1847/ 1995, p. 416.

[2335] A. R. W. Chow, J.-B. Pingault and J. R. Baldwin, "Early risk factors for joint trajectories of bullying victimization and perpetration," *European Child and Adolescent Psychiatry,* pp. 1-9, 2022.

[2336] H. A. Ball, L. Arseneault, A. Taylor, B. Maughan, A. Caspi and T. E. Moffitt, "Genetic and environmental influences on victims, bullies and bully-victims in childhood," *Journal of Child Psychology and Psychiatry,* vol. 49, no. 1, p. 104–112, 2008.

[2337] S. A. M. Veldkamp, D. I. Boomsma, E. L. de Zeeuw, C. E. M. van Beijsterveldt, M. Bartels, C. V. Dolan and E. van Bergen, "Genetics and environmental influences on different forms of bullying perpetration, bullying victimization, and their co-occurrence," *Behavior Genetics,* vol. 49, pp. 432-443, 2019.

[2338] UNESCO, UN, "School violence and bullying: Global status and trends, drivers and consequences," United Nations Educational, Scientific, and Cultural Organization, Paris, 2018.

[2339] J. M. Hettema, M. C. Neale and K. S. Kendler, "A review and meta-analysis of the genetic epidemiology of anxiety disorders," *The American Journal of Psychiatry,* vol. 158, no. 10, pp. 1568-1578, 2001.

[2340] Wang, R., Lifelines Cohort Study, Hartman. C. A., Snieder, H., "Stress-related exposures amplify the effects of genetic susceptibility on depression and anxiety," *Translational Psychiatry,* vol. 13, p. 27, 2023.

[2341] D. Schwartz, "Subtypes of victims and aggressors in children's peer groups," *Journal of Abnormal Child Psychology,* vol. 28, no. 2, p. 181–192, 2000.

[2342] K. A. Fanti, P. J. Frick and S. Georgiou, "Linking callous-unemotional traits to instrumental and non-instrumental forms of aggression," *Journal of Psychopathology and Behavioral Assessment,* vol. 31, no. 4, pp. 285-298, 2009.

[2343] E. Viding, E. Simmonds, K. V. Petrides and N. Frederickson, "The contribution of callous-unemotional traits and conduct problems to bullying in early adolescence," *Journal of Child Psychology and Psychiatry Volume,* vol. 50, no. 4, pp. 471-481, 2009.

[2344] M. Dargis and M. Koenigs, "Two subtypes of psychopathic criminals differ in negative affect and history of childhood abuse," *Psychological Trauma,* 2017.

[2345] A. F. Bettencourt, L. K. Clary, N. Ialongo and R. J. Musci, "Long-term consequences of bullying involvement in first grade," *Journal of School Psychology,* vol. 97, pp. 63-76, 2023.

[2346] M. E. Solberg, D. Olweus and I. M. Endresen, "Bullies and victims at school: Are they the same pupils?," *British Journal of Educational Psychology,* vol. 77, no. 2, pp. 441-464, 2007.

[2347] N. R. Crick and M. A. Bigbee, "Relational and overt forms of peer victimization: A multiinformant approach," *Journal of Consulting and Clinical Psychology,* vol. 66, no. 2, p. 337–347, 1998.

[2348] E. V. E. Hodges and D. G. Perry, "Personal and interpersonal antecedents and consequences of victimization by peers," *Journal of Personality and Social Psychology,* vol. 74, no. 4, p. 677–685, 1999.

[2349] D. Olweus, "The revised Olweus bully/victim questionnaire," Mimeo, Bergen, 1996.

[2350] M. E. Solberg and D. Olweus, "Prevalence estimation of school bullying with the Olweus Bully/Victim Questionnaire," *Aggressive Behavior,* vol. 29, no. 3, pp. 239-268, 2003.

[2351] D. Olweus, Bullying at School: What We Know and What We Can Do, First ed., Oxford: Wiley-Blackwell Publishers, 1993, p. 160.

[2352] D. Olweus and K. Breivik, "Chapter 90: Plight of victims of school bullying: The opposite of well-being," in *Handbook of Child Well-Being: Theories, Methods and Policies in Global Perspective*, A. Ben-Arieh, F. Casas, I. Frønes and J. E. Korbin, Eds., Springer, 2014, p. 2626.

[2353] K. E. O'Connor, A. D. Farrell, W. Kliewer and S. J. Lepore, "Social and emotional adjustment across aggressor/victim subgroups: Are aggressive-victims distinct?," *Journal of Youth and Adolescence,* vol. 48, pp. 2222-2240, 2019.

[2354] D. Schwartz, L. J. Proctor and D. H. Chien, "The aggressive victim of bullying: Emotional and behavioural dysregulation as a pathway to victimization by peers. pp. 147–174," in *Peer Harassment in School: The Plight of the Vulnerable and Victimized*, New York, Guildford Press, 2001, p. 440.

[2355] E. Espino, A. Guarini and L. D. R. R. Menabò, "Why are some victims also bullies? The role of peer relationship management and anger regulation in traditional bullying," *Youth and Society,* vol. 55, no. 6, pp. 1056-1078, 2022.

[2356] R. Desmarais, Joint Trajectories of Bullying Victimization and Perpetration: Investigating the Role of the COMT Gene, Ottawa, Ontario: University of Ottawa, 2023.

[2357] C. Marlow, G. S. and K. L. Mulvey, "Adolescents' expectations for types of victim retaliation following direct bullying," *Journal of Youth and Adolescence,* vol. 52, p. 533–546, 2023.

[2358] M. A. Schiele, M. G. Gottschalk and K. Domschke, "The applied implications of epigenetics in anxiety, affective and stress-related disorders - A review and synthesis on psychosocial stress, psychotherapy and prevention," *Clinical Psychology Review,* vol. 77, p. 101830, 2020.

[2359] M.-K. Koskinen and I. Hovatta, "Genetic insights into the neurobiology of anxiety," *Trends in Neurosciences,* vol. 46, no. 4, pp. 318-331, 2023.

[2360] B. W. Haas, M. M. Filkowski, R. N. Cochran, L. Denison, A. Ishak, S. Nishitani and A. K. Smith, "OXT and sociability," *Proceedings of the National Academy of Sciences,* vol. 113, no. 27, pp. E3816-E3823, 2016.

[2361] J. Greek, Director, *Babylon 5: A Voice in the Wilderness, Part 2.* [Film]. USA/ Canada: Babylonian Productions; Warner Bros. Television, 1994.

[2362] T. Ein-Dor, W. Verbeke, M. Mokry and P. Vrtička, "Epigenetic modification of the oxytocin and glucocorticoid receptor genes is linked to attachment avoidance in young adults," *Attachment and Human Development,* vol. 20, no. 4, pp. 439-454, 2018.

[2363] L. Bevilacqua, S. Doly, J. Kaprio, Q. Yuan, R. Tikkanen, T. Paunio, Z. Zhou, J. Wedenoja, L. Maroteaux, S. Diaz, M. Virkkunen, A. Belmer, C. A. Hodgkinson, L. Dell'osso, J. Suvisaari, E. Coccaro, R. Rose, L. Peltonen, M. Virkkunen and Go, "A population-specific HTR2B stop codon predisposes to severe impulsivity," *Nature,* pp. 1061-1066, 2010.

[2364] A. Delprato, B. Bonheur, M. Algéo, A. Murillo, E. Dhawan, L. Lu, R. W. Williams and W. E. Crusio, "A quantitative trait locus on chromosome 1 modulates intermale aggression in mice," *Genes, Brain and Behavior,* vol. 0, no. 0, p. e12469., 2018.

[2365] L. J. Siever, "Neurobiology of aggression and violence," *American Journal of Psychiatry,* vol. 165, no. 4, pp. 429-442, 2008.

[2366] X. Lin, Y. Cao, L. Ji and W. Zhang, "Inhibitory control mediates the interaction between serotonin transporter gene (5-HTTLPR) and peer victimization on adolescent depressive symptoms," *Science Reports,* vol. 11, no. 1, p. 14640, 2021.

[2367] D. P. Jacobsen, M. B. Nielsen, S. Einasen and J. Gjerstad, "Negative social acts and pain: evidence of a workplace bullying and 5-HTT genotype interaction," *Scandinavian Journal of Work, Environment, and Health,* vol. 44, no. 3, pp. 283-290, 2018.

[2368] C. Bursztein and A. Apter, "Adolescent suicide," *Current Opinion in Psychiatry,* vol. 22, no. 1, pp. 1-6, 2009.

[2369] K. A. Pavlov, D. A. Chistiakov and V. P. Chekhonin, "Genetic determinants of aggression and impulsivity in humans," *Journal of Applied Genetics,* vol. 53, no. 1, pp. 61-82, 2012.

[2370] R. Waltes, A. G. Chiocchetti and C. M. Freitag, "he neurobiological basis of human aggression: A review on genetic and epigenetic mechanisms," *Medical Genetics B: Neuropsychiatric Genetics,* vol. 171, no. 5, pp. 650-675, 2016.

[2371] G. N. Erjavec, L. Tudor, M. N. Perkovic, J. Podobnik, K. D. Curkovic, M. Curkovic, D. S. Strac, M. Cusek, M. Bortolato and N. Pivac, "Serotonin 5-HT2A receptor polymorphisms are associated with irritability and aggression in conduct disorder," *Progress in Neuro-Psychopharmacology and Biological Psychiatry,* vol. 117, p. 110542, 2022.

[2372] R. Upthegrove, "Bullying, victimisation, and psychosis," *The Lancet Psychiatry,* vol. 2, no. 7, pp. 574-576, 2015.

[2373] C. Vrijen, I. M. Nolte, A. J. Oldehinkel, R. Veenstra and T. Kretschmer, "Genetic confounding in bullying research: Causal claims revisited," *Development and Psychopathology,* pp. 1-12, 2023.

[2374] I. Ouellet-Morin, C. C. Y. Wong, A. Danese, C. M. Pariante, A. S. Papadopoulos and L. Arseneault, "Increased serotonin transporter gene (SERT) DNA methylation is associated with bullying victimization and blunted cortisol response to stress in childhood: a longitudinal study of discordant monozygotic twins," *Psychological Medicine,* vol. 43, no. 9, pp. 1813-1823, 2013.

[2375] E. Maier and V. Grinevich, "Social neuroscience: How we learn to avoid the bully," *Current Biology,* vol. 34, no. 8, pp. PR320-R322, 2024.

[2376] A. K. Beery, "Antisocial oxytocin: Complex effects on social behavior," *Current Opinion in Behavioral Sciences,* vol. 6, pp. 174-182, 2015.

[2377] R. Feldman, "The neurobiology of hatred: Tools of Dialogue© intervention for youth reared amidst intractable conflict impacts brain, behaviour, and peacebuilding attitudes," *Acta Paediatrica,* vol. 112, no. 4, pp. 603-616, 2023.

[2378] C. D. Navarrete, D. M. T. Fessler and S. J. Eng, "Elevated ethnocentrism in the first trimester of pregnancy," *Evolution of Human Behavior,* vol. 28, no. 1, pp. 60-65, 2007.

[2379] M. Q. Steinman and B. C. Trainor, "Sex differences in the effects of social defeat on brain and behavior in the California mouse: Insights from a monogamous rodent," *Seminars in Cell and Developmental Biology,* vol. 61, no. 1, pp. 92-98, 2017.

[2380] D. LoParo and I. D. Waldman, "The oxytocin receptor gene (OXTR) is associated with autism spectrum disorder: a meta-analysis," *Molecular Psychiatry,* vol. 20, no. 5, pp. 640-646, 2015.

[2381] V. Gigliucci, B. M., S. F., P. C., B. A., S. F. and C. B. Muscatelli F., "Oxytocin receptors in the Magel2 mouse model of autism: Specific region, age, sex and oxytocin treatment effects," *Frontiers in Neuroscience,* vol. 14, no. 17, p. 1026939, 2023.

[2382] E. Stafflinger, K. K. Hansen, F. Hauser and C. J. P. Gimmelikhuijzen, "Cloning and identification of an oxytocin/vasopressin-like receptor and its ligand from insects," *Proceedings of the National Academy of Sciences,* vol. 105, no. 9, pp. 3262-3267, 2008.

[2383] D. R. Nässel and C. Wegener, "A comparative review of short and long neuropeptide F signaling in invertebrates: Any similarities to vertebrate neuropeptide Y signaling?," *Peptides,* vol. 32, no. 6, pp. 1335-1355, 2011.

[2384] H. A. Dierick and R. J. Greenspan, "Serotonin and neuropeptide F have opposite modulatory effects on fly aggression," *Nature Genetics,* vol. 39, p. 678–682, 2007.

[2385] K. Eaton, F. R. Sallee and R. Sah, "Relevance of neuropeptide Y (NPY) in psychiatry," *Current Topics in Medical Chemistry,* vol. 7, no. 17, p. 1645–1659, 2007.

[2386] L. B. Comeras, N. Hörmer, P. M. Bethuraj and R. O. Tasan, "NPY released from GABA neurons of the dentate gyrus specially reduces contextual fear without affecting cued or trace fear," *Frontiers in Synaptic Neuroscience,* vol. 13, p. 635726, 2021.

[2387] J. S. Womersley, L. I. Martin, L. van der Merwe, S. Seedat and S. M. J. Hemmings, "Hypothalamic-pituitary-adrenal axis variants and childhood trauma influence anxiety sensitivity in South African adolescents," *Metabolic Brain Disease,* vol. 33, pp. 601-613, 2018.

[2388] L. H. Vähätalo, S. T. Ruohonen, L. Ailanen and E. Savontaus, "Neuropeptide Y in noradrenergic neurons induces obesity in transgenic mouse models," *Neuropeptides,* vol. 55, pp. 31-37, 2016.

[2389] M. A. Raghanti, E. N. Miller, D. N. Jones, H. N. Smith, E. L. Munger, M. K. Edler, K. A. Phillips, W. D. Hopkins, P. R. Hof, C. C. Sherwood and C. O. Lovejoy, "Hedonic eating, obesity, and addiction result from increased neuropeptide Y in the nucleus accumbens during human brain evolution," *Proceedings of the National Academy of Sciences,* vol. 120, no. 38, p. e2311118120, 2023.

[2390] J. Sommer, H. Ehnis, T. Seitz, J. Schneider, A. B. Wild, S. Moceri, C. Buechler, A. Bozec, G. F. Weber, S. Merkel, R. Beckervordersandforth, A. Steinkasserer, R. Schüle, J. Trebicka, A. Hartmann, A. Bosserhoff and S. von Hörsten, "our-and-a-Half LIM-Domain Protein 2 (FHL2) Induces Neuropeptide Y (NPY) in Macrophages in Visceral Adipose Tissue and Promotes Diet-Induced Obesity," *International Journal of Molecular Sciences,* vol. 24, no. 19, p. 14943, 2023.

[2391] S. X. Zhang, A. Kim, J. C. Madara, P. K. Zhu, L. F. Christenson, A. Lutas, P. N. Kalugin, P. S. Sunkavalli, Y. Jin, A. Pal, L. Tian, B. B. Lowell and M. L. Andermann, "Stochastic neuropeptide signals compete to calibrate the rate of satiation," *Nature,* vol. 637, no. 8044, pp. 137-144, 2025.

[2392] M. Heilig and R. Murison, "Intracerebroventricular neuropeptide Y protects against stress-induced gastric erosion in the rat," *European Journal of Pharmacology,* vol. 137, no. 1, pp. 127-129, 1987.

[2393] M. Nakazato, N. Murakami, Y. Date, M. Kojima, H. Matsuo, K. Kangawa and S. Matsukura, "A role for ghrelin in the central regulation of feeding," *Nature,* vol. 409, no. 6817, pp. 194-198, 2001.

[2394] A. J. Tomiyama, "Stress and Obesity," *Annual Review of Psychology,* vol. 70, pp. 703-718, 2019.

[2395] Y. Chen, R. A. Essner, S. Kosar, O. H. Miklller, Y.-C. Lin, S. Mesgarzadeh and Z. A. Kinght, "Sustained NPY signaling enables AgRP neurons to drive feeding," *eLife,* vol. 8, p. e46348, 2019.

[2396] M. J. Krashes, S. Koda, C.-P. Ye, S. C. Rogan, A. C. Adams, D. S. Cusher, E. Maratos-Flier, B. L. Roth and B. B. Lowell, "Rapid, reversible activation of AgRP neurons drives feeding behavior in mice," *Journal of Clinical Investigation,* vol. 121, no. 4, pp. 1424-1428, 2011.

[2397] T. P. Chacko, J. T. Toole, S. Richman, G. L. Spink Jr., M. J. Reinhard, R. C. Brewster, M. E. Costanzo and G. Broderick, "Mapping the network biology of metabolic response to stress in posttraumatic stress disorder and obesity," *Frontiers in Psychology,* vol. 13, p. 941019, 2022.

[2398] C. Singh, J. Rihel and D. A. Prober, "Neuropeptide Y regulates sleep by modulating noradrenergic signaling," *Current Biology,* vol. 27, no. 24, pp. 3796-3811.e5, 2017.

[2399] M. Heilig, B. Söderpalm and J. A. W. E. Engel, "Centrally administered neuropeptide Y (NPY) produces anxiolytic-like effects in animal anxiety models," *Psychopharmacology,* vol. 98, pp. 524-529, 1989.

[2400] A. Dzaja, M. A. Dalal, H. Himmerich, M. Uhr, T. Pollmächer and A. Schuld, "Sleep enhances nocturnal plasma ghrelin levels in healthy subjects," *American Journal of Physiology: Endocrinology and Metabolism,* vol. 286, no. 6, pp. E963-E967, 2004.

[2401] J. Kornhuber and I. Zoicas, "Neuropeptide Y reduces expression of social fear via simultaneous activation of Y1 and Y2 receptors," *Journal of Psychopharmacology,* vol. 33, no. 12, p. 1533–1539, 2019.

[2402] R. O. Tasan, D. Verma, J. Wood, G. Lach, B. Hörmer, T. C. M. de Lima, H. Herzog and G. Sperk, "The role of Neuropeptide Y in fear conditioning and extinction," *Neuropeptides,* vol. 55, pp. 111-126, 2016.

[2403] P. Broqua, J. G. Wettstein, M. N. Rocher, B. Gauthier-Martin and J. L. Junien, "Behavioral effects of neuropeptide Y receptor agonists in the elevated plus-maze and fear-potentiated startle procedures," *Behavioral Pharmacology,* vol. 6, no. 3, p. 215–222, 1995.

[2404] M. A. Cortes, A. F. Bartley, Q. Li, T. R. Davis, S. E. Cunningham, M. A. Garner, P. J. Perez, A. C. Harvey, A. K. Gross and L. E. Dobrunz, "Modulation of temporoammonic-CA1 synapses by neuropeptide Y is through Y1 receptors in mice," *Neuropeptides,* vol. 110, p. 102504, 2025.

[2405] J. Kornhuber and I. Zoicas, "Valence-dependent effects of neuropeptide Y on the expression of conditioned fear and anxiety-like behavior: Involvement of the bed nucleus of the stria terminalis," *Neuropharmacology,* vol. 246, p. 109847, 2024.

[2406] J. H. Urban, "Neuropeptide Y and amygdala circuitry: Modulation of stress-related behavior," *Handbook of Behavioral Neuroscience,* vol. 26, pp. 141-159, 2020.

[2407] J. Lu, S. Li, H. Li, T. Mou, L. Zhou, B. Huang, M. Huang and Y. Xu, "Changes in plasma NPY, IL-1β and hypocretin in people who died by suicide," *Neuropsychiatric Diseases and Treatment,* vol. 15, pp. 2893-2900, 2019.

[2408] S. Sabol, S. Hu and D. Hamer, "A functional polymorphism in the monoamine oxidase A gene promoter," *Human Genetics,* vol. 103, no. 3, p. 273–279, 1998.

[2409] Y. Hu, Y. Liu, C. Zhou, H. Li, J. Fan and Z. Ma, "Effects of food quantity on aggression and monoamine levels of juvenile pufferfish (Takifugu rubripes)," *Fish Physiology and Biochemistry,* vol. 47, no. 6, pp. 1983-1993, 2021.

[2410] N. James and A. Bell, "Minimally invasive brain injections for viral-mediated transgenesis: New tools for behavioral genetics in sticklebacks," *PLoS One,* vol. 16, no. 5, p. e0251653, 2021.

[2411] A. A. Ahmed, W. Ma, Y. Ni, Q. Zhou and R. Zhao, "Embryonic exposure to corticosterone modifies aggressive behavior through alterations of the hypothalamic pituitary adrenal axis and the serotonergic system in the chicken," *Hormones and Behavior,* vol. 65, no. 2, pp. 97-105, 2014.

[2412] M. Wysocki, S. Preuss, P. Stratz and J. Bennewitz, "Investigating gene expression differences in two chicken groups with variable propensity to feather pecking," *Animal Genetics,* vol. 44, no. 6, pp. 773-777, 2013.

[2413] Q. Chu, T. Liang, L. Fu, H. Li and B. Zhou, "Behavioural genetic differences between Chinese and European pigs," *Journal of Genetics,* vol. 96, no. 4, pp. 707-715, 2017.

[2414] P. G. Eusebi, N. Sevane, O. Cortés, E. Contreras, J. Cañon and S. Dunner, "Aggressive behavior in cattle is associated with a polymorphism in the MAOA gene promoter," *Animal Genetics,* vol. 51, no. 1, pp. 14-21, 2019.

[2415] G. Lühken, K. Glenske, H. Brandt and G. Erhardt, "Genetic variation in monoamine oxidase A and analysis of association with behaviour traits in beef cattle," *Journal of Animal Breeding and Genetics,* vol. 127, no. 5, pp. 411-418, 2010.

[2416] P. G. Eusebi, N. Sevane, T. O'Rourke, M. Pizarro and C. D. S. Boeckx, "Age Effects Aggressive Behavior: RNA-Seq Analysis in Cattle with Implications for Studying Neoteny Under Domestication," *Behavior Genetics,* vol. 52, no. 2, pp. 141-153, 2022.

[2417] J. W. Eo, H.-E. Lee, G.-H. Nam, Y.-J. Kwon, Y. Chio, B.-H. Choi, J.-W. Huh, M. Kim, S.-E. Lee, S. Bohyun and H.-S. Kim, "Association of DNA methylation and monoamine oxidase A gene expression in the brains of different dog breeds," *Gene,* vol. 580, no. 2, pp. 177-182, 2016.

[2418] L. Kerkerian, O. Bosler, G. Pelletier and A. Nieoullon, "Striatal neuropeptide Y neurons are under the influence of the nigrostriatal dopaminergic pathway: immunohistochemical evidence," *Neuroscience Letters,* vol. 66, no. 1, pp. 106-112, 1986.

[2419] N. N. Kudryavtseva, D. A. Smagin, I. L. Kovalenko, A. G. Galyamina, G. B. Vishnivetskaya, V. Babenko and Y. L. Orlov, "[Serotonergic genes in the development of anxiety/depression-like state and pathology of aggressive behavior in male mice: RNA-seq data] (English abstract)," *Molekuliarnaia Biologiia,* vol. 51, no. 2, pp. 288-300, 2017.

[2420] H. Chen, D. S. Pine, M. Ernst, E. Gorodetsky, S. Kasen, K. Gordon, D. Goldman and P. and Cohen, "The MAOA gene predicts happiness in women," *Progress in Neuro-Psychopharmacology & Biological Psychiatry,* vol. 40, no. 1, pp. 122-125, 2013.

[2421] J. W. Buckholtz and A. Meyer-Lindenberg, "MAOA and the neurogenetic architecture of human aggression," *Trends in Neurosciences,* vol. 31, no. 3, pp. 120-129, 2008.

[2422] S. Bissière, N. Plachta, D. Hoyer, H.-R. Olpe, A. A. Grace and J. F. Cryan, "The Rostral anterior cingulate cortex modulates the efficiency of amygdala-dependent fear learning," *Biological Psychiatry,* vol. 63, no. 9, pp. 821-831, 2008.

[2423] T. B. Franklin, B. A. Silva, Z. Perova, L. Marrone, M. E. Masferrer, Y. Zhang, A. Kaplan, L. Greetham, V. Verrechia, A. Halman, S. Pagella, A. L. Vyssotski, A. Illarionova, V. Grinevich, T. Branco and C. T. Gross, "Prefrontal cortical control of a brainstem social behavior circuit," *Nature Neuroscience,* vol. 20, pp. 260-270, 2017.

[2424] X. Liu, J. Li, J. Gao, Z. Zhou, F. Meng, G. Pan and B. Luo, "Association of medial prefrontal cortex connectivity with consciousness level and its outcome in patients with acquired brain injury," *Journal of Neuroscience,* vol. 42, pp. 160-166, 2017.

[2425] A. L. Byrd and S. B. Manuck, "MAOA, childhood maltreatment, and antisocial behavior: Meta-analysis of a gene-environment interaction," *Biological Psychiatry,* vol. 75, no. 1, pp. 9-17, 2014.

[2426] G. Frazzetto, G. Di Lorenzo, V. Carola, L. Proietti, E. Sokolowska, A. Siracusano, C. Gross and A. Troisi, "Early trauma and increased risk for physical aggression during adulthood: The moderating role of MAOA genotype," *PLoS One,* vol. 2, no. 5, p. e486, 2007.

[2427] N. Weder, B. Z. Yang, H. Douglas-Palumberi, J. Massey, J. H. Krystal, J. Gelernter and J. Kaufman, "MAOA genotype, maltreatment, and aggressive behavior: The changing impact of genotype at varying levels of trauma," *Biological Psychiatry,* vol. 65, no. 5, p. 417–424, 2009.

[2428] C. A. Galán, D. E. Choe, E. E. Forbes and D. S. Shaw, "The interaction between monoamine oxidase A and punitive discipline in the development of antisocial behavior: Mediation by maladaptive social information processing," *Developmental Psychopathology,* vol. 29, no. 4, pp. 1235-1252, 2017.

[2429] A. Caspi, J. McClay, T. E. Moffitt, J. Mill, J. Martin, I. W. Craig, A. Taylor and R. Poulton, "Role of genotype in the cycle of violence in maltreated children," *Science,* vol. 297, no. 5582, pp. 851-854, 2002.

[2430] C. Mariz, O. S. Cruz and D. Moreira, "The influence of environmental and genetic factors on the development of psychopathy: A systematic review," *Aggression and Violent Behavior,* vol. 62, p. 101715, 2022.

[2431] D. Checknita, M. Bendre, T. J. Ekström, E. Comasco, J. Tiihonen, S. Hodgins and K. W. Nilsson, "Monoamine oxidase A genotype and methylation moderate the association of maltreatment and aggressive behaviour," *Brain Research Bulletin,* vol. 382, p. 112476, 2023.

[2432] A. Ziomkiewicz-Wichary, "Serotonin and Dominance (pp 7053–7057)," in *Encyclopedia of Evolutionary Psychological Science*, T. K. Shackelford and V. A. Weekes-Shackelford, Eds., Cham, Springer, 2016, p. 8679.

[2433] P. Gong, J. Zhang, J. Liu, L. He and W. Guo, "Bright side of the MAOA-uVNTR on trait and situational forgiveness," *Psychoneuroendocrinology,* vol. 151, p. 106057, 2023.

[2434] V. Mertins, A. B. Schote and J. Meyer, "Variants of the Monoamine Oxidase A gene (MAOA) predict free-riding behavior in women in a strategic public goods experiment," *Journal of Neuroscience, Psychology, and Economics,* vol. 6, no. 2, pp. 97-114, 2013.

[2435] T. F. Denson, C. Dobson-Stone, R. Ronay, W. von Hippel and M. M. Schira, "A functional polymorphism of the MAOA gene is associated with neural responses to induced anger control," *Journal of Cognitive Neuroscience,* vol. 26, no. 7, pp. 1418-1427, 2014.

[2436] R. McDermott, D. Tingley, J. Cowden, G. Frazzetto and D. D. P. Johnson, "Monoamine oxidase A gene (MAOA) predicts behavioral aggression following provocation," *Proceedings of the National Academy of Sciences,* vol. 106, no. 7, pp. 2118-2123, 2009.

[2437] K. Sawyer, "Biophysiological risk-factors for political violence," *Journal of Conflict Resolution,* vol. 67, no. 5, pp. 809-827, 2022.

[2438] R. Boyle and R. Haynes, Power Play: Sport, the Media and Popular Culture.Pearson Education, Second ed., Edinburgh: Edinburgh University Press, 2009, p. 264.

[2439] D. Rowe, "The global love-match. Sport and television (pp 565-582)," in *Television: Critical Concepts in Media and Cultural Studies*, First ed., T. Miller, Ed., London, Routledge, 2002, p. 2048.

[2440] M. Hardin, J. E. Dodd, J. Chance and K. Walsdorf, "Sporting images in Black and White: Race in newspaper coverage of the 2000 Olympic Games," *Howard Journal of Communications,* vol. 15, no. 4, pp. 211-227, 2004.

[2441] J. Ter Wal, L. D'Haenens and J. Koeman, "(Re)presentation of ethnicity in EU and domestic news: A qualitative analysis," *Media, Culture and Society,* vol. 27, no. 6, pp. 937-950, 2005.

[2442] M. C. Duncan, "A hermeneutic of spectator sport: The 1976 and 1984 Olympic Games," *Quest,* vol. 38, no. 1, pp. 50-77, 1986.

[2443] C. T. Higgs and K. H. Weiller, "Gender bias and the 1992 Summer Olympic Games: An analysis of television coverage," *Journal of Sport and Social Issues,* vol. 18, no. 3, pp. 234-246, 1994.

[2444] M. C. Duncan, "Representation and the gun that points backwards," *Journal of Sport and Social Issues,* vol. 17, pp. 42-46, 1993.

[2445] M. C. Duncan and C. A. Hasbrook, "Denial of power in televised women's sports," *Sociology of Sport Journal,* vol. 5, no. 1, pp. 1-21, 1988.

[2446] K. L. Bissell and A. M. Duke, "Bump, set, spike: An analysis of commentary and camera angles of women's beach volleyball during the 2004 Summer Olympics (pp 35-53)," in *Investigating the Use of Sex in Media Promotion and Advertising,* First ed., Milton Park, Routledge, 2019, p. 192.

[2447] J. E. Korbin, "Children, childhoods, and violence," *Annual Review of Anthropology,* vol. 31, no. 1, pp. 431-446, 2003.

[2448] G. Margolin and E. B. Gordis, "The Effects of Family and Community Violence on Children," *Annual Review of Psychology,* vol. 51, pp. 445-479, 2000.

[2449] D. L. Espelage, "Bullying, sexual violence, and suicide in education (pp 323- 328)," in *The Wiley Handbook on Violence in Education: Forms, Factors, and Preventions*, Hoboken, NJ: Wiley-Blackwell, 2018, p. 640.

[2450] J. Levi, "Bullying and the laws pertaining to it," *GPSOLO,* vol. 34, no. 1, 2017.

[2451] E. Payne and M. J. Smith, "Violence against LGBTO students: Punishing and marginalizing difference pp 393-416," in *The Wiley Handbook on Violence in Education: Forms, Factors, and Preventions*, Hoboken, NJ: Wiley-Blackwell, 2018, p. 640.

[2452] S. H. Sherwood, A. Amadori, S. T. Russell and S. Ioverno, "School outness, bias-based bullying, and bias-based cyberbullying across Europe," *Child Development,* vol. 96, no. 5, pp. 1730-1748, 2025.

[2453] V. Rawlings, "'It's not bullying,' 'It's just a joke': Teacher and student discursive manoeuvres around gendered violence," *British Educational Research Journal,* vol. 45, no. 4, pp. 698-716, 2019.

[2454] P. A. Newman, S. Fantus, M. R. Woodford and M.-J. Rwigema, "'Pray that God will change you': The religious social ecology of bias-based bullying targeting sexual and gender minority youth—a qualitative study of service providers and educators," *Journal of Adolescent Research,* 2017.

[2455] Y. Wang, H. Yu, R. Li, A. Wilson, S. Wang, J. Drescher and R. Chen, "The victim-bully cycle of sexual minority school adolescents in China: prevalence and the association of mood problems and coping strategies," *Epidemiology and Psychiatric Sciences,* vol. 29, p. e179, 2020.

[2456] I. Warwick, P. Aggleton and N. Douglas, "Playing it safe: Addressing the emotional and physical health of lesbian and gay pupils in the U.K.," *Journal of Adolescence,* vol. 24, no. 1, pp. 129-140, 2001.

[2457] A. J. White, R. Magrath and B. Thomas, "The experiences of lesbian, gay and bisexual students and staff at a Further Education college in South East England," *British Educational Research Journal,* vol. 44, no. 3, pp. 480-495, 2018.

[2458] M. E. Eisenberg, A. L. Gower, C. Brown, Y.-S. Nam, G. N. Rider and M. Ramirez, "'It was never really bullying, but...': Reflections of Adolescents and Parents on Experiences of Microaggressions in School," *International Journal of Bullying Prevention,* vol. 4, p. 285–296, 2022.

[2459] S. M. Swearer, R. K. Turner, J. E. Givens and W. S. Pollack, "'You're so gay!': Do different forms of bullying matter for adolescent males?," *School Psychology Review,* vol. 37, no. 2, pp. 160-173, 2008.

[2460] M. L. Ybarra, K. J. Mitchell, J. G. Kosciw and J. D. Korchmaros, "Understanding linkages between bullying and suicidal ideation in a national sample of LGB and heterosexual youth in the United States," *Prevention Science,* vol. 16, no. 3, pp. 451-462, 2015.

[2461] CBS, "Lawmaker: Rejection Of Anti-Bullying Bill Is 'Disgusting'," 23 5 2012. [Online]. Available: https://chicago.cbslocal.com/2012/05/23/lawmaker-rejection-of-anti-bullying-bill-is-disgusting/. [Accessed 27 6 2018].

[2462] M. Moen and R. Otwell, "Anti-bullying measure fails in Illinois legislature. Government," 23 5 2012. [Online]. Available: http://disqus.com/forums/npr-wnij/anti_bullying_measure_fails_in_illinois_legislature/trackback/. [Accessed 3 6 2012].

[2463] V. Harlan, Director, *Jud Süß.* [Film]. Nazi Germany: Terra Film, 1940.

[2464] F. Hippler, Director, *Der Ewige Jude.* [Film]. Nazi Germany: Deutsche Film Gesellschaft, 1940.

[2465] J. J. Heslen, "Neurocognitive hacking: A new capability in cyber conflict?," *Politics and the Life Sciences,* vol. 39, no. 1, pp. 87-100, 2020.

[2466] R. Epstein and J. Friedman, Directors, *The Celluloid Closet.* [Film]. USA: Channel Four Films; HBO Pictures, 1996.

[2467] V. Russo, The Celluloid Closet, New York: Harper and Row, 1987, p. 368.

[2468] H. Wessel, Composer, *Horst-Wessel-Lied/ Die Fahne hoch.* [Sound Recording]. 1929.

[2469] W. A. Donohue, "The identity trap: The language of genocide," *Journal of Language and Social Psychology,* vol. 31, no. 1, pp. 13-29, 2012.

[2470] S. Bikindi, Composer, *Twasezereye Ingoma Ya Cyami.* [Sound Recording].

[2471] S. Bikindi, "Nanga Abahutu," [Online]. Available: https://genocidearchiverwanda.org.rw/index.php/Nanga_Abahutu.

[2472] *International Crime Tribunal for Rwanda: Prosecutor v. Simon Bikindi,* 2008.

[2473] K. Lovelace, N. Thrasher and K. M. Allison, Composers, *Try That in a Small Town.* [Sound Recording]. BBR. 2023.

[2474] K. Trumpener, "Béla Bartók and the rise of comparative ethnomusicology: Nationalism, race purity, and the legacy of the Austro-Hungarian empire. pp 403-434," in *Music and the Racial Imagination (Chicago Studies in Ethnomusicology),* Chicago, University of Chicago Press, 2000, p. 703.

[2475] N. M. Baur, N. P. Kalmoe and E. B. Russell, "Candidate aggression and gendered voter evaluations," *Political Psychology,* vol. 43, no. 1, pp. 23-43, 2021.

[2476] M. M. Joo, B. Bolte, N. Huynh, V. Yadav and B. Mukherjee, "Right-wing populist leaders, nationalist rhetoric, and dispute initiation in international politics," *Journal of Conflict Resolution,* p. 00220027241247041, 2024.

[2477] P. Bull and P. Wells, "Adversarial discourse in prime minister's questions," *Journal of Language and Social Psychology,* vol. 31, no. 1, pp. 30-48, 2011.

[2478] G. d. P. di Bernardone and P. Robinson, The Writings of St. Francis of Assisi, Philadelphia: Dolphin Press/ Evinity Publishing Inc, 1905, p. 255 .

[2479] A. Frisen and S. Bjarnelind, "Health-related quality of life and bullying in adolescence," *Acta Paediatrica,* vol. 99, no. 4, pp. 597-603, 2010.

[2480] I. Gobina, A. Zaborskis, I. Pudule, I. Kalnins and A. Villerusa, "Bullying and subjective health among adolescents at schools in Latvia and Lithuania," *International Journal of Public Health,* vol. 53, no. 5, pp. 272-276, 2008.

[2481] P. Schneider, M. Pätz, R. Spanagel and M. Schneider, "Adolescent social rejection alters pain processing in a CB1 receptor dependent manner," *European Neuropsychopharmacology,* vol. 26, no. 7, pp. 1201-1212, 2016a.

[2482] P. Schneider, L. Bindila, C. Schmahl, M. Bohus, A. Meyer-Lindenberg, B. Lutz, R. Spanagel and M. Schneider, "Adverse social experiences in adolescent rats result in enduring effects on social competence, pain sensitivity and endocannabinoid signaling," *Frontiers in Behavioral Neuroscience,* vol. 10, p. 203, 2016b.

[2483] J. L. Lakin, T. L. Chartrand and R. M. Arkin, "I am too just like you: Nonconscious mimicry as an automatic behavioral response to social exclusion," *Psychological Science,* vol. 19, no. 8, p. 816–822, 2008.

[2484] L. Hou, X. Gu and G. Din, "From an identity process theory perspective: a daily investigation of why and when ostracism triggers ingratiation," *The Journal of Social Psychology,* pp. 1-15, 2022.

[2485] P. Martín and A. Guelbenzu, "Aggravated solitary confinement in Turkey. A form of institutionalised torture," *Torture,* vol. 34, no. 1, pp. 135-140, 2024.

[2486] J. F. Benenson, H. Markovits, M. E. Thompson and R. W. Wrangham, "Under threat of social exclusion, females exclude more than males," *Psychological Science,* vol. 22, no. 4, pp. 538-544, 2011.

[2487] M. Biron, "Effective and ineffective support: How different sources of support buffer the short- and long-term effects of a working day," *European Journal of Work and Organizational Psychology,* vol. 22, no. 2, pp. 150-164, 2013.

[2488] J. B. Nezlek, E. D. Wesselmann, L. Wheeler and K. D. Williams, "Ostracism in everyday life: The effects of ostracism on those who ostracize," *The Journal of Social Psychology,* vol. 155, no. 5, pp. 432-451, 2015.

[2489] Y. Al-Saggaf and S. B. O'Donnell, "Phubbing: Perceptions, reasons behind, predictors, and impacts," *Human Behavior and Emerging Technologies,* vol. 1, no. 2, pp. 132-140, 2019.

[2490] T. R. Nuñez and T. Radtke, "Is socially disruptive smartphone use detrimental to well-being? A systematic meta-analytic review on being phubbed," *Behaviour and Information Technology,* pp. 1-29, 2023.

[2491] J. J. Paul and A. H. N. Cillessen, "Dynamics of peer victimization in early adolescence," *Journal of Applied School Psychology,* vol. 19, no. 2, pp. 25-43, 2003.

[2492] D. M. Doumas, A. Midgett and M. Peck, "Gender differences in defending behavior among elementary school students trained in bullying bystander program: Is self-esteem a moderator?," *Journal of Applied School Psychology,* pp. 1-24, 2022.

[2493] K. D. Rudolph, H. V. Skymba, H. H. Modi, M. M. Davis and W. Y. Sze, "Biological embedding of peer experiences: The contribution of peer adversity to stress regulation (pp 111–151)," in *Biosocial Interplay During Elementary School*, P. van Lier and K. Deater-Deckard, Eds., Cham, Springer, 2022, p. 111.

[2494] E. D. Barker, L. Arseneault, M. Brendgen, N. Fontaine and B. Maughan, "Joint development of bullying and victimization in adolescence: Relations to delinquency and self-harm," *Journal of the American Academy of Child & Adolescent Psychiatry,* vol. 49, no. 7, pp. 1030-1038, 2008.

[2495] I. C. Muñoz and P. J. Frick, "Callous-unemotional traits and their implication for understanding and treating aggressive and violent youths," *Criminal Justice and Behavior,* vol. 39, no. 6, pp. 794-813, 2012.

[2496] N. Carey, The Epigenetics Revolution: How Modern Biology Is Rewriting Our Understanding of Genetics, Disease, and Inheritance, New York: Columbia University Press, 2013, p. 352.

[2497] T. R. Nansel, W. M. Craig, M. D. Overpeck, G. Saluja and W. J. Ruan, "Cross-national consistency in the relationship between bullying behaviours and psychosocial adjustment," *Archives of Pediatrics and Adolescent Medicine,* vol. 158, no. 8, p. 730–736, 2004.

[2498] K. Rigby, "Psychosocial functioning in families of Australian adolescent schoolchildren involved in bully/victim problems," *Journal of Family Therapy,* vol. 16, no. 2, pp. 173-187, 1994.

[2499] S. N. Georgiou, M. Ioannou and P. Stavrinides, "Parenting styles and bullying at school: The mediating role of locus of control," *International Journal of School and Educational Psychology,* vol. 5, no. 4, pp. 226-242, 2016.

[2500] D. I. Grama, R. D. Georgescu, I. M. Cosa and A. Dobrean, "Parental risk and protective factors associated with bullying victimization in children and adolescents: A systematic review and meta-analysis," *Clinical Child and Family Psychology Review,* vol. 27, p. 627–657, 2024.

[2501] T. E. Moffitt, "The new look of behavioral genetics in developmental psychopathology: Gene–environment interplay in antisocial behaviors," *Psychological Bulletin,* vol. 131, no. 4, p. 533–554, 2005.

[2502] S. H. Rhee and I. D. Waldman, "Genetic and environmental influences on antisocial behaviour: A meta-analysis of twin and adoption studies," *Psychological Bulletin,* vol. 28, no. 2, p. 181–192, 2002.

[2503] P. M. Adamczyk, A. Shaw, I. M. Morella and L. More, "Neurobiology, molecular pathways, and environmental influences in antisocial traits and personality disorders," *Neuropharmacology,* vol. 269, p. 110322, 2025.

[2504] T. J. Polderman, B. Benyamin, C. A. de Leeuw, P. F. Sullivan, A. van Bochoven, P. M. Visscher and D. Posthuma, "Meta-analysis of the heritability of human traits based on fifty years of twin studies," *Nature Genetics,* vol. 47, no. 7, pp. 702-709, 2015.

[2505] D. A. Mason and P. J. Frick, "The heritability of antisocial behavior: A meta-analysis of twin and adoption studies," *Journal of Psychopathology and Behavioral Assessment,* vol. 16, p. 301–323, 1994.

[2506] K. C. Jacobson, C. A. Prescott and K. S. Kendler, "Sex differences in the genetic and environmental influences on the development of antisocial behavior," *Development and Psychopathology,* vol. 14, no. 2, p. 395– 416, 2002.

[2507] F. Razoux, H. Russig, T. Mueggler, C. Baltes, K. Dikaiou, M. Rudin and I. M. Mansuy, "Transgenerational disruption of functional 5-HT1AR-induced connectivity in the adult mouse brain by traumatic stress in early life," *Molecular Psychiatry,* vol. 22, no. 4, pp. 519-526, 2016.

[2508] S. M. J. Hemmings, L. I. Martin, L. van der Merwe, R. Benecke, K. Domschke and S. Seedat, "Serotonin transporter variants play a role in anxiety sensitivity in South African adolescents," *The World Journal of Biological Psychiatry,* vol. 17, no. 1, pp. 66-75, 2016.

[2509] S. D. Calkins, K. L. Gill, M. C. Johnson and C. L. Smith, "Emotional reactivity and emotional regulation strategies as predictors of social behaviour with peers during toddlerhood," *Social Development,* vol. 8, no. 3, p. 310–334, 1999.

[2510] J. Sutton, P. K. Smith and J. Sweetenham, "Social cognition and bullying: Social inadequacy or skilled manipulation," *British Journal of Developmental Psychology,* vol. 17, no. 4, pp. 435-450, 1999.

[2511] Y. Wang, S. Shang, W. Xie, S. Hong, Z. Liu and Y. Su, "The relation between aggression and theory of mind in children: A meta-analysis," *Developmental Science,* vol. 26, no. 2, p. e13310, 2023.

[2512] S. H. Parade, K. K. Ridout, R. Seifer, D. A. Armstrong, C. J. Marsit, M. A. McWilliams and A. R. Tyrka, "Methylation of the glucocorticoid receptor gene promoter in preschoolers: Links with internalizing behavior problems," *Child Development,* vol. 87, no. 1, p. 86–97, 2016.

[2513] A. Nyembwe, Y. Zhao, B. A. Caceres, D. Belsky, C. P. Ryan, B. Taylor, M. T. Morrison, L. Prescott, S. Potts-Thompson, A. Aziz, F. Aruleba, E. Matute-Arcos, O. Williams, C. Crusto and J. Y. Taylor, "Discrimination, coping and DNAm accelerated aging among African American mothers of the InterGEN study," *Epigenomes,* vol. 269, p. 110322, 2025.

[2514] K. Gapp, A. Corcoba, G. van Steenwyk, I. M. Mansuy and J. M. Duarte, "Brain metabolic alterations in mice subjected to postnatal traumatic stress and in their offspring," *Journal of Cerebral Blood Flow and Metabolism,* vol. 37, no. 7, pp. 2423-2432, 2017.

[2515] Y. S. Yap, P. Patrizio, L. Cimmino, K. Sdrimas and A. G. Telonis, "The small non-coding RNA profile of human and mouse sperm," *Noncoding RNA,* vol. 11, no. 1, p. 15, 2025.

[2516] E. T. Cánepa and B. G. Berardino, "Epigenetic mechanisms linking early-life adversities and mental health," *Biochemical Journal,* vol. 418, no. 10, pp. 615-642, 2024.

[2517] J. P. Curley, C. L. Jensen, R. Mashoodh and F. A. Champagne, "Social influences on neurobiology and behavior: epigenetic effects during development," *Psychoneuroendocrinology,* vol. 36, no. 3, pp. 352-371, 2011.

[2518] B. Labonté, M. Suderman, G. Maussion, L. Navaro, V. Yerko, I. Mahar, A. Bureau, N. Mechawar, M. Szyf, M. J. Meaney and G. Turecki, "Genome-wide epigenetic regulation by early-life trauma," *Journal of the American Medical Association Psychiatry,* vol. 69, no. 7, pp. 722-731, 2012.

[2519] D. T. Denhardt, "Effect of stress on human biology: Epigenetics, adaptation, inheritance, and social significance," *Journal of Cellular Physiology,* vol. 233, no. 3, pp. 1975-1984, 2017.

[2520] L. C. Houtepen, C. H. Vinkers, T. Carrillo-Roa, M. Hiemstra, P. A. van Lier, W. S. Meeus, C. M. Heim, C. B. Nemeroff, J. Mill, L. C. Schalkwyk, M. P. Creyghton, R. S. Kahn, M. Joëls, B. E. B. and M. P. M. Boks, "Genome-wide DNA methylation levels and altered cortisol stress reactivity following childhood trauma in humans," *Nature Communications,* vol. 7, p. 10967, 2016.

[2521] I. C. G. Weaver, N. Cervoni, F. A. Champagne, A. C. D'Alessio, S. Sharma, J. R. Seckl, S. Dymov, M. Szyf and M. J. Meaney, "Epigenetic programming by maternal behavior," *Nature Neuroscience,* vol. 7, no. 8, p. 847–854, 2004.

[2522] A. K. Smith, S. Katrinli, D. O. Cobb, E. G. Goff, M. Simmond, G. M. Christensen, T. Prusisz, S. N. Garth, M. Brashear, A. Hüls, E. J. Wolf, E. J. Trapido, A. L. Rung, N. R. Nugent and E. S. Peters, "Epigenetic age acceleration and disparities in posttraumatic stress in women in southeast Louisiana: NIMHD social epigenomics program," *JAMA Network Open,* vol. 7, no. 7, p. e2421884, 2024.

[2523] M. T. Brown, M. Mutambudzi and N.-W. Chen, "Relationship between accelerated biological aging, race, perceived discrimination, and limitations in activities of daily living," *International Journal of Geriatrics and Gerontology,* vol. 9, no. 1, p. 199, 2025.

[2524] A. Elbasheir, S. Katrinli, B. E. Kearney, R. A. Lanius, N. G. Harnett, S. E. Carter, T. D. Ely, B. Bradley, C. F. Gillespie, J. S. Stevens, A. Lori, S. J. H. van Rooij, A. Powers, T. Jovanovic, A. K. Smith and N. Fani, "Racial discrimination, neural connectivity, and epigenetic aging among black women," *JAMA Network Open,* vol. 7, no. 6, p. e2416588, 2024.

[2525] C. W. Kuzawa and E. Sweet, "Epigenetics and the embodiment of race: Developmental origins of US racial disparities in cardiovascular health," *American Journal of Human Biology,* vol. 21, no. 1, pp. 2-15, 2009.

[2526] R. Yehuda and A. Lehrner, "Intergenerational transmission of trauma effects: Putative role of epigenetic mechanisms," *World Psychiatry,* vol. 17, no. 3, pp. 243-257, 2018.

[2527] E. K. Adam, J. A. Heissel, K. H. Zeiders, J. A. Richeson, E. C. Rossa, K. B. Ehrlich, D. J. Levy, M. Kemeny, A. B. Brodish, O. Malanchuk, S. C. Peck, T. E. Fuller-Rowell and J. S. Eccles, "Developmental histories of perceived racial discrimination and diurnal cortisol profiles in adulthood: A 20-year prospective study," *Psychoneuroendocrinology,* vol. 62, pp. 279-291, 2015.

[2528] T. W. McDade, C. Ryan, M. J. Jones, J. L. MacIsaac, A. M. Morin, J. M. Meyer, J. B. Borja, G. E. Miller, M. S. Kobor and C. W. Kuzawa, "Social and physical environments early in development predict DNA methylation of inflammatory genes in young adulthood," *Proceedings of the National Academy of Sciences USA,* vol. 114, no. 29, pp. 7611-7616, 2017.

[2529] M. Weinstock, "Alterations induced by gestational stress in brain morphology and behaviour of the offspring," *Progress in Neurobiology,* vol. 65, no. 5, pp. 427-451, 2001.

[2530] P. B. Adkins-Jackson, C. H. Tejera, D. Cotton-Samuel, C. L. Foster, L. L. Brown, K. T. Watson, T. N. Ford, T. Bragg, B. B. Wondimu and J. J. Manly, ""Rest of the folks are tired and weary": The impact of historical lynchings on biological and cognitive health for older adults racialized as Black," *Social Science and Medicine,* vol. 364, p. 117537, 2025.

[2531] G. S. Malhi, P. Das, T. Outhred, E. Bell, D. Gessler, R. Bryant and Z. Mannie, "Significant age by childhood trauma interactions on grey matter volumes: A whole brain VBM analysis," *Bipolar Disorders,* pp. 209-220, 2023.

[2532] A. M. Letkiewicz, J. D. Spring, A. Weinberg and S. A. Shankman, "Childhood trauma predicts blunted error monitoring in adulthood: An event-related potential study," *Cognitive, Affective, and Behavioral Neuroscience,* vol. 23, pp. 427-439, 2023.

[2533] B. Labonté, M. Suderman, G. Maussion, J. P. Lopez, L. Navarro-Sánchez, V. Yerko, N. Mechawar, S. M., M. J. Meaney and G. and Turecki, "Genome-wide methylation changes in the brains of suicide completers," *American Journal of Psychiatry,* vol. 170, no. 5, pp. 511-520, 2013.

[2534] P. O. McGowan, A. Sasaki, A. C. D'Alessio, S. L. B. Dymov, M. Szyf, G. Turecki and M. J. and Meaney, "Epigenetic regulation of the glucocorticoid receptor in human brain associates with childhood abuse," *Nature Neuroscience,* vol. 12, p. 342–348, 2009.

[2535] M. Khalil, A. Kalyoncu and A. Bellon, "Genetics of suicide," *Genes,* vol. 16, no. 4, p. 428, 2025.

[2536] M. Arancibia, M. Manterola, U. Ríos, P. R. Moya, J. Moran-Kneer and M. L. Bustamante, "The rs1360780 variant of FKBP5: Genetic variation, epigenetic regulation, and behavioral phenotypes," *Genes,* vol. 16, no. 3, p. 325, 2025.

[2537] P. O. McGowan, A. Sasaki, T. C. T. Huang, A. Unterberger, M. Suderman, C. Ernst, M. J. Meaney, G. Turecki and M. Szyf, "Promoter-wide hypermethylation of the ribosomal RNA gene promoter in the suicide brain," *PLOS One,* vol. 3, no. 5, p. e2085, 2008.

[2538] T. Sarkar, N. Patro and I. K. Patro, "Cumulative multiple early life hits- a potent threat leading to neurological disorders," *Brain Research Bulletin,* vol. 147, pp. 58-68, 2019.

[2539] C. M. Marsters, D. Nesan, R. Rar, N. Klenin, Q. J. Pittman and D. M. Kurrasch, "Embryonic microglia influence developing hypothalamic glial populations," *Journal of Neuroinflammation,* vol. 17, no. 1, p. 146, 2020.

[2540] J. M. Rosin, S. Sinha, J. Biernaskie and D. M. Kurrasch, "A subpopulation of embryonic microglia respond to maternal stress and influence nearby neural progenitors," *Developmental Cell,* vol. 56, no. 9, pp. 1326-1345, 2021.

[2541] E. Haroon, A. H. Miller and G. Sanacora, "Inflammation, glutamate, and glia: A trio of trouble in mood disorders," *Neuropsychopharmacology,* vol. 42, no. 1, pp. 193-215, 2017.

[2542] G. Warhaftig, D. Almeida and G. Turecki, "Early life adversity across different cell- types in the brain," *Neuroscience and Behavioral Reviews,* vol. 148, p. 105113, 2023.

[2543] S. G. Torres-Platas, C. Hercher, M. A. Davoli, G. Maussion, B. Labonté, G. Turecki and N. Mechawar, "Astrocytic hypertrophy in anterior cingulate white matter of depressed suicides," *Neuropsychopharmacology,* vol. 36, no. 3, p. 2650–2658, 2011.

[2544] Y. G. Byun, N.-S. Kim, G. Kim, Y.-S. Jeon, J. B. Choi, C.-W. Park, K. Kim, H. Jang, J. Kim, E. Kim, Y.-M. Han, K.-J. Yoon, S.-H. Lee and W.-S. Chung, "Stress induces behavioral abnormalities by increasing expression of phagocytic receptor MERTK in astrocytes to promote synapse phagocytosis," *Immunity,* vol. 56, no. 9, pp. 2105-2120.e13, 2023.

[2545] S. Z. Zhang, Q. Q. Wang, Q. Q. Yang, H. Y. Gu, Y. Q. Yin, Y. D. Li, J. C. Hou, R. Chen, Q. Q. Sun, Y. F. Sun, G. Hu and J. W. Zhou, "NG2 glia regulate brain innate immunity via TGF-β2/TGFBR2 axis," *BCM,* vol. 15, p. 204, 2019.

[2546] F. Birkey, M. Kloc, M. H. I. Chivali, M. Wilson, D. Christoffel, T. Chen, M. A. Frohman, J. K. Robinson, S. J. Russo, A. Maffei and A. Aguirre, "Genetic and stress-induced loss of NG2 glia triggers emergence of depressive-like behaviors through reduced secretion of FGF2," *Neuron,* vol. 88, no. 5, pp. 941-956, 2015.

[2547] S. C. Lin and D. E. Bergles, "Synaptic signaling between neurons and glia," *Glia,* vol. 47, no. 3, pp. 290-298, 2004.

[2548] X. Zhang, Y. Liu, X. Hong, X. Li, C. K. Meshul, C. Moore, Y. Yang, Y. Han, W.-G. Li, X. Qi, H. Lou, S. Duan, T.-L. Xu and X. Tong, "NG2 glia-derived GABA release tunes inhibitory synapses and contributes to stress-induced anxiety," *Nature Communications,* vol. 12, no. 1, p. 5740, 2012.

[2549] A. L. Moon, N. E. Clifton, N. Wellard, K. L. Thomas, J. Hall and N. M. Brydges, "Social interaction following prepubertal stress alters prefrontal gene expression associated with cell signalling and oligodendrocytes," *Translational Psychiatry,* vol. 12, no. 1, p. 561, 2022.

[2550] P. E. Lutz, A. Tanti, A. Gasecka, B.-B. S., K. J. J., Y. Zhou, G. G. Chen, M. Wakid, M. Shaw, D. Almeida, M.-A. Chay, J. Yang, V. Larivière, M.-N. M'Boutchou, L. C. van Kempen, V. Yerko, J. Prud'homme, M. A. Davoli and Vaillancourt, "Association of a history of child abuse with impaired myelination in the anterior cingulate cortex: Convergent epigenetic, transcriptional, and morphological evidence," *American Journal of Psychiatry,* vol. 174, no. 12, pp. 1185-1194, 2017.

[2551] A. G. Kokkosis, M. M. Madeira, M. M. R. and S. E. Tsirka, "Chronic stress disrupts the homeostasis and progeny progression of oligodendroglial lineage cells, associating immune oligodendrocytes with prefrontal cortex hypomyelination," *Molecular Psychiatry,* vol. 27, no. 6, pp. 2833-2848, 2022.

[2552] G. Poggi, J. Albiez and C. R. Pryce, "Effects of chronic social stress on oligodendrocyte proliferation-maturation and myelin status in prefrontal cortex and amygdala in adult mice," *Neurobiology of Stress,* vol. 18, p. 100451, 2022.

[2553] M. Flanigan, H. Aleyasin, A. Takahashi, S. A. Golden and S. J. Russo, "An emerging role for the lateral habenula in aggressive behavior," *Pharmacology, Biochemistry, and Behavior,* vol. 162, pp. 79-86, 2017.

[2554] P. Guan, C. Huang, Q. Lan, S. Huang, P. Zhou and C. Zhang, "Activation of ventral tegmental area dopaminergic neurons ameliorates anxiety-like behaviors in single prolonged stress-induced PTSD model rats," *Neurochemistry International,* vol. 161, p. 105424, 2022.

[2555] A. Teissier, C. le Magueresse, J. Olusakin, B. L. S. A. da Costa, A. M. De Stasi, A. Bacci, Y. I. Kawasawa, V. A. Vaidya and P. Gaspar, "Early-life stress impairs postnatal oligodendrogenesis and adult emotional behaviour through activity-dependent mechanisms," *Molecular Psychiatry,* vol. 25, no. 6, pp. 1159-1174, 2020.

[2556] G. Treccani, H. Yigit, T. Linger, V. Schleuβner, F. Mey, M. A. van der Kooij, W. M., D. P. Herzog, M. Linke, M. Fricke, M. J. Schmeisser, G. Wegener, T. Mittmann, J. Trotter and M. M. B., "Early life adversity targets the transcriptional signature of hippocampal NG2+ glia and affects voltage gated sodium (Nav) channels properties," *Neurobiological Stress,* vol. 15, p. 100338, 2021.

[2557] M. Munyeshyaka and R. D. Fields, "Oligodendroglia are emerging players in several forms of learning and memory," *Communications Biology,* vol. 5, no. 1, p. 1148, 2022.

[2558] K. Q. Fan, Y. Y. Li, H. L. Wang, X. T. Mao, J. X. Guo, F. Wang, L. J. Huang, Y. N. Li, X. Y. Ma, Z. J. Gao, W. Chen, D. D. Qian, W. J. Xue, Q. Cao, L. Zhang, L. Shen, L. Zhang, C. Tong, J. Y. Zhong, W. Lu, L. Lu, K. M. Ren and G. Zhong, "Stress-induced metabolic disorder in peripheral CD4+ T cells leads to anxiety-like behavior," *Cell,* vol. 179, no. 4, pp. 864-879.e19, 2019.

[2559] J. P. Curley, R. Mashoodh and F. A. Champagne, "Epigenetics and the origins of paternal effects," *Hormones and Behavior,* vol. 59, no. 3, pp. 306-314, 2011.

[2560] C. Gregg, J. Zhang, B. Weissbourd, S. Luo, G. P. Schroth, D. Haig and C. Dulac, "High Resolution Analysis of Parent-of-Origin Allelic Expression in the Mouse Brain," *Science,* vol. 329, no. 5992, pp. 643-648, 2010b.

[2561] R. Mott, W. Yuan, P. Kaisaki, X. Gan, J. Cleak, A. Edwards, A. Baud and J. Flint, "The architecture of parent-of-origin effects in mice," *Cell,* vol. 156, no. 1-2, p. 332–342, 2014.

[2562] C. Gregg, J. Zhang, J. E. Butler, D. Haig and C. Dulac, "Sex-Specific Parent-of-Origin Allelic Expression in the Mouse Brain," *Science,* vol. 329, no. 5992, pp. 682-685, 2010a.

[2563] A. Gimelbrant, J. Hutchinson, B. R. Thompson and A. Chess, "Widespread monoallelic expression on human autosomes," *Science,* vol. 318, no. 5853, pp. 1136-1140, 2007.

[2564] E. M. O'Learly and P. J. Bonthuis, "Mom genes and dad genes: genomic imprinting in the regulation of social behaviors," *Epigenomics,* vol. 17, no. 8, pp. 555-573, 2025.

[2565] A. S. Garfield, M. Cowley, F. M. Smith, K. Moorwood, J. E. Stewart-Cox, K. Gilroy, S. Baker, J. Xia, J. W. Dalley, L. D. Hurst, L. S. Wilkinson, A. R. Isles and A. Ward, "Distinct physiological and behavioural functions for parental alleles of imprinted Grb10," *Nature,* pp. 534-538, 2011.

[2566] M. J. Chomsky, Director, *Star Trek: And the Children Shall Lead.* [Film]. Paramount Studios, 1968.

[2567] S. Tzu, The Art of War, Project Gutenberg, 1910/2021.

[2568] L. R. Huesmann, "An information processing model for the development of aggression," *Aggressive Behavior,* vol. 14, pp. 13-24, 1988.

[2569] S. Stavrou, N. C. Nicolaides, E. Critselis, C. Darviri, E. Charmandari and G. P. Chrousos, "Paediatric stress: from neuroendocrinology to contemporary disorders," *European Journal of Clinical Investigation,* vol. 47, no. 3, pp. 262-269, 2017.

[2570] R. H. Mulder, E. Walton, A. Neumann, L. C. Houtepen, J. F. Felix, M. J. Bakermans-Kranenburg, M. Suderman, H. Tiemeier, M. H. van Ijzendoorn, C. L. Relton and C. A. M. Cecil, "Epigenomics of being bullied: changes in DNA methylation following bullying exposure," *Epigenetics,* vol. 17, no. 6-7, pp. 750-764, 2020.

[2571] R. Shouval, "Payoff expectancies between contingent serial moves of victim and aggressor in three hypothetical conflict situations," *Aggressive Behavior,* vol. 17, no. 3, pp. 155-169, 1991.

[2572] S. Bailey-Shah, "Expert: Suppressing kids' play-fighting may erode self-esteem," Portland, 2013.

[2573] N. Gibbs, "The growing backlash against overparenting," *Time,* 20 11 2009.

[2574] K. G. Odenweller, M. Booth-Butterfield and K. Weber, "Investigating helicopter parenting, family environments, and relational outcomes for millennials," *Communication Studies,* vol. 65, no. 4, pp. 407-425, 2014.

[2575] V. Darlow, J. M. Norvilitis and P. Schuetze, "The relationship between helicopter parenting and adjustment to college," *Journal of Child and Family Studies,* p. 2291–2298, 2017.

[2576] T. Garner, The Relation of Helicopter Parenting to Depressive Symptomatology in Emerging Adults (M.S. Thesis), Mobile, AL: University of South Alabama, 2017, p. 58.

[2577] L. M. Padilla-Walker and L. J. Nelson, "Black Hawk down?: Establishing helicopter parenting as a distinct construct from other forms of parental control during emerging adulthood," *Sociological Spectrum,* vol. 31, no. 4, pp. 399-418, 2012.

[2578] K. Mrnjaus, "The child's right to play?!," *Croatian Journal of Education,* vol. 16, no. 1, pp. 217-233, 2014.

[2579] L. Wray-Lake, A. C. Crouther and S. M. McHale, "Developmental patterns in decision-making autonomy across middle childhood and adolescence: European American parents' perspectives," *Child Development,* vol. 81, no. 2, pp. 636-651, 2010.

[2580] C. Friedersdorf, "Working mom arrested for letting her 9-year-old play alone at park," *The Atlantic,* 15 7 2014.

[2581] M. Thernstrom, "The anti-helicopter parent's plea: Let our kids play!," *New York Times Magazine,* 19 10 2016.

[2582] E. G. Porter, "Tort liability in the age of the helicopter parent," *Alabama Law Review,* vol. 64, p. 533, 2012.

[2583] Wikipedia, "Zero tolerance," 12 7 2023. [Online]. Available: https://en.wikipedia.org/wiki/Zero_tolerance. [Accessed 14 7 2018].

[2584] M. French-Marcelin and S. Hinger, Bullies in Blue: The Origins and Consequences of School Policing, New York: American Civil Liberties Union, 2017, p. 53.

[2585] F. Scani, *Expelled for Possession of a Butter Knife,* ABC NEWS, 2007.

[2586] N. Dreier, "Girl suspended for using child-sized butter knife to cut peach at school," *Dayton Daily News,* 8 November 2016.

[2587] M. Seidon, "Girl suspended from middle school for cutting peach with child butter knife," 14 November 2016. [Online]. Available: https://www.local10.com/education/2016/11/14/girl-suspended-from-middle-school-for-cutting-peach-with-child-butter-knife/. [Accessed 4 June 2023].

[2588] J. Moore, "Girl suspended from school for cutting peach with child's butter knife," 19 November 2016. [Online]. Available: https://www.wjhl.com/news/girl-suspended-from-school-for-cutting-peach-with-childs-butter-knife/. [Accessed 11 August 2025].

[2589] D. Moye, "5-year-old suspended for pretending stick is a gun on playground," *Huffington Post,* 29 March 2017.

[2590] T. Starnes, "'Stick gun' gets 5-year-old suspended from North Carolina school," 29 March 2017. [Online]. Available: https://www.foxnews.com/opinion/stick-gun-gets-5-year-old-suspended-from-north-carolina-school. [Accessed 3 June 2023].

[2591] United States Education Department Office for Civil Rights, "Data Collection, 2017-18 State and National Estimations," United States Department of Education, Washington DC, released 2021.

[2592] K.-A. K. Henry, R. M. Catagnus, A. K. Griffith and Y. A. Garcia, "Ending the school-to-prison pipeline: Perception and experience with zero-tolerance policies and interventions to address racial inequality," *Behavior Analysis in Practice,* vol. 15, no. 4, pp. 1254-1263, 2021.

[2593] J. C. Williams, A. Andreou, T. Lemelle, E. E. Coates and J. Q. Bostic, "Editorial: Racial battle fatigue: The toll of policing Black students," *Journal of the American Academy of Child and Adolescent Psychiatry,* vol. 61, no. 8, pp. 977-979, 2022.

[2594] A. Flitner, S. McQuillin, M. Kornbluh and D. Thompson, "Spotlighting racism in schools: Teacher mentors and the mediating effect of school safety," *American Journal of Community Psychology,* vol. 73, no. 1-2, pp. 133-143, 2024.

[2595] S. Pownall, "A, B. C. D. STPP: How School Discipline Feeds the School-to-Prison Pipeline," New York Civil Liberties Union, New York, 2013.

[2596] I. Lambie and I. Randell, "The impact of incarceration on juvenile offenders," *Clinical Psychology Review,* vol. 33, no. 3, pp. 448-459, 2013.

[2597] B. Justice, "Hobbling: The effects of proactive policing and mass imprisonment on children's education," *Annual Review of Law and Social Science,* vol. 17, pp. 31-51, 2021.

[2598] S. S. Taylor, "School disturbance laws: What they are, how they are used, and how they impact students," *SAGE Open,* vol. 14, no. 2, 2024.

[2599] Annie E. Casey Foundation, "At Onset of the COVID-19 Pandemic, Dramatic and Rapid Reductions in Youth Detention," 23 April 2020a. [Online]. Available: https://www.aecf.org/blog/youth-detention-admissions-remain-low-but-releases-stall-despite-covid-19/. [Accessed 6 June 2023].

[2600] Annie E. Casey Foundation, "Youth Detention Admissions Remain Low, But Releases Stall Despite COVID-19," 9 July 2020b. [Online]. Available: https://www.aecf.org/blog/youth-detention-admissions-remain-low-but-releases-stall-despite-covid-19.

[2601] A. D. Pellegrini, "Elementary-school children's rough-and-tumble play and social competence," *Developmental Psychology,* vol. APA, no. 24, pp. 802-806, 1988.

[2602] A. D. Pellegrini, "Elementary school children's rough-and-tumble play," *Early Childhood Research Quarterly,* vol. 4, no. 2, pp. 245-260, 1989.

[2603] S. M. Pellis, V. C. Pellis and J. R. Ham, "Play fighting revisited: its design features and how they shape our understanding of its mechanisms and functions," *Frontiers in Ethology,* vol. 3, p. 1362052, 2024.

[2604] P. E. Smaldino, E. Palagi, G. M. Burghardt and S. M. Pellis, "The evolution of two types of play," *Behavioral Ecology,* vol. 30, no. 5, pp. 1388-1397, 2019.

[2605] E. Palagi, "Not just for fun! Social play as a springboard for adult social competence in human and non-human primates," *Behavioral Ecology and Sociobiology,* vol. 72, p. 90, 2018.

[2606] D. Ellingworth, R. Barrett, A. Partington and J. Radcliffe, "Utilising play-based approaches in higher education to facilitate graduates transition into professional practice," *Journal of Hospitality, Leisure, Sport and Tourism Education,* vol. 36, p. 100557, 2025.

[2607] M. Francesconi, A. Galotti, K. Ludmann and E. Palagi, "Late bloomers in play: social isolation and playful skills in an adult bonobo," *Primates,* pp. 1-7, 2025.

[2608] K. Vékony, F. Prónik and P. Pongrácz, "Personalized dominance – a questionnaire-based analysis of the associations among personality traits and social rank of companion dogs," *Applied Animal Behaviour Science,* vol. 247, p. 105544, 2022.

[2609] E. Palagi, "Adult play and the evolution of tolerant and cooperative societies," *Neuroscience and Biobehavioral Reviews,* vol. 148, p. 105124, 2023.

[2610] R. Stark and S. M. Pellis, "Male Long Evans rats reared with a Fisher-344 peer during the juvenile period show deficits in social competency: a role for play," *International Journal of Play,* vol. 9, no. 1, pp. 76-91, 2020.

[2611] R. A. Stark and S. M. Pellis, "Using the 'stranger test' to assess social competency in adult female Long Evans rats reared with a Fischer 344 partner," *Behavioural Processes,* vol. 192, p. 104492, 2021.

[2612] S. M. Pellis, V. C. Pellis, J. R. Ham and R. A. Stark, "Play fighting and the development of the social brain: The rat's tale," *Neuroscience and Behavioral Reviews,* vol. 145, p. 105037, 2023.

[2613] M. A. Cooper, J. A. Grizzell, C. J. Whitten and G. M. Burghardt, "Comparing the ontogeny, neurobiology, and function of social play in hamsters and rats," *Neuroscience and Biobehavioral Reviews,* vol. 147, p. 105102, 2023.

[2614] E. J. M. Achterberg and L. J. M. J. Vanderschuren, "The neurobiology of social play behaviour: Past, present and future," *Neuroscience and Biobehavioral Reviews,* vol. 152, p. 105319, 2023.

[2615] H. C. Bell, S. M. Pellis and B. Kolb, "Juvenile peer play experience and the development of the orbitofrontal and medial prefrontal cortices," *Behavioural Brain Research,* vol. 207, no. 1, pp. 7-13, 2010.

[2616] S. M. Pellis, V. C. Pellis, B. T. Himmler and T. Brett, "How play makes for a more adaptable brain: A comparative and neural perspective," *American Journal of Play,* vol. 7, no. 1, pp. 73-98, 2014.

[2617] H. C. Bell, D. McCaffrey, M. L. Forgie, B. Kolb and S. M. Pellis, "The role of the medial prefrontal cortex in the play fighting in rats," *Behavioral Neuroscience,* vol. 123, no. 6, pp. 1158-1168, 2009.

[2618] R. A. Stark, B. Brinkman, R. L. Gibb, A. N. Iwaniuk and S. M. Pellis, "Atypical play experiences in the juvenile period has an impact on the development of the medial prefrontal cortex in both male and female rats," *Behavioural Brain Research,* vol. 439, p. 114222, 2023.

[2619] B. T. Himmler, S. M. Pellis and B. Kolb, "Juvenile play experience primes neurons in the medial prefrontal cortex to be more responsive to later experiences," *Neuroscience Letters,* vol. 556, pp. 42-45, 2013.

[2620] S. M. Pellis, B. T. Himmler, S. M. Himmler and V. C. Pellis, "Chapter 12- Rough-and-tumble play and the development of the social brain: What do we know, how do we know it, and what do we need to know (pp 315-337)," in *Neurobiology of Brain and Behavioral Development*, R. Gibb and B. Kolb, Eds., London, Academic Press, 2017, p. 496.

[2621] L. Ahnert, L. Teufl, N. Ruiz, B. Piskernik, B. Supper, S. Remiorz, A. Gesing and K. Nowacki, "Father-child playing during the preschool years and child internalizing behaviors: between robustness and vulnerability," *Infant Mental Health Journal,* vol. 38, no. 6, pp. 743-756, 2017.

[2622] S. Anderson, W. Qiu and S. J. Wheeler, "The quality of father-child rough-and-tumble play and toddlers' aggressive behavior in China," *Infant Mental Health Journal,* vol. 38, no. 6, pp. 726-742, 2017.

[2623] J. Stgeorge and E. Freeman, "Measurement of father-child rough-and-tumble play and its relations to child behavior," *Infant Mental Health Journal,* vol. 38, no. 6, pp. 709-725, 2017.

[2624] D. Paquette and J. M. StGeorge, "Proximate and ultimate mechanisms of human father-child rough-and-tumble play," *Neuroscience and Biobehavioural Reviews,* vol. 149, p. 105151, 2023.

[2625] C. K. Broadfoot, C. Lenell, C. A. Kelm-Nelson and M. R. Ciucci, "Effects of social isolation on 50-kHz ultrasonic vocalizations, affective state, cognition, and neurotransmitter concentrations in the ventral tegmental and locus coeruleus of adult rats," *Behavioural Brain Research,* vol. 473, p. 114157, 2023.

[2626] A. Menashe-Grinberg and N. Atzaba-Poria, "Mother-child and father-child play interactions: The importance of parental playfulness as a moderator of the links between parental behavior and child negativity," *Infant Mental Health Journal,* vol. 28, no. 6, pp. 772-784, 2017.

[2627] S. M. Pellis and V. C. Pellis, "What is play fighting and what is it good for?," *Learning and Behavior,* vol. 45, pp. 355-366, 2017.

[2628] M. Wenner, "The serious need for play," *Scientific American,* p. 8, 2009.

[2629] A. S. Lillard, M. D. Lerner, E. J. Hopkins, R. A. Dore, E. D. Smith and C. M. Palmquist, "The impact of pretend play on children's development: A review of the evidence," *Psychological Bulletin,* vol. 139, no. 1, pp. 1-34, 2013.

[2630] S. Feshbach, "The catharsis hypothesis and some consequences of interaction with aggressive and neutral play objects," *Journal of Personality,* vol. 24, no. 4, pp. 449-462, 1956.

[2631] P. Mussen and E. Rutherford, "Effects of aggressive cartoons on children's aggressive play," *Journal of Abnormal and Social Psychology,* vol. 62, no. 2, pp. 461-464, 1961.

[2632] K. K. Fehr and S. W. Russ, "Aggression in pretend play and aggressive behavior in the classroom," *Early Education and Development,* vol. 24, no. 3, pp. 332-345, 2013.

[2633] A. Keary and E. Keary, "The Dwarfs and the Fairies," in *Young Folks' Shelf of Books: The Junior Classics*, vol. 2: Folk Tales and Myths, W. Patten, Ed., New York, P. F. Collier & Son/ Project Gutenberg, 1912/ 2018.

[2634] E. Palagi, "Social play in bonobos (Pan paniscus) and chimpanzees (Pan troglodytes): Implications for natural social systems and interindividual relationships," *American Journal of Physical Anthropology,* vol. 129, no. 3, pp. 418-426, 2005.

[2635] K. Koops, "Animal behavior: Chimpanzee play and the evolutionary roots of cooperation," *Current Biology,* vol. 35, no. 3, pp. R71-R73, 2025.

[2636] S. M. Pellis, E. Castañeda, M. M. McKenna, L. T. Tran-Nguyen and I. Q. Whishaw, "The role of the striatum in organizing sequences of play fighting in neonatally dopamine-depleted rats," *Neuroscience Letters,* vol. 158, no. 1, pp. 13-15, 1993.

[2637] S. Uchida, K. Hara, A. Kobayashi, K. Otsuki, H. Yamagata, T. Hobara, T. Suzuki, N. Miyata and Y. Watanabe, "Epigenetic status of Gdnf in the ventral striatum determines susceptibility and adaptation to daily stressful events," *Neuron,* vol. 69, no. 2, pp. 359-372, 2011.

[2638] M. D. Murphy and E. A. Heller, "Convergent actions of stress and stimulants via epigenetic regulation of neural circuitry (Review)," *Trends in Neuroscience,* vol. 45, no. 12, pp. 955-967, 2022.

[2639] M. A. Bermudez and W. Schultz, "Responses of amygdala neurons to positive reward-predicting stimuli depend on background reward (contingency) rather than stimulus-reward pairing (contiguity)," *Journal of Neurophysiology,* vol. 103, no. 3, pp. 1158-1170, 2010.

[2640] N. C. Ferrara and M. Opendak, "Amygdala circuit transitions supporting developmentally-appropriate social behavior," *Neurobiology of Learning and Memory,* vol. 201, p. 107762, 2023.

[2641] S. A. Fleary, R. W. Heffer, E. L. McKyer and D. A. Newman, "Dopamine signals for reward value and risk: Basic and recent data," *Behavioral and Brain Functions, Article 24,* vol. 6, 2010.

[2642] Y. Li, G. Vanni-Mercier, J. Isnard, F. Mauguière and J. C. Dreher, "The neural dynamics of reward value and risk coding in the human orbitofrontal cortex," *Brain,* vol. 139, no. 4, pp. 1295-1309, 2016.

[2643] M. O'Neill and W. Schultz, "Coding of reward risk by orbitofrontal neurons is mostly distinct from coding of reward value," *Neuron,* vol. 68, no. 4, pp. 789-800, 2010.

[2644] W. Schultz, "Dopamine signals for reward value and risk: Basic and recent data," *Behavioral and Brain Functions,* vol. 6, no. 24, 2010.

[2645] I. L. Elias, I. K. Succi, M. D. Schaffler, W. Foster, M. A. Gradwell, M. Bohic, A. Fushiki, L. L. Ejoh, R. Schwark, R. Frazer, B. Bistis, J. E. Burke, V. Saltz, J. E. Boyce, A. Jhumka, R. M. Costa, V. E. Abraira and I. Abdus-Saboor, "Touch neurons underlying dopaminergic pleasurable touch and sexual receptivity," *Cell,* vol. 186, no. 3, pp. 577-590, 2023.

[2646] V. Pando-Naude, T. E. Matthews, A. Højlund, S. Jakobsen, K. Østergaard, E. Johnsen, E. A. Garza-Villarreal, M. A. G. Witek, V. Penhune and P. Vuust, "Dopamine dysregulation in Parkinson's disease flattens the pleasurable urge to move to musical rhythms," *European Journal of Neuroscience,* vol. 59, no. 1, pp. 101-118, 2024.

[2647] C. J. Burke, P. N. Tobler, M. Baddeley and W. Schultz, "Neuronal mechanisms of observational learning," *Proceedings of the National Academy of Science USA,* pp. 14431-14436, 2010.

[2648] K. M. Witt, D. N. Harper and B. A. Ellenbroek, "The role of the dopamine D1 receptor in anticipatory pleasure and social play," *Neuropharmacology,* vol. 264, p. 110225, 2025.

[2649] W. Schultz, "Multiple dopamine functions at different time courses," *Annual Review of Neuroscience,* vol. 30, pp. 259-288, 2007.

[2650] P. N. Tobler, C. D. Fiorillo and W. Schultz, "Adaptive coding of reward value by dopamine Neurons," *Science,* vol. 307, no. 5715, pp. 1642-1645, 2005.

[2651] P. N. Tobler, J. P. O'Doherty, R. Dolan and W. Schultz, "Reward value coding distinct from risk attitude-related uncertainty coding in human reward systems," *Journal of Neurophysiology,* vol. 97, no. 2, pp. 1621-1632, 2007.

[2652] A. Noritake and M. Isoda, "The macaque medial prefrontal cortex simultaneously represents self and others' reward prediction error," *Cell Reports,* vol. 44, no. 3, p. 115358, 2025.

[2653] S. M. Pellis, "Conservative motor systems, behavioral modulation and neural plasticity," *Behavioural Brain Research,* vol. 2014, no. 1, pp. 25-29, 2010.

[2654] S. M. Pellis, V. C. Pellis and H. C. Bell, "The function of play in the development of the social brain," *American Journal of Play,* vol. 2, no. 3, pp. 278-296, 2010.

[2655] S. M. Siviy and J. Panksepp, "In search of the neurobiological substrates for social playfulness in mammalian brains," *Neuroscience and Biobehavioral Reviews,* vol. 35, no. 9, pp. 1821-1830, 2011.

[2656] W. W. Beatty and K. B. Costello, "Naloxone and play fighting in juvenile rats," *Pharmacology Biochemistry and Behavior,* vol. 17, no. 5, pp. 905-907, 1982.

[2657] T. Yamasaki, Y. Kiyokawa, A. Munetomo and Y. Takeuchi, "Naloxone increases conditioned fear responses during social buffering in male rats," *European Journal of Neuroscience,* vol. 59, no. 12, pp. 3256-3272, 2024.

[2658] J. Panksepp, J. Jalowiec, F. G. DeEskinazi and P. Bishop, "Opiates and play dominance in juvenile rats," *Behavioral Neuroscience,* vol. 99, no. 3, pp. 441-453, 1985.

[2659] H. G. Guard, J. D. Newman and R. L. Roberts, "Morphine administration selectively facilitates social play in common marmosets," *Developmental Psychobiology,* vol. 41, no. 1, p. 37–49, 2002.

[2660] S. M. Pellis and M. M. McKenna, "Intrinsic and extrinsic influences on play fighting in rats: Effects of dominance, partner's playfulness, temperament and neonatal exposure to testosterone propionate," *Behavioural Brain Research,* vol. 50, no. 1, pp. 135-145, 1992.

[2661] S. M. Pellis and M. M. McKenna, "What do rats find rewarding in play fighting? — An analysis using drug-induced non-playful partners," *Behavioural Brain Research,* vol. 68, no. 1, pp. 65-73, 1995.

[2662] D. H. Thor and W. R. J. Holloway, "Scopolamine blocks play fighting behavior in juvenile rats," *Physiology & Behavior,* vol. 30, no. 4, pp. 545-549, 1983.

[2663] D. H. Thor and W. R. J. Holloway, "Social play in juvenile rats during scopolamine withdrawal," *Physiology & Behavior,* vol. 32, no. 2, pp. 217-220, 1984.

[2664] L. I. Wilson, R. A. Bierley and W. W. Beatty, "Cholinergic agonists suppress play fighting in juvenile rats," *Pharmacology Biochemistry and Behavior,* vol. 24, no. 5, pp. 1157-1159, 1986.

[2665] W. W. Beatty, A. M. Dodge, L. J. Dodge, K. White and J. Panksepp, "Psychomotor stimulants, social deprivation and play in juvenile rats," *Pharmacology Biochemistry and Behavior,* vol. 16, no. 3, pp. 417-422, 1982.

[2666] W. W. Beatty, K. B. Costello and S. L. Berry, "Suppression of play fighting by amphetamine: Effects of catecholamine antagonists, agonists and synthesis inhibitors," *Pharmacology Biochemistry and Behavior,* vol. 20, no. 5, pp. 747-755, 1984.

[2667] E. F. Field and S. M. Pellis, "Differential effects of amphetamine on the attack and defense components of play fighting in rats," *Physiology & Behavior,* vol. 56, no. 2, pp. 325-330, 1994.

[2668] S. M. Pellis and V. C. Pellis, "Play fighting of rats in comparative perspective: A schema for neurobehavioral analyses," *Neuroscience & Biobehavioral Reviews,* vol. 23, no. 1, pp. 87-101, 1998.

[2669] A. Foroud and S. M. Pellis, "The development of roughness in the play fighting of rats: A Laban movement analysis perspective," *Developmental Psychobiology,* vol. 42, no. 1, pp. 35-43, 2003.

[2670] A. Bundy and H. Kuhaneck, "The occupation of play," *American Journal of Occupational Therapy,* vol. 78, no. 4, p. 7804170010, 2024.

[2671] L. L. Sharpe, "Play fighting does not affect subsequent fighting success in wild meerkats," *Animal Behaviour,* vol. 69, no. 5, pp. 1023-1029, 2005.

[2672] S. Y. Cheng and Y. Delville, "Play fighting and corticotropin-releasing hormone in the lateral septum of golden hamsters," *Neuroscience,* vol. 169, no. 1, pp. 236-245, 2010.

[2673] N. Dedic, C. Kühne, K. S. Gomes, J. Hartmann, K. J. Ressler, M. V. Schmidt and J. M. Deussing, "Deletion of CRH from GABAergic forebrain neurons promotes stress resilience and dampens stress-induced changes in neuronal activity," *Frontiers in Neuroscience,* vol. 13, 2019.

[2674] S. Mondoloni, P. Molina, S. Lecca, C.-H. Wu, L. Michel, D. Osypenko, F. Cachin, M. Flanigan, M. Congiu, A. L. Lalive, T. Kash, F. Deng, Y. Li and M. Mameli, "Serotonin release in the habenula during emotional contagion promotes resilience," *Science,* vol. 385, no. 6713, pp. 1081-1086, 2024.

[2675] E. Elharrar, G. Warhaftig, O. Issler, Y. Sztainberg, Y. Dikshtein, R. Zahut, L. Redlus, A. Chen and G. Yadid, "Overexpression of Corticotropin-Releasing Factor Receptor Type 2 in the bed nucleus of the stria terminalis improves posttraumatic stress disorder-like symptoms in a model of incubation of fear," *Biological Psychiatry,* vol. 74, no. 11, pp. 827-836, 2013.

[2676] S. Maul, I. Giegling, C. Fabbri, F. Corponi, A. Serretti and D. Rujescu, "Genetics of resilience: Implications from genome-wide association studies and candidate genes of the stress response system in posttraumatic stress disorder and depression," *American Journal of Medical Genetics Part B,* vol. 77, no. 94, pp. 77-94, 2020.

[2677] M. Ryan and R. Ryznar, "The molecular basis of resilience: A narrative review," *Frontiers in Psychiatry,* vol. 13, 2022.

[2678] T. D. Wager, M. L. Davidson, B. L. Hughes, M. A. Lindquist and K. N. Ochsner, "Prefrontal-subcortical pathways mediating successful emotional regulation," *Neuron,* vol. 59, no. 6, pp. 1037-1050, 2008.

[2679] S. M. Mooney and E. I. Varlinskaya, "Acute prenatal exposure to ethanol and social behavior: Effects of age, sex, and timing of exposure," *Behavioural Brain Research,* vol. 216, no. 1, pp. 358-364, 2011.

[2680] A. H. Metwalli, A. Pross, E. Desfilis, A. Abellán and L. Medina, "Mapping of corticotropin-releasing factor, receptors and binding protein mRNA in the chicken telencephalon throughout development," *Journal of Comparative Neurology,* vol. 531, no. 14, pp. 1389-1424, 2023.

[2681] R. Chudoba and J. Dabrowska, "Distinct populations of corticotropin-releasing factor (CRF) neurons mediate divergent yet complementary defensive behaviors in response to a threat," *Neuropharmacology,* vol. 228, p. 109416, 2023.

[2682] J. Frasnelli, B. Schuster and T. Hummel, "Interactions between olfaction and the trigeminal system: what can be learned from olfactory loss," *Cerebral Cortex,* vol. 17, no. 10, pp. 2268-2275, 2007.

[2683] C. E. Barrett, M. E. Modi, B. C. Zhang, H. Walum, K. Inoue and L. J. Young, "Neonatal melanocortin receptor agonist treatment reduces play fighting and promotes adult attachment in prairie voles in a sex-dependent manner," *Neuropharmacology,* vol. 85, pp. 357-366, 2014.

[2684] Z. V. Johnson and L. J. Young, "Neurobiological mechanisms of social attachment and pair bonding," *Current Opinion in Behavioral Sciences,* vol. 3, pp. 38-44, 2015.

[2685] F. van Winden, M. Stallen and K. R. Ridderinkhof, "On the nature, modeling, and neural bases of social ties," in *Neuroeconomics, Vol. 20, Advances in Health Economics and Health Services*, vol. 20, Emerald Group Publishing Limited, 2008, pp. 125 - 159.

[2686] P. Smith and J. M. StGeorge, "Play fighting (rough-and-tumble play) in children: developmental and evolutionary perspectives," *International Journal of Play,* vol. 12, no. 1, pp. 113-126, 2023.

[2687] C. L. Masten, N. I. Eisenberger, L. A. Borofsky, J. H. Pfeifer, K. McNealy, J. C. Mazziotta and M. Dapretto, "Neural correlates of social exclusion during adolescence: Understanding the distress of peer rejection," *Scan,* vol. 4, no. 2, pp. 143-157, 2009.

[2688] S. M. Pellis, E. F. Field, L. K. Smith and V. C. Pellis, "Multiple differences in the play fighting of male and female rats. Implications for the causes and functions of play," *Neuroscience & Biobehavioral Reviews,* vol. 2, no. 1, pp. 105-120, 1997.

[2689] S. M. Pellis, V. C. Pellis and J. R. Ham, "The rough-and-tumble play of rats as a natural behavior suitable for studying the social brain," *Frontiers in Behavioral Neuroscience,* vol. 16, 2022.

[2690] C. L. Marley, T. M. Pollard, R. A. Barton and S. E. Street, "A systematic review of sex differences in rough and tumble play across non-human mammals," *Behavioral Ecology and Sociobiology,* vol. 76, p. 158, 2022.

[2691] S. M. Pellis and V. C. Pellis, "Rough-and-tumble play and the development of the social brain," *Current Directions in Psychological Science,* vol. 16, no. 2, pp. 95-98, 2007.

[2692] A. Carraro, E. Gobbi and A. Moè, "Brief report: Play fighting to curb self-reported aggression in young adolescents," *Journal of Adolescence,* vol. 37, no. 8, pp. 1303-1307, 2014.

[2693] D. Einon and M. Portegal, "Enhanced defense in adult rats deprived of playfighting experience as juveniles," *Aggressive Behavior,* vol. 17, no. 1, pp. 27-40, 1991.

[2694] N. Asensio, E. Zandonà, J. C. Dunn and J. Cristóbal-Azkarate, "Socioecological correlates of social play in adult mantled howler monkeys," *Animal Behaviour,* vol. 186, pp. 219-229, 2022.

[2695] S. Nunes and M. P. M. Montemayor, "Multiple benefits of juvenile play: A ground squirrel's perspective," *Neuroscience and Biobehavioral Reviews,* vol. 147, p. 105099, 2023.

[2696] I. Q. Whishaw, C. J. Burke and S. M. Pellis, "Does play shape hand use skill in rats?," *Experimental Brain Research,* vol. 239, no. 6, pp. 1895-1909, 2021.

[2697] J. Hailman, Optical Signals: Animal Communication and Light, Bloomington: Indiana University Press, 1977, p. 416.

[2698] M. Bekoff, "The communication of play intention: Are play signals functional?," *Semiotica,* vol. 15, no. 3, pp. 231-239, 1975.

[2699] V. Maglieri, A. Zanoli, F. Mastrandrea and E. Palagi, "Don't stop me now, I'm having such a good time! Czechoslovakian wolfdogs renovate the motivation to play with a bow," *Current Zoology,* vol. 69, no. 1, pp. 50-58, 2023.

[2700] M. Spinka, R. C. Newberry and M. Bekoff, "Mammalian play: Training for the unexpected," *Quarterly Review of Biology,* vol. 76, no. 2, pp. 141-168, 2001.

[2701] M. Davila-Ross and E. Palagi, "Laughter, play faces and mimicry in animals: evolution and social functions," *Philosophical Transactions of the Royal Society B,* vol. 377, no. 1863, p. 20210177, 2022.

[2702] V. Maglieri, A. Zanoli, F. Mastrandrea and E. Palagi, "The relaxed open mouth is a true signal in dogs: demonstrating Tinbergen's ritualization process," *Animal Behaviour,* vol. 188, pp. 65-74, 2022.

[2703] S. Bagnato, L. Pedruzzi, J. Goracci and E. Palagi, "The interconnection of hierarchy, affiliative behaviours, and social play shapes social dynamics in Maremmana beef cattle," *Applied Animal Behaviour Science,* vol. 260, p. 105868, 2023.

[2704] M. Francesconi, L. Pedruzzi, S. Bagnato, J. Goracci, A. Ripamonti, M. Mele and E. Palagi, "Social play and affiliation as possible coping strategies in a group of Maremmana beef cattle," *Journal of Ethology,* vol. 42, pp. 41-52, 2023.

[2705] S. M. Pellis, D. P. O'Brien, V. C. Pellis, P. Teitelbaum, D. L. Wolgin and S. Kennedy, "Escalation of feline predation along a gradient from avoidance through 'play' to killing," *Behavioral Neuroscience,* vol. 102, no. 5, pp. 760-777, 1988.

[2706] N. Gajdoš-Kmecová, B. Peťková, J. Kottferová, V. Halls, C. Haddon, L. S. de Assis and D. S. Mills, "An ethological analysis of close-contact inter-cat interactions determining if cats are playing, fighting, or something in between," *Science Reports,* vol. 13, p. 92, 2023.

[2707] C. J. Reinhart, D. C. McIntyre, G. A. Metz and S. M. Pellis, "Play fighting between kindling-prone (FAST) and kindling-resistant (SLOW) rats," *Journal of Comparative Psychology,* vol. 120, no. 1, pp. 19-30, 2006.

[2708] G. W. Martins, G. S. Chagas, F. C. Patrício, A. O. Lira, J. Fragoso, A. B. Santos and C. G. Leandro, "Can the frequency of defensive posture adoption influence the expression of active behavior of adult rats?," *Behavioural Processes,* vol. 177, p. 104151, 2020.

[2709] H. Kamitakahara, M.-H. Monfils, M. L. Forgie, B. Kolb and S. M. Pellis, "The modulation of play fighting in rats: Role of the motor cortex," *Behavioral Neuroscience,* vol. 121, no. 1, pp. 164-176, 2007.

[2710] A. Foroud, I. Q. Whishaw and S. M. Pellis, "Experience and cortical control over the pubertal transition to rougher play fighting in rats," *Behavioural Brain Research,* vol. 149, no. 1, pp. 69-76, 2004.

[2711] A. Ilany, A. Barocas, L. Koren, M. Kam and E. Geffen, "Structural balance in the social networks of a wild mammal," *Animal Behaviour,* vol. 85, no. 6, pp. 1397-1405, 2013.

[2712] J. B. Silk, "The natural history of social bonds," *Annals of the New York Academy of Sciences,* vol. 1546, no. 1, pp. 90-99, 2025.

[2713] J. R. Ham and S. M. Pellis, "Groups of familiar male rats form unstable partner preferences when play fighting during the juvenile period," *iScience,* vol. 28, no. 6, p. 112562, 2025.

[2714] J. Massen, E. Sterck and H. de Vos, "Close social associations in animals and humans: functions and mechanisms of friendship," *Behaviour,* vol. 147, no. 11, pp. 1379-1412, 2010.

[2715] S. Nunes, E.-M. Muecke, Z. Sanchez, R. R. Hoffmeier and L. T. Lancaster, "Play behavior and motor development in juvenile Belding's ground squirrels (Spermophilus beldingi)," *Behavioral Ecology and Sociobiology,* vol. 56, no. 2, pp. 97-105, 2004.

[2716] C. M. Evans and P. F. Brain, "Some studies on endocrine influences on aggressive behavior in the golden hamster (Mesocricetus auratus Waterhouse)," *Progress in Brain Research,* vol. 41, pp. 473-480, 1974.

[2717] R. F. Guerra and C. R. Nunes, "Effects of litter size on maternal care, body weight and infant development in golden hamsters (Mesocricetus auratus)," *Behavioural Processes,* vol. 55, no. 3, pp. 127-142, 2001.

[2718] M. L. Vieira and E. Otta, "Play fighting in juvenile golden hamsters (Mesocricetus auratus): Effects of litter size and analysis of social interaction among males," *Behavioural Processes,* vol. 43, no. 3, pp. 265-273, 1998.

[2719] R. F. Guerra, M. L. Vieira, E. Takase and S. Gasparetto, "Sex differences in the play fighting activity of golden hamster infants," *Physiology & Behavior,* vol. 52, no. 1, pp. 1-5, 1992.

[2720] E. B. Bauer and B. B. Smuts, "Cooperation and competition during dyadic play in domestic dogs, Canis familiaris," *Animal Behaviour,* vol. 73, no. 3, pp. 489-499, 2007.

[2721] S. M. Pellis and G. M. Burghardt, "Play and exploration," *APA Handbook of Comparative Psychology: Basic Concepts, Methods, Neural Substrate, and Behavior,* pp. 699-722, 2017.

[2722] M. Bekoff, "Play signals as punctuation: The structure of social play in canids," *Behaviour,* vol. 132, no. 5-6, pp. 419-429, 1995.

[2723] C. Llamazares-Martín, C. Scopa, F. Guillén-Salazar and E. Palagi, "Relaxed Open Mouth reciprocity favours playful contacts in South American sea lions (Otaria flavescens)," *Behavioural Processes,* pp. 87-95, 2017.

[2724] N. W. Owens, "A comparison of aggressive play and aggression in free-living baboons, Papio anubis," *Animal Behaviour,* vol. 23, no. 4, pp. 757-765, 1975.

[2725] C. Douglas, Composer, *Kung Fu Fighting.* [Sound Recording]. 20th Century Fox . 1974.

[2726] S. M. Pellis and V. C. Pellis, The Playful Brain: Venturing to the Limits of Neuroscience, London, England: One World Publications, Oxford University Press, 2009, p. 336 .

[2727] A. D. Pellegrini, The Role of Play in Human Development, New York, NY: Oxford University Press, 2009, p. 288 .

[2728] S. M. Pellis and V. C. Pellis, "'I am going to groom you': Multiple forms of play fighting in gray mouse lemurs (Microcebus murinus)," *Journal of Comparative Psychology,* vol. 132, no. 1, pp. 6-15, 2018.

[2729] V. Maglieri, F. Vantaggio, C. Pilenga, M. Böye, A. Lemasson, L. Favaro and E. Palagi, "Smiling underwater: Exploring playful signals and rapid mimicry in bottlenose dolphins," *iScience,* vol. 27, no. 10, p. 110966, 2024.

[2730] unknown, "The Agincourt Song," [Online]. Available: http://folk-lyrics.co.uk/Lyrics/TheAgincourtSong. [Accessed 12 7 2018].

[2731] R. O. Deaner and B. A. Smith, "Sex difference in sport across 50 societies," *Cross-Cultural Research,* vol. 47, no. 3, pp. 268-309, 2012.

[2732] R. O. Deaner, D. C. Geary, D. A. Puts, S. A. Ham, J. Kruger, E. W. B. Fles and T. Grandis, "A sex difference in the predisposition for physical competition: males play sports much more than females even in the contemporary US," *PLOS One,* vol. 7, no. 11, p. e49168, 2012.

[2733] R. W. Wrangham, "Hypotheses for the evolution of reduced reactive aggression in the context of human self-domestication," *Frontiers in Psychology,* vol. 10, p. 1914, 2019.

[2734] M. N. Muller and R. W. Wrangham, "The reproductive ecology of male hominoids (pp 397-427)," in *Reproductive Ecology and Human Evolution*, P. T. Ellison, Ed., Oxford, Aldine, 2001, p. 478.

[2735] J. Archer, "Sex difference in aggression in real-world settings: A meta-analytic review," *Review of General Psychology,* vol. 8, no. 4, pp. 291-322, 2004.

[2736] L. Ellis, S. Hershberger, E. Field, S. Wersinger, S. Pellis, D. Geary, C. Palmer, K. Hoyenga, A. Hetsroni and K. Karadi, Sex Differences: Summarizing More than a Century of Scientific Research, New York: Psychology Press: Taylor and Francis, 2008, p. 990.

[2737] M. Bollig, "Ethnic conflicts in North-West Kenya: Pokot-Turkana Raiding 1969-1984," *Zeitschrift für Ethnologie / Journal of Social and Cultural Anthropology,* vol. 114, pp. 73-90, 1990.

[2738] L. Glowacki, M. L. Wilson and R. W. Wrangham, "The evolutionary anthropology of war," *Journal of Economic Behavior and Organization,* vol. 178, pp. 963-982, 2020.

[2739] S. Mathew and R. Boyd, "The cost of cowardice: punitive sentiments towards free riders in Turkana raids," *Evolution and Human Behavior,* vol. 35, no. 1, pp. 58-64, 2014.

[2740] D. W. Sellen and D. J. Hruschka, "Extracted-food resource-defense polygyny in native Western North American societies at contact," *Current Anthropology,* vol. 45, no. 5, pp. 707-714, 2004.

[2741] A. Keith, "Dira canam: Marriage and war in Ovid's Metamorphoses Dira canam: Marriage and war in Ovid's Metamorphoses (pp 143-162)," in *Au-delà de l'épithalame: Le mariage dans la littérature latine (IIIe s. av. - VIe s. ap. J.-C.),* L. G. Milic and A. Stoehr-Monjou, Eds., Turnhout, Brepols, 2022, p. 584.

[2742] C. Dunn, "Abduction and Forced Marriage: Rape, Abduction, and Adultery, 1100–1500 (pp 82-97)," in *Stolen Women in Medieval England*, First ed., Cambridge, Cambridge University Press, 2012, p. 274.

[2743] United Nations Committee on the Elimination of Discrimination Against Women, "Inquiry concerning Kyrgyzstan under article 8 of the Optional Protocol to the Convention on the Elimination of All Forms of Discrimination Against Women: Follow-up Report Submitted by Kyrgyzstan (CEDAW/c/0p.8/kgz/3, English Translation)," United Nations, Washington DC, 2019.

[2744] United Nations Committee on the Elimination of Discrimination Against Women., "Inquiry concerning Kyrgyzstan Under Article 8 of the Optional Protocol to the Convention on the Elimination of All Forms of Discrimination Against Women (CEDAW/c/p[.8/kgz/1, English Translation)," United Nations, New York, 2018.

[2745] R. S. Walker and D. H. Bailey, "Body counts in lowland South American violence," *Evolution and Human Behavior,* vol. 34, no. 1, pp. 29-34, 2013.

[2746] S. Beckerman, P. I. Erickson, J. Yost and K. Long, "Life histories, blood revenge, and reproductive success among the Waorani of Ecuador," *Proceedings of the National Academy of Sciences,* vol. 106, no. 20, pp. 8134-8139, 2009.

[2747] S. J. Macfarlan, P. I. Erickson, J. Yost, J. Regalado, L. Jaramillo and S. Beckerman, "Bands of brothers and in-laws: Waorani warfare, marriage and alliance formation," *Proceedings of the Royal Society B,* vol. 285, no. 1890, p. 20181859, 2018.

[2748] J. S. Eccles and R. D. Harold, "Gender differences in sport involvement: Applying the Eccles' expectancy-value model," *Journal of Applied Sport Psychology,* vol. 3, no. 1, pp. 7-35, 1991.

[2749] M. S. Sugiyama, M. Mendoza, F. White and L. Sugiyama, "Coalitional play fighting and the evolution of coalitional intergroup aggression," *Human Nature,* pp. 1-26, 2018.

[2750] K. M. Kniffin and D. Palacio, "Trash-talking and trolling," *Human Nature,* pp. 1-17, 2018.

[2751] N. Barbaro, J. K. Mogilski, T. K. Shackelford and M. N. Pham, "Men's interest in allying with a previous combatant for future group combat," *Human Nature,* pp. 1-9, 2018.

[2752] M. P. Lombardo, "On the evolution of sport," *Evolutionary Psychology,* vol. 10, no. 1, 2012.

[2753] K. Ficek, P. Cieszczyk, M. Kaczmarczyk, A. Maciejewska-Karłowska, M. Sawczuk, J. Cholewinski, A. Leonska-Duniec, M. Stepien-Slodkowska, A. Zarebska, N. K. Stepto, D. J. Bishop and N. Eynon, "Gene variants within the COL1A1 gene are associated with reduced anterior cruciate ligament injury in professional soccer players," *Journal of Science and Medicine in Sports,* vol. 16, no. 5, pp. 396-400, 2013.

[2754] C. Pickering and J. Kiely, "ACTN3: More than just a gene for speed," *Frontiers in Physiology,* vol. 8, p. 1080, 2017.

[2755] N. Yang, D. G. MacArthur, J. P. Gulbin, A. G. Hahn, A. H. Beggs, S. Easteal and K. North, "ACTN3 genotype is associated with human elite athletic performance," *American Journal of Human Genetics,* vol. 73, no. 3, pp. 627-631, 2003.

[2756] National Center for Biotechnology, "CDH13 cadherin 13 [Homo sapiens (human)] - Gene - NCBI," 8 7 2018. [Online]. Available: https://www.ncbi.nlm.nih.gov/gene/1012#. [Accessed 12 7 2018].

[2757] O. Rivero, M. M. Sselten, S. Sich, S. Popp, L. Bacmeister, E. Amendola, M. Negwer, D. Schubert, F. Proft, D. Kiser, A. G. Schmitt, C. Gross, S. M. Kolk, T. Sterkalova, D. van den Hove, T. J. Resink, N. N. Kasri and K. P. Lesch, "Cadherin-13, a risk gene for ADHD and comorbid disorders, impacts GABAergic function in hippocampus and cognition," *Translational Psychiatry,* vol. 5, p. e655, 2015.

[2758] B. Mossink, J.-R. van Rhijn, S. Wang, K. Linda, M. R. Vitale, J. E. M. Zöller, E. J. H. van Hugte, J. Bak, A. H. A. Verboven, M. Selten, M. Negwer, B. L. Latour, I. van der Werf, J. M. Keller, T. M. K. Gunnewiek and C. Schoenmaker, "Cadherin-13 is a critical regulator of GABAergic modulation in human stem-cell-derived neuronal networks," *Molecular Psychiatry,* vol. 17, pp. 1-18, 2021.

[2759] J. L. Nuzzo, "Sex differences in skeletal muscle fiber types: A meta-analysis," *Clinical Anatomy,* vol. 37, no. 1, pp. 81-91, 2023.

[2760] Á. M. Serrano, A. Cazorla-Rincón, J. Villafañe, Á. González-de-la-Flor, C. Rodríguez-Anadón, M. Bravo-Aguilar, F. García-Sanz and C. Romero-Morales, "Sex-related differences in core muscle morphology and activation in amateur CrossFit® athletes: An ultrasound imaging study," *Journal of Bodywork and Movement Therapies,* vol. 44, pp. 489 - 493 , 2025.

[2761] M. Apostolou and M. Zacharia, "The evolution of sports: Exploring parental interest in watching sports," *Evolutionary Psychological Science,* vol. 1, no. 3, p. 155–162, 2015.

[2762] M. Apostolou, "The athlete and the spectator inside the man: A cross-cultural investigation of the evolutionary origins of athletic behavior," *Cross-Cultural Research,* vol. 49, no. 2, pp. 151-173, 2014.

[2763] M. A. Gómez, L. Lorenzo, S.-J. Ibañez and J. Sampaio, "Ball possession effectiveness in men's and women's elite basketball according to situational variables in different game periods," *Journal of Sports Sciences,* vol. 31, no. 13, pp. 1578-1587, 2013.

[2764] K. R. Ford, G. D. Myer, H. E. Toms and T. E. Hewett, "Gender differences in the kinematics of unanticipated cutting in young athletes," *Medicine and Science in Sports & Exercise,* vol. 37, no. 1, pp. 124-129, 2004.

[2765] M. L. Beaulieu, M. Lamontagne and L. Xu, "Gender differences in time-frequency analysis of unanticipated cutting maneuvers," *Medicine & Science in Sports & Exercise,* vol. 40, no. 10, pp. 1795-1804, 2008.

[2766] V. Thibault, M. Guillaume, G. Berthelot, N. El Helou, K. Schaal, L. Quinquis, H. Nassif, M. Tafflet, S. Escolano, O. Hermine and J.-F. Toussaint, "Women and men in sport performance: The gender gap has not evolved since 1983," *Journal of Sports Medicine & Science,* vol. 9, no. 2, p. 214–223, 2010.

[2767] W. W. Beatty, "Hormonal organization of sex differences in play fighting and spatial behavior," *Progress in Brain Research,* vol. 61, pp. 315-330, 1984.

[2768] R. M. Fagan, Animal Play Behaviour, Oxford: Oxford University Press, 1981, p. 702.

[2769] M. F. Stevenson and T. B. Poole, "Playful interactions in family groups of the common marmoset (Callithrix jacchus jacchus)," *Animal Behaviour,* vol. 30, no. 3, pp. 886-900, 1982.

[2770] V. A. Davies and E. D. Kemble, "Social play behaviours and insect predation in northern grasshopper mice (Onychomys leucogaster)," *Behavioural Processes,* vol. 8, no. 2, pp. 197-204, 1983.

[2771] N. Caine and G. Mitchell, "A review of play in the genus Macaca: Social correlates," *Primates,* vol. 20, no. 4, pp. 535-546, 1979.

[2772] P. K. Smith, "Does play matter? Functional and evolutionary aspects of animal and human play," *Behavioral and Brain Sciences,* vol. 5, no. 1, pp. 139-155, 1982.

[2773] S. M. Pellis and A. N. Iwaniuk, "Evolving a playful brain: A levels of control approach," *International Journal of Comparative Psychology,* vol. 17, no. 1, pp. 90-116, 2004.

[2774] D. Fry, "Chapter 4: Rough-and-tumble social play in humans. (pp 54-88)," in *The Nature Of Play: Great Apes And Humans*, New York, The Guilford Press, 2004, p. 308.

[2775] P. K. Smith, "Social and pretend play in children (pp 173-212)," in *The Nature of Play: Great Apes and Humans* , New York, Guilford Press, 2005, p. 308.

[2776] L. Carol, Alice in Wonderland, Huntington, WV: Empire Books, 1865/ 2011, p. 108.

[2777] J. Decety, K. J. Michalska, Y. Akitsuki and B. B. Lahey, "Atypical empathic responses in adolescents with aggressive conduct disorder: A functional MRI investigation," *Biological Psychology,* vol. 80, no. 2, pp. 203-211, 2009.

[2778] R. Zhang, R. J. Blair, K. S. Blair, M. Dobbertin, J. Elowsky, J. Bashford-Largo, A. J. Dominguez, M. Hatch and S. Bajaj, "Reduced grey matter volume in adolescents with conduct disorder: a region-of-interest analysis using multivariate generalized linear modeling," *Discover Mental Health,* vol. 3, no. 25, 2023.

[2779] J. R. Dugré and S. Potvin, "Investigating the impact of lumping heterogeneous conduct problems: aggression and rule-breaking rely on distinct spontaneous brain activity," *European Child and Adolescent Psychiatry,* vol. 34, no. 3, p. 1207–1219, 2024b.

[2780] N. Goulter and M. M. Moretti, "Network structures of callous-unemotional traits and conduct disorder symptoms among at-risk youth: Multi-informant reports and gender differences," *Research on Child and Adolescent Psychopathology,* vol. 49, pp. 1179-1196, 2021.

[2781] N. R. Crick and C. Zahn-Waxler, "The development of psychopathology in females and males: Current progress and future challenges," *Developmental and Psychopathology,* vol. 15, no. 3, pp. 719-742, 2003.

[2782] D. Cho, B. R. L. Zatto and W. L. G. Hoglund, "Forms of peer victimization in adolescence: Covariation with symptoms of depression," *Developmental Psychology,* vol. 58, no. 2, p. 392–404, 2022.

[2783] J. Decety and K. J. Michalska, "Neurodevelopmental changes in the circuits underlying empathy and sympathy from childhood to adulthood," *Developmental Science,* vol. 13, no. 6, pp. 886-899, 2009.

[2784] G. M. Mitchison, J. M. Liber, D. K. Hannesdottir and U. Njardvik, "Emotion dysregulation, ODD and conduct problems in a sample of five and six-year-old children," *Child Psychiatry and Human Development,* vol. 51, pp. 71-79, 2020.

[2785] N. D. Thompson, S. L. Kjaervik, V. Blondell and L. E. Hazlett, "The interplay between fear reactivity and callous-unemotional traits predicting reactive and proactive aggression," *Children,* vol. 11, no. 4, p. 379, 2024.

[2786] A. Bernhard, K. Ackermann, A. Martinelli, A. G. Chiocchetti, L. Vllasaliu, K. onzález-Madruga, M. Batachelor, N. M. Raschle, H. Oldenhof, L. M. C. Jansen, G. Kohls, K. Konrad, A. Popma, C. Stadler, G. Fairchild and C. M. Freitag, "Neuroendocrine stress response in female and male youths with conduct disorder and associations with early adversity," *Journal of the American Academy of Child and Adolescent Psychiatry,* vol. 61, no. 5, pp. 698-710, 2022.

[2787] S. Fecteau, A. Pascual-Leone and H. Théoret, "Psychopathy and the mirror neuron system: Preliminary findings from a non-psychiatric sample," *Psychiatric Research,* vol. 150, no. 2, pp. 137-144, 2008.

[2788] A. C. Gallup, M. E. Kret, O. T. Eldakar, J. Folz and J. J. M. Massen, "People that score high on psychopathic traits are less likely to yawn contagiously," *Scientific Reports,* vol. 11, no. 1, p. 23779, 2021.

[2789] C. Ward, R. Trisko and B. B. Smuts, "Third-party interventions in dyadic play between littermates of domestic dogs, Canis lupus familiaris," *Animal Behaviour,* vol. 78, no. 5, pp. 1153-1160, 2009.

[2790] D. Finkelhor, "The international epidemiology of child sexual abuse," *Child Abuse & Neglect,* vol. 18, no. 5, pp. 409-417, 1994.

[2791] N. Pereda, G. Guilera, M. Forns and J. Gómez-Benito, "The prevalence of child sexual abuse in community and student samples: A meta-analysis," *Clinical Psychology Review,* vol. 29, no. 4, pp. 328-338, 2009a.

[2792] N. Pereda, G. Guilera, M. Forns and J. Gómez-Benito, "The international epidemiology of child sexual abuse: A continuation of Finkelhor (1994)," *Child Abuse & Neglect,* vol. 33, no. 6, pp. 331-342, 2009b.

[2793] C. A. Plummera and W. Njugunab, "Cultural protective and risk factors: Professional perspectives about child sexual abuse in Kenya," *Child Abuse & Neglect,* vol. 33, no. 8, pp. 524-532, 2009.

[2794] M. H. van IJzendoorn, E. M. Euser, P. F. Prinzie, Juffer, J. and M. Bakermans-Kranenburg, "Elevated risk of child maltreatment in families with stepparents but not with adoptive parents," *Child Maltreat,* vol. 14, no. 4, pp. 369-375, 2009.

[2795] S. D. Levitt and S. J. Dubner, Freakonomics [Revised and Expanded]: A Rogue Economist Explores the Hidden Side of Everything, rough-cut ed., New York, NY: William Morrow, 2006, p. 320.

[2796] V. Hugo, Les Misérables, New York: Penguin Classics, 1862/1982, p. 1232.

[2797] P. Lamarque, "On the distance between literary narratives and real-life narratives (pp 117–132)," in *Narrative and Understanding Persons*, Cambridge, Cambridge University Press, 2007, p. 236.

[2798] C. Pan and M. Zhao, "How does upward social comparison impact the delay discounting: The chain mediation of belief in a just world and relative deprivation," *Psychology in the Schools,* vol. 61, no. 1, pp. 7-16, 2024.

[2799] L. V. Linhares, A. R. R. Torres and C. R. Pereira, "Live by the sword, die by the sword: Measuring belief in a just world with popular sayings," *Personality and Individual Differences,* vol. 195, p. 111673, 2022.

[2800] M. B. Schofield, B. L. H. Roberts, C. A. Harvey, I. S. Baker and G. Crouch, "Tales from the dark side: The dark tetrad of personality, supernatural, and scientific belief," *Journal of Humanistic Psychology,* vol. 62, no. 2, pp. 298-315, 2022.

[2801] H. Kaplan, "Belief in a just world, religiosity and victim blaming," *Archive for the Psychology of Religion,* vol. 34, no. 3, pp. 397-409, 2012.

[2802] C. J. M. White and A. Norenzayan, "Chapter 1 - Belief in karma: How cultural evolution, cognition, and motivations shape belief in supernatural justice," *Advances in Experimental Social Psychology,* vol. 60, pp. 1-63, 2023.

[2803] M. Burt, "Cultural myths and supports for rape," *Journal of Personality and Social Psychology,* vol. 38, no. 2, pp. 217-230, 1980.

[2804] B. A. Kopper, "Gender, gender identity, rape myth acceptance, and time of initial resistance on the perception of acquaintance rape blame and avoidability," *Sex Roles,* pp. 81-93, 1996.

[2805] A. Furnham, "Belief in a just world: Research progress over the past decade," *Personality and Individual Differences,* vol. 34, no. 5, p. 795–817, 2003.

[2806] G. Andrade, "New thought in the aftermath of the COVID-19 pandemic," in *The New Thought Movement in Healthcare: History, Uses, and Abuses*, Champs, Palgrave Macmillan, 2025, p. 129.

[2807] P. Boyer and E. Chantland, "Victims of misfortune are blamed for imposing costs on others: Testing a cooperation-dilemma factor in victim-blame," *Human Nature,* 2025.

[2808] P. Chobthamkit, R. M. Sutton, A. K. Uskul and T. Chaleeraktrakoon, "Personal versus general belief in a just world, karma, and well-being: Evidence from Thailand and the UK," *Social Justice Research,* vol. 35, pp. 296-317, 2022.

[2809] P. Larionov and K. Mudło-Głagolska, "Polska adaptacja Skali Personalnej i Ogólnej Wiary w Sprawiedliwy Świat. (Polish adaptation of the personal and general belief in a just world scales)," *Annales Universitatis Mariae Curie-Skłodowska. Sectio J, Paedagogia-Psychologia,* vol. 34, no. 2, pp. 245-264, 2021.

[2810] A. M. Smith and R. R. Ortiz, "#MeToo social media engagement and perceived hypersensitivity in the workplace," *Communication Studies,* vol. 72, no. 4, pp. 531-546, 2021.

[2811] B. Ashdown, J. Hackathorn, R. Antenucci, J. Mings and S. Dimitrakopoulou, "Perceptions of sexual harassment in the United States: Empathy predicts derogation of victims and perpetrators," *Sexuality and Culture,* vol. 28, p. 575–609, 2024.

[2812] M. Stubbs-Richardson, N. E. Rader and A. G. Cosby, "Tweeting rape culture: Examining portrayals of victim blaming in discussions of sexual assault cases on Twitter," *Feminism and Psychology,* vol. 28, no. 1, pp. 90-108, 2018.

[2813] M. Catlin and K. C. Scherr, "The deleterious effect of victimization on just world beliefs," *Journal of Interpersonal Violence,* vol. 37, no. 17-18, p. NP16013–NP16036, 2022.

[2814] N. Li, M. B. Augustyn and G. M. Pinchevsky, "Prior sexual victimization and rape myth acceptance: Direct victimization, vicarious victimization, and gender," *Victims and Offenders,* p. 1–27, 2025.

[2815] K. J. Thomas and W. M. Mucherah, "How fair is my world? Development of just world beliefs among Kenyan students," *Journal of Adolescence,* vol. 49, pp. 244-253, 2016.

[2816] E. García-Sánchez, I. Correia, C. R. Pereira, G. B. Willis, R. Rodríguez-Bailón and J. Vala, "How fair is economic inequality? Belief in a just world and the legitimation of economic disparities in 27 European countries," *Personality and Social Psychology Bulletin,* vol. 48, no. 3, pp. 382-395, 2022.

[2817] A. V. Komatsu and S. Kühn, "The effect of the physical environment on adolescents' sense of justice," *Journal of Environmental Psychology,* vol. 103, p. 102582, 2025.

[2818] G. Yu, F. Zhao, H. Wang and S. Li, "Subjective social class and distrust among Chinese college students: The mediating roles of relative deprivation and belief in a just world," *Current Psychology,* vol. 39, pp. 2221-2230, 2025.

[2819] J. Oldmeadow and S. T. Fiske, "System-justifying ideologies moderate status = competence stereotypes: Roles for belief in a just world and social dominance orientation," *European Journal of Social Psychology,* vol. 37, no. 6, pp. 1135-1148, 2007.

[2820] Z. Rubin and L. A. Peplau, "Who believes in a just world?," *Journal of Social Issues,* vol. 31, no. 3, pp. 65-89, 1975.

[2821] C. L. Kleinke and C. Meyer, "Evaluation of rape victim by men and women with high and low belief in a just world," *Psychology of Women Quarterly,* vol. 14, no. 3, pp. 343-353, 1990.

[2822] J. E. Grusec, M. Davidov and L. Lundell, "Prosocial and helping behavior," *Blackwell Handbook of Childhood Social Development,* pp. 457-474, 2002.

[2823] Y. Guo, X. M. J. Chen, Y. Li and C. Hommeny, "How belief in a just world leads to prosocial behaviours: The role of communal orientation," *Personality and Individual Differences,* vol. 195, p. 111642, 2022.

[2824] S. B. Özdemir, S. Y. T. Cucurachi and M. Özdemir, "Understanding the association between moral disengagement and ethnic victimization: roles of bystanders in class," *European Journal of Developmental Psychology,* vol. 21, no. 4, pp. 644-662, 2023.

[2825] G. A. Haynes and J. M. Olson, "Coping with threats to just-world beliefs: Derogate, blame, or help?," *Journal of Applied Social Psychology,* vol. 36, pp. 664-682, 2006.

[2826] M. J. Lerner and D. T. Miller, "Just world research and the attribution process: Looking back and ahead," *Psychological Bulletin,* vol. 85, no. 5, pp. 1030-1051, 1978.

[2827] D. Aderman, S. S. Brehm and L. B. Katz, "Empathic observation of an innocent victim: The just world revisited," *Journal of Personality and Social Psychology,* vol. 29, no. 3, pp. 342-347, 1974.

[2828] K. Bocian, R. D. Szekely-Copîndean, K. Myslinska-Szarek and B. Wojciszke, "Reevaluating the beautiful is moral stereotype by examining the impact of personal liking and belief in a just world," *Scientific Reports,* vol. 15, p. 12387, 2025.

[2829] J. H. Langlois, L. Kalakanis, A. J. Rubenstein, A. Larson, M. Hallam and M. Smoot, "Maxims or myths of beauty? A meta-analytic and theoretical review," *Psychological Bulletin,* vol. 126, no. 3, pp. 390-423, 2000.

[2830] A. A. Fultz, M. D. Stosic and F. J. Bernieri, "Nonverbal expressivity, physical attractiveness, and liking: First impression to established relationship," *Journal of Nonverbal Behavior,* vol. 48, p. 189–211, 2024.

[2831] J. M. Binnall, Twenty Million Angry Men: The Case for Including Convicted Felons in Our Jury System, First ed., Oakland, California: University of California Press, 2021, p. 288.

[2832] K. C. Gerbasi, M. Zuckerman and H. T. Reis, "Justice needs a new blindfold: A review of mock jury research," *Psychological Bulletin,* vol. 82, no. 4, pp. 323-345, 1977.

[2833] M. J. Lerner and E. Agar, "The consequences of perceived similarity: Attraction and rejection, approach and avoidance," *Journal of Experimental Research in Personality,* vol. 6, no. 1, pp. 69-75, 1972.

[2834] M. J. Lerner and C. H. Simmons, "Observer's reaction to the 'innocent victim': Compassion or rejection?," *Journal of Personality and Social Psychology,* vol. 4, no. 2, pp. 203-210, 1966.

[2835] L. B. Myers, "The importance of the repressive coping style: Findings from 30 years of research," *Anxiety Stress Coping,* vol. 23, no. 1, pp. 3-17, 2010.

[2836] D. F. Stramiccia, A.-K. Meyer, K. M. Rischer, J. M. Fawcett and R. G. Benoit, "Memory suppression and its deficiency in psychological disorders: A focused meta-analysis," *Journal of Experimental Psychology: General,* vol. 150, no. 5, p. 828–850, 2021.

[2837] C. L. Hafer and L. Gosse, "Predicting alternative strategies for preserving a belief in a just world: The case of repressive coping style," *European Journal of Social Psychology,* vol. 41, no. 6, p. 730–739, 2011.

[2838] A. Kudrnáč, M. A. Eger and M. Hjerm, "Scapegoating immigrants in times of personal and collective crises: Results from a Czech panel study," *International Migration Review,* vol. 58, no. 2, pp. 573-592, 2023.

[2839] C. J. Sung, "Economic slowdowns and international conflict," *Journal of Peace Research,* vol. 61, no. 2, pp. 180-196, 2022.

[2840] J. Gordils, A. J. Elliot, S. Toprakkiran and J. P. Jamieson, "The effects of COVID-19 on perceived intergroup competition and negative intergroup outcomes," *Journal of Social Psychology,* vol. 161, no. 4, pp. 419-434, 2021.

[2841] B. Savun and C. Gineste, "From protection to persecution: Threat environment and refugee scapegoating," *Journal of Peace Research,* vol. 56, no. 1, pp. 88-102, 2018.

[2842] P. Bertin, O. Ionescu, R. Green, K. Abts, J. Rogenhofer, S. Delouvée, V. Yzerbyt and O. Klein, "Conspiracy blaming in the aftermath of group relative deprivation: The moderating role of national narcissism," *European Journal of Social Psychology,* vol. 54, no. 6, pp. 1393-1415, 2024.

[2843] D. Bhanot, T. Singh, S. K. Verma and S. Sharad, "Stigma and discrimination during COVID-19 pandemic," *Frontiers in Public Health,* vol. 8, p. 577018, 2021.

[2844] N. Keady-Tabbal and I. Mann, "Weaponizing rescue: Law and the materiality of migration management in the Aegean," *Leiden Journal of International Law,* vol. 36, no. 1, p. 61–82, 2022.

[2845] G. A. Pormbescu, D. Moynihan, J. Anastasopoulos and A. L. Olsen, "When blame avoidance backfires: Responses to performance framing and outgroup scapegoating during the COVID-19 pandemic," *Governance,* vol. 36, no. 3, pp. 779-803, 2023.

[2846] J. S. Sun, "Strategic state violence and migration in conflict," *American Journal of Political Science,* pp. 1-18, 2024.

[2847] I. Salehyan and B. Savun, "Strategic humanitarianism: Host states and refugee policy," *Annual Review of Political Science,* vol. 27, p. 107–125, 2024.

[2848] M. J. Lerner and L. Montada, "The two forms of belief in a just world: Some thoughts on why and how people care about justice. pp 247-266," in *Responses to Victimizations and Belief in a Just World*, New York, Plenum Publishing Co., 1998, p. 278.

[2849] I. Pichon and V. Saroglou, "Religion and helping: Impact of target thinking styles and just-world beliefs," *Archiv für Religionspsychologie / Archive for the Psychology of Religions,* vol. 31, no. 2, pp. 215-236, 2009.

[2850] K. Stroebe, T. Postmes, S. Täuber, A. Stegeman and M.-S. John, "Belief in a just what? Demystifying just world beliefs by distinguishing sources of justice," *Plos One,* vol. 10, no. 3, p. e0120145, 2015.

[2851] L. C. Torres-Vega, J. Ruiz and M. Moya, "Dangerous world view and perceived sociopolitical control: Two mechanisms to understand trust in authoritarian political leaders in economically threatening contexts," *Frontiers in Psychology,* vol. 12, p. 603116, 2021.

[2852] J.-E. Lönnqvist, Z. P. Szabó and L. Kelemen, "'The new state that we are building': Authoritarianism and system-justification in an illiberal democracy," *Frontiers in Psychology,* vol. 12, p. 703280, 2021.

[2853] M. A. Hogg, J. R. Adelman and R. D. Blag, "Religion in the face of uncertainty: An uncertainty-identity theory account of religiousness," *Personality and Social Psychology Review,* vol. 14, no. 1, pp. 72-83, 2010.

[2854] J. C. Jackson, D. Dillion, B. Bastian, J. Watts, W. Buckner, N. DiMaggio and K. Grey, "Supernatural explanations across 114 societies are more common for natural than social phenomena," *Nature Human Behavior,* vol. 7, pp. 707-717, 2023.

[2855] M. J. van Deursen, A. R. D. Pope and R. H. Warner, "Just world maintenance patterns among intrinsically and extrinsically religious individuals," *Personality and Individual Differences,* vol. 52, no. 6, p. 755–758, 2012.

[2856] D. T. Miller, "Altruism and threat to a belief in a just world," *Journal of Experimental Social Psychology,* vol. 13, no. 2, p. 113–124, 1977.

[2857] J. L. Preston and A. Baimel, "Towards a psychology of religion and the environment [Review]," *Current Opinion in Psychology,* vol. 40, pp. 145-149, 2021.

[2858] H. Lui, "Meritocracy as authoritarian co-optation: Political selection and upward mobility in China," *American Political Science Review,* p. 1–17, 2023.

[2859] A. García-Sierra, "The dark side of meritocratic beliefs: Is believing in meritocracy detrimental to individuals from low socioeconomic backgrounds?," *Social Justice Research,* vol. 36, pp. 385-409, 2023.

[2860] A. Batruch, J. Jetten, H. Van de Werfhorst, C. Darnon and F. Butera, "Belief in school meritocracy and the legitimization of social and income inequality," *Psychological and Personality Science,* vol. 14, no. 5, pp. 621-635, 2023.

[2861] C. Rauscher, E. Louvet and D. Bourguignon, "'To succeed at work, you just need to work hard': How beliefs in professional meritocracy serve to blame unemployed people," *Journal of Community and Applied Social Psychology,* vol. 34, no. 1, p. e2762, 2024.

[2862] A. Alesina, E. Glaser and B. Sacerdote, "Why doesn't the US have a European-type welfare state?," *Brookings Papers on Economic Activity,* vol. 2001, no. 2, pp. 187-277, 2001.

[2863] A. Furnham, "Just world beliefs in twelve societies," *Journal of Social Psychology,* vol. 133, no. 3, pp. 317-329, 1991.

[2864] Z. Rubin and L. A. Peplau, "Belief in a just world and reactions to another's lot: A study of participants in the national draft lottery," *Journal of Social Issues,* vol. 29, no. 4, pp. 73-93, 1973.

[2865] I. A. Betawi, "Children for integrity: The influence of moral stories on pre-schoolers' integrity values," *Clinical Child Psychology and Psychiatry,* vol. 28, no. 2, 2022.

[2866] P. M. Cooper, "Teaching young children self-regulation through children's books," *Early Childhood Education Journal,* vol. 34, p. 315–322, 2007.

[2867] V. A. Umscheid, C. E. Smith, F. Warneken, S. A. Gelman and H. M. Wellman, "What makes Voldemort tick? Children's and adults' reasoning about the nature of villains," *Cognition,* vol. 233, p. 105357, 2023.

[2868] M. Nicholson, "Narrative Matters: Is obedience good? The classics may surprise us," *Child and Adolescent Mental Health,* vol. 30, no. 1, pp. 89-91, 2025.

[2869] C. Xueqin, Dream of the Red Chamber (Hónglóu Mèng), Chicago, NY: Centaur Editions, 2016/ca 1760, p. 329.

[2870] P. K. Nayar, "The human rights torture novel: Unmade subjects, unmaking worlds," *Orbis Litterarum,* vol. 72, no. 4, pp. 318-347, 2017.

[2871] A. A. Raney, "The role of morality in emotional reactions to and enjoyment of media entertainment," *Journal of Media Psychology: Theories, Methods, and Applications,* vol. 23, no. 1, pp. 18-23, 2011.

[2872] J. Best, "Reading literary fiction: More empathy, but at what possible cost?," *North American Journal of Psychology,* vol. 22, no. 2, pp. 269-288, 2020.

[2873] N. Buttrick, E. C. Westgate and S. Oishi, "Reading literary fiction is associated with a more complex worldview," *Personality and Social Psychology Bulletin,* p. 01461672221106059, 2022.

[2874] M. Appel, "Fictional narratives cultivate just-world beliefs," *Journal of Communication,* vol. 58, no. 1, p. 62–83, 2008.

[2875] E. Castano, J. Zanella, F. Saedi, L. Zunshine and L. Ducceschi, "On the complexity of literary and popular fiction," *Empirical Studies of the Arts,* vol. 0, no. 0, 2023.

[2876] J. E. Black and J. L. Barnes, "Fiction and morality: Investigating the associations between reading exposure, empathy, morality, and moral judgement," *Psychology of Popular Media,* vol. 10, no. 3, pp. 149-164, 2021.

[2877] D. Smith, P. Schlaepfer, K. Major, M. Dyble, A. E. Page, J. Thompson, N. Chaudhary, G. D. Salali, R. Mace, L. Astete, M. Ngales, L. Vinicius and A. B. Migliano, "Cooperation and the evolution of hunter-gatherer storytelling," *Nature Communications,* vol. 8, p. 1853, 2017.

[2878] A. Nowell, "Oral storytelling and knowledge transmission in Upper Paleolithic children and adolescents," *Journal of Archaeological Methods and Theory,* vol. 30, pp. 9-31, 2023.

[2879] B. Boyd, "The evolution of stories: from mimesis to language, from fact to fiction," *Wiley Interdisciplinary Reviews. Cognitive Science,* vol. 9, no. 1, p. e1444, 2018.

[2880] C. Taylor, Sources of the Self: The Making of the Modern Identity, Cambridge, MA: Harvard University Press, 1989, p. 624.

[2881] S. Vice, "Literature and the narrative self," *Philosophy,* vol. 78, no. 2, p. 93–108, 2003.

[2882] A. A. Raney, "Moral judgment as a predictor of enjoyment of crime drama," *Media Psychology,* vol. 4, no. 4, pp. 305-322, 2002.

[2883] A. A. Raney and J. Bryant, "Moral judgment in crime drama: An integrated theory of enjoyment," *Journal of Communication,* vol. 52, no. 2, pp. 402-415, 2002.

[2884] P. Highsmith, The Talented Mr. Ripley, New York: Coward-McCann, 1955, p. 252.

[2885] C. Kepnes, You, New York: Atria/Emily Bestler Books, 2014, p. 424.

[2886] T. Harris, Hannibal, New York: Delacorte Press, 1999, p. 484.

[2887] J. Lindsay, Darkly Dreaming Dexter, New York: Orion Publishing, 2004, p. 304.

[2888] A. A. Raney, "The psychology of disposition-based theories of media enjoyment (pp 137-150)," in *Psychology of Entertainment*, Mahwah, Erlbaum, 2006, p. 472.

[2889] D. Zillmann, "Mechanisms of emotional involvement with drama," *Poetics,* vol. 23, no. 1-2, p. 33–51, 1994.

[2890] D. Zillmann and J. R. Cantor, "Directionality of transitory dominance as a communication variable affecting humor appreciation," *Journal of Personality and Social Psychology,* vol. 24, no. 2, p. 191–198, 1972.

[2891] C. Dickens, The Complete Works of Charles Dickens: Novels, Short Stories, Plays, Poetry, Essays, Articles, Speeches, Travel Sketches & Letters (Illustrated), e-Artnow, 2017, p. 25083.

[2892] G. Puccini, L. Illica and G. Giacosa, Composers, *Madam Butterfly.* [Sound Recording]. 1904.

[2893] M. Grizzard, C. J. Francemone, R. Frazer, K. Fitzgerald, C. K. Monge and C. Henry, "A comprehensive experimental test of the affective disposition theory of drama," *Journal of Communication,* p. jqac053, 2023.

[2894] K. Fitzgerald, C. J. Francemone and M. Grizzard, "Memorable, meaningful, pleasurable: An exploratory examination of narrative character deaths," *Omega- Journal of Death and Dying,* vol. 86, no. 2, p. 553–575, 2020.

[2895] S. Collins, The Hunger Games, New York: Scholastic Press, 2008, p. 384.

[2896] J. K. Rowling, Harry Potter and the Philosopher's Stone, London: Bloomsbury, 1997, p. 223.

[2897] H. Dong-hyuk and K. Ji-yong, Directors, *The Squid Games [TV Series].* [Film]. South Korea: Siren Pictures Inc, 2021-2025.

[2898] M. MacLaren, A. Bernstein and V. Gilligan, Directors, *Breaking Bad.* [Film]. High Bridge Entertainment; Gran Via Productions ; Sony Pictures Television, 2008-2013.

[2899] K. van Krieken, A. de Graaf and E. Das, "Can we distinguish mixed from ambiguous emotions and morality?," *Cognitive Neuroscience: Current Debates, Research and Reports,* vol. 14, no. 2, pp. 68-69, 2023.

[2900] M. Grizzard, C. J. Francemone, K. Fitzgerald, J. Huang and C. Ahn, "Interdependence of narrative characters: Implications for media theories," *Journal of Communication,* vol. 70, no. 2, p. 274–301, 2020.

[2901] M. B. Vaage, "On punishment and why we enjoy it in fiction: Lisbeth Salander of the Millennium Trilogy and Eli in Let the Right One In as Scandinavian Avengers," *Poetics Today,* vol. 40, no. 3, pp. 543-557, 2019.

[2902] J. Huang, "Loving Shuang Ju: Chinese audiences' entertainment experience of retribution narratives," *Psychology of Popular Media,* vol. 13, no. 3, pp. 447-456, 2024.

[2903] R. Weber, F. R. Hopp, A. Eden, J. T. Fisher and H.-E. Lee, "Vicarious punishment of moral violations in naturalistic drama narratives predicts cortical synchronization," *NeuroImage,* vol. 292, p. 120613, 2024.

[2904] G. Lucas, Director, *Star Wars.* [Film]. 20th Century Fox: Lucas Films Ltd, 1977.

[2905] Circa24, Silent Consent, Chicago: Prairie House Gallery Press, 2022, p. 426.

[2906] H. Piper, "Happy Valley: Compassion, evil and exploitation in an ordinary 'Trouble Town.'," in *Social Class and Television Drama in Contemporary Britain*, D. Forrest and B. Johnson, Eds., London, Palgrave Macmillan, 2017, p. 271.

[2907] I. Correia and J. Vala, "When will a victim be secondarily victimized? The effect of observer's belief in a just world, victim's innocence and persistence of suffering," *Social Justice Research,* vol. 16, no. 4, pp. 379-400, 2003.

[2908] I. Correia, J. Vala and P. Aguiar, "Victim's innocence, social categorization, and the threat to the belief in a just world," *Journal of Experimental Social Psychology,* vol. 43, no. 1, p. 31–38, 2007.

[2909] A. Christie, Murder on the Orient Express: A Hercule Poirot Mystery, reprint ed., New York: Harper, 1934/2011, p. 336.

[2910] A. A. Raney, "Expanding disposition theory: Reconsidering character liking, moral evaluation, and enjoyment," *Communication Theory,* vol. 14, no. 4, pp. 348-368, 2004.

[2911] E. S. Gardner, The Case of the Velvet Claw, White Plains, NY: Reader's Digest, 1933/2017, p. 261.

[2912] S. Grafton, "A" is for Alibi, New York: Henry Holt and Company, 1982, p. 274.

[2913] S. Grafton, "B" is for Burglar, New York: Henry Holt and Company, 1985, p. 229.

[2914] S. Grafton, "V" is for Vengeance, New York: G. P. Putnam's Sons, 2011, p. 448.

[2915] R. Cook, Coma, New York, NY: Little Brown and Co., 1977, p. 306.

[2916] R. Cook, Viral, New York: G. P. Putnam's Sons, 2021, p. 432.

[2917] J. Grisham, A Time to Kill, Bonners Ferry, Idaho: Wynwood Press, 1989a, p. 672.

[2918] Circa24, Endured by the Gods, Chicago: Prairie House Gallery Books, 2024, p. 365.

[2919] I. Correia, S. V. Kamble and C. Dalbert, "Belief in a just world and well-being of bullies, victims and defenders: A study with Portuguese and Indian students," *Anxiety, Stress, and Coping,* vol. 22, no. 5, pp. 497-508, 2009.

[2920] N. Frye, Anatomy of Criticism: Four Essays, Princeton, NJ: Princeton University Press, 1957/2000, p. 400.

[2921] B. Herrmann, C. Thöni and S. Gächter, "Antisocial punishment across societies," *Science,* vol. 319, no. 5868, pp. 1362-1367, 2008.

[2922] M. Salahshour, "Evolution of prosocial punishment in unstructured and structured populations and in the presence of antisocial punishment," *PLOS One,* vol. 16, no. 8, p. e0254860, 2021.

[2923] A. C. Doyle, "The Adventure of the Blue Carbuncle (Location 5694)," in *The Complete Sherlock Holmes, Heirloom Collection*, Amazon Digital Services, 1892/2018a, p. 2509 KB .

[2924] A. C. Doyle, "The Adventure of the Speckled Band (Location 6024)," in *The Complete Sherlock Holmes, Heirloom Collection*, Amazon Digital Services LLC, 1892/2018b.

[2925] J. Scalzi, The Last Colony, New York City, New York: Tor Books, 2007, p. 320.

[2926] H. B. Piper, Little Fuzzy, Halcyon Press Ltd, 1962/2010, p. 118 .

[2927] J. Cameron, Director, *Avatar.* [Film]. USA: Lightstorm Entertainment; Dune Entertainment; Indigenous Media, 2009.

[2928] V. Swarup, Q & A, New York: Scribner, 2008/2005, p. 336.

[2929] J. London, The Call of the Wild, New York: Macmillan, 1903, p. 232.

[2930] C. Dickens, David Copperfield, Project Gutenberg, 1869/1996.

[2931] M. Gershman, J. A. Contner, D. Solomon and M. Grabiak, Directors, *Buffy the Vampire Slayer.* [Film]. 2003.

[2932] K. Kesey, One Flew Over the Cuckoo's Nest, New York: Viking Press, 1962, p. 320.

[2933] M. Waters, Director, *Mean Girls.* [Film]. USA.2004.

[2934] D. Contandriopoulos, N. Stake-Doucet and J. Schilling, "Fake kindness, caring and symbolic violence," *Nursing Ethics,* vol. 3, no. 6, pp. 1041-1049, 2024.

[2935] E. L. F. Schipper and A. Mukherji, "Misguided negative adaptation narratives are hurting the poor: Measurement and maladaptation are being weaponized," *Science,* vol. 386, no. 6722, pp. 624-628, 2024.

[2936] S. Mohammed, E. Peter, T. Killackey and J. Maciver, "The 'nurse as hero' discourse in the COVID-19 pandemic: A poststructural discourse analysis," *International Journal of Nursing Studies,* vol. 117, p. 103887, 2021.

[2937] A. Bailey, "On Anger, Silence, and Epistemic Injustice," *Royal Institute of Philosophy Supplement,* vol. 84, pp. 93-115, 2018.

[2938] K. Melino, B. Bell and K. Freborg, "Deconstructing professionalism as code for White (power): Authenticity as resistance in nursing," *Nursing Philosophy,* vol. 26, no. 1, p. 70002, 2024.

[2939] C. Biddle and E. Hufnagel, "Navigating the 'Danger Zone': Tone policing and the bounding of civility in the practice of student voice," *American Journal of Education,* vol. 125, no. 4, pp. 487-520, 2019.

[2940] C. Dickens, "CHAPTER LVII. In which the Pickwick Club is finally dissolved, and everything concluded to the satisfaction of everybody," in *The Pickwick Papers*, Project Gutenberg, 1836/ 2009.

[2941] J. D. Velleman, "Narrative explanation. The philosophical review," *The Philosophical Review,* vol. 112, no. 1, p. 1–25, 2003.

[2942] A. Almeida, I. Correia and S. Marinho, "Moral disengagement, normative belief of peer group, and attitudes regarding roles in bullying," *Journal of School Violence,* vol. 9, no. 1, pp. 23-26, 2010.

[2943] P. Aguiar, J. Vala, I. Correia and C. Pereira, "Justice in our world and in that of others: Belief in a just world and reactions to victims," *Social Justice Research,* vol. 21, no. 1, pp. 50-68, 2008.

[2944] L. Goldenberg and J. P. Forgas, "Can happy mood reduce the just world bias? Affective influences on blaming the victim," *Journal of Experimental Social Psychology,* vol. 48, no. 1, p. 239–243, 2012.

[2945] N. Friedman, Form and Meaning in Fiction, Athens, GA: University of Georgia Press, 1982, p. 432.

[2946] T. Hardy, Tess of the D'Urbervilles, T. Dolin, Ed., London: Penguin Classics, 1891/2003, p. 592.

[2947] T. Hardy, The Mayor of Casterbridge, K. Wilson, Ed., London: Penguin Classics, 1886/ 2004, p. 448.

[2948] F. S. Fitzgerald, The Great Gatsby, Independently published, 2022/1925, p. 114.

[2949] M. Bal, Narratology: Introduction to the Theory of Narrative, Fourth Edition, Toronto: University of Toronto Press, 2017, p. 206.

[2950] B. Richardson, Unlikely Stories: Causality and the Nature of Modern Narrative, Newark, DE: University of Delaware Press, 1997, p. 219.

[2951] S. Knobloch-Westerwick and C. Keplinger, "Murder for pleasure: Impacts of plot complexity and need for cognition on mystery enjoyment," *Journal of Media Psychology: Theories, Methods, and Applications,* vol. 20, no. 3, pp. 117-128, 2008.

[2952] P. Brooks, Reading for the Plot: Design and Intention in Narrative, Cambridge, MA: Harvard University Press, 1992, p. 392.

[2953] L. J. Davis, Resisting Novels: Ideology and Fiction, New York: Routledge, 1987, p. 264.

[2954] F. Kermode, The Sense of an Ending: Studies in the Theory of Fiction, Oxford: Oxford University Press, 2000, p. 150.

[2955] Sophocles, Oedipus the King, Chicago: University of Chicago Press, ca 239 BCE./ 2010, p. 88.

[2956] Y.-H. Lee, Q. Xu, X. Yu, H. He and Q. Xu, "Seeking justice in video games: Belief in a just world and video game preferences," *Psychology of Popular Media,* 2024.

[2957] I. Correia and C. Dalbert, "School bullying," *European Psychologist,* vol. 16, no. 4, pp. 248-254, 2008.

[2958] C. L. Fox, T. Elder, J. Gater and E. Johnson, "The association between adolescents' beliefs in a just world and their attitudes to victims of bullying," *British Journal of Educational Psychology,* vol. 80, no. 2, pp. 183-198, 2010.

[2959] M. Donat, M. Knigge and C. Dalbert, "Being a good or a just teacher: Which experiences of teachers' behavior can be more predictive of school bullying?," *Aggressive Behavior,* vol. 44, no. 1, pp. 29-39, 2018.

[2960] D.-P. Jones, "The sayings of Secretary Henry," *New York Times,* p. 337, 28 10 1973.

[2961] I. Prusa, "Heroes Beyond Good and Evil: Theorizing Transgressivity in Japanese and Western Fiction," *Electronic Journal of Contemporary Japanese Studies,* vol. 16, no. 1, pp. 1-30, 2016.

[2962] R. Graves, I, Claudius: From the Autobiography of Tiberius Claudius Born 10 B.C. Murdered and Deified A.D. 54, Reissue ed., New York: Vintage, 1934/ 1989, p. 468.

[2963] S. Pütz, Vampire und ihre Opfer. Der Blutsauger als literarische Figur, Bielefeld: Aisthesis Verlag, 1993, p. 208.

[2964] J. Nußbaumer, The Vampire in Literature: A Comparison of Bram Stoker's Dracula and Anne Rice's Interview with the Vampire, Albany, NY: Anchor Academic Publishing, 2013, p. 44.

[2965] C. McHugh and J. Wylie, "Moral judgments," in *Encyclopedia of Heroism Studies*, Cham, Springer, 2023, pp. 1-7.

[2966] A. A. Raney, H. Schmid, J. Niemann and M. Ellensohn, "Testing Affective Disposition Theory: A Comparison of the Enjoyment of Hero and Antihero Narratives," in *Annual Meeting of the International Communication Association, Marriott, Chicago*, Chicago, 2009-05-21.

[2967] K. M. Krakowiak and M. Tsay, "The role of moral disengagement in the enjoyment of real and fictional characters," *International Journal of Arts and Technology,* vol. 4, no. 1, pp. 90-101, 2011.

[2968] A. M. J. de Jonge, Fiction Reading and the Moral Shift (Ph.D. dissertation), Amsterdam: Vrije Universiteit Amsterdam, 2024, p. 166.

[2969] T. Andreyeva, R. M. Puhl and K. D. Brownell, "Changes in perceived weight discrimination among Americans, 1995–1996 through 2004–2006," *Obesity,* vol. 16, no. 5, pp. 1129-1134, 2012.

[2970] C. S. Crandall, "Prejudice against fat people: Ideology and self-interest," *Journal of Personality and Social Psychology,* vol. 66, no. 5, pp. 882-894, 1994.

[2971] S. A. Flanagan, "Obesity: The last bastion of prejudice," *Obesity Surgery,* vol. 6, no. 5, pp. 430-437, 1996.

[2972] M. Brandon and M. S. Himmelstein, "The role of masculinity and weight stigma in men's body dissatisfaction," *Psychology of Men and Masculinities,* p. e0000528, 2025.

[2973] I. B. Rosenberg, "Height discrimination in employment," *Utah Law Review,* vol. 3, pp. 907-954, 2009.

[2974] G. Lasco, "Beyond 'heightism' and 'height premium': An anthropology and sociology of human stature," *Sociology Compas,* vol. 18, no. 2, p. e13178, 2024.

[2975] F. J. Marco-Gracia, "Revisiting the effect of height on wages in a historical context: the case of the city of Zaragoza (Spain), 1924," *Labor History,* vol. 66, no. 1, p. 105–127, 2025.

[2976] A. C. Downs and S. K. Harrison, "Embarrassing age spots or just plain ugly? Physical attractiveness stereotyping as an instrument of sexism on American television commercials," *Sex Roles,* vol. 13, no. 1-2, pp. 9-19, 1985.

[2977] S. Daníelsdóttira, K. S. O'Brienb and A. Ciaoc, "Anti-fat prejudice reduction: A review of published studies," *Obesity Facts,* vol. 3, pp. 47-58, 2010.

[2978] E. B. Kramer, E. S. Peitri and A. D. Bryan, "Reducing anti-fat bias toward the self and others: a randomized controlled trial," *Journal of Eating Disorders,* vol. 12, p. 46, 2024.

[2979] L. C. Carter, A. C. Campbell and S. Muncer, "The Dark Triad personality: Attractiveness to women," *Personality and Individual Differences,* vol. 56, pp. 57-61, 2014.

[2980] M. Lyons, "Mating Strategies (Mate Competition, Mate Attraction, and Established Relationships) (pp. 81-110)," in *The Dark Triad of Personality: Narcissism, Machiavellianism, and Psychopathy in Everyday Life*, First ed., Cambridge, MA: Academic Press, 2019, p. 219.

[2981] A. C. Davis, H. Brittain, S. Arnocky and T. Vaillancourt, "Longitudinal Associations Between Primary and Secondary Psychopathic Traits, Delinquency, and Current Dating Status in Adolescence," *Evolutionary Psychology,* vol. 20, no. 1, 2022.

[2982] R. Yao and J. Yang, "Exploring the appeal of villainous characters in film-induced tourism: perceived charismatic leadership and justice sensitivity," *Humanities and Social Sciences Communications,* vol. 11, p. 267, 2024.

[2983] A. W. Eaton, "Robust Immoralism," *Aesthetics and Art Criticism,* vol. 70, no. 3, pp. 281-292, 2012.

[2984] D. R. Feinberg, B. C. Jones, A. C. Little, D. M. Burt and D. I. Perrett, "Manipulations of fundamental and formant frequencies affect the attractiveness of human male voices," *Animal Behaviour,* vol. 69, p. 561–568, 2005.

[2985] D. R. Feinberg, B. C. Jones, M. J. Law Smith, F. R. Moore, L. M. DeBruine, R. E. Cornwell, S. G. Hiller and D. I. Perrett, "Menstrual cycle, trait estrogen level, and masculinity preferences in the human voice," *Hormones and Behavior,* vol. 49, no. 2, pp. 215-222, 2006.

[2986] D. A. Puts, "Mating context and menstrual phase affect women's preferences for male voice pitch," *Evolution and Human Behavior,* vol. 26, p. 388–397, 2005.

[2987] D. A. Puts, "Cyclic variation in women's preferences for masculine traits: Potential hormonal causes," *Human Nature,* vol. 17, no. 1, p. 114–127, 2006.

[2988] C. L. Apicella and D. R. Feinberg, "Voice pitch alters mate-choice-relevant perception in hunter-gatherers," *Proceedings of the Royal Society B,* vol. 276, no. 1659, 2009.

[2989] C. L. Apicella, D. Feinberg and F. W. Marlowe, "Voice pitch predicts reproductive success in male hunter-gatherers," *Biological Letters,* vol. 3, no. 6, pp. 628-684, 2007.

[2990] D. A. Puts, A. K. Hill, D. H. Bailey, R. S. Walker, D. Rendall, J. R. Wheatley, L. L. M. Welling, K. Dawood, R. Cárdenas, R. P. Burriss, N. Jablonski, M. D. Shriver, D. Weiss, A. R. Lameira, C. L. Apicella, M. J. Owren and Barelli, "Sexual selection on male vocal fundamental frequency in humans and other anthropoids," *Proceedings of the Royal Society B,* vol. 283, no. 1829, 2016.

[2991] J. van Ginneken, "Voice and Speech (pp. 117-136)," in *The Profile of Political Leaders: Archetypes of Ascendancy*, First ed., London, Palgrave Macmillan (Springer), 2016, p. 108.

[2992] S. Evans, N. Neave and D. Wakelin, "Relationships between vocal characteristics and body size and shape in human males: An evolutionary explanation for a deep male voice," *Biological Psychology,* vol. 72, no. 2, p. 160–163, 2006.

[2993] B. C. Jones, D. R. Feinberg, C. D. Watkins, C. L. Fincher, A. C. Little and L. M. DeBruine, "Pathogen disgust predicts women's preferences for masculinity in men's voices, faces, and bodies," *Behavioral Ecology,* vol. 24, no. 2, pp. 373-379, 2012.

[2994] Q. Ma, L. Cheng, W. Qiu and G. Pei, "Role effect on beauty premium: Female as proposer may gain more fairness," *Psych Journal,* vol. 11, no. 5, pp. 691-706, 2022.

[2995] T. S. Rosenblat, "The beauty premium: Physical attractiveness and gender in dictator games," *Negotiation Journal,* vol. 24, no. 4, pp. 465-481, 2008.

[2996] M. Paquette-Smith, H. Buckler and E. K. Johnson, "How sociolinguistic factors shape children's subjective impressions of teacher quality," *Quarterly Journal of Experimental Psychology,* vol. 76, no. 3, pp. 485-496, 2023.

[2997] T. J. Bouchard, D. T. Lykken, M. McGue, M. L. Segal and A. Tellegen, "Sources of human psychological differences - the Minnesota study of twins reared apart," *Science,* vol. 250, no. 4978, pp. 223-226, 1990.

[2998] S. M. van den Berg, D. Posthuma and D. I. Boomsma, "A longitudinal genetic study of vocabulary knowledge in adults," *Twin Research,* vol. 7, no. 3, pp. 284-291, 2004.

[2999] J. R. Coffey, C. L. Shafto, J. C. Geren and J. Snedeker, "The effects of maternal input on language in the absence of genetic confounds: Vocabulary development in internationally adopted children," *Child Development,* vol. 93, no. 1, pp. 237-253, 2021.

[3000] F. M. Menger, "Molecular Lamarckism: On the evolution of human intelligence," *The Journal of New Paradigm Research,* vol. 73, no. 2, pp. 89-103, 2017.

[3001] B. Sauce and L. D. Matzel, "The paradox of intelligence: Heritability and malleability coexist in hidden gene-environment interplay," *Psychological Bulletin,* vol. 144, no. 1, pp. 26-47, 2018.

[3002] D. M. Buss, M. Abbott, A. Angleitner, A. Asherian, A. Biaggio, A. Blanco-Villasenor, M. Bruchon-Schweitzer, H.-Y. Ch'U, J. Czapinski, B. Deraad, B. Ekehammar, N. E. Lohamy, M. Fioravanti and K.-S. Yang, "International preferences in selecting mates: A study of 37 cultures," *Journal of Cross-Cultural Psychology,* vol. 21, no. 1, pp. 5-47, 1990.

[3003] R. Fisman, S. S. Iyengar, E. Kamenica and I. Simonson, "Gender differences in mate selection: Evidence from a speed dating experiment," *Quarterly Journal of Economics,* vol. 121, pp. 6673-679, 2006.

[3004] B. Pawlowski and S. Koziel, "The impact of traits offered in personal advertisements on response rates," *Evolution and Human Behavior,* vol. 23, no. 2, pp. 139-149, 2002.

[3005] G. F. Miller, "How mate choice shaped human nature: A review of sexual selection and human evolution (pp 87-130)," in *Handbook of Evolutionary Psychology*, Mahwah, NJ: Lawrence Erlbaum Associates Inc, 1998, p. 680.

[3006] G. Miller, The Mating Mind, New York: Random House, 2001, p. 528.

[3007] J. Rosenberg and R. J. Tunney, "Human vocabulary use as display," *Evolutionary Psychology,* vol. 6, no. 3, 2008.

[3008] T. Van der Zanden and A. P. Schouten, "Creativity, expectation violations, and impression formation: Effects of novelty and appropriateness in online dating profile texts," *Media Psychology,* vol. 27, no. 6, pp. 842-868, 2024.

[3009] L. Richie, Composer, *Hello.* [Sound Recording]. Motown. 1984.

[3010] J. Legend, Composer, *All of Me.* [Sound Recording]. Columbia Records.. 2013.

[3011] C. Broadus, P. Williams and C. Hugo, Composers, *Beautiful.* [Sound Recording]. Priority (Capitol). 2022.

[3012] A. P. Møller and R. Thronhill, "Bilateral symmetry and sexual selection: A meta-analysis," *The American Naturalist,* vol. 151, no. 2, pp. 174-192, 1998.

[3013] K. Grammer and R. Thornhill, "Human (Homo sapiens) facial attractiveness and sexual selection: The role of symmetry and averageness," *Journal of Comparative Psychology,* vol. 108, no. 3, pp. 233-242, 1994.

[3014] J. Goller, H. Leder, H. Cursiter and R. Jenkins, "Anchoring effects in facial attractiveness," *Perception,* vol. 47, no. 10-11, pp. 1043-1053, 2018.

[3015] T. C. Bates, "Fluctuating asymmetry and intelligence," *Intelligence,* vol. 35, no. 1, pp. 41-46, 2007.

[3016] M. D. Prokosch, R. A. Yeo and G. F. Miller, "Intelligence tests with higher g-loadings show higher correlations with body symmetry: Evidence for a general fitness factor mediated by developmental stability," *Intelligence,* vol. 33, no. 2, pp. 203-213, 2005.

[3017] B. Fink, B. Weege, J. T. Manning and R. Trivers, "Body symmetry and physical strength in human males," *American Journal of Human Biology,* vol. 26, no. 5, pp. 697-700, 2014.

[3018] R. M. Montoya, R. S. Horton and J. Kirchner, "Is actual similarity necessary for attraction? A meta-analysis of factual and perceived similarity," *Journal of Social and Personal Relationships,* vol. 25, no. 6, pp. 889-922, 2008.

[3019] Y. Z. Foo, L. W. Simmons and G. Rhodes, "Predictors of facial attractiveness and health in humans," *Scientific Reports,* vol. 7, p. 39731, 2017.

[3020] J. Li, D. He, W. Zhang, R. Huang and H. X, "The effects of moral behavior on facial attractiveness," *Psychology Research and Behavior Management,* vol. 16, p. 1521—1532, 2023.

[3021] G. Rhodes, "The evolutionary psychology of facial beauty," *Annual Review of Psychology,* vol. 57, pp. 199-226, 2006.

[3022] B. Fink, N. Neave, J. T. Manning and K. Grammer, "Facial symmetry and judgements of attractiveness, health and personality," *Personality and Individual Differences,* vol. 41, no. 3, pp. 491-499, 2006.

[3023] R. H. A. H. Jacobs, R. Renken and F. W. Cornelissen, "Neural correlates of visual aesthetics--beauty as the coalescence of stimulus and internal state," *PLoS One,* vol. 7, no. 2, p. e31248, 2012.

[3024] H. Chuan-Peng, Y. Huang, S. B. Eickhoff, K. Peng and J. Sui, "Seeking the 'beauty center' in the brain: A meta-analysis of fMRI studies of beautiful human faces and visual art," *Cognitive, Affective and Behavioral Neuroscience,* vol. 20, no. 6, pp. 1200-1215, 2020.

[3025] R. Yu, D. Mobbs, B. Seymour and A. J. Calder, "Insula and striatum mediate the default bias," *Journal of Neuroscience,* vol. 30, no. 44, p. 14702–14707, 2010.

[3026] H. Li, Q. Qang, G. Li and D. Cai, "Do looks matter for hosts on the peer-to-peer sharing accommodation market?," *Annals of Tourism Research,* vol. 98, p. 103510, 2023.

[3027] Z. Suo, Q. Guo and J. Shen, "Revisiting the effect of trustworthy face and attractive appearance on trust and trustworthiness behavior," *Applied Economics,* p. 1–21, 2024.

[3028] S. Weber, K. Geukes, M. Leckelt and M. D. Back, "The attractiveness of narcissists: Hard work or natural beauty?," *Self and Identity,* vol. 20, no. 2, pp. 235-267, 2021.

[3029] N. S. Holtzman and M. J. Strube, "People with dark personalities tend to create a physically attractive veneer," *Social Psychology and Personality Science,* vol. 4, no. 4, pp. 461-467, 2013.

[3030] K. J. Brazil, C. J. Dias and A. E. Forth, "Successful and selective exploitation in psychopathy: Convincing others and gaining trust," *Personality and Individual Differences,* vol. 170, no. 15, p. 110394, 2021.

[3031] J. Górniak, M. Zajenkowski, K. Szymaniak and P. K. Jonason, "Kindness or intelligence? Angry men are perceived as less intelligent by their female romantic partners," *Evolutionary Psychology,* vol. 22, no. 3, 2024.

[3032] S. W. Gangestad and D. M. Buss, "Pathogen prevalence and human mate preferences," *Ethology and Sociobiology,* vol. 14, no. 2, pp. 89-96, 1993.

[3033] J. L. Nedelec and K. M. Beaver, "Physical attractiveness as a phenotypic marker of health: An assessment using a nationally representative sample of American adults," *Evolution and Human Behavior,* vol. 35, no. 6, pp. 456-463, 2014.

[3034] C. I. Workman, S. Humphries, F. Hartung, G. K. Aguirre, J. W. Kable and A. Chatterjee, "Morality is in the eye of the beholder: the neurocognitive basis of the 'anomalous-is-bad' stereotype," *Annals of the New York Academy of Sciences,* vol. 1494, no. 1, pp. 3-17, 2021.

[3035] D. E. Han and S. M. Laurent, "Beautiful seems good, but perhaps not in every way: Linking attractiveness to moral evaluation through perceived vanity," *Journal of Personality and Social Psychology,* vol. 124, no. 2, pp. 264-286, 2023.

[3036] D. S. Hamermesh and J. E. Biddle, "Beauty and the labor market," *American Economic Review,* vol. 84, no. 5, p. 1174 – 1194, 1994.

[3037] N. Ambady and R. Rosenthal, "Half a minute: Predicting teacher evaluations from thin slices of nonverbal behavior and physical attractiveness," *Journal of Personality and Social Psychology,* vol. 64, no. 3, pp. 431-441, 1993.

[3038] H. Alkusari, N. D. Gupta and N. Etcoff, "In the eye of the promoter? How faculty ratings of attractiveness matter for junior academic careers," *Journal of Economic Behavior and Organization,* pp. 645-659, 2024.

[3039] G. Hale, T. Regev and Y. Rubinstein, "Do looks matter for an academic career in economics?," *Journal of Economic Behavior and Organization,* vol. 215, pp. 406-420, 2023.

[3040] T. Heffernan, "Sexism, racism, prejudice, and bias: a literature review and synthesis of research surrounding student evaluations of courses and teaching," *Assessment & Evaluation in Higher Education,* vol. 47, no. 1, pp. 144-154, 2022.

[3041] C. Goulão, J. A. Lacomba, F. Lagos and D.-O. Rooth, "Weight, attractiveness, and gender when hiring: A field experiment in Spain," *Journal of Economic Behavior and Organization,* vol. 218, pp. 132-145, 2024.

[3042] K. Bocian, R. D. Szekely-Copîndean, K. Myslinska-Szarek and B. Wojciszke, "Reevaluating the beautiful is moral stereotype by examining the impact of personal liking and belief in a just world," *Scientific Reports,* vol. 15, p. 12387, 2025.

[3043] S. W. Gangestad, L. A. Merriman and M. E. Thompson, "Men's oxidative stress, fluctuating asymmetry and physical attractiveness," *Animal Behaviour,* vol. 80, no. 6, pp. 1005-1013, 2010.

[3044] S. G. Young, D. F. Sacco and K. Hugenberg, "Vulnerability to disease is associated with a domain-specific preference for symmetrical faces relative to symmetrical non-face stimuli," *European Journal of Social Psychology,* vol. 41, no. 5, pp. 558-563, 2011.

[3045] L. M. DeBruine, B. C. Jones, J. R. Crawford, L. L. M. Welling and A. C. Little, "The health of a nation predicts their mate preferences: Cross-cultural variation in women's preferences for masculinized male faces," *Proceedings of the Royal Society B,* vol. 277, no. 1692, pp. 2405-2410, 2010.

[3046] A. C. Little, L. M. DeBruine and B. C. Jones, " Exposure to visual cues of pathogen contagion changes preferences for masculinity and symmetry in opposite-sex faces," *Proceedings of the Royal Society B,* vol. 278, no. 1714, p. 2032–2039, 2011.

[3047] M. Brown and D. F. Sarcco, "How and when crowd salience activates pathogen-avoidant motives," *Evolutionary and Behavioral Sciences,* vol. 16, no. 1, pp. 23-27, 2022.

[3048] A. J. Lee and B. P. Zietsch, "Experimental evidence that women's mate preferences are directly influenced by cues of pathogen prevalence and resource scarcity," *Biology Letters,* vol. 7, no. 6, pp. 892-895, 2011.

[3049] J. F. Lame Deer, "Top 15 quotes by John Fire Lame Deer," 14 7 2018. [Online]. Available: http://www.azquotes.com/author/22546-John_Fire_Lame_Deer. [Accessed 14 7 2018].

[3050] A. H. G. Rickover, "'Energy resources and our future' - remarks by Admiral Hyman Rickover delivered in 1957," *Energy Bulletin,* 1957/2006.

[3051] U. S. Department of Commerce, "Greenhouse gases continued to increase rapidly in 2022," 5 April 2023. [Online]. Available: https://www.noaa.gov/news-release/greenhouse-gases-continued-to-increase-rapidly-in-2022. [Accessed 5 July 2023].

[3052] S. Nordebo, M. F. Naeem and P. Tans, "Estimating the short-time rate of change in the trend of the Keeling curve," *Scientific Reports,* vol. 10, p. 21222, 2020.

[3053] R. Friberg, "Evidence II: The simple model of global atmospheric CO2 concentration 2 and the global atmospheric CO2 emission transfer constant," *ResearchGate,* 2024.

[3054] D. Soeder, "The Anomaly (pp 21-32)," in *Energy Futures: The Story of Fossil Fuel, Greenhouse Gas, and Climate Change*, Cham, Springer, 2022, p. 302.

[3055] E. Winiarska, M. Jutel and M. Zemelka-Wiacek, "The potential impact of nano- and microplastics on human health: Understanding human health risks," *Environmental Research,* vol. 251, no. (Pt 2), p. 118535, 2024.

[3056] R. E. Zurub, Y. Cariaco, M. G. Wade and S. A. Bainbridge, "Microplastics exposure: implications for human fertility, pregnancy and child health," *Frontiers in Endocrinology,* vol. 14, p. 1330396, 2024.

[3057] A. J. Nihart, M. A. Garcia, E. E. Hayek, R. Liu, M. Olewine, J. D. Kingston, E. F. Castillo, R. R. Gullapilli, T. Howarad, B. Bleske, J. Scott, J. Gonzalez-Estrella, J. M. Gross, M. Spilde, N. L. Adolphi, D. F. Gallego and H. S. Jarrell, "Bioaccumulation of microplastics in decedent human brains," *Nature Medicine,* vol. 31, no. 4, pp. 1114-1119, 2025.

[3058] N. Alziny, F. M. Elgarhy, A. M. Mohamed, H. Y. Shalaby, N. Nounou, O. Soliman and A. Abdelnaser, "The impact of microplastics in food and drugs on human health: A review of the MENA region," *International Journal of Environmental Research and Public Health,* vol. 22, no. 3, p. 380, 2025.

[3059] S. J. Heo, N. Moon and J. H. Kim, "A systematic review and quality assessment of estimated daily intake of microplastics through food," *Reviews on Environmental Health,* vol. 40, no. 2, pp. 371-392, 2024.

[3060] M. Rossi, "Can Lebanon's cedars outlive climate change and a pesky insect?," 20 April 2019. [Online]. Available: https://www.aljazeera.com/news/2019/4/20/can-lebanons-cedars-outlive-climate-change-and-a-pesky-insect#:~:text=Only%2017%20square%2Dkilometres%2C%200.4%20percent%20of%20the,of%20Nature%27s%20red%20list%20of%20threatened%20species.. [Accessed 19 November 2024].

[3061] R. Dahl, Charlie and the Chocolate Factory, London: George Allen & Unwin, 1964.

[3062] L. Lehmann, K. Bargum and M. Reuter, "An evolutionary analysis of the relationship between spite and altruism," *Journal of Evolutionary Biology,* vol. 19, no. 5, pp. 1507-1516, 2006.

[3063] J. Grisham, A Time to Kill, Bonners Ferry, ID: Wynewood Press, 1989, p. 515.

[3064] D. B. Bowser, Artist, *Sic Semper Tyrannis - 22th Regt. U.S. Colored Troops.* [Art]. United States Library of Congress, circa 1860-1870.

[3065] M. Beilmann, "Dropping out because of the others: bullying among the students of Estonian vocational schools," *British Journal of Sociology of Education,* vol. 38, no. 8, pp. 1139-1151, 2015.

[3066] A. Barnett, "A circadian rhythm of mating type reversals in Paramecium multimicronucleatum, syngen 2 and its genetic control," *Journal of Cell Physiology,* vol. 67, no. 2, pp. 239-270, 1966.

[3067] C. Borghi, M. Veronesi, E. Cosentino, A. F. G. Cicero, F. Kuria, A. Dormi and E. Ambrosioni, "Interaction between serum cholesterol levels and the renin-angiotensin system on the new onset of arterial hypertension in subjects with high-normal blood pressure," *Journal of Hypertension,* vol. 25, no. 10, pp. 2051-2057, 2007.

[3068] *The People of the State of Illinois, Plaintiff-Appellee, v. Patrick SYKES, Defendant-Appellant,* 2003.

[3069] W. E. Butler, "Crime in the Soviet Union," *British Journal of Criminology,* vol. 32, no. 2, pp. 144-159, 1992.

[3070] J. Banerjee, R. K. Layek, S. K. Sasmal and D. Ghosh, "Delayed evolutionary model for public goods competition with policing in phenotypically variant bacterial biofilms," *European Letters,* vol. 126, p. 18002, 2019.

[3071] V. Barone, "Bronx Zoo apologizes for putting African man on display in 1906," *New York Post,* 30 July 2020.

[3072] M. Bogen, "All the ways hiring algorithms can introduce bias," *Harvard Business Review,* 6 May 2019.

[3073] C. Boberg, Writer, *How Great Thou Art.* [Performance]. 1885.

[3074] A. Burgess, A Clockwork Orange, London: William Heinemann, 1962, p. 192.

[3075] S. P. Bradish, Old Norse Stories, New York: American Book Company, 1900, p. 240.

[3076] W. B. Boykin, "Runaway," *Columbus Democrat,* p. 4, 18 August 1838.

[3077] D. G. Blanchflower and A. Bryson, "The adult consequences of being bullied in childhood," *Social Science and Medicine,* p. 116690, 2024.

[3078] C. R. Browning, Ordinary Men: Reserve Police Battalion 101 and the Final Solution in Poland, New York: Harper Collins, 1992, p. 231.

[3079] I. Arita, "Discovery of forgotten variola specimens at the National Institutes of Health in the USA," *Expert Review of Anti-infective Therapy,* vol. 12, no. 12, 2014.

[3080] S. Ahmer, A. W. Yousafzai, N. Bhutto, S. Alam, A. Sarangzai and A. Iqbal, "Bullying of medical students in Pakistan: A cross-sectional questionnaire," *PLoS ONE,* vol. 3, p. e3889, 2008.

[3081] C. Achebe, Things Fall Apart, London: Penguin, 1994, p. 209.

[3082] S. Abril and C. Gómez, "Factors triggering queen executions in the Argentine ant," *Scientific Reports,* vol. 9, p. 10427, 2019.

[3083] D. Abnett, First and Only, London: Games Workshop, 1999/ 2013, p. 416.

[3084] L. F. Abbott, D. D. Bock, E. M. Callaway, W. Denk, C. Dulac, L. A. Fairhall, I. Fiete, K. M. Harris, M. Helmstaedter, V. Jain, N. Kasthuri, Y. LeCun, J. W. Lichtman, P. B. Littlewood, L. Luo, J. H. R. Maunsell, R. C. Reid and B. R. Rosen, "The mind of a mouse," *Cell,* vol. 182, no. 6, pp. 1372-1376, 2020.

[3085] P.-C. Chou, Y.-C. Huang and S. Yu, "Mechanisms of epigenetic inheritance in post-traumatic stress disorder," *Life,* vol. 14, no. 1, p. 98, 2024.

[3086] X. Chen, Y. Jiang, J. Wang, Y. Liu, M. Xiao, C. Song, Y. Bai, N. Y. Han and F. Han, "Synapse impairment associated with enhanced apoptosis in post-traumatic stress disorder," *Synapse,* vol. 74, no. 2, p. e22134, 2020.

[3087] B. Chakraborty, *Disabled Ga. Girl Gang-Raped,* Marietta, Georgia: ABC News, 2000.

[3088] W. R. Chae, J. Nübel, J. Baumert, S. M. Gould and C. Otte, "Association of depression and obesity with C-reactive protein in Germany: A large nationally representative study," *Brain, Behavior, and Immunity,* vol. 103, pp. 223-231, 2022.

[3089] M. Catalini, J. Carr Smyth and B. Shipkowski, "WATCH: Trump amplifies false racist rumor about Ohio's Haitian immigrants in debate," 11 September 2024. [Online]. Available: https://www.pbs.org/newshour/politics/watch-trump-amplifies-false-racist-rumor-about-ohios-haitian-immigrants-in-debate. [Accessed 15 June 2025].

[3090] K. M. (. b. Cassidy, "Full Text of HB5290," 8 February 2012. [Online]. Available: https://ilga.gov/legislation/97/HB/PDF/09700HB5290lv.pdf.

[3091] E. A. Caspar, Just Following Orders: Atrocities and the Brain Science of Obedience, Cambridge: Cambridge University Press, 2024, p. 261.

[3092] A. J. Calder, J. Kean, F. Manes, N. Antoun and A. W. Young, "Impaired recognition and experience of disgust following brain injury," *Nature Neuroscience,* vol. 3, pp. 1077-1078, 2000.

[3093] A. Cutway, "George Zimmerman says he retweeted photo of Trayvon Martin's body by accident," *Orlando Sentinel,* 6 October 2015.

[3094] R. Smith, L. Tolhurst and M. Dempsey, Composers, *Boys Don't Cry.* [Sound Recording]. Fiction. 1979.

[3095] R. Cook, G. Bird, C. Catmur, C. Press and C. Heyes, "Mirror neurons: from origin to function," *The Behavioral and Brain Sciences,* vol. 37, no. 2, pp. 177-192, 2014.

[3096] M. Connaughton, O. Mitchell, E. Cullen, M. O'Connor, T. Banaschewski, G. J, A. L. W. Bokde, R. Brühl, S. Desrivières, H. Flor, H. Garavan, P. Gowland, A. Grigis, A. Heinz, H. Lemaitre, J.-L. Martinot and M.-L. Paillère Martinot, "Bullying and Early Brain Development: A Longitudinal Structural Magnetic Resonance Imaging Study from Adolescence to Early Adulthood," *bioRxiv Preprint Server,* 2024.

[3097] D. K. Enigk, M. E. Thompson, Z. P. Machanda, R. W. Wrangham and M. N. Muller, "Competitive ability determines coalition participation and partner selection during maturation in wild male chimpanzees (Pan troglodytes schweinfurthii)," *Behavioral Ecology and Sociobiology,* vol. 74, 2020.

[3098] J. Emerson, "Study: Chicago splitting from Illinois would be economic disaster," 13 July 2021. [Online]. Available: https://khqa.com/news/local/siuc-study-chicago-splitting-from-illinois-would-be-economic-disaster.

[3099] C. R. Ember, "Residential variation among hunter-gatherers," *Behavior Science Research,* vol. 10, no. 3, pp. 199-227, 1975.

[3100] N. Elias, The Civilizing Process: Sociogenetic and Psychogenetic Investigations Revised Edition (Über den Prozeß der Zivilisation), Hoboken, NJ: Blackwell Publishing, 2000/1934, p. 592.

[3101] M. Eisner, "From swords to words: Does macro-level change in self-control predict long-term variation in levels of homicide?," *Crime and Justice,* vol. 43, no. 1, pp. 65-134, 2014.

[3102] S. K. Egan and D. G. Perry, "Does low self-regard invite victimization?," *Developmental Psychology,* p. 299–309, 1998.

[3103] K. M. Edwards, "Intimate partner violence and the rural-urban-suburban divide: Myth or reality? A critical review of the literature," *Trauma, Violence, and Abuse,* vol. 16, no. 3, pp. 359-373, 2015.

[3104] B. N. Dulca, A. K. Bourdon, C. T. Clinard, M. B. Muvvala, S. R. Campagna and M. A. Cooper, "Metabolomics reveals distinct neurochemical profiles associated with stress resilience.," *Neurobiology of Stress,* vol. 7, pp. 103-112, 2017a.

[3105] F. Dostoyevsky, The House of the Dead; or, Prison Life in Siberia, 2011.

[3106] K. Domschke, "A methylome-based signature of stress: Association with accelerated epigenetic aging and alcohol use," *Biological Psychiatry,* vol. 93, no. 4, pp. 296-297, 2023a.

[3107] S. DeWeerdt, "How to map the brain," *Nature Outlook,* vol. 571, pp. S6-S8, 24 July 2019.

[3108] A. L. Des Forges, Leave No One to Tell the Story: Genocide in Rwanda, New York: Human Rights Watch, 1999b, p. 789.

[3109] F. B. M. de Waal, Chimpanzee Politics: Power and Sex Among Apes, New York: Harper and Row, 1982.

[3110] B. B. de Mesquita and A. Smith, The Dictator's Handbook: Why Bad Behavior Is Almost Always Good Politics, New York: PublicAffairs, 2011b, p. 352.

[3111] C. K. W. De Dreu, L. L. Greer, G. A. van Kleef, S. Shalvi and M. J. J. Handgraaf, "Oxytocin promotes human ethnocentrism," *Proceedings of the National Academy of Sciences of the USA,* vol. 108, no. 4, pp. 1252-1266, 2011.

[3112] R. Ginzburg, 100 Years of Lynchings, Baltimore, MD: Black Classic Press, 1988/1996, p. 270.

[3113] J. Gillray, Artist, *The Antisaccharrites or John Bull (etching).* [Art]. 1792.

[3114] Giambologna, *The Sabines (Sculpture),* New York, New York: Stacy, George, ca. 1870.

[3115] B. Gegenhuber, M. V. Wu, R. Bronstein and J. Tollkuhn, "Gene regulation by gonadal hormone receptors underlies brain sex differences," *Nature,* vol. 606, pp. 153-159, 2022.

[3116] N. Ganti and S. Baek, "Why people stand by," *Journal of Student Research,* vol. 10, no. 1, 2021.

[3117] C. A. Frye and A. A. Walf, "Progesterone enhances learning and memory of aged wildtype and progestin receptor knockout mice," *Neuroscience Letters,* vol. 472, no. 1, pp. 38-42, 2010.

[3118] R. Freedman, We Will Not Be Silent: The White Rose Student Resistance Movement That Defied Adolf Hitler, New York: Clarion Books, 2016, p. 112.

[3119] J. A. Fox and M. W. Zawitz, "Homicide Trends in the United States," [Online]. Available: https://bjs.ojp.gov/content/pub/pdf/htius.pdf. [Accessed 11 March 2023].

[3120] S. Footnotes, "Title Footnote 2," *Journal 2,* pp. 4-9, 2000.

[3121] F. Footnote, "Title 1," *Journal 1,* pp. 1-3, 1900.

[3122] K. Follett, Pillars of the Earth, Reissue ed., New York: Viking, 1989/2016, p. 992.

[3123] E. Fantino, S. Gaitan, A. Kennelly and S. Stolarz-Fantino, "How reinforcer type affects choice in economic games," *Behavioural Processes,* vol. 75, no. 2, pp. 107-114, 2007.

[3124] T. Husband, Ed., Propaganda: Truth and Lies in Times of Conflict, London: Arcturus Publishing Limited, 2014, p. 192.

[3125] V. Hugo, Oeuvres Complètes de Victor Hugo, Paris: A. Lacroix, Verboeckhoven & Cie., 1862, p. 1462.

[3126] B. Hölldobler and E. O. Wilson, Journey to the Ants: A Story of Scientific Exploration, Cambridge, Massachusetts: Belknap Press, 1994, p. 228.

[3127] W. Hogarth, Artist, *Credulity, Superstition and Fanaticism: A Medley.* [Art]. 1761/62.

[3128] E. Hoffer, The True Believer: Thoughts on the Nature of Mass Movements, New York: Harper & Brothers, 1951, p. 176.

[3129] A. Hoeffler, "What are the costs of violence?," *Politics, Philosophy and Economics,* vol. 16, no. 4, pp. 422-445, 2017.

[3130] M. Hobson, "Quotes Come Early This Year," 6 December 2021. [Online]. Available: https://sports.yahoo.com/quotes-come-early-140041948.html. [Accessed 23 April 2023].

[3131] J. P. Hobbs, Dear General: Eisenhower's Wartime Letters to Marshall, First ed., Baltimore, MD: Johns Hopkins University Press, 1999/1971, p. 276.

[3132] G. Hinge, "Consolidation of Indian Legislation Volume II: Indian Acts and Amendments, 1868-1975," Government of Canada, 1927.

[3133] T. R. Hettler and L. H. Cohen, "Intrinsic religiousness as a stress moderator for adult protestant churchgoers," *Journal of Community Psychology,* vol. 26, no. 6, pp. 597-609, 1998.

[3134] S. Herculano-Houzel, C. Watson and G. Paxinos, "Distribution of neurons in functional areas of the mouse cerebral cortex reveals quantitatively different cortical zones," *Frontiers in Neuroanatomy,* vol. 7, p. 35, 2013.

[3135] E. Hemmingsson, P. Nowicka, S. Ulijaszek and T. I. A. Sørensen, "The social origins of obesity within and across generations," *Obesity,* vol. 24, no. 1, p. e13514, 2023.

[3136] T. Hardy, Jude the Obscure, CreateSpace, 1895/ 2016, p. 230.

[3137] Hammurabi, *Code of Hammurabi,* 1700 BCE.

[3138] R. Hames, "Pacifying hunter-gatherers," *Human Nature,* vol. 30, pp. 155-175, 2019a.

[3139] F. Hafez, "Opinion | Why Is the Far Right Obsessed With Burning the Koran?," 24 January 2023. [Online]. Available: https://bridge.georgetown.edu/today_islamophobia/today-in-islamophobia-january-25-2023/.

[3140] Y. Guan, X. Chen, B. Zhao, Y. Shi and F. Han, "What happened in the hippocampal axon in a rat model of posttraumatic stress disorder," *Cellular and Molecular Neurobiology,* vol. 24, no. 3, pp. 723-737, 2022.

[3141] J. Grisham, A Time to Kill, Bonners Ferry, Idaho: Wynwood Press, 1989b, p. 672.

[3142] N. M. Grebe, J. P. Hirwa, T. S. Stoinski, L. Vigilant and S. Rosenbaum, "Mountain gorillas maintain strong affiliative biases for maternal siblings despite high male reproductive skew and extensive exposure to paternal kin," *Elife,* vol. 11, p. e80820, 2022.

[3143] F. D. Gray, The Tuskegee Syphilis Study: The Real Story and Beyond, 1st ed., Montgomery, AL: River City Publishers, 1998, p. 175.

[3144] R. Graves and J. Pulman, *I Claudius, Episode 7: Reign of Terror,* 1976.

[3145] M. Grant, Cleopatra, New York, New York: Barnes and Noble Books, 1972/1992, p. 301.

[3146] J. Goodall, The Chimpanzees of Gombe: Patterns of Behavior, Cambridge, MA: Belknap Press of Harvard University Press, 1986, p. 673.

[3147] W. Golding, Lord of the Flies, London: Faber and Faber, 1954, p. 224.

[3148] S. Z. Goldenberg and G. Wittemyer, "Orphaned female elephant social bonds reflect lack of access to mature adults," *Scientific Reports,* vol. 7, no. 1, pp. 1-7, 2017.

[3149] J. D. Johnson and M. S. Gazzaniga, "Interhemisphere imitation in split-brain monkeys," *Experimental Neurology,* vol. 79, no. 2, pp. 206-212, 1970.

[3150] X. Jiang, R. Sanford and M. D. Pell, "Neural systems for evaluating speaker (Un)believability," *Human Brain Mapping,* vol. 38, no. 7, pp. 3732-3749, 2017.

[3151] Y. Jia, Y. Han, X. Wang and F. Han, "Role of apoptosis in the Post-traumatic stress disorder model-single prolonged stressed rats," *Psychoneuroendocrinology,* vol. 95, pp. 97-105, 2018.

[3152] K. Jensen, "Punishment and Spite, the dark side of cooperation," *Philosophical Transactions of the Royal Society B,* vol. 365, no. 1553, pp. 2635-2650, 2010b.

[3153] M. D. Jennions and D. W. Macdonald, "Cooperative breeding in mammals," *Trends in Ecology and Evolution,* vol. 9, no. 3, pp. 89-93, 1994.

[3154] C. Konishi, S. Hymel, T. K. Y. Wong and T. Waterhouse, "School climate and bystander responses to bullying," *Psychology in the Schools,* vol. 58, no. 8, 2021.

[3155] M. Kivimäki, A. Bartolomucci and I. Kawachi, "The multiple roles of life stress in metabolic disorders," *Nature Reviews Endocrinology,* vol. 19, pp. 10-27, 2023b.

[3156] H. Kissinger, Interviewee, *Viewpoint: The Day After.* [Interview]. 20 November 1987.

[3157] J. Kirkpatrick, Turnabout: New Help for the Woman Alcoholic, Fort Lee, New Jersey: Barricade, 1977, p. 190.

[3158] M. L. J. King, *Letter From a Birmingham Jail,* 1963.

[3159] M. L. King, "I Have a Dream," 28 August 1963. [Online]. Available: https://www.archives.gov/files/social-media/transcripts/transcript-march-pt3-of-3-2602934.pdf. [Accessed 26 July 2025].

[3160] I. Kim, D. Burgess and S. S. Chatters, "Perceived school fairness and willingness to report bias-based bullying among youth during COVID-19 Pandemic," *Journal of School Health,* vol. 1, 2022.

[3161] A. Kennedy and A. M. Brausch, "Emotion dysregulation, bullying, and suicide behaviors in adolescents," *Journal of Affective Disorders Reports,* vol. 15, p. 100715, 2024.

[3162] Kelly, "The third culture," *Science,* vol. 279, no. 5353, pp. 992-993, 1998.

[3163] M. H. Kärcher and F. L. W. Ratnieks, "Killing and replacing queen-laid eggs: low cost of worker policing in the honeybee," *The American Naturalist,* vol. 184, no. 1, pp. 110-118, 2014.

[3164] M. Kämmerle, H.-F. Unterrainer, P. Dahmen-Wassenberg, A. Fink and H.-P. Kapfhammer, "Dimensions of religious/spiritual well-being and the dark triad of personality," *Psychopathology,* vol. 47, pp. 297-302, 2014.

[3165] R. Luscombe, "DeSantis sends migrants to Martha's Vineyard, causing 'humanitarian situation'," 15 September 2022. [Online]. Available: https://www.theguardian.com/us-news/2022/sep/15/ron-desantis-migrants-marthas-vineyard. [Accessed 27 February 2023].

[3166] N. Luna, "McDonald's president who made $7.4 million last year says proposal to pay fast-food workers $22 an hour is 'costly and job-destroying'," 26 January 2023. [Online]. Available: https://www.msn.com/en-us/money/companies/mcdonald-s-president-who-made-7-4-million-last-year-says-proposal-to-pay-fast-food-workers-22-an-hour-is-costly-and-job-destroying/ar-AA16Mc7D. [Accessed 28 January 2023].

[3167] D. Lukas and E. Huchard, "The evolution of infanticide by females in mammals," *Philosophical Transactions of the Royal Society B,* vol. 374, no. 1780, p. 20180075, 2019.

[3168] R. Lowell, Lowell's Poems: A Selection, J. Raban, Ed., London: Faber and Faber, 1974, p. 192.

[3169] G. Lopez, "Here are the members of Congress who voted against protecting gay people from hate crimes," 13 June 2016. [Online]. Available: https://www.vox.com/2016/6/12/11912076/orlando-florida-mass-shooting-gay-hate-crime-law.

[3170] R. Wise and J. Robbins, Directors, *West Side Story.* [Film]. Mirisch Pictures; Steven Arts Productions, 1961.

[3171] J. H. Lee, S. K. Nam, A.-R. Kim, B. Kim, M. Y. Lee and S. M. Lee, "Resilience: A meta-analytic approach," *Journal of Counseling and Development,* vol. 91, no. 3, pp. 269-279, 2013.

[3172] L. L. Leape, "Unnecessary surgery," *Health Services Research,* vol. 24, no. 3, pp. 351-407, 1989.

[3173] C. Lauper, Composer, *Girls Just Want To Have Fun.* [Sound Recording]. Rock Plant Studio. 1983.

[3174] J. Lang, "Explaining Genocide: Hannah Arendt and the Social-Scientific Concept of Dehumanization," in *The Anthem Companion to Hannah Arendt*, New York, Anthem Press, 2017, pp. 175-196.

[3175] V. Kumar, R. Talwar, D. Raut and S. Singh, "Psychosocial determinants of tobacco use among school going adolescents in Delhi, India," *Journal of Addiction,* vol. 1, p. 170941, 2014.

[3176] N. K. Kulick, S. Cheves, C. Chaves-Cordero, R. Lopez, S. R. Morales, L. M. Fedigan and K. M. Jack, "Female-committed infanticide followed by juvenile-enacted cannibalism in wild white-faced capuchins," *Primates,* vol. 62, no. 6, pp. 1037-1043, 2021b.

[3177] A. V. Kukekova, J. L. Johnson, X. Xiang, S. Feng, S. Liu, H. M. Rando, A. V. Kharlamova, Y. Herbeck, N. A. Serdyukova, Z. Xiong, V. Beklemischeva, K.-P. Koepfli, R. G. Gulevich, A. V. Vladimirova, J. P. Hekman, P. L. Perelman and Graphodatsky,

"Red fox genome assembly identifies genomic regions associated with tame and aggressive behaviours," *Nature Ecology and Evolution,* vol. 2, p. 1479–1491, 2018.

[3178] S. Kühn, D. T. Kugler, K. Schmalen, M. Weichenberger, C. Witt and J. Gallinat, "Does playing violent video games cause aggression? A longitudinal intervention study," *Molecular Psychiatry,* pp. http://dx.doi.org/10.1038/s41380-018-0031-7, 2018.

[3179] T. T. Kristensen, L. L. Christensen, J. Frystyk, D. Glintborg, G. T'Sjoen, K. K. Roessler and M. S. Anderson, "Effects of testosterone therapy on constructs related to aggression in transgender men: A systematic review," *Hormones and Behavior,* vol. 128, p. 104912, 2021.

[3180] C. Musanabaganwa, A. H. Wani, J. Donglasan, S. Fatumo, S. Jansen, J. Mutabaruka, E. Rutembesa, A. Uwineza, E. J. Hermans, B. Roozendaal, D. E. Wildman, L. Mutesa and M. Uddin, "Leukocyte methylomic imprints of exposure to the genocide against the Tutsi in Rwanda: a pilot epigenome-wide analysis," *Epigenomics,* vol. 14, no. 1, pp. 11-25, 2022b.

[3181] F. Murphy, The Long Haul: A Trucker's Tales of Life on the Road, New York: W. W. Norton, 2017, p. 256.

[3182] L. S. Morris, J. G. McCall, D. Charney and J. W. Murrough, "The role of the locus coeruleus in the generation of pathological anxiety," *Brain Neuroscience Advances,* vol. 4, p. 2398212820930321, 2020.

[3183] F. B. Mohamed, S. H. Faro, N. J. Gordon, M. Platek, H. Ahmad and J. M. Williams, "Brain mapping of deception and truth telling about an ecologically valid situation: functional MR imaging and polygraph investigation--initial experience," *Radiology,* vol. 238, no. 2, pp. 679-688, 2006.

[3184] S. Mitchell, Gilgamesh: A New English Version, New York: Free Press, 2004.

[3185] J. S. Mintert and A. G. T. T. Tran, "Bullying and personality," in *The Wiley Encyclopedia of Personality and Individual Differences: Clinical, Applied, and Cross-Cultural Research*, John Wiley & Sons Ltd, 2020, pp. 523-531.

[3186] O. Matilainen, P. M. Quirós and J. Auwerx, "Mitochondria and epigenetics - Crosstalk in homeostasis and stress," *Trends in Cell Biology,* vol. 27, no. 6, pp. 453-463, 2017.

[3187] J. Mathias and G. R. Yabsley, "Conspiracy of legislation: The suppression of Indian rights in Canada," *British Columbian Studies,* vol. 89, pp. 34-45, 1991.

[3188] J. W. Mason, E. L. Giller, T. R. Kosten and L. Harkness, "Elevation of urinary norepinephrine/cortisol ratio in posttraumatic stress disorder," *Journal of Nervous and Mental Disorders,* vol. 176, no. 8, pp. 498-502, 1988.

[3189] A. H. Maslow, "A theory of human motivation," *Psychological Review,* vol. 50, no. 4, p. 370–396, 1943.

[3190] N. M. Malamuth, T. Addison and M. Koss, "Pornography and sexual aggression: Are there reliable effects and can we understand them?," *Annual Review of Sex Research,* vol. 11, no. 1, pp. 26-91, 2012.

[3191] S. Malaivijitnond, Y. Hamada, B. Suryobroto and O. Takenaka, "Female long-tailed macaques with scrotum-like structures," *American Journal of Primatology,* vol. 69, no. 7, pp. 721-735, 2007.

[3192] A. Pusey, J. Williams and J. Goodall, "The influence of dominance rank on the reproductive success of female chimpanzees," *Science,* vol. 277, no. 5327, p. 828–831, 1997.

[3193] T. Pratchett, Discworld: Equal Rites, vol. 3, London: Harper, 1987/ 2004, p. 256.

[3194] R. Platten, Composer, *Fight Song.* [Sound Recording]. 2015.

[3195] F. Pistrucci, Artist, *The rape of the Sabine women.* [Art]. British Museum, 1830-1835.

[3196] B. Picart, Artist, *Susanna's Accusers Stoned.* [Art]. 1775.

[3197] M. Picard, Der Mensch und das Wort, Erlenbach-Zürich: E. Rentsch, 1955, p. 201.

[3198] F. Petty and A. Sherman, "Regional aspects of the prevention of learned helplessness by desipramine," *Life Sciences,* vol. 26, no. 17, pp. 1447-1452, 1980.

[3199] N. Perroud, E. Rutembesa, A. Paoloni-Giacobino, J. Mutabaruka, L. Mutesa, L. Stenz, A. Malafosse and F. Karege, "The Tutsi genocide and transgenerational transmission of maternal stress: epigenetics and biology of the HPA axis," *The World Journal of Biological Psychiatry,* vol. 15, no. 4, pp. 334-345, 2014.

[3200] J. C. Penagos-Corzo, M. C. van-Hasselt, D. Escobar, R. A. Vázquez-Roque and G. Flores, "Mirror neurons and empathy-related regions in psychopathy: Systematic review, meta-analysis, and a working model," *Social Neuroscience,* vol. 17, no. 5, pp. 462-479, 2022.

[3201] L. Peeples, "What the data say about police brutality and racial bias — and which reforms might work," *Nature (news feature),* vol. 583, pp. 22-24, 26 May 2021 (update).

[3202] A. Pascual-Sanchez, N. Hickey, A. Mateu, M. Martinez-Herves, T. Kramer and D. Nicholls, "Personality traits and self-esteem in traditional bullying and cyberbullying," *Personality and Individual Differences,* vol. 177, p. 110809, 2021.

[3203] S. A. Park, S. Goïame, D. A. O'Connor and J.-C. Dreher, "Integration of individual and social information for decision-making in groups of different sizes," *PLoS Biology,* vol. 15, no. 6, p. e2001958, 2017.

[3204] P. Panjiar, "Hindutva is nothing but Brahminism," *Outlook,* 3 February 2022.

[3205] C. Painter, Artist, *Ajax and Cassandra.* [Art]. Musée du Louvre, circa 440 and circa 430 BCE.

[3206] M. Simon, Director, *Axis of Evil Comedy Tour.* [Film]. USA: Image Entertainment, 2008.

[3207] R. Noë, F. B. M. de Waal and J. A. R. A. M. van Hooff, "Types of dominance in a chimpanzee colony," *Folia Primatologica,* vol. 34, no. 1-2, pp. 90-110, 1980.

[3208] S. Noble, "Google Has a Striking History of Bias Against Black Girls," *Time Magazine,* 26 March 2018.

[3209] T. Nishida, N. Corp, M. Hamai, T. Hasegawa, M. Hiraiwa-Hasegawa, K. Hosaka, K. D. Hunt, N. Itoh, K. Kawanaka, A. Matsumoto-Oda, J. C. Mitani, M. Nakamura, K. Norikoshi, T. Sakamaki, L. Turner, S. Uehara and K. Zamma, "Demography, female life history, and reproductive profiles among the chimpanzees of Mahale," *American Journal of Primatology,* vol. 59, no. 3, pp. 99-121, 2003.

[3210] T. Nishida, Chimpanzees of the Lakeshore: Natural History and Culture at Mahale, 1st ed., Cambridge: Cambridge University Press, 2012, p. 358.

[3211] P. Newkirk, Spectacle: The Astonishing Life of Ota Benga by Pamela Newkirk(1905-06-18), Amistad (Harper Collins), 2015/2016, p. 336.

[3212] A. Nelmes, "An open letter to the ranga-haters of Australia (by a proud ginger)," 2015. [Online]. Available: https://www.nine.com.au/entertainment/viral/ginger-pride/eca34e43-814f-47cb-be0f-c105097ebc9f.

[3213] J. Nalson, An Impartial Collection of the Great Affairs of State from the Beginning of the Scotch Rebellion in the Year MDCXXXIX to the Murther of King Charles I. (p vii), London: Printed for S. Mearne, T. Dring, B. Tooke, T. Sawbrige, and C. Mearne, 1683.

[3214] M. Nall, M. Corbett, J. McLoughlin, J. Petrosko, D. Garcia and J. Karibo, "Impact of short-term oral steroid use upon children's school achievement and behavior," *Annals of Allergy,* vol. 69, no. 3, p. 218–220, 1992.

[3215] H. Ryu, "Social relationships in female bonobos," in *Bonobos and People at Wamba: 50 Years of Research*, Singapore, Springer, 2023, pp. 267-284.

[3216] *H.R.55 Emmett Till Antilynching Act,* 2023.

[3217] J. M. Ruiz, M. Quintela and R. Barreiro, "Tributyltin and imposex: no uncertainty shown," *Marine Ecological Progress Series,* vol. 170, pp. 293-294, 1998.

[3218] S. Rosenbaum, V. Vecellio and T. Stoinski, "Observations of severe and lethal coalitionary attacks in wild mountain gorillas," *Science Reports,* vol. 6, no. 1, pp. 1-8, 2016.

[3219] V. Rojas, "The gender of Latinidad: Latinas speak about Hispanic television," *The Communications Review,* vol. 7, no. 2, pp. 125-153, 2004.

[3220] K. Robb, "Epstein, the Headlines, and the Children We Ignore," 25 July 2025. [Online]. Available: https://verdict.justia.com/2025/07/25/epstein-the-headlines-and-the-children-we-ignore. [Accessed 25 July 2025].

[3221] K. Rigby, "Consequences of bullying in schools," *The Canadian Journal of Psychiatry,* vol. 48, no. 9, pp. 583-590, 2003.

[3222] A. Rice, Queen of the Damned, New York: Knopf, 1988, p. 448.

[3223] K. Reilly, "Here Are All the Times Donald Trump Insulted Mexico," 31 August 2016. [Online]. Available: https://time.com/4473972/donald-trump-mexico-meeting-insult/. [Accessed 27 February 2023].

[3224] S. Reardon, "'Forgotten' NIH smallpox virus languishes on death row," *Nature,* vol. 514, p. 544, 2014.

[3225] C. Rayner and H. Hoel, "A summary review of literature relating to workplace bullying," *Journal of Community & Applied Social Psychology,* vol. 7, no. 3, pp. 181-191, 1998b.

[3226] S. Ratcliffe, Ed., Enemies, Fifth ed., Oxford: Oxford University Press, 2017.

[3227] F. Range, L. Horn, Z. Viranyi and L. Huber, "The absence of reward induces inequity aversion in dogs," *Proceedings of the National Academy of Sciences,* vol. 106, no. 1, pp. 340-345, 2009.

[3228] D. M. Smith, Endless Holocausts: Mass Death in the History of the United States Empire, New York: Monthly Review Press, 2023, p. 504.

[3229] P. Shrimpton, Conscience before Conformity: Hans and Sophie Scholl and the White Rose resistance in Nazi Germany, Leominster Motors: Gracewing Publishing, 2018, p. 328.

[3230] M. Shermer, The Believing Brain: From Ghosts and Gods to Politics and Conspiracies---How We Construct Beliefs and Reinforce Them as Truths, New York: St. Martin's Press, 2012, p. 385.

[3231] K. I. Shepherd, Why I Am Not a Hindu: A Sudra Critique of Hindutva Philosophy, Culture and Political Economy, First ed., New Delhi: Sage, 2019, p. 192.

[3232] M. Forman, Director, *Amadeus.* [Film]. USA: Orion Pictures, 1984.

[3233] S. J. Shabel, C. D. Proulx, J. Piriz and R. Malinow, "Mood regulation. GABA/glutamate co-release controls habenula output and is modified by antidepressant treatment," *Science,* vol. 345, no. 6203, pp. 1494-1498, 2014.

[3234] R. Shabad, J. Soboroff and Z. Richards, "Sen. Alex Padilla is forcibly removed from DHS Secretary Kristi Noem's news conference in Los Angeles," 12 June 2025. [Online]. Available: https://www.nbcnews.com/politics/congress/sen-alex-padilla-forcibly-removed-dhs-sec-kristi-noems-press-conferenc-rcna212688. [Accessed 16 June 2025].

[3235] J. Seabrook, "Suffering Souls," *The New Yorker,* p. 10, 10 November 2008.

[3236] M. Schindler, "On Dr. Eduard Bloch, Hitler's Family Physician (Who Happened to Be Jewish)," 18 October 2021. [Online]. Available: https://lithub.com/on-dr-eduard-bloch-hitlers-family-physician-who-happened-to-be-jewish/.

[3237] M. A. Schiele, M. G. Gottschalk and K. Domschke, "The applied implications of epigenetics in anxiety, affective and stress-related disorders - A review and synthesis on psychosocial stress, psychotherapy and prevention," *Clinical Psychology Review,* vol. 77, 2020.

[3238] R. M. Sapolsky, Behave: The Biology of Humans at Our Best and Worst, New York: Penguin Books, 2018, p. 800.

[3239] C. Sagan, Interviewee, *Viewpoint: On The Day After.* [Interview]. 20 11 1983.

[3240] A. Szyk, Artist, *Poland fights Nazi dragon, Polish war relief. US WW2 Poster.* [Art]. United States Government, 1943.

[3241] O. Suleiman, "Islam beyond phobia: Quelling the obsession with burning the Quran," 31 January 2023. [Online]. Available: https://religionnews.com/2023/01/31/quelling-the-obsession-with-burning-the-quran/.

[3242] Y. Zhang, Director, *Raise the Red Lantern (Da hong denglong gaogao gua).* [Film]. People's Republic of China.1991.

[3243] P. Stockley and J. Bro-Jørgensen, "Female competition and its evolutionary consequences in mammals," *Biological Reviews,* vol. 86, no. 2, pp. 341-366, 2011.

[3244] S. D. Stelwagen, B. D. Opell and K. G. Short, "Temperature mediates the effect of humidity on the viscoelasticity of glycoprotein glue within the droplets of an orb-weaving spider's prey capture threads," *Journal of Experimental Biology,* vol. 217, no. 9, p. 1563–1569, 2014.

[3245] R. Sriwoon and N. Poonsud, "Growth and reproductive of Babylonia areolata Link 1807 in Songkhla and Pattani, the lower part of Gulf of Thailand," *Songklanakarin Journal of Science Technology,* vol. 41, no. 6, pp. 1404-1412, 2019.

[3246] L. Spytska, "Symptoms and main differences between a psychopath and a sociopath," *The Journal of Nervous and Mental Disease,* vol. 212, no. 1, pp. 52-56, 2024.

[3247] A. Soldati, P. Fedurek, C. Crockford, S. Adue, J. W. Akankwasa, C. Asiimwe, J. Asua, G. Atayo, B. Chandia, E. Feymann, C. Fryns, G. Muhumuza, D. Taylor, K. Zuberbühler and C. Hobaiter, "Dead-infant carrying by chimpanzee mothers in the Budongo Forest," *Primates,* vol. 63, pp. 497-508, 2022.

[3248] A. E. Vaughn, "The association between offender socioeconomic status and victim-offender relationship in rape offences," *Psychology, Evolution, and Gender,* vol. 3, no. 2, pp. 121-136, 2001.

[3249] R. R. Varma, "Birth of Krishna by Raja Ravi Varma (1890)," 23 April 2011. [Online]. Available: https://upload.wikimedia.org/wikipedia/commons/8/8f/Raja_Ravi_Varma%2C_Birth_of_Krishna_%281890%29.jpg.

[3250] unknown, Artist, *Ota Benga at Bronx Zoo.* [Art]. Library of Congress, ca 1906; published ca 1915-1916.

[3251] unknown, Artist, *Flag, announcing lynching, flown from the window of the NAACP headquarters on 69 Fifth Ave., New York City.* [Art]. United States Library of Congress, 1936.

[3252] S. Tzu, The Art of War, New York: Oxford University Press, n.d./1971, p. 187.

[3253] J. M. Tybur, L. Fan, B. C. Jones, I. J. Holzleitner, A. J. Lee and L. M. DeBruine, "Re-evaluating the relationship between pathogen avoidance and preferences for facial symmetry and sexual dimorphism: A registered report," *Evolution and Human Behavior,* vol. 43, no. 3, pp. 212-223, 2022.

[3254] D. A. Trump, "Truth Social," 14 October 2024. [Online]. Available: https://truthsocial.com/@realDonaldTrump/posts/113309607102307550.

[3255] M. Travis, Composer, *Sixteen Tons.* [Sound Recording]. Capitol Americana. 1947.

[3256] N. Tokuyama, K. Toda, M.-L. Poirret, B. Iyokango, B. Bakaa and S. Ishizuka, "Two wild female bonobos adopted infants from a different social group at Wamba," *Nature: Scientific Reports,* vol. 11, p. 4967, 2021.

[3257] E. L. Thomas, "Field study of the social behavior of the black lizard, Ctenosaura pectinata," *American Museum Novitates,* vol. 1493, pp. 24-26, 1951.

[3258] C. Thomas, "Just one maximum compensation payout in Magdalene scheme since expansion," *The Journal,* 11 February 2019.

[3259] A. B. Taylor and M. L. Goldsmith, Gorilla Biology: A Multidisciplinary Perspective, 1st ed., Cambridge: Cambridge University Press, 2003, p. 530.

[3260] H. Tao, S. Guo, T. Ge, K. M. Kendrick, Z. Xue, Z. Liu and J. Feng, "Depression uncouples brain hate circuit," *Molecular Psychiatry,* vol. 18, pp. 101-111, 2013.

[3261] B. Wyatt-Brown, The Shaping of Southern Culture: Honor, Grace, and War, 1760s-1880s, Chapel Hill: NC, 2001, p. 412.

[3262] J. S. Womersley, J. Nothling, S. Toikumo, S. Malan-Müller, L. L. v. d. Heuvel, N. W. McGregor, S. Seedat and S. M. J. Hemmings, "Childhood trauma, the stress response and metabolic syndrome: A focus on DNA methylation," *European Journal of Neuroscience,* vol. 55, no. 9-10, pp. 2253-2296, 2022.

[3263] R. Wollast, E. Puvia, P. Bernard, P. Tevichapong and O. Klein, "How Sexual Objectification Generates Dehumanization in Western and Eastern Cultures," *Social Psychology,* vol. 53, no. 3, 2022.

[3264] M. L. Wilson, C. Boesch, B. Fruth, T. Furuichi, I. C. Gilby, C. Hashimoto, C. L. Hobaiter, G. Hohmann, N. Itoh, K. Koops, J. N. Lloyd, T. Matsuzawa, J. C. Mitani, D. C. Mjungu, D. Morgan, M. N. Muller, R. Mundry, M. Nakamura and J. Pruetz, "Lethal aggression in Pan is better explained by adaptive strategies than human impacts," *Nature,* vol. 513, pp. 414-417, 2014.

[3265] B. Wilson, Alcoholics Anonymous: The Big Book, reprint ed., Mincola, New York: Ixia Press/ Dover, 1939/2019.

[3266] I. Wilkerson, Caste: The Origins of Our Discontents, New York, New York: Random House, 2020, p. 496.

[3267] O. Wild, The Picture of Dorian Gray, Mineola, NY: Dover, 1891/ 1993, p. 176.

[3268] W. White, Rope and Faggot: A Biography of Judge Lynch, Notre Dame , Indiana: Notre Dame University, 1969/2002, p. 272.

[3269] S. White, R. Acierno, K. J. Ruggiero, K. C. Koenen, D. G. Kilpatrick, S. Galea, J. Gelernter, V. Williamson, O. McMichael, V. I. Vladimirov and A. B. Amstadter, "Association of CRHR1 variants and posttraumatic stress symptoms in hurricane exposed adults," *Journal of Anxiety Disorders,* vol. 27, no. 7, pp. 678-683, 2013.

[3270] A. White, Klansmen: Guardians of Liberty, Zarephath, New Jersey: Pillar of Fire, 1926, p. 90.

[3271] D. P. Watts, "Social relationships of immigrant and resident female mountain gorillas. I. Male-female relationships," *American Journal of Primatology,* vol. 28, no. 3, pp. 159-181, 1992.

[3272] C. Wahlestedt, J. A. Blendy, K. J. Kellar, M. Heilig, E. Widerlöv and R. Ekman, "Electroconvulsive shocks increase the concentration of neocortical and hippocampal neuropeptide Y (NPY)-like immunoreactivity in the rat," *Brain Research,* vol. 507, no. 1, pp. 65-68, 1990.

[3273] G. T. Viki, P. Chiroro and D. Abrams, "Hostile sexism, type of rape, and self-reported rape proclivity within a sample of Zimbabwean males," *Violence Against Women,* vol. 12, no. 8, pp. 789-800, 2006.

[3274] W. Vickery, J. S. Brown and G. J. FitzGerald, "Spite: altruism's evil twin," *OIKOS,* vol. 102, no. 2, pp. 413-416, 2003.

[3275] H. Vervaecke and L. Van Elsacker, "Sexual competition in a group of captive bonobos (Pan paniscus)," *Primates,* vol. 41, no. 1, p. 109–115, 2000a.

[3276] H. Vervaecke, G. J. De Vries and L. Van Elsacker, "Dominance and its behavioral measures in a captive group of bonobos (Pan paniscus)," *International Journal of Primatology,* vol. 21, no. 1, pp. 47-68, 2000a.

[3277] C. L. Vernier, J. J. Krupp, K. Marcus, A. Hefetz, J. D. Levine and Y. Ben-Shahar, "The cuticular hydrocarbon profiles of honey bee workers develop via a socially-modulated innate process," *eLife,* vol. 8, p. e41855, 2019.

[3278] É. Zola, Au Bohheur des Dames, Paris: G. Charpentier, 1883.

[3279] S. M. T. Ziegler and M. J. Morrier, "Increasing social interactions of preschool children with autism through cooperative outdoor play," *Hammill Institute on Disabilities,* vol. 56, no. 1, 2022.

[3280] Z. Zheng, J. Lauritzen, E. Perlman, C. Robinson, M. Nichols, D. Milkie, O. Torrens, J. Price, C. Fisher and N. Sharifi, "A complete electron microscopy volume of the brain of adult Drosophila melanogaster," *Cell,* vol. 174, no. 3, pp. 730-743.e22, 2018.

[3281] J. Kander and F. Ebb, Composers, *The Cell Block Tango.* [Sound Recording]. 2002.

[3282] X. Zang, X. Lu, Y. Chi, Y. Jiang, C. Wang, S. A. Al-Farraj, A. Vallesi and F. Gao, "Timing and characteristics of nuclear events during conjugation and genomic exclusion in Paramecium multimicronucleatum," *Marine Life Science & Technology,* vol. 4, pp. 317-328, 2022.

[3283] L. R. S. Zanette, S. D. L. Miller, C. M. A. Faria, T. J. Huggins, W. C. Jordan and A. F. G. Bourke, "Reproductive conflict in bumblebees and the evolution of worker policing," *Evolution,* vol. 66, no. 12, pp. 3765-3777, 2012.

[3284] K. Yoshizawa, R. L. Ferreira, C. Lienhard and Y. Kamimura, "Why did a female penis evolve in a small group of cave insects?," *BioEssays,* vol. 41, no. 6, p. 1900005, 2019.

[3285] K. Yoshizawa, R. L. Ferreira, I. Yao, C. Lienhard and Y. Kamimura, "Independent origins of female penis and its coevolution with male vagina in cave insects (Psocodea: Prionoglarididae)," *Biology Letters,* vol. 14, no. 11, 18 November 2018.

[3286] M. P. Yavropoulou, G. C. Tsokos, G. P. Chrousos and P. P. Sfikakis, "Protracted stress-induced hypocortisolemia may account for the clinical and immune manifestations of Long COVID," *Clinical Immunology,* vol. 245, p. 109133, 2022.

[3287] *Zanesville Signal,* p. 4, 25 September 1946.

[3288] United States Geological Survey, "USGS 09379500 SAN JUAN RIVER NEAR BLUFF, UT," 20 05 2023. [Online]. Available: https://nwis.waterdata.usgs.gov/nwis/peak/?site_no=09379500&agency_cd=USGS.

[3289] New York Times, "Troop Deaths and Injuries in Ukraine War Near 500,000, U.S. Officials Say," 18 August 2023. [Online]. Available: https://www.nytimes.com/2023/08/18/us/politics/ukraine-russia-war-casualties.html. [Accessed 16 July 2024].

[3290] *This Is the Enemy.* [Art]. US National Archives, ca 1941-1945.

[3291] *The Jolly Darkie Target Game.* [Art]. ca 1890.

[3292] Federal Reserve Bank of St. Louis, "Student Loans Owned and Scrutinized (SLOAS)," 7 December 2022. [Online]. Available: https://fred.stlouisfed.org/series/SLOAS. [Accessed 7 January 2023].

[3293] United States Senate, "S.909 - Matthew Shepard Hate Crimes Prevention Act," 28 April 2009. [Online]. Available: https://www.congress.gov/bill/111th-congress/senate-bill/909/text. [Accessed 19 April 2023].

[3294] Federal Reserve Bank of New York, "Quarterly Report on Household Debt and Credit," Federal Reserve Bank of New York, New York, 2019.

[3295] New Books in German, "New Books in German," 2008. [Online]. Available: https://www.new-books-in-german.com/recommendations/hitlers-noble-jew-the-life-of-doctor-eduard-bloch/#sharethisrecommendation.

[3296] Smithsonian Design Museum, "Massacre of the Innocents by Etienne Delaune," 9 December 2017. [Online]. Available: https://commons.wikimedia.org/wiki/File:Print,_Massacre_of_the_Innocents_(CH_18099823).jpg.

[3297] "List of Egyptian Hieroglyphs," 23 June 2023. [Online]. Available: https://en.wikipedia.org/wiki/List_of_Egyptian_hieroglyphs. [Accessed 4 July 2023].

[3298] Harvard Medical School, "Lifetime prevalence DSM-IV/WMH-CIDI disorders by sex and cohort," 19 July 2007a. [Online]. Available: https://www.hcp.med.harvard.edu/ncs/ftpdir/table_ncsr_LTprevgenderxage.pdf. [Accessed 29 May 2023].

[3299] State of Illinois, "Illinois Anti-Bullying Laws & Policies: Components of State Anti-Bullying Laws and Regulations," 2023. [Online]. Available: https://www.stopbullying.gov/resources/laws/illinois. [Accessed 29 March 2023].

[3300] United States Government, "Federal Laws [on bullying]," 2023. [Online]. Available: https://www.stopbullying.gov/resources/laws/federal. [Accessed 29 March 2023].

[3301] The Guardian, "Factchecking Trump's debate claims: from abortion to Project 2025," 11 September 2024. [Online]. Available: https://www.theguardian.com/us-news/article/2024/sep/10/harris-trump-debate-fact-check-updates. [Accessed 15 June 2025].

[3302] 157th Infantry Association (Felix L. Sparks, Secretary), *Dachau and its Liberation,* 1989.

[3303] Supreme Judicial Court of Massachusetts, Middlesex, "Commonwealth v. Garabedian," 25 February 1987. [Online]. Available: https://case-law.vlex.com/vid/commonwealth-v-garabedian-890380592.

[3304] "Bhagavad Gita," [Online]. Available: https://asitis.com/2/34.html.

[3305] C. Boehm, Hierarchy in the Forest: The Evolution of Egalitarian Behavior, Cambridge: Harvard University Press, 2001, p. 304.

[3306] C. Boehm, Hierarchy in the Forest, Cambridge: Harvard University Press, 1999b, p. 304.

[3307] Z. Lang, "Reexamination of individual knowledge and common behavior rules: A cross-disciplinary view based on empirical evidences," *Munich Personal RePEc Archive,* 2009.

[3308] A. Bolton, "Lawmakers take in $608 per hour for the time they work in DC," 22 September 2014. [Online]. Available: https://thehill.com/homenews/news/218579-time-in-dc-for-lawmakers-works-out-to-608-per-hour/.

[3309] T. G. Townsend, T. B. Neilands, A. J. Thomas and T. R. Jackson, "I'm no Jezebel; I am young, gifted and Black: Identity, sexuality, and Black girls," *Psychology of Women,* vol. 34, no. 3, pp. 273-285, 2010.

[3310] B. Sepers, K. J. F. Verhoeven and K. van Oers, "Early developmental carry-over effects on exploratory behaviour and DNA methylation in wild great tits (Parus major)," *Evolutionary Applications,* vol. 17, no. 3, p. e13664, 2024.

[3311] R. M. Dawes, J. McTavish and H. Shaklee, "Behavior, communication, and assumptions about other people's behavior in a commons dilemma situation," *Journal of Personality and Social Psychology,* vol. 35, no. 1, pp. 1-11, 1977.

[3312] P. Dawe, Artist, *The Bostonians Paying the Excise-man, or Tarring and Feathering.* [Art]. Library of Congress, ca 1774.

[3313] M.-N. Tsai, C.-L. Wu, L.-P. Tseng, C.-P. An and H.-C. Chen, "Extraversion is a mediator of gelatophobia: A study of autism spectrum disorder and the Big Five," *Frontiers in Psychology,* vol. 9, p. 150, 2018.

[3314] C. Chen, A.-Y. Hung, Y.-T. Fan, S. Tan, H. Hong and Y. Cheng, "Linkage between pain sensitivity and empathic response in adolescents with autism spectrum conditions and conduct disorder symptoms," *Autism Research,* vol. 10, no. 2, pp. 267-275, 2017.

[3315] J. Silverstein, "George Zimmerman goes on depraved Twitter rant after retweeting picture of Trayvon Martin's corpse," *New York Daily News,* 28 September 2015.

[3316] S. Bauer, "George Zimmerman posted a photo of Trayvon Martin's dead body," *Mother Jones,* 28 September 2015.

[3317] Luigi Mangione's New York Legal Defense Team, "Home Page," 13 August 2025. [Online]. Available: https://www.luigimangioneinfo.com/. [Accessed 13 August 2025].

[3318] W. Shakespeare, "The Tragical History of Hamlet, Prince of Denmark," in *The Complete Shakespeare,* Garden City, NJ: Garden City Publishing, 1936.

[3319] N. S. Holtzman, A. A. Augustine and A. L. Senne, "Are pro-social or socially aversive people more physically symmetrical? Symmetry in relation to over 200 personality variables," *Journal of Research in Personality,* vol. 45, no. 6, pp. 687-691, 2011.

[3320] J. D. Miller, A. Dir, B. Gentile, L. Wilson, L. R. Pryor and W. K. Campbell, "Searching for a vulnerable dark triad: Comparing factor 2 psychopathy, vulnerable narcissism, and borderline personality disorder," *Journal of Personality,* vol. 78, no. 5, pp. 1529-1564, 2010.

[3321] S. R. Curtis and D. N. Jones, "Understanding what makes dark traits 'vulnerable': A distinction between indifference and hostility," *Personality and Individual Differences,* vol. 160, p. 109941, 2020.

[3322] J. S. Mill, "Inaugural Address at University of St. Andrews: 1867 February 1 (p 664)," in *Littell's Living Age,* vol. 92, Boston, MA: T.H. Carter & Company, 1868, p. 830.

[3323] N. Heym, F. Kibowski, C. A. J. Bloxsom, A. Blanchard, A. Harper, L. Wallace, J. Firth and A. Sumich, "The dark empath: Characterising dark traits in the presence of empathy," *Personality and Individual Differences,* vol. 169, p. 110172, 2021.

[3324] A. Mason, What's Wrong with Lookism?: Personal Appearance, Discrimination, and Disadvantage, Oxford: Oxford University Press, 2023, p. 252.

[3325] E. A. Caspar, K. Loumpa, C. Keysers and V. Gazzola, "Obeying orders reduces vicarious brain activation towards victims' pain," *NeuroImage,* vol. 222, no. 15, p. 117251, 2020a.

[3326] B. Chapais, "Alliances as a means of competition in primates: Evolutionary, developmental, and cognitive aspects," *American Journal of Physical Anthropology,* vol. 38, no. S21, pp. 115-136, 1995b.

[3327] J. M. Darley and B. Latane, "Bystander intervention in emergencies: Diffusion of responsibility," *Journal of Personality and Social Psychology,* vol. 8, no. 4, pt 1, p. 377–383, 1968b.

[3328] F. Hagenmuller, W. Rössler, J. Endrass, A. Rossegger and H. Haker, "[Impaired resonance in offenders with psychopathic traits] Empathische Resonanzfähigkeit bei Straftätern mit psychopathischen Persönlichkeitszügen]," *Neuropsychiatrie,* vol. 26, pp. 65-71, 2012b.

[3329] L. T. Harris and S. T. Fiske, "Dehumanized perception: A psychological means to facilitate atrocities, torture, and genocide?," *Zeitschrift für Psychologie,* vol. 219, no. 3, pp. 175-181, 2011b.

[3330] G. King, Devil in the Grove, First ed., Harper, 2012b, p. 448.

[3331] G. King, Devil in the Grove: Thurgood Marshall, the Groveland Boys, and the Dawn of a New America, New York: Harper, 2012, p. 448.

[3332] M. Kivimäki, A. Bartolomucci and I. Kawachi, "The multiple roles of life stress in metabolic disorders," *Nature Reviews Endocrinology,* vol. 19, pp. 10-27, 2023c.

[3333] C. Longhi, L. A. Martinez-Vaquero, F. W. Vázquez and V. Trianni, "Strong leaders don't cheat: an evolutionary appraisal of population heterogeneity and leadership," in *ALIFE 2024: Proceedings of the 2024 Artificial Life Conference*, 2024b.

[3334] A.-H. Luukkonen, P. Rasanen, H. Hakko and K. Riala, "Bullying behavior in relation to psychiatric disorders and physical health among adolescents: A clinical cohort of 508 underage inpatient adolescents in Northern Finland," *Psychiatry Research Volume,* vol. 178, no. 1, pp. 166-170, 2010b.

[3335] Q. Ma, L. Cheng, W. Qiu and G. Pei, "Role effect on beauty premium: Female as proposer may gain more fairness," *PsyCh Journal,* vol. 11, no. 5, pp. 691-706, 2022b.

[3336] C. McGettrick, K. O'Donnell, M. O'Rourke, J. M. Smith and M. Steed, Ireland and the Magdalene Laundries: A Campaign for Justice, London: I. B. Tauris, 2021b, p. 304.

[3337] L. R. Moscovice, P. H. Douglas, L. Martinez-Iñigo, M. Surbeck, L. Vigilante and G. Hohmann, "Stable and fluctuating social preferences and implications for cooperation among female bonobos at LuiKotale, Salonga National Park, DRC," *American Journal of Biological Anthropology,* vol. 163, no. 1, pp. 158-172, 2017b.

[3338] E. Palagi, "Not just for fun! Social play as a springboard for adult social competence in human and non-human primates," *Behavioral Ecology and Sociobiology,* vol. 72, p. 90, 2018b.

[3339] D. J. Rankin and M. Taborsky, "Assortment and the evolution of generalized reciprocity," *Evolution,* vol. 63, no. 7, pp. 1913-1922, 2009b.

[3340] R. B. Reddy, L. Samuni, V. Städele, L. Vigilant and M. Surbeck, "Maternal conflict intervention is more frequent in chimpanzee compared to bonobo development," *Animal Behaviour,* vol. 208, pp. 127-136, 2024b.

[3341] J. K. Rowling, Harry Potter and the Philosopher's Stone, Bloomsbury, 1997b, p. 223.

[3342] Z. Rubin and L. A. Peplau, "Who believes in a just world?," *Journal of Social Issues,* vol. 21, no. 3, pp. 65-89, 2025.

[3343] S. Sathe and R. Kümmerli, "Antagonistic interactions subdue inter-species green-beard cooperation in bacteria," *Journal of Evolutionary Biology,* vol. 33, no. 9, pp. 1245-1255, 2020b.

[3344] J. M. Smith, Ireland's Magdalen Laundries and the Nation's Architecture of Containment, Notre Dame, Indiana: Notre Dame Press, 2007b, p. 312.

[3345] T. B. Smith, M. E. McCullough and J. Poll, "Religiousness and depression: Evidence for a main effect and the moderating influence of stressful life events," *Psychological Bulletin,* vol. 103, no. 1, pp. 614-636, 2003b.

[3346] A. Sourander, D. Gyllenberg, A. Burnstein Klomek, L. Sillanmäki, A. M. Ilola and K. Kumpulainen, "Association of Bullying Behavior at 8 Years of Age and Use of Specialized Services for Psychiatric Disorders by 29 Years of Age," *Journal of the American Medical Association Psychiatry,* vol. 73, no. 2, pp. 195-165, 2016b.

[3347] J. S. Sun, "Strategic state violence and migration in conflict (preprint)," *American Journal of Political Science,* 2024b.

[3348] A. Surget and C. Belzung, "Adult hippocampal neurogenesis shapes adaptation and improves stress response: a mechanistic and integrative perspective," *Molecular Psychiatry,* vol. 27, pp. 403-421, 2022b.

[3349] H. Vervaecke, J. Stevens and L. Van Elsacker, "Interfering with others: Female-female reproductive competition in Pan Paniscus," in *Sexual Selection and Reproductive Competition in Primates: New Perspectives and Directions*, C. B. Jones, Ed., Norman, OK: American Society of Primatologists, 2003b, pp. 231-256.

[3350] S. A. West and A. Gardner, "Altruism, spite, and greenbeards," *Science,* vol. 327, no. 5971, pp. 1341-1344, 2010.

[3351] J. Rich, Director, *All in the Family: Lionel Moves Into the Neighborhood.* [Film]. US.1971.

[3352] A. M. Mutwill, H. Schielzeth, T. D. Zimmermann, S. H. Richter, S. Kaiser and N. Sachser, "Individuality meets plasticity: Endocrine phenotypes across male dominance rank acquisition in guinea pigs living in a complex social environment," *Hormones and Behavior,* vol. 131, p. 104967, 2021.

[3353] C. Becker and K.-R. Rose, Unterm Hakenkreuz : Tostedt und seine Aussendörfer in der Zeit vom aufkommenden Nationalsozialismus bis zum Ende des Dritten Reiches (p3-4), 1996, p. 128.

[3354] M. Chan, "He flipped his middle finger at a war memorial. Then he was shot dead," 9 September 2025. [Online]. Available: https://www.nbcnews.com/news/us-news/flipped-middle-finger-war-memorial-was-shot-dead-rcna228826. [Accessed 10 September 2025].

[3355] unknown, "Gestapo photos of Hans Scholl, taken after his capture on February 18, 1943.jpg," 18 February 1943. [Online]. Available: https://commons.wikimedia.org/wiki/File:Gestapo_photos_of_Hans_Scholl,_taken_after_his_capture_on_February_18,_1943.jpg. [Accessed 16 April 2023].

[3356] M. Twain, The Tragedy of Pudd'nhead Wilson, Hartford, CT: American Publishing Company, 1894, p. 337.

[3357] T. W. Scott, G. Wild and A. Gardner, "Kin-discriminating partner choice promotes the evolution of helping," *Evolution,* vol. 79, no. 6, pp. 962-972, 2025.

[3358] D. Bunsha, Scarred: Experiment's with Violence in Gujrat, Penguin Global, 2007, p. 332.

[3359] E. S. Brown and P. A. Chandle, "Mood and cognitive changes during systemic corticosteroid therapy," *Primary Care Companion to the Journal of Clinical Psychiatry,* vol. 3, no. 1, pp. 17-21, 2001.

[3360] J. H. Brunvand, "The eaten pets," in *Too Good to Be True: The Colossal Book of Urban Legends*, New York, W. W. Norton & Company, 2011, p. 480.

[3361] S. A. Dagbovie-Mullins, "Pigtails, ponytails, and getting tail: The infantilization and hyper-sexualization of African American females in popular culture," *The Journal of Popular Culture,* vol. 46, no. 4, pp. 745-771, 2013.

[3362] N. Cross, A. Kiefner, J. Rossi, J. Borushok, N. Hinman, J. Burmeister and R. A. Carels, "Facial attractiveness, weight status, and personality trait attribution: The role of attractiveness in weight stigma," *Obesity Research and Clinical Practice,* vol. 11, no. 4, pp. 377-388, 2017.

[3363] G. Cordoni, E. Collarini, M. Gioia and I. Norscia, "Play fighting versus real fighting in piglets (Sus scrofa): Similar patterns, different structure," *Behavioural Processes,* vol. 203, p. 104778, 2022.

[3364] M. A. Cooper, J. A. Grizzell, C. J. Whitten and G. M. Burghardt, "Comparing the ontogeny, neurobiology, and function of social play in hamsters and rats," *Neuroscience and Biobehavioral Reviews,* vol. 147, p. 105102, 2023.

[3365] L. B. Comeras, N. Hörmer, P. M. Bethuraj and R. O. Tasan, "NPY released from GABA neurons of the dentate gyrus specially reduces contextual fear without affecting cued or trace fear," *Frontiers in Synaptic Neuroscience,* vol. 13, p. 635726, 2021.

[3366] R. S. Clein, M. R. Warren and J. P. Neunuebel, "Mice employ a bait-and-switch escape mechanism to de-escalate social conflict," *PLoS Biology,* vol. 22, no. 10, p. e3002496, 2024.

[3367] D. Evon, "Did a young girl disconnect her grandfather's life support to plug in her phone's charger?," 1 November 2017. [Online]. Available: https://www.snopes.com/fact-check/girl-disconnect-life-support-plug-in-charger/. [Accessed 19 January 2023].

[3368] I. Ettekal and G. W. Ladd, "Development of aggressive-victims from childhood through adolescence: Associations with emotion dysregulation, withdrawn behaviors, moral disengagement, peer rejection, and friendships," *Development and Psychopathology,* vol. 32, no. 1, 2019.

[3369] D. A. Evans, A. V. Stempel, R. Vale and T. Branco, "Cognitive control of escape behavior," *Trends in Cognitive Science,* vol. 23, no. 4, pp. 334-348, 2019.

[3370] K. Eriksson, P. Strimling, P. A. Andersson, M. Aveyard, M. Bauer, V. Gritskov, T. Kiyonari, D. M. Kuhlman, A. T. Maitner, Z. Manesi, L. S. Peperkoorn, M. Rizwan, A. W. Stivers, Q. Tian, P. A. M. Van Lange, I. Vartanova, J. Wu and T. Yamagishi, "Cultural universals and cultural differences in meta-norms about peer punishment," *Management and Organization Review,* vol. 13, no. 4, pp. 851-870, 2017.

[3371] G. Dorantes-Argandar, C. R. Barrales-Díaz, E. S. Ferrero-Pastor, M. D. Madrigal-Arroyo, M. A. Maegli-Quezada, L. P. Marroquín-Gracias, M. Tortosa-Pérez and P. A. Villalobos-Ruano, "Aggressive driving, rule-breaking, and driver stress in Spanish-speaking countries," *Transportation Research Interdisciplinary Perspectives,* vol. 21, p. 100907, 2003.

[3372] L. M. Diamond, J. A. Dickenson and K. L. Blair, "Menstrual cycle changes in daily sexual motivation and behavior among sexually diverse cisgender women," *Archives of Sexual Behavior,* vol. 51, no. 1, pp. 577-588, 2022.

[3373] J. Diamond, The World Until Yesterday: What Can We Learn from Traditional Societies?, New York: Viking, 2013, p. 512.

[3374] J. Defreyne, B. Kreukels, G. T'Sjoen, A. Staphorsius, M. D. Heijer, G. Heylens and E. Elaut, "No correlation between serum testosterone levels and state-level anger intensity in transgender people: Results from the European Network for the Investigation of Gender Incongruence," *Hormones and Behavior,* vol. 110, pp. 29-39, 2019.

[3375] J. Decety and C. Holvoet, "The emergence of empathy: A developmental neuroscience perspective," *Developmental Review,* vol. 62, p. 100999, 2021.

[3376] Í. B. de Castro, A. O. de Meielles, H. Matthews-Cascon, C. d. A. Rocha-Barreira, P. Penchaszadeh and G. Bigatti, "Imposex in endemic Volutid from northeast Brazil (Mollusca: Gastropoda)," *Brazilian Archives of Biology and Technology,* vol. 51, no. 5, pp. 1065-1069, 2008.

[3377] C. Gençoglu, E. Sahin and N. Topkaya, "General self-efficacy and forgiveness of self, others, and situations as predictors of depression, anxiety, and stress in university students," *Educational Sciences: Theory and Practice,* vol. 18, no. 3, pp. 605-626, 2018.

[3378] C. J. Gee and L. M. Leith, "Aggressive behavior in professional ice hockey: A cross-cultural comparison of North American and European born NHL players," *Psychology of Sport and Exercise,* vol. 8, no. 4, pp. 567-583, 2007.

[3379] T. Furuichi and C. Hashimoto, "Why female bonobos have a lower copulation rate during estrus than chimpanzees," in *Behavioural Diversity in Chimpanzees and Bonobos*, 1st ed., L. Marchant, C. Boesch and G. Hohmann, Eds., Cambridge, Cambridge University Press, 2002, pp. 256-167.

[3380] A. Furnham, "Belief in a just world: Research progress over the past decade," *Personality and Individual Differences,* vol. 34, no. 5, pp. 795-817, 2003.

[3381] D. A. Frederick, J. C. Alley, S. Semeyna and J. R. Garcia, "Mating strategy variation by sexual orientation (pp 286-314)," in *The Oxford Handbook of Human Mating*, Oxford, Oxford University Press, 2023, p. 848.

[3382] K. R. Foster and F. L. Ratnieks, "Convergent evolution of worker policing by egg eating in the honeybee and common wasp," *Proceedings of the Royal Society B,* vol. 268, pp. 169-174, 2001.

[3383] D. Fatherree, "Trashing their rights: Alabama town uses 'debtors' prison' for people who fall behind on their garbage bills," 31 March 2023. [Online]. Available: https://www.splcenter.org/news/2023/03/31/alabama-town-debtors-prison-garbage-bills. [Accessed 2 April 2023].

[3384] K. Hurme, K. Gonzalez, M. Halvorsen, B. Foster, D. Moore and B. D. Chepko-Sade, "Environmental enrichment for dendrobatid frogs," *Journal of Applied Animal Welfare Science,* vol. 6, no. 4, pp. 285-299, 2003.

[3385] S. B. Hrdy, Father Time: A Natural History of Men and Babies, First ed., Princeton: Princeton University Press, 2024, p. 432.

[3386] M. J. Holmes, K. Tan, Z. Wang, B. P. Oldroyd and M. Beekman, "Why acquiesce? Worker reproductive parasitism in the Eastern honeybee (Apis cerana)," *Journal of Evolutionary Biology,* vol. 27, no. 5, pp. 939-949, 2014.

[3387] A. F. Hill, "Testimony of Anita F. Hill, Professor of Law, University of Oklahoma, Norman, Oklahoma," in *Complete Transcripts of the Clarence Thomas-Anita Hill Hearings: October 11, 12, 13, 1991*, N. Totenberg, Ed., Chicago, Academy Chicago Publishers, 2005, p. 496.

[3388] M. V. Harris-Perry, Sister Citizen: Shame, Stereotypes, and Black Women in America, Reprint ed., New Haven: Yale University Press, 2013, p. 400.

[3389] A. F. d. C. Hamilton, "Reflecting on the mirror neuron system in autism: A systematic review of current theories," *Developmental Cognitive Neuroscience,* vol. 3, pp. 91-105, 2013.

[3390] B. Hamann, Hitlers Edeljude. Das Leben des Armenarztes Eduard Bloch (Hitler's Noble Jew: The Life of Doctor Eduard Bloch), Munich: Piper Verlag GmbH, 2008.

[3391] A. L. Webber and T. Rice, Composers, *Simon Zealotes/Poor Jerusalem.* [Sound Recording]. Olympic Studios. 1970.

[3392] T. Grzyb and D. Dolinski, "You and I are alike, so I will hold back – The effect of directed empathy on the behavior of participants of Stanley Milgram's obedience paradigm," *Acta Psychologica,* vol. 234, p. 103859, 2023.

[3393] T. Gruber, K. B. Potts, C. Krupenye, M.-R. Byrne, C. Mackworth-Young, W. C. McGrew, V. Reynolds and K. Zuberbühler, "The influence of ecology on chimpanzee (Pan troglodytes) cultural behavior: A case study of five Ugandan chimpanzee communities," *Journal of Comparative Psychology,* vol. 126, no. 4, p. 446–457, 2012.

[3394] C. T. Gross and N. S. Canteras, "The many paths to fear," *Nature Reviews,* vol. 13, pp. 651-658, 2012.

[3395] P. B. Gray, T. S. McHale and J. M. Carré, "A review of human male field studies of hormones and behavioral reproductive effort," *Hormones and Behavior,* vol. 91, pp. 52-67, 2017.

[3396] S. B. Johnson, A. W. Riley, D. A. Granger and J. Riis, "The science of early life toxic stress for pediatric practice and advocacy," *Pediatrics,* vol. 131, no. 2, pp. 319-327, 2017.

[3397] M. Jog and M. Watve, "Role of parasites and commensals in shaping host behaviour," *Current Science,* vol. 89, no. 7, pp. 1184-1191, 2005.

[3398] X. Jiang and M. D. Pell, "On how the brain decodes vocal cues about speaker confidence," *Cortex,* vol. 66, pp. 9-34, 2015.

[3399] X. Jiang, R. Sanford and M. D. Pell, "Neural architecture underlying person perception from in-group and out-group voices," *NeuroImage,* vol. 181, pp. 582-597, 2018.

[3400] C. Jiang and J. Shi, "The long arm of childhood bullying victimization: Associations with social isolation, gender differences and depression in later life," *Journal of Interpersonal Violence,* vol. 39, no. 13-14, pp. 1-23, 2024.

[3401] R. Jewkes, Y. Sikweyiya, R. Morrell and K. Dunkle, "Gender inequitable masculinity and sexual entitlement in rape perpetration South Africa: Findings of a cross-sectional study," *PLoS One,* vol. 6, no. 12, p. e29590, 2011.

[3402] M. Iqbal, H. Bardwell and D. Hammond, "Estimating the global economic cost of violence: Methodology improvements and estimate updates," *Defence and Peace Economics,* vol. 23, no. 4, pp. 403-426, 2021.

[3403] M. Koenigs, L. Young, R. Adolphs, D. Tranel, F. Cushman, M. Hauser and A. Damasio, "Damage to the prefrontal cortex increases utilitarian moral judgements," *Nature,* vol. 446, no. 7138, pp. 908-111, 2007.

[3404] A. M. U. D. Khanday, Q. R. Khan and S. T. Rabani, "Identifying propaganda from online social networks during COVID-19 using machine learning techniques," *International Journal of Information Technology,* vol. 13, pp. 115-122, 2021.

[3405] A. Khan, E. Fleva and T. Qazi, "Role of self-esteem and general self-efficacy in teachers' efficacy in primary schools," *Psychology,* vol. 6, no. 1, pp. 117-124, 2015.

[3406] L. Keller, L. Passera and J.-P. Suzzoni, "Queen execution in the Argentine ant, Iridomyrmex humilis," *Physiological Entomology,* vol. 14, no. 2, pp. 157-163, 1989.

[3407] M. Kamatini, M. Ito, Y. Miyazaki and J. I. Kawahara, "Effects of masks worn to protect against Covid-19 on the perception of facial attractiveness," *i-Perception,* vol. 12, no. 3, 2021.

[3408] M. Locker, "You may now call a police horse 'gay' in the U.K.," *Time,* 16 January 2013.

[3409] A. Lippert, D. Baker, G. Hawk, N. Gongora and J. Golding, "Whom would you help? The impact of perpetrator and victim gender on bystander behavior during a sexual assault," *Violence Against Women,* vol. 0, no. 0, 2024.

[3410] G. Li, C. Lu, M. Yin, P. Wang, P. Zhang, J. Wu, W. Wang, D. Wang, M. Wang, J. Liu, X. Lin, J.-X. Zhang, Z. Wang, Y. Yu and Y.-F. Zhang, "Neural substrates for regulating self-grooming behavior in rodents (啮齿动物自我梳理行为调控的神经基质)," *Journal of Zhejiang University-SCIENCE B,* vol. 25, pp. 841-856, 2024.

[3411] T. Li, Q. Xiong, R. Shi and L. Yi, "Autistic and non-autistic children's pain perception is modulated by their first-hand pain sensitivity and theory of mind," *Research on Child and Adolescent Psychopathology,* vol. 52, pp. 1173-1182, 2024.

[3412] S.-B. Lee, S. Yerim and Y. Myung-Sook, "The Effects of relative deprivation of youth on suicide: The serial mediation effect of prospects and social isolation," *Health and Social Welfare Review,* vol. 42, no. 2, pp. 369-389, 2022.

[3413] S. K. Murnen, C. Wright and G. Kaluzny, "If 'Boys will be boys,' then girls will be victims? A meta-analytic review of the research that relates masculine ideology to sexual aggression," *Sex Roles,* vol. 46, pp. 359-375, 2002.

[3414] K. Murayama and M. N. J. C. Zhou, "A cross-cultural examination of the psychometric properties of responses to the achievement goal questionnaire," *Educational and Psychological Measurement,* vol. 69, no. 2, pp. 266-286, 2009.

[3415] M. Monday, "Donald Trump: Mexico is bringing drugs, crime and rapists to the US," 25 June 2015. [Online]. Available: https://www.usatoday.com/videos/news/2015/06/25/29292957/. [Accessed 27 February 2023].

[3416] S. Milgram, "Liberating effects of group pressure," *Journal of Personal and Social Psychology,* vol. 1, no. 2, pp. 127-234, 1965.

[3417] T. M. Milewski, W. Lee, C. F. A. and C. J. P., "Behavioural and physiological plasticity in social hierarchies," *Philosophical Transactions of the Royal Society B,* vol. 377, no. 1845, p. 20200443, 2022.

[3418] T. Migliaccio, A. Rivas, B. Rivas and R. Stark, "Bullying within academia: A cultural and structural analysis," *Learning, Culture and Social Interaction,* vol. 44, p. 100783, 2024.

[3419] A. H. Metwalli, A. Pross, E. Desfilis, A. Abellán and L. Medina, "Mapping of corticotropin-releasing factor, receptors and binding protein mRNA in the chicken telencephalon throughout development," *Journal of Comparative Neurology,* vol. 531, no. 14, pp. 1389-1424, 2023.

[3420] D. Merskin, "Three faces of Eva: Perpetuation of the hot-Latina stereotype in Desperate Housewives," *Howard Journal of Communications,* vol. 18, no. 2, pp. 133-151, 2007.

[3421] Y. Mekawi and N. R. Todd, "Focusing the lens to see more clearly: Overcoming definitional challenges and identifying new directions in racial microaggressions research," *Perspectives on Psychological Science,* vol. 16, no. 5, pp. 879-1098, 2021.

[3422] B. McLaughlin, N. S. Rodriguez, J. A. Dunn and J. Martinez, "Stereotyped identification: How identifying with fictional Latina characters increases acceptance and stereotyping," *Mass Communication and Society,* vol. 21, no. 5, pp. 585-605, 2018.

[3423] B. S. McEwen, "Redefining neuroendocrinology: Epigenetics of brain-body communication over the life course," *Frontiers in Neuroendocrinology,* vol. 49, pp. 8-30, 2018.

[3424] A. A. Mathé, S. Gruber, P. A. Jimenez and E. Theodorsson, "Effects of electroconvulsive stimuli and MK-801 on neuropeptide Y, neurokinin A, and calcitonin gene-related peptide in rat brain," *Neurochemical Research,* vol. 22, pp. 629-636, 1997.

[3425] D. E. Mastro and E. B. Behm-Morawitz, "Latino representations on primetime television," *Journalism and Mass Communication Quarterly,* vol. 82, no. 1, pp. 110-130, 2005.

[3426] R. C. Matheson, "Button, Button. (pp 129-137)," in *The Best of Richard Matheson*, London, Penguin Classics, 1970/2017, p. 432.

[3427] N. S. Marzuki, F. P. Idris, H. D. Kartapradja, A. R. Harahap and J. R. L. Batubara, "Characterising SRD5A2 gene variants in 37 Indonesian patients with 5-alpha-reductase type 2 deficiency," *International Journal of Endocrinology,* vol. 2019, p. 7676341, 2019.

[3428] *COMMONWEALTH vs. DAVID GARABEDIAN,* 1987.

[3429] M. C. Madsen, M. den Heijer, C. Pees, N. R. Biermasz and L. E. H. Bakker, "Testosterone in men with hypogonadism and transgender males: A systematic review comparing three different preparations," *Endocrine Connections,* vol. 11, no. 8, p. e220112, 2022.

[3430] W. A. Pridemore, "Demographic, temporal, and spatial patterns of homicide rates in Russia," *European Sociological Review,* vol. 19, no. 1, p. 41–59, 2003.

[3431] D. Pontier and E. Natoli, "Infanticide in rural male cats (Felis catus L.) as a reproductive mating tactic," *Aggressive Behavior,* vol. 25, no. 6, pp. 445-449, 1999b.

[3432] B. W. Penninx, D. S. Pine, E. A. Holmes and A. Reif, "Anxiety disorders (review)," *Lancet,* vol. 397, no. 10277, pp. 915-927, 2021.

[3433] T. O. Patton, "Reflections of a Black woman professor: racism and sexism in academia," *Howard Journal of Communications,* vol. 15, no. 3, pp. 185-200, 2004.

[3434] S. A. Park, S. Goïame, D. A. O'Connor and J.-C. Dreher, "Integration of individual and social information for decision-making in groups of different sizes," *PLoS Biology,* vol. 15, no. 6, p. e2001958, 2017.

[3435] N. O'Sullivan, "Magdalene compensation snub is 'rejection of Laundry women'," *Irish Post,* 2 August 2013.

[3436] C. Nwokafor, L. I. Serova, R. J. Nahvi, J. McCloskey and E. L. Sabban, "Activation of NPY receptor subtype 1 by [D-His26]NPY is sufficient to prevent development of anxiety and depressive like effects in the single prolonged stress rodent model of PTSD," *Neuropeptides,* vol. 80, p. 102001, 2020.

[3437] S. Nunes and M. P. M. Montemayor, "Multiple benefits of juvenile play: A ground squirrel's perspective," *Neuroscience and Biobehavioral Reviews,* vol. 147, p. 105099, 2023.

[3438] S. U. Noble, Algorithms of Oppression: How Search Engines Reinforce Racism, New York: NYU Press, 2018b, p. 248.

[3439] C. Njue, I. Askew and J. Chege, "Non-consensual sexual experiences of young people in Kenya: boys as perpetrators and victims (pp 139-157)," in *Sex without Consent: Young People in Developing Countries*, First ed., S. Jejeebhoy, I. Shah and S. Thapa, Eds., New York, Zed Books, 2005, p. 384.

[3440] Robespierre, "On Political Morality," 5 February 1794. [Online]. Available: https://revolution.chnm.org/d/413. [Accessed 12 September 2025].

[3441] E. Roca-Campos, E. Duque, O. Ríos and M. Ramis-Salas, "The zero violence brave club: A successful intervention to prevent and address bullying in schools," *Frontiers in Psychiatry,* vol. 12, p. 601424, 2021.

[3442] M. I. Rivera-Hechem, C. Rodríguez-Sickert, R. A. Guzmán, T. Ramírez-Parada, F. Benavides, V. Ladaeta, M. Aspé-Sánchez and G. M. Repetto, "No association between genetic variants in MAOA, OXTR, and AVPR1a and cooperative strategies," *PLOS One,* vol. 15, no. 12, p. 0244189, 2020.

[3443] S. M. Rich and F. J. Ayala, "The recent origin of allelic variation in antigenic determinants of Plasmodium falciparum," *Genetics,* vol. 150, no. 1, p. 515–517, 1998.

[3444] V. Reynolds, "Some behavioral comparisons between the chimpanzee and the mountain gorilla in the wild," *American Anthropologist,* vol. 67, no. 3, pp. 691-706, 1965.

[3445] F. L. W. Ratnieks, "Reproductive harmony via mutual policing by workers in eusocial Hymenoptera," *The American Naturalist,* vol. 132, no. 2, pp. 217-236, 1988.

[3446] W. Qing and Z. Xuebo, "Rational resistance amidst gentle breeze and mild rain: Examining online collective behavior among the Chinese public using the elaborated social identity model," *PLoS One,* vol. 19, no. 5, p. e0303603, 2024.

[3447] J. C. Simon and U. Herberlein, "Social hierarchy is established and maintained with distinct acts of aggression in male Drosophila melanogaster," *Journal of Experimental Biology,* vol. 223, no. 24, p. jeb232439, 2020.

[3448] A. D. Sherman and F. Petty, "Neurochemical basis of the action of antidepressants on learned helplessness," *Behavioral Neural Biology,* vol. 30, no. q, pp. 119-134, 1980.

[3449] R. P. Shaw and Y. Wong, "Preface," in *Genetic Seeds of Warfare: Evolution, Nationalism, and Patriotism*, Milton Park, Routledge, 2020, p. 288.

[3450] W. Shakespeare, The Merchant of Venice, Project Gutenberg, 1998/ca 1596.

[3451] J. M. Setchell, T. Smith, E. J. Wickings and L. A. Knapp, "Stress, social behaviour, and secondary sexual traits in a male primate," *Hormones and Behavior,* vol. 58, no. 5, pp. 720-728, 2010.

[3452] U. Scholz, B. G. Doña, S. Shonali and R. Schwarzer, "Is general self-efficacy a universal construct? Psychometric findings from 25 countries," *European Journal of Psychological Assessment,* vol. 18, no. 3, pp. 242-251, 2002.

[3453] H. Scholl and S. Scholl, At the Heart of the White Rose: Letters and Diaries of Hans and Sophie Scholl, Translated edition ed., I. Jens, Ed., Walden, New York: Plough Publishing House, 2017, p. 381.

[3454] T. H. Sawyer, M. S. Carroll and D. P. Connaughton, "Bullying and legal liability," *Journal of Physical Education, Recreation and Dance,* vol. 81, no. 7, pp. 51-53, 2010.

[3455] K. Sawada, "Simultaneous hermaphroditism in fishes (pp 39-63)," in *Hermaphroditism and Mating Systems in Fish*, T. Kuwamura, K. Sawada, T. Sunobe, Y. Sakai and T. Kadota, Eds., Singapore, Springer, 2023, p. 259.

[3456] A. C. Samson, O. Huber and W. Ruch, "Teasing, ridiculing and the relation to the fear of being laughed at in individuals with Asperger's Syndrome," *Journal of Autism and Developmental Disorders,* vol. 41, pp. 475-483, 2011.

[3457] T. Saler, "When it comes to unemployment compensation, things are much different in 2021," *Milwaukee Journal Sentinel,* 9 June 2021.

[3458] J. Swift, Gulliver's Travels, London: Benjamin Motte/ Project Gutenberg, 1726-7 (1997).

[3459] M. Surbeck and G. Hohmann, "Affiliations, aggressions and an adoption: Male–male relationships in wild bonobos (pp 35-46)," in *Bonobos: Unique in Mind, Brain, and Behavior*, B. Hare and S. Yamamoto, Eds., Oxford, Oxford University Press, 2017, p. 384.

[3460] J. Sun, J. Hao and Y. Liu, "Short-term effects of competitive video games on aggression: An event-related potential study," *Brain Sciences,* vol. 13, no. 6, p. 904, 2023.

[3461] S. L. Stewart, J. N. vanDyke and J. W. Poss, "Examining the mental health presentations of treatment-seeking transgender and gender nonconforming (TGNC) youth," *Child Psychiatry & Human Development,* vol. 32, no. 1, 2021.

[3462] K. Stephen, "Sexualized bodies (pp 29-45)," in *Real Bodies*, 2002 ed., M. Evans and E. Lee, Eds., Red Glove, 2002, p. 2012.

[3463] O. van Veen, Artist, *La pauurete ne vit pas tousiours a la vertu (Poverty does not always harm to virtue).* [Art]. 1646.

[3464] M. A. van Noordwijk, "From maternal investment to lifetime maternal care (pp 321-342)," in *The Evolution of Primate Societies*, First ed., J. C. Mitani, J. Call, P. M. Kappeler, R. A. Palombit and J. B. Silk, Eds., Chicago, University of Chicago Press, 2012, p. 730.

[3465] G. Toner, "The Transmission of 'Tochmarc Emire'," *Ériu,* vol. 49, p. 71–88, 1998.

[3466] A. E. Thigpen, D. L. Davis, T. Gautier, J. Imperato-McGinley and D. W. Russell, "The molecular basis of steroid 5α-reductase deficiency in a large Dominican kindred," *The New England Journal of Medicine,* vol. 327, no. 17, pp. 1216-1219, 1992.

[3467] D. Terburg, B. Morgan and J. van Honk, "The testosterone-cortisol ratio: A hormonal marker for proneness to social aggression," *International Journal of Law and Psychiatry,* vol. 32, no. 4, pp. 216-223, 2009.

[3468] T. Takahashi, "Economic decision-making in the ultimatum game by smokers," *Neuroendocrinology Letters,* vol. 28, no. 5, pp. 659-661, 2007.

[3469] J. Xu, W. Troop-Gordon and K. D. Rudolph, "Within-person reciprocal associations between peer victimization and need for approval," *Developmental Psychology,* vol. 58, no. 10, p. 1999–2011, 2022.

[3470] C.-L. Wu, C.-P. An, L.-P. Tseng, H.-C. Chen, Y.-C. Chan, S.-L. Cho and M.-L. Tsai, "Fear of being laughed at with relation to parent attachment in individuals with autism," *Disorders,* vol. 10, pp. 116-123, 2015.

[3471] I. L. C. M. Wrigley, E. Mascheroni, S. Bonichini and R. Montirosso, "Epigenetic protection: maternal touch and DNA-methylation in early life," *Current Opinion in Behavioral Science,* vol. 43, pp. 111-117, 2022.

[3472] J. C. Williams, A. Andreou, T. Lemelle, E. E. Coates and J. Q. Bostic, "Editorial: Racial battle fatigue: The toll of policing Black students," *Journal of the American Academy of Child and Adolescent Psychiatry,* vol. 61, no. 8, pp. 977-979, 2022.

[3473] D. J. Wilkinson, L. S. Bearup and T. Soprach, "Youth gang rape in Phnom Penh (pp 158-168)," in *Sex without Consent: Young People in Developing Countries*, S. Jejeebhoy, I. Shah and S. Thapa, Eds., New York, Zed Books, 2005, p. 384.

[3474] D. Watts and A. Pusey, "Behavior of juvenile and adolescent great apes," in *Juvenile Primates: Life History, Development, and Behavior*, First ed., M. E. Pereira and L. A. Fairbanks, Eds., Chicago, University of Chicago Press, 2002, pp. 148-167.

[3475] F. L. Wang, K. E. Bountress, K. Lemery-Chalfant, M. N. Wilson and D. S. Shaw, "A polygenic risk score enhances risk prediction for adolescents' antisocial behavior over the combined effect of 22 extra-familial, familial, and individual risk factors in the context of the family check-up," *Prevention Science,* vol. 24, pp. 739-751, 2023.

[3476] N. Wagner, M. Rieger and K. Voorvelt, "Gender, ethnicity and teaching evaluations: Evidence from mixed teaching teams," *Economics of Education Review,* vol. 54, no. 1, pp. 79-94, 2016.

[3477] M. Verderame, V. Migliaccio and R. Scudiero, "Role of estrogen receptors, P450 aromatase, PCNA and p53 in high-fat-induced impairment of spermatogenesis in rats," *Comptes Rendus Biologies,* vol. 341, no. 7-8, pp. 371-379, 2018.

[3478] W. J. Zielinski and J. G. Vandenbergh, "Increased survivorship of testosterone-treated female house mice, Mus musculus, in high-density field conditions," *Animal Behaviour,* vol. 42, no. 6, pp. 955-967, 1991.

[3479] W. Zhao, Q. Liu, X. Zhang, X. Song, Z. Zhang, P. Qing, X. Liu, S. Zhu, W. Yang and K. M. Kendrick, "Differential responses in the mirror neuron system during imitation of individual emotional facial expressions and association with autistic traits," *NeuroImage,* vol. 277, p. 120263, 2023.

[3480] J. X. Zhang and R. Schwarzer, "Measuring optimistic self-beliefs: a Chinese adaptation of the general self-efficacy scale," *Psychologia: An International Journal of Psychology in the Orient,* vol. 38, no. 3, pp. 174-181, 1995.

[3481] L. Yuxin, L. Chen, L. Xiaoxia, L. Yue, L. Junjie, L. Youzhu, Z. Huiliang and L. Qicai, "Research progress on the relationship between obesity-inflammation-aromatase axis and male infertility," *Oxidative Medicine and Cellular Longevity,* vol. 2021, p. 6612796, 2021.

[3482] R. Yehuda, S. Brand and R.-K. Yang, "Plasma neuropeptide Y concentrations in combat exposed veterans: Relationship to trauma exposure, recovery from PTSD, and coping," *Biological Psychiatry,* vol. 59, no. 7, pp. 660-663, 2006.

[3483] X. Yang, Y. Fang, H. Chen, T. Zhang, X. Yin, J. Man, L. Yang and M. Lu, "Global, regional and national burden of anxiety disorders from 1990 to 2019: results from the Global Burden of Disease Study 2019," *Epidemiology and Psychiatric Sciences,* vol. 30, p. E36, 2021.

[3484] N. Yamawaki, R. Darby and A. Queiroz, "The moderating role of ambivalent sexism: The influence of power status on perception of rape victim and rapist," *Journal of Social Psychology,* vol. 147, no. 1, pp. 41-56, 2007.

[3485] Canadian Parliament, "The Indian Act, 1927 (Indian Act, Chapter 98: An Act Respecting Indians)," 1927. [Online]. Available: http://kopiwadan.ca/wp-content/uploads/2017/01/4_1_1927-_AN-ACT-RESPECTIING-THE-INDIANS-1927.pdf. [Accessed 22 May 2023].

[3486] Harvard Medical School, "Table 2. 12-month prevalence of DSM-IV/WMH-CIDI disorders by sex and cohort," 19 July 2007. [Online]. Available: https://www.hcp.med.harvard.edu/ncs/ftpdir/table_ncsr_LTprevgenderxage.pdf. [Accessed 29 May 2023].

[3487] State of Massachusetts, "Massachusetts law about bullying and cyberbullying," 2023. [Online]. Available: https://www.mass.gov/info-details/massachusetts-law-about-bullying-and-cyberbullying. [Accessed 29 March 2023].

[3488] Southern Poverty Law Center, "After Election Day: The Trump Effect: The Impact of the 2016 Presidential Election on Our Nation's Schools," Southern Poverty Law Center, Montgomery AL, 2016.

[3489] M. Abavisani, N. Faraji, N. Ebadpour, P. Kesharwani and A. Sahebkar, "Beyond digestion: Exploring how the gut microbiota modulates human social behaviors," *Neuroscience,* vol. 565, pp. 52-62, 2025.

[3490] D. H. Abbott, "Behaviourally mediated suppression of reproduction in female primates," *Journal of Zoology,* vol. 213, no. 3, pp. 455-470, 1987.

[3491] P. Foot, "The problem of abortion and the doctrine of the double effect (pp 19-32)," in *Virtues and Vices: And Other Essays in Moral Philosophy*, Reprint ed., Oxford, Clarendon Press, 1967/ 2003, p. 232.

[3492] C. Musanabaganwa, A. H. Wani, J. Donglasan, S. Fatumo, S. Jansen, J. Mutabaruka, E. Rutembesa, A. Uwineza, E. J. Hermans, B. Roozendaal, D. E. Wildman and L. U. M. Mutesa, "Leukocyte methylomic imprints of exposure to the genocide against the Tutsi in Rwanda: a pilot epigenome-wide analysis," *Epigenomics,* vol. 14, no. 1, pp. 11-25, 2022c.

[3493] M. Sundvik, H. Puttonen, S. Semenova and P. Panula, "The bullies are the leaders of the next generation: Inherited aminergic neurotransmitter system changes in socially dominant zebrafish, Danio rerio," *Behavioural Brain Research,* vol. 49, no. 9, p. 113309, 2021c.

[3494] S. Anderson, "Blaming immigrants for eating pets is an old American urban legend," *Forbes,* 15 September 2024.

[3495] B. E. Sheese and W. G. Graziano, "Deciding to defect: The effects of video-game violence on cooperative behavior," *Psychological Science,* vol. 16, no. 5, pp. 354-357, 2005.

[3496] D. Andres, T. H. Clutton-Brock, L. E. B. Kruuk, J. M. Pemberton, K. V. Stopher and K. E. Ruckstuhl, "Sex differences in the consequences of maternal loss in a long-lived mammal, the red deer (Cervus elaphus)," *Behavioral Ecology and Sociobiology,* vol. 67, pp. 1249-1258, 2013.

[3497] M. L. East and H. Hofer, "Male spotted hyenas (Crocuta crocuta) queue for status in social groups dominated by females," *Behavioral Ecology,* vol. 12, no. 5, p. 558–568, 2001.

[3498] B. W. Dunlop, B. O. Rothbaum, E. B. Binder, E. Duncan, P. D. Harvey, T. Jovanovic, M. E. Kelley, B. Kinkead, M. Kutner, D. V. Iosifescu, S. J. Mathew, T. C. Neylan, C. D. Kilts, C. B. Nemeroff and H. S. Mayberg, "Evaluation of a corticotropin releasing hormone type 1 receptor antagonist in women with posttraumatic stress disorder: study protocol for a randomized controlled trial," *Trials,* vol. 15, p. 240, 2014.

[3499] J. R. Dugré and S. Potvin, "Mapping the brain network of conduct disorder: Heterogeneous fMRI findings converge on a common brain circuit," *MedRxiv,* (2024a; preprint).

[3500] B. L. Fredrickson and T.-A. Roberts, "Objectification theory: Toward understanding women's lived experiences and mental health risks," *Psychology of Women,* vol. 21, no. 2, pp. 173-206, 1997.

[3501] M. Farhan, M. A. Javed, M. H. K. Khattak and Z. Sheikh, "The role of religiosity and locus of control on psychological well-being among Muslim undergraduate students: The moderating role of resilience," *Remittances Review,* vol. 9, no. 1, pp. 3467-3513, 2024.

[3502] A. Hodges and I. B. Prine, "Personal interview with 'Aunt' Adeline Hodges, 3 Frye Street, Mobile Alabama: Hongry for pun'kin pie," Federal Writers' Project of the Works Progress Administration for the State of Alabama, Mobile AL, n.d..

[3503] S. Lee, P. K. Smith and C. P. Monks, "Meaning and usage of a term for bullying-like phenomena in South Korea: A lifespan perspective," *Journal of Language and Social Psychology,* vol. 31, no. 3, pp. 342-349, 2012.

[3504] J. Lee, Y.-S. Kwak, Y.-J. Kim, E.-J. P. E. J. Kim, Y. Shin, B.-H. Lee, S. H. Lee, H. Y. Jung, I. Lee, J. I. Hwang, D. Kim and S. I. Lee, "Transgenerational transmission of trauma: Psychiatric evaluations of offspring of former 'Comfort Women,' Survivors of the Japanese military sexual slavery during World War II," *Psychiatric Investigation,* vol. 16, no. 3, pp. 249-253, 2019.

[3505] T. Merz, "'Kick a Ginger Kid Day' leads to attacks on schoolchildren," 18 October 2013. [Online]. Available: https://www.telegraph.co.uk/education/educationnews/10387636/Kick-a-Ginger-Kid-Day-leads-to-attacks-on-schoolchildren.html. [Accessed 30 May 2023].

[3506] K. Putnam, R. Shoemaker, F. Yiannikouris and L. A. Cassis, "The renin-angiotensin system: a target of and contributor to dyslipidemias, altered glucose homeostasis, and hypertension of the metabolic syndrome," *American Journal of Physiology: Heart and Circulatory Physiology,* vol. 302, no. 6, p. H1219–H1230, 2012.

[3507] S. N. Schmeltzer, J. P. Herman and R. Sah, "Neuropeptide Y (NPY) and posttraumatic stress disorder (PTSD): A translational update," *Experimental Neurology,* vol. 284 (pt B), pp. 196-210, 2016.

[3508] unknown, One Thousand and One Nights, vol. 10, The Burton Club, 1866.

[3509] C. Anacker, V. M. Luna, G. S. Stevens, A. Millette, R. Shores, J. C. Jimenez, B. Chen and R. Hen, "Hippocampal neurogenesis confers stress resilience by inhibiting the ventral dentate gyrus," *Nature,* vol. 559, pp. 98-102, 2018.

[3510] R. Alexander, The Biology of Moral Systems (Evolutionary Foundations of Human Behavior Series), reprint ed., New York: Aldine, 1987, p. 324.

[3511] A. R. Ekrich, At Day's Close: Night in Times Past, First ed., New York: W.W. Norton, 2006, p. 480.

[3512] A. Rosenberg, *Letter from the Reichsminister for the Occupied Eastern Territories to Gauleiter Fritz Sauckel. Document 018 PS,* 1942.

[3513] J. Knebel, Evolution of Cooperation in Bacterial Biofilms: A Stochastic Model for the Growth and Sinking of Bacterial Mats (master's thesis), Munich: Ludwig Maximilians University, 2011, p. 113.

[3514] B. Madley, "Reexamining the American genocide debate: Meaning, historiography, and new methods," *American Historical Review,* vol. 120, no. 1, pp. 98-139, 2015.

[3515] A. Powell, S. Shennan and M. G. Thomas, "Late Pleistocene demography and the appearance of modern human behavior," *Science,* vol. 324, no. 5932, pp. 1298-1301, 2009b.

[3516] F. L. W. Ratnieks, K. R. Foster and T. Wenseleers, "Conflict Resolution in Insect Societies," *Annual Review of Entomology,* vol. 51, pp. 581-608, 2006.

[3517] S. A. Surguladze, W. El-Hage and T. Dalgleish, "Depression is associated with increased sensitivity to signals of disgust: A functional magnetic resonance imaging study," *Journal of Psychiatric Research,* vol. 44, no. 14, pp. 894-902, 2010.

[3518] Amnesty International, "USA: Pattern of brutality and cruelty-- war crimes at Abu Ghraib," 7 May 2004. [Online]. Available: https://www.amnesty.org/en/wp-content/uploads/2021/09/amr510772004en.pdf. [Accessed 12 May 2023].

[3519] W. C. Cockerham, "Theoretica approaches to research on the social determinants of obesity," *American Journal of Preventive Medicine,* vol. 63, no. 1, S1, pp. S8-S17, 2022.

ABOUT THE AUTHOR

Ptera Hunter, PhD. teaches biology at Northeastern Illinois University. She received her Doctorate in Zoology from the University of Toronto. She has taught classes on environmental biology, animal behavior, evolutionary biology, parasitology, and animal diversity. She has published a book on insects, *The Unabashedly Totally Incomplete Book about Bugs*, a book on deceit in the natural world, *The Wisdom of Loki*, and several novels, short stories, and journal articles. Writing is her avocation, but teaching is her life.

Made in United States
Orlando, FL
06 November 2026

90299268R00380